AF361026

Vitamin D in the New Decade: Facts, Controversies, and Future Perspectives for Daily Clinical Practice

Vitamin D in the New Decade: Facts, Controversies, and Future Perspectives for Daily Clinical Practice

Editors

Spyridon N. Karras
Pawel Pludowski

MDPI • Basel • Beijing • Wuhan • Barcelona • Belgrade • Manchester • Tokyo • Cluj • Tianjin

Editors
Spyridon N. Karras
Aristotle University of
Thessaloniki
Greece

Pawel Pludowski
The Children's Memorial
Health Institute
Warsaw
Poland

Editorial Office
MDPI
St. Alban-Anlage 66
4052 Basel, Switzerland

This is a reprint of articles from the Special Issue published online in the open access journal *Nutrients* (ISSN 2072-6643) (available at: https://www.mdpi.com/journal/nutrients/special_issues/Vitamin_D_in_the_New_Decade).

For citation purposes, cite each article independently as indicated on the article page online and as indicated below:

LastName, A.A.; LastName, B.B.; LastName, C.C. Article Title. *Journal Name* **Year**, *Volume Number*, Page Range.

ISBN 978-3-0365-5105-0 (Hbk)
ISBN 978-3-0365-5106-7 (PDF)

Contents

About the Editors

Spyridon N. Karras

Spyridon N. Karras was born in Thessaloniki and graduated from the Medical School of the Aristotle University of Thessaloniki. He acquired his specialization in endocrinology in 2010. His PhD thesis (completed in 2015) focused on the pleiotropic effects of vitamin D during pregnancy. This research endeavor led to the development of several global collaborations in the field of vitamin D research and was the initial basis for the foundation of the Mediterranean Experts Meeting on vitamin D. He has published more than 130 full papers in peer-reviewed international journals, with the vast majority relating to extraskeletal vitamin D behaviors. According to Expertscape and other ranking sites, he ranks in the top 20 positions in the field worldwide. He is a regular reviewer of more than 100 international journals and has been a lecturer at national and international conferences in the field, receiving several awards for his research activities. His current position is as a faculty member at the Laboratory of Biochemistry of the Medical School of the Aristotle University of Thessaloniki, Greece.

Pawel Pludowski

Pawel Pludowski works as an Associate Professor at the Department of Biochemistry, Radioimmunology, and Experimental Medicine at the Children's Memorial Health Institute in Warsaw. Pawel Pludowski is the Chairman of the European Vitamin D Association (EVIDAS). He has published over 250 peer-reviewed articles and chapters in books (till this date, about 70 are indexed in Pubmed). He is a doctor habil. of medical sciences, and a professor of an institute. He is a specialist in clinical densitometry (certified clinical densitometrist, CCD) and vitamin D metabolism. His other peripheral qualifications include the following: disturbances of calcium–phosphate metabolism, disturbances of bone tissue metabolism, body composition and biological development, basics of statistics, GLP, and research methodologies. His research work has been devoted primarily to densitometry (DXA and pQCT) as a noninvasive method for assessing bone mass and body composition during growth and maturation. Other areas of interest are the metabolism of vitamin D and the importance of vitamin D for human health.

Review

Evidence that Vitamin D Supplementation Could Reduce Risk of Influenza and COVID-19 Infections and Deaths

William B. Grant [1,*], Henry Lahore [2], Sharon L. McDonnell [3], Carole A. Baggerly [3], Christine B. French [3], Jennifer L. Aliano [3] and Harjit P. Bhattoa [4]

[1] Sunlight, Nutrition, and Health Research Center, P.O. Box 641603, San Francisco, CA 94164-1603, USA
[2] 2289 Highland Loop, Port Townsend, WA 98368, USA; hlahore@vitamindwiki.com.
[3] GrassrootsHealth, Encinitas, CA 92024, USA; Sharon@grassrootshealth.org (S.L.M.); carole@grassrootshealth.org (C.A.B.); Christine@grassrootshealth.org (C.B.F.); jen@grassrootshealth.org (J.L.A.)
[4] Department of Laboratory Medicine, Faculty of Medicine, University of Debrecen, Nagyerdei Blvd 98, H-4032 Debrecen, Hungary; harjit@med.unideb.hu
[*] Correspondence: wbgrant@infionline.net; Tel.: +1-415-409-1980

Received: 12 March 2020; Accepted: 31 March 2020; Published: 2 April 2020

Abstract: The world is in the grip of the COVID-19 pandemic. Public health measures that can reduce the risk of infection and death in addition to quarantines are desperately needed. This article reviews the roles of vitamin D in reducing the risk of respiratory tract infections, knowledge about the epidemiology of influenza and COVID-19, and how vitamin D supplementation might be a useful measure to reduce risk. Through several mechanisms, vitamin D can reduce risk of infections. Those mechanisms include inducing cathelicidins and defensins that can lower viral replication rates and reducing concentrations of pro-inflammatory cytokines that produce the inflammation that injures the lining of the lungs, leading to pneumonia, as well as increasing concentrations of anti-inflammatory cytokines. Several observational studies and clinical trials reported that vitamin D supplementation reduced the risk of influenza, whereas others did not. Evidence supporting the role of vitamin D in reducing risk of COVID-19 includes that the outbreak occurred in winter, a time when 25-hydroxyvitamin D (25(OH)D) concentrations are lowest; that the number of cases in the Southern Hemisphere near the end of summer are low; that vitamin D deficiency has been found to contribute to acute respiratory distress syndrome; and that case-fatality rates increase with age and with chronic disease comorbidity, both of which are associated with lower 25(OH)D concentration. To reduce the risk of infection, it is recommended that people at risk of influenza and/or COVID-19 consider taking 10,000 IU/d of vitamin D_3 for a few weeks to rapidly raise 25(OH)D concentrations, followed by 5000 IU/d. The goal should be to raise 25(OH)D concentrations above 40–60 ng/mL (100–150 nmol/L). For treatment of people who become infected with COVID-19, higher vitamin D_3 doses might be useful. Randomized controlled trials and large population studies should be conducted to evaluate these recommendations.

Keywords: acute respiratory distress syndrome (ARDS); ascorbic acid; cathelicidin; coronavirus; COVID-19; cytokine storm; influenza; observational; pneumonia; prevention; respiratory tract infection; solar radiation; treatment; UVB; vitamin C; vitamin D

---

1. Introduction

The world is now experiencing its third major epidemic of coronavirus (CoV) infections. A new CoV infection epidemic began in Wuhan, Hubei, China, in late 2019, originally called 2019-nCoV [1] and renamed COVID-19 by the World Health Organization on February 11, 2020. Previous CoV epidemics include severe acute respiratory syndrome (SARS)-CoV, which started in China in 2002 [2], and the ongoing Middle East respiratory syndrome (MERS)-CoV in the Middle East, first reported in 2012 [3]. Those epidemics all began with animal-to-human infection. The direct cause of death is generally due to ensuing severe atypical pneumonia [4,5].

Seasonal influenza has a high health burden. According to one recent estimate, 389,000 (uncertainty range 294,000–518,000) respiratory deaths were associated with influenza during the period 2002–2011 [6]. According to the U.S. Center for Disease Control and Prevention, during the period 2010–2019, annual symptomatic illness affected between 9 and 45 million people, resulting in between 4 and 21 million medical visits, 140,000–810,000 hospitalizations, and 23,000–61,000 deaths (https://www.cdc.gov/flu/about/burden/).

This review is a narrative one. Searches were made in PubMed.gov and scholar.google.com for publications regarding influenza, CoVs, COVID-19, and pneumonia with respect to epidemiology, innate and adaptive immune response, vitamin D, 25-hydroxyvitamin D (25(OH)D), and parathyroid hormone.

2. Vitamin D and Mechanisms to Reduce Microbial Infections

The general metabolism and actions of vitamin D are well known [7]. Vitamin D_3 is produced in the skin through the action of UVB radiation reaching 7-dehydrocholesterol in the skin, followed by a thermal reaction. That vitamin D_3 or oral vitamin D is converted to 25(OH)D in the liver and then to the hormonal metabolite, $1,25(OH)_2D$ (calcitriol), in the kidneys or other organs as needed. Most of vitamin D's effect arises from calcitriol entering the nuclear vitamin D receptor, a DNA binding protein that interacts directly with regulatory sequences near target genes and that recruits chromatin active complexes that participate genetically and epigenetically in modifying transcriptional output [8]. A well-known function of calcitriol is to help regulate serum calcium concentrations, which it does in a feedback loop with parathyroid hormone (PTH), which itself has many important functions in the body [7].

Several reviews consider the ways in which vitamin D reduces the risk of viral infections [9–17].

Vitamin D has many mechanisms by which it reduces the risk of microbial infection and death. A recent review regarding the role of vitamin D in reducing the risk of the common cold grouped those mechanisms into three categories: physical barrier, cellular natural immunity, and adaptive immunity [16]. Vitamin D helps maintain tight junctions, gap junctions, and adherens junctions (e.g., by E-cadherin) [18]. Several articles discussed how viruses disturb junction integrity, increasing infection by the virus and other microorganisms [19–21].

Vitamin D enhances cellular innate immunity partly through the induction of antimicrobial peptides, including human cathelicidin, LL-37, by 1,25-dihdroxyvitamin D [22,23], and defensins [24]. Cathelicidins exhibit direct antimicrobial activities against a spectrum of microbes, including Gram-positive and Gram-negative bacteria, enveloped and nonenveloped viruses, and fungi [25]. Those host-derived peptides kill the invading pathogens by perturbing their cell membranes and can neutralize the biological activities of endotoxins [26]. They have many more important functions, as described therein. In a mouse model, LL-37 reduced influenza A virus replication [27]. In another laboratory study, $1,25(OH)_2D$ reduced the replication of rotavirus both in vitro and in vivo by another process [28]. A clinical trial reported that supplementation with 4000 IU/d of vitamin D decreased dengue virus infection [29].

Vitamin D also enhances cellular immunity, in part by reducing the cytokine storm induced by the innate immune system. The innate immune system generates both pro-inflammatory and anti-inflammatory cytokines in response to viral and bacterial infections, as observed in COVID-19 patients [30]. Vitamin D can reduce the production of pro-inflammatory Th1 cytokines, such as tumor necrosis factor α and interferon γ [31]. Administering vitamin D reduces the expression of pro-inflammatory cytokines and increases the expression of anti-inflammatory cytokines by macrophages ([17] and references therein).

Vitamin D is a modulator of adaptive immunity [16,32]; 1,25(OH)$_2$D$_3$ suppresses responses mediated by the T helper cell type 1 (Th1), by primarily repressing production of inflammatory cytokines IL-2 and interferon gamma (INFγ) [33]. Additionally, 1,25(OH)$_2$D$_3$ promotes cytokine production by the T helper type 2 (Th2) cells, which helps enhance the indirect suppression of Th1 cells by complementing this with actions mediated by a multitude of cell types [34]. Furthermore, 1,25(OH)$_2$D$_3$ promotes induction of the T regulatory cells, thereby inhibiting inflammatory processes [35].

Serum 25(OH)D concentrations tend to decrease with age [36], which may be important for COVID-19 because case-fatality rates (CFRs) increase with age [37]. Reasons include less time spent in the sun and reduced production of vitamin D as a result of lower levels of 7-dehydrocholesterol in the skin [38]. In addition, some pharmaceutical drugs reduce serum 25(OH)D concentrations by activating the pregnane-X receptor [39]. Such drugs include antiepileptics, antineoplastics, antibiotics, anti-inflammatory agents, antihypertensives, antiretrovirals, endocrine drugs, and some herbal medicines. Pharmaceutical drug use typically increases with age.

Vitamin D supplementation also enhances the expression of genes related to antioxidation (glutathione reductase and glutamate–cysteine ligase modifier subunit) [40]. The increased glutathione production spares the use of ascorbic acid (vitamin C), which has antimicrobial activities [41,42], and has been proposed to prevent and treat COVID-19 [43]. Moreover, a former director of the Center for Disease Control and Prevention, Dr. Tom Frieden, proposed using vitamin D to combat the COVID-19 pandemic on 23 March 2020 (https://www.foxnews.com/opinion/former-cdc-chief-tom-frieden-coronavirus-risk-may-be-reduced-with-vitamin-d).

3. Discussion

3.1. Seasonal Influenza

Influenza virus affects the respiratory tract by direct viral infection or by damage to the immune system response. The proximate cause of death is usually from the ensuing pneumonia. Patients who develop pneumonia are more likely to be < 5 years old, > 65 years old, white, and nursing home residents, to have chronic lung or heart disease and a history of smoking, and to be immunocompromised [44].

Seasonal influenza infections generally peak in winter [45]. Cannell et al. hypothesized that the winter peak was due in part to the conjunction with the season when solar UVB doses, and thus 25(OH)D concentrations, are lowest in most mid- and high-latitude countries [46], extended in [47]. Mean serum 25(OH)D concentrations in north and central regions of the United States are near 21 ng/mL in winter and 28 ng/mL in summer, whereas in the south region, they are near 24 ng/mL in winter and 28 ng/mL in summer [48]. In addition, the winter peak of influenza also coincides with weather conditions of low temperature and relative humidity that allow the influenza virus to survive longer outside the body than under warmer conditions [49–51].

Ecological studies suggest that raising 25(OH)D concentrations through vitamin D supplementation in winter would reduce the risk of developing influenza. Table 1 presents results from randomized controlled trials (RCTs) investigating how vitamin D supplementation affects risk of influenza. The RCTs included confirmed that the respiratory tract infection was indeed derived from influenza. Only two RCTs reported beneficial effects: one among schoolchildren in Japan [52], the other among infants in China [53]. An RCT in Japan that reported no beneficial effect did not measure baseline 25(OH)D concentration [54] and included many participants who had been vaccinated against influenza (M. Urashima; private communication). The two most recent RCTs included participants with above average mean baseline 25(OH)D concentrations [55,56]. A comprehensive review of the role of vitamin D and influenza was published in 2018 [15]. It concluded that the evidence of vitamin D's effects on the immune system suggest that vitamin D should reduce the risk of influenza, but that more studies are required to evaluate that possibility. Large population studies would also be useful, in which vitamin D supplementation is also related to changes in serum 25(OH)D concentration.

Table 1. Results of vitamin D randomized controlled trials (RCTs) on risk of influenza.

Country	Population	Baseline 25(OH)D (ng/mL)	Vitamin D Dose (IU/d)	Influenza Cases in Vitamin D, Placebo Arms	Outcome	Ref
Japan	Schoolchildren aged 6–15 yrs	N/A	0, 1200	Type A: 18/167; 31/167. If not taking vitamin D before enrollment: 8/140; 22/140. Type B: 39/167; 28/167	Type A: RR = 0.58 (95% CI, 0.34 to 0.99); if not taking vitamin D before enrollment, RR = 0.36 (95% CI, 0.17 to 0.79); no effect for Type B	[52]
Japan	High school students, including many vaccinated against influenza	N/A	0, 2000	20/148; 12/99	Type A, RR = 1.11 (95% CI, 0.57 to 2.18)	[54]
China	Infants, 3–12 mos	17	400, 1200		Diff. in influenza A viral load, high vs. low vitamin D on day 4 of illness: 1.3 ± 0.5 vs. 4.5 ± 1.4 × 10^6 copies/mL	[53]
Japan	223 patients with IBD, mean age 45 yrs	23–24	0, 500	8/115; 6/108	RR = 1.25 (95% CI, 0.45 to 3.49)	[55]
Vietnam	Children aged 3–17 yrs	26	0, 14,000 /wk	50/650; 43/650	HR = 1.18 (95% CI, 0.79 to 1.78)	[56]

Note: 95% confidence interval (95% CI); day (d); hazard ratio (HR); inflammatory bowel disease (IBD); months (mos); not available (N/A); relative risk (RR); upper respiratory tract infection (URTI); week (wk); years (yrs).

An observational study conducted in Connecticut on 198 healthy adults in the fall and winter of 2009–2010 examined the relationship between serum 25(OH)D concentration and incidence of acute RTIs (ARTIs) [57]. Only 17% of people who maintained 25(OH)D >38 ng/mL throughout the study developed ARTIs, whereas 45% of those with 25(OH)D < 38 ng/mL did. Concentrations of 38 ng/mL or more were associated with a significant ($p < 0.0001$) twofold reduction in risk of developing ARTIs and with a marked reduction in the percentage of days ill. Eight influenza-like illnesses (ILIs) occurred, seven of which were the 2009 H1N1 influenza.

3.2. Clinical and Epidemiological Findings Regarding COVID-19

The first step in developing a hypothesis is to outline the epidemiological and clinical findings regarding the disease of interest and their relationship with 25(OH)D concentrations. From the recent journal literature, it is known that COVID-19 infection is associated with the increased production of pro-inflammatory cytokines [58], C-reactive protein [30], increased risk of pneumonia [58], sepsis [59], acute respiratory distress syndrome [59], and heart failure [59]. CFRs in China were 6%–10% for those with cardiovascular disease, chronic respiratory tract disease, diabetes, and hypertension [37]. Two regions hard hit by COVID-19 are regions of high air pollution in China [60] and northern Italy [61].

The possible roles of vitamin D for the clinical and epidemiological characteristics of diseases associated with the increased risk of COVID-19 CFR are given in Table 2. Most of the beneficial effects of vitamin D given in Table 2 are from observational studies of disease incidence or prevalence with respect to serum 25(OH)D concentrations. RCTs comparing outcomes for participants treated or given a placebo are preferred to establish causality related to health outcomes. However, most vitamin D RCTs have not reported that vitamin D supplementation reduced the risk of disease [62,63]. Reasons for the lack of agreement between observational studies and RCTs seems to be due to several factors, including enrolling participants with relatively high 25(OH)D concentrations and using low vitamin D doses and not measuring baseline and achieved 25(OH)D concentrations. Previous studies proposed that RCTs of nutrients such as vitamin D be based on nutrient status, such as 25(OH)D concentration, seeking to enroll participants with low values, supplementing them with enough agent to raise the concentration to values associated with good health, and measuring achieved concentrations as well as cofactors such as vitamin C, omega-3 fatty acids, and magnesium [64,65],. Two recently completed RCTs reported significantly reduced incidence in the secondary analyses for cancer [66] and diabetes mellitus [67].

Table 2. How vitamin D is related to the clinical and epidemiological findings for incidence and case-fatality rates.

Characteristics	Relation to 25(OH)D	Reference
Clinical		
Severe cases associated with pneumonia	Inverse correlation for CAP	[68,69]
Increased production of pro-inflammatory cytokines such as IL-6	Inverse correlation	[70,71]
Increased CRP	Inverse correlation	[72,73]
Increased risk of sepsis	Inverse correlation	[74,75]
Risk of ARDS	Inverse correlation	[76,77]
Risk of heart failure	Inverse correlation	[78,79]
Risk of diabetes mellitus	Inverse correlation	[67,80]
Epidemiological		
Began in December 2019 in China, spread mainly to northern midlatitude countries	Low 25(OH)D values in winter	[48,81]
Males have higher incidence and much higher CFRs than females	Smoking reduces 25(OH)D	[82]
CFR increases with age	Chronic disease rates increase with age; vitamin D plays a role in reducing risk of chronic diseases	[83]
Higher CFR for diabetics	Diabetics may have lower 25(OH)D	[84]
Higher CFR for diabetics	Lower 25(OH)D associated with increased risk of incidence	[85]
Higher CFR for hypertension	Lower 25(OH)D may be associated with increased risk of incidence	[86]
Higher CFR for cardiovascular disease	Lower 25(OH)D associated with increased risk of incidence and death	[87]
Higher CFR for chronic respiratory disease	For COPD patients, 25(OH)D inversely correlated with risk, severity, and exacerbation	[88]
Found at higher rates in regions with elevated air pollution	Air pollution associated with lower 25(OH)D concentrations	[89]

Note: 25-hydroxyvitamin D ((25(OH)D); acute respiratory distress syndrome (ARDS); community-acquired pneumonia (CAP); case-fatality rate (CFR); interleukin 6 (IL-6); chronic obstructive pulmonary disease (COPD); C-reactive protein (CRP); vitamin D deficiency (VDD).

Table 3 lists some findings for vitamin D supplementation in reducing the clinical effects of COVID-19 infection found from treating other diseases.

Table 3. How vitamin D supplementation is related to the clinical and epidemiological findings for treatment.

Clinical Characteristics	Findings from Vitamin D Supplementation Trials	Reference
Treatment of CAP with vitamin D	Did not significantly result in complete resolution. Baseline 25(OH)D was 20 ng/ml. Achieved 25(OH)D in the treatment arm was 40 ng/mL.	[90]
Increased production of pro-inflammatory cytokines such as IL-6	Reduces concentration of IL-6	[11]
Increased CRP	Reduces CRP in diabetic patients	[91]
Increased risk of sepsis	No reduction in mortality rate found for adults with sepsis supplemented with vitamin D. Most trials included participants with 25(OH)D <20 ng/mL; vitamin D_3 doses between 250 and 600 thousand IU.	[92]
Risk of ARDS	Vitamin D deficiency contributes to development of ARDS	[77,93]

Acute respiratory distress syndrome (ARDS); community-acquired pneumonia (CAP); case-fatality rate (CFR); interleukin 6 (IL-6); chronic obstructive pulmonary disease (COPD); C-reactive protein (CRP); vitamin D deficiency (VDD).

A possible reason for the monotonic increase in CFR with increasing age could be that the presence of chronic diseases increases with age. For example, the global prevalence of diabetes mellitus increases from about 1% below the age of 20 years, to ~10% at 45 years and to 19% at 65 years, decreasing to 14% by 95 years [94]. Invasive lung cancer incidence rates for females in the United States in 2015 increased from 1.1/100,000 for those aged 30–34 years, to 51.0/100,000 for those aged 50–54 years, 204.1/100,000 for those aged 65–79 years, and 347.3 for those aged 75–79 years [95]. Several studies report that people with chronic diseases have lower 25(OH)D concentrations than healthy people. A study in Italy reported that male chronic obstructive pulmonary disease patients had mean 25(OH)D concentrations of 16 (95% CI, 13–18) ng/mL, whereas female patients had concentrations of 13 (95% CI, 11–15) ng/ml [96]. A study in South Korea reported that community-acquired pneumonia (CAP) patients had a mean 25(OH)D concentration at admission of 14 ± 8 ng/mL [97]. A study in Iran reported that hypertensive patients had lower 25(OH)D concentrations than control subjects: males, 13 ± 11 vs. 21 ± 11 ng/mL; females, 13 ± 10 vs. 20 ± 11 ng/mL [98].

Another factor that affects immune response with age is reduced 1,25-dihydroxyvitamin D ($1,25(OH)_2D$, or calcitriol), the active vitamin D metabolite, with increased age. Parathyroid hormone (PTH) concentration increases with age. A U.S. study was based on 312,962 paired serum PTH and 25(OH)D concentration measurements from July 2010 to June 2011. For participants with 20-ng/mL 25(OH)D concentration, PTH increased from 27 pg/mL for those <20 years to 54 pg/mL for those >60 years [99]. Serum calcitriol concentrations are inversely related to PTH concentrations. In a study conducted in Norway on patients with a mean age of 50 (SD, 21) years, calcitriol decreased from 140 pmol/L for those aged 20–39 years to 98 pmol/L for those >80 years despite an increase in serum 25(OH)D from 24 ng/mL for those 20–39 years to 27 ng/mL for those >80 years [100].

The seasonality of many viral infections is associated with low 25(OH)D concentrations, as a result of low UVB doses owing to the winter in temperate climates and the rainy season in tropical climates—such as respiratory syncytial virus (RSV) infection [101,102],. This is the case for influenza [45, 46], and SARS-CoV [103]. However, MERS showed a peak in the April–June quarter [104], probably affected by both Hajj pilgrims gathering and the fact that 25(OH)D concentrations show little seasonal variation in the Middle East [105]. In the tropics, seasonality is related more to rainy periods (low UVB doses), for example, for influenza [106].

Considerable indirect evidence is inferred from effects found for other enveloped viruses. Table 4 presents the findings from various studies.

Table 4. Findings regarding the associations and effects of vitamin D on enveloped viral infections.

Virus	Vitamin D Effect	Reference
Dengue	Vitamin D mechanisms discussed	[107]
Dengue	Inverse association between 25(OH)D concentration and progression of disease state	[108]
Dengue	Vitamin D supplementation trial with 1000 and 4000 IU/d. 4000 IU/d resulted in higher resistance to DENV-2 infection. MDDCs from those supplemented with 4000 IU/d showed decreased mRNA expression of TLR3, 7, and 9; downregulation of IL-12/IL-8 production; and increased IL-10 secretion in response to DENV-2 infection	[29]
Hepatitis C	1,25-hydroxyvitamin-D3-24-hydroxylase, encoded by CYP24A1 gene, is a key enzyme that neutralizes 1,25(OH)$_2$D. This study found that alleles of CYP24A1 had different effects on risk of chronic hepatitis C infection.	[109]
CHB	25(OH)D concentrations were lower in CHB patients than that of healthy controls and inversely correlated with HBV viral loads	[110]
KSHV	Found that cathelicidin significantly reduced KSVH by disrupting the viral envelope.	[111]
HIV-1	Review of 29 clinical studies of vitamin D supplementation showed there was a decrease in inflammation. In 3 of 7 studies, CD4+ T cell count increased, but effect on viral load was inconclusive since most patients were on cART.	[112]
H9N2 influenza	In a lung epithelial cell study, calcitriol treatment prior to and post infection with H9N2 influenza significantly decreased expression of the influenza M gene, IL-6, and IFN-β in A549 cells, but did not affect virus replication.	[113]
RSV	Demonstrated that the human cathelicidin LL-37 has effective antiviral activity against RSV in vitro and prevented virus-induced cell death in epithelial cultures,	[114]
RSV	Performed a laboratory study that identified the mechanism by which vitamin D reduced risk of RSV.	[28]
RSV	Found that the T-allele of the vitamin D receptor has a lower prevalence in African populations and runs parallel to the lower incidence of RSV-associated severe ALRI in African children, 1 year.	[115]
Rotaviral diarrhea	Found serum 25(OH)D <20 ng/mL associated with an odds ratio of 6.3 (95% CI, 3.6 to 10.9) for rotaviral diarrhea	[116]

Note: acute respiratory tract infection (ALRI); combination Antiretroviral Therapy (cART); chronic hepatitis B (CHB); dengue virus-2 (DENV-2). Human immunodeficiency virus 1 (HIV-1); Kaposi's sarcoma-associated herpesvirus (KSHV); monocyte-derived dendritic cells (MDDCs); respiratory syncytial virus (RSV).

One way that CoVs injure the lung epithelial cells and facilitate pneumonia is through increased production of Th1-type cytokines as part of the innate immune response to viral infections, giving rise to the cytokine storm. A laboratory cell study reported that interferon γ is responsible for acute lung injury during the late phase of the SARS-CoV pathology [117].

Pro-inflammatory cytokine storms from CoV infections have resulted in the most severe cases for SARS-CoV [118] and MERS-CoV [119]. However, COVID-19 infection also initiated increased secretion of the Th2 cytokines (e.g., interleukins 4 and 10) that suppress inflammation, which differs from SARS-CoV infection [30].

3.3. Pneumonia

An example of the role of vitamin D in reducing the risk of death from pandemic respiratory tract infections is found in a study of CFRs resulting from the 1918–1919 influenza pandemic in the United States [120]. The U.S. Public Health Service conducted door-to-door surveys of 12 communities from

New Haven, Connecticut, to San Francisco, California, to ascertain incidence and CFRs. The canvasses were made as soon as possible after the autumn 1918 wave of the epidemic subsided in each locality. A total of 146,203 people, 42,920 cases, and 730 deaths were found. As shown in their Table 25, fatality rates averaged 1.70 per 100 influenza cases but averaged 25.5 per 100 cases of pneumonia. The percentage of influenza complicated by pneumonia was 6.8%. The pneumonia CFR (excluding Charles County, MD, because of inconsistencies in recording cause of death) was 28.8 per 100 for whites and 39.8 per 100 for "coloreds". As shown in Table 23, "coloreds" in the southeastern states had between a 27% and 80% higher rate of pneumonia compared to whites. As discussed in an ecological study using those CFR data, communities in the southwest had lower CFR than those in the northeast because of higher summertime and wintertime solar UVB doses [121]. Previous work suggested that higher UVB doses were associated with higher 25(OH)D concentrations, leading to reductions in the cytokine storm and the killing of bacteria and viruses that participate in pneumonia. African Americans had much higher mortality rates than white Americans for the period 1900–1948 [122]. The reasons CFRs were higher for "coloreds" than whites may include that they have higher rates of chronic diseases, are more likely to live in regions impacted by air pollution, and that with darker skin pigmentation, blacks have lower 25(OH)D concentrations. A clinical trial involving postmenopausal women living on Long Island, NY with mean baseline 25(OH)D concentration 19 ± 8 ng/mL found that supplementation with 2000 IU/d resulted in significantly fewer upper respiratory tract infections, including influenza, than a placebo or supplementation with 800 IU/d [123]. See, also, references in [11]. An analysis of serum 25(OH)D concentrations by race for 2001–2004 indicated mean 25(OH)D concentrations for people over 40 years: non-Hispanic whites, ~25–26 ng/mL; non-Hispanic blacks, 14–17 ng/mL; Mexican–Americans, 18–22 ng/mL [124]. A reason proposed for the higher mortality rates in some communities during the 1918–1919 influenza pandemic was that they were near to coal-fired electricity generating plants [125]. Recent studies have confirmed that air pollution, from combustion sources, increases the risk of influenza [126,127]. The highest concentration of these plants is in the northeast, where solar UVB doses are lowest.

A high-dose (250,000 or 500,000 IU) vitamin D_3 trial in ventilated intensive care unit patients in Georgia with mean a baseline 25(OH)D concentration of 20–22 ng/mL reported that hospital length of stay was reduced from 36 (SD, 19) days in the control group to 25 (SD, 14) days in the 250,000-IU group [25(OH)D = 45 ± 20 ng/mL] and 18 (SD, 11) days in the 500,000-IU group [25(OH)D = 55 ± 14 ng/mL]; $p = 0.03$ [93]. In a follow-on pilot trial involving 30 mechanically ventilated critically ill patients, 500,000 IU of vitamin D_3 supplementation significantly increased hemoglobin concentrations and lowered hepcidin concentrations, improving iron metabolism and the blood's ability to transport oxygen [128].

4. Recommendations

4.1. Hospital-Acquired Infections

Hospitals are a source of RTIs for both patients and medical personnel. For example, during the SARS-CoV epidemic, a woman returned to Toronto from Hong Kong with SARS-CoV in 2003 and went to a hospital. The disease was transmitted to other people, leading to an outbreak among 257 people in several Greater Toronto Area hospitals [129]. During the 2014–2015 influenza season, 36% of health care workers in a German hospital developed influenza infection [130].

Working in a hospital dealing with COVID-19 patients is associated with increased risk of COVID-19 infection. For example, 40 of 138 hospitalized COVID-19 patients in Wuhan in the Zhongnan Hospital from 1 to 28 January were medical staff, and 17 more were infected while in the hospital [58]. It was announced on February 14, 2020, that more than 1700 Chinese health workers were infected by COVID-19 and six had died (https://www.huffpost.com/entry/chinese-health-workers-infected-by-virus_n_5e46a0fec5b64d860fc97c1b).

Vitamin D supplementation to raise serum 25(OH)D concentrations can help reduce hospital-associated infections [131]. Concentrations of at least 40–50 ng/mL (100–125 nmol/L) are indicated on the basis of observational studies [132,133]. During the COVID-19 epidemic, all people in the hospital, including patients and staff, should take vitamin D supplements to raise 25(OH)D concentrations as an important step in preventing infection and spread. Trials on that hypothesis would be worth conducting.

4.2. Proposed Actions

The data reviewed here supports the role of higher 25(OH)D concentrations in reducing risk of infection and death from ARTIs, including those from influenza, CoV, and pneumonia. The peak season for ARTIs is generally when 25(OH)D concentrations are lowest. Thus, vitamin D_3 supplementation should be started or increased several months before winter to raise 25(OH)D concentrations to the range necessary to prevent ARTIs. Studies reviewed here generally reported that 25(OH)D concentrations of 20–30 ng/mL reduced the risk of ARTIs [134]. One reason for that result may be that the studies included few participants with higher 25(OH)D concentrations. However, one observational study reported that 38 ng/mL was the appropriate concentration to reduce the risk of CAP [57]. Although the degree of protection generally increases as 25(OH)D concentration increases, the optimal range appears to be in the range of 40–60 ng/mL (100–150 nmol/l). To achieve those levels, approximately half the population could take at least 2000–5000 IU/d of vitamin D_3 [135]. Various loading doses have been studied for achieving a 25(OH)D concentration of 30 ng/mL. For example, one study used a weekly or fortnightly dose totaling 100,000–200,000 IU over 8 weeks (1800 or 3600 IU/d) [136]. However, to achieve 40–60 ng/mL would take higher loading doses. A trial involving Canadian breast cancer patients with bone metastases treated with bisphosphonates but without comorbid conditions reported that doses of 10,000 IU/d of vitamin D_3 over a four-month period showed no adverse effects, but did unmask two cases of primary hyperparathyroidism [137]. A study involving 33 participants, including seven taking 4000 IU/d of vitamin D_3 and six who took 10,000 IU/d of vitamin D_3 for 8 weeks, reported that 25(OH)D concentrations increased from 20 ± 6 to 39 ± 9 for 4000 IU/d and from 19 ± 4 to 67 ± 3 for 10,000 IU/d and improved gut microbiota with no adverse effects [138]. Thus, from the literature, it is reasonable to suggest taking 10,000 IU/d for a month, which is effective in rapidly increasing circulating levels of 25(OH)D into the preferred range of 40–60 ng/mL. To maintain that level after that first month, the dose can be decreased to 5000 IU/d [135,139,140]. When high doses of vitamin D are taken, calcium supplementation should not be high to reduce risk of hypercalcemia.

A recent review suggested using vitamin D loading doses of 200,000–300,000 IU in 50,000-IU capsules to reduce the risk and severity of COVID-19 [43].

The efficacy and safety of high-dose vitamin D supplementation has been demonstrated in a psychiatric hospital in Cincinnati, Ohio [141]. The age range was from 18 to 90 years. Half of the patients were black, and nearly half were white. All patients entering since 2011 were offered supplementation of 5000 or 10,000 IU/d vitamin D_3. For 36 patients who received 5000 IU/d for 12 months or longer, mean serum 25(OH)D concentration rose from 24 to 68 ng/mL, whereas for the 78 patients who received 10,000 IU/d, mean concentrations increased from 25 to 96 ng/mL. No cases of vitamin D–induced hypercalcemia were reported. This article includes a brief review of other high-dose vitamin D studies, including the fact that vitamin D doses of 60,000–600,000 IU/d were found to treat and control such diseases as asthma, rheumatoid arthritis, rickets, and tuberculosis in the 1930s and 1940s. Those doses are much higher than the 10,000–25,000 IU/d of vitamin D_3 that can be made from solar UVB exposure [142]. However, after reports of hypercalcemia associated with use of supra-physiological doses of vitamin D surfaced, e.g., [143], high-dose vitamin D supplementation fell out of favor.

A recent article on a high-dose vitamin D supplementation trial in New Zealand involving 5110 participants reported that, over a median of 3.3 years, monthly supplementation with 100,000 IU of vitamin D_3 did not affect the incidence rate of kidney stone events or hypercalcemia [144].

Unfortunately, most countries do not have guidelines supporting vitamin D supplementation doses and desirable serum 25(OH)D concentrations that would deal with wintertime RTIs. Guidelines for many countries consider 20 ng/mL (50 nmol/L) adequate. According to the statement from the European Society for Clinical and Economic Aspects of Osteoporosis, Osteoarthritis, and Musculoskeletal Diseases, "attainment of serum 25-hydroxyvitamin D levels well above the threshold desired for bone health cannot be recommended based on current evidence, since safety has yet to be confirmed" [145]. This statement, published in 2017, is no longer correct since a number of vitamin D supplementation studies have reported that long-term vitamin D supplementation has health benefits without adverse health effects, e.g., 2000 IU/d for cancer risk reduction [66,146] and 4000 IU/d for reduced progression from prediabetes to diabetes [67].

A recent review on the status of vitamin D deficiency worldwide stated that because of inadequate evidence from clinical trials, "a 25(OH)D level of >50 nmol/L or 20 ng/mL is, therefore, the primary treatment goal, although some data suggest a benefit for a higher threshold" [147]. A companion article in the same issue of the journal stated, "although 20 ng/mL seems adequate to reduce risk of skeletal problems and ARTIs, concentrations above 30 ng/mL have been associated with reduced risk of cancer, type 2 diabetes mellitus, and adverse pregnancy and birth outcomes" [148]. However, on the basis of the findings in several studies discussed here, as well as recommendations for breast and colorectal cancer prevention [149], the desirable concentration should be at least 40–60 ng/mL.

The U.S. Institute of Medicine issued vitamin D and calcium guidelines in 2011 [150]. The institute recommended vitamin D supplementation of 600 IU/d for people younger than 70 years, 800 IU/d for those older than 70 years, and a serum 25(OH)D concentration of 20 ng/mL (50 nmol/L) or higher. That recommendation was based on the effects of vitamin D for bone health. The institute recognized that no studies had reported adverse effects of supplementation with less than 10,000 IU/d of vitamin D, but set the upper intake level at 4000 IU/d, partly out of concerns stemming from observational studies that found U-shaped 25(OH)D concentration–health outcome relationships. However, later investigation determined that most reports of J- or U-shaped relationships were from observational studies that did not measure serum 25(OH)D concentrations and that the likely reason for those relationships was a result of enrolling some participants who had started taking vitamin D supplements shortly before enrolling [151].

Moreover, in 2011, the Endocrine Society recommended supplementation of 1000–4000 IU/d of vitamin D and a serum 25(OH)D concentration of 30 ng/mL or higher [152]. Those guidelines were for patients. It appears that anyone with chronic disease should be considered in that category. The U.S. Institute of Medicine noted that no adverse effects of vitamin D supplementation had been reported for daily doses <10,000 IU/d [150].

Measuring serum 25(OH)D concentration would be useful to determine baseline and achieved 25(OH)D concentrations. A recent article recommended testing for groups of people who were likely to have low concentrations and could benefit from higher concentrations, such as pregnant women, the obese, people with chronic diseases, and the elderly [148]. Part of the rationale for testing was to increase awareness of actual 25(OH)D concentrations and the benefits of higher concentrations. In addition, increases in 25(OH)D concentration with respect to vitamin D supplementation depend on various personal factors, including genetics, digestive system health, weight, and baseline 25(OH)D concentration. For about half the population, taking 5000 IU/d of vitamin D_3 or 30,000–35,000 IU/wk would raise 25(OH)D concentration to 40 ng/mL. Taking 6235–7248 IU/d as proposed to ensure that 97.5% of the population has concentrations >20 ng/mL [153] would not exceed the 10,000-IU/d threshold.

Vitamin D supplementation is required for many individuals to reach 25(OH)D concentrations above 30 ng/mL, especially in winter [154]. However, vitamin D fortification of basic foods such as dairy and flour products [83,155] can raise serum 25(OH)D concentrations of those members of various populations with the lowest concentrations by a few ng/mL. Doing so can result in reduced risk of ARTIs for individuals with extreme vitamin D deficiency [134,156]. However, for greater benefits, daily or weekly vitamin D supplementation is recommended [134], as is the annual determination of serum 25(OH)D concentration for those with health risks [148].

Magnesium supplementation is recommended when taking vitamin D supplements. Magnesium helps activate vitamin D, which in turn helps regulate calcium and phosphate homeostasis to influence the growth and maintenance of bones. All the enzymes that metabolize vitamin D seem to require magnesium, which acts as a cofactor in the enzymatic reactions in the liver and kidneys [157]. The dose of magnesium should be in the range of 250–500 mg/d, along with twice that dose of calcium.

The hypothesis that vitamin D supplementation can reduce the risk of influenza and COVID-19 incidence and death should be investigated in trials to determine the appropriate doses, serum 25(OH)D concentrations, and the presence of any safety issues. The RCT on vitamin D supplementation for ventilated ICU patients conducted in Atlanta, Georgia, is a good model [93].

A recent review stated: "Although contradictory data exist, available evidence indicates that supplementation with multiple micronutrients with immune-supporting roles may modulate immune function and reduce the risk of infection. Micronutrients with the strongest evidence for immune support are vitamins C and D and zinc. Better design of human clinical studies addressing dosage and combinations of micronutrients in different populations are required to substantiate the benefits of micronutrient supplementation against infection." [17].

Author Contributions: Conceptualization, W.B.G., H.L. and C.A.B.; methodology, W.B.G., H.L., C.A.B.; writing—original draft preparation, W.B.G., C.A.B., H.P.B; writing—review and editing, W.B.G., H.L., S.L.M., C.A.B., C.B.F., J.L.A., and H.P.B.; supervision, W.B.G., C.A.B. All authors have read and agreed to the published version of the manuscript. Authorship must be limited to those who have contributed substantially to the work reported.

Funding: No funding was received for this study.

Conflicts of Interest: W.B.G receives funding from Bio-Tech Pharmacal, Inc. (Fayetteville, AR). H.L. sells vitamin D supplements. GrassrootsHealth works with various supplement suppliers to test the efficacy of their products in various custom projects. These suppliers may be listed as sponsors of GrassrootsHealth. H.P.B. has no conflicts of interest to declare.

References

1. Zhu, N.; Zhang, D.; Wang, W.; Li, X.; Yang, B.; Song, J.; Zhao, X.; Huang, B.; Shi, W.; Lu, R.; et al. A Novel Coronavirus from Patients with Pneumonia in China, 2019. *N. Engl. J. Med.* **2020**. [CrossRef] [PubMed]

2. Zhong, N.S.; Zheng, B.J.; Li, Y.M.; Poon, L.L.M.; Xie, Z.H.; Chan, K.H.; Li, P.H.; Tan, S.Y.; Chang, Q.; Xie, J.P.; et al. Epidemiology and cause of severe acute respiratory syndrome (SARS) in Guangdong, People's Republic of China, in February, 2003. *Lancet* **2003**, *362*, 1353–1358. [CrossRef]

3. Assiri, A.; McGeer, A.; Perl, T.M.; Price, C.S.; Al Rabeeah, A.A.; Cummings, D.A.; Alabdullatif, Z.N.; Assad, M.; Almulhim, A.; Makhdoom, H.; et al. Hospital outbreak of Middle East respiratory syndrome coronavirus. *N. Engl. J. Med.* **2013**, *369*, 407–416. [CrossRef] [PubMed]

4. Song, Z.; Xu, Y.; Bao, L.; Zhang, L.; Yu, P.; Qu, Y.; Zhu, H.; Zhao, W.; Han, Y.; Qin, C. From SARS to MERS, Thrusting Coronaviruses into the Spotlight. *Viruses* **2019**, *11*, 59. [CrossRef]

5. Yin, Y.; Wunderink, R.G. MERS, SARS and other coronaviruses as causes of pneumonia. *Respirology* **2018**, *23*, 130–137. [CrossRef]

6. Paget, J.; Spreeuwenberg, P.; Charu, V.; Taylor, R.J.; Iuliano, A.D.; Bresee, J.; Simonsen, L.; Viboud, C. Global mortality associated with seasonal influenza epidemics: New burden estimates and predictors from the GLaMOR Project. *J. Glob. Health* **2019**, *9*, 020421. [CrossRef]

7. Holick, M.F. Vitamin D deficiency. *N. Engl. J. Med.* **2007**, *357*, 266–281. [CrossRef]

8. Pike, J.W.; Christakos, S. Biology and Mechanisms of Action of the Vitamin D Hormone. *Endocrinol. Metab. Clin.* **2017**, *46*, 815–843. [CrossRef]

9. Beard, J.A.; Bearden, A.; Striker, R. Vitamin D and the anti-viral state. *J. Clin. Virol.* **2011**, *50*, 194–200. [CrossRef]

10. Hewison, M. An update on vitamin D and human immunity. *Clin. Endocrinol.* **2012**, *76*, 315–325. [CrossRef]

11. Greiller, C.L.; Martineau, A.R. Modulation of the immune response to respiratory viruses by vitamin D. *Nutrients* **2015**, *7*, 4240–4270. [CrossRef] [PubMed]

12. Wei, R.; Christakos, S. Mechanisms Underlying the Regulation of Innate and Adaptive Immunity by Vitamin D. *Nutrients* **2015**, *7*, 8251–8260. [CrossRef] [PubMed]

13. Coussens, A.K. The role of UV radiation and vitamin D in the seasonality and outcomes of infectious disease. *Photochem. Photobiol. Sci.* **2017**, *16*, 314–338. [CrossRef]

14. Lang, P.O.; Aspinall, R. Vitamin D Status and the Host Resistance to Infections: What It Is Currently (Not) Understood. *Clin. Ther.* **2017**, *39*, 930–945. [CrossRef] [PubMed]

15. Gruber-Bzura, B.M. Vitamin D and Influenza-Prevention or Therapy? *Int. J. Mol. Sci.* **2018**, *19*, 2419. [CrossRef] [PubMed]

16. Rondanelli, M.; Miccono, A.; Lamburghini, S.; Avanzato, I.; Riva, A.; Allegrini, P.; Faliva, M.A.; Peroni, G.; Nichetti, M.; Perna, S. Self-Care for Common Colds: The Pivotal Role of Vitamin D, Vitamin C, Zinc, and Echinacea in Three Main Immune Interactive Clusters (Physical Barriers, Innate and Adaptive Immunity) Involved during an Episode of Common Colds-Practical Advice on Dosages and on the Time to Take These Nutrients/Botanicals in order to Prevent or Treat Common Colds. *Evid. Based Complement. Alternat. Med.* **2018**, *2018*, 5813095. [CrossRef]

17. Gombart, A.F.; Pierre, A.; Maggini, S. A Review of Micronutrients and the Immune System-Working in Harmony to Reduce the Risk of Infection. *Nutrients* **2020**, *12*, 236. [CrossRef]

18. Schwalfenberg, G.K. A review of the critical role of vitamin D in the functioning of the immune system and the clinical implications of vitamin D deficiency. *Mol. Nutr. Food Res.* **2011**, *55*, 96–108. [CrossRef]

19. Kast, J.I.; McFarlane, A.J.; Globinska, A.; Sokolowska, M.; Wawrzyniak, P.; Sanak, M.; Schwarze, J.; Akdis, C.A.; Wanke, K. Respiratory syncytial virus infection influences tight junction integrity. *Clin. Exp. Immunol.* **2017**, *190*, 351–359. [CrossRef]

20. Chen, Y.; Leng, K.; Lu, Y.; Wen, L.; Qi, Y.; Gao, W.; Chen, H.; Bai, L.; An, X.; Sun, B.; et al. Epidemiological features and time-series analysis of influenza incidence in urban and rural areas of Shenyang, China, 2010–2018. *Epidemiol. Infect.* **2020**, *148*, e29. [CrossRef]

21. Rossi, G.A.; Fanous, H.; Colin, A.A. Viral strategies predisposing to respiratory bacterial superinfections. *Pediatr. Pulmonol.* **2020**. [CrossRef] [PubMed]

22. Liu, P.T.; Stenger, S.; Li, H.; Wenzel, L.; Tan, B.H.; Krutzik, S.R.; Ochoa, M.T.; Schauber, J.; Wu, K.; Meinken, C.; et al. Toll-like receptor triggering of a vitamin D-mediated human antimicrobial response. *Science* **2006**, *311*, 1770–1773. [CrossRef]

23. Adams, J.S.; Ren, S.; Liu, P.T.; Chun, R.F.; Lagishetty, V.; Gombart, A.F.; Borregaard, N.; Modlin, R.L.; Hewison, M. Vitamin d-directed rheostatic regulation of monocyte antibacterial responses. *J. Immunol.* **2009**, *182*, 4289–4295. [CrossRef] [PubMed]

24. Laaksi, I. Vitamin D and respiratory infection in adults. *Proc. Nutr. Soc.* **2012**, *71*, 90–97. [CrossRef] [PubMed]

25. Herr, C.; Shaykhiev, R.; Bals, R. The role of cathelicidin and defensins in pulmonary inflammatory diseases. *Expert Opin. Biol. Ther.* **2007**, *7*, 1449–1461. [CrossRef] [PubMed]

26. Agier, J.; Efenberger, M.; Brzezinska-Blaszczyk, E. Cathelicidin impact on inflammatory cells. *Cent. Eur. J. Immunol.* **2015**, *40*, 225–235. [CrossRef]

27. Barlow, P.G.; Svoboda, P.; Mackellar, A.; Nash, A.A.; York, I.A.; Pohl, J.; Davidson, D.J.; Donis, R.O. Antiviral activity and increased host defense against influenza infection elicited by the human cathelicidin LL-37. *PLoS ONE* **2011**, *6*, e25333. [CrossRef]

28. Zhao, Y.; Ran, Z.; Jiang, Q.; Hu, N.; Yu, B.; Zhu, L.; Shen, L.; Zhang, S.; Chen, L.; Chen, H.; et al. Vitamin D Alleviates Rotavirus Infection through a Microrna-155-5p Mediated Regulation of the TBK1/IRF3 Signaling Pathway In Vivo and In Vitro. *Int. J. Mol. Sci.* **2019**, *20*. [CrossRef]

29. Martinez-Moreno, J.; Hernandez, J.C.; Urcuqui-Inchima, S. Effect of high doses of vitamin D supplementation on dengue virus replication, Toll-like receptor expression, and cytokine profiles on dendritic cells. *Mol. Cell. Biochem.* **2020**, *464*, 169–180. [CrossRef]

30. Huang, C.; Wang, Y.; Li, X.; Ren, L.; Zhao, J.; Hu, Y.; Zhang, L.; Fan, G.; Xu, J.; Gu, X.; et al. Clinical features of patients infected with 2019 novel coronavirus in Wuhan, China. *Lancet* **2020**. [CrossRef]

31. Sharifi, A.; Vahedi, H.; Nedjat, S.; Rafiei, H.; Hosseinzadeh-Attar, M.J. Effect of single-dose injection of vitamin D on immune cytokines in ulcerative colitis patients: A randomized placebo-controlled trial. *APMIS* **2019**, *127*, 681–687. [CrossRef] [PubMed]

32. Cantorna, M.T. Mechanisms underlying the effect of vitamin D on the immune system. *Proc. Nutr. Soc.* **2010**, *69*, 286–289. [CrossRef] [PubMed]

33. Lemire, J.M.; Adams, J.S.; Kermani-Arab, V.; Bakke, A.C.; Sakai, R.; Jordan, S.C. 1,25-Dihydroxyvitamin D3 suppresses human T helper/inducer lymphocyte activity in vitro. *J. Immunol.* **1985**, *134*, 3032–3035. [PubMed]
34. Cantorna, M.T.; Snyder, L.; Lin, Y.D.; Yang, L. Vitamin D and 1,25(OH)2D regulation of T cells. *Nutrients* **2015**, *7*, 3011–3021. [CrossRef] [PubMed]
35. Jeffery, L.E.; Burke, F.; Mura, M.; Zheng, Y.; Qureshi, O.S.; Hewison, M.; Walker, L.S.; Lammas, D.A.; Raza, K.; Sansom, D.M. 1,25-Dihydroxyvitamin D3 and IL-2 combine to inhibit T cell production of inflammatory cytokines and promote development of regulatory T cells expressing CTLA-4 and FoxP3. *J. Immunol.* **2009**, *183*, 5458–5467. [CrossRef] [PubMed]
36. Vasarhelyi, B.; Satori, A.; Olajos, F.; Szabo, A.; Beko, G. Low vitamin D levels among patients at Semmelweis University: Retrospective analysis during a one-year period. *Orv. Hetil.* **2011**, *152*, 1272–1277. [CrossRef] [PubMed]
37. Novel, C.P.E.R.E. The epidemiological characteristics of an outbreak of 2019 novel coronavirus diseases (COVID-19) in China. *Zhonghua Liu Xing Bing Xue Za Zhi* **2020**, *41*, 145–151. [CrossRef]
38. MacLaughlin, J.; Holick, M.F. Aging decreases the capacity of human skin to produce vitamin D3. *J. Clin. Invest.* **1985**, *76*, 1536–1538. [CrossRef]
39. Grober, U.; Kisters, K. Influence of drugs on vitamin D and calcium metabolism. *Dermatoendocrinol* **2012**, *4*, 158–166. [CrossRef]
40. Lei, G.S.; Zhang, C.; Cheng, B.H.; Lee, C.H. Mechanisms of Action of Vitamin D as Supplemental Therapy for Pneumocystis Pneumonia. *Antimicrob. Agents Chemother.* **2017**, *61*. [CrossRef]
41. Mousavi, S.; Bereswill, S.; Heimesaat, M.M. Immunomodulatory and Antimicrobial Effects of Vitamin C. *Eur. J. Microbiol. Immunol.* **2019**, *9*, 73–79. [CrossRef] [PubMed]
42. Colunga Biancatelli, R.M.L.; Berrill, M.; Marik, P.E. The antiviral properties of vitamin C. *Expert Rev. Anti Infect. Ther.* **2020**, *18*, 99–101. [CrossRef] [PubMed]
43. Wimalawansa, S.J. Global epidemic of coronavirus–COVID-19: What we can do to minimze risksl. *Eur. J. Biomed. Pharm. Sci.* **2020**, *7*, 432–438.
44. Kalil, A.C.; Thomas, P.G. Influenza virus-related critical illness: Pathophysiology and epidemiology. *Crit. Care* **2019**, *23*, 258. [CrossRef]
45. Hope-Simpson, R.E. The role of season in the epidemiology of influenza. *J. Hyg.* **1981**, *86*, 35–47. [CrossRef]
46. Cannell, J.J.; Vieth, R.; Umhau, J.C.; Holick, M.F.; Grant, W.B.; Madronich, S.; Garland, C.F.; Giovannucci, E. Epidemic influenza and vitamin D. *Epidemiol. Infect.* **2006**, *134*, 1129–1140. [CrossRef]
47. Cannell, J.J.; Zasloff, M.; Garland, C.F.; Scragg, R.; Giovannucci, E. On the epidemiology of influenza. *Virol. J.* **2008**, *5*, 29. [CrossRef]
48. Kroll, M.H.; Bi, C.; Garber, C.C.; Kaufman, H.W.; Liu, D.; Caston-Balderrama, A.; Zhang, K.; Clarke, N.; Xie, M.; Reitz, R.E.; et al. Temporal relationship between vitamin D status and parathyroid hormone in the United States. *PLoS ONE* **2015**, *10*, e0118108. [CrossRef]
49. Lowen, A.C.; Mubareka, S.; Steel, J.; Palese, P. Influenza virus transmission is dependent on relative humidity and temperature. *PLoS Pathog.* **2007**, *3*, 1470–1476. [CrossRef]
50. Shaman, J.; Kohn, M. Absolute humidity modulates influenza survival, transmission, and seasonality. *Proc. Natl. Acad. Sci. USA* **2009**, *106*, 3243–3248. [CrossRef]
51. Shaman, J.; Pitzer, V.E.; Viboud, C.; Grenfell, B.T.; Lipsitch, M. Absolute humidity and the seasonal onset of influenza in the continental United States. *PLoS Biol.* **2010**, *8*, e1000316. [CrossRef]
52. Urashima, M.; Segawa, T.; Okazaki, M.; Kurihara, M.; Wada, Y.; Ida, H. Randomized trial of vitamin D supplementation to prevent seasonal influenza A in schoolchildren. *Am. J. Clin. Nutr.* **2010**, *91*, 1255–1260. [CrossRef] [PubMed]
53. Zhou, J.; Du, J.; Huang, L.; Wang, Y.; Shi, Y.; Lin, H. Preventive Effects of Vitamin D on Seasonal Influenza A in Infants: A Multicenter, Randomized, Open, Controlled Clinical Trial. *Pediatr. Infect. Dis. J.* **2018**, *37*, 749–754. [CrossRef] [PubMed]
54. Urashima, M.; Mezawa, H.; Noya, M.; Camargo, C.A., Jr. Effects of vitamin D supplements on influenza A illness during the 2009 H1N1 pandemic: A randomized controlled trial. *Food Funct.* **2014**, *5*, 2365–2370. [CrossRef] [PubMed]

55. Arihiro, S.; Nakashima, A.; Matsuoka, M.; Suto, S.; Uchiyama, K.; Kato, T.; Mitobe, J.; Komoike, N.; Itagaki, M.; Miyakawa, Y.; et al. Randomized Trial of Vitamin D Supplementation to Prevent Seasonal Influenza and Upper Respiratory Infection in Patients With Inflammatory Bowel Disease. *Inflamm. Bowel Dis.* **2019**, *25*, 1088–1095. [CrossRef] [PubMed]

56. Loeb, M.; Dang, A.D.; Thiem, V.D.; Thanabalan, V.; Wang, B.; Nguyen, N.B.; Tran, H.T.M.; Luong, T.M.; Singh, P.; Smieja, M.; et al. Effect of Vitamin D supplementation to reduce respiratory infections in children and adolescents in Vietnam: A randomized controlled trial. *Influenza Other Respir. Viruses* **2019**, *13*, 176–183. [CrossRef] [PubMed]

57. Sabetta, J.R.; DePetrillo, P.; Cipriani, R.J.; Smardin, J.; Burns, L.A.; Landry, M.L. Serum 25-hydroxyvitamin d and the incidence of acute viral respiratory tract infections in healthy adults. *PLoS ONE* **2010**, *5*, e11088. [CrossRef]

58. Wang, D.; Hu, B.; Hu, C.; Zhu, F.; Liu, X.; Zhang, J.; Wang, B.; Xiang, H.; Cheng, Z.; Xiong, Y.; et al. Clinical Characteristics of 138 Hospitalized Patients With 2019 Novel Coronavirus-Infected Pneumonia in Wuhan, China. *JAMA* **2020**. [CrossRef]

59. Zhou, F.; Yu, T.; Du, R.; Fan, G.; Liu, Y.; Liu, Z.; Xiang, J.; Wang, Y.; Song, B.; Gu, X.; et al. Clinical course and risk factors for mortality of adult inpatients with COVID-19 in Wuhan, China: A retrospective cohort study. *Lancet* **2020**. [CrossRef]

60. He, Q.; Gu, Y.; Zhang, M. Spatiotemporal trends of PM2.5 concentrations in central China from 2003 to 2018 based on MAIAC-derived high-resolution data. *Environ. Int.* **2020**, *137*, 105536. [CrossRef]

61. Longhin, E.; Holme, J.A.; Gualtieri, M.; Camatini, M.; Ovrevik, J. Milan winter fine particulate matter (wPM2.5) induces IL-6 and IL-8 synthesis in human bronchial BEAS-2B cells, but specifically impairs IL-8 release. *Toxicol. In Vitro* **2018**, *52*, 365–373. [CrossRef] [PubMed]

62. Autier, P.; Mullie, P.; Macacu, A.; Dragomir, M.; Boniol, M.; Coppens, K.; Pizot, C.; Boniol, M. Effect of vitamin D supplementation on non-skeletal disorders: A systematic review of meta-analyses and randomised trials. *Lancet Diabetes Endocrinol.* **2017**, *5*, 986–1004. [CrossRef]

63. Rejnmark, L.; Bislev, L.S.; Cashman, K.D.; Eiriksdottir, G.; Gaksch, M.; Grubler, M.; Grimnes, G.; Gudnason, V.; Lips, P.; Pilz, S.; et al. Non-skeletal health effects of vitamin D supplementation: A systematic review on findings from meta-analyses summarizing trial data. *PLoS ONE* **2017**, *12*, e0180512. [CrossRef] [PubMed]

64. Heaney, R.P. Guidelines for optimizing design and analysis of clinical studies of nutrient effects. *Nutr. Rev.* **2014**, *72*, 48–54. [CrossRef] [PubMed]

65. Grant, W.B.; Boucher, B.J.; Bhattoa, H.P.; Lahore, H. Why vitamin D clinical trials should be based on 25-hydroxyvitamin D concentrations. *J. Steroid Biochem. Mol. Biol.* **2018**, *177*, 266–269. [CrossRef]

66. Manson, J.E.; Cook, N.R.; Lee, I.M.; Christen, W.; Bassuk, S.S.; Mora, S.; Gibson, H.; Gordon, D.; Copeland, T.; D'Agostino, D.; et al. Vitamin D Supplements and Prevention of Cancer and Cardiovascular Disease. *N. Engl. J. Med.* **2019**, *380*, 33–44. [CrossRef]

67. Pittas, A.G.; Dawson-Hughes, B.; Sheehan, P.; Ware, J.H.; Knowler, W.C.; Aroda, V.R.; Brodsky, I.; Ceglia, L.; Chadha, C.; Chatterjee, R.; et al. Vitamin D Supplementation and Prevention of Type 2 Diabetes. *N. Engl. J. Med.* **2019**, *381*, 520–530. [CrossRef]

68. Lu, D.; Zhang, J.; Ma, C.; Yue, Y.; Zou, Z.; Yu, C.; Yin, F. Link between community-acquired pneumonia and vitamin D levels in older patients. *Z. Gerontol. Geriatr.* **2018**, *51*, 435–439. [CrossRef]

69. Zhou, Y.F.; Luo, B.A.; Qin, L.L. The association between vitamin D deficiency and community-acquired pneumonia: A meta-analysis of observational studies. *Medicine* **2019**, *98*, e17252. [CrossRef]

70. Manion, M.; Hullsiek, K.H.; Wilson, E.M.P.; Rhame, F.; Kojic, E.; Gibson, D.; Hammer, J.; Patel, P.; Brooks, J.T.; Baker, J.V.; et al. Vitamin D deficiency is associated with IL-6 levels and monocyte activation in HIV-infected persons. *PLoS ONE* **2017**, *12*, e0175517. [CrossRef]

71. Dalvi, S.M.; Ramraje, N.N.; Patil, V.W.; Hegde, R.; Yeram, N. Study of IL-6 and vitamin D3 in patients of pulmonary tuberculosis. *Indian J. Tuberc.* **2019**, *66*, 337–345. [CrossRef] [PubMed]

72. Poudel-Tandukar, K.; Poudel, K.C.; Jimba, M.; Kobayashi, J.; Johnson, C.A.; Palmer, P.H. Serum 25-hydroxyvitamin d levels and C-reactive protein in persons with human immunodeficiency virus infection. *AIDS Res. Hum. Retrovir.* **2013**, *29*, 528–534. [CrossRef] [PubMed]

73. Zhang, M.; Gao, Y.; Tian, L.; Zheng, L.; Wang, X.; Liu, W.; Zhang, Y.; Huang, G. Association of serum 25-hydroxyvitamin D3 with adipokines and inflammatory marker in persons with prediabetes mellitus. *Clin. Chim. Acta* **2017**, *468*, 152–158. [CrossRef] [PubMed]

74. Zhou, W.; Mao, S.; Wu, L.; Yu, J. Association Between Vitamin D Status and Sepsis. *Clin. Lab.* **2018**, *64*, 451–460. [CrossRef] [PubMed]

75. Li, Y.; Ding, S. Serum 25-Hydroxyvitamin D and the risk of mortality in adult patients with Sepsis: A meta-analysis. *BMC Infect. Dis.* **2020**, *20*, 189. [CrossRef] [PubMed]

76. Thickett, D.R.; Moromizato, T.; Litonjua, A.A.; Amrein, K.; Quraishi, S.A.; Lee-Sarwar, K.A.; Mogensen, K.M.; Purtle, S.W.; Gibbons, F.K.; Camargo, C.A., Jr.; et al. Association between prehospital vitamin D status and incident acute respiratory failure in critically ill patients: A retrospective cohort study. *BMJ Open Respir. Res.* **2015**, *2*, e000074. [CrossRef]

77. Dancer, R.C.; Parekh, D.; Lax, S.; D'Souza, V.; Zheng, S.; Bassford, C.R.; Park, D.; Bartis, D.G.; Mahida, R.; Turner, A.M.; et al. Vitamin D deficiency contributes directly to the acute respiratory distress syndrome (ARDS). *Thorax* **2015**, *70*, 617–624. [CrossRef]

78. Hou, Y.M.; Zhao, J.Y.; Liu, H.Y. Impact of serum 25-hydroxyvitamin D on cardiac prognosis in Chinese patients with heart failure. *Br. J. Nutr.* **2019**, *122*, 162–171. [CrossRef]

79. Aparicio-Ugarriza, R.; Salguero, D.; Mohammed, Y.N.; Ferri-Guerra, J.; Baskaran, D.J.; Mirabbasi, S.A.; Rodriguez, A.; Ruiz, J.G. Is vitamin D deficiency related to a higher risk of hospitalization and mortality in veterans with heart failure? *Maturitas* **2020**, *132*, 30–34. [CrossRef]

80. McDonnell, S.L.; Baggerly, L.L.; French, C.B.; Heaney, R.P.; Gorham, E.D.; Holick, M.F.; Scragg, R.; Garland, C.F. Incidence rate of type 2 diabetes is >50% lower in GrassrootsHealth cohort with median serum 25-hydroxyvitamin D of 41 ng/mL than in NHANES cohort with median of 22 ng/mL. *J. Steroid Biochem. Mol. Biol.* **2016**, *155*, 239–244. [CrossRef]

81. Xie, Z.; Xia, W.; Zhang, Z.; Wu, W.; Lu, C.; Tao, S.; Wu, L.; Gu, J.; Chandler, J.; Peter, S.; et al. Prevalence of Vitamin D Inadequacy Among Chinese Postmenopausal Women: A Nationwide, Multicenter, Cross-Sectional Study. *Front. Endocrinol.* **2019**, *9*, 782. [CrossRef] [PubMed]

82. Brot, C.; Jorgensen, N.R.; Sorensen, O.H. The influence of smoking on vitamin D status and calcium metabolism. *Eur. J. Clin. Nutr.* **1999**, *53*, 920–926. [CrossRef]

83. Pilz, S.; Marz, W.; Cashman, K.D.; Kiely, M.E.; Whiting, S.J.; Holick, M.F.; Grant, W.B.; Pludowski, P.; Hiligsmann, M.; Trummer, C.; et al. Rationale and Plan for Vitamin D Food Fortification: A Review and Guidance Paper. *Front. Endocrinol.* **2018**, *9*, 373. [CrossRef] [PubMed]

84. Chen, X.; Wu, W.; Wang, L.; Shi, Y.; Shen, F.; Gu, X.; Jia, Z. Association Between 25-Hydroxyvitamin D and Epicardial Adipose Tissue in Chinese Non-Obese Patients with Type 2 Diabetes. *Med. Sci. Monit.* **2017**, *23*, 4304–4311. [CrossRef] [PubMed]

85. Lucato, P.; Solmi, M.; Maggi, S.; Bertocco, A.; Bano, G.; Trevisan, C.; Manzato, E.; Sergi, G.; Schofield, P.; Kouidrat, Y.; et al. Low vitamin D levels increase the risk of type 2 diabetes in older adults: A systematic review and meta-analysis. *Maturitas* **2017**, *100*, 8–15. [CrossRef]

86. Qi, D.; Nie, X.L.; Wu, S.; Cai, J. Vitamin D and hypertension: Prospective study and meta-analysis. *PLoS ONE* **2017**, *12*, e0174298. [CrossRef]

87. Gholami, F.; Moradi, G.; Zareei, B.; Rasouli, M.A.; Nikkhoo, B.; Roshani, D.; Ghaderi, E. The association between circulating 25-hydroxyvitamin D and cardiovascular diseases: A meta-analysis of prospective cohort studies. *BMC Cardiovasc. Disord.* **2019**, *19*, 248. [CrossRef]

88. Zhu, M.; Wang, T.; Wang, C.; Ji, Y. The association between vitamin D and COPD risk, severity, and exacerbation: An updated systematic review and meta-analysis. *Int. J. Chron. Obstruct. Pulmon. Dis.* **2016**, *11*, 2597–2607. [CrossRef]

89. Hoseinzadeh, E.; Taha, P.; Wei, C.; Godini, H.; Ashraf, G.M.; Taghavi, M.; Miri, M. The impact of air pollutants, UV exposure and geographic location on vitamin D deficiency. *Food Chem. Toxicol.* **2018**, *113*, 241–254. [CrossRef]

90. Slow, S.; Epton, M.; Storer, M.; Thiessen, R.; Lim, S.; Wong, J.; Chin, P.; Tovaranonte, P.; Pearson, J.; Chambers, S.T.; et al. Effect of adjunctive single high-dose vitamin D3 on outcome of community-acquired pneumonia in hospitalised adults: The VIDCAPS randomised controlled trial. *Sci. Rep.* **2018**, *8*, 13829. [CrossRef]

91. Mirzavandi, F.; Talenezhad, N.; Razmpoosh, E.; Nadjarzadeh, A.; Mozaffari-Khosravi, H. The effect of intramuscular megadose of vitamin D injections on E-selectin, CRP and biochemical parameters in vitamin D-deficient patients with type-2 diabetes mellitus: A randomized controlled trial. *Complement. Ther. Med.* **2020**, *49*, 102346. [CrossRef] [PubMed]

92. Amrein, K.; Papinutti, A.; Mathew, E.; Vila, G.; Parekh, D. Vitamin D and critical illness: What endocrinology can learn from intensive care and vice versa. *Endocr. Connect.* **2018**, *7*, R304–R315. [CrossRef] [PubMed]

93. Han, J.E.; Jones, J.L.; Tangpricha, V.; Brown, M.A.; Brown, L.A.S.; Hao, L.; Hebbar, G.; Lee, M.J.; Liu, S.; Ziegler, T.R.; et al. High Dose Vitamin D Administration in Ventilated Intensive Care Unit Patients: A Pilot Double Blind Randomized Controlled Trial. *J. Clin. Transl. Endocrinol.* **2016**, *4*, 59–65. [CrossRef] [PubMed]

94. Cho, N.H.; Shaw, J.E.; Karuranga, S.; Huang, Y.; da Rocha Fernandes, J.D.; Ohlrogge, A.W.; Malanda, B. IDF Diabetes Atlas: Global estimates of diabetes prevalence for 2017 and projections for 2045. *Diabetes Res. Clin. Pract.* **2018**, *138*, 271–281. [CrossRef]

95. Henley, S.J.; Gallaway, S.; Singh, S.D.; O'Neil, M.E.; Buchanan Lunsford, N.; Momin, B.; Richards, T.B. Lung Cancer Among Women in the United States. *J. Womens Health* **2018**, *27*, 1307–1316. [CrossRef]

96. Malinovschi, A.; Masoero, M.; Bellocchia, M.; Ciuffreda, A.; Solidoro, P.; Mattei, A.; Mercante, L.; Heffler, E.; Rolla, G.; Bucca, C. Severe vitamin D deficiency is associated with frequent exacerbations and hospitalization in COPD patients. *Respir. Res.* **2014**, *15*, 131. [CrossRef]

97. Kim, H.J.; Jang, J.G.; Hong, K.S.; Park, J.K.; Choi, E.Y. Relationship between serum vitamin D concentrations and clinical outcome of community-acquired pneumonia. *Int. J. Tuberc. Lung Dis.* **2015**, *19*, 729–734. [CrossRef]

98. Naghshtabrizi, B.; Borzouei, S.; Bigvand, P.; Seifrabiei, M.A. Evaluation of the Relationship between Serum 25-Hydroxy Vitamin D and Hypertension in Hamadan, Iran-A Case Control Study. *J. Clin. Diagn. Res.* **2017**, *11*, LC01–LC03. [CrossRef]

99. Valcour, A.; Blocki, F.; Hawkins, D.M.; Rao, S.D. Effects of age and serum 25-OH-vitamin D on serum parathyroid hormone levels. *J. Clin. Endocrinol. Metab.* **2012**, *97*, 3989–3995. [CrossRef]

100. Christensen, M.H.; Lien, E.A.; Hustad, S.; Almas, B. Seasonal and age-related differences in serum 25-hydroxyvitamin D, 1,25-dihydroxyvitamin D and parathyroid hormone in patients from Western Norway. *Scand. J. Clin. Lab. Invest.* **2010**, *70*, 281–286. [CrossRef]

101. Nam, H.H.; Ison, M.G. Respiratory syncytial virus infection in adults. *BMJ* **2019**, *366*, l5021. [CrossRef] [PubMed]

102. Paynter, S.; Ware, R.S.; Sly, P.D.; Weinstein, P.; Williams, G. Respiratory syncytial virus seasonality in tropical Australia. *Aust. N. Z. J. Public Health* **2015**, *39*, 8–10. [CrossRef] [PubMed]

103. Feng, X.; Guo, T.; Wang, Y.; Kang, D.; Che, X.; Zhang, H.; Cao, W.; Wang, P. The vitamin D status and its effects on life quality among the elderly in Jinan, China. *Arch. Gerontol. Geriatr.* **2016**, *62*, 26–29. [CrossRef] [PubMed]

104. Nassar, M.S.; Bakhrebah, M.A.; Meo, S.A.; Alsuabeyl, M.S.; Zaher, W.A. Global seasonal occurrence of middle east respiratory syndrome coronavirus (MERS-CoV) infection. *Eur. Rev. Med. Pharmacol. Sci.* **2018**, *22*, 3913–3918. [CrossRef] [PubMed]

105. Grant, W.B.; Fakhoury, H.M.A.; Karras, S.N.; Al Anouti, F.; Bhattoa, H.P. Variations in 25-Hydroxyvitamin D in Countries from the Middle East and Europe: The Roles of UVB Exposure and Diet. *Nutrients* **2019**, *11*, 2065. [CrossRef] [PubMed]

106. Moura, F.E.; Perdigao, A.C.; Siqueira, M.M. Seasonality of influenza in the tropics: A distinct pattern in northeastern Brazil. *Am. J. Trop. Med. Hyg.* **2009**, *81*, 180–183. [CrossRef] [PubMed]

107. Arboleda, J.F.; Urcuqui-Inchima, S. Vitamin D-Regulated MicroRNAs: Are They Protective Factors against Dengue Virus Infection? *Adv. Virol.* **2016**, *2016*, 1016840. [CrossRef]

108. Villamor, E.; Villar, L.A.; Lozano, A.; Herrera, V.M.; Herran, O.F. Vitamin D serostatus and dengue fever progression to dengue hemorrhagic fever/dengue shock syndrome. *Epidemiol. Infect.* **2017**, *145*, 2961–2970. [CrossRef]

109. Fan, H.Z.; Zhang, R.; Tian, T.; Zhong, Y.L.; Wu, M.P.; Xie, C.N.; Yang, J.J.; Huang, P.; Yu, R.B.; Zhang, Y.; et al. CYP24A1 genetic variants in the vitamin D metabolic pathway are involved in the outcomes of hepatitis C virus infection among high-risk Chinese population. *Int. J. Infect. Dis.* **2019**, *84*, 80–88. [CrossRef]

110. Hu, Y.C.; Wang, W.W.; Jiang, W.Y.; Li, C.Q.; Guo, J.C.; Xun, Y.H. Low vitamin D levels are associated with high viral loads in patients with chronic hepatitis B: A systematic review and meta-analysis. *BMC Gastroenterol.* **2019**, *19*, 84. [CrossRef]

111. Brice, D.C.; Toth, Z.; Diamond, G. LL-37 disrupts the Kaposi's sarcoma-associated herpesvirus envelope and inhibits infection in oral epithelial cells. *Antivir. Res.* **2018**, *158*, 25–33. [CrossRef] [PubMed]

112. Alvarez, N.; Aguilar-Jimenez, W.; Rugeles, M.T. The Potential Protective Role of Vitamin D Supplementation on HIV-1 Infection. *Front. Immunol.* **2019**, *10*, 2291. [CrossRef] [PubMed]

113. Gui, B.; Chen, Q.; Hu, C.; Zhu, C.; He, G. Effects of calcitriol (1, 25-dihydroxy-vitamin D3) on the inflammatory response induced by H9N2 influenza virus infection in human lung A549 epithelial cells and in mice. *Virol. J.* **2017**, *14*, 10. [CrossRef] [PubMed]

114. Currie, S.M.; Findlay, E.G.; McHugh, B.J.; Mackellar, A.; Man, T.; Macmillan, D.; Wang, H.; Fitch, P.M.; Schwarze, J.; Davidson, D.J. The human cathelicidin LL-37 has antiviral activity against respiratory syncytial virus. *PLoS ONE* **2013**, *8*, e73659. [CrossRef]

115. Laplana, M.; Royo, J.L.; Fibla, J. Vitamin D Receptor polymorphisms and risk of enveloped virus infection: A meta-analysis. *Gene* **2018**, *678*, 384–394. [CrossRef]

116. Bucak, I.H.; Ozturk, A.B.; Almis, H.; Cevik, M.O.; Tekin, M.; Konca, C.; Turgut, M.; Bulbul, M. Is there a relationship between low vitamin D and rotaviral diarrhea? *Pediatr. Int.* **2016**, *58*, 270–273. [CrossRef]

117. Theron, M.; Huang, K.J.; Chen, Y.W.; Liu, C.C.; Lei, H.Y. A probable role for IFN-gamma in the development of a lung immunopathology in SARS. *Cytokine* **2005**, *32*, 30–38. [CrossRef]

118. Wong, C.K.; Lam, C.W.; Wu, A.K.; Ip, W.K.; Lee, N.L.; Chan, I.H.; Lit, L.C.; Hui, D.S.; Chan, M.H.; Chung, S.S.; et al. Plasma inflammatory cytokines and chemokines in severe acute respiratory syndrome. *Clin. Exp. Immunol.* **2004**, *136*, 95–103. [CrossRef]

119. Mahallawi, W.H.; Khabour, O.F.; Zhang, Q.; Makhdoum, H.M.; Suliman, B.A. MERS-CoV infection in humans is associated with a pro-inflammatory Th1 and Th17 cytokine profile. *Cytokine* **2018**, *104*, 8–13. [CrossRef]

120. Britten, R. The incidence of epidemic influenza, 1918-19. *Public Health Rep.* **1932**, *47*, 303–339. [CrossRef]

121. Grant, W.B.; Giovannucci, E. The possible roles of solar ultraviolet-B radiation and vitamin D in reducing case-fatality rates from the 1918–1919 influenza pandemic in the United States. *Dermatoendocrinol* **2009**, *1*, 215–219. [CrossRef] [PubMed]

122. Feigenbaum, J.J.; Muller, C.; Wrigley-Field, E. Regional and Racial Inequality in Infectious Disease Mortality in U.S. Cities, 1900–1948. *Demography* **2019**, *56*, 1371–1388. [CrossRef] [PubMed]

123. Aloia, J.F.; Li-Ng, M. Re: Epidemic influenza and vitamin D. *Epidemiol. Infect.* **2007**, *135*, 1095–1096, author reply 1097–1098. [CrossRef] [PubMed]

124. Ginde, A.A.; Liu, M.C.; Camargo, C.A., Jr. Demographic differences and trends of vitamin D insufficiency in the US population, 1988–2004. *Arch. Intern. Med.* **2009**, *169*, 626–632. [CrossRef]

125. Clay, K.; Lewis, J.; Severnini, E. What explains cross-city variation in mortality during the 1918 influenza pandemic? Evidence from 438 U.S. cities. *Econ. Hum. Biol.* **2019**, *35*, 42–50. [CrossRef] [PubMed]

126. Liu, X.X.; Li, Y.; Qin, G.; Zhu, Y.; Li, X.; Zhang, J.; Zhao, K.; Hu, M.; Wang, X.L.; Zheng, X. Effects of air pollutants on occurrences of influenza-like illness and laboratory-confirmed influenza in Hefei, China. *Int. J. Biometeorol.* **2019**, *63*, 51–60. [CrossRef]

127. Croft, D.P.; Zhang, W.; Lin, S.; Thurston, S.W.; Hopke, P.K.; van Wijngaarden, E.; Squizzato, S.; Masiol, M.; Utell, M.J.; Rich, D.Q. Associations between Source-Specific Particulate Matter and Respiratory Infections in New York State Adults. *Environ. Sci. Technol.* **2020**, *54*, 975–984. [CrossRef]

128. Smith, E.M.; Jones, J.L.; Han, J.E.; Alvarez, J.A.; Sloan, J.H.; Konrad, R.J.; Zughaier, S.M.; Martin, G.S.; Ziegler, T.R.; Tangpricha, V. High-Dose Vitamin D3 Administration Is Associated With Increases in Hemoglobin Concentrations in Mechanically Ventilated Critically Ill Adults: A Pilot Double-Blind, Randomized, Placebo-Controlled Trial. *JPEN J. Parenter. Enter. Nutr.* **2018**, *42*, 87–94. [CrossRef]

129. Centers for Disease Control and Prevention. Update: Severe acute respiratory syndrome–Toronto, Canada, 2003. *MMWR Morb. Mortal. Wkly. Rep.* **2003**, *52*, 547–550.

130. Hagel, S.; Ludewig, K.; Moeser, A.; Baier, M.; Loffler, B.; Schleenvoigt, B.; Forstner, C.; Pletz, M.W. Characteristics and management of patients with influenza in a German hospital during the 2014/2015 influenza season. *Infection* **2016**, *44*, 667–672. [CrossRef]

131. Youssef, D.A.; Ranasinghe, T.; Grant, W.B.; Peiris, A.N. Vitamin D's potential to reduce the risk of hospital-acquired infections. *Derm. Endocrinol.* **2012**, *4*, 167–175. [CrossRef] [PubMed]

132. Quraishi, S.A.; Bittner, E.A.; Blum, L.; Hutter, M.M.; Camargo, C.A., Jr. Association between preoperative 25-hydroxyvitamin D level and hospital-acquired infections following Roux-en-Y gastric bypass surgery. *JAMA Surg.* **2014**, *149*, 112–118. [CrossRef] [PubMed]

133. Laviano, E.; Sanchez Rubio, M.; Gonzalez-Nicolas, M.T.; Palacian, M.P.; Lopez, J.; Gilaberte, Y.; Calmarza, P.; Rezusta, A.; Serrablo, A. Association between preoperative levels of 25-hydroxyvitamin D and hospital-acquired infections after hepatobiliary surgery: A prospective study in a third-level hospital. *PLoS ONE* **2020**, *15*, e0230336. [CrossRef] [PubMed]

134. Martineau, A.R.; Jolliffe, D.A.; Hooper, R.L.; Greenberg, L.; Aloia, J.F.; Bergman, P.; Dubnov-Raz, G.; Esposito, S.; Ganmaa, D.; Ginde, A.A.; et al. Vitamin D supplementation to prevent acute respiratory tract infections: Systematic review and meta-analysis of individual participant data. *BMJ* **2017**, *356*, i6583. [CrossRef] [PubMed]

135. Heaney, R.P.; Davies, K.M.; Chen, T.C.; Holick, M.F.; Barger-Lux, M.J. Human serum 25-hydroxycholecalciferol response to extended oral dosing with cholecalciferol. *Am. J. Clin. Nutr.* **2003**, *77*, 204–210. [CrossRef] [PubMed]

136. van Groningen, L.; Opdenoordt, S.; van Sorge, A.; Telting, D.; Giesen, A.; de Boer, H. Cholecalciferol loading dose guideline for vitamin D-deficient adults. *Eur. J. Endocrinol.* **2010**, *162*, 805–811. [CrossRef]

137. Amir, E.; Simmons, C.E.; Freedman, O.C.; Dranitsaris, G.; Cole, D.E.; Vieth, R.; Ooi, W.S.; Clemons, M. A phase 2 trial exploring the effects of high-dose (10,000 IU/day) vitamin D(3) in breast cancer patients with bone metastases. *Cancer* **2010**, *116*, 284–291. [CrossRef]

138. Charoenngam, N.; Shirvani, A.; Kalajian, T.A.; Song, A.; Holick, M.F. The Effect of Various Doses of Oral Vitamin D3 Supplementation on Gut Microbiota in Healthy Adults: A Randomized, Double-blinded, Dose-response Study. *Anticancer Res.* **2020**, *40*, 551–556. [CrossRef]

139. Ekwaru, J.P.; Zwicker, J.D.; Holick, M.F.; Giovannucci, E.; Veugelers, P.J. The importance of body weight for the dose response relationship of oral vitamin D supplementation and serum 25-hydroxyvitamin D in healthy volunteers. *PLoS ONE* **2014**, *9*, e111265. [CrossRef]

140. Shirvani, A.; Kalajian, T.A.; Song, A.; Holick, M.F. Disassociation of Vitamin D's Calcemic Activity and Non-calcemic Genomic Activity and Individual Responsiveness: A Randomized Controlled Double-Blind Clinical Trial. *Sci. Rep.* **2019**, *9*, 17685. [CrossRef]

141. McCullough, P.J.; Lehrer, D.S.; Amend, J. Daily oral dosing of vitamin D3 using 5000 TO 50,000 international units a day in long-term hospitalized patients: Insights from a seven year experience. *J. Steroid Biochem. Mol. Biol.* **2019**, *189*, 228–239. [CrossRef] [PubMed]

142. Holick, M.F. Environmental factors that influence the cutaneous production of vitamin D. *Am. J. Clin. Nutr.* **1995**, *61*, 638S–645S. [CrossRef] [PubMed]

143. Howard, J.E.; Meyer, R.J. Intoxication with vitamin D. *J. Clin. Endocrinol. Metab.* **1948**, *8*, 895–910. [CrossRef] [PubMed]

144. Malihi, Z.; Lawes, C.M.M.; Wu, Z.; Huang, Y.; Waayer, D.; Toop, L.; Khaw, K.T.; Camargo, C.A.; Scragg, R. Monthly high-dose vitamin D supplementation does not increase kidney stone risk or serum calcium: Results from a randomized controlled trial. *Am. J. Clin. Nutr.* **2019**, *109*, 1578–1587. [CrossRef]

145. Cianferotti, L.; Bertoldo, F.; Bischoff-Ferrari, H.A.; Bruyere, O.; Cooper, C.; Cutolo, M.; Kanis, J.A.; Kaufman, J.M.; Reginster, J.Y.; Rizzoli, R.; et al. Vitamin D supplementation in the prevention and management of major chronic diseases not related to mineral homeostasis in adults: Research for evidence and a scientific statement from the European society for clinical and economic aspects of osteoporosis and osteoarthritis (ESCEO). *Endocrine* **2017**, *56*, 245–261. [CrossRef]

146. Lappe, J.; Watson, P.; Travers-Gustafson, D.; Recker, R.; Garland, C.; Gorham, E.; Baggerly, K.; McDonnell, S.L. Effect of Vitamin D and Calcium Supplementation on Cancer Incidence in Older Women: A Randomized Clinical Trial. *JAMA* **2017**, *317*, 1234–1243. [CrossRef]

147. Amrein, K.; Scherkl, M.; Hoffmann, M.; Neuwersch-Sommeregger, S.; Kostenberger, M.; Tmava Berisha, A.; Martucci, G.; Pilz, S.; Malle, O. Vitamin D deficiency 2.0: An update on the current status worldwide. *Eur. J. Clin. Nutr.* **2020**. [CrossRef]

148. Grant, W.B.; Al Anouti, F.; Moukayed, M. Targeted 25-hydroxyvitamin D concentration measurements and vitamin D3 supplementation can have important patient and public health benefits. *Eur. J. Clin. Nutr.* **2020**. [CrossRef]

149. Garland, C.F.; Gorham, E.D.; Mohr, S.B.; Garland, F.C. Vitamin D for cancer prevention: Global perspective. *Ann. Epidemiol.* **2009**, *19*, 468–483. [CrossRef]

150. Ross, A.C.; Manson, J.E.; Abrams, S.A.; Aloia, J.F.; Brannon, P.M.; Clinton, S.K.; Durazo-Arvizu, R.A.; Gallagher, J.C.; Gallo, R.L.; Jones, G.; et al. The 2011 report on dietary reference intakes for calcium and vitamin D from the Institute of Medicine: What clinicians need to know. *J. Clin. Endocrinol. Metab.* **2011**, *96*, 53–58. [CrossRef]

151. Grant, W.B.; Karras, S.N.; Bischoff-Ferrari, H.A.; Annweiler, C.; Boucher, B.J.; Juzeniene, A.; Garland, C.F.; Holick, M.F. Do studies reporting 'U'-shaped serum 25-hydroxyvitamin D-health outcome relationships reflect adverse effects? *Derm. Endocrinol.* **2016**, *8*, e1187349. [CrossRef] [PubMed]

152. Holick, M.F.; Binkley, N.C.; Bischoff-Ferrari, H.A.; Gordon, C.M.; Hanley, D.A.; Heaney, R.P.; Murad, M.H.; Weaver, C.M.; Endocrine, S. Evaluation, treatment, and prevention of vitamin D deficiency: An Endocrine Society clinical practice guideline. *J. Clin. Endocrinol. Metab.* **2011**, *96*, 1911–1930. [CrossRef] [PubMed]

153. Veugelers, P.J.; Pham, T.M.; Ekwaru, J.P. Optimal Vitamin D Supplementation Doses that Minimize the Risk for Both Low and High Serum 25-Hydroxyvitamin D Concentrations in the General Population. *Nutrients* **2015**, *7*, 10189–10208. [CrossRef] [PubMed]

154. Pludowski, P.; Holick, M.F.; Grant, W.B.; Konstantynowicz, J.; Mascarenhas, M.R.; Haq, A.; Povoroznyuk, V.; Balatska, N.; Barbosa, A.P.; Karonova, T.; et al. Vitamin D supplementation guidelines. *J. Steroid Biochem. Mol. Biol.* **2018**, *175*, 125–135. [CrossRef]

155. Grant, W.B.; Boucher, B.J. A Review of the Potential Benefits of Increasing Vitamin D Status in Mongolian Adults through Food Fortification and Vitamin D Supplementation. *Nutrients* **2019**, *11*, 2452. [CrossRef]

156. Camargo, C.A., Jr.; Ganmaa, D.; Frazier, A.L.; Kirchberg, F.F.; Stuart, J.J.; Kleinman, K.; Sumberzul, N.; Rich-Edwards, J.W. Randomized trial of vitamin D supplementation and risk of acute respiratory infection in Mongolia. *Pediatrics* **2012**, *130*, e561–e567. [CrossRef]

157. Uwitonze, A.M.; Razzaque, M.S. Role of Magnesium in Vitamin D Activation and Function. *J. Am. Osteopath Assoc.* **2018**, *118*, 181–189. [CrossRef]

 *nutrients*

Article

Circulating Levels of Muscle-Related Metabolites Increase in Response to a Daily Moderately High Dose of a Vitamin D3 Supplement in Women with Vitamin D Insufficiency—Secondary Analysis of a Randomized Placebo-Controlled Trial

Lise Sofie Bislev [1,2,*], Ulrik Kræmer Sundekilde [3,4], Ece Kilic [3], Trine Kastrup Dalsgaard [3,4], Lars Rejnmark [1,2] and Hanne Christine Bertram [3,4]

[1] Department of Endocrinology and Internal Medicine, Aarhus University Hospital, 8200 Aarhus N, Denmark; lars.rejnmark@rm.dk
[2] Department of Clinical Medicine, Aarhus University, 8200 Aarhus N, Denmark
[3] Department of Food Science, Aarhus University, Agro Food Park 48, 8200 Aarhus N, Denmark; uksundekilde@food.au.dk (U.K.S.); ecekilic90@gmail.com (E.K.); trine.dalsgaard@food.au.dk (T.K.D.); hannec.bertram@food.au.dk (H.C.B.)
[4] iFOOD, Centre for Innovative Food Research, Aarhus University, 8200 Aarhus N, Denmark
* Correspondence: lise.sofie@auh.rm.dk; Tel.: +45-2091-4277

Received: 30 March 2020; Accepted: 1 May 2020; Published: 4 May 2020

Abstract: Recently, we demonstrated negative effects of vitamin D supplementation on muscle strength and physical performance in women with vitamin D insufficiency. The underlying mechanism behind these findings remains unknown. In a secondary analysis of the randomized placebo-controlled trial designed to investigate cardiovascular and musculoskeletal health, we employed NMR-based metabolomics to assess the effect of a daily supplement of vitamin D3 (70 µg) or an identically administered placebo, during wintertime. We assessed the serum metabolome of 76 postmenopausal, otherwise healthy, women with vitamin D (25(OH)D) insufficiency (25(OH)D < 50 nmol/L), with mean levels of 25(OH)D of 33 ± 9 nmol/L. Compared to the placebo, vitamin D3 treatment significantly increased the levels of 25(OH)D (−5 vs. 59 nmol/L, respectively, $p < 0.00001$) and 1,25(OH)$_2$D (−10 vs. 59 pmol/L, respectively, $p < 0.00001$), whereas parathyroid hormone (PTH) levels were reduced (0.3 vs. −0.7 pmol/L, respectively, $p < 0.00001$). Analysis of the serum metabolome revealed a significant increase of carnitine, choline, and urea and a tendency to increase for trimethylamine-N-oxide (TMAO) and urinary excretion of creatinine, without any effect on renal function. The increase in carnitine, choline, creatinine, and urea negatively correlated with muscle health and physical performance. Combined with previous clinical findings reporting negative effects of vitamin D on muscle strength and physical performance, this secondary analysis suggests a direct detrimental effect on skeletal muscle of moderately high daily doses of vitamin D supplements.

Keywords: vitamin D; secondary hyperparathyroidism; skeletal muscle; metabolomics; postmenopausal women

---

1. Introduction

The prevalence of vitamin D insufficiency is high, especially during wintertime [1]. As low levels of 25-hydroxyvitamin D (25(OH)D) are associated with adverse skeletal and non-skeletal health outcomes and as a correction of vitamin D deficiency by supplementation is cheap and feasible, studies elucidating the effects of treating vitamin D insufficiency are obviously of major public health interest.

It is well described that low levels of 25(OH)D may elevate the levels of parathyroid hormone (PTH), causing secondary hyperparathyroidism (SHPT) [2]. High PTH levels are associated with adverse health outcomes independently of low 25(OH)D levels [2–4]. Vitamin D insufficiency is pragmatically defined as the level of 25(OH)D below which PTH increases, and SHPT has been suggested as the best marker of vitamin D insufficiency [2,5]. Vitamin D supplementation normalizes the levels of 25(OH)D and PTH, and it has been suggested that individuals with SHPT are more prone to adverse effects of low 25(OH)D levels.

Findings from cross-sectional and cohort studies suggest an inverse association between vitamin D status and adverse health outcomes, whereas data from randomized clinical trials (RCTs) are less conclusive or even report negative effects of vitamin D supplementation on musculoskeletal health [6,7]. The discrepancy between observational studies and RCTs is an indisputable fact [8,9]. So far, many RCTs have been criticized for including participants with a replete vitamin D status, thereby not reflecting the findings from observational studies [9,10].

Metabolomics is a post-genomic advanced method of analysis. Through an explorative approach, metabolomics seeks to characterize and quantify as many metabolites as possible, which constitute the so-called metabolome [11]. The metabolome provides an expression of an individual's metabolic state, and studies suggest that the metabolome may predict individuals' different responses to interventions [12].

Taking into account the large number of publications on the role of vitamin D in metabolic health, metabolomics studies are sparse, and RCTs investigating the effect of vitamin D supplements on the metabolome almost non-existing [13].

In the present study, designed to investigate the cardiovascular and musculoskeletal effects of vitamin D supplementation, we applied a nuclear magnetic resonance (NMR)-based approach, to study metabolic changes in otherwise healthy, postmenopausal women with vitamin D insufficiency and relatively high levels of PTH, randomized to a daily oral supplement of cholecalciferol (vitamin D3) of 70 µg (2800 IU) or a similarly administered placebo for 12 weeks during wintertime. Using this explorative method, we hypothesized that a normalization of plasma 25(OH)D associated with a decrease of PTH levels changes the metabolome, providing knowledge of the underlying metabolic pathways involved.

2. Materials and Methods

The study was an investigator-initiated parallel group, single-center, randomized double-blinded placebo-controlled trial.

The Danish Data Protection Agency (1-16-02-492-14), the Danish Health Authority (2014-003645-10), the Regional Committee on Biomedical Research Ethics (1-10-72-326-14), and the Danish Health Data Authority (FSEID-00001274) approved the project. The local unit for Good Clinical Practice at Aarhus University Hospital monitored the study. Clinicaltrials.gov: #NCT02572960.

The recruitment of participants has previously been reported in detail [14]. Briefly, a total of 81 healthy postmenopausal women with SHPT (PTH > 6.9 pmol/L) and 25(OH)D levels <50 nmol/L were recruited from the area nearby Aarhus University Hospital, Denmark. Inclusion criteria involved subjects who had not received any treatment with antihypertensives, diuretics, systemic glucocorticoids, nonsteroidal anti-inflammatory drugs, lithium, or anti-osteoporotic drugs. The study was conducted at latitude 56° N during wintertime (between November and April) to prevent cutaneous synthesis of cholecalciferol. Informed consent was obtained from all individual participants included in the study.

The study design is depicted in Figure 1.

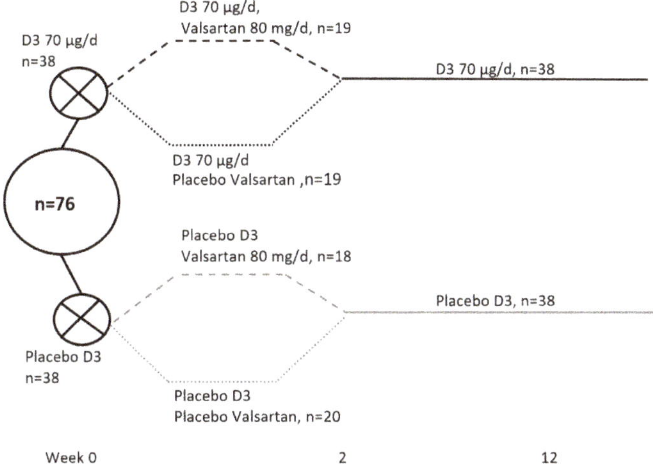

Figure 1. Study design: All women were included from November to February to avoid cutaneous vitamin D synthesis (latitude 56° N).

The participants received a daily supplement of 70 μg (2800 IU) of cholecalciferol or a similarly administered placebo for 12 weeks. For the first two weeks, the design was 2 × 2 factorial with an angiotensin II receptor blocker (valsartan, 80 mg per day) or similar placebo in order to study the response of PTH to the treatment in the presence of a blockade of the renin–angiotensin–aldosterone system. These findings have previously been reported, showing no impact of the angiotensin II receptor blocker on PTH levels [15].

The overall compliance, as assessed by pill-count, was 99.2%.

As previously reported, the normalization of vitamin D/PTH levels had no effect on most markers of cardiovascular disease (CVD), quality of life, or body composition as assessed by dual energy X ray absorptiometry [14,16], but improved bone microarchitecture and estimated bone strength [17]. Contrary to what expected, the moderately high dose of vitamin D impaired muscle strength (as assessed by hand grip strength and knee flexion) and physical performance (as assessed by the Timed Up and Go test, TUG) [16].

For the present study, samples from five participants (placebo group, $n = 3$, vitamin D group, $n = 2$) were missing, leaving 76 participants for the metabolomics analysis, as shown in Figure 1.

Participants were included based on a biochemical screening indicating plasma 25(OH)D concentration below 50 nmol/L, plasma calcium and creatinine levels below the upper normal limit, and PTH levels above the upper limit of the normal reference range (i.e., >6.9 pmol/L), thereby excluding patients with hypercalcemic primary hyperparathyroidism or other causes of SHPT than vitamin D insufficiency [14].

The samples reported in the study were collected at baseline and at the end of study as fasting blood samples drawn in the morning after an overnight fast.

Metabolomics analyses and analyses of 25(OH)D and 1,25(OH)$_2$D were conducted on serum samples, whereas the rest of the analyses were conducted on plasma samples collected in tubes containing lithium heparin. All blood samples were centrifuged at 4000 rpm at 5 °C for 10 min and subsequently stored at −80 °C. To minimize the intra-individual variability, all women rested while lying down for at least 30 min prior to the collection of blood samples.

The total plasma levels of 25(OH)D (25(OH)D2 + 25(OH)D3) and 1,25-dihydroxyvitamin (1,25(OH)$_2$D) were quantified using isotope dilution liquid chromatography–tandem mass spectrometry (LC–MS/MS), which is the gold standard for 25(OH)D measurements [18]. Using a second-generation immunoassay on an automated immunoanalyzer (Roche Diagnostics GmbH, Mannheim, Germany), plasma intact PTH was measured in duplicate.

Plasma glucose and lipid profile, as well as measurements of 24 h urine electrolytes were consecutively analyzed using standard laboratory procedures at the Department of Clinical Biochemistry, Aarhus University Hospital Denmark. Participants were verbally informed and received a written instruction from the laboratory prior to urine collection to ensure high quality of the measurements.

Muscle strength was assessed as maximum voluntary isometric muscle strength, with an adjustable dynamometer chair (Good Strength, Metitur Ltd., Jyvaskyla, Finland) [16]. Using a stopwatch, the TUG test provided the time to stand up, walk 3 m as fast as possible in a straight line, and immediately return to the chair [16].

NMR analyses were conducted in October 2019. Prior to the NMR analyses, serum samples were thawed at room temperature, vortexed for 30 s, and filtered using spin filters with a 10 kDa cutoff (13,000× g for 90 min at 4 °C; Amicon Ultra-0.5 mL 10K, Merck, Darmstadt, Germany). Prior to use, the filters were washed 3 times with 0.5 mL milliQ H_2O. A volume of 400 µL of filtrate was transferred to a 5 mm NMR tube (VWR International, Herlev, Denmark) with 100 µL phosphate buffer (pH 7.4, 300 mM) and 100 µL D_2O containing 0.05 wt. % sodium salt (TSP) (Sigma-Aldrich, Søborg, Denmark). All NMR spectra were acquired at 298 K on a Bruker Avance III 600 spectrometer operating at a ^{1}H frequency of 600.13 MHz and equipped with a 5 mm TXI probe (Bruker BioSpin, Rheinstetten, Germany). A one-dimensional (1D) nuclear Overhauser enhancement spectroscopy (NOESY)-presat pulse sequence (noesypr1d) with water suppression was applied, and a total of 64 scans were collected into 32 K data points spanning a spectral width of 12.15 ppm, and relaxation delay was 5 s. Baseline and phase correction of the spectra were done manually using TopSpin 3.0 (Bruker BioSpin). Assignment and quantification of the ^{1}H NMR signals was performed by using Chenomx NMR Suite version 8.1 (Chenomx Inc., Edmonton, AB, Canada). The concentration of metabolites was calculated based on the known glucose concentration.

Data distribution was tested using QQ plots and histograms. Normally distributed data were analyzed with parametric tests, and non-normally distributed data with non-parametric tests. Baseline data are reported as means with standard deviation, medians with interquartile (25th, 75th percentiles) range (IQR), or numbers with percentages. The effects of treatment are reported as absolute changes from baseline. Significance was tested using a test for two independent samples or, Mann–Whitney U test, as appropriate. Pearson correlation coefficient (r) was used to assess associations between changes in the serum metabolome and changes in plasma levels of 25(OH)D, as well as changes in fat and lean mass, strength in handgrip, knee flexion at 60°, and the TUG test. We considered a two-tailed p value <0.05 as statistically significant. SPSS version 26 was used for the statistical analyses.

3. Results

3.1. Baseline Characteristics and Effects of Vitamin D Supplementation on 25(OH)D, 1,25(OH)$_2$D, and PTH Levels

Baseline characteristic are reported in Table 1. The randomization was well balanced.

Table 1. Baseline characteristics. Data are reported as mean ± SD, median with interquartile (25th, 75th percentiles) range (IQR) or numbers (%). HDL, high-density lipoprotein, LDL, low-density lipoprotein.

	Vitamin D (n = 38)	Placebo (n = 38)	p-Value
Age and body composition			
Age (years)	64.5 [61.0; 68.25]	65.5 [62.0; 68.25]	0.56
Body weight (Kg)	75.3 [67.3; 90.3]	70.4 [65.0; 78.2]	0.17
Height (cm)	166.2 ± 4.7	165.1 ± 6.0	0.39
Body mass index (Kg/m^2)	27.3 [23.3; 32.0]	26.8 [23.6; 28.8]	0.42

Table 1. *Cont.*

	Vitamin D (*n* = 38)	Placebo (*n* = 38)	*p*-Value
Appendicular lean mass index (Kg/m^2)	10.8 [10.0; 12.1]	10.7 [10.1; 11.5]	0.56
Fat mass index (Kg/m^2)	18.7 [14.1; 23.8]	17.9 [12.5; 20.2]	0.16
Indices of bone health			
Calcium intake (mg/day)	850 [700; 950]	700 [650; 1075]	0.87
History of fracture in adulthood *n* (%)	13 (34)	7 (20)	0.12
Smoking status*n*(%)			0.22
Never	23 (61)	22 (68)	
Current	1 (3)	5 (13)	
Former	14 (37)	11 (29)	
Use of medication			
Any *n* (%)	13 (34)	13 (34)	0.60
Indices of cardiovascular health			
Systolic 24 h blood pressure (mmHg)	129 [125; 146]	128 [118; 135]	0.14
Diastolic 24 h blood pressure (mmHg)	75 [68; 83]	74 [68; 79]	0.31
Total cholesterol (mmol/L)	5.2 [4.8; 5.9]	5.5 [5.0; 6.4]	0.12
HDL (mmol/L)	1.8 ± 0.4	1.8 ± 0.4	0.76
LDL (mmol/L)	3.0 ± 0.8	3.3 ± 1.0	0.14
Triglycerides (mmol/L)	1.1 [0.7; 1.3]	0.9 [0.7; 1.3]	0.65
Arterial stiffness (m^2/s)	9.7 ± 1.7	9.1 ± 1.3	0.10

Baseline plasma levels of vitamin D, PTH, renal function, and electrolytes, as well as responses to treatment are depicted in Table 2.

Table 2. Baseline levels of 25(OH)D, 1,25(OH)$_2$D, PTH, and electrolytes. Baseline data are reported as mean ± SD or median with IQR (25%–75% percentile). The mean of the entire group is reported at baseline, as there was no difference between the groups in any of the measurements. Changes are reported as means ± SD. Significant results are shown in bold.

			Changes (Δ)		
	Ref.Range	Baseline, *n* = 76	Vitamin D, *n* = 38	Placebo, *n* = 38	*p*-Value
Plasma					
25(OH)D (nmol/L)	50–160	33 ± 9	58.5 ± 16.3	−4.5 ± 6.3	**<0.00001**
1.25(OH)2D (pmol/L)	60–180	53 ± 14	18.5 ± 15.2	−9.6 ± 9.9	**<0.00001**
PTH (pmol/L)	1.6–6.9	6.1 ± 1.3	-0.69 ± 0.79	0.28 ± 0.80	**<0.00001**
Ca^{2+} (mmol/L)	1.18–1.32	1.25 ± 0.04	0.00 ± 0.04	−0.01 ± 0.03	0.20
Magnesium (mmol/L)	0.7–1.1	0.88 ± 0.06	-0.01 ± 0.04	0.01 ± 0.04	0.24
Phosphate (mmol/L)	0.76–1.41	1.00 ± 0.14	0.06 ± 0.11	0.04 ± 0.12	0.52
eGRF	>60 mL/min	82.4 [73.1; 90.7]	−2.18 [−5.45; 4.21]	−1.15 [−5.30; 1.76]	0.94
Urine					
Creatinine (mmol/24 h)	6–15	10.3 ± 1.9	0.33 ± 1.53	−0.37 ± 1.67	0.06

Abbreviations: 25(OH)D, 25-hydroxy vitamin D, 1,25(OH)$_2$D, 1,25dihydroxy vitamin D, PTH, parathyroid hormone, Ca^{2+}, ionized calcium, eGFR, estimated glomerular filtration rate. Significant results are shown in bold.

At baseline, the plasma levels of 25(OH)D were 33 nmol/L in the total group of women (*n* = 76). At the end of the study, the levels of 25(OH) had increased significantly to 90 nmol/L (95%, CI: 86 to 95) in the vitamin D group as compared with 30 nmol/L (95% CI: 28 to 33) in the control group. Within the vitamin D group, 25(OH)D and 1,25(OH)$_2$D increased, and PTH decreased (p_{all} < 0.001). Similarly, within the placebo group, 25(OH)D and 1,25(OH)$_2$D decreased significantly (p_{all} < 0.001), whereas PTH increased (*p* = 0.02, data not shown).

3.2. Changes in Muscle-Related Metabolites

The urinary 24 h excretion of creatinine tended to increase in the vitamin D group (0.33 ± 1.53) as compared with the placebo group (−0.37 ± 1.67), *p* = 0.06.

Compared with the placebo, vitamin D supplementation significantly increased the serum levels of carnitine to 6.0 μmol/L (95% CI: −1.1 to 13) vs. −5.5 μmol/L (−13 to 2.0), $p = 0.03$), choline to −0.00 μmol/L (−2.2 to 2.1) vs. −4.1 μmol/L (−6.7 to −1.6), $p = 0.02$, and urea to 45 μmol/L (24 to 66 vs. 13 μmol/L (−7.0 to 34, $p = 0.03$), whereas trimethylamine-N-oxide (TMAO) tended to increase, reaching 6.3 μmol/L (1.5 to 11) vs. 0.6 μmol/L (−2.7 to 4.0, $p = 0.05$), Table 3.

Table 3. Significance of the changes in metabolites level observed in fasting serum after a 12-week intervention with vitamin D supplementation (70 μg/d) compared to placebo.

Metabolites	Baseline, μmol/L, $n = 76$	Changes (Δ), μmol/L		p-Value
		Vitamin D, $n = 38$	Placebo $n = 38$	
Hydroxybutyrate	38 [22; 90]	−0.34 [−55; 25]	2.1 [−34; 29]	0.58
Acetate	16 [9; 23]	−2.0 [−11; 2.8]	−0.33 [−12; 2.7]	0.92
Acetoacetate	18 [10; 29]	1.4 [−14; 14]	2.4 [−7.8; 10]	0.93
Acetone	7.1 [5.6; 11]	−1.3 [−7.4; 1.3]	−1.0 [−3.1; 1.8]	0.33
Alanine	210 ± 57	8.3 [−15; 47]	7.1 [−30; 38]	0.26
Betaine	16 [9; 23]	3.3 (−0.9 to 5.5)	0.3 (−2.7 to 3.3)	0.36
Carnitine	77 [65; 86]	6.0 (−1.1 to 13)	−5.5 (−13 to 2.0)	**0.03**
Choline	15 ± 6	−0.00 (−2.2 to 2.1)	−4.1 (−6.7 to −1.6)	**0.02**
Citrate	93 [82; 108]	−0.42 (−8.9 to 8.0)	−1.1 (−7.8 to 5.7)	0.91
Creatine	23 [18; 31]	3.5 (−0.44 to 7.5)	4.2 (−0.76 to 9.1)	0.84
Creatinine	49 ± 11	8.6 (3.4 to 14)	5.3 (1.7 to 8.9)	0.30
Dimethylamine	1.5 [1.0; 4.3]	0.3 [0.0; 0.8]	0.1 [−0.30; 0.5]	0.25
Formate	1.3 [0.8; 2.1]	0.0 (−0.24 to 0.3)	−0.01 (−0.29 to 0.3)	0.91
Glucose	5500 [5125; 6000]	0.0 [−225; 300]	−100 [−325; 100]	0.31
Glutamate	48 [35; 73]	−2.1 [−23; 18]	−14 [−28; 7.0]	0.15
Glutamine	470 [410; 520]	64 [14; 130]	60 [15; 127]	0.82
Glycerol	390 [320; 530]	0.0 [−180; 80]	−34 [−160; 71]	0.65
Glycine	69 ± 22	6.4 (−0.30 to 13)	2.9 (−5.7 to 11)	0.52
Isoleucine	67 [59; 80]	4.6 [−6.1; 19]	4.4 [−10; 12]	0.19
Lactate	750 [650; 970]	77 (−31 to 180)	10.7 (−92 to 110)	0.37
Leucine	150 [140; 170]	12 [−8.2; 36]	−2.5 [−20; 24]	0.17
Lysine	260 [240; 280]	13 [−30; 60]	6.2 [−8.3; 46]	0.83
Methionine	19 [17; 26]	−1.8 [−8.6; 2.1]	−2.7 [−7.6; 0.8]	0.58
OPhosphocholine	19 [16; 24]	−1.2 (−2.7 to 0.4)	−1.2 (−2.9 to 0.5)	0.98
Ornithine	83 [61; 110]	5.9 (−11 to 23)	5.0 (−8.0 to 18)	0.93
Phenylalanine	57 [51; 65]	4.7 (−0.1 to 9.6)	0.4 (−3.5 to 4.4)	0.17
Proline	150 [110; 280]	52 (27 to 77)	25 (2 to 47)	0.11
Pyruvate	12 [7; 18]	4.5 (1.7 to 7.4)	4.9 (1.9 to 7.9)	0.87
Succinate	3.6 [2.6; 6.6]	−0.25 [−2.9; 0.6]	−0.9 [−4.7; 1.4]	0.87
Threonine	83 [74; 91]	8.2 [−8.3; 17]	6.8 [−2.6; 18]	0.89
TMAO	36 [31; 42]	6.3 (1.5 to 11)	0.6 (−2.7 to 4.0)	**0.05**
Tyrosine	59 ± 15	3.0 [−5.5; 12]	1.0 [−4.2; 9.0]	0.32
Urea	180 [150; 220]	45 (24 to 66)	13 (−7.0 to 34)	**0.03**
Valine	240 [200; 280]	14 [−5.0; 46]	0.9 [−21; 31]	0.11
τMethylhistidine	110 [100; 120]	3.6 [−13; 30]	4.8 [−13; 30]	0.34

The metabolites were quantified by ^{1}H NMR spectroscopy. Except from choline (vitamin D, 14 ± 16 vs. placebo 17 ± 6.2, $p = 0.02$), none of the data at baseline differed between groups when stratified by treatment allocation. The mean ± standard deviation or median (25th, 75th percentiles) for the whole group is reported. Changes were calculated as individual post-intervention values minus baseline values for each metabolite, and data are reported as median (25, 75 percentiles) or mean with 95% confidence intervals. Abbreviation: TMAO, trimethylamine N-oxide. Significant results are shown in bold.

3.3. The Effect of Valsartan on the Metabolome

No significant interactions of valsartan with the metabolome were found when the drug was given either alone or in combination with vitamin D (data not shown). In a secondary analysis reporting the effect of valsartan (plus/minus vitamin D) vs. placebo, valsartan (plus/minus vitamin D) did not affect any of the measures of the metabolome (data not shown).

3.4. Nutrient Intake and Physical Activity

No differences in estimated daily calcium intake were found (Table 1). There were no differences in the intake of major sources of vitamin D (egg and fish) between the two groups, and no differences in the intake of fruit and vegetables (data not shown). The estimated physically activity did not differ between groups, as previously reported [4].

3.5. Correlations between Muscle-Related Metabolites, Body Composision, Musle Strength, and Physical Performance

As reported in Table 4, changes in total fat mass correlated positively with changes in the levels of carnitine (r = 0.29, *p* = 0.01) and urea (r = 0.25, *p* = 0.03). Moreover, changes in the levels of carnitine (r = 0.29, *p* = 0.01), choline (r = 0.23, *p* = 0.04), and urea (r = 0.26, *p* = 0.02) correlated positively with changes in the TUG test, i.e., increases in these metabolites were associated with a longer time spent on performing the test. Changes in handgrip strength were negatively correlated with changes in choline levels (r = −0.25, *p* = 0.05) and excreted creatinine (r = −0.25, *p* = 0.04), i.e., increases in serum choline and excreted creatinine were associated with a decreased handgrip strength.

Table 4. Correlations between changes in the levels of 25-hydroxyvitamin D, carnitine, choline, and urea, as well as 24 h renal excretion of creatinine and previoulys reported significant markers of muscle health [16] and body composition (*n* = 76). A positive correlation at the TUG test means spending longer time performing the test (worse performance).

Changes (Δ)	Total Fat Mass		TUG		Handgrip Strength		Knee Flexion 60°	
	r	*p*-Value	r	*p*-Value	r	*p*-Value	R	*p*-Value
25(OH)D, nmol/L	-	-	-	-	−0.27	0.03	−0.29	0.02
Carnitine, mmol/L	0.29	0.01	0.29	0.01	-	-	-	-
Choline, mmol/L	-	-	0.23	0.04	−0.25	0.04	-	-
Urea, mmol/L	0.25	0.03	0.26	0.02	-	-	-	-
Urine creatinine, mmol/day	-	-	-	-	−0.26	0.04	-	-

Abbreviations: r: Pearson correlation coefficient, 25(OH)D: 25-hydroxyvitamin D, TUG: Time Up and Go test, knee flexion 60°: maximum voluntary muscle strength with the knee flexed 60° from the fully extended leg.

Bivariate correlation analysis showed no correlation between any of 25(OH)D, choline, carnitine, TMAO, excreted creatinine, and urea and total lean mass/appendicular lean mass index, as assessed by dual-energy X-ray absorptiometry, data not shown.

4. Discussion

In this exploratory study, we investigated changes in the human metabolome in response to the normalization of vitamin D levels with a daily moderately high dose supplement of vitamin D during wintertime. Vitamin D supplementation effectively normalized 25(OH)D levels. Compared to placebo, vitamin D supplementation significantly increased the serum levels of carnitine, choline, and urea and tended to increase the serum levels of TMAO and those of creatinine excreted in urine.

Carnitine and choline are two essential nutrients. The major sources of these nutrients are animal products, especially red meat [19,20]. Carnitine is required for energy production, as carnitine acts as a transporter of long-chain fatty acids into the mitochondria to be oxidized and produce energy [19]. Within the body, carnitine is accumulated in the cardiac and skeletal muscles. The content of carnitine in skeletal muscle is about 70-fold higher than in plasma [19]. Supplementation with carnitine is proposed to play a role in muscle health, and supplements are widely used among athletes to enhance performance [19,21]. Choline is required to produce acetylcholine and is used at the neuromuscular junction. Choline deficiency is associated with muscle damage [20]. As with carnitine, skeletal muscle contains a large quantity of choline [22,23].

Choline and carnitine are metabolized by gut microorganisms to produce trimethylamine (TMA), which is subsequently absorbed by the gut and oxidized by flavin-monooxygenases (FMOs) in the liver to produce TMAO [19,20] (Figure 2).

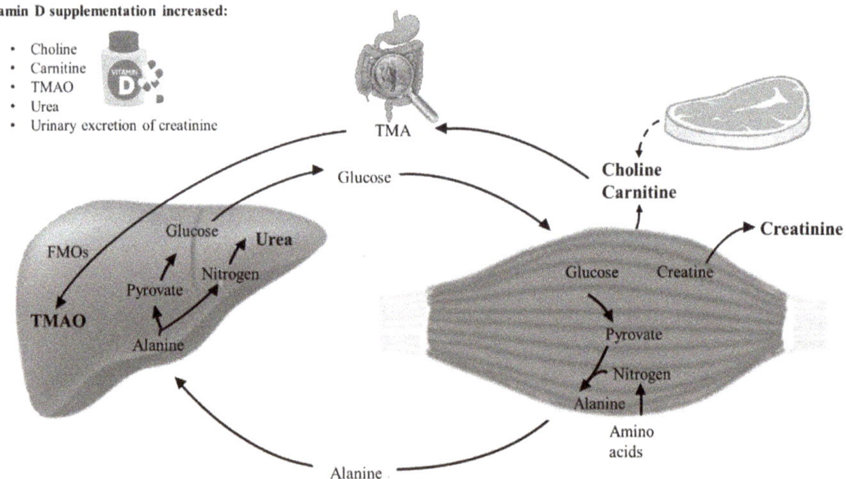

Figure 2. Schematic illustration of the metabolic pathways of the significant and border-significant findings on choline, carnitine, creatinine, TMAO, and urea. Choline and carnitine are nutrients normally ingested through protein-rich diets. TMAO is generated from the hepatic oxidation of trimethylamine (TMA), formed by the gut microbiota from carnitine and choline. In the body, high concentrations of choline and carnitine are present in skeletal muscle. In the glucose–alanine cycle, amino groups and carbons from skeletal muscle are transported to the liver. In the liver, alanine is converted to pyruvate and nitrogen. Nitrogen enters the urea cycle, and pyruvate is used to produce glucose [24]. Creatinine is a waste product of a non-enzymatic degradation of creatine phosphate, serving as a reserve of high-energy phosphates in skeletal muscle. Together with previous clinical findings on muscle strength and physical performance, the data suggest that the increase in choline, carnitine, creatinine, TMAO, and urea, all waste products originating from muscle catabolism, may be caused by a direct toxic effect on skeletal muscle.

TMAO is mainly known as a waste product of carnitine and choline metabolism [25]. TMAO has received attention as a consequence of a proposed negative effect on cardiovascular health [26], although not all studies support this observation [27]. Intriguingly, the POUNDS Lost trial suggests a positive relationship between circulating TMAO and bone mineralization [28]. Although the 12 weeks of vitamin D supplementation did not affect bone mineral density in our study, we observed improved bone health in terms of increased trabecular thickness and estimated bone strength at the tibia [17].

Blood urea is a product of protein catabolism, and in the urea cycle, nitrogen from the muscles are converted to ammonia and, via liver enzymes, to water-soluble urea, which can be excreted by the kidneys (Figure 2). Creatinine is a breakdown product of creatine from muscle and protein metabolism [29]. We observed that 24 h urinary excretion of creatinine tended ($p = 0.06$) to increase in the vitamin D group.

A possible explanation for the increase in serum carnitine and choline is an increased intestinal absorption. A reduced use and/or degradation is also plausible.

Our findings need to be considered in relation to our clinical findings on cardiovascular and musculoskeletal health [14,16,17]. Overall, there was no effect of vitamin D supplementation on measures of cardiovascular health. Cardiac and/or smooth muscle cells also contain the examined muscle-related metabolites. Thus, we cannot rule out that the metabolites derive from cardiac and/or smooth muscle cells, but in a post hoc analysis no associations were found between the muscle-related metabolites and blood pressure, arterial stiffness, or cardiac conductivity. In contrast, we previously reported a detrimental effect on muscle health and physical performance [16]. Carnitine, choline, TMAO, urea, and creatinine derive from the muscles [29] (Figure 2), and the increase in carnitine, choline, creatinine, and urea correlated negatively with the findings on muscle strength/performance, suggesting that these findings can be ascribed to changes in skeletal muscle.

The amount of carnitine, choline, TMAO, urea, and creatinine are dependent on protein intake, the body's capacity to catabolize protein, and their adequate excretion by the renal system. There was no between-group difference in renal function, physical activity, or estimated intake of different nutrients [16].

In recent years, an increased number of studies have demonstrated adverse effects of higher dosages of vitamin D. Negative effects are mainly reported on muscle strength and risk of falls [16,30–36]. A recent study with vitamin D3, 70 µg per day, suggested a negative effect on lean body mass [30].

Previous studies have reported an increased risk of falls in response to vitamin D supplementation [31–33,36]. Orthostatic hypotension due to decreased activity of the renin–angiotensin–aldosterone system has been suggested, but in this study, markers of this system were not affected by vitamin D supplementation [14].

To the best of our knowledge, higher dosages of vitamin D supplementation has not been reported to impair postural stability [16,30,37,38], and it therefore seems most likely that the increased risk of falls is attributable to an impaired muscle strength/function.

The mechanisms behind the studies reporting negative effects of vitamin D have not yet been fully elucidated. Vitamin D receptors are almost ubiquitously expressed in human tissues. Over-expression of vitamin D receptors and inadequate differentiation of muscle fibers are reported in response to active vitamin D in supra-physiological dosages [39,40]. Elevated levels of creatine kinase (which converts creatine to creatinine) [40] as well as fat infiltration are also reported [41].

Together with the existing data, this explorative study suggests a direct detrimental effect on skeletal muscles causing a leak of muscle products to the blood stream and subsequently to urine (creatinine).

That vitamin D could heal myopathies was a clinical observation before it was possible to measure 25(OH)D levels, and threshold levels are largely based on findings from observational studies. In general, randomized clinical trials have largely failed to demonstrate any effect of vitamin D. Possibly, vitamin D has divergent effects on different tissues (e.g., an increased risk of falls and thereby increased risk of fractures despite an improved bone health), which overall counterbalance each other.

The present study has several strengths as well as limitations. Most importantly, the well-balanced randomized placebo-controlled design conducted during wintertime in women with low levels of 25(OH)D and relatively high PTH levels leaves a unique study group not previously investigated with respect to NMR-based metabolomics. The NMR-metabolomics data were not a pre-planned endpoint, and there was no a priori hypothesis. As the study is exploratory, it is important to establish that the study is hypothesis-generating rather than hypothesis-testing. We did not adjust for multiple testing as this is a rather conservative approach, which lowers the chance of detecting potential associations. This is, on the other hand, a major limitation, and we cannot rule out that some of the findings are type I errors.

The three months duration of the intervention is relatively short.

The initial factorial design is a limitation. There was no interaction between vitamin D and valsartan in the reported outcomes, and valsartan did not affect any of the measures of the metabolome, but as the half-life of valsartan is 6–9 h, a potential effect of valsartan on the metabolome in the reported 12-week measures is unlikely [42].

The estimated intake of vitamin D was not calculated, and neither was estimated intake of meat. Finally, PTH level at baseline was substantial lower than expected. This has previously been discussed in details [14].

Unfortunately, we did not use different dosages of vitamin D or assess muscle health with biopsies. The dose used to treat vitamin D insufficiency was larger than recommended in most guidelines, and our results do not allow for conclusions on lower dosages of vitamin D [19]. Detrimental effects on muscle health are not reported with dosages below 20 µg/day [16]. In contrast, dosages at 70 µg/day are commonly used to treat insufficiency.

At present, it is unknown whether it is the rapid increase in 25(OH)D or the levels of 25(OH)D at the end of the study that cause potential adverse effects on skeletal muscle and/or falls. "Very high dose bolus studies" [33,36] has been reproduced by studies using moderately high daily dosages of vitamin D3 [16,30,32], also in participants with vitamin D insufficiency [16,32]. The fact that adverse effects have been reported also for levels of 25(OH)D within the reference range suggests that a rapid increase in 25(OH)D is associated with adverse effects on skeletal muscles and/or falls. The mechanism behinds those findings needs to be established.

In 2011, the upper tolerance limit was increased from 50 to 100 µg/day based on the lack of occurrence of hypercalcemia. Data on falls and muscle health from 2010 on (mainly from 2015) suggest a reduction of the upper tolerance limit [16,30–36].

5. Conclusions

Normalization of 25(OH)D levels with a moderately high daily dose of vitamin D supplementation during wintertime causes changes in the metabolome in terms of increased serum levels of carnitine, choline, and urea and a tendency towards increased serum levels of TMAO and urinary creatinine. This study suggests a potential detrimental effect of vitamin D supplements on skeletal muscles, with leak of muscle products to the circulation.

Author Contributions: Conceptualization, L.S.B., L.R., and H.C.B.; methodology, H.C.B.; validation, L.S.B.; formal analysis, L.S.B. and U.K.S.; investigation, L.S.B., E.K., and U.K.S.; resources, L.R. and H.C.B.; data curation, L.S.B., writing—original draft preparation, L.S.B.; writing—review and editing, L.S.B., U.K.S., E.K., T.K.D., L.R., H.C.B.; visualization, L.B.; supervision, U.K.S., L.R., H.C.B.; project administration, L.S.B.; funding acquisition, H.C.B. All authors have read and agreed to the published version of the manuscript.

Funding: The present study was part of the DFORT project, J-nr. 4105-00008B supported by Innovation Fund Denmark.

Acknowledgments: We thank all included women for their participation.

Conflicts of Interest: L.S.B., U.K.S., E.K., T.K.D., L.R., and H.C.B. declare that they have no conflict of interest and no competing financial interests.

References

1. Hansen, L.; Tjønneland, A.; Køster, B.; Brot, C.; Andersen, R.; Cohen, A.S.; Frederiksen, K.; Olsen, A. Vitamin D Status and Seasonal Variation among Danish Children and Adults: A Descriptive Study. *Nutritients* **2018**, *10*, 1801. [CrossRef] [PubMed]
2. Lips, P. Vitamin D Deficiency and Secondary Hyperparathyroidism in the Elderly: Consequences for Bone Loss and Therapeutic Implications. *Endocr. Rev.* **2001**, *22*, 477–501. [CrossRef]
3. Rødbro, L.L.; Bislev, L.S.; Sikjær, T.; Rejnmark, L. Bone metabolism, density, and geometry in postmenopausal women with vitamin D insufficiency: A cross-sectional comparison of the effects of elevated parathyroid levels. *Osteoporos. Int.* **2018**, *29*, 2211–2218. [CrossRef] [PubMed]
4. Bislev, L.S.; Rødbro, L.L.; Sikjaer, T.; Rejnmark, L. Effects of elevated PTH levels on muscle health, postural stability and quality of life in vitamin D insufficient healthy women: A cross-sectional study. *Calcif. Tissue Int.* **2019**, *105*, 642–650. [CrossRef] [PubMed]
5. Bouillon, R. Comparative analysis of nutritional guidelines for vitamin D. *Nat. Rev. Endocrinol.* **2017**, *13*, 466–479. [CrossRef] [PubMed]
6. Sanders, K.M.; Seibel, M.J. Therapy: New findings on vitamin D3 supplementation and falls—When more is perhaps not better. *Nat. Rev. Endocrinol.* **2016**, *12*, 190–191. [CrossRef]
7. Reid, I.R. High-Dose Vitamin D: Without Benefit but Not Without Risk. *J. Intern. Med.* **2018**, *284*, 694–696. [CrossRef]

8. Grübler, M.R.; März, W.; Pilz, S.; Grammer, T.B.; Trummer, C.; Müllner, C.; Schwetz, V.; Pandis, M.; Verheyen, N.; Tomaschitz, A.; et al. Vitamin-D concentrations, cardiovascular risk and events—A review of epidemiological evidence. *Rev. Endocr. Metab. Disord.* **2017**, *18*, 259–272. [CrossRef]

9. Rejnmark, L.; Bislev, L.S.; Cashman, K.D.; Eiríksdottir, G.; Gaksch, M.; Grübler, M.; Grimnes, G.; Gudnason, V.; Lips, P.; Pilz, S.; et al. Non-skeletal health effects of vitamin D supplementation: A systematic review on findings from meta-analyses summarizing trial data. *PLoS ONE* **2017**, *12*, e0180512. [CrossRef]

10. Jorde, R.; Grimnes, G. Vitamin D and health: The need for more randomized controlled trials. *J. Steroid Biochem. Mol. Biol.* **2015**, *148*, 269–274. [CrossRef]

11. Tokarz, J.; Haid, M.; Cecil, A.; Prehn, C.; Artati, A.; Möller, G.; Adamski, J. Endocrinology Meets Metabolomics: Achievements, Pitfalls, and Challenges. *Trends Endocrinol. Metab.* **2017**, *28*, 705–721. [CrossRef] [PubMed]

12. Kaddurah-Daouk, R.; Weinshilboum, R. Metabolomic signatures for drug response phenotypes: Pharmacometabolomics enables precision medicine. *Clin. Pharmacol. Ther.* **2015**, *98*, 71–75. [CrossRef] [PubMed]

13. O'Sullivan, A.; Gibney, M.J.; Connor, A.O.; Mion, B.; Kaluskar, S.; Cashman, K.D.; Flynn, A.; Shanahan, F.; Brennan, L. Biochemical and metabolomic phenotyping in the identification of a vitamin D responsive metabotype for markers of the metabolic syndrome. *Mol. Nutr. Food Res.* **2011**, *55*, 679–690. [CrossRef] [PubMed]

14. Bislev, L.S.; Langagergaard Rødbro, L.; Bech, J.N.; Pedersen, E.B.; Kjaergaard, A.D.; Ladefoged, S.A.; Rolighed, L.; Sikjaer, T.; Rejnmark, L. The effect of vitamin D3 supplementation on markers of cardiovascular health in hyperparathyroid, vitamin D insufficient women: A randomized placebo-controlled trial. *Endocrine* **2018**, *62*, 82–194. [CrossRef]

15. Bislev, L.S.; Langagergaard Rødbro, L.; Nørgaard Bech, J.; Bjerregaard Pedersen, E.; Rolighed, L.; Sikjaer, T.; Rejnmark, L. Effects of treatment with an angiotensin 2 receptor blocker and/or vitamin D3 on parathyroid hormone and aldosterone: A randomized, placebo-controlled trial. *Clin. Endocrinol.* **2018**, *89*, 656–666. [CrossRef]

16. Bislev, L.S.; Langagergaard Rødbro, L.; Rolighed, L.; Sikjaer, T.; Rejnmark, L. Effects of Vitamin D3 Supplementation on Muscle Strength, Mass, and Physical Performance in Women with Vitamin D Insufficiency: A Randomized Placebo-Controlled Trial. *Calcif. Tissue Int.* **2018**, *103*, 483–493. [CrossRef]

17. Bislev, L.S.; Langagergaard Rødbro, L.; Rolighed, L.; Sikjaer, T.; Rejnmark, L. Bone Microstructure in Response to Vitamin D3 Supplementation: A Randomized Placebo-Controlled Trial. *Calcif. Tissue Int.* **2018**, *104*, 160–170. [CrossRef]

18. Roth, H.J.; Schmidt-Gayk, H.; Weber, H.; Niederau, C. Accuracy and clinical implications of seven 25-hydroxyvitamin D methods compared with liquid chromatography-tandem mass spectrometry as a reference. *Ann. Clin. Biochem.* **2008**, *45*, 153–159. [CrossRef]

19. Fielding, R.; Riede, L.; Lugo, J.P.; Bellamine, A. L-carnitine supplementation in recovery after exercise. *Nutrients* **2018**, *10*, 349. [CrossRef]

20. Wiedeman, A.M.; Barr, S.I.; Green, T.J.; Xu, Z.; Innis, S.M.; Kitts, D.D. Dietary choline intake: Current state of knowledge across the life cycle. *Nutrients* **2018**, *10*, 1513. [CrossRef]

21. Peeling, P.; Binnie, M.J.; Goods, P.S.R.; Sim, M.; Burke, L.M. Evidence-based supplements for the enhancement of athletic performance. *Int. J. Sport Nutr. Exerc. Metab.* **2018**, *28*, 178–187. [CrossRef] [PubMed]

22. Consolo, S.; Romano, H.; Scozzesi, C.; Bonetti, A.; Ladinsky, H. Acetylcholine and Choline Contents of Rat Skeletal Muscle Determined by a Radioenzymatic Microassay: Effects of Drugs. *Dyn. Cholinergic Funct.* **1986**, *30*, 583–592.

23. Fayad, L.M.; Salibi, N.; Wang, X.; Machado, A.J.; Jacobs, M.A.; Ouwerkerk, R.; Okollie, B.; David, A.; Eng, J.; Barker, P.B. Quantification of Muscle Choline Concentration by Proton MR Spectroscopy at 3 Tesla: Technical Feasibility. 2010, 194, W73–W79. *Am. J. Roentgenol.* **2010**, *194*, W73–W79. [CrossRef] [PubMed]

24. Felig, P. The glucose-alanine cycle. *Metabolism* **1973**, *22*, 179–207. [CrossRef]

25. Ufnal, M.; Anna, Z.; Ryszard, O. TMAO: A small molecule of great expectations. *Nutrition* **2015**, *31*, 1317–1323. [CrossRef]

26. Koeth, R.A.; Wang, Z.; Levison, B.S.; Buffa, J.A.; Org, E.; Sheehy, B.T.; Britt, E.B.; Fu, X.; Wu, Y.; Li, L.; et al. Intestinal microbiota metabolism of l-carnitine, a nutrient in red meat, promotes atherosclerosis. *Nat. Med.* **2013**, *19*, 576–585. [CrossRef]

27. Aldana-Hernández, P.; Leonard, K.A.; Zhao, Y.Y.; Curtis, J.M.; Field, C.J.; Jacobs, R.L. Dietary Choline or Trimethylamine N-oxide Supplementation Does Not Influence Atherosclerosis Development in Ldlr-/- and Apoe-/- Male Mice. *J. Nutr.* **2020**, *150*, 249–255. [CrossRef]

28. Zhou, T.; Heianza, Y.; Chen, Y.; Li, X.; Sun, D.; DiDonato, J.A.; Pei, X.; LeBoff, M.S.; Bray, G.A.; Sacks, F.M.; et al. Circulating gut microbiota metabolite trimethylamine N-oxide (TMAO) and changes in bone density in response to weight loss diets: The Pounds lost trial. *Diabetes Care* **2019**, *42*, 1365–1371. [CrossRef]

29. Salazar, J.H. Overview of Urea and Creatinine. *Lab. Med.* **2014**, *45*, e19–e20. [CrossRef]

30. Grove-Laugesen, D.; Cramon, P.K.; Malmstroem, S.; Ebbehoj, E.; Watt, T.; Hansen, K.W.; Rejnmark, L. Effects of supplemental vitamin D on muscle performance and quality of life in Graves' disease. A randomized clinical trial. *Thyroid* **2020**, *30*, 661–671. [CrossRef]

31. Ginde, A.A.; Blatchford, P.; Breese, K.; Zarrabi, L.; Linnebur, S.A.; Wallace, J.I.; Schwartz, R.S. High-Dose Monthly Vitamin D for Prevention of Acute Respiratory Infection in Older Long-Term Care Residents: A Randomized Clinical Trial. *J. Am. Geriatr. Soc.* **2017**, *65*, 496–503. [CrossRef] [PubMed]

32. Smith, L.M.; Gallagher, J.C.; Suiter, C. Medium doses of daily vitamin D decrease falls and higher doses of daily vitamin D3 increase falls: A randomized clinical trial. *J. Steroid Biochem. Mol. Biol.* **2017**, *173*, 317–322. [CrossRef] [PubMed]

33. Bischoff-Ferrari, H.A.; Dawson-Hughes, B.; John Orav, E.; Staehelin, H.B.; Meyer, O.W.; Theiler, R.; Dick, W.; Willett, W.C.; Egli, A. Monthly high-dose Vitamin D treatment for the prevention of functional decline a randomized clinical trial. *JAMA Intern. Med.* **2016**, *176*, 175–183. [CrossRef] [PubMed]

34. Mason, C.; Tapsoba, J.D.; Duggan, C.; Imayama, I.; Wang, C.-Y.; Korde, L.; McTiernan, A. Effects of Vitamin D 3 Supplementation on Lean Mass, Muscle Strength, and Bone Mineral Density During Weight Loss: A Double-Blind Randomized Controlled Trial. *J. Am. Geriatr. Soc.* **2016**, *64*, 769–778. [CrossRef] [PubMed]

35. Uusi-Rasi, K.; Patil, R.; Karinkanta, S.; Kannus, P.; Tokola, K.; Lamberg-Allardt, C.; Sievänen, H. Exercise and vitamin din fall prevention among older women a randomized clinical trial. *JAMA Intern. Med.* **2015**, *175*, 703–711. [CrossRef]

36. Sanders, K.M.; Stuart, A.L.; Williamson, E.J.; Simpson, J.A.; Kotowicz, M.A.; Young, D.; Nicholson, G.C. Annual high-dose oral vitamin D and falls and fractures in older women: A randomized controlled trial. *JAMA* **2010**, *303*, 1815–1822. [CrossRef]

37. Burt, L.A.; Gabel, L.; Billington, E.O.; Hanley, D.A.; Boyd, S.K. Postural Balance Effects Associated with 400, 4000 or 10,000 IU Vitamin D3 Daily for Three Years: A Secondary Analysis of a Randomized Clinical Trial. *Nutrients* **2020**, *12*, 527. [CrossRef]

38. Grimnes, G.; Emaus, N.; Cashman, K.D.; Jorde, R. The effect of high-dose vitamin D supplementation on muscular function and quality of life in postmenopausal women—A randomized controlled trial. *Clin. Endocrinol.* **2017**, *87*, 20–28. [CrossRef]

39. Camperi, A.; Pin, F.; Costamagna, D.; Penna, F.; Menduina, M.L.; Aversa, Z.; Zimmers, T.; Verzaro, R.; Fittipaldi, R.; Caretti, G.; et al. Vitamin D and VDR in cancer cachexia and muscle regeneration. *Oncotarget* **2017**, *8*, 21778–21793. [CrossRef]

40. Owens, D.J.; Sharples, A.P.; Polydorou, I.; Alwan, N.; Donovan, T.; Tang, J.; Fraser, W.D.; Cooper, R.G.; Morton, J.P.; Stewart, C.; et al. A systems-based investigation into vitamin D and skeletal muscle repair, regeneration, and hypertrophy. *Am. J. Physiol.-Endocrinol. Metab.* **2015**, *309*, E1019–E1031. [CrossRef]

41. Debruin, D.A.; Timpani, C.A.; Lalunio, H.; Rybalka, E.; Goodman, C.A.; Hayes, A. Exercise May Ameliorate the Detrimental Side Effects of High Vitamin D Supplementation on Muscle Function in Mice. *J. Bone Miner. Res.* **2020**. [CrossRef] [PubMed]

42. Dina, R.; Jafari, M. Angiotensin II-receptor antagonists: An overview. *Am. J. Health Pharm.* **2000**, *57*, 1231–1241. [CrossRef] [PubMed]

 *nutrients*

Article

High-Dose Vitamin D Supplementation Improves Microcirculation and Reduces Inflammation in Diabetic Neuropathy Patients

Tatiana Karonova [1,2,*], Anna Stepanova [2], Anna Bystrova [1,2] and Edward B. Jude [3]

[1] Almazov National Medical Research Centre, Institute of Endocrinology, 2 Akkuratova str., 197341 St. Petersburg, Russia; bystrova@inbox.ru

[2] Internal Medicine Department, Pavlov First Saint Petersburg State Medical University, 6-8 L.Tolstoy str., 197022 St. Petersburg, Russia; annstepanova12@gmail.com

[3] Tameside Hospital NHS Foundation Trust, Ashton Under Lyne OL69RW, UK; Edward.jude@tgh.nhs.uk

* Correspondence: karonova@mail.ru; Tel.: +79-213-106-041

Received: 30 June 2020; Accepted: 14 August 2020; Published: 20 August 2020

Abstract: We assessed the effect of different doses of vitamin D supplementation on microcirculation, signs and symptoms of peripheral neuropathy and inflammatory markers in patients with type 2 diabetes (T2DM). Sixty-seven patients with T2DM and peripheral neuropathy (34 females) were randomized into two treatment groups: Cholecalciferol 5000 IU and 40,000 IU once/week orally for 24 weeks. Severity of neuropathy (NSS, NDS scores, visual analogue scale), cutaneous microcirculation (MC) parameters and inflammatory markers (ILs, CRP, TNFα) were assessed before and after treatment. Vitamin D deficiency/insufficiency was detected in 78% of the 62 completed subjects. Following treatment with cholecalciferol 40,000 IU/week, a significant decrease in neuropathy severity (NSS, $p = 0.001$; NDS, $p = 0.001$; VAS, $p = 0.001$) and improvement of cutaneous MC were observed ($p < 0.05$). Also, we found a decrease in IL-6 level (2.5 pg/mL vs. 0.6 pg/mL, $p < 0.001$) and an increase in IL-10 level (2.5 pg/mL vs. 4.5 pg/mL, $p < 0.001$) after 24 weeks of vitamin D supplementation in this group. No changes were detected in the cholecalciferol 5000 IU/week group. High-dose cholecalciferol supplementation of 40,000 IU/week for 24 weeks was associated with improvement in clinical manifestation, cutaneous microcirculation and inflammatory markers in patients with T2DM and peripheral neuropathy.

Keywords: diabetes; neuropathy; microcirculation; 25(OH)D; vitamin D; inflammatory markers

---

1. Introduction

It is well known that vitamin D deficiency along with type 2 diabetes mellitus (T2DM) is a modern pandemic [1,2]. The development of microvascular complications in T2DM worsens both the prognosis and the patients' quality of life. There is increasing evidence of a possible contribution of vitamin D deficiency to the pathogenesis of diabetes and its complications [3]. Large-scale studies have shown 40% increased risk of developing diabetes in individuals with a reduced 25(OH)D (25-hydroxy vitamin D) level [4], as well as 24% decrease in diabetes risk for every 25 nmol/L increase in 25(OH)D concentration [5]. However, some studies found no association between diabetes risk and vitamin D status [6]. Thus, a recent interventional prospective study demonstrated no decrease in the risk of T2DM development in patients with prediabetes after two-year treatment with 4000 IU of vitamin D per day [7]. However, some experts suggested that 4000 IU is not a sufficient supplementation dose for patients with already existing impaired glucose metabolism, and besides, most study participants had normal basal 25(OH)D levels [8]. These results do not exclude the presence of pleiotropic vitamin D effects on insulin secretion, insulin resistance and adipocytokine system [9–12], and the possibility of influencing the development of microvascular diabetic complications [3]. Along with immune-mediated mechanisms, microcirculation deterioration in patients with diabetes has been found to play an important role in the pathogenesis of microvascular complications, including peripheral neuropathy (DPN) [12,13].

Vitamin D deficiency is also believed to play a role in the progression of DPN [14–16]. One study showed that vitamin D supplementation in patients with T2DM and DPN resulted in the pain decrease and reduction or withdrawal of semisynthetic opioids, and that an increase in 25(OH)D by 1 ng/mL was associated with the decrease in neuropathy severity and increase in impulse conduction frequency along nerve fibres by 2.2% and 3.4%, respectively [15]. Another study demonstrated a relationship between serum 25(OH)D levels and the severity of neuropathy in T2DM, where the greatest changes were found in patients with 25(OH)D levels of less than 16 ng/mL [16]. Possible association of vitamin D deficiency with DPN was revealed by other investigators, but relationships between 25(OH)D level and DPN remain unclear [17]. The Cochrane systematic review demonstrated no convincing evidence regarding vitamin D effectiveness in chronic painful conditions [18]. Some studies found high levels of IL-13 and IL-17 in patients with T2DM and DPN, and negative correlations between these interleukins and 25(OH)D levels [19].

Thus, the correction of vitamin D deficiency in patients with T2DM is becoming increasingly attractive for the prevention and treatment of microvascular complications. However, the question of the required vitamin D dose and the treatment duration remains highly debatable. According to some studies, the daily dose of vitamin D for pleiotropic effects should exceed the dose recommended for prophylaxis [20,21]. The inconsistency of the evidence dictates the need for further research in this field.

The aim of this study was to assess the effect of therapy with different doses of cholecalciferol for 24 weeks on parameters of microcirculation, clinical manifestations of peripheral neuropathy and inflammatory markers in patients with T2DM.

2. Materials and Methods

2.1. Study Population

We conducted a prospective randomized trial in patients with T2DM and DPN. Ninety-eight patients with T2DM and DPN were screened for the study from January 2018 to January 2019. Patients were selected based on the following inclusion criteria: (i) Males and females with T2DM aged 18 to 65 years; (ii) diabetes duration ≥5 years; (iii) HbA1c <9%; (iv) stable hypoglycaemic, hypotensive and hypolipidemic therapy; and (v) neurological deficit of 4 points or more according to the neuropathy disability score (NDS). Exclusion criteria were as follows: Current and former smokers, obliterating atherosclerosis, diabetic foot or Charcot osteoarthropathy, inflammatory joint diseases, B12 deficiency, vitamin D supplementation, treatment with tricyclic antidepressants, anticonvulsants, opiates or nonsteroidal anti-inflammatory drugs. The patient's decision against further participation in the trial, failure to appear at the scheduled time and any acute inflammatory/infectious disease during the trial were withdrawal criteria.

Sixty-seven patients (34 females, median age 56 (49; 61) years) were enrolled into the study. Patients were randomized using the even/odd method into two cholecalciferol treatment groups: Group I (*n* = 34) 5000 IU once weekly and Group II (*n* = 33) 40,000 IU once weekly, taken orally for 24 weeks. Three patients refused to participate in the study after randomization. Two patients developed upper respiratory tract infection and were withdrawn from the study soon after randomization. Thus, 62 patients completed the study (31 patients from each group; Figure 1).

The trial was performed at the First Pavlov State Medical University and Almazov National Medical Research Centre, St. Petersburg, Russia, and it was conducted in compliance with the principles of the Declaration of Helsinki. Each patient gave written informed consent before enrolment. The study was approved by the local ethics committee.

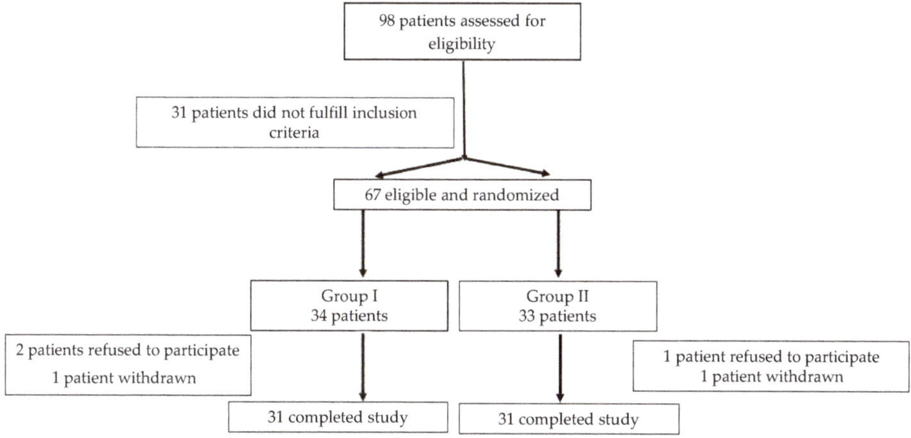

Figure 1. Flowchart showing patient randomization and disposition.

2.2. Data Collection

Patient demographics (gender, age, height, body weight, calculated body mass index (BMI)) and blood pressure (BP), anamnesis (diabetes duration, complications, concomitant diseases, medications, smoking and alcohol intake) and anthropometric data were assessed at baseline.

Neuropathy assessment was done using standard tests: NSS (neuropathic symptomatic score) [22], NDS (neuropathic disability score) [23] and VAS (visual analogue scale, to measure painful symptoms) [24].

Laboratory tests were performed before and 24 weeks after cholecalciferol treatment. Blood samples were taken from the antecubital vein in the morning after an overnight fast (not less than 12 h after the last meal) and centrifuged at 4000 rpm and serum was stored at −20 °C until analysis.

Serum total cholesterol (TC, reference values 3.5–5.0 mmol/L) and C-reactive protein (CRP, reference values 0.00–5.00 mg/L) levels were evaluated on automatic biochemical analyser (COBAS INTEGRA 400 plus, Roche Diagnostics GmbH, Mannheim, Germany). Determination of HbA1c (reference values 4.0–6.0%) was carried out on a Bio-Rad D-10 Chemistry Analyzer (Bio-Rad Diagnostics, Hercules, USA). Serum 25(OH)D level was measured by chemiluminescent immunoassay with commercial laboratory kits and control kits (Abbott Laboratories, Waukegan, USA) using an Architect i2000 analyser (Abbott, Abbott Park, IL, USA). Vitamin D deficiency was defined as serum 25(OH)D level < 20 ng/mL, insufficiency—from 20 ng/mL to 30 ng/mL and adequate vitamin D level > 30 ng/mL [25]. The level of parathyroid hormone (PTH, reference values 15.0–65.0 pg/mL) was estimated using chemiluminescent immunoassay on microparticles (Architect i2000, Abbott, Abbott Park, IL, USA). Serum interleukins (IL) and tumour necrosis factor-α (TNFα) were determined by enzyme-linked immunosorbent assay (Bio-Rad 680 Microplate Reader, Hercules, USA) using the appropriate sets of reagents for enzyme immunoassay to determine the concentration of IL-1β (reference values 0–5.0 pg/mL), IL-6 (reference values 0–7.0 pg/mL), IL-10 (reference values 0–9.1 pg/mL) and TNFα (reference values 0–8.21 pg/mL) (Vector-Best, Novosibirsk, Russia).

Skin microcirculation (MC) was assessed at baseline and after 24 weeks of vitamin D therapy by the laser Doppler flowmetry (LDF) method (LAKK-M complex, LAZMA LLC, Moscow, Russia) using standard functional tests (occlusal and orthostatic). LDF measurements were carried out at room temperature of 24 °C. Basal MC was evaluated on the plantar surface of the big toe of the right lower limb in supine position after 15-min rest, during which the test area was not covered [26]. The average MC parameters measured in perfusion (pf) units were automatically calculated: M—average perfusion value, σ—average blood flow modulation, Kv—coefficient of variation (%). Post occlusal (Mbase—average value of MC before occlusion (pf unit); Moccl—indicator of MC in the process of occlusion ("biological zero") (pf unit); Mmax—maximum value of MC during the postocclusal hyperemia (pf unit); RCB—reserve of capillary blood flow (the ratio of Mmax to Mbase, %) and orthostatic LDF tests (Mbase average value of MC before orthostasis (pf unit), Mmin (pf unit) —minimal decrease in blood flow, and the degree of decrease in blood flow (DDB) (%)) were performed for each diabetic patient. In healthy subjects, RCB ranged from +80 to +150%, and the normal decrease in the level of MC during the postural test reached 30–45%.

2.3. Study Objective

The primary outcome was to evaluate the effect of high and low dose of vitamin D on skin microcirculation after 24 weeks of treatment. The secondary outcomes were change from baseline in plasma interleukins and TNFα at 24 weeks. Other secondary outcomes were change in clinical and symptom scores for neurological status (NDS, VAS and NSS), all assessed at baseline and at 24 weeks.

2.4. Statistical Analysis

Statistical data processing was carried out using the licensed software package SAS 9.4 (SAS, Buckinghamshire, UK). All data points for both treatment groups were collected, hence data imputation was not implemented in this study. Results are presented as median and interquartile range [IQR, Q25; Q75]. Comparison of the indicators in the groups before and after treatment was performed by Wilcoxon T-test. Parameters of the two treatment groups were compared using Mann–Whitney U-test. Analysis of clinical and laboratory data was evaluated using the Pearson χ^2 criterion. The relationship between the indicators was assessed by the Spearman rank correlation method. Statistical significance was defined as $p < 0.05$.

Clinical Trial Registration: URL: https://clinicaltrials.gov. Unique Identifier: NCT04377399.

3. Results

The basic characteristics of the study participants are presented in Table 1. The median age of the study participants was 56 (49; 61) (range 36 to 65) years, BMI—30.2 (28.3; 32.7) kg/m^2, serum 25(OH)D—17.2 (10.2; 27.9) ng/mL, HbA1c—7.9 (7.2; 8.4)%. Both treatment groups were matched for age, gender, diabetes duration, BMI, HbA1c, neuropathy severity, comorbidities and concomitant medications. Glucose-lowering and concomitant therapy was stable throughout the study period.

At the beginning of the study most patients had vitamin D deficiency/insufficiency (25 patients (79.7%) from group I; 24 patients (77.4%) from group II). After 24 weeks of cholecalciferol intake, an increase in serum 25(OH)D was observed in both groups. Thus, all patients taking 40,000 IU per week reached 25(OH)D levels of ≥30 ng/mL after 24 weeks of treatment, while only 15 patients (48.4%) from group I (5000 IU weekly) achieved a normal vitamin D value.

After 24 weeks of treatment, a significant decrease in BMI ($p = 0.001$), HbA1c level ($p = 0.004$), serum IL-6 ($p < 0.001$) and an increase in serum IL-10 ($p < 0.001$) were found in patients taking 40,000 IU of cholecalciferol per week. Weight loss of more than 5% was seen in 19 patients (61%) from this group. No significant changes of any of the above parameters were observed in group I. Both groups showed no significant changes in total cholesterol, PTH, IL-1β, TNFα and CRP after 24 weeks of treatment. Baseline and follow-up values of investigated parameters are presented in Table 2.

At the end of the study, a negative correlation between the level of 25(OH)D and HbA1c ($r = -0.388$, $p = 0.031$) and positive correlation of HbA1c with IL-6 ($r = 0.426$, $p = 0.017$) and IL-10 ($r = -0.391$, $p = 0.030$) was observed in Group II. Also, the correlation analysis revealed an interlink between the severity of neurological deficit and HbA1c level ($r = -0.352$, $p = 0.003$).

Table 1. Baseline clinical characteristics of type 2 diabetic patients according to randomization.

Characteristics	5000 IU/Week, $n = 31$ (Group I)	40,000 IU/Week, $n = 31$ (Group II)	p
Males, n (%)/Females, n (%)	15 (48.4)/16 (51.6)	16 (51.6)/15 (48.4)	0.800
Age, years	57 (48; 62)	55 (52; 60)	0.756
Body mass index, kg/m^2	30 (28.3; 31.8)	31 (29.5; 32.)7	0.155
Obesity, n (%)	21 (68)	20 (65)	0.789
Duration of type 2 diabetes, years	6 (5; 8.5)	7 (5; 11)	0.733
Diabetic peripheral neuropathy, n (%)	31 (100)	31 (100)	1.000
Neuropathic symptomatic score, points	5 (4; 6)	5 (4; 6)	0.799
Neuropathic dysfunctional score, points	8 (7; 9)	8 (7; 9)	0.857
Visual analog scale, mm	50 (40; 60)	50 (42.5; 55)	0.744
Diabetic retinopathy, n (%)	21(68)	24(77)	0.394
Diabetic nephropathy, n (%)	11(35)	9(29)	0.584
Arterial hypertension, n (%)	23 (74)	25 (81)	0.544
Coronary heart disease, n (%)	17 (55)	15 (48)	0.701
Insulin, n (%)	11 (35)	9 (29)	0.587
Metformin, n (%)	29 (94)	25 (81)	0.130
Sulfonylureas, n (%)	4 (13)	5 (16)	0.719
DPP-4 inhibitors, n (%)	5(16)	5(16)	1.000
SGLT-2 inhibitors, n (%)	1 (3)	3(10)	0.302
GLP-1R agonists, n (%)	-	1(3)	0.314
ACE inhibitors/ARB, n (%)	23 (74)	25 (81)	0.544
Calcium channel blockers, n (%)	5 (16)	7 (22)	0.521
β-adrenergic receptor blockers, n (%)	21 (68)	23 (74)	0.576
Diuretics, n (%)	14 (45)	11 (35)	0.438
Statins, n (%)	15 (48)	16 (52)	0.800

Data are presented as median, interquartile range [Q25; Q75] and percentages (%); DPP-4—Dipeptidyl-peptidase-4; SGLT-2—sodium-glucose transport protein 2; GLP-1R—Glucagon-Like Peptide-1 Receptor; ACE—angiotensin converting enzyme; ARB—angiotensin II receptor blockers.

Table 2. Anthropometric and laboratory parameters at baseline and after 24-week treatment.

Parameters	5000 IU/Week, $n = 31$ (Group I)		40,000 IU/Week, $n = 31$ (Group II)	
	Baseline	After 24 Weeks	Baseline	After 24 Weeks
BMI, kg/m^2	30 (28.3; 31.8)	30 (28.4; 31.8)	31 (29.5; 32.7)	28,7 (25.4; 30.4) **,#
25(OH)D, ng/mL	18.8 (10.7; 27.4)	26.9 (20; 34.6) *	16.2 (8.7; 25.3)	71.6 (54.8; 88.3) **,##
HbA1c, %	7.9 (7.1; 8.3)	7.9 (7.2; 8.4)	7.9 (7.1; 8.5)	7.4 (6.5; 7.7) *,#
PTH, pg/mL	34.5 (24.3; 45.7)	28.6 (23.4; 40.4)	32.8 (23.5; 45.2)	26.6 (19.2; 34.6)
TC, mmol/L	4.9 (4.1; 6.1)	5.3 (4.1; 6.3)	5.5 (4.5; 6.5)	5.4 (4.7; 6.1)
TNFα pg/mL	2.0 (2.0; 2.0)	2.0 (2.0; 2.0)	2.0 (2.0; 2.0)	2.0 (2.0; 2.0)
CRP ml/L	1.4 (0.7; 2.0)	1.4 (0.8; 2.1)	1.5 (1.1; 2.0)	2.0 (0.8; 3.0)
IL-1β pg/mL	1.0 (1.0; 1.0)	1.0 (1.0; 1.0)	1.0 (1.0; 1.0)	1.0 (1.0; 1.0)
IL-6 pg/mL	1.9 (1.3; 3.1)	2.3 (1.3; 3.1)	2.5 (1.5; 4.1)	0.6 (0.5; 0.8) **,##
IL-10 pg/mL	3.3 (2.5; 4.8)	3.5 (2.5; 5.0)	2.5 (2.5; 3.6)	4.5 (3.5; 5.7) **,#

Data are presented as median and interquartile range (Q25; Q75); p value: * $p < 0.05$, ** $p < 0.001$—compared with previous results in the same group; # $p < 0.05$, ## $p < 0.001$—between groups at baseline and after 24 weeks of therapy; BMI—body mass index; 25(OH)D—25-hydroxyvitamin D; HbA1c—glycated hemoglobin; PTH—parathyroid hormone; TC—total cholesterol; TNFα—tumor necrosis factor α; CRP—C-reactive protein; IL-1β—interleukin 1β; IL-6—interleukin-6; IL-10—interleukin-10.

Baseline parameters of MC (M, σ, and Kv) did not differ between the treatment groups (p_{MI-II} = 0.08; $p_{σI-II}$ = 0.08; p_{KvI-II} = 0.74). After 24 weeks of treatment, a significant difference between the initial and final Kv ($p < 0.001$) was found only in Group II. This increase in Kv reflects an improvement in microcirculation in patients taking 40,000 IU cholecalciferol per week. Correlation analysis revealed a significant relationship between final levels of 25(OH)D and Kv ($r = 0.51$; $p = 0.04$) in patients from Group II. No associations and significant changes were detected in Group I (Table 2). The postural test demonstrated a significant increase in DDB after 24 weeks of treatment ($p < 0.001$) that was detected only in Group II. After 24 weeks of treatment, M_{max} increased significantly in both groups of patients ($p = 0.012$; $p = 0.003$). There were no differences between the initial and final RCB in Group I ($p = 0.056$), but a significant increase in RCB was found in Group II ($p < 0.001$). Indicators of skin microcirculation before and after 24 weeks of vitamin D therapy are presented in Table 3.

Table 3. Microcirculation parameters at baseline and after 24-week treatment.

Parameters	5000 IU/Week, $n = 31$ (Group I)		40,000 IU/Week, $n = 31$ (Group II)	
	Baseline	**After 24 Weeks**	**Baseline**	**After 24 Weeks**
M, pf units	7.41 ± 3.97	7.16 ± 4.26 [#]	6.01 ± 1.89	7.01 ± 2.46 [*,#]
σ, pf units	1.11 ± 0.57	1.05 ± 0.56 [#]	0.85 ± 0.57	1.81 ± 1.14 [*,#]
Kv *, %	17.68 ± 10.14	18.89 ± 10.83 [#]	16.65 ± 10.99	27.96 ± 16.38 [*,#]
Δ Kv, %		+6.8%		+68.3%
		Postural Test		
M_{base}, pf unit	7.75 ± 1.8	7.78 ± 2.3 [#]	6.69 ± 1.51	7.97 ± 2.13 [*,#]
M_{min}, pf unit	6.10 ± 1.52	6.13 ± 2.26 [#]	5.36 ± 1.47	5.07 ± 1.72 [*,#]
DDB, %	24.82 ± 9.27	23.87 ± 9.1 [#]	23.4 ± 12.68	51.88 ± 36.71 [**,#]
Δ DDB, %		−3.8%		+121.7%
		Occlusal Test		
M_{base} pf unit	7.10 ± 1.72	6.74 ± 1.75 [#]	6.49 ± 2.10	7.54 ± 2.89 [*,#]
M_{max} pf unit	9.73 ± 2.25	8.97 ± 3.60 [#]	9.59 ± 3.15	14.57 ± 3.63 [*,#]
RCB, %	40.85 ± 20.31	35.79 ± 17.10 [#]	48.57 ± 18.56	106.8 ± 44.8 [**,#]
Δ RCB, %		−12.4%		+120%

Data are presented as median and interquartile range (Q25; Q75); p value: * $p < 0.05$—compared with previous results in the same group; ** $p < 0.01$—compared with previous results in the same group [#] $p < 0.05$—between groups after 24 weeks of therapy; M—average perfusion value; σ—average blood flow modulation; Kv—coefficient of variation (%); M_{base}—average value of microcirculation before orthostasis or occlusion, M_{min}—minimal decrease in blood flow; pf units—perfusion units; DDB—the degree of decrease in blood flow (%); M_{max}—maximum value of microcirculation during the postocclusal hyperaemia; RCB—reserve of capillary blood flow (the ratio of M_{max} to M_{base}, %); Δ—delta between baseline and 24 weeks parameters in the same group.

Initially, all patients had neurological deficit of more than 4 points according to the neuropathy disability score (NDS). The median severity of neurological deficit was 8, which corresponds to moderately severe diabetic neuropathy. No differences in neuropathy manifestation evaluated by NSS and VAS were observed between the groups. After 24 weeks of treatment, patients from Group II (40,000 IU/week) demonstrated a significant decrease in neurological deficit (NDS points decreased from 8 to 6, $p = 0.001$), reduction of pain severity assessed by VAS (from 50 (42.5; 55) mm to 47 (37.5; 51) mm, $p = 0.001$), and significant decrease in neuropathic symptomatic score points (from 5 (4; 6) to 4 (4; 5), $p = 0.001$). No changes were found in Group I (5000 IU/week).

4. Discussion

Vitamin D deficiency is widespread throughout the world, and patients with obesity, prediabetes, gestational diabetes and T2DM constitute a high-risk group [9,27–29]. Given the presence of obesity in most subjects with impaired glucose metabolism, prophylactic doses of vitamin D for this population should be significantly higher than for individuals with normal body weight [25,27,30]. Our study revealed a very high prevalence of vitamin D deficiency/insufficiency in patients with T2DM, which is consistent with previously reported data [6,17]. After six months, all patients taking 40,000 IU of cholecalciferol per week achieved normal vitamin D levels, while only half of the patients receiving

5000 IU weekly reached normal 25(OH)D concentration. This finding suggests the need to prescribe higher doses of vitamin D for patients with T2DM.

Interestingly enough, vitamin D supplementation has been reported to be associated with body weight reduction, decrease in HbA1c and insulin resistance and improvement in insulin sensitivity [28,29]. Also, patients with higher baseline vitamin D levels have a greater degree of weight loss than those with lower baseline 25(OH)D level [31]. After 24 weeks of treatment, our study found a negative correlation between serum 25(OH)D and BMI in patients receiving 40,000 IU of cholecalciferol weekly, which was 5714 IU/day. We also found a decrease in HbA1c level in patients in group II (40,000 IU/week), though no change in diabetes treatment was introduced. Whether it was an independent vitamin D effect or mediated through body weight reduction remains to be determined. Our results support previously demonstrated correlation between increase in serum 25(OH)D and decrease in HbA1c in patients with T2DM [5,20].

Another fact we know is that chronic microvascular complications in T2DM lead to early disability and significantly increase the cost of treatment [32]. Vitamin D deficiency has been shown to affect diabetic complications by influencing glucose metabolism and inflammatory process [3,33]. Pleiotropic effect of vitamin D on inflammation has been found to play an important role in DPN development, and it is of great scientific and practical interest [34]. Some studies have shown higher TNFα and lower IL-10 levels associated with increased HbA1c in patients with T2DM and DPN than in patients with impaired glucose tolerance and healthy controls [35].

Regarding CRP, its concentration is considered to be a surrogate marker of inflammation, and its increase in T2DM has been also discussed [36]. Thus, in patients with metabolic syndrome, vitamin D therapy resulted in significant IL-6 reduction but did not change CRP concentration [37]. The REGARDS study showed an association between low serum 25(OH)D and increase in IL-6 and CRP levels, and found no associations with IL-10 [38]. Meta-analysis of 20 randomized clinical trials demonstrated lower levels of CRP and TNFα and no differences in IL-6 concentration in patients taking vitamin D therapy compared to the control group [39]. At the same time, active forms of vitamin D have been shown to reduce TNFα and IL-6 production and stimulate IL-10 production by immune cells [40]. The results of our study appeared to be consistent with the previously reported data concerning association between 25(OH)D levels and markers of inflammation, but a significant decrease in IL-6 and increase in IL-10 were revealed only in patients who received high-dose vitamin D therapy (40,000 IU per week) and reached normal 25(OH)D levels. Our findings suggest that normalization of serum 25(OH)D with high-dose cholecalciferol treatment affects inflammatory markers. It is known that immune cells have vitamin D receptors and can participate in the final stage of hydroxylation and in calcitriol formation [40]. An increase in 25(OH)D concentration with cholecalciferol treatment may contribute to activation of calcitriol synthesis, which, in turn, may influence release of proinflammatory and anti-inflammatory cytokines [39].

As for neuropathy, we found a decrease in neurological deficit and pain severity after 24 weeks of treatment with 40,000 IU of cholecalciferol weekly. We found no correlation between values of pain scales with serum 25(OH)D and ILs but found a correlation with HbA1c level. It can be assumed that the effect of cholecalciferol on peripheral nervous system in patients with T2DM and DPN was most likely mediated by improvement in metabolic parameters rather than resulting from direct vitamin D action.

Our study demonstrated significant improvement in skin microcirculation parameters only in patients receiving 40,000 IU of cholecalciferol per week. This effect can be explained by the direct protective action of vitamin D on endothelial cells through specific receptors [41], or it can be mediated by improvement in metabolic parameters and inflammatory status associated with high-dose therapy [42].

Study Limitations

There is lack of data on 1,25-dihydroxyvitamin D (calcitriol) levels, which implements the main pleiotropic effects of vitamin D. There are several methods available to measure 25(OH)D levels. In this study, we used Abbott chemiluminescent immunoassay, which is the method available in our centre. Our study was an open-label one, so the possible effect of information about therapy on outcomes should be considered. We found a relationship between 25(OH)D and only some inflammatory markers, which makes further research in this area necessary. Most patients included in the study had vitamin D deficiency or insufficiency, so the effect of cholecalciferol therapy on peripheral neuropathy in patients with T2DM and normal 25(OH)D remains to be investigated. In addition, we want to highlight that since there is no gold standard for the assessment of microcirculation in diabetic patients, we chose Doppler flowmetry with two functional tests to perform this.

5. Conclusions

The study demonstrated that high-dose cholecalciferol therapy (40,000 IU/week) for 24 weeks resulted in 25(OH)D normalization and was associated with reduction in neuropathy severity, as well as improvement in skin microcirculation and cytokines profile (decrease in proinflammatory IL-6 and increase in anti-inflammatory IL-10), in patients with T2DM and DPN. Our findings suggest that vitamin D deficiency may be a modifiable factor which affects diabetic peripheral neuropathy and requires timely identification and correction with cholecalciferol at doses of more than 5000 IU/day. Further studies are needed to clarify the treatment duration and determine the optimal dose of vitamin D for patients with T2DM and DPN.

Author Contributions: T.K.—concept, design of work, randomization, analysis of results, writing text; A.S.—screened and selected the patients, collected material, analyzing results, analyzing literature, wrote the first draft; A.B.—analysis of results, writing text; E.B.J.—concept, design of work, analysis of results, writing text. All authors made a significant contribution to the research and preparation of the article, read and approved the final version of the article before publication. All authors have read and agreed to the published version of the manuscript.

Funding: The study was supported by the Russian Science Foundation (grant number 17-75-30052 "Development of personalized therapy for obesity and type 2 diabetes in order to reduce cardiovascular risks").

References

1.	Pop-Busui, R.; Boulton, A.J.; Feldman, E.L.; Bril, V.; Freeman, R.; Malik, R.A.; Sosenko, J.M.; Ziegler, D. Diabetic neuropathy: A position statement by the American Diabetes Association. *Diabetes Care* **2017**, *40*, 136–154. [CrossRef]

2.	Cashman, K.D.; Dowling, K.G.; Škrabáková, Z.; Gonzalez-Gross, M.; Valtueña, J.; De Henauw, S.; Moreno, L.; Damsgaard, C.T.; Michaelsen, K.F.; Christian, M.; et al. Vitamin D deficiency in Europe: Pandemic? *Am. J. Clin. Nutr.* **2016**, *103*, 1033–1044. [CrossRef]

3.	Qu, G.B.; Wang, L.L.; Tang, X.; Wu, W.; Sun, Y.H. The association between vitamin D level and diabetic peripheral neuropathy in patients with type 2 diabetes mellitus: An update systematic review and meta-analysis. *J. Clin. Transl. Endocrinol.* **2017**, *9*, 25–31. [CrossRef]

4.	Liu, E.; Meigs, J.B.; Pittas, A.G.; Economos, C.D.; McKeown, N.M.; Booth, S.L.; Jacques, P.F. Predicted 25-hydroxyvitamin D score and incident type 2 diabetes in the Framingham Offspring Study. *Am. J. Clin. Nutr.* **2010**, *91*, 1627–1633. [CrossRef]

5.	Gagnon, C.; Lu, Z.X.; Magliano, D.J.; Dunstan, D.W.; Shaw, J.E.; Zimmet, P.Z.; Sikaris, K.; Grantham, N.; Ebeling, P.R.; Daly, R.M. Serum 25-hydroxyvitamin D, calcium intake, and risk of type 2 diabetes after 5 years: Results from a national, population-based prospective study (the Australian Diabetes, Obesity and Lifestyle study). *Diabetes Care* **2011**, *34*, 1133–1138. [CrossRef]

6. Moreira-Lucas, T.S.; Duncan, A.M.; Rabasa-Lhoret, R.; Vieth, R.; Gibbs, A.L.; Badawi, A.; Wolever, T.M. Effect of vitamin D supplementation on oral glucose tolerance in individuals with low vitamin D status and increased risk for developing type 2 diabetes (EVIDENCE): A double-blind, randomized, placebo-controlled clinical trial. *Diabetes Obes. Metab.* **2017**, *19*, 133–141. [CrossRef]

7. Pittas, A.G.; Dawson-Hughes, B.; Sheehan, P.; Ware, J.H.; Knowler, W.C.; Aroda, V.R.; Brodsky, I.; Ceglia, L.; Chadha, C.; Chatterjeeet, R.; et al. Vitamin D Supplementation and Prevention of Type 2 Diabetes. *N. Engl. J. Med.* **2019**, *381*, 520–530. [CrossRef]

8. Davidson, M.B.; Duran, P.; Lee, M.L.; Friedman, T.C. High-Dose Vitamin D Supplementation in People with Prediabetes and Hypovitaminosis D. *Diabetes Care* **2012**, *36*, 260–266. [CrossRef]

9. Muñoz-Garach, A.; García-Fontana, B.; Muñoz-Torres, M. Vitamin D Status, Calcium Intake and Risk of Developing Type 2 Diabetes: An Unresolved Issue. *Nutrients* **2019**, *11*, 642. [CrossRef]

10. Jamwal, S.; Sharma, S. Vascular endothelium dysfunction: A conservative target in metabolic disorders. *Inflamm. Res.* **2018**, *67*, 391–405. [CrossRef]

11. Magrinelli, F.; Briani, C.; Romano, M.; Ruggero, S.; Toffanin, E.; Triolo, G.; Peter, G.C.; Praitano, M.; Lauriola, F.; Zanette, G.; et al. The Association between Serum Cytokines and Damage to Large and Small Nerve Fibers in Diabetic Peripheral Neuropathy. *J. Diabetes Res.* **2015**, *2015*, 1–7. [CrossRef]

12. Fuchs, D.; Dupon, P.P.; Schaap, L.A.; Draijer, R. The association between diabetes and dermal microvascular dysfunction non-invasively assessed by laser Doppler with local thermal hyperemia: A systematic review with meta-analysis. *Cardiovasc. Diabetol.* **2017**, *16*. [CrossRef]

13. Strain, W.D.; Paldanius, P.M. Diabetes, cardiovascular disease and the microcirculation. *Cardiovasc. Diabetol.* **2018**, *17*, 57. [CrossRef]

14. Pop-Busui, R.; Ang, L.; Holmes, C.; Gallagher, K.; Feldman, E.L. Inflammation as a Therapeutic Target for Diabetic Neuropathies. *Curr. Diabetes Rep.* **2016**, *16*, 29. [CrossRef]

15. Alamdari, A.; Mozafari, R.; Tafakhori, A.; Faghihi-Kashani, S.; Hafezi-Nejad, N.; Sheikhbahaei, S.; Naderi, N.; Ebadi, M.; Esteghamati, A. An inverse association between serum vitamin D levels with the presence and severity of impaired nerve conduction velocity and large fiber peripheral neuropathy in diabetic subjects. *Neurol. Sci.* **2015**, *36*, 1121–1126. [CrossRef]

16. He, R.; Hu, Y.; Zeng, H.; Zhao, J.; Zhao, J.; Chai, Y.; Lu, F.; Liu, F.; Jia, W. Vitamin D deficiency increases the risk of peripheral neuropathy in Chinese patients with Type 2 diabetes. *Diabetes Metab. Res. Rev.* **2017**, *33*, 2820. [CrossRef]

17. Celikbilek, A.; Gocmen, A.Y.; Tanik, N.; Borekci, E.; Adam, M.; Celikbilek, M.; Suher, M.; Delibas, N. Decreased serum vitamin D levels are associated with diabetic peripheral neuropathy in a rural area of Turkey. *Acta Neurol. Belg.* **2015**, *115*, 47–52. [CrossRef]

18. Straube, S.; Derry, S.; Straube, C.; Moore, R.A. Vitamin D for the treatment of chronic painful conditions in adults. *Cochrane Database Syst. Rev.* **2015**, *5*. [CrossRef]

19. Bilir, B.; Tulubas, F.; Bilir, B.E.; Atile, N.S.; Kara, S.P.; Yildirim, T.; Gumustas, S.A.; Topcu, B.; Kaymaz, O.; Aydin, M. The association of vitamin D with inflammatory cytokines in diabetic peripheral neuropathy. *J. Phys. Ther. Sci.* **2016**, *28*, 2159–2163. [CrossRef]

20. Randhawa, F.A.; Mustafa, S.; Khan, D.M.; Hamid, S. Effect of Vitamin D supplementation on reduction in levels of HbA1 in patients recently diagnosed with type 2 Diabetes Mellitus having asymptomatic Vitamin D deficiency. *Pak. J. Med. Sci.* **2017**, *33*, 881–885. [CrossRef]

21. Basit, A.; Basit, K.A.; Fawwad, A.; Shaheen, F.; Fatima, N.; Petropoulos, I.N.; Alam, U.; Malik, R.A. Vitamin D for the treatment of painful diabetic neuropathy. *BMJ Open Diabetes Res. Care* **2016**, *4*, e000148. [CrossRef] [PubMed]

22. Grant, I.A.; O'Brien, P.; Dyck, P.J. Neuropathy tests and normative results. In *Diabetic Neuropathy*, 2nd ed.; Dyck, P.J., Thomas, P.K., Eds.; Saunders: Philadelphia, PA, USA, 1999; pp. 123–141.

23. Gries, A.; Cameron, N.E. Severity and Staging of Diabetic Polyneuropathy. *Textb. Diabet. Neuropathy* 2003. [CrossRef]

24. Hawker, G.A.; Mian, S.; Kendzerska, T.; French, M. Measures of adult pain: Visual Analog Scale for Pain (VAS Pain), Numeric Rating Scale for Pain (NRS Pain), McGill Pain Questionnaire (MPQ), Short-Form McGill Pain Questionnaire (SF-MPQ), Chronic Pain Grade Scale (CPGS), Short Form-36 Bodily Pain Scale (SF-36 BPS), and Measure of Intermittent and Constant Osteoarthritis Pain (ICOAP). *Arthritis Care Res.* **2011**, *63*, 240–252. [CrossRef]

25. Pigarova, E.A.; Rozhinskaia, L.Y.; Belaia, J.E.; Dzeranova, L.K.; Karonova, T.L.; Ilyin, A.V.; Melnichenko, G.A.; Dedov, I.I. Russian Association of Endocrinologists recommendations for diagnosis, treatment and prevention of vitamin D deficiency in adults. *Probl. Endocrinol.* **2016**, *62*, 60–84. [CrossRef]

26. Karnafel, W.; Juskowa, J.; Maniewski, R.; Liebert, A.; Jasik, M.; Zbieć, A. Microcirculation in the diabetic foot as measured by a multichannel laser Doppler instrument. *Med. Sci. Monit.* **2002**, *8*, MT137–MT144.

27. Karonova, T.; Andreeva, A.; Nikitina, I.; Belyaeva, O.; Mokhova, E.; Galkina, O.; Vasilyeva, E.; Grineva, E. Prevalence of Vitamin D deficiency in the North-West region of Russia: A cross-sectional study. *J. Steroid Biochem. Mol. Biol.* **2016**, 230–234. [CrossRef]

28. Maddaloni, E.; Cavallari, I.; Napoli, N.; Conte, C. Vitamin D and Diabetes Mellitus. *Vitam. D Clin. Med.* **2018**, *50*, 161–176. [CrossRef]

29. Lips, P.; Eekhoff, M.; van Schoor, N.; Oosterwerff, M.; de Jongh, R.; Krul-Poel, Y.; Simsek, S. Vitamin D and type 2 diabetes. *J. Steroid Biochem. Mol. Biol.* **2017**, *173*, 280–285. [CrossRef]

30. Pludowski, P.; Karczmarewicz, E.; Bayer, M.; Carter, G.; Chlebna-Sokół, D.; Czech-Kowalska, J.; Dębski, R.; Decsi, T.; Dobrzańska, A.; Franek, E. Practical guidelines for the supplementation of vitamin D and the treatment of deficits in Central Europe-recommended vitamin D intakes in the general population and groups at risk of vitamin D deficiency. *Endokrynol. Pol.* **2013**, *64*, 319–327. [CrossRef]

31. Earthman, C.P.; Beckman, L.M.; Masodkar, K.; Sibley, S.D. The link between obesity and low circulating 25-hydroxyvitamin D concentrations: Considerations and implications. *Int. J. Obes.* **2012**, *36*, 387–396. [CrossRef]

32. Chapman, D.; Foxcroft, R.; Dale-Harris, L.; Ronte, H.; Bidgoli, F.; Bellary, S. Insights for Care: The Healthcare Utilisation and Cost Impact of Managing Type 2 Diabetes-Associated Microvascular Complications. *Diabetes Ther.* **2019**, *10*, 575–585. [CrossRef] [PubMed]

33. Calton, E.K.; Keane, K.N.; Newsholme, P.; Soares, M.J. The Impact of Vitamin D Levels on Inflammatory Status: A Systematic Review of Immune Cell Studies. *PLoS ONE* **2015**, *10*, e0141770. [CrossRef] [PubMed]

34. Stepanova, A.P.; Karonova, T.L.; Bystrova, A.A.; Bregovsky, V.B. Role of vitamin D deficiency in type 2 diabetes mellitus and diabetic neuropathy development. *Diabetes Mellitus.* **2018**, *21*, 301–306. (In Russian) [CrossRef]

35. Zeng, J.; Xu, Y.; Shi, Y.; Jiang, C. Inflammation role in sensory neuropathy in Chinese patients with diabetes/prediabetes. *Clin. Neurol. Neurosurg.* **2018**, *166*, 136–140. [CrossRef]

36. Mazidi, M.; Toth, P.P.; Banach, M. C-reactive Protein Is Associated with Prevalence of the Metabolic Syndrome, Hypertension, and Diabetes Mellitus in US Adults. *Angiology* **2017**, *69*, 438–442. [CrossRef]

37. Salekzamani, S.; Bavil, A.S.; Mehralizadeh, H.; Jafarabadi, M.A.; Ghezel, A.; Gargari, B.P. The effects of vitamin D supplementation on proatherogenic inflammatory markers and carotid intima media thickness in subjects with metabolic syndrome: A randomized double-blind placebo-controlled clinical trial. *Endocrine* **2017**, *57*, 51–59. [CrossRef]

38. Jackson, J.L.; Judd, S.E.; Panwar, B.; Howard, V.J.; Wadley, V.G.; Jenny, N.S.; Guti, O.M. Associations of 25-hydroxyvitamin D with markers of inflammation, insulin resistance and obesity in black and white community-dwelling adults. *J. Clin. Transl. Endocrinol.* **2016**, *5*, 21–25. [CrossRef]

39. Mousa, A.; Naderpoor, N.; Teede, H.; Scragg, R.; de Courten, B. Vitamin D supplementation for improvement of chronic low-grade inflammation in patients with type 2 diabetes: A systematic review and meta-analysis of randomized controlled trials. *Nutr. Rev.* **2018**, *76*, 380–394. [CrossRef]

40. Colotta, F.; Jansson, B.; Bonelli, F. Modulation of inflammatory and immune responses by vitamin D. *J. Autoimmun.* **2017**, *85*, 78–97. [CrossRef]

41. Vitamin D Attenuates Endothelial Dysfunction in Uremic Rats and Maintains Human Endothelial Stability. *J. Am. Heart Assoc.* **2018**, *7*, e02709. [CrossRef]

42. Gil, Á.; Plaza-Diaz, J.; Mesa, M.D. Vitamin D: Classic and Novel Actions. *Ann. Nutr. Metab.* **2018**, *72*, 87–95. [CrossRef] [PubMed]

Article

Vitamin D Doses from Solar Ultraviolet and Dietary Intakes in Patients with Depression: Results of a Case-Control Study

Haitham Jahrami [1,2,*], Nicola Luigi Bragazzi [3], William Burgess Grant [4], Hala Shafeeq Mohamed AlFarra [1], Wafa Shafeeq Mohamed AlFara [1], Shahla Mashalla [1] and Zahra Saif [1]

[1] Ministry of Health, Manama, Bahrain, P.O. Box 12 Manama, Bahrain; hfara@health.gov.bh (H.S.M.A.); wshafiq@health.gov.bh (W.S.M.A.); smaki@health.gov.bh (S.M.); zsaif@health.gov.bh (Z.S.)
[2] College of Medicine and Medical Sciences, Arabian Gulf University, P.O. Box 26671 Manama, Bahrain
[3] Laboratory for Industrial and Applied Mathematics (LIAM), Department of Mathematics and Statistics, York University, ON M3J 1P3, Canada; robertobragazzi@gmail.com
[4] Sunlight, Nutrition, and Health Research Center, P.O. Box 641603, San Francisco, CA 94164-1603, USA; wbgrant@infionline.net
* Correspondence: hjahrami@health.gov.bh; Fax: +97-317-270-637

Received: 3 August 2020; Accepted: 24 August 2020; Published: 26 August 2020

Abstract: The purpose of this study to estimate cumulative vitamin D doses from solar ultraviolet and dietary intakes in patients with depression and compare it to healthy controls. Using a case-control research design, a sample of 96 patients with depression were age- and sex-matched with 96 healthy controls. Dietary vitamin D dose was estimated from diet analysis. Vitamin D-weighted ultraviolet solar doses were estimated from action spectrum conversion factors and geometric conversion factors accounting for the skin type, the fraction of body exposed, and age factor. Patients with depression had a lower dose of vitamin D (IU) per day with 234, 153, and 81 per day from all sources, sunlight exposure, and dietary intake, respectively. Controls had a higher intake of vitamin D (IU) per day with 357, 270, and 87 per day from all sources, sunlight exposure, and dietary intake, respectively. Only 19% and 30% met the minimum daily recommended dose of ≥400 IU per day for cases and controls, respectively. The sensitivity, specificity, percentage correctly classified and receiver operating characteristic (ROC) Area for the estimated vitamin D against serum vitamin D as reference were 100%, 79%, 80%, and 89%. Physical activity level was the only predictor of daily vitamin D dose. Vitamin D doses are lower than the recommended dose of ≥400 IU (10 mcg) per day for both cases with depression and healthy controls, being much lower in the former.

Keywords: 25OHD; mood disorders; UVB; vitamin D analogs; vitamin D supplementation

---

1. Introduction

Depression is a universal mental illness that affects a large proportion of any community [1]. A recent meta-analysis showed that the estimated point-, 12-months, and lifetime-prevalence rates of depression are 12.9%, 7.2%, and 10.8%, respectively [1]. The illness affects 350 million persons worldwide and is considered a leading cause of disability [2]. The first line of treatment for major depression is pharmacotherapy [3]. Recent network meta-analyses show that drug pharmacotherapy demonstrates minimal difference from placebo [4,5]. Other modalities such as electroconvulsive therapy and psychotherapy also showed similar results compared to sham procedures [6–8]. Dietary and lifestyle approaches hold potential as a novel intervention for the management of symptoms of depression [9]. They can be used in support of pharmacotherapy for severe cases. Therefore, understanding the specific components of dietary and lifestyle interventions that improve mental health are needed.

The association between depression and the status of vitamin D from lack of sun exposure is well established and was first described two thousand years ago [10]. Results from epidemiological studies shows that vitamin D deficiency is associated with an 8%–14% increase in depression [11] and a 50% increase in suicide [12]. In the past 10 years, an increasing body of literature has linked vitamin D to the pathophysiology of depression [13]. This comes from three lines of evidence; first, the presence of vitamin D receptors in various parts of the cortex and limbic system [14]; second, the important modulatory role that vitamin D plays in regulating immunoinflammatory pathways that are relevant to the pathophysiology of depression [15,16]; third, lower serum vitamin D levels in depressed patients compared to controls [13,17,18]. The reasons for the difference in serum vitamin D between cases with depression and controls remained unclear.

Vitamin D deficiency for patients with depression as well as a healthy population has become an important community health concern. Previous research focused on either laboratory approaches of measuring vitamin D serum 25-hydroxyvitamin D 25(OH)D [19,20], or focused on establishing an association between diet or dietary supplements and vitamin D [21]. No previous work was established to cover the estimation of solar ultraviolet (UV) doses of patients with depression and vitamin D3 production. Accordingly, the current study was designed to estimate how much vitamin D3 is acquired from diet and produced from everyday outdoor ultraviolet type B doses in Bahrain (26 °N) for cases with depression in comparison to age- and sex-matched controls. We hypothesize that healthy controls acquire higher daily vitamin D3 doses from both dietary and sunlight exposure compared to cases. We also hypothesize that the severity of depressive symptoms is associated with the level of vitamin D3 acquired.

2. Materials and Methods

2.1. Study Design and Setting

The current research utilized the guidelines of the strengthening the reporting of observational studies in epidemiology (STROBE) statement [22]. The study took place between March and December 2019.

Cases with depression were recruited from the outpatient clinics of the general adult psychiatry services of the Psychiatric Hospital, Ministry of Health, Manama, Kingdom of Bahrain. The Psychiatric Hospital, Bahrain, is the national center for mental illness in Bahrain. The hospital registry shows that there are about 750 cases with depression only without another psychiatric morbidity. Controls were recruited from local health centers during regular non-emergency visits, and routine/investigation visits. The local health centers are the primary healthcare clinics belonging to the Ministry of Health, Bahrain.

2.2. Participants

Cases—We included cases with depression (major depressive disorder, single episode, unspecified depressive disorder). Diagnosis was made using the International Statistical Classification of Diseases, 10th Revision. We included adults aged between 20–60 years who were diagnosed in the past six months or more using the ICD-10 criteria. We excluded: women who are pregnant or lactating; the coexistence of any other psychiatric disorder, e.g., eating disorder, generalized anxiety disorder, etc.; or those who were dieting, taking dietary supplements, or enrolled in lifestyle experimental studies.

Controls—We included controls, defined as individuals free from a known history of mental illness including depression. Controls were achieved by matching each case with depression with a person from the local care centers. Age match was on the basis of year of birth. We excluded: women who are pregnant or lactating, positive history of psychiatric disorder, those who were dieting, taking dietary supplements, or enrolled in lifestyle experimental studies.

2.3. Sample Size and Sampling Techniques

Using a matched case-control design, we estimated the sample needed for our research to be 75 patients and 75 controls. Sample size calculations are based on a z test, with a 1:1 ratio design assuming the difference in vitamin D3 intake by 33% based on previous research [23]. The sample size was estimated for the two-sided test with error probabilities of alpha = 0.05 and 80% power (beta = 0.20). To further increase the statistical power, we aimed to include 95–100 patients in each group.

Probability sampling techniques were used for recruiting cases and controls. The sample of depression cases (n = 96) was selected using a simple random sampling technique from the case registry. Similarly, controls (n = 96) were selected using simple random sampling after matching.

2.4. Data Collection Procedure

Data were collected using structured forms and included sociodemographic and anthropometric variables, medical history, and comprehensive lifestyle assessment. The anthropometric measurements included weight, height, and body composition analysis. Weight was measured using electronic scales with rod height attachment. During measurements, individuals were advised to stand straight, without footwear, and keep on only light clothes. Body composition analysis (BCA) was completed using a bioelectrical impedance system (The InBody 230 model: MW160, Seoul/Korea). BCA involved fat mass, and body fat percentage. Body mass index (BMI) (kg/m^2) was classified corresponding to the World Health Organization (WHO) categories of underweight (<18.5), normal (18.5–24.9), overweight (25.0–29.9), or obese (≥30) [24].

The electronic medical record was accessed to obtain data available in the past six months from the interview on serum vitamin D, and no special request was made to collect a new blood sample. Vitamin D was analyzed as 25(OH)D using a chemiluminescent immunoassay in our study. This method (in Ministry of Health, Bahrain laboratories) has a correlation coefficient with the high performance liquid chromatography assay of 0.92.

For cases with depression, the Beck Depression Inventory-II (BDI-II) was used to assess the severity of symptoms. The BDI-II is a sum score of all 21 items of the scale; each item is evaluated on a 4 points (0–3) Likert scale [25]. The following algorithm has been used to interpret the BDI-II: minimal depression = 0–13, mild depression = 14–19, moderate depression = 20–28, and severe depression = 29–63. We used the validated Arabic version of the BDI-II in our study [26].

A quantitative food frequency questionnaire (covering 102 foods distributed on 38 items/groups) was used [27]. Participants were requested to report the frequency of consuming a standard serving of a specific food item in six categories (1 time/day, ≥2 times/day, 1–2 times/week, 3–6 times/week, 1–3 times/month, rarely, or never). Special attention was given to vitamin D rich food including fatty fish, liver, meat, cheese, eggs, dairy products, and foods fortified with vitamin D such as juices and cereals [28]. The responses on the frequent consumption of a specific serving size were standardized using visual aids to determine a standard unit for portions. Dietary intake assessed using the FFQ was analyzed using nutrition and fitness software (ESHA Food Processor SQL, version 10.1.1, Salem, OR, USA). ESHA was used to estimate a gross mean of daily vitamin D3 intake from food. We also obtained data on current smoking history and physical activity. Individuals were considered to be physically active when they met the target of 150 min of moderate-intensity (or 75 min of vigorous-intensity) per week [29].

Solar ultraviolet doses and vitamin D3 production were estimated using the approach described by Godar and colleagues [30]. To do that, we obtained information on the following: Fitzpatrick skin type scale, duration and timing of direct exposure to sunlight per day, the fraction of body exposed, age factor, action spectrum conversion factors (ASCF), and geometric conversion factors (GCF).

The Fitzpatrick skin type scale is utilized to evaluate the reaction of different types of skin to ultraviolet light [31]. Type I (scores 0–6—pale white) easily burns, does not tan. Type II (scores 7–13—white) typically burns, tans slightly. Type III (scores 14–20—light brown) mild burn, tans consistently. Type IV (scores 21–27—moderate brown) minimally burns, tans. Type V (scores

28–34—dark brown) infrequently burns, tans easily. Type VI (scores 35–36—dark brown or black) never burns.

Scattered duration and timing of direct exposure to sunlight per day were obtained by asking the participants to estimate the average time spent on outdoor activities with an emphasis on the proportion being exposed to direct sunlight. This was used to calculate Standard Erythemal Dose (SED) [32]. The solar zenith angle was not considered in our research.

The fraction of body exposed is the body surface exposed to sunlight. The following standard fractions were used: sun on arms and hands only (short-sleeved shirt, head is covered) = 11%; sun on face, neck, arms, and hands (same like before, but no head cover) = 18%; sun on face, neck, arms, hands, and lower legs (wearing shorts and shirt, no head cover) = 32%; sun on the top half of body (stripped up to waist) = 53%; sun on whole body except for one-piece bathing costume (ladies) = 73%; sun on whole body except for swimming costume = 88%; sun on whole body = 100% [30].

Age factor encompasses the ability of an adult to synthesize vitamin D3. The ability to produce vitamin D3 is decreased as human age due to decreased 7-dehydrocholesterol in the skin. The following age factor conversion was used: 0–20 years (100% or 1.0), 22–40 years (83% or 0.83), 41–59 years (66% or 0.66), and 60+ years (49% or 0.49) [30,33].

The action spectrum conversion factors are the differences between wavelength contributions approximated by the erythemal action spectrum and the previtamin D action spectrum toward previtamin D3 production. ASCFs for Bahrain (26 °N) were compensated with values latitude 30 °N as follows: 1.110 for summer, 1.061 for fall, 0.910 for winter, and 1.065 for spring, respectively [34].

The standard vitamin D dose, which represents a horizontal plane or planar doses, is converted to whole-body doses using geometric conversion factors based on a full-cylinder model representing the human body. GCF for Bahrain (26 °N) is 0.580 during the summer and spring and 0.644 during the winter and fall [34]. The daily estimate of synthesized vitamin D3 per day was estimated using the following equations:

- Estimate vitamin D3 (IU) per day = Vitamin D Dose (VDD) × (4900 IU) × skin type factor × fraction of body exposed × age factor.
- Standard Vitamin D Dose (SVD) = Standard Erythemal Dose (SED/day) × Action Spectrum Conversion Factor (ASCF).
- Vitamin D Dose (VDD) = Standard Vitamin D Dose (SVD) × Geometric Conversion Factors (GCF).

To convert vitamin D from IU to mc: 1 IU is approximated to be the biological equivalent of 0.025 mcg cholecalciferol or ergocalciferol [35].

2.5. Ethical Considerations

This research was approved by the Secondary Healthcare Research Ethics Committee in the Ministry of Health, Bahrain (No.2018/REC/EF023). Before the start of data collection, informed consent was requested and secured from each person included.

2.6. Statistical Analyses

Descriptive statistics were used to a provide summary of the demographic characteristics, health status, and daily vitamin D from diet and sunlight exposure. The arithmetic mean and standard deviation (SD) were utilized for continuous variables, and the count and percentage for categorical variables. A daily dose of vitamin D < 400 IU, serum levels < 30 nmol/L was considered as deficient, levels between 30 nmol/L and 50 nmol/L (≥30, <50) were classified as vitamin D insufficiency, and optimal levels were ≥50 nmol/L. Sensitivity, specificity, percentage correctly classified, and receiver operating characteristic (ROC) Area were calculated for the estimated intake of vitamin D using 25(OH)D as reference. Multiple linear regression analysis was performed to assess the association between the dose of vitamin D per day and selected predictors. A statistically significant result was *p*-value < 0.05. All analyses were executed using Stata 16.1 programming [36].

3. Results

This study involved 192 participants: 96 patients with depression and 96 age- and sex-matched controls. The mean age was approximately 43 years, with 60% being female sex. Table 1 shows the characteristics of the study participants. The results generally show that patients with depression are more likely to be unemployed, single, and overweight or obese. During the study, all patients were on active pharmacological treatments, 42% were on selective serotonin reuptake inhibitors, 35% were on serotonin and norepinephrine reuptake inhibitors, 12% were on tricyclic antidepressants, and the remaining 11% were on others or combined antidepressants therapy.

Table 1. Sociodemographic and anthropometric characteristics of the study participants.

Variable *	Cases, *n* = 96	Controls, *n* = 96	*p*-Value **
Sex			
Male	37 (39%)	37 (39%)	1.0
Female	59 (61%)	59 (61%)	
Job-status			
Employed	27 (28%)	69 (72%)	0.001
Unemployed	69 (72%)	27 (28%)	
Marital status			
Single	48 (50%)	23 (24%)	0.001
Married	48 (50%)	73 (76%)	
BMI classification			
Underweight	4 (4%)	2 (2%)	
Normal	26 (27%)	34(35%)	0.25
Overweight	30 (31%)	35 (37%)	
Obese	36 (38%)	25 (26%)	
Current tobacco smoker	37%	10%	0.001
Beck Depression Inventory-II (BDI-II)			
Mild	13 (13%)		
Moderate	40 (42%)	Not applicable	-
Severe	43 (45%)		
Age (year)	44 ± 13	43 ± 15	0.4
Weight (kg)	76 ± 19	75 ± 17	0.63
Height (cm)	163 ± 10	165 ± 10	0.13
BMI (kg/m^2)	29 ± 7	28 ± 6	0.11
Body fat percentage (%)	35 ± 12	33 ± 10	0.09
Total body water percentage (%)	36 ± 6	36 ± 7	0.91
Body surface area (m^2)	2 ± 0.2	2 ± 0.2	0.98
Lean mass (kg)	49 ± 8	49 ± 8	0.68
Fat mass (kg)	28 ± 13	26 ± 10	0.31
Serum 25(OH)D (nmol/L) *** [a,b]	35 ± 7	38 ± 6	0.01

* Frequency count and (%) OR Mean ± SD; ** Independent samples *t*-test or Pearson's Chi-Squared; *** [a] To convert to ng/mL, divide by 2.5, [b] data available for 43 cases and 50 controls.

Table 2 shows the vitamin D status of the study participants. The daily dose of vitamin D is approximately 260 IU (7 mcg) per day for the entire participants (*n* = 192), with 212 IU (5 mcg) per day acquired from sunlight exposure and 84 IU (2 mcg) per day from dietary intake. Only 47 (25%) met the minimum daily recommended dose of ≥400 IU (10 mcg) per day. Patients with depression had a lower intake of vitamin D per day with 234 IU (6 mcg), 153 (4 mcg), and 81 (2 mcg) per day from all sources, sunlight exposure, and dietary intake, respectively. Controls had a higher intake of vitamin D per day with 357 IU (9 mcg), 270 (7 mcg), and 87 (2 mcg) per day from all sources, sunlight exposure, and dietary intake, respectively. Intake of vitamin D from the diet was equal between the two groups $p = 0.5$, but intake from sunlight exposure and cumulative daily intake of vitamin D was statistically significant for the favor of controls $p = 0.001$ and $p = 0.001$, respectively. Serum 25(OH)D for cases with depression and controls were 35 ± 7 nmol/L (ng/mL) and 38 ± 6 nmol/L (ng/mL), respectively. The difference was statistically significant $p = 0.01$. Recent research in Bahrain showed that controls have a mean serum of 39.95 nmol/L [37]. The proportions of persons at the cutoff 25(OH)D ≥ 35 nmol/L were 56% and 76%, and at cutoff 25(OH)D ≥ 40 nmol/L were 21% and 46% for cases and controls, respectively. The difference was significant at both cutoffs points $p = 0.04$ and $p = 0.01$, respectively. See Table 2.

Table 2. Vitamin D status of the study participants.

* Variable	Cases, $n = 96$				Controls, $n = 96$				p-Value **
	Mean	SD	SE	95%CI	Mean	SD	SE	95%CI	
Vitamin D intake from diet per day (IU)	81	65	7	68–94	87	66	7	74–101	0.50
Vitamin D synthesis from sunlight per day (IU)	153	206	21	111–195	270	260	27	218–323	0.001
Vitamin D per day (IU)	234	275	23	189–280	357	275	28	301–413	0.001
Share of Vitamin D from diet per day	35%				25%				0.11
Share of Vitamin D from sunlight exposure per day	65%				75%				0.11
Compliance with the recommended minimum daily intake (400 IU per day)	18 (19%)				29 (30%)				0.048
Vitamin D according to 25(OH)D Optimal—≥50 nmol/L Insufficient—≥30 <50 nmol/L Deficient—<30 nmol/L	1 (2%) 34 (79%) 8 (19)				3 (6%) 44 (88%) 3 (6%)				0.13
Serum 25(OH)D ≥ 30 nmol/L	35 (83%)				47 (94%)				0.06
Serum 25(OH)D ≥ 35 nmol/L	24 (56%)				38 (76%)				0.04
Serum 25(OH)D ≥ 40 nmol/L	9 (21%)				23 (46%)				0.01
Serum 25(OH)D ≥ 45 nmol/L	3 (7%)				4 (8%)				0.90
Serum 25(OH)D ≥ 50 nmol/L	1 (2%)				3 (6)				0.40

* Frequency count and (%) OR Mean ± SD; ** Independent samples *t*-test or Pearson's Chi-Squared.

The relationship between serum 25(OD)D and daily vitamin D dose from dietary intake and solar ultraviolet B is presented in Figures 1 and 2, respectively.

The sensitivity, specificity, percentage correctly classified, and ROC Area for the estimated vitamin D against the 25(OH)D as reference were 100%, 79%, 80%, and 89%.

Figure 3 illustrates the intake of vitamin D among patients with depression according to symptoms of severity. Figure 4 illustrates serum 25(OH)D among patients with depression according to symptoms of severity.

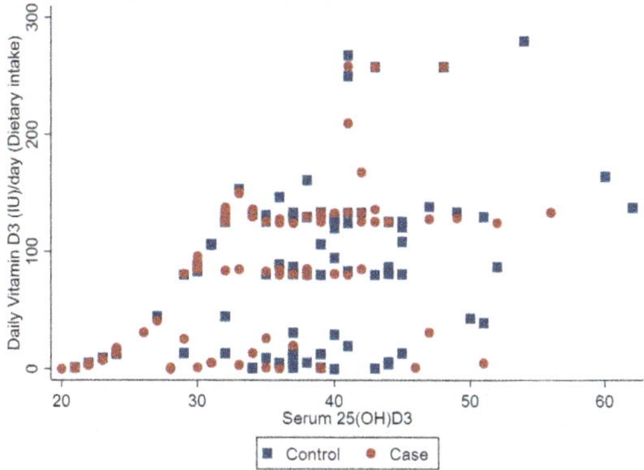

Figure 1. The association between serum vitamin D 25(OH)D and daily vitamin D from dietary intake.

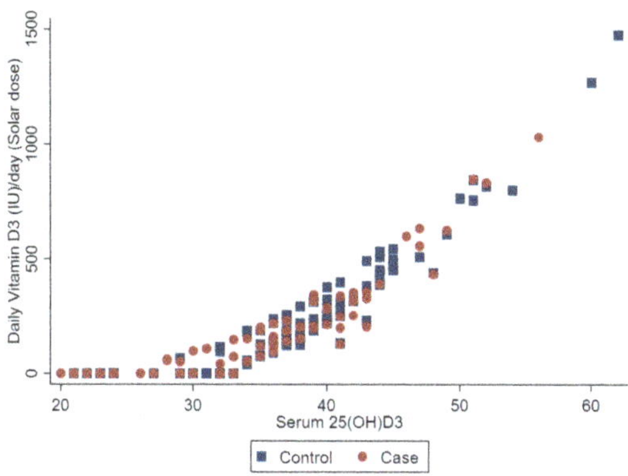

Figure 2. The association between serum vitamin D 25(OH)D and daily vitamin D from solar ultraviolet B.

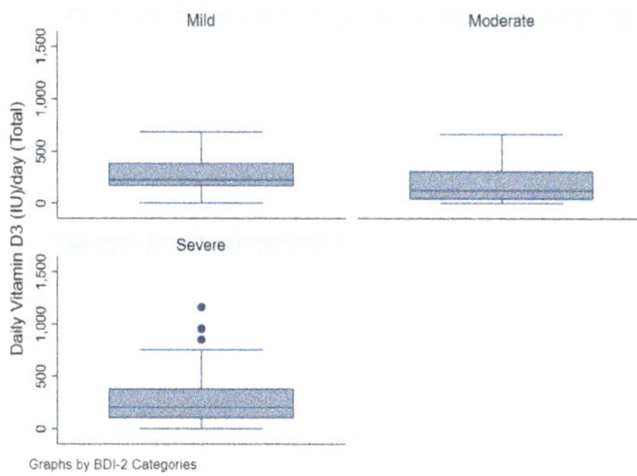

Figure 3. Vitamin D dose (IU/day) of patients with depression according to symptoms severity.

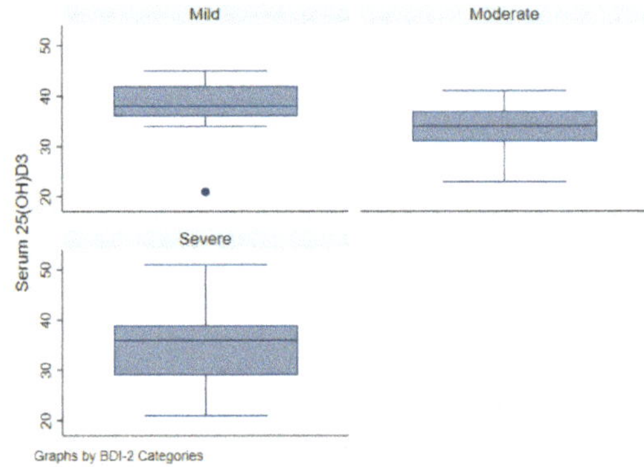

Figure 4. Serum vitamin D 25(OH)D of patients with depression according to symptoms severity.

One-way analysis of variance (ANOVA) revealed that the mean daily dose of vitamin D for patients with depression did not significantly differ according to symptoms severity as measured by the BDI-II with $p = 0.15$. Patients with mild, moderate, and severe symptoms had a daily dose of 268 IU (7 mcg), 181 IU (5 mcg), and 275 IU (7 mcg) accordingly.

Multiple linear regression analysis showed that the only predictor for vitamin D doses per day is physical activity for both cases with depression and controls $p = 0.001$. Detailed results are presented in Table 3.

Table 3. Association * between total vitamin D doses and selected predictors.

Cases with depression ($n = 96$)		
	Outcome variable: Daily vitamin D Dose	
Explanatory Variables	β	*p*-Value
Education level	60	0.12
Smoking	1	1
Physical activity	318	0.001 *
Controls ($n = 96$)		
	Outcome variable: Daily vitamin D Dose	
Explanatory Variables	β	*p*-Value
Education level	111	0.08
Smoking	−52	0.60
Physical activity	267	0.001 *

* Multiple linear regression analysis—Adjusting for age, sex, caloric intake, social status, and job status.

4. Discussion

To the authors' best knowledge this is the first study to measure vitamin D doses from solar ultraviolet and dietary intakes in patients with depression. The major finding of this study is that: patients with depression have significantly lower doses of vitamin D compared to age- and sex-matched healthy controls. While dietary intakes of vitamin D are equal in both groups, patients with depression appeared to have statistically significantly less vitamin D from solar ultraviolet B. The proportion of patients with depression meeting the daily recommended dose of vitamin D is less than one out of five. The daily dose of vitamin D did not vary significantly among patients with depression according to symptoms of severity.

A recent laboratory-based study found a very high prevalence of vitamin D deficiency among patients with mental illness with only 18% showing adequate levels of vitamin D [19]. A meta-analysis of fourteen observational studies with approximately 31,500 patients revealed that lower vitamin D levels were found in patients with depression compared to healthy individuals [20]. Our results are consistent with previous research, which suggest that generally 20% of patients with depression have lowered vitamin D and increased vitamin D deficiency.

A low 25(OH)D in depressed patients can be also attributed to antidepressants drug use. Previous research found that antidepressants use, especially tricyclic antidepressants, appeared significantly associated with lower vitamin D [38].

Previous research demonstrated an association between adequate diet and sensible sun exposure to vitamin D deficiency among patients with depression [19]. Our findings suggest that sun exposure plays a more important role in explaining vitamin D deficiency in both patients with depression and healthy controls. It is well documented that low vitamin D can be linked with many health problems including neuropsychiatric disorders [20,39–42]. Specifically, observational and experimental studies showed a relationship between low levels 25(OH)D and depression [27,42–44].

The low doses of vitamin D for solar ultraviolet can be explained by the fact that adults with depression and depressive disorders engage in low levels of physical activity and poor lifestyle behavior [45,46]. Thus, because lower levels of vitamin D may precipitate mental disorders [13,47], a reestablishment of adequate levels may improve mental wellbeing and offer a feasibly adjunct treatment option. This is especially true if it is offered as part of a comprehensive lifestyle intervention that includes an outdoor physical activity with solar light exposure. Recent research showed that vitamin D and exercise have independent desirable influence on mood. Thus, the active engagement in outdoor activities under the sunlight can neutralize the vitamin D deficiency problem and the severity of mood disorders [48]. Sun avoidance inventory (SAI) can be used to examine outlooks towards sun avoidance attitudes in the context of vitamin D deficiency. In our study, we excluded participants who are taking dietary supplements; however, vitamin D exposure through supplementation should be also included in measures of overall vitamin D exposure.

This is the first research to estimate vitamin D doses from solar ultraviolet and dietary intakes in patients with depression using a rigorous approach and using a case-control methodology. Another strength is that we compared the estimated vitamin D doses against serum 25(OH)D. We focused on outpatients with depression to eliminate the role of hospital-based restricted diets and inpatients closed wards policy; however, future studies are needed to compare inpatients vs. outpatients.

5. Conclusions

The present study showed that about 80% of patients with depression and 70% of controls do not receive adequate daily doses of vitamin D. Effective detection and interventions on adequate vitamin D levels in patients with depression might prove to be an easy and cost-effective intervention to improve long-term health outcomes.

Author Contributions: H.J. and Z.S. contributed to the conception and design of the work. H.J. and Z.S. coordinated data collection, data entry, and data cleaning. H.J. and N.L.B. performed data analyses. H.J. and W.B.G. drafted the manuscript. All authors (H.J., N.L.B., W.B.G., H.S.M.A., W.S.M.A., S.M., Z.S.) were engaged in writing the paper. All authors have read and agreed to the published version of the manuscript.

Funding: This research received no specific grant from any funding agency in the public, commercial, or not-for-profit sectors.

Conflicts of Interest: The authors have no conflicts of interest to declare.

Disclosure: WBG receives funding from Bio-Tech Pharmacal, Inc. (Fayetteville, AR, USA). The other authors have no conflicts of interest to declare. The subject matter of this paper will not have any direct bearing on that work, nor has that activity exerted any influence on this project.

References

1. Lim, G.Y.; Tam, W.W.; Lu, Y.; Ho, C.S.; Zhang, M.W.; Ho, R.C. Prevalence of Depression in the Community from 30 Countries between 1994 and 2014. *Sci. Rep.* **2018**, *8*, 2861. [CrossRef]
2. Shadrina, M.; Bondarenko, E.A.; Slominsky, P.A. Genetics Factors in Major Depression Disease. *Front. Psychiatry* **2018**, *9*, 334. [CrossRef]
3. Kern, D.M.; Cepeda, M.S.; Defalco, F.; Etropolski, M. Treatment patterns and sequences of pharmacotherapy for patients diagnosed with depression in the United States: 2014 through 2019. *BMC Psychiatry* **2020**, *20*, 1–10. [CrossRef]
4. Kirsch, I. Placebo Effect in the Treatment of Depression and Anxiety. *Front. Psychiatry* **2019**, *10*, 407. [CrossRef]
5. Bracken, P.; Thomas, P.; Timimi, S.; Asen, E.; Behr, G.; Beuster, C.; Bhunnoo, S.; Browne, I.; Chhina, N.; Double, D.; et al. Psychiatry beyond the current paradigm. *Br. J. Psychiatry* **2012**, *201*, 430–434. [CrossRef]
6. Jelovac, A.; Kolshus, E.; McLoughlin, D.M. Relapse Following Successful Electroconvulsive Therapy for Major Depression: A Meta-Analysis. *Neuropsychopharmacology* **2013**, *38*, 2467–2474. [CrossRef]
7. Van Bronswijk, S.; Moopen, N.; Beijers, L.; Ruhe, H.G.; Peeters, F. Effectiveness of psychotherapy for treatment-resistant depression: A meta-analysis and meta-regression. *Psychol. Med.* **2019**, *49*, 366–379. [CrossRef] [PubMed]
8. Cuijpers, P.; Noma, H.; Karyotaki, E.; Vinkers, C.H.; Cipriani, A.; Furukawa, T.A. A network meta-analysis of the effects of psychotherapies, pharmacotherapies and their combination in the treatment of adult depression. *World Psychiatry* **2020**, *19*, 92–107. [CrossRef] [PubMed]
9. Firth, J.; Marx, W.; Dash, S.; Carney, R.; Teasdale, S.B.; Solmi, M.; Stubbs, B.; Schuch, F.B.; Carvalho, A.F.; Jacka, F.; et al. The Effects of Dietary Improvement on Symptoms of Depression and Anxiety: A Meta-Analysis of Randomized Controlled Trials. *Psychosom. Med.* **2019**, *81*, 265–280. [CrossRef] [PubMed]
10. Spedding, S. Vitamin D and depression: A systematic review and meta-analysis comparing studies with and without biological flaws. *Nutrients* **2014**, *6*, 1501–1518. [CrossRef] [PubMed]
11. Kjærgaard, M.; Joakimsen, R.; Jorde, R. Low serum 25-hydroxyvitamin D levels are associated with depression in an adult Norwegian population. *Psychiatry Res.* **2011**, *190*, 221–225. [CrossRef] [PubMed]
12. Umhau, J.C.; George, D.T.; Heaney, R.P.; Lewis, M.D.; Ursano, R.J.; Heilig, M.; Hibbeln, J.R.; Schwandt, M.L. Low vitamin D status and suicide: A case-control study of active duty military service members. *PLoS ONE* **2013**, *8*, e51543. [CrossRef]
13. Milaneschi, Y.; Hoogendijk, W.; Lips, P.; Heijboer, A.C.; Schoevers, R.; van Hemert, A.M.; Beekman, A.T.F.; Smit, J.H.; Penninx, B.W.J.H. The association between low vitamin D and depressive disorders. *Mol. Psychiatry* **2014**, *19*, 444–451. [CrossRef] [PubMed]
14. Lardner, A.L. Vitamin D and hippocampal development-the story so far. *Front. Mol. Neurosci.* **2015**, *8*, 58. [CrossRef] [PubMed]
15. Kaufmann, F.N.; Costa, A.P.; Ghisleni, G.; Diaz, A.P.; Rodrigues, A.L.S.; Peluffo, H.; Kaster, M.P. NLRP3 inflammasome-driven pathways in depression: Clinical and preclinical findings. *Brain Behav. Immun.* **2017**, *64*, 367–383. [CrossRef]
16. Muthuramalingam, A.; Menon, V.; Rajkumar, R.P.; Negi, V.S. Is Depression an Inflammatory Disease? Findings from a Cross-Sectional Study at a Tertiary Care Center. *Indian J. Psychol. Med.* **2016**, *38*, 114–119. [CrossRef]
17. Zhu, D.-M.; Zhao, W.; Zhang, B.; Zhang, Y.; Yang, Y.; Zhang, C.; Wang, Y.; Zhu, J.; Yu, Y. The Relationship Between Serum Concentration of Vitamin D, Total Intracranial Volume, and Severity of Depressive Symptoms in Patients With Major Depressive Disorder. *Front. Psychiatry* **2019**, *10*, 322. [CrossRef]
18. Dana-Alamdari, L.; Kheirouri, S.; Noorazar, S.G. Serum 25-Hydroxyvitamin D in Patients with Major Depressive Disorder. *Iran. J. Public Health* **2015**, *44*, 690–697.
19. Cuomo, A.; Maina, G.; Bolognesi, S.; Rosso, G.; Beccarini Crescenzi, B.; Zanobini, F.; Goracci, A.; Facchi, E.; Favaretto, E.; Baldini, I.; et al. Prevalence and Correlates of Vitamin D Deficiency in a Sample of 290 Inpatients With Mental Illness. *Front. Psychiatry* **2019**, *10*, 167. [CrossRef]
20. Anglin, R.E.S.; Samaan, Z.; Walter, S.D.; McDonald, S.D. Vitamin D deficiency and depression in adults: Systematic review and meta-analysis. *Br. J. Psychiatry* **2013**, *202*, 100–107. [CrossRef]

21. Sarris, J.; Murphy, J.; Mischoulon, D.; Papakostas, G.I.; Fava, M.; Berk, M.; Ng, C.H. Adjunctive Nutraceuticals for Depression: A Systematic Review and Meta-Analyses. *Am. J. Psychiatry* **2016**, *173*, 575–587. [CrossRef] [PubMed]

22. Von Elm, E.; Altman, D.G.; Egger, M.; Pocock, S.J.; Gøtzsche, P.C.; Vandenbroucke, J.P.; Initiative, S. The Strengthening the Reporting of Observational Studies in Epidemiology (STROBE) statement: Guidelines for reporting observational studies. *PLoS Med.* **2007**, *4*, e040296. [CrossRef] [PubMed]

23. Al-Haddad, F.A.; Al-Mahroos, F.T.; Al-Sahlawi, H.S.; Al-Amer, E. The impact of dietary intake and sun exposure on vitamin D deficiency among couples. *Bahrain Med. Bull.* **2014**, *158*, 1–5. [CrossRef]

24. WHO. Body Mass Index—BMI. Available online: https://www.euro.who.int/en/health-topics/disease-prevention/nutrition/a-healthy-lifestyle/body-mass-index-bmi (accessed on 1 July 2020).

25. Upton, J. Beck Depression Inventory (BDI). In *Encyclopedia of Behavioral Medicine*; Gellman, M.D., Turner, J.R., Eds.; Springer: New York, NY, USA, 2013; pp. 178–179. [CrossRef]

26. Naja, S.; Al-Kubaisi, N.; Chehab, M.; Al-Dahshan, A.; Abuhashem, N.; Bougmiza, I. Psychometric properties of the Arabic version of EPDS and BDI-II as a screening tool for antenatal depression: Evidence from Qatar. *BMJ Open* **2019**, *9*, e030365. [CrossRef] [PubMed]

27. Jahrami, H.; Saif, Z.; AlHaddad, M.; Faris, M.e.A.-I.; Hammad, L.; Ali, B. Assessing dietary and lifestyle risk behaviours and their associations with disease comorbidities among patients with depression: A case-control study from Bahrain. *Heliyon* **2020**, *6*, e04323. [CrossRef] [PubMed]

28. Grant, W.B.; Fakhoury, H.M.A.; Karras, S.N.; Al Anouti, F.; Bhattoa, H.P. Variations in 25-Hydroxyvitamin D in Countries from the Middle East and Europe: The Roles of UVB Exposure and Diet. *Nutrients* **2019**, *11*, 2065. [CrossRef]

29. WHO. Physical Activity and Adults. Available online: https://www.who.int/dietphysicalactivity/publications/recommendations18_64yearsold/en/ (accessed on 1 July 2020).

30. Godar, D.E.; Pope, S.J.; Grant, W.B.; Holick, M.F. Solar UV Doses of Young Americans and Vitamin D3 Production. *Environ. Health Perspect.* **2012**, *120*, 139–143. [CrossRef]

31. Fitzpatrick, T.B. The Validity and Practicality of Sun-Reactive Skin Types I Through VI. *Arch. Dermatol.* **1988**, *124*, 869–871. [CrossRef]

32. Silva, A.A. Erythemal dose rate under noon overcast skies. *Photochem. Photobiol. Sci.* **2013**, *12*, 777–786. [CrossRef]

33. Godar, D.E.; Pope, S.J.; Grant, W.B.; Holick, M.F. Solar UV Doses of Adult Americans and Vitamin D3 Production. *Derm. Endocrinol.* **2011**, *3*, 243–250. [CrossRef]

34. Pope, S.J.; Holick, M.F.; Mackin, S.; Godar, D.E. Action spectrum conversion factors that change erythemally weighted to previtamin D3-weighted UV doses. *Photochem. Photobiol.* **2008**, *84*, 1277–1283. [CrossRef] [PubMed]

35. USDA. Dietary Supplement Ingredient Database. Available online: https://dietarysupplementdatabase.usda.nih.gov/ingredient_calculator/help.php (accessed on 1 July 2020).

36. *STATA, M.P. 16.1*; StataCorp: College Station, TX, USA, 2020.

37. Farid, E.; Jaradat, A.A.; Al-Segai, O.; Hassan, A.B. Prevalence of Vitamin D Deficiency in Adult Patients with Systemic Lupus Erythematosus in Kingdom of Bahrain. *Egypt. J. Immunol.* **2017**, *24*, 1–8. [PubMed]

38. Voshaar, R.C.O.; Derks, W.J.; Comijs, H.C.; Schoevers, R.A.; de Borst, M.H.; Marijnissen, R.M. Antidepressants differentially related to 1,25-(OH)2 vitamin D3 and 25-(OH) vitamin D3 in late-life depression. *Transl. Psychiatry* **2014**, *4*, e383. [CrossRef] [PubMed]

39. Grant, W.B.; Soles, C.M. Epidemiologic evidence supporting the role of maternal vitamin D deficiency as a risk factor for the development of infantile autism. *Dermatoendocrinology* **2009**, *1*, 223–228. [CrossRef]

40. Itzhaky, D.; Amital, D.; Gorden, K.; Bogomolni, A.; Arnson, Y.; Amital, H. Low serum vitamin D concentrations in patients with schizophrenia. *Isr. Med. Assoc. J.* **2012**, *14*, 88–92.

41. Annweiler, C.; Schott, A.M.; Rolland, Y.; Blain, H.; Herrmann, F.R.; Beauchet, O. Dietary intake of vitamin D and cognition in older women: A large population-based study. *Neurology* **2010**, *75*, 1810–1816. [CrossRef]

42. Brouwer-Brolsma, E.M.; Feskens, E.J.; Steegenga, W.T.; de Groot, L.C. Associations of 25-hydroxyvitamin D with fasting glucose, fasting insulin, dementia and depression in European elderly: The SENECA study. *Eur. J. Nutr.* **2013**, *52*, 917–925. [CrossRef]

43. Jaddou, H.Y.; Batieha, A.M.; Khader, Y.S.; Kanaan, S.H.; El-Khateeb, M.S.; Ajlouni, K.M. Depression is associated with low levels of 25-hydroxyvitamin D among Jordanian adults: Results from a national population survey. *Eur. Arch. Psychiatry Clin. Neurosci.* **2012**, *262*, 321–327. [CrossRef]
44. Föcker, M.; Antel, J.; Ring, S.; Hahn, D.; Kanal, Ö.; Öztürk, D.; Hebebrand, J.; Libuda, L. Vitamin D and mental health in children and adolescents. *Eur. Child Adolesc. Psychiatry* **2017**, *26*, 1043–1066. [CrossRef]
45. Schuch, F.; Vancampfort, D.; Firth, J.; Rosenbaum, S.; Ward, P.; Reichert, T.; Bagatini, N.C.; Bgeginski, R.; Stubbs, B. Physical activity and sedentary behavior in people with major depressive disorder: A systematic review and meta-analysis. *J. Affect. Disord.* **2017**, *210*, 139–150. [CrossRef]
46. Vancampfort, D.; Firth, J.; Schuch, F.B.; Rosenbaum, S.; Mugisha, J.; Hallgren, M.; Probst, M.; Ward, P.B.; Gaughran, F.; De Hert, M.; et al. Sedentary behavior and physical activity levels in people with schizophrenia, bipolar disorder and major depressive disorder: A global systematic review and meta-analysis. *World Psychiatry Off. J. World Psychiatric Assoc. WPA* **2017**, *16*, 308–315. [CrossRef] [PubMed]
47. McCann, J.C.; Ames, B.N. Is there convincing biological or behavioral evidence linking vitamin D deficiency to brain dysfunction? *FASEB J.* **2008**, *22*, 982–1001. [CrossRef] [PubMed]
48. Irandoust, K.; Taheri, M. The Effect of Vitamin D supplement and Indoor Vs Outdoor Physical Activity on Depression of Obese Depressed Women. *Asian J. Sports Med.* **2017**, *8*, 1. [CrossRef]

Article

Association of Vitamin D Status with Lower Limb Muscle Strength in Professional Basketball Players: A Cross-Sectional Study

Do Kyung Kim [1], Geon Park [1], Liang-Tseng Kuo [2,3,*,†] and Won-Hah Park [1,*,†]

[1] Department of Sports Medicine Center, Samsung Medical Center, Sungkyunkwan University School of Medicine, Seoul 03063, Korea; hrmax1@naver.com (D.K.K.); Geon2.park@samsung.com (G.P.)

[2] Sports Medicine Center, Department of Orthopaedic Surgery, Chang Gung Memorial Hospital, Chiayi 613, Taiwan

[3] School of Medicine, College of Medicine, Chang Gung University, Taoyuan 333, Taiwan

* Correspondence: light71829@gmail.com (L.-T.K.); pk90007@naver.com (W.-H.P.); Tel.: +886-5-3621000 (L.-T.K.); +82-2-3410-3847 (W.-H.P.)

† The two authors contributed equally.

Received: 2 August 2020; Accepted: 2 September 2020; Published: 5 September 2020

Abstract: Vitamin D deficiency in athletes may play a role in influencing fracture risk and athletic performance. This study aimed to examine the vitamin D status of basketball players and determine its correlation with muscle strength. We included 36 male professional basketball players (mean age, 22.6 ± 3.2 years) categorized by vitamin D status. We examined the muscle strength of knee extension/flexion and ankle dorsiflexion/plantarflexion using an isokinetic dynamometer. Eleven (30.5%), fifteen (41.7%), and ten (27.8%) players had deficient (<20 ng/mL), insufficient (20–32 ng/mL), and sufficient vitamin D levels (>32 ng/mL), respectively. In the dominant side, there were no significant correlations of vitamin D level with knee extension/flexion strength ($r = 0.134$, $p = 0.436$; $r = -0.017$, $p = 0.922$, respectively), or with plantarflexion/dorsiflexion ankle strength ($r = -0.143$, $p = 0.404$; $r = 1.109$, $p = 0.527$, respectively). Moreover, the isokinetic lower limb strengths were not significantly different between the three groups in all settings (all $p > 0.05$). In conclusion, professional basketball players had a high prevalence of vitamin D insufficiency. Though it may not be associated with muscle strength, maintaining adequate vitamin D levels by micronutrients monitoring, regular dietician consultation, and supplementation is still a critically considerable strategy to enhance young athletes' health.

Keywords: vitamin D insufficiency; muscle strength; basketball; athletes

1. Introduction

Vitamin D is an essential hormone for calcium and phosphate metabolism and, hence, influences bone homeostasis and muscle function [1]. The synthesis of vitamin D relies on skin exposure to ultraviolet radiation B (UVB) in sunlight [2]. UVB exposure is moderate since the latitude of South Korea ranges from 33° N to 38° N; however, vitamin D deficiency is common in Korea [3]. Due to seasonal variation of UVB exposure, the vitamin D level of Koreans was lower in winter and spring [3]. According to previous studies, risk factors for vitamin D deficiency in Korea included living in urban areas, lacking exercise, working indoors, and being younger (20–49 years), especially for those who used sunscreen daily [3,4]. Prevention of skin aging and maintenance of youthful skin were critical factors associated with sunscreen use in young Koreans, regardless of sex [4].

Vitamin D deficiency can result in muscle loss and weakness [5]. Severe vitamin D deficiency can cause myopathy, accompanied by muscle weakness, amyotrophy, and muscle pain [6,7]. Vitamin D is essential for muscle function. Vitamin D may enhance muscle function by synthesizing protein in muscles and optimizing muscle growth while improving nerve-muscle function [8,9]. Optimal serum 25-hydroxyvitamin D (25(OH)D) helps to optimize overall performance for athletes and enhance muscle contraction [10]. For athletes, vitamin D is important not only for exercise performance but also for recovery [11]. Previous literature revealed that an insufficient vitamin D level was associated with an increased incidence of muscle damage [12], decreased athletic ability [13–15], and lower restorative ability after training and competition [11,16–18].

Vitamin D is essential to musculoskeletal health and exercise performance; however, vitamin D deficiency is not uncommon in the general population. According to a Korean national database study [3,19], the prevalence of vitamin D deficiency (<20 ng/mL) in 2014 was 75.2% in males and 82.5% in females. However, the cut-off value for vitamin D deficiency in Korea was higher than that of western countries [20,21]. Furthermore, the scene of vitamin D deficiency was also common among athletes. In a study on 279 NBA players, 79.3% of players had vitamin D insufficiency or deficiency, with 90 having vitamin D deficiency (<20 ng/mL) (32.3%) and 131 having insufficiency (20–32 ng/mL) (47.0%) [22]. According to a recent meta-analysis, 44–67% of the athletes were vitamin D inadequate (<32 ng/mL) [12].

Research on the effect of vitamin D on athletes has been gathering interest due to its potential role in improving athletic performance since enhanced muscle function is essential in boosting performance and reducing injuries for athletes [23,24]. However, studies on the effects of vitamin D for athletes are conflicting [14,25–27]. Some studies reported that vitamin D supplements in athletes with insufficient levels of vitamin D could increase quadricep strength and enhance vertical jump and sprint performance [14,25]. In contrast, other studies showed that vitamin D levels were not associated with muscle strength and motor ability [26,27]. Therefore, the association between vitamin D level and muscular performance in athletes remains uncertain. Our previous work showed that vitamin D insufficiency was common in Korean elite volleyball players, although their shoulder muscle strengths were not affected by low vitamin D status [28]. Little information about the vitamin D status and its association with lower limb muscle strength in professional basketball players is available.

Thus, we performed this study to investigate the vitamin D status of professional basketball players who participate in one of the most popular indoor sports and to evaluate the relationship between vitamin D concentration and extension/flexion strength of knees and plantar/dorsiflexion strength of ankles, both of which are critical components in jumping motions during a basketball game.

2. Materials and Methods

2.1. Participants and Demographics

We enrolled healthy male professional basketball players from the Samsung Thunders in the Korean Basketball League (KBL) by using the convenience sampling technique in this cross-sectional study from January 2015 to June 2017. The participants were routinely medically evaluated, healthy athletes and cleared for participation by an orthopedic specialist. All participants in this study regularly underwent training, including team-specific training supervised by coaching staff five times a week for at least four hours per day. During the two months of pre-season training and the regular season, the players were provided with a controlled diet under a nutritionist's supervision. The controlled diet met the basic requirements for micronutrients including calcium and vitamin D. No other additional supplements including omega-3s and vitamin D were given. We excluded players who had undergone major knee and ankle surgeries and those already taking vitamin D supplements. All research procedures were reviewed and approved by the bioethical committee of the University of Sungkyunkwan. The study conformed to the tenets of the Declaration of Helsinki for medical research involving human subjects (IRB file No: 2020-04-199). All participants received a clear explanation of the study, including the risks and benefits of participation, and they provided written informed consent for testing and data analysis before the beginning of the study.

2.2. Assessments

All participants received an examination for serum vitamin D levels and muscle strength of their lower limbs in April (off-season period). The participants were instructed to fast overnight for at least 8 h and were abstained from any vigorous physical activity and exercise at least 24 h before the assessments to avoid the confounding effects of post-exertional muscle fatigue. After arriving in the laboratory, the fasting serum samples (3 mL) were taken from the antecubital vein area of the arm and collected into a tube. The collected blood samples were clotted for 30 min at room temperature (20–22 °C), and centrifuged at 3000 rpm (revolutions per minute) for 15 min. The separated serum samples were stored at −70 °C until analysis. All archived samples of the discovery cohort were sent to the Department of Laboratory Medicine at the Samsung Medical Center, and vitamin D levels were determined from a single baseline serum sample. Serum levels of 25(OH)D2 and D3 were determined by high-performance liquid chromatography-tandem mass spectrometry detection (Euroimmun AG, Lübeck, Germany). All assays met reproducibility requirements of ≤20% coefficient of variation (CV) and were acceptable for clinical use. Total vitamin D levels were quantitated using calibration curves constructed from the mass chromatogram.

There is no universally accepted standard definition for vitamin D deficiency, insufficiency, or sufficiency. In our study, vitamin D levels were defined as deficient at <20 ng/mL, insufficient at 20–32 ng/mL, and sufficient at >32 ng/mL, which were described by Fishman et al. [22]. We used an isokinetic dynamometer (CSMI Medical Solutions, MA, USA) to evaluate the knee and ankle muscle strengths of all participants. This testing protocol was conducted on the dominant and non-dominant legs of each subject. The dominant leg was determined by which hand is dominant. Musculoskeletal physiotherapists performed standardized testing under the supervision of one of the authors. After warm-up using a stationary bike for 10 min, the participants performed three submaximal familiarization trials. Thereafter, they underwent maximal concentric tests. The participants were given verbal support to encourage maximal effort. Concentric knee extension/flexion peak torques were measured at angular velocities of 60°/s. Next, the participants performed ankle dorsiflexion/plantarflexion muscle strength tests at a speed of 30°/s. The maximum peak torque (Nm) for each velocity was also recorded.

2.3. Statistics

Statistical analysis was performed to evaluate the correlation between player parameters and vitamin D levels using Pearson correlation coefficients. We also calculated the concentric extension/flexion and dorsiflexion/plantarflexion muscle strengths of the dominant side and analyzed the isokinetic strength with respect to vitamin D level using one-way analysis of variance (ANOVA). The level of statistical significance was set at 0.05. One-way ANOVA and a post-hoc Bonferroni test was used to analyze the data. Statistical analysis was conducted using SPSS ver 18.0 (SPSS Inc., Chicago, IL, USA)

3. Results

We included 36 participants in this study. The mean age of the athletes was 22.6 ± 3.0 years. The mean vitamin D level was 24.7 ± 7.2 ng/mL. Regarding vitamin D levels, there were 11 (30.5%), 15 (41.7%), and 10 (27.8%) players who were deficient, insufficient, and sufficient, respectively. Twenty-six players (72.2%) were either vitamin D deficient or insufficient (Table 1).

Correlation between vitamin D and knee and ankle strengths revealed no significant findings. There was no significant bivariate correlation between vitamin D and extension/flexion knee strength of the dominant side at 60 deg/sec ($r = 0.134, p = 0.436; r = -0.017, p = 0.922$, respectively). Similarly, we found no significant correlation between vitamin D level and isokinetic ankle plantarflexion/dorsiflexion strength of the dominant side at 30 deg/sec ($r = -0.143, p = 0.404$ and $r = 0.109, p = 0.527$, respectively)

(Table 2). In the non-dominant side, no significant correlations were noted between vitamin D level and knee or ankle isokinetic strengths (all $p > 0.05$, Table 2).

Table 1. Players' demographics by vitamin D status.

Variable	Vitamin D Level			p Value [a]
	Deficiency (<20 ng/mL)	Insufficiency (20–32 ng/mL)	Sufficiency (>32 ng/mL)	
No. players (%)	11 (30.5)	15 (41.7)	10 (27.8)	
Vitamin D (ng/mL)	16.4 ± 3.2	24.6 ± 2.6	33.9 ± 1.4	<0.001 *
Age (yr)	22.3 ± 3.2	22.3 ± 3.3	23.2 ± 2.4	0.731
Height (cm)	187.5 ± 7.2	188.5 ± 5.5	192.8 ± 8.9	0.198
Weight (kg)	82.8 ± 7.8	83.5 ± 7.1	85.6 ± 8.8	0.697
Body mass index (kg/m^2)	23.3 ± 1.4	23.4 ± 1.5	22.1 ± 1.5	0.098

Values are presented as mean ± standard deviation. [a] p value for between group comparisons, and the significance level was set as 0.05. * $p < 0.05$.

Table 2. Correlation coefficients (r) between vitamin D level and other characteristics.

Characteristics	Vitamin D (ng/mL)	p Value
Age (yr)	0.045	0.796
Height (cm)	0.227	0.184
Weight (kg)	0.077	0.656
BMI (kg/m^2)	−0.295	0.080
Dominant side		
Knee		
Extension	0.134	0.436
Flexion	−0.017	0.922
Ankle		
Plantarflexion	−0.143	0.404
Dorsiflexion	0.109	0.527
Non-Dominant side		
Knee		
Extension	−0.058	0.737
Flexion	−0.056	0.748
Ankle		
Plantarflexion	−0.014	0.934
Dorsiflexion	0.028	0.871

Abbreviation: BMI, body mass index.

The participants were divided into three groups by the level of vitamin D to examine maximal muscle strength differences according to vitamin D levels. There were no significant differences in knee and ankle maximal concentric isokinetic muscle strengths between the three groups, neither dominant nor non-dominant sides (Tables 3 and 4, Figures 1 and 2).

Table 3. Comparison of maximal isokinetic knee strength by vitamin D level.

Knee Strength at 60°/s (Nm)	Vitamin D Level			*p* Value [a]
	Deficiency (<20 ng/mL)	Insufficiency (20–32 ng/mL)	Sufficiency (>32 ng/mL)	
Dominant				
Extension	165.3 ± 33.0	172.3 ± 22.7	166.2 ± 20.5	0.753
Flexion	93.5 ± 11.012214	99.9 ± 16.0	92.2 ± 14.1	0.340
Non-Dominant				
Extension	160.1 ± 30.3	173.3 ± 25.2	154.2 ± 21.5	0.182
Flexion	89.0 ± 10.1	98.0 ± 20.5	88.8 ± 18.6	0.316

Values are presented as mean ± standard deviation. [a] *p*-value for between-group comparisons, and significance level was set as 0.05.

Table 4. Comparison of maximal isokinetic ankle strength by vitamin D level.

Ankle Strength at 30°/s (Nm)	Vitamin D level			*p* Value
	Deficiency (<20 ng/mL)	Insufficiency (20–32 ng/mL)	Sufficiency (>32 ng/mL)	
Dominant				
Plantarflexion	81.4 ± 14.7	88.3 ± 22.4	78.5 ± 20.3	0.445
Dorsiflexion	27.3 ± 6.2	30.0 ± 7.1	30.3 ± 13.8	0.694
Non-Dominant				
Plantarflexion	70.8 ± 13.6	85.2 ± 27.8	78.7 ± 19.4	0.277
Dorsiflexion	27.7 ± 4.6	29.3 ± 10.2	28.4 ± 11.2	0.901

Values are presented as mean ± standard deviation.

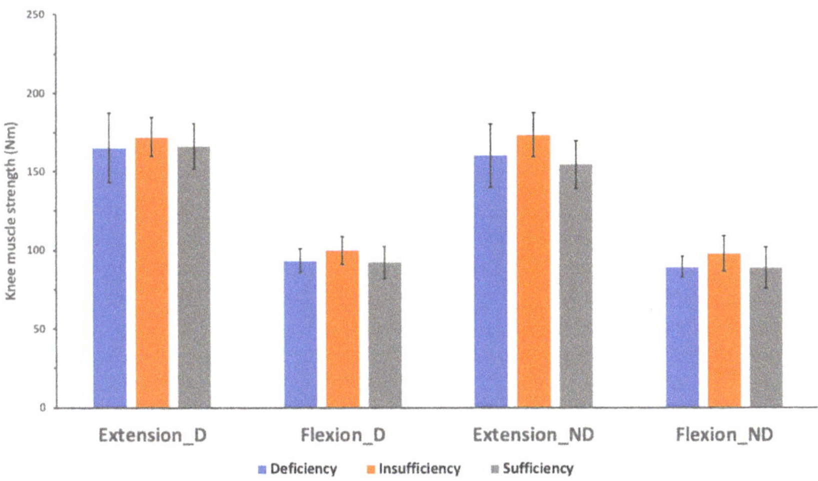

Figure 1. Isokinetic knee muscle strength at 60°/s according to vitamin D status. The isokinetic knee strengths were not significantly different across the three groups in all settings (Extension_D, knee extension at dominant side; Flexion_D, knee flexion at dominant side; Extension_ND, knee extension at non-dominant side; Flexion_ND, knee flexion at non-dominant side).

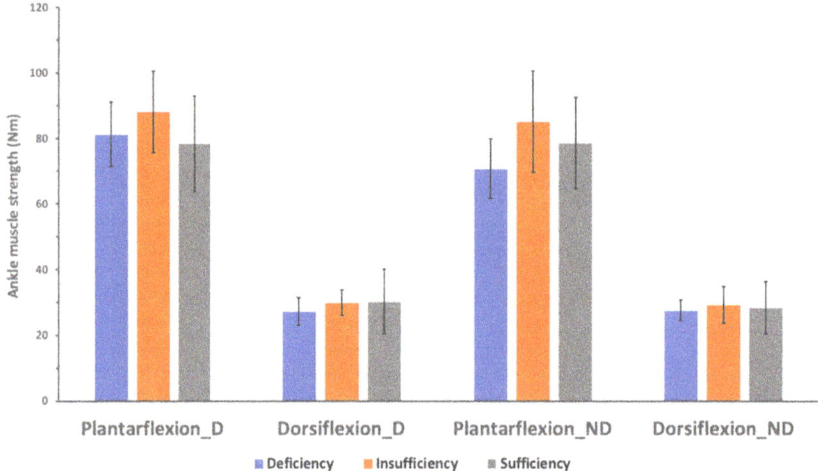

Figure 2. Isokinetic ankle muscle strength at 30°/s according to vitamin D status. The isokinetic ankle strengths were not significantly different across the three groups in all settings (Plantarflexion_D, ankle plantarflexion at dominant side; Dorsiflexion_D, ankle dorsiflexion at dominant side; Plantarflexion_ND, ankle plantarflexion at non-dominant side; Dorsiflexion_D, ankle dorsiflexion at non-dominant side).

4. Discussion

The present study examined the serum concentration of vitamin D and its effect on Korean professional basketball players' muscle strength. The most important finding of this study was that most of the basketball players were either vitamin D insufficient or deficient, although the vitamin D status did not significantly affect the lower limb muscle strength.

In a study on NBA players, 79.3% had vitamin D insufficiency (20~32 ng/mL) (47.0%) or deficiency (<20ng/mL) (32.3%) [22]. Similar to our results, the findings of this study showed that the majority of professional basketball players lack vitamin D. According to a meta-analysis, which surveyed 2,313 professional athletes playing various sports, 44–67% of the athletes were vitamin D inadequate (<32 ng/mL) [12]. The vitamin D synthesis relied on skin exposure to UVB radiation in sunlight [2]. Therefore, the choice of an indoor or outdoor training environment influences sun exposure, ultimately affecting vitamin D synthesis. Emerging evidence has indicated that athletes who train outdoors have higher vitamin D levels than those who train indoors or avoid peak daylight hours, regardless of latitude or season [29]. In this study, the professional basketball team trained five times a week (for more than five hours), but the athletes spent most of their training time at an indoor gym. That may be why most of the participants in this study had vitamin D deficiency or insufficiency. Previous literature also shows that basketball players have a relatively lower vitamin D level than other outdoor athletes [26,30].

There may not be a universal cut-off value for optimal vitamin D status. A blood 25(OH)D concentration below 10 ng/mL or 12 ng/mL is considered the lower limit of vitamin D status and an indicator of risk of vitamin D deficiency [20,21]. The World Health Organization (WHO) has also defined vitamin D "insufficiency" as a serum 25(OH)D below 20 ng/mL and "deficiency" as a serum 25(OH)D below 10 ng/mL [20]. In this study, we took a recommendation from the American Nutrition Society as a reference. It recommended that vitamin D deficiency be defined as a 25(OH)D level of 20 ng/mL or less [31], which was echoed in the consensus for optimal serum 25(OH)D concentrations from Central Europe, including Poland, Hungary, Belarus, Estonia, Czech Republic, and Ukraine [32]. This cut-off level was also utilized in various previous studies [3,7,19,28,33], including the Korean national survey [3,19] and our previous study [28].

Vitamin D is essential for athletes because it reduces injury rates, is useful in skeletal muscle repair and remodeling [34], and aids in efficient muscle recovery before vigorous-intensity exercises [18,34,35].

Moreover, vitamin D affects muscle tissue, primarily by increasing the size and quantity of type II (fast-twitch) muscle fibers [34,36,37]. Close et al. [14] reported that increasing vitamin D intake for eight weeks in athletes decreased 10 m sprint times and enhanced exercise abilities, such as vertical jumps [14,38]. Wyon et al. also found a significant increase in some of the muscle strength measurements [39]. However, in a study on vitamin D levels and lower limb muscle strength in isokinetic exercise among professional soccer players, there was no association between lower limb muscle strength and vitamin D levels [27].

Furthermore, Todd et al. found that the prevalence of vitamin D deficiency can be resolved by oral supplementation, but muscle function and respiratory function did not improve after that [40,41]. Brännström et al. found no significant correlation between these parameters, including jump and sprint performance, and vitamin D levels [42]. Other studies also reported that the associations of muscle strength and physical performance with vitamin D in athletes could not be adequately explained [36,43]. The results of research examining the association between serum vitamin D concentration and muscle strength and function in athletes have been contradictory. Moreover, not accounting for physical activity level or body composition change [14,39] or lacking a suitable priori power calculation [27,42] made interpretation of the above findings difficult. Since the current study was a cross-sectional design, we could not imply a dose-response of vitamin D on muscle function. The difference of results in the present study from the other cross-sectional Polish study indicating a positive relationship between vitamin D and muscle function [43] might come from the difference of baseline vitamin D levels from different exposure amounts of UVB in different latitudes. Meanwhile, the training environments in different kinds of sports might also play a role.

We measured the muscle strength of knee extension/flexion and ankle plantarflexion/dorsiflexion using isokinetic equipment to evaluate the muscles typically utilized while jumping, a significant movement in basketball. While the application of constant angular velocities does not necessarily characterize performance in sports, the evaluation of muscle strength using isokinetic equipment is valid, reliable, and widely used in the assessment of muscle function in athletics [44].

The findings of the current cross-sectional study did not find a relationship between vitamin D deficiency and impaired lower limb strength in professional basketball players. There were two explanations for these findings. The first is the appropriateness of the isokinetic test. Some studies have shown that vitamin D deficiency results in a reduction in type II muscle fibers [45]. Type II fibers produce faster muscle contractions and provide greater strength than do type I fibers; therefore, the performance of explosive and nimble movements such as sprinting, jumping, and turning is closely related to the contraction of type II muscle fibers [46].

However, accurate functional assessment of fast-twitch fibers (type II), used for vertical jumps, may not have been achieved since the evaluation of lower extremity strength by using isokinetic equipment in the current study was more consistent with the evaluation of type I fibers than that of type II fibers. As a result, there is insufficient data to assess young and healthy basketball players' athletic abilities with only isokinetic test results.

The second one is the physical well-preparedness of the participants. The participants included in this study were elite athletes who were already highly skilled and may have little room for muscle strength improvement. The effect of long-term training may potentially overcome the negative effect of vitamin D deficiency on muscle. Meanwhile, the current assessment tools may be unable to detect the tiny difference, and thus, sensitive and standardized measurement techniques for athletes are urgently needed.

There were still some limitations to this study. First, the external validity of the findings was limited. The results of the current investigation would be difficult to apply in other settings, since only male Korean basketball players were included within the analysis. Second, this is a cross-sectional study where the vitamin D levels were only examined one time. Third, other potential confounding factors for calcium and vitamin D levels, including diet, degree of sun exposure, individual training hours/modality, and sunlight practice, were not completely controlled. Although we only included young, healthy athletes and excluded those with diseases or medications that might affect vitamin D, bias should be considered while applying the findings of the current investigation. Fourth, a convenience sample was used in this study. Although we enrolled almost all players on a single professional basketball team, the condition of the participants may not be representative of players in other groups. A future study involving multiple teams may be needed to validate the present findings.

5. Conclusions

The current study showed that more than two-thirds of young professional basketball players had vitamin D insufficiency or deficiency. Although this is not associated with lower limb muscle weakness, maintaining an adequate vitamin D level by micronutrients monitoring, regular dietician consultation, and supplementation is still a critically considerable strategy to enhance the young athletes' health and performance. Future study investigating the effect of vitamin D on athletic performance should be performed under more critical control of the potential confounders, such as sunlight exposure and dietary intake.

Author Contributions: Conceptualization, D.K.K. and G.P.; methodology, D.K.K.; investigation and data curation, D.K.K. and G.P.; analysis and interpretation, D.K.K. and L.-T.K.; writing—original draft preparation, G.P. and D.K.K.; writing—review and editing, L.-T.K.; supervision, W.-H.P. All authors have read and agreed to the published version of the manuscript.

Funding: This research received no external funding. The Article Processing Charge was funded by the Samsung Medical Center.

Conflicts of Interest: The authors declare no conflict of interest.

References

1. Ceglia, L.; Harris, S.S. Vitamin D and its role in skeletal muscle. *Calcif. Tissue Int.* **2013**, *92*, 151–162. [CrossRef] [PubMed]
2. Holick, M.F. Sunlight and vitamin D for bone health and prevention of autoimmune diseases, cancers, and cardiovascular disease. *Am. J. Clin. Nutr.* **2004**, *80*, 1678S–1688S. [CrossRef] [PubMed]
3. Choi, H.S.; Oh, H.J.; Choi, H.; Choi, W.H.; Kim, J.G.; Kim, K.M.; Kim, K.J.; Rhee, Y.; Lim, S.K. Vitamin D insufficiency in Korea—A greater threat to younger generation: The Korea National Health and Nutrition Examination Survey (KNHANES) 2008. *J. Clin. Endocrinol. Metab.* **2011**, *96*, 643–651. [CrossRef]
4. Park, G.Y.; Oh, S.; Park, H.S.; Cho, S.; Yoon, H.S. Desire for a youthful look is associated with daily sunscreen use: Results of a questionnaire-based study. *Photodermatol. Photoimmunol. Photomed.* **2015**, *31*, 227–230. [CrossRef]
5. Visser, M.; Deeg, D.J.; Lips, P. Low vitamin D and high parathyroid hormone levels as determinants of loss of muscle strength and muscle mass (sarcopenia): The Longitudinal Aging Study Amsterdam. *J. Clin. Endocrinol. Metab.* **2003**, *88*, 5766–5772. [CrossRef] [PubMed]
6. Prabhala, A.; Garg, R.; Dandona, P. Severe myopathy associated with vitamin D deficiency in western New York. *Arch. Intern. Med.* **2000**, *160*, 1199–1203. [CrossRef] [PubMed]
7. Plotnikoff, G.A.; Quigley, J.M. Prevalence of severe hypovitaminosis D in patients with persistent, nonspecific musculoskeletal pain. *Mayo Clin. Proc.* **2003**, *78*, 1463–1470. [CrossRef] [PubMed]
8. Bouillon, R.; Bischoff-Ferrari, H.; Willett, W. Vitamin D and health: Perspectives from mice and man. *J. Bone Miner. Res.* **2008**, *23*, 974–979. [CrossRef]
9. Grimaldi, A.S.; Parker, B.A.; Capizzi, J.A.; Clarkson, P.M.; Pescatello, L.S.; White, M.C.; Thompson, P.D. 25(OH) vitamin D is associated with greater muscle strength in healthy men and women. *Med. Sci. Sports Exerc.* **2013**, *45*, 157–162. [CrossRef]
10. Zhang, L.; Quan, M.; Cao, Z.B. Effect of vitamin D supplementation on upper and lower limb muscle strength and muscle power in athletes: A meta-analysis. *PLoS ONE* **2019**, *14*, e0215826. [CrossRef]
11. Dahlquist, D.T.; Dieter, B.P.; Koehle, M.S. Plausible ergogenic effects of vitamin D on athletic performance and recovery. *J. Int. Soc. Sports Nutr.* **2015**, *12*, 33. [CrossRef]

12. Farrokhyar, F.; Tabasinejad, R.; Dao, D.; Peterson, D.; Ayeni, O.R.; Hadioonzadeh, R.; Bhandari, M. Prevalence of vitamin D inadequacy in athletes: A systematic-review and meta-analysis. *Sports Med.* **2015**, *45*, 365–378. [CrossRef] [PubMed]

13. Girgis, C.M.; Baldock, P.A.; Downes, M. Vitamin D, muscle and bone: Integrating effects in development, aging and injury. *Mol. Cell. Endocrinol.* **2015**, *410*, 3–10. [CrossRef] [PubMed]

14. Close, G.L.; Russell, J.; Cobley, J.N.; Owens, D.J.; Wilson, G.; Gregson, W.; Fraser, W.D.; Morton, J.P. Assessment of vitamin D concentration in non-supplemented professional athletes and healthy adults during the winter months in the UK: Implications for skeletal muscle function. *J. Sports Sci.* **2013**, *31*, 344–353. [CrossRef]

15. Close, G.L.; Leckey, J.; Patterson, M.; Bradley, W.; Owens, D.J.; Fraser, W.D.; Morton, J.P. The effects of vitamin D(3) supplementation on serum total 25[OH]D concentration and physical performance: A randomised dose-response study. *Br. J. Sports Med.* **2013**, *47*, 692–696. [CrossRef] [PubMed]

16. Garcia, L.A.; King, K.K.; Ferrini, M.G.; Norris, K.C.; Artaza, J.N. 1,25(OH)2vitamin D3 stimulates myogenic differentiation by inhibiting cell proliferation and modulating the expression of promyogenic growth factors and myostatin in C2C12 skeletal muscle cells. *Endocrinology* **2011**, *152*, 2976–2986. [CrossRef]

17. Stratos, I.; Li, Z.; Herlyn, P.; Rotter, R.; Behrendt, A.K.; Mittlmeier, T.; Vollmar, B. Vitamin D increases cellular turnover and functionally restores the skeletal muscle after crush injury in rats. *Am. J. Pathol.* **2013**, *182*, 895–904. [CrossRef]

18. Barker, T.; Schneider, E.D.; Dixon, B.M.; Henriksen, V.T.; Weaver, L.K. Supplemental vitamin D enhances the recovery in peak isometric force shortly after intense exercise. *Nutr. Metab.* **2013**, *10*, 69. [CrossRef]

19. Park, J.H.; Hong, I.Y.; Chung, J.W.; Choi, H.S.H. Vitamin D status in South Korean population. *Medicine* **2018**, *97*, e11032. [CrossRef]

20. Lamberg-Allardt, C.; Brustad, M.; Meyer, H.E.; Steingrimsdottir, L. Vitamin D–a systematic literature review for the 5th edition of the Nordic Nutrition Recommendations. *Food Nutr. Res.* **2013**, *57*, 22671. [CrossRef]

21. IOM (Institute of Medicine). *Dietary Reference Intakes for Calcium and Vitamin D*; The National Academies Press: Washington, DC, USA, 2011.

22. Fishman, M.P.; Lombardo, S.J.; Kharrazi, F.D. Vitamin D Deficiency among Professional Basketball Players. *Orthop. J. Sports Med.* **2016**, *4*, 2325967116655742. [CrossRef] [PubMed]

23. Hootman, J.M.; Dick, R.; Agel, J. Epidemiology of collegiate injuries for 15 sports: Summary and recommendations for injury prevention initiatives. *J. Athl. Train* **2007**, *42*, 311–319. [PubMed]

24. Abrams, G.D.; Feldman, D.; Safran, M.R. Effects of Vitamin D on Skeletal Muscle and Athletic Performance. *J. Am. Acad. Orthop. Surg.* **2018**, *26*, 278–285. [CrossRef] [PubMed]

25. Wyon, M.A.; Koutedakis, Y.; Wolman, R.; Nevill, A.M.; Allen, N. The influence of winter vitamin D supplementation on muscle function and injury occurrence in elite ballet dancers: A controlled study. *J. Sci. Med. Sport* **2014**, *17*, 8–12. [CrossRef] [PubMed]

26. Ksiazek, A.; Zagrodna, A.; Dziubek, W.; Pietraszewski, B.; Ochmann, B.; Slowinska-Lisowska, M. 25(OH)D3 Levels Relative to Muscle Strength and Maximum Oxygen Uptake in Athletes. *J. Hum. Kinet.* **2016**, *50*, 71–77. [CrossRef] [PubMed]

27. Hamilton, B.; Whiteley, R.; Farooq, A.; Chalabi, H. Vitamin D concentration in 342 professional football players and association with lower limb isokinetic function. *J. Sci. Med. Sport* **2014**, *17*, 139–143. [CrossRef]

28. Kim, D.K.; Park, G.; Kuo, L.T.; Park, W.H. The Relationship between Vitamin D Status and Rotator Cuff Muscle Strength in Professional Volleyball Athletes. *Nutrients* **2019**, *11*, 2768. [CrossRef]

29. Valtueña, J.; Dominguez, D.; Til, L.; González-Gross, M.; Drobnic, F. High prevalence of vitamin D insufficiency among elite Spanish athletes the importance of outdoor training adaptation. *Nutr. Hosp.* **2014**, *30*, 124–131.

30. Willis, K.S.; Peterson, N.J.; Larson-Meyer, D.E. Should we be concerned about the vitamin D status of athletes? *Int. J. Sport Nutr. Exerc. Metab.* **2008**, *18*, 204–224. [CrossRef]

31. Bischoff-Ferrari, H.A.; Giovannucci, E.; Willett, W.C.; Dietrich, T.; Dawson-Hughes, B. Estimation of optimal serum concentrations of 25-hydroxyvitamin D for multiple health outcomes. *Am. J. Clin. Nutr.* **2006**, *84*, 18–28. [CrossRef]

32. Pludowski, P.; Karczmarewicz, E.; Bayer, M.; Carter, G.; Chlebna-Sokół, D.; Czech-Kowalska, J.; Dębski, R.; Decsi, T.; Dobrzańska, A.; Franek, E.; et al. Practical guidelines for the supplementation of vitamin D and the treatment of deficits in Central Europe—Recommended vitamin D intakes in the general population and groups at risk of vitamin D deficiency. *Endokrynol. Pol.* **2013**, *64*, 319–327. [CrossRef] [PubMed]

33. Holick, M.F. Vitamin D status: Measurement, interpretation, and clinical application. *Ann. Epidemiol.* **2009**, *19*, 73–78. [CrossRef] [PubMed]

34. Owens, D.J.; Sharples, A.P.; Polydorou, I.; Alwan, N.; Donovan, T.; Tang, J.; Fraser, W.D.; Cooper, R.G.; Morton, J.P.; Stewart, C.; et al. A systems-based investigation into vitamin D and skeletal muscle repair, regeneration, and hypertrophy. *Am. J. Physio. Endocrinol. Metab.* **2015**, *309*, E1019–E1031. [CrossRef] [PubMed]

35. Barker, T.; Henriksen, V.T.; Martins, T.B.; Hill, H.R.; Kjeldsberg, C.R.; Schneider, E.D.; Dixon, B.M.; Weaver, L.K. Higher serum 25-hydroxyvitamin D concentrations associate with a faster recovery of skeletal muscle strength after muscular injury. *Nutrients* **2013**, *5*, 1253–1275. [CrossRef] [PubMed]

36. Farrokhyar, F.; Sivakumar, G.; Savage, K.; Koziarz, A.; Jamshidi, S.; Ayeni, O.R.; Peterson, D.; Bhandari, M. Effects of Vitamin D Supplementation on Serum 25-Hydroxyvitamin D Concentrations and Physical Performance in Athletes: A Systematic Review and Meta-analysis of Randomized Controlled Trials. *Sports Med.* **2017**, *47*, 2323–2339. [CrossRef]

37. Ogan, D.; Pritchett, K. Vitamin D and the athlete: Risks, recommendations, and benefits. *Nutrients* **2013**, *5*, 1856–1868. [CrossRef]

38. Maroon, J.C.; Mathyssek, C.M.; Bost, J.W.; Amos, A.; Winkelman, R.; Yates, A.P.; Duca, M.A.; Norwig, J.A. Vitamin D profile in National Football League players. *Am. J. Sports Med.* **2015**, *43*, 1241–1245. [CrossRef]

39. Wyon, M.A.; Wolman, R.; Nevill, A.M.; Cloak, R.; Metsios, G.S.; Gould, D.; Ingham, A.; Koutedakis, Y. Acute Effects of Vitamin D3 Supplementation on Muscle Strength in Judoka Athletes: A Randomized Placebo-Controlled, Double-Blind Trial. *Clin. J. Sport Med.* **2016**, *26*, 279–284. [CrossRef]

40. Todd, J.; Madigan, S.; Pourshahidi, K.; McSorley, E.; Laird, E.; Healy, M.; Magee, P. Vitamin D Status and Supplementation Practices in Elite Irish Athletes: An Update from 2010/2011. *Nutrients* **2016**, *8*, 485. [CrossRef]

41. Todd, J.J.; McSorley, E.M.; Pourshahidi, L.K.; Madigan, S.M.; Laird, E.; Healy, M.; Magee, P.J. Vitamin D3 supplementation using an oral spray solution resolves deficiency but has no effect on VO2 max in Gaelic footballers: Results from a randomised, double-blind, placebo-controlled trial. *Eur. J. Nutr.* **2017**, *56*, 1577–1587. [CrossRef]

42. Brännström, A.; Yu, J.G.; Jonsson, P.; Åkerfeldt, T.; Stridsberg, M.; Svensson, M. Vitamin D in relation to bone health and muscle function in young female soccer players. *Eur. J. Sport Sci.* **2017**, *17*, 249–256. [CrossRef] [PubMed]

43. Książek, A.; Dziubek, W.; Pietraszewska, J.; Słowińska-Lisowska, M. Relationship between 25(OH)D levels and athletic performance in elite Polish judoists. *Biol. Sport* **2018**, *35*, 191–196. [PubMed]

44. Croisier, J.L.; Ganteaume, S.; Binet, J.; Genty, M.; Ferret, J.M. Strength imbalances and prevention of hamstring injury in professional soccer players: A prospective study. *Am. J. Sports Med.* **2008**, *36*, 1469–1475. [CrossRef] [PubMed]

45. Ceglia, L. Vitamin D and skeletal muscle tissue and function. *Mol. Aspects Med.* **2008**, *29*, 407–414. [CrossRef]

46. Dzik, K.P.; Kaczor, J.J. Mechanisms of vitamin D on skeletal muscle function: Oxidative stress, energy metabolism and anabolic state. *Eur. J. Appl. Physiol.* **2019**, *119*, 825–839. [CrossRef]

Article

25-Hydroxycholecalciferol Concentration Is Associated with Protein Loss and Serum Albumin Level during the Acute Phase of Burn Injury

Andrzej Krajewski [1], Krzysztof Piorun [1], Dominika Maciejewska-Markiewicz [2,*],
Marta Markowska [3], Karolina Skonieczna-Żydecka [2], Ewa Stachowska [2], Zofia Polakowska [4],
Maciej Mazurek [1] and Małgorzata Szczuko [2]

[1] West Pomeranian Center of Treating Severe Burns and Plastic Surgery in Gryfice, 72-300 Gryfice, Poland;
krajewski1407@gmail.com (A.K.); krzysztof.piorun@gmail.com (K.P.); maciek.j.mazurek@gmail.com (M.M.)
[2] Department of Human Nutrition and Metabolomics, Pomeranian Medical University in Szczecin,
71-460 Szczecin, Poland; karzyd@pum.edu.pl (K.S.-Ż.); ewast@pum.edu.pl (E.S.);
malgorzata.szczuko@pum.edu.pl (M.S.)
[3] Clinic of Plastic, Endocrine and General Surgery, Pomeranian Medical University in Szczecin, 72-010 Police,
Poland; markowskamh@gmail.com
[4] Department of Dermatology and Venereology, Pomeranian Medical University in Szczecin, 71-010 Police,
Poland; zpolakowska@gmail.com
* Correspondence: dmaciejewska.pum@gmail.com; Tel.: +48-91-441-48-09

Received: 23 July 2020; Accepted: 9 September 2020; Published: 11 September 2020

Abstract: Background: Burned patients have an increased need for vitamin D supply related to the maintenance of calcium–phosphate homeostasis and the regulation of cell proliferation/differentiation. This study aimed to analyze the concentration of 25-hydroxycholecalciferol and its relationship with severe condition after burn injury. Methods: 126 patients were enrolled in the study. Patients were qualified due to thermal burns—over 10% of total body surface area. On the day of admission, the following parameters were assessed: 25-hydroxycholecalciferol concentration, total protein concentration, albumin concentration, aspartate transaminase activity, alanine transaminase activity, albumin concentration, creatinine concentration, c-reactive protein concentration, procalcitonin concentration, and interleukin-6 concentration. Results: Almost all patients (92%) in the study group had an improper level of vitamin D (<30 ng/mL), with the average of 11.6 ± 10.7 ng/mL; 17.5% of patients had levels of vitamin D below the limit of determination—under 3 ng/mL. The study showed that there are several factors which correlated with vitamin D concentration during the acute phase of burn injury, including: total protein ($r = 0.42$, $p < 0.01$), albumin, ($r = 0.62$, $p < 0.01$), percentage of body burns ($r = 0.36$, $p < 0.05$), aspartate aminotransferase ($r = 0.21$, $p < 0.05$), and c-reactive protein ($r = 0.22$, $p < 0.05$). We did not find any significant correlation between vitamin D concentration and body mass index. Conclusions: The burn injury has an enormous impact on the metabolism and the risk factors of the deficiency for the general population (BMI) have an effect on burned patients. Our study showed that concentration of 25-hydroxycholecalciferol is strongly correlated with serum albumin level, even more than total burn surface area and burn degrees as expected. We suspect that increased supplementation of vitamin D should be based on albumin level and last until albumin levels are balanced.

Keywords: vitamin D; burns; albumin; total protein; burn body surface

---

1. Introduction

Burns are one of the most serious injuries, which often include multi-organ dysfunction. Every year, about 1% of Polish people (both children and adults) suffer from various types of burns [1]. The stress and metabolic response associated with burn injury are linked to bone demineralization. These specific conditions promote large production of glucocorticoids that decrease the number of osteoblasts and block osteoclastogenesis. Moreover, interleukin (IL) 1-β and IL-6, produced in inflammatory conditions, increase osteoclastogenic bone resorption which leads to bone loss [2]. Due to such large metabolic changes, burned patients have an increased need for vitamin D supply related to the maintenance of calcium–phosphate homeostasis [3] and the regulation of cell proliferation/differentiation [4]. Extensive burns can lead to the dysfunction of many organs, which have a big impact on vitamin D biotransformation. Liver and kidney failures are responsible for insufficient conversion of cholekalcyferol to its active metabolites. Liver dysfunction may also result in impaired production of vitamin D-binding protein [4]. Vitamin D plays a very important role in the healing of dermal wounds. In vitro studies have shown that 25-hydroxycholecalciferol had a positive effect on the regulation of the transforming growth factor beta (TGFβ). TGFβ affects many processes related to the development and regulation of cell growth, such as wound healing and scar formation. The role of the beta-transforming agent in the treatment of thermal injury wounds can be associated with the stimulation of fibroblast proliferation, myofibroblast differentiation, and collagen synthesis. Vitamin D deficiency accompanying patients after extensive burns may have a negative impact on the healing process and prolong treatment and convalescence [5].

The aim of our study was to analyze the level of vitamin D and its relationship with severe condition during the acute phase of burn injury.

2. Materials and Methods

2.1. Patients

One hundred twenty-six patients with burn injuries were enrolled in the study. Participants were patients of the Western Pomeranian Center for the Treatment of Burns Injuries and Plastic Surgery in Poland. Patients were qualified due to thermal burns—over 10% of the total body surface area (TBSA); 88% of patients involved in the project meet the major burn criteria. According to the classification used in our hospital, major burn needs to include: ≥25% TBSA, or ≥20% in adults over 40 years old, or ≥10% TBSA with full-thickness burn, or all burn injuries complicated by major trauma/inhalation injury; 12% of patients met the moderate burn criteria (10–20% partial-thickness burn). According to their medical history from admission to the unit, none of the included patients suffered from chronic kidney disease. The protocol used in our hospital includes no albumin administration during the first 24 h of burn injury. Instead of albumin patients were given Ringer's lactate and fresh frozen plasma. The protocol of the study has been accepted by the local bioethical committee at the Pomeranian Medical University in Szczecin (KB-0012/143/16). Every participant signed a consent to take part in the study and was informed about its course, benefits, and potential side effects. Patient characteristics are shown in Tables 1 and 2.

Table 1. Burn degree among patients enrolled in the study.

Burn Degree	Amount of Patients ($n = 126$)
I/II	9 (7%)
II	13 (10.3%)
II/III	61 (48.5%)
III	35 (27.8%)
III/IV	7 (5.6%)
IV	1 (0.8%)

Table 2. Patient characteristics.

Patient Characteristics	Mean	±SD
Age [years]	49.06	17.53
BMI [kg/m^2]	24.59	3.76
Percentage of body burns [%]	21.11	20.80
Day after burn [day]	1.66	5.06
Phosphate [mmol/L]	1.18	0.22
Calcium [mmol/L]	2.29	0.26

2.2. Vitamin D and Other Biochemical Parameters Measurements

On the day of admission, the following parameters were assessed: 25-hydroxycalciferol concentration (vitamin D status predictor), total protein concentration, albumin concentration, aspartate transaminase activity, alanine transaminase (ALT) activity, albumin concentration, creatinine concentration, c-reactive protein (CRP) concentration, procalcitonin concentration, and IL-6 concentration. All measurements were performed in a commercial certificated laboratory in the Hospital. The 25-hydroxycholecalciferol measurement was based on validated automatic immunochemical method. Serum was used as basic material for all analysis.

2.3. Statistical Analysis

The statistical analysis was performed using the "R 3.0.2" program. In order to check the normal distribution, the Shapiro–Wilk test was used. The distribution did not deviate from the norm, and parametric tests were used in the calculations. The results are presented as mean values and standard deviation (SD). In order to estimate the correlations, the Pearson's correlation test was used. To estimate the connection between burn degree and concentration of vitamin D, the Poisson regression was used. The values of $p < 0.05$ were considered as statistically significant. To control type I errors, the false discovery rate (FDR) approach was used. The calculations were performed using the p.adjust function of the stats package in R 4.0.2. Multiple regression was used to assess the relationship between albumin, total protein, and vitamin D concentration. The values being at the threshold of statistical significance were established at $p < 0.055$ and the statistical tendency from $p = 0.06$ to $p = 0.1$. In reference to the results which were not statistically significant, the abbreviation NS (not significant) was used instead of p.

3. Results

Almost all patients (92%) in the study group had an improper level of 25-hydroxycholecalciferol (<30 ng/mL), with the average of 11.6 ± 10.7 ng/mL; 17.5% of patients had a level below the limit of determination—under 3 ng/mL. Poisson regression showed that there is a statistical tendency between 25-hydroxycholecalciferol concentration and burn degree ($p = 0.08$). The average concentration in particular subgroups is shown in Table 3.

Table 3. 25-hydroxycholecalciferol concentration according to burn degree.

25-Hydroxycholecalciferol Concentration [ng/mL]	Mean	±SD
Whole cohort	11.6	10.7
Superficial	18.2	13.5
Superficial partial thickness	13.08	9.9
Superficial deep dermal	11.7	13.4
Full thickness	12.8	11.7
Full thickness with catastrophic	8.45	8.3
Catastrophic	7.4	6.2

Pearson's test showed a significant correlation between body mass index (BMI), total protein, albumin, percentage of body burns, ALT, CRP, and vitamin D concentration (Table 4). The most significant correlations are shown in Figure 1.

Table 4. Correlation between 25-hydroxycholecalciferol concentration and biochemical parameters.

25-Hydroxycholecalciferol [ng/mL] vs., $n = 126$	r	p Value	FDR * p Value
BMI [kg/m^2]	0.18	NS	0.43
Total protein [g/dL]	0.42	$p < 0.01$	$p < 0.01$
Albumin [g/dL]	0.62	$p < 0.01$	$p < 0.01$
Percentage of body burns [%]	(−) 0.36	$p < 0.05$	0.10
AST [U/L]	(−) 0.21	$p < 0.05$	0.07
ALT [U/L]	(−) 0.08	NS	0.43
CRP [mg/L]	(−) 0.22	$p < 0.05$	0.07
IL-6 [pg/mL]	(−) 0.16	NS	0.20
Creatinine [mg/dL]	(−) 0.18	$p < 0.055$ [#]	0.10

[#] the verge of significance. * false discovery rate. The false discovery rate (FDR); AST: aspartate aminotransferase; ALT: alanine transaminase; CRP: c-reactive protein; NS: not statistically significant.

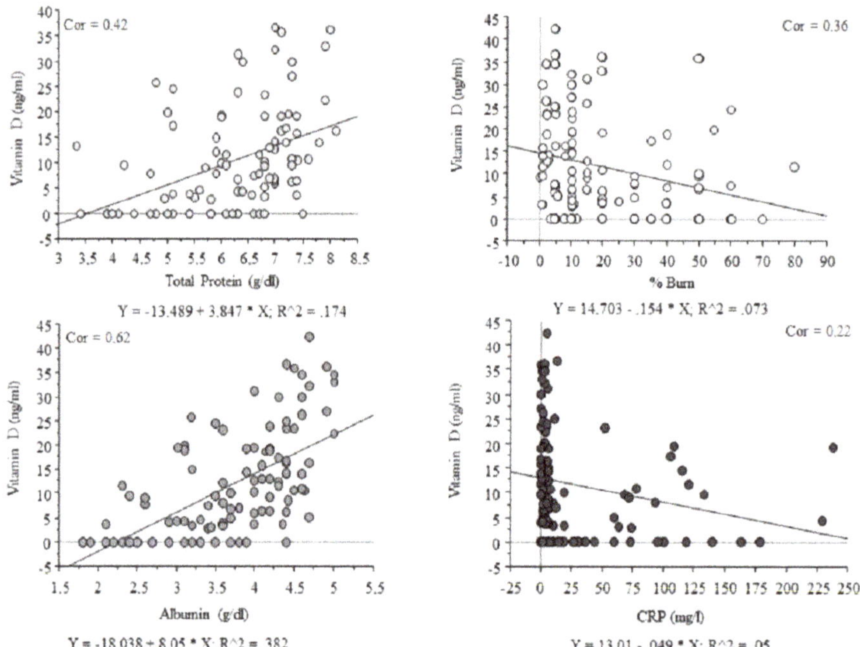

Figure 1. 25-hydroxycholecalciferol concentration and biochemical status; CRP: c-reactive protein.

As we demonstrated that total protein and albumin concentration were significantly correlated (y = 2.393 + 1.069x; x2 = 0.7), we decided to conduct a multiple regression analysis to demonstrate that only albumin was significantly associated with vitamin D concentration (b = 7.9, SE = 1.68, t = 4.69, $p < 0.0001$). We also performed analysis of the correlation of percentage of body burns with albumin/protein. The results are presented in Table 5.

Table 5. Correlation between percentage of body burns and protein parameters.

Percentage of Body Burns [%]	r	p Value
Total protein [g/dL]	(−) 0.58	$p < 0.01$
Albumin [g/dL]	(−) 0.62	$p < 0.01$

4. Discussion

Deficiencies of minerals and vitamins in burned patients are a serious clinical challenge either during hospitalization, or while in outpatient care. In many cases, the recommended supplementation is not sufficient for deficiencies and patients cannot reach the proper level of vitamin D. Dickerson et al. reported that 76% of critically ill patients after traumatic injury were vitamin D deficient and severely deficient [6]. Similar observations were made by Alizedeh et al., where 74% patient had an improper level of vitamin D [7].

Our study revealed that 92% of burned patients had an improper concentration of 25-hydroxycholecalciferol (average concentration: 11.6 ± 10.7 ng/mL) and almost 20% of them had a level below the limit of the quantification (<3 ng/mL). Płudowski et al. revealed that 89.9% of the Polish population is vitamin D-deficient, with 18.0 ± 9.6 ng/mL of average concentration of 25-hydroxycholecalciferol [8]. A lower concentration of vitamin D before admission can have a big impact on the concentration of vitamin D deficient after a burn injury and can be an additional factor of such low level during the acute phase of a burn injury.

The need of supplementation in burned patients is well known and described in many medical protocols [9]. Unfortunately, past research suggested that universal supplementation does not significantly improve concentration of Vitamin D in serum. There is no recommendation for a sufficient dose or time of increased supplementation for these group of patients [10]. Vitamin D regulates many crucial metabolic processes which are critical for burned patients' convalescence [11,12]. Appropriate supplementation should be implemented immediately after admission to the hospital. There is a great need to identify the factors that have the biggest influence on the concentration of vitamin D, and at the same time are analyzed during routine admission to the hospital.

There are multiple factors that are associated with low level of circulating vitamin D. Many of them are connected with poor prognosis in critically ill patients, among others: organ failure [13], sepsis, and short- or long-term mortality. Several meta-analyses have revealed that very low 25-hydroxycholecalciferol concentration is associated with the higher incidence of either infection or sepsis, and greater mortality in these groups of patients [14,15]. Our study showed that there are several factors which correlated with serum vitamin D concentration during the acute phase of burn injury (Figure 1), including: serum total protein ($r = 0.42$), serum albumin, ($r = 0.62$), percentage of body burns ($r = 0.36$), AST ($r = 0.21$), and CRP ($r = 0.22$). However, we did not see a significant relationship between BMI and concentration of 25-hydroxycholecalciferol, which are considered to be one of the most important factors of vitamin D deficiency in the general population [16]. We can assume that the burn injury has an enormous impact on the metabolism and the risk factors of the deficiency for the general population have a negligible effect on burned patients.

Evaluation of the vitamin D level in burned patients is a difficult issue, related to the acute phase development, which is associated with decreased levels of vitamin D binding protein (VDBP). The amount of protein that can bind 25-hydroxycholecalciferol, significantly decreases, and the tested "free amount" of the active form can be falsified. Reduced protein synthesis persists for several months after burns; therefore, the results obtained can be false [17]. However, VDBP levels increase after the acute phase of thermal injury [18], but albumin concentration may recover after six months or more [19]. Due to such multifactorial problems, interpretation of 25-hydroxycholecalciferol concentration in the diagnosis of vitamin D deficiency remains challenging [20]. We have to remember that, during hospital admission, the analysis of VDBP protein or vitamin D status is not included in the standard analysis.

Our study reveals that serum protein level, mostly albumin, strongly correlated (Figure 1) with serum vitamin D status during the acute phase of burn injury. We also noticed a trend and a weaker correlation with either the burn degree or TBSA and vitamin D concentration. Therefore, these two factors are strongly associated with the plasma protein (and albumin) concentration [21]. The appropriate level of protein is essential for the maintenance of plasma colloid oncotic pressure and responsible for the transport of various substances in the blood stream including: hormones, drugs and vitamins, such as vitamin D [22]. Albumin is one of the most important proteins synthesized by the liver and has several relevant functions [23]. Because of its long half-life and the fact that serum level depends on many factors, albumin is a reliable marker of mortality and morbidity in hospitalized patients [24,25]. The acute period of burn injury is associated with severe conditions, such as: increase of free-radical oxidation and higher vascular permeability in the burned wounds, which significantly decreases the level of albumin [24]. Our study revealed that albumin strongly correlated with 25-hydroxycholecalciferol concentration ($r = 0.62$) and the multiple regression confirm that thesis. A majority (85–90%) of 25-hydroxycholecalciferol D is bound to VDBP and 10–15% to albumin [25]. The acute phase of burn injury is associated with higher vascular permeability in the burned wounds, which significantly decreases the level of all proteins, including albumin and VDBP. Yonemura et al. revealed that the active form of vitamin D (calcitriol) is associated with albumin level in patients with end-stage renal disease. Moreover, supplementation with an active form of vitamin D tends to normalize low serum albumin concentrations [26]. We can hypothesize that albumin can be a good predictor of vitamin D status, especially that the concentration of 25-hydroxycholecalciferol is not measured in standard analysis during admission to hospital [27]. However, long-term studies are needed to confirm the usefulness of albumin as a factor reflecting the need for vitamin D supplementation. It should be highlighted that burn injury decreases vitamin D synthesis in the skin, therefore patients need to be supplemented permanently [28].

5. Conclusions

Burn injuries have an enormous impact on the metabolism in burned patients. On the other hand, the risk factors of deficiency for the general population (e.g., BMI) have a negligible effect on burned patients. Our study shows that the concentration of 25-hydroxycholecalciferol is strongly correlated with serum albumin levels, even more than TBSA and burn degrees, as expected. Albumin can be a good predictor of vitamin D status, especially since the concentration of 25-hydroxycholecalciferol is not measured in standard analysis during admission to hospital. We presume that this direction of vitamin D diagnostic should be tested in randomized clinical trials.

Author Contributions: A.K.—principal investigator, study design; K.P. and M.M. (Marta Markowska)—data collection; K.S.-Ż.—statistical analysis; E.S.—study design, co-investigator; Z.P.—writing; M.M. (Maciej Mazurek)—writing; M.S.—writing; D.M.-M.—study design, co-investigator, data collection, writing. All authors have read and agreed to the published version of the manuscript.

Funding: The project is financed by the program of the Minister of Science and Higher Education under the name "Regional Initiative of Excellence" in 2019–2022 project number 002 / RID / 2019/20 amount of financing 12 000 000 PLN.

Conflicts of Interest: All authors declare no conflict of interest.

References

1. Chrzanowska-Wąsik, M.; Chemperek, E.; Sokołowski, D.; Goniewicz, M.; Badnarz, K.; Rzońca, P. Burn analysis in adult patients hospitalized in the East Centre of Burn Treatment and Reconstructive Surgery in Łęczna. *J. Educ. Health Sport* **2017**, *7*, 410–419.
2. Klein, G.L. The interaction between burn injury and vitamin D metabolism and consequences for the patient. *Curr. Clin. Pharmacol.* **2008**, *3*, 204–210. [CrossRef]
3. Lips, P. Interaction between vitamin D and calcium. Scand. *J. Clin. Lab. Investig.* **2012**, *243*, 60–64.

4. Klein, G.L.; Herndon, D.N.; Rutan, T.C.; Sherrard, D.J.; Coburn, J.W.; Langman, C.B.; Thomas, M.L.; Haddad, J.G., Jr.; Cooper, C.W.; Miller, N.L.; et al. Bone disease in burn patients. *J. Bone Miner. Res.* **1993**, *8*, 337–345. [CrossRef]

5. Ding, J.; Kwan, P.; Ma, Z.; Iwashina, T.; Wang, J.; Shankowsky, H.A.; Tredget, E.E. Synergistic effect of vitamin D and low concentration of transforming growth factor beta 1, a potential role in dermal wound healing. *Burns* **2016**, *42*, 1277–1286. [CrossRef]

6. Dickerson, R.N.; Van Cleve, J.R.; Swanson, J.M.; Maish, G.O.; Minard, G.; Croce, M.A.; Brown, R.O. Vitamin D deficiency in critically ill patients with traumatic injuries. *Burns Trauma* **2016**, *4*, 1–9. [CrossRef]

7. Alizadeh, N.; Khalili, H.; Mohammadi, M.; Abdollahi, A. Serum Vitamin D levels at admission predict the length of intensive care unit stay but not in-hospital mortality of critically ill surgical patients. *J. Res. Pharm. Pract.* **2015**, *4*, 193.

8. Płudowski, P.; Ducki, C.; Konstantynowicz, J.; Jaworski, M. Vitamin D status in Poland. *Pol. Arch Med. Wewn* **2016**, *126*, 530–539. [CrossRef]

9. Rech, M.A.; Hidalgo, D.C.; Larson, J.; Zavala, S.; Mosier, M. Vitamin D in burn-injured patients. *Burns* **2019**, *45*, 32–41. [CrossRef]

10. Rousseau, A.-F.; Losser, M.-R.; Ichai, C.; Berger, M.M. ESPEN endorsed recommendations: Nutritional therapy in major burns. *Clin. Nutr.* **2013**, *32*, 497–502. [CrossRef]

11. Matyjaszek-Matuszek, B.; Lenart-Lipińska, M.; Woźniakowska, E. Clinical implications of vitamin D deficiency. *Menopause Rev.* **2015**, *14*, 75. [CrossRef] [PubMed]

12. Prietl, B.; Treiber, G.; Pieber, T.R.; Amrein, K. Vitamin D and immune function. *Nutrients* **2013**, *5*, 2502–2521. [CrossRef] [PubMed]

13. Jeschke, M.G. The Hepatic Response to Thermal Injury: Is the Liver Important for Postburn Outcomes? *Mol. Med.* **2009**, *15*, 337–351. [CrossRef] [PubMed]

14. Al-Tarrah, K.; Hewison, M.; Moiemen, N.; Lord, J.M. Vitamin D status and its influence on outcomes following major burn injury and critical illness. *Burns Trauma* **2018**, *6*, s41038-018-0113-4. [CrossRef]

15. Herndon, D.N.; Tompkins, R.G. Support of the metabolic response to burn injury. *Lancet* **2004**, *363*, 1895–1902. [CrossRef]

16. Lagunova, Z.; Porojnicu, A.C.; Lindberg, F.; Hexeberg, S.; Moan, J. The dependency of vitamin D status on body mass index, gender, age and season. *Anticancer. Res.* **2009**, *29*, 3713–3720. [CrossRef]

17. Klein, G.L. Burns: Where Has All the Calcium (and Vitamin D) Gone? *Adv. Nutr.* **2011**, *2*, 457–462. [CrossRef]

18. Rousseau, A.-F.; Foidart-Desalle, M.; Ledoux, D.; Remy, C.; Croisier, J.-L.; Damas, P.; Cavalier, E. Effects of cholecalciferol supplementation and optimized calcium intakes on vitamin D status, muscle strength and bone health: A one-year pilot randomized controlled trial in adults with severe burns. *Burns* **2015**, *41*, 317–325. [CrossRef]

19. Klein, G.L.; Herndon, D.N.; Chen, T.C.; Kulp, G.; Holick, M.F. Standard multivitamin supplementation does not improve vitamin D insufficiency after burns. *J. Bone Miner. Metab.* **2009**, *27*, 502–506. [CrossRef]

20. Quraishi, S.A.; Camargo, C.A. Vitamin D in acute stress and critical illness. *Curr. Opin. Clin. Nutr. Metab. Care* **2012**, *15*, 625–634. [CrossRef]

21. Lehnhardt, M.; Jafari, H.J.; Druecke, D.; Steinstraesser, L.; Steinau, H.U.; Klatte, W.; Homann, H.H. A qualitative and quantitative analysis of protein loss in human burn wounds. *Burns* **2005**, *31*, 159–167. [CrossRef]

22. Caironi, P.; Gattinoni, L. The clinical use of albumin: The point of view of a specialist in intensive care. *Blood Transfus.* **2009**, *7*, 259–267.

23. Lyons, O.; Whelan, B.; Bennett, K.; O'Riordan, D.; Silke, B. Serum albumin as an outcome predictor in hospital emergency medical admissions. *Eur. J. Intern. Med.* **2010**, *21*, 17–20. [CrossRef]

24. Pérez-Guisado, J.; de Haro-Padilla, J.M.; Rioja, L.F.; DeRosier, L.C.; de la Torre, J.I. Serum albumin levels in burn people are associated to the total body surface burned and the length of hospital stay but not to the initiation of the oral/enteral nutrition. *Int. J. Burn. Trauma* **2013**, *3*, 159–163.

25. Robinson, R. Low serum albumin and total lymphocyte count as predictors of 30 day hospital readmission in patients 65 years of age or older. *PeerJ* **2015**, *3*, e1181. [CrossRef]

26. Bikle, D.; Bouillon, R.; Thadhani, R.; Schoenmakers, I. Vitamin D metabolites in captivity? Should we measure free or total 25(OH)D to assess vitamin D status? *J. Steroid Biochem. Mol. Biol.* **2017**, *173*, 105–116. [CrossRef]

27. Yonemura, K.; Fujimoto, T.; Fujigaki, Y.; Hishida, A. Vitamin D deficiency is implicated in reduced serum albumin concentrations in patients with end-stage renal disease. *Am. J. Kidney Dis.* **2000**, *36*, 337–344. [CrossRef]

28. Klein, G.L.; Chen, T.C.; Holick, M.F.; Langman, C.B.; Price, H.; Celis, M.M.; Herndon, D.N. Synthesis of vitamin D in skin after burns. *Lancet* **2004**, *363*, 291–292. [CrossRef]

Review

Vitamin D and Autoimmune Thyroid Disease—Cause, Consequence, or a Vicious Cycle?

Inês Henriques Vieira [1,*], Dírcea Rodrigues [1,2] and Isabel Paiva [1]

[1] Endocrinology Department of Coimbra Hospital and University Centre, Praceta Professor Mota Pinto, 3004-561 Coimbra, Portugal; dircearodrigues@chuc.min-saude.pt (D.R.); isapaiva@chuc.min-saude.pt (I.P.)

[2] Faculty of Medicine of the University of Coimbra, R. Larga 2, 3000-370 Coimbra, Portugal

* Correspondence: 11285@chuc.min-saude.pt

Received: 17 July 2020; Accepted: 9 September 2020; Published: 11 September 2020

Abstract: Vitamin D is a steroid hormone traditionally connected to phosphocalcium metabolism. The discovery of pleiotropic expression of its receptor and of the enzymes involved in its metabolism has led to the exploration of the other roles of this vitamin. The influence of vitamin D on autoimmune disease—namely, on autoimmune thyroid disease—has been widely studied. Most of the existing data support a relationship between vitamin D deficiency and a greater tendency for development and/or higher titers of antibodies linked to Hashimoto's thyroiditis, Graves' disease, and/or postpartum thyroiditis. However, there have also been some reports contradicting such relationships, thus making it difficult to establish a unanimous conclusion. Even if the existence of an association between vitamin D and autoimmune thyroid disease is assumed, it is still unclear whether it reflects a pathological mechanism, a causal relationship, or a consequence of the autoimmune process. The relationship between vitamin D's polymorphisms and this group of diseases has also been the subject of study, often with divergent results. This text presents a review of the recent literature on the relationship between vitamin D and autoimmune thyroid disease, providing an analysis of the likely involved mechanisms. Our thesis is that, due to its immunoregulatory role, vitamin D plays a minor role in conjunction with myriad other factors. In some cases, a vicious cycle is generated, thus contributing to the deficiency and aggravating the autoimmune process.

Keywords: Vitamin D; autoimmune thyroid disease; Vitamin D receptor; Graves' disease; Hashimoto thyroiditis

1. Introduction

The term vitamin D (VitD) encompasses a group of steroid compounds, namely VitD2 (ergocalciferol) and VitD3 (cholecalciferol) [1].

Its main functions are the regulation of phosphocalcium metabolism and the promotion of bone homeostasis. However, the discovery of the widespread expression of the VitD receptor (VDR) and the enzymes responsible for its metabolism suggests the pleiotropic role of this vitamin and its influence in several diseases [2,3]. An immunomodulatory role is evident and its influence on the development of autoimmune diseases (AID) has been proposed. Autoimmune thyroid disease (AITD) is the most common organ-specific AID [3] and several studies have been carried out to explore the role of VitD in its development and course, as well as the possible impact of supplementation.

The aim of this review is to analyze the most recent evidence on the relationship between VitD and AITD.

2. Materials and Methods

A search was conducted in Pubmed using the Medical Subject Headings (MESH) terms "vitamin D" and "thyroid disease" for publications from January 2009 to July 2020. Articles with full text in English,

Portuguese, or Spanish (n = 205) were selected based on their title and/or abstract. Articles focusing on nodular thyroid disease (benign or malignant), parathyroid disease, or otherwise not referring to autoimmune thyroid disease were excluded at this stage. Additional articles were excluded after reading the full text if they did not relate to the study matter or if the information provided was redundant. The bibliographies of the publications thus selected were also analyzed, with the inclusion of additional relevant articles published in the same time interval. Further research was conducted to provide context and to answer particular questions which emerged upon reading the selected articles or during the peer-review process (Figure 1).

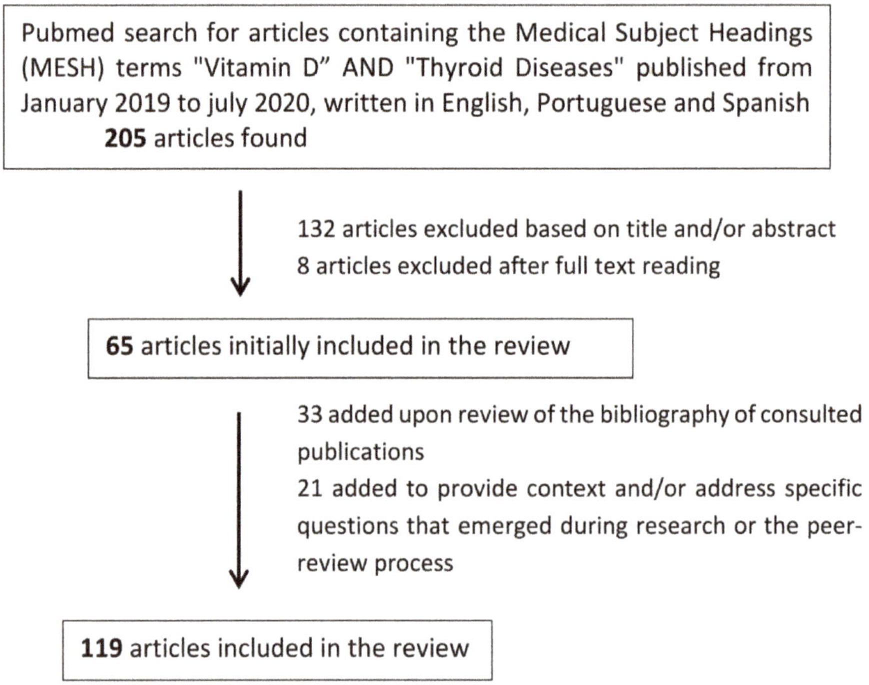

Pubmed search for articles containing the Medical Subject Headings (MESH) terms "Vitamin D" AND "Thyroid Diseases" published from January 2019 to july 2020, written in English, Portuguese and Spanish
205 articles found

132 articles excluded based on title and/or abstract

8 articles excluded after full text reading

65 articles initially included in the review

33 added upon review of the bibliography of consulted publications

21 added to provide context and/or address specific questions that emerged during research or the peer-review process

119 articles included in the review

Figure 1. Literature search process.

3. Metabolism and Functions of Vitamin D

In humans, VitD3 is produced in the skin under the action of ultraviolet light on 7-desidroxicolesterol [1,4,5]. Additionally, it can be obtained nutritionally, predominantly from fish oil and eggs [4,5]. In fungi and plants, VitD2 is synthesized from ergosterol [5].

There are three essential steps in the metabolism of this vitamin, which are carried out by cytochrome P450 oxidases: 25-hydroxylation, which produces 25(OH)D (calcidiol); 1α- hydroxylation, which generates 1,25(OH)$_2$D (calcitriol); and 24-hydroxylation, which inactivates 25(OH)D and 1,25(OH)$_2$D (preferentially), preventing the accumulation of toxic levels [4–6]; see Figure 2.

25(OH)D has little biological activity [2], but is the main circulating form, being considered to best reflect an organism's reserves. As such, its quantification is widely used to assess the levels of VitD [7]. Conversion to 1,25(OH)$_2$D requires the action of 1α-hydroxylase (CYP27B1). Although the majority of human cells express this enzyme, levels of 1,25(OH)$_2$D seem to reflect its activity in the cells of the proximal tubules of the kidney [7]. In these cells, its activity is stimulated by parathyroid hormone (PTH) and is inhibited by fibroblast growth factor-23 (FGF23) and by 1,25(OH)$_2$D itself [5].

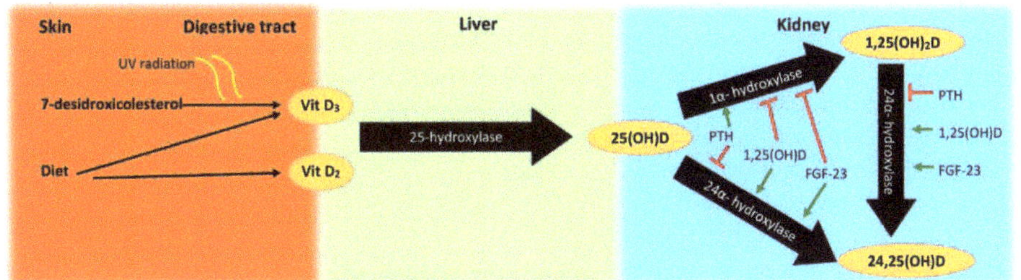

Figure 2. Schematic representation of vitamin D metabolism. UV, ultraviolet; Vit D₂, vitamin D₂; VitD₃, vitamin D₃; PTH, parathyroid hormone; FGF-23, fibroblast growth factor 23.

CYP24A1 is the only established 24-hydroxylase, which has an inverse regulation from the kidney's 1α-hydroxylase, as it is induced by 1,25(OH)$_2$D and FGF23 [5].

Most other human cells include 1α-hydroxylase and VDR, but seem to essentially regulate the levels of 1,25(OH)$_2$D on a tissue level [7], which may be subject to different regulatory mechanisms than those in renal cells [5].

The main function of 1,25(OH)$_2$D is to increase calcium absorption from the intestines and, along with PTH, it contributes to maintaining serum calcium levels. When there is a low VitD status, PTH levels tend to rise, in order to compensate for impaired intestinal calcium absorption [8].

VitD exerts most of its effects by binding to the nuclear receptor VDR, which dimerizes with the retinoid X receptor; this heterodimer binds to VitD-responsive genes [9]. Rapid actions, independent of gene transcription [10], which modulate intracellular calcium levels and several signaling pathways have also been described. Thus, this compound can directly or indirectly influence up to 5% of the human genome. A randomized controlled clinical trial evaluated gene expression in the white blood cells of eight adults after daily supplementation with 400 and 2000 UI of VitD3. There was a differential expression of ≥291 genes involved in functions such as cell proliferation and differentiation, immune function, and DNA repair in a continuous manner with increasing levels of 25(OH)D [11].

Therefore, it is not surprising that, in addition to having been implicated in several skeletal diseases, the hypothesis has been raised regarding the association of VitD with neoplasms, cardiovascular disease, metabolic diseases, infections, AID, and neurocognitive dysfunction [12]. However, a clear role has not been definitively established for any of these conditions [13].

4. Vitamin D and Immune Modulation

The immune system defends the organism against what is recognized as non-self. Failure to recognize the body's cells as the self generates autoimmune phenomena, which may be physiological (elimination of unnecessary cells) or pathological (AID) [14].

Given the immunomodulatory role of VitD, its relationship with AID has been extensively explored. Evidence of associations between VitD deficiency and several AIDs has been presented; namely, multiple sclerosis, systemic sclerosis, systemic lupus erythematosus, Sjögren's syndrome, mixed connective tissue disease, rheumatoid arthritis, antiphospholipid syndrome, type 1 diabetes mellitus, AITD, celiac disease, and primary biliary cirrhosis [14]. In a population-based longitudinal study, Skaaby et al. observed a decreased risk of AID in general, and thyrotoxicosis in particular, with each increment of 4 ng/mL (10 nmol/L) on the level of 25(OH)D (hazard ratios of 0.94 and 0.83, respectively) [15]. Additionally, birth month can influence the risk of developing AID, most likely in relation to exposure to ultraviolet radiation [14].

In general, VitD tends to activate the innate immune response and to regulate the adaptive immune response [5,6,14]. VitD appears to have the ability to stimulate the differentiation of monocytes into macrophages and the production of antibacterial substances by these cells, promoting an initial response [9,16], but also helps to avoid excessive innate responses and consequent tissue damage [16].

Concerning the regulation of adaptative immunity, the result of antigen presentation to T cells differs when performed by immature or mature dendritic cells (Figure 3), promoting tolerance or an immune response, respectively [9]. Physiological levels of 1,25(OH)$_2$D inhibit the maturation of dendritic cells and maintain a more tolerogenic phenotype [17]. As dendritic cells become more mature, they express more 1α-hydroxylase and less VDR. As a consequence, mature antigen-presenting dendritic cells can be relatively insensitive to the action of 1,25(OH)$_2$D, allowing for the induction of an initial T response. However, they synthesize 1,25(OH)$_2$D, which acts on a paracrine level on immature dendritic cells and prevents their excessive proliferation [16]. Dendritic cells generated with the use of biologically active forms of VitD have high immunoregulatory capacity [18] while maintaining cell mobility [19].

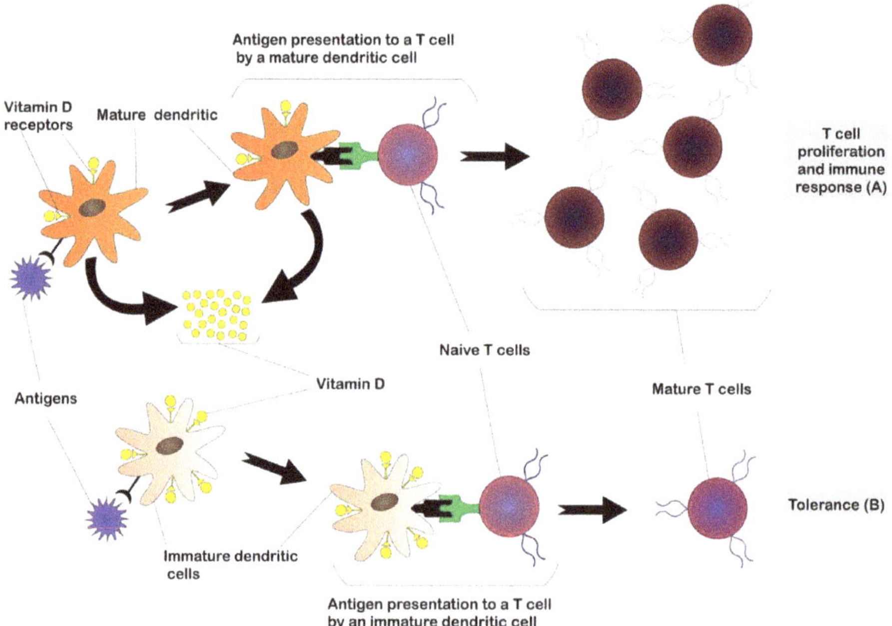

Figure 3. Influence of vitamin D in activation of adaptive immunity. The different results of antigen presentation to T cells by mature vs. immature dendritic cells, leading to immune response (**A**) or tolerance (**B**), respectively, are depicted. Vitamin D inhibits the maturation of dendritic cells, maintaining a more tolerogenic phenotype. Mature dendritic cells have less vitamin D receptor (VDR) but synthesize 1,25(OH)$_2$D, which acts on a paracrine level on immature dendritic cells and prevents their excessive proliferation.

VitD also plays a role in the regulation of adaptive immunity. B and T lymphocytes have low VDR expression at rest and higher expression when activated. An ability to synthetize VitD was also detected, which plays a regulatory role, acting in an autocrine/paracrine fashion [16].

Previous studies have led to the conclusion that VitD has a role in promoting the change from Th1 to Th2 phenotype, limiting the damage induced by the cellular immune response [17]. However, it has been found that, in vivo, the effects of VitD on T cells are more complex [16,17].

In T cells, 1,25(OH)$_2$D inhibits the proliferation of Th17 (linked to organ-specific autoimmunity, inflammation, and tissue damage), appears to induce regulatory T cells (Treg), which have a suppressive role in the proliferation of T cells, and helps direct T cells to tissues. FoxP3 is important in Treg cell development, and VitD levels have been found to be associated with FoxP3 expression in 32 children with chronic autoimmune thyroiditis. An increase in FoxP3 expression has been observed after VitD supplementation [20]. Additionally, VitD can reduce the production of cytokines by CD8+ T cells and regulate their proliferation after specific stimuli, although a significant effect has not been shown in animal models [16].

In B cells, $1,25(OH)_2D$ has a direct and indirect regulatory role (through T helper cells), seeming to inhibit their differentiation and the production of immunoglobulins [16].

5. Vitamin D and Autoimmune Thyroid Disease

AITD is the most common AID, with a prevalence of around 5% [21,22].

Autoimmunity requires an autoantigen to which the individual is normally tolerant and a process which leads to breaking that tolerance [23]. The potential autoantigens in the thyroid are the thyroid stimulating hormone (TSH) receptor (TSH-R), thyroid peroxidase (TPO), and thyroglobulin (Tg). Autoimmunity to these antigens leads to the creation of anti-thyroid antibodies. Anti-TPO and anti-Tg are usually associated with chronic autoimmune thyroiditis/Hashimoto thyroiditis (HT), and TSH-R (TRABs) with Graves' disease (GD) [24].

Both GD and HT are characterized by lymphocytic infiltration of the thyroid parenchyma. In GD, the infiltration is mild, such that the gland remains intact but TRABs play a central role in stimulating the gland's function and growth. In HT, the lymphocytic infiltrate causes the destruction of the follicles, which may lead to hypothyroidism [21]. In the thyroid tissue, the recruitment of Th1 lymphocytes may increase the production of interferon-y and tumor necrosis factor-α, which stimulate the secretion of CXCL10 by thyroid cells and create a positive feedback, thus initiating and perpetuating the autoimmune process [22].

B cells are found in secondary lymphoid follicles in the thyroid tissue and produce antibodies spontaneously, making the thyroid the probable main source of autoantibodies in AITD [21].

AITD has a multifactorial etiology, influenced by genetic factors (e.g., polymorphisms of TSH-R, Tg, human leukocyte antigens, and other genes associated with the immune response) [24], environmental factors (e.g., radiation, iodine, smoking habits, infections, selenium, drugs, stress, and dietary habits) [21,24,25], and endogenous factors (e.g., body mass index, adipokines, estrogens, selective X chromosome inactivation, microquimerism, glucocorticoids [21], and potentially the gastrointestinal microbiome) [26–28]. Given the immunomodulatory role of VitD, its relationship with AITD has been extensively studied in recent years [29].

5.1. Data on Vitamin D and Thyroid Function

A role in the modulation of the hypothalamic–pituitary–thyroid axis has been proposed for VitD, both at the pituitary [30] and thyroid gland levels [31]. Previous studies have reported the presence of VDR in murine thyrotropic cells [31]. A strong molecular homology between VDR and thyroid hormone has been demonstrated, as well as the presence of VDR in murine follicular thyroid cells. The incubation of these cells with $1,25(OH)_2D$ inhibited the uptake of iodine and cell growth [32].

Barchetta et al. studied the seasonality of TSH levels in euthyroid adults and found a strong inverse correlation between this hormone and 25(OH)D, with TSH levels being highest in autumn–winter and 25(OH)D levels being highest in spring–summer [31]. The relationship between the season of birth and risk of AITDs has also been evaluated, with inconsistent results. No impact of birth month in GD and discretely higher birth rates in autumn in HT females were reported in a study with data from Europe (mostly from the U.K.) [33]. A higher risk of autoimmune thyroiditis in subjects born in summer [34] and no relationship between month of birth and GD [35] were described in Danish register-based studies. A higher frequency of birth in spring was noticed in Greek children with HT [36]. Seasonality of birth month may be related to VitD levels (higher frequency of deficit in the end of winter, beginning of spring), but also may relate to viral exposure and other factors which vary in different regions and years [36].

Mackawy et al. also found an inverse relationship between VitD levels and TSH values, with a high prevalence of hypovitaminosis D and hypocalcemia in patients with hypothyroidism [37]. Two population-based studies corroborated these data in young people [38], as well as in middle-aged and elderly men with negative anti-thyroid antibodies [39]. A study performed in Korea revealed that iodine excess was associated with thyroid dysfunction only in VitD-deficient individuals [40].

In patients with AITD, Vondra et al. found a positive relationship between 25(OH)D levels and the fT4/fT3 ratio, which disappeared after supplementation with cholecalciferol. The authors speculated that the decreased ratio may be a compensatory adaptation to VitD deficiency [41].

5.2. Data on Vitamin D Levels and Autoimmune Thyroid Disease

Most data on VitD and AITD have come from cross-sectional studies and tend to support the existence of an association.

Kivity et al. reported an association between VitD deficiency, defined as 25(OH)D < 10 ng/mL (~25 nmol/L), and a higher frequency of AITD (mainly HT) and the presence of thyroid antibodies, in general [42]. Unal et al. found lower levels of 25(OH)D in individuals with AITD, with the GD group registering lower levels than those with HT and an inverse correlation between the levels of 25(OH)D and antithyroid antibody titers [43]. Another cross-sectional study examined 25(OH)D levels in 140 people with AITD versus 70 controls and found lower levels in the study group. Higher levels of 25(OH)D had a weak correlation with lower TRABs, but were not associated with anti-TPO/Tg titers [44]. In a meta-analysis in 2015, Wang et al. reported lower levels of 25(OH)D and higher prevalence of deficiency in individuals with AITD vs. controls. In sub-group analysis, the relationship remained when HT and GD patients were analyzed separately [45].

A role of VitD has also been proposed in polycystic ovary syndrome (PCOS); these patients had a high prevalence of AITD, making it plausible that there was a pathophysiological association. Muscogiuri et al. evaluated 50 women with PCOS and found lower 25(OH)D levels in those who also had AITD [46].

However, there are also data that contradict the presence of an association between VitD and AITD. D'Aurizio et al. did not find a statistically significant difference in the levels of VitD in AITD patients when compared to healthy controls [47]. Effraimidis et al. compared euthyroid individuals without anti-thyroid antibodies and with a family history of AITD (used as a marker for genetic pre-disposition) versus individuals without anti-thyroid antibodies and with no family history of AITD. The authors found higher levels of 25(OH)D in those with a family history. In a longitudinal analysis by the same authors, individuals who developed de novo anti-TPO antibodies were compared with control subjects, with no statistically significant difference in the levels of 25(OH)D or 1,25(OH)$_2$D at baseline, nor at the time of seroconversion [48].

A study comparing pre-/post-menopausal women and men with AITD found an association of AITD and VitD levels only in pre-menopausal women. These data raise the possibility of an interaction between VitD and estrogens in the development of AITD. 17-β estradiol may play a protective role by suppressing the transcription of CYP24A1, increasing VDR biosynthesis, inducing greater binding, and internalizing D-binding protein to T cells and macrophages [49]. The results from an analysis of the 6th Korean National Health and Nutrition Study Examination Survey corroborate this hypothesis, with lower VitD levels in anti-TPO-positive women (but not men) and an association of lower VitD with thyroid dysfunction exclusively in TPO-positive pre-menopausal woman [50].

5.2.1. Data in Hashimoto's Thyroiditis/Chronic Autoimmune Thyroiditis

There is evidence supporting a relationship between vitamin D and HT. Tamer et al. identified lower 25(OH)D levels in individuals with HT versus control subjects, with a tendency for a higher prevalence of deficiency in patients with hypothyroidism than in those in euthyroidism [51]. Studies in other populations corroborated the association between lower levels of 25(OH)D and the risk of HT, namely

Bozkurt et al. 2013 [52]; Mansournia et al. 2014 [53]; Vondra et al. 2015 [7]; Maciejewski et al. 2015 [54]; Kim D et al. 2016 [55]; Giovinazzo et al. 2017 [56]; Ke et al. 2017 [57]; and Pergola et al. 2018 [58].

There are also data supporting this relationship at age extremes. A higher prevalence of AITD and anti-TPO titers in association with 25(OH)D < 20 ng/mL (~50 nmol/L) was found in individuals over 65 years of age. It should be noted, however, that the AITD group was older and had higher creatinine levels [59]. In pediatric patients with HT vs. healthy controls, a higher prevalence of VitD deficiency was also found [60–63]. However, in an analysis of pediatric patients with type 1 DM with vs. without HT, 25(OH)D levels < 20 ng/mL were found in both groups, with no difference between the two [64].

The relationship with antibody titers is characterized by more inconsistent data. Bozkurt et al. reported a correlation between 25(OH)D deficiency severity, duration of HT, thyroid volume, and antibody titers [52]. An inverse correlation between 25(OH)D and anti-TPO was also verified by Giovinazzo et al. in recently diagnosed euthyroid HT patients vs. control subjects [56]; by Arslan et al. in healthy subjects with moderate–severe 25(OH)D deficiency [65]; and by Shin et al. in individuals with AITD [66]. Goswami et al. detected only a weak correlation between the levels of 25(OH)D and anti-TPO [67]. Wang et al. found a negative correlation between the levels of 25(OH)D and anti-Tg, but not anti-TPO [68]. Ke et al. found no relationship with thyroid function, antibody titers, and serum cytokines in a group with HT [57]. Sönmenzgöz et al. found no correlation between the levels of 25(OH)D and anti-TPO in a pediatric population [60]. An absence of correlation between the levels of 25(OH)D, anti-TPO, and anti-Tg was also observed in two population-based studies in Thailand [38] and China [39].

The results obtained by Yasmeh et al. contradict most of the published data, indicating higher levels of 25(OH)D in women with HT vs. controls (no difference in males) and a positive correlation between levels of 25(OH)D and anti-TPO titers only in males [69].

VitD may also affect disease manifestations: Xu et al. reported a highly significant correlation between mild cognitive impairment (defined as a Montreal Cognitive Assessment score < 26) and 25(OH)D deficiency in adult patients with HT, in both univariate and multivariate analyses [70].

The effect of this steroid hormone may depend on its interaction with other factors. For instance, there exist data suggesting that adequate levels of 25(OH)D allow an anti-inflammatory and immunomodulatory effect of simvastatin, with a consequent reduction in the levels of anti-TPO and anti-Tg [71,72].

5.2.2. Data in Graves' Disease

Data on the relationship between VitD and GD are more scarce. Misharin et al. investigated the response to TRABs induction by immunizing two BALB/c and C57BL/6 murine strains receiving VitD-sufficient or -depleted diet. BALB/c strains are more susceptible to disease induction and displayed a reduced ability to convert 25(OH)D to 1,25(OH)$_2$D when compared C57BL/6 strains. The authors found that BALB/c mice had a slightly different immune response, depending on the diet administered; however, the main difference was the greater probability of developing persistent hyperthyroidism [73].

Several studies have reported lower levels of serum 25(OH)D in GD patients [74–77]; however, there were important differences in the results. The study by Zhang et al. reported an association between VitD levels and TRAB titers [75], while the remaining studies did not support such an association [74,76,77]. Yasuda et al. described an association with greater thyroid volume [74]; conversely, Mangaraj et al. found no differences in glandular volume between VitD-deficient and non-deficient GD patients [77]. Two metanalyses from 2015 reported a greater probability of 25(OH)D deficiency in individuals with GD [45,78].

Levels of 25(OH)D may be important in the response to treatment, with lower levels being associated with a lower likelihood of remission [79] and higher recurrence rate [80] when anti-thyroid drug therapy is used. Contrary to these findings, Planck et al. found no association between VitD levels at baseline and relapse within 1 year of completion of a 18 month anti-thyroid drug cycle [76]. Serum levels of 25(OH)D < 20 ng/mL were also identified as an independent risk factor for therapeutic failure with radioactive iodine [81]. Furthermore, cases of symptomatic hypocalcemia have been reported following GD treatment; not only surgical, but also after radioiodine [82] and with methimazole [83]. In both cases, low 25(OH)D levels and high compensatory 1,25(OH)$_2$D levels were reported, and prior VitD deficiency was appointed as a possible contributing cause [82,83]. However, in the case following radioiodine therapy, PTH was inappropriately normal and prior hypoparathyroidism, although unlikely, could not be excluded [82].

5.2.3. Data on Postpartum Thyroiditis (PPT)

Analyses performed on women with PPT also identified a relationship between lower levels of 25(OH)D and development of the disease [44,84]. Regarding anti-thyroid antibody titers, the results differ: while Krysiak et al. reported a negative correlation with the levels of 25(OH)D [84], Ma et al. found no relationship [44].

The inability to reach clear conclusions is partly due to limitations in the design of the studies, which were mostly cross-sectional with limited samples, heterogeneous populations, different latitudes and seasonality of blood sampling, variable criteria for defining AITD, and different cutoffs for defining insufficiency and deficiency of 25(OH)D. It is also necessary to take into account the possible interaction with several factors influencing the association (e.g., age, body mass index, ethnicity, other hormone levels, and so on).

5.3. Polymorphisms of Genes Associated with Vitamin D and AITD

An association has been hypothesized between polymorphisms of genes involved in the function and metabolism of VitD and AITD.

The most widely studied polymorphisms in this context are those of the VDR gene. This gene is located on chromosome 12q13.11 and contains 14 exons and about 75 kilobases. Several single nucleotide polymorphisms (SNP) have been identified in this gene, some of which have been associated with a risk of AITD [85]. The four main SNPs which have been described are Fok1 (rs10735870), BsmI (rs1544410), ApaI (rs7975232), and TaqI (rs731236); the latter three are in linkage disequilibrium with each other [56].

The results of association studies of VDR polymorphisms with AITD are inconsistent, even when meta-analyses were used to obtain a higher statistical power [85–88]. Table 1 summarizes the main data of the four meta-analyses on this subject published within the time frame reviewed in this text.

Table 1. Meta-analyses summarizing the association between VDR polymorphisms and AITD.

Autor	N Included Studies	PMF	Population (Cases/Controls)	Main Results
Zhou H., Xu C., and Gu C., 2009 (data from 2000–2008) [86]	Nine on the relationship between VDR-PMF relationship with GD	ApaI	1820/1866	Increased risk of GD in Asians (OR 1.31) No statistical association in Caucasians
		BsmI	1815/2066	Increased risk of GD in Asians (OR 1.58) No statistical association in Caucasians
		TaqI	1348/1175	No statistical association in Caucasians
		FoxI	1662/1840	Increased risk of GD in Asians (OR 1.68) No statistical association in Caucasians
Feng M. et al. 2012 (data up to 08/2012) [87]	Eight on the relationship between VDR-PMF with AITD	ApaI	1009/1080	No statistical association
		BsmI	1158/1049	Risk decreased B allele vs. b (OR 0.801)
		TaqI	1211/1184	Risk decreased t allele vs. T (OR 0.854)
		FoxI	739/924	No statistical association

Table 1. *Cont.*

Autor	N Included Studies	PMF	Population (Cases/Controls)	Main Results
Gao X. and Yu Y., 2017 (data until 08/2017) [85]	Two on the relationship between VDR-PMF with AITD	ApaI	3544/3117 [1]	Increased risk in Africans (OR 3.62) [1] No statistical association in general
		BsmI	3636/3373 [1]	Reduced risk in Europeans (OR 0.79) [1] and Africans (OR 0.42) [1] Increased risk in Asians (OR 1.41) [1]
		TaqI	2950/2254 [1]	Reduced risk of HT in the African population (OR 0.33) [1]
		FoxI	3174/2836 [1]	Reduced risk of HT in the Asian population (OR 0.65) [1]
Veneti S. et al. 2019 (data up to 12/2018) [88]	Ten on the relationship between VDR-PMF relationship with GD	ApaI	2533/2474	No statistical association
		BsmI	2536/2576	No statistical association in general Risk decreased in Asians (OR 0.67), but increased in Caucasians (OR1.31) of subtype bb
		TaqI	2380/2235	Increased risk of GD with TT (OR 1.42)
		FoxI	2587/2603	No statistical association

[1] Dominant model. Abbreviations: AITD, autoimmune thyroid disease; GD, Graves' disease; HT, Hashimoto thyroiditis; OR, odds ratio; PMF, polymorphism; VDR, vitamin D receptor.

Genome-wide association studies (GWAs) have shown that the genes encoding D-binding protein and CYP2R1 are associated with circulating VitD levels. Polymorphisms in these genes may be associated with treatment unresponsiveness in GD [89].

The somewhat divergent results of the polymorphism studies may be due, at least in part, to limited sample sizes, as the effect of each susceptibility locus is limited.

6. Relevance of Supplementation

The multitude of data suggesting a relationship between low levels of 25(OH)D and AITD have generated interest in the investigation of the use of VitD supplements in the prevention/treatment of this group of conditions.

Most recent results (Table 2) support the benefit of supplementation in individuals with AITD, which is generally higher in the presence of a deficiency, both in HT [90–94] and in TPP [84,95]. Three of the studies mentioned below also analyzed PTH and calcium levels at baseline and after supplementation, showing some degree of tendency towards the normalization of high PTH and low calcium levels [41,91,95].

Table 2. Prospective studies on AITD and VitD supplementation.

Authors Study Type	Number of Subjects and Intervention	Results	Effect on Ca^{2+}/PTH
Chaudhary S. et al. 2016 [91] Open-label RCT	One hundred and two AITD subjects randomized to receive cholecalciferol 6000 IU + calcium 500 mg/d (G1) or only calcium (G2) Positive response defined as a decrease ≥ 25% in anti-TPO titers.	Response in 68% of G1 vs. 44% of G2 Only significant in those with TSH ≤ 10 mUI/mL.	Higher PTH in those with lower 25(OH)D$_2$, no statistically significant difference in Ca^{2+} and P$^-$ levels. PTH reduction after supplementation.
Krysiak R. et al. 2016 [95] Longitudinal, Case–Control trial	Thirty-eight PPT vs. 21 healthy postpartum women. VitD supplementation in the subjects with PPT: -4000 IU/day if deficiency [25(OH)D < 20 ng/mL] -2000 IU/day or no supplement for the remaining patients	Lower baseline 25(OH)D levels in those with PPT. After supplementation of VitD according to baseline values→reduction in anti-TPO titers, with a more marked effect in those with deficiency at baseline.	Higher PTH and lower Ca^{2+} in those with PPT. Significant PTH reduction in those with a deficiency of 25(OH)D.
Simsek Y. et al. 2016 [96] Longitudinal, RCT	Eighty-two AITD patients -46 were supplemented with VitD 1000 IU/day for 1 month -36 were not supplemented	Reduction in anti-TPO and anti-Tg titers only in the supplementation group.	

Table 2. *Cont.*

Authors Study Type	Number of Subjects and Intervention	Results	Effect on Ca^{2+}/PTH
Krysiak R. et al. 2017 [92] Longitudinal, Case–Control trial	Thirty-two women with HT, euthyroid, or with sub-clinical hypothyroidism and 25(OH) > 30 ng/mL -18 were supplemented with VitD 2000 UI/day for 6 months -16 were not supplemented	At baseline: inverse correlation of 25(OH)D with antibody titers with non-significant difference between groups. At 6 months: reduction in antibody titers (mainly anti-TPO) in relation to the increase in 25(OH)D only statistically significant in those with sub-clinical hypothyroidism (vs. euthyroidism) and dependent on baseline antibody titers.	
Krysiak R. et al. 2019 [93] Non-randomized	Thirty-two men with AITD in euthyroidism -20 supplemented with VitD 4000 IU/day -17 with selenomethionine 200 µg/day	Similar reduction in anti-TPO and anti-Tg titers in both groups. Greater effect of VitD on antibody titers in those with 25(OH)D < 30 ng/mL (~75 nmol/L) at baseline.	
Mazokopakis E. et al. 2015 [90] Non-randomized	From a group of 218 HT, the 186 with 25(OH) < 30 ng/mL were supplemented with cholecalciferol 1200–4000 IU/day.	Negative correlation between baseline 25(OH)D and anti-TPO. Significant decrease in anti-TPO after 4 months of supplementation.	No statistically significant difference in Ca^{2+} and P$^-$ at baseline or after supplementation.
Vondra K. et al. 2017 [41] Non-randomized	Thirty-seven women with AITD were supplemented with 4300 IU/day of cholecalciferol for 3 months.	Positive relationship between fT4/fT3 ratio in patients with AITD and 25(OH)D deficiency which disappeared after supplementation with cholecalciferol.	Correlation with higher PTH and lower Ca^{2+} at baseline. Normalization after supplementation.

Legend: AITD, autoimmune thyroid disease; anti-Tg, anti-thyroglobulin; anti-TPO, anti-thyroid peroxidase; fT3, free triiodothyronine; fT4, free thyroxine; G1, group 1; G2, group 2; PPT, post-partum thyroiditis; PTH, parathyroid hormone; RCT, randomized controlled trial; VitD, vitamin D; TSH, thyroid stimulating hormone.

In a systematic review and meta-analysis, Wang et al. concluded that supplementation with VitD appeared to significantly reduce levels of anti-TPO (for treatments ≥6 months) and anti-Tg, with no reported serious adverse effects [97]. More recently, Koehler et al. retrospectively analyzed 933 patients with autoimmune thyroiditis and found a greater reduction in anti-TPO levels in a 58-patient sub-group that had an improvement in their initially insufficient VitD level (<30 ng/mL) vs. a control group that maintained a VitD level below the threshold. The difference between the groups, however, was not statistically significant [98].

Other factors may influence the effect of VitD supplementation on HT. Testosterone replacement in testosterone-deficient men has been associated with a more pronounced reduction in anti-TPO/-Tg titers and increased thyroid secretory capacity (SPINA-GT index) with VitD supplementation (vs. testosterone-naïve men) [99]. Selenomethionine supplementation has also been shown to enhance the effect of VitD on these parameters in 47 HT women [100].

Supplementation may also have a preventive component. A group of 11,017 participants in a wellness program were supplemented with VitD for over a year, aiming to reach physiological levels defined as 25(OH)D > 40 ng/mL (100 nmol/L). It was found that concentrations of 25(OH)D ≥ 50 ng/mL (125 nmol/L) reduced the risk of hypothyroidism by 30% (from 0.4%–44 cases/11,017 participants—to 0.28%—31 cases) and elevated antibody titers by 32%. Increased levels of 25(OH)D in patients with hypothyroidism have been associated with improved thyroid function [101].

Some recent studies evaluated the effects of VitD supplementation and outcomes in GD. Supplementation may delay the onset, but does not seem to prevent disease recurrence [102]. This intervention may have beneficial effects on cardiovascular outcomes (as suggested by a reduction in pulse wave velocity), which are limited to patients with VitD deficiency [103]. Conversely, VitD supplementation may be detrimental to muscle strength recovery [104].

It should be noted, however, that supplementation with excessive doses of 25(OH)D may be harmful. A possible increased risk of fractures has been reported with high-dose 25(OH)D supplementation [105]. In a large retrospective study, an association between 25(OH)D and mortality in the form of an inverted J-curve was suggested, with the lowest risk for serum levels between 20 and 24 ng/mL [106]. Therefore, it is important to emphasize that, indeed, some undesirable effects of attaining levels above the physiologic range may exist.

Given the paucity of data in this regard, a logical approach is to aim for VitD levels within the reference ranges suggested by international guidelines. The Institute of Medicine considers 20 ng/mL to be sufficient for most of the general population [107]. The Endocrine Society Guidelines, focused on individuals with risk of VitD deficiency, identify an optimal level of 25(OH)D > 30 ng/mL and that values up to 100 ng/mL (250 nmol/L) are safe (as they do not cause hypercalcemia) [108].

7. What Is the Nature of the Relationship between Vitamin D Levels and Autoimmune Thyroid Disease?

Although there exists some inconsistency in the results of the studies carried out so far, most of the data are consistent with the presence of an association between vitamin D and AITD. However, there are several possible interpretations for this association.

The most commonly cited explanation is the decrease in the immunomodulatory role of $1,25(OH)_2D$, in patients with deficiency, contributing to the development of AID. However, the data obtained to date are mostly resultant from cross-sectional studies, which do not allow for the establishment of causal effects. It is, therefore, essential to evaluate alternative explanatory models.

Some authors have raised the possibility that the various data favoring the involvement of VitD in AITD reflect a consequence, rather than a cause, of the disease. AID may lead to VitD deficiency by causing incapacitation and lower sunlight exposure, malabsorption, and the use of corticosteroids [42,109]. In hyperthyroidism, there may be accelerated bone turnover [32]. Kozai et al. found marked decreases in $1,25(OH)_2D$ and CYP27B1 expression in rats with T3-induced hyperthyroidism [110]. In HT, the increase in fat mass caused by hypothyroidism could contribute to the deficiency [111]. Botello et al. studied 88 patients with long-term HT and found a positive correlation between 25(OH)D levels, fT4, and (contrary to expectations) Th17 and TNFα. The authors hypothesized that low levels of fT4 are predictors of a deficiency of 25(OH)D and that the long evolution of the disease and treatment of hypothyroidism are related to a decrease in cytotoxic immune response, regardless of the levels of 25(OH)D [112]. The coexistence of AITD with other AID, such as celiac disease, also deserves consideration. Celiac disease leads to malabsorption with a deficiency of several nutrients [113], including VitD [114], and it is associated with an increased risk of developing other AIDs [113,114]. The presence of biopsy-proven celiac disease in patients with AITD is small, around 1.6% according to a recent meta-analysis (although there may be some underdiagnosis) [115]; therefore, it cannot fully explain the reported lower values of VitD in all AITD patients. However, it is likely to contribute to this association in patients in which both diseases coexist. A group of HT patients with positive transglutaminase antibodies and no symptoms of celiac disease were divided, receiving gluten-free vs. gluten-containing diets. The former group, but not the second one, experienced a reduction in antibody titers and an increase in VitD levels [116]. However, the possibility of VitD deficiency being exclusively a consequence of AID seems unlikely, given that this relationship has been found in several studies, independently of factors such as age, body mass index, thyroid function tests (i.e., presence of hyper-, hypo-, or euthyroidism) and the presence or absence of other AIDs. Furthermore, in a study that evaluated patients with GD and 25(OH)D insufficiency, no statistically significant difference was found in the values of 25(OH)D at baseline and 1 to 2 years after hyperthyroidism therapy (with achievement of euthyroidism) [117]. Therefore, contrary to what would be expected if low levels of VitD were a consequence of the autoimmune disease, treating the autoimmune disease does not improve VitD status.

Another possibility is that the lower levels of 25(OH)D in AID are the result of a pathophysiological mechanism involved in the development of the disease; that is, VDR dysfunction caused by chronic infection by intra-phagocytic microorganisms [111]. This dysfunction could lead to lower production of the antimicrobial peptides that would usually result from activation of VDR. VDR dysfunction could also lead to lesser expression of 24-hydroxylase, with a consequent increase in $1,25(OH)_2D$ levels. Excess $1,25(OH)_2D$ has the ability to displace ligands of nuclear receptors such as α-thyroid, glucocorticoids, and androgens, which can lead to glandular dysfunction [118]. Elevated levels of $1,25(OH)_2D$ further bind to the pregnane X receptor and inhibit the synthesis of 25(OH)D in the liver. In this context, the various data pointing towards a relationship between AID and VitD deficiency may be explained by the fact that the metabolite usually measured is 25(OH)D [119]. This is a counterintuitive hypothesis, with some theoretical background but with little data to support or contradict it directly, as $1,25(OH)_2D$ is rarely quantified. However, some of the above-mentioned studies on VitD supplementation reported elevated PTH and normal/slightly low calcium values, associated with a deficiency of 25(OH)D at baseline with a tendency towards normalization after VitD supplementation [41,91,95]. This does not support the possibility that there is an increase in $1,25(OH)_2D$ in AITD concealed by the quantification of 25(OH)D. Although it may be argued that PTH level elevation and lowering of calcium levels may be explained by VDR dysfunction, it is unlikely that such alterations were susceptible to correction by VitD supplementation, as it would not correct the primary mechanism. The fact that VitD supplementation has shown some beneficial effects on autoimmunity parameters is also against this hypothesis.

Analyzing the current evidence, we conclude that, although a direct and marked contribution of VitD levels alone in the pathogenesis of AITD is unlikely, given the marked inconsistency of the data, a minor contribution is probable, as the existence of an association has been supported by the majority of the studies cited above (refer to Section 5.2. Data on vitamin D levels and autoimmune thyroid disease). Therefore, it is plausible that the levels of VitD, the polymorphisms of its receptor [85–88], and the enzymes that govern its metabolism [89] influence its regulatory capacity and, thus, it likely plays a small, yet significant, role in the development and course of AITD. It is likely that this contribution depends upon a multiplicity of other factors, such as age and gender, sex hormones [49,99], and micronutrients [100]. Genetic, epigenetic, and other endogenous and environmental factors which contribute to the predisposition to AITD may also influence this correlation, explaining some of the inconsistency in the results obtained in different populations. The above-mentioned consequences of AITD (e.g., incapacitation, lower sunlight exposure, obesity in hypothyroidism, and increased bone turnover in hyperthyroidism) and, in some cases, the coexistence of other AID may generate a vicious cycle and contribute to the observed relationship.

8. Discussion and Conclusions

Several questions can be raised regarding the relationship between VitD and AITD, the first one being whether such a relationship actually exists. With respect to this matter, although there is some inconsistency in the results of the studies carried out to date, most of the data point toward an association between lower VitD levels and increased risk of developing the disease and/or higher antibody titers and/or more difficulty in its treatment, especially for vitamin D deficiency. Polymorphisms in genes associated with VitD function/metabolism also appear to have some influence on the risk of AITD.

The second question concerns the exact nature of this relationship. We propose that VitD plays a small, yet significant, role in the pathogenesis of AITD, which may only be apparent when other factors that contribute to its expression are gathered. After the onset of AITD, its consequences may generate a vicious cycle, contributing to aggravation of the deficiency.

The third question, with more immediate implications on clinical practice, is the role of VitD supplementation on the prevention and/or treatment of AITD, as well as whether a supraphysiological level would be desirable. At present, there is a paucity of data establishing the exemption from harm and the presence of benefit of obtaining supraphysiological levels of 25(OH)D. There are even data suggesting possible associations with increased fracture and mortality risks. Therefore, a sensitive approach is to aim for a 25(OH)D level within the reference ranges suggested in international guidelines.

In the future, more data from investigations with a larger number of individuals, a more global scope, and involving year-round evaluations of VitD levels are necessary, in order to provide more uniform and consistent answers to these questions.

Author Contributions: Conceptualization and methodology, I.H.V., D.R. and I.P.; writing—original draft preparation, I.H.V.; review and editing, I.H.V., D.R. and I.P.; supervision, D.R. and I.P. All authors have read and agreed to the published version of the manuscript.

Funding: This research received no external funding.

Acknowledgments: The authors which to thank Elisabete Santos for her initial language corrections.

Conflicts of Interest: The authors declare no conflict of interest.

References

1. Kmiec, P.; Sworczak, K. Vitamin D in thyroid disorders. *Exp. Clin. Endocrinol. Diabetes* **2015**, *123*, 386–393.
2. Muscogiuri, G.; Mitri, J.; Mathieu, C.; Badenhoop, K.; Tamer, G.; Orio, F.; Mezza, T.; Vieth, R.; Colao, A.; Pittas, A. Mechanisms in Endocrinology: Vitamin D as a potential contributor in endocrine health and disease. *Eur. J. Endocrinol.* **2014**, *171*, R101–R110. [CrossRef] [PubMed]
3. Kim, D. The Role of Vitamin D in Thyroid Diseases. *Int. J. Mol. Sci.* **2017**, *18*, 1949. [CrossRef] [PubMed]
4. Altieri, B.; Muscogiuri, G.; Barrea, L.; Mathieu, C.; Vallone, C.V.; Mascitelli, L.; Bizzaro, G.; Altieri, V.M.; Tirabassi, G.; Balercia, G.; et al. Does vitamin D play a role in autoimmune endocrine disorders? A proof of concept. *Rev. Endocr. Metab. Disord.* **2017**, *18*, 335–346. [CrossRef]
5. Bikle, D.D. Vitamin D metabolism, mechanism of action, and clinical applications. *Chem. Biol.* **2014**, *21*, 319–329. [CrossRef] [PubMed]
6. Nettore, I.C.; Albano, L.; Ungaro, P.; Colao, A.; Macchia, P.E. Sunshine vitamin and thyroid. *Rev. Endocr. Metab. Disord.* **2017**, *18*, 347–354. [CrossRef]
7. Vondra, K.; Stárka, L.; Hampl, R. Vitamin D and Thyroid Diseases. *Physiol. Res.* **2015**, S95–S100. [CrossRef]
8. Martins, J.S.; Palhares, M.D.O.; Teixeira, O.C.M.; Ramos, M.G. Vitamin D Status and Its Association with Parathyroid Hormone Concentration in Brazilians. *J. Nutr. Metab.* **2017**, *2017*, 1–5. [CrossRef]
9. Aranow, C. Vitamin D and the immune system. *J. Investig. Med.* **2011**, *59*, 881–886. [CrossRef]
10. Wierzbicka, J.; Piotrowska, A.; Żmijewski, M.A. The renaissance of vitamin D. *Acta Biochim. Pol.* **2014**, *61*, 679–686. [CrossRef]
11. Hossein-Nezhad, A.; Spira, A.; Holick, M.F. Influence of Vitamin D Status and Vitamin D3 Supplementation on Genome Wide Expression of White Blood Cells: A Randomized Double-Blind Clinical Trial. *PLoS ONE* **2013**, *8*, e58725. [CrossRef] [PubMed]
12. Płudowski, P.; Holick, M.F.; Pilz, S.; Wagner, C.L.; Hollis, B.W.; Grant, W.B.; Shoenfeld, Y.; Lerchbaum, E.; Llewellyn, D.J.; Kienreich, K.; et al. Vitamin D effects on musculoskeletal health, immunity, autoimmunity, cardiovascular disease, cancer, fertility, pregnancy, dementia and mortality—A review of recent evidence. *Autoimmun. Rev.* **2013**, *12*, 976–989. [CrossRef] [PubMed]
13. Theodoratou, E.; Tzoulaki, I.; Zgaga, L.; Ioannidis, A.J.P. Vitamin D and multiple health outcomes: Umbrella review of systematic reviews and meta-analyses of observational studies and randomised trials. *BMJ* **2014**, *348*, g2035. [CrossRef] [PubMed]
14. Rosen, Y.; Daich, J.; Soliman, I.; Brathwaite, E.; Shoenfeld, Y. Vitamin D and autoimmunity. *Scand. J. Rheumatol.* **2016**, *45*, 439–447. [CrossRef]
15. Skaaby, T.; Husemoen, L.L.N.; Thuesen, B.; Linneberg, A. Prospective population-based study of the association between vitamin D status and incidence of autoimmune disease. *Endocrine* **2015**, *50*, 231–238. [CrossRef]
16. Hewison, M. Vitamin D and the Immune System: New Perspectives on an Old Theme. *Endocrinol. Metab. Clin. N. Am.* **2010**, *39*, 365–379. [CrossRef]
17. Kamen, D.L.; Tangpricha, V. Vitamin D and molecular actions on the immune system: Modulation of innate and autoimmunity. *J. Mol. Med.* **2010**, *88*, 441–450. [CrossRef]
18. Nikolic, T.; Roep, B.O. Regulatory Multitasking of Tolerogenic Dendritic Cells—Lessons Taken from Vitamin D3-Treated Tolerogenic Dendritic Cells. *Front. Immunol.* **2013**, *4*, 113. [CrossRef]

19. Ferreira, G.B.; Gysemans, C.A.; Demengeot, J.; Da Cunha, J.P.M.C.M.; Vanherwegen, A.-S.; Overbergh, L.; Van Belle, T.L.; Pauwels, F.; Verstuyf, A.; Korf, H.; et al. 1,25-Dihydroxyvitamin D3 Promotes Tolerogenic Dendritic Cells with Functional Migratory Properties in NOD Mice. *J. Immunol.* **2014**, *192*, 4210–4220. [CrossRef]

20. Şıklar, Z.; Karataş, D.; Doğu, F.; Hacıhamdioğlu, B.; Ikincioğulları, A.; Berberoğlu, M. Regulatory T Cells and Vitamin D Status in Children with Chronic Autoimmune Thyroiditis. *J. Clin. Res. Pediatr. Endocrinol.* **2016**, *8*, 276–281. [CrossRef]

21. Orgiazzi, J. Thyroid autoimmunity. *Presse Medicale* **2012**, *41*, e611–e625. [CrossRef] [PubMed]

22. Antonelli, A.; Ferrari, S.M.; Corrado, A.; Di Domenicantonio, A.; Fallahi, P. Autoimmune thyroid disorders. *Autoimmun. Rev.* **2015**, *14*, 174–180. [CrossRef] [PubMed]

23. Merrill, S.J.; Minucci, S.B. Thyroid Autoimmunity: An Interplay of Factors. In *Vitamins and Hormones*; Academic Press: Cambridge, MA, USA, 2018; Volume 106, pp. 129–145. [CrossRef]

24. Wiersinga, W.M. Thyroid Autoimmunity. In *Paediatric Thyroidology*; Karger Publishers: Basel, Switzerland, 2014; pp. 139–157. [CrossRef]

25. Wiersinga, W.M. Clinical Relevance of Environmental Factors in the Pathogenesis of Autoimmune Thyroid Disease. *Endocrinol. Metab.* **2016**, *31*, 213–222. [CrossRef] [PubMed]

26. Köhling, H.L.; Plummer, S.F.; Marchesi, J.R.; Davidge, K.S.; Ludgate, M. The microbiota and autoimmunity: Their role in thyroid autoimmune diseases. *Clin. Immunol.* **2017**, *183*, 63–74. [CrossRef]

27. Virili, C.; Fallahi, P.; Antonelli, A.; Benvenga, S.; Centanni, M. Gut microbiota and Hashimoto's thyroiditis. *Rev. Endocr. Metab. Disord.* **2018**, *19*, 293–300. [CrossRef]

28. Zhao, F.; Feng, J.; Li, J.; Zhao, L.; Liu, Y.; Chen, H.; Jin, Y.; Zhu, B.; Wei, Y. Alterations of the Gut Microbiota in Hashimoto's Thyroiditis Patients. *Thyroid* **2018**, *28*, 175–186. [CrossRef]

29. Chiara, M.; Caputo, M.; Bisceglia, A.; Samà, M.T.; Zavattaro, M.; Gianluca, A.; Pagano, L.; Prodam, F.; Paolo, M. Immunomodulatory Effects of Vitamin D in Thyroid Diseases. *Nutrients* **2020**, *12*, 1444. [CrossRef]

30. Kucukler, F.K.; Şimşek, Y.; Gorkem, U.; Dogan, B.A.; Guler, S. Relationship between gestational transient thyrotoxicosis and vitamin D. *Turk. J. Med. Sci.* **2016**, *46*, 1374–1378. [CrossRef]

31. Barchetta, I.; Baroni, M.; Leonetti, F.; De Bernardinis, M.; Bertoccini, L.; Fontana, M.; Mazzei, E.; Fraioli, A.; Cavallo, M.G. TSH levels are associated with vitamin D status and seasonality in an adult population of euthyroid adults. *Clin. Exp. Med.* **2015**, *15*, 389–396. [CrossRef]

32. Muscogiuri, G.; Tirabassi, G.; Bizzaro, G.; Orio, F.; Paschou, S.A.; Vryonidou, A.; Balercia, G.; Shoenfeld, Y.; Colao, A. Vitamin D and thyroid disease: To D or not to D? *Eur. J. Clin. Nutr.* **2015**, *69*, 291–296. [CrossRef]

33. Hamilton, A.; Newby, P.R.; Carr-Smith, J.D.; Disanto, G.; Allahabadia, A.; Armitage, M.; Brix, T.H.; Chatterjee, K.; Connell, J.M.; Hegedüs, L.; et al. Impact of Month of Birth on the Development of Autoimmune Thyroid Disease in the United Kingdom and Europe. *J. Clin. Endocrinol. Metab.* **2014**, *99*, E1459–E1465. [CrossRef] [PubMed]

34. Thvilum, M.; Brandt, F.; Brix, T.H.; Hegedüs, L. Month of birth is associated with the subsequent diagnosis of autoimmune hypothyroidism. A nationwide Danish register-based study. *Clin. Endocrinol.* **2017**, *87*, 832–837. [CrossRef] [PubMed]

35. Sivalingam, S.; Thvilum, M.; Brix, T.H.; Hegedüs, L.; Brandt, F. No link between season of birth and subsequent development of Graves' disease or toxic nodular goiter: A nationwide Danish register-based study. *Endocr. Connect.* **2018**, *7*, 1090–1095. [CrossRef] [PubMed]

36. Kyrgios, I.; Giza, S.; Tsinopoulou, V.R.; Maggana, I.; Haidich, A.-B.; Galli-Tsinopoulou, A. Seasonality of month of birth in children and adolescents with autoimmune thyroiditis: A continuing conundrum. *J. Pediatr. Endocrinol. Metab.* **2018**, *31*, 1123–1131. [CrossRef] [PubMed]

37. Mackawy, A.M.H.; Al-Ayed, B.M.; Al-Rashidi, B.M. Vitamin D Deficiency and Its Association with Thyroid Disease. *Int. J. Health Sci.* **2013**, *7*, 267–275. [CrossRef]

38. Chailurkit, L.-O.; Aekplakorn, W.; Ongphiphadhanakul, B. High Vitamin D Status in Younger Individuals Is Associated with Low Circulating Thyrotropin. *Thyroid* **2013**, *23*, 25–30. [CrossRef]

39. Zhang, Q.; Wang, Z.; Sun, M.; Cao, M.; Zhu, Z.; Fu, Q.; Gao, Y.; Mao, J.; Li, Y.; Shi, Y.; et al. Association of High Vitamin D Status with Low Circulating Thyroid-Stimulating Hormone Independent of Thyroid Hormone Levels in Middle-Aged and Elderly Males. *Int. J. Endocrinol.* **2014**, *2014*, 1–6. [CrossRef]

40. Kim, M.; Song, E.; Oh, H.-S.; Park, S.; Kwon, H.; Jeon, M.J.; Shong, Y.K.; Kim, T.-Y.; Kim, W.G.; Kim, W.B. Vitamin D deficiency affects thyroid autoimmunity and dysfunction in iodine-replete area: Korea national health and nutrition examination survey. *Endocrine* **2017**, *58*, 332–339. [CrossRef]

41. Vondra, K.; Bílek, R.; Matucha, P.; Salátová, M.; Vosátková, M.; Stárka, L.; Hampl, R. Vitamin D Supplementation Changed Relationships, Not Levels of Metabolic-Hormonal Parameters in Autoimmune Thyroiditis. *Physiol. Res.* **2017**, *26*, S409–S417. [CrossRef]

42. Kivity, S.; Agmon-Levin, N.; Zisappl, M.; Shapira, Y.; Nagy, E.V.; Dankó, K.; Szekanecz, Z.; Langevitz, P.; Shoenfeld, Y. Vitamin D and autoimmune thyroid diseases. *Cell. Mol. Immunol.* **2011**, *8*, 243–247. [CrossRef]

43. Unal, A.D.; Tarcin, O.; Parildar, H.; Cigerli, O.; Eroglu, H.; Demirag, N.G. Vitamin D deficiency is related to thyroid antibodies in autoimmune thyroiditis. *Cent. Eur. J. Immunol.* **2014**, *39*, 493–497. [CrossRef] [PubMed]

44. Ma, J.; Wu, D.; Li, C.; Fan, C.; Chao, N.; Liu, J.; Li, Y.; Wang, R.; Miao, W.; Guan, H.; et al. Lower Serum 25-Hydroxyvitamin D Level is Associated With 3 Types of Autoimmune Thyroid Diseases. *Medicine* **2015**, *94*, e1639. [CrossRef] [PubMed]

45. Wang, J.; Lv, S.; Chen, G.; Gao, C.; He, J.; Zhong, H.; Xu, Y. Meta-Analysis of the Association between Vitamin D and Autoimmune Thyroid Disease. *Nutrients* **2015**, *7*, 2485–2498. [CrossRef] [PubMed]

46. Muscogiuri, G.; Palomba, S.; Caggiano, M.; Tafuri, D.; Colao, A.; Orio, F. Low 25 (OH) vitamin D levels are associated with autoimmune thyroid disease in polycystic ovary syndrome. *Endocrine* **2016**, *53*, 538–542. [CrossRef] [PubMed]

47. D'Aurizio, F.; Villalta, D.; Metus, P.; Doretto, P.; Tozzoli, R. Is vitamin D a player or not in the pathophysiology of autoimmune thyroid diseases? *Autoimmun. Rev.* **2015**, *14*, 363–369. [CrossRef]

48. Effraimidis, G.; Badenhoop, K.; Tijssen, J.G.P.; Wiersinga, W.M. Vitamin D deficiency is not associated with early stages of thyroid autoimmunity. *Eur. J. Endocrinol.* **2012**, *167*, 43–48. [CrossRef]

49. Choi, Y.M.; Kim, W.G.; Kim, T.-Y.; Bae, S.J.; Kim, H.-K.; Jang, E.K.; Jeon, M.J.; Han, J.M.; Lee, S.H.; Baek, J.H.; et al. Low Levels of Serum Vitamin D3 Are Associated with Autoimmune Thyroid Disease in Pre-Menopausal Women. *Thyroid* **2014**, *24*, 655–661. [CrossRef]

50. Kim, C.-Y.; Lee, Y.J.; Choi, J.H.; Lee, S.Y.; Lee, H.Y.; Jeong, D.H.; Choi, Y.J. The Association between Low Vitamin D Status and Autoimmune Thyroid Disease in Korean Premenopausal Women: The 6th Korea National Health and Nutrition Examination Survey, 2013-2014. *Korean J. Fam. Med.* **2019**, *40*, 323–328. [CrossRef]

51. Tamer, G.; Arik, S.; Tamer, I.; Coksert, D. Relative Vitamin D Insufficiency in Hashimoto's Thyroiditis. *Thyroid* **2011**, *21*, 891–896. [CrossRef]

52. Bozkurt, N.; Karbek, B.; Uçan, B.; Sahin, M.; Cakal, E.; Özbek, M.; Delibasi, T. The Association between Severity of Vitamin D Deficiency and Hashimoto's Thyroiditis. *Endocr. Prat.* **2013**, *19*, 479–484. [CrossRef]

53. Mansournia, N.; Mansournia, M.A.; Saeedi, S.; Dehghan, J. The association between serum 25OHD levels and hypothyroid Hashimoto's thyroiditis. *J. Endocrinol. Investig.* **2014**, *37*, 473–476. [CrossRef] [PubMed]

54. Maciejewski, A.; Wójcicka, M.; Roszak, M.; Losy, J.; Lacka, K. Assessment of Vitamin D Level in Autoimmune Thyroiditis Patients and a Control Group in the Polish Population. *Adv. Clin. Exp. Med.* **2015**, *24*, 801–806. [CrossRef] [PubMed]

55. Kim, H. Low vitamin D status is associated with hypothyroid Hashimoto's thyroiditis. *Hormones* **2016**. [CrossRef] [PubMed]

56. Giovinazzo, S.; Vicchio, T.M.; Certo, R.; Alibrandi, A.; Palmieri, O.; Campennì, A.; Cannavò, S.; Trimarchi, F.; Ruggeri, R.M. Vitamin D receptor gene polymorphisms/haplotypes and serum 25(OH)D3 levels in Hashimoto's thyroiditis. *Endocrine* **2017**, *55*, 599–606. [CrossRef] [PubMed]

57. Ke, W.; Sun, T.; Zhang, Y.; He, L.; Wu, Q.; Liu, J.; Zha, B. 25-Hydroxyvitamin D serum level in Hashimoto's thyroiditis, but not Graves' disease is relatively deficient. *Endocr. J.* **2017**, *64*, 581–587. [CrossRef] [PubMed]

58. De Pergola, G.; Triggiani, V.; Bartolomeo, N.; Giagulli, V.A.; Anelli, M.; Masiello, M.; Candita, V.; De Bellis, D.; Silvestris, F. Low 25 Hydroxyvitamin D Levels are Independently Associated with Autoimmune Thyroiditis in a Cohort of Apparently Healthy Overweight and Obese Subjects. *Endocr. Metab. Immune Disord. Drug Targets* **2018**, *18*, 646–652. [CrossRef]

59. Muscogiuri, G.; Mari, D.; Prolo, S.; Fatti, L.M.; Cantone, M.C.; Garagnani, P.; Arosio, B.; Di Somma, C.; Vitale, G. 25 Hydroxyvitamin D Deficiency and Its Relationship to Autoimmune Thyroid Disease in the Elderly. *Int. J. Environ. Res. Public Health* **2016**, *13*, 850. [CrossRef]

60. Sönmezgöz, E.; Ozer, S.; Yılmaz, R.; Bilge, S.; Önder, Y.; Bütün, I. Hypovitaminosis D in Children with Hashimotos Thyroiditis. *Rev. Méd. Chile* **2016**, *144*, 611–616. [CrossRef]

61. Evliyaoğlu, O.; Acar, M.; Özcabı, B.; Erginoz, E.; Bucak, F.; Ercan, O.; Kucur, M. Vitamin D Deficiency and Hashimoto's Thyroiditis in Children and Adolescents: A Critical Vitamin D Level for This Association? *J. Clin. Res. Pediatr. Endocrinol.* **2015**, *7*, 128–133. [CrossRef]

62. Metwalley, K.A.; Farghaly, H.S.; Sherief, T.; Elderwy, A. Vitamin D status in children and adolescents with autoimmune thyroiditis. *J. Endocrinol. Investig.* **2016**, *39*, 793–797. [CrossRef]

63. Çamurdan, O.M.; Döğer, E.; Bideci, A.; Celik, N.; Cinaz, P. Vitamin D status in children with Hashimoto thyroiditis. *J. Pediatr. Endocrinol. Metab.* **2012**, *25*, 467–470. [CrossRef] [PubMed]

64. Demir, K.; Keskin, M.; Kör, Y.; Karaoglan, M.; Bülbül, Ö.G. Autoimmune thyroiditis in children and adolescents with type 1 diabetes mellitus is associated with elevated IgG4 but not with low vitamin D. *Hormones* **2014**, *13*, 361–368. [CrossRef] [PubMed]

65. Arslan, M.S.; Topaloglu, O.; Uçan, B.; Karakose, M.; Karbek, B.; Tutal, E.; Çaliskan, M.; Ginis, Z.; Cakal, E.; Sahin, M.; et al. Isolated Vitamin D Deficiency Is Not Associated with Nonthyroidal Illness Syndrome, but with Thyroid Autoimmunity. *Sci. World J.* **2015**, *2015*, 1–5. [CrossRef] [PubMed]

66. Shin, D.Y.; Kim, K.J.; Kim, D.; Hwang, S.; Lee, E.J. Low Serum Vitamin D Is Associated with Anti-Thyroid Peroxidase Antibody in Autoimmune Thyroiditis. *Yonsei Med. J.* **2014**, *55*, 476–481. [CrossRef]

67. Goswami, R.; Marwaha, R.K.; Gupta, N.; Tandon, N.; Sreenivas, V.; Tomar, N.; Ray, D.; Kanwar, R.; Agarwal, R. Prevalence of vitamin D deficiency and its relationship with thyroid autoimmunity in Asian Indians: A community-based survey. *Br. J. Nutr.* **2009**, *102*, 382–386. [CrossRef]

68. Wang, X.; Zynat, J.; Guo, Y.; Osiman, R.; Tuhuti, A.; Zhao, H.; Abdunaimu, M.; Wang, H.; Jin, X.; Xing, S. Low Serum Vitamin D Is Associated with Anti-Thyroid-Globulin Antibody in Female Individuals. *Int. J. Endocrinol.* **2015**, *2015*, 1–6. [CrossRef]

69. Yasmeh, J.; Farpour, F.; Rizzo, V.; Kheradnam, S.; Sachmechi, I. Hashimoto thyroiditis not associated with vitamin D deficiency. *Endocr. Pract.* **2016**, *22*, 809–813. [CrossRef]

70. Xu, J.; Zhu, X.-Y.; Sun, H.; Xu, X.-Q.; Xu, S.-A.; Suo, Y.; Cao, L.-J.; Zhou, Q.; Yu, H.-J.; Cao, W.-Z. Low vitamin D levels are associated with cognitive impairment in patients with Hashimoto thyroiditis. *BMC Endocr. Disord.* **2018**, *18*, 87. [CrossRef]

71. Krysiak, R.; Szkróbka, W.; Okopień, B. Moderate-dose simvastatin therapy potentiates the effect of vitamin D on thyroid autoimmunity in levothyroxine-treated women with Hashimoto's thyroiditis and vitamin D insufficiency. *Pharmacol. Rep.* **2018**, *70*, 93–97. [CrossRef]

72. Krysiak, R.; Szkróbka, W.; Okopień, B. The Relationship between Statin Action on Thyroid Autoimmunity and Vitamin D Status: A Pilot Study. *Exp. Clin. Endocrinol. Diabetes* **2019**, *6*, 23–28. [CrossRef]

73. Misharin, A.V.; Hewison, M.; Chen, C.-R.; Lagishetty, V.; Aliesky, H.A.; Mizutori, Y.; Rapoport, B.; McLachlan, S.M. Vitamin D Deficiency Modulates Graves' Hyperthyroidism Induced in BALB/c Mice by Thyrotropin Receptor Immunization. *Endocrinology* **2009**, *150*, 1051–1060. [CrossRef] [PubMed]

74. Yasuda, T.; Okamoto, Y.; Hamada, N.; Miyashita, K.; Takahara, M.; Sakamoto, F.; Miyatsuka, T.; Kitamura, T.; Katakami, N.; Kawamori, D.; et al. Serum vitamin D levels are decreased and associated with thyroid volume in female patients with newly onset Graves' disease. *Endocrine* **2012**, *42*, 739–741. [CrossRef] [PubMed]

75. Zhang, H.; Liang, L.; Xie, Z. Low Vitamin D Status is Associated with Increased Thyrotropin-Receptor Antibody Titer in Graves Disease. *Endocr. Pract.* **2015**, *21*, 258–263. [CrossRef] [PubMed]

76. Planck, T.; Shahida, B.; Malm, J.; Manjer, J. Vitamin D in Graves Disease: Levels, Correlation with Laboratory and Clinical Parameters, and Genetics. *Eur. Thyroid. J.* **2018**, *7*, 27–33. [CrossRef]

77. Choudhury, A.K.; Mangaraj, S.; Swain, B.M.; Sarangi, P.K.; Mohanty, B.K.; Baliarsinha, A.K. Evaluation of vitamin D status and its impact on thyroid related parameters in new onset Graves' disease—A cross-sectional observational study. *Indian J. Endocrinol. Metab.* **2019**, *23*, 35–39. [CrossRef]

78. Xu, M.-Y.; Cao, B.; Yin, J.; Wang, D.-F.; Chen, K.-L.; Lu, Q.-B. Vitamin D and Graves' Disease: A Meta-Analysis Update. *Nutrients* **2015**, *7*, 3813–3827. [CrossRef]

79. Yasuda, T.; Okamoto, Y.; Hamada, N.; Miyashita, K.; Takahara, M.; Sakamoto, F.; Miyatsuka, T.; Kitamura, T.; Katakami, N.; Kawamori, D.; et al. Serum vitamin D levels are decreased in patients without remission of Graves' disease. *Endocrine* **2013**, *43*, 230–232. [CrossRef]

80. Ahn, H.Y.; Chung, Y.J.; Cho, B.Y. Serum 25-hydroxyvitamin D might be an independent prognostic factor for Graves disease recurrence. *Medicine* **2017**, *96*, e7700. [CrossRef]

81. Li, X.; Wang, G.; Lu, Z.; Chen, M.; Tan, J.; Fang, X. Serum 25-hydroxyvitamin D predict prognosis in radioiodine therapy of Graves' disease. *J. Endocrinol. Investig.* **2015**, *38*, 753–759. [CrossRef]

82. Komarovskiy, K.; Raghavan, S. Hypocalcemia Following Treatment with Radioiodine in a Child with Graves' Disease. *Thyroid* **2012**, *22*, 218–222. [CrossRef]

83. Miyashita, K.; Yasuda, T.; Kaneto, H.; Kuroda, A.; Kitamura, T.; Otsuki, M.; Okamoto, Y.; Hamada, N.; Matsuhisa, M.; Shimomura, I. A Case of Hypocalcemia with Severe Vitamin D Deficiency following Treatment for Graves' Disease with Methimazole. *Case Rep. Endocrinol.* **2013**, *2013*, 1–4. [CrossRef] [PubMed]

84. Krysiak, R.; Kowalska, B.; Okopień, B. Serum 25-Hydroxyvitamin D and Parathyroid Hormone Levels in Non-Lactating Women with Post-Partum Thyroiditis: The Effect ofl-Thyroxine Treatment. *Basic Clin. Pharmacol. Toxicol.* **2015**, *116*, 503–507. [CrossRef] [PubMed]

85. Gao, X.-R.; Yu, Y. Meta-Analysis of the Association between Vitamin D Receptor Polymorphisms and the Risk of Autoimmune Thyroid Disease. *Int. J. Endocrinol.* **2018**, *2018*, 1–15. [CrossRef] [PubMed]

86. Zhou, H.; Xu, C.; Gu, M. Vitamin D receptor (VDR) gene polymorphisms and Graves' disease: A meta-analysis. *Clin. Endocrinol.* **2009**, *70*, 938–945. [CrossRef]

87. Feng, M.; Li, H.; Chen, S.-F.; Li, W.-F.; Zhang, F.-B. Polymorphisms in the vitamin D receptor gene and risk of autoimmune thyroid diseases: A meta-analysis. *Endocrine* **2013**, *43*, 318–326. [CrossRef]

88. Veneti, S.; Anagnostis, P.; Adamidou, F.; Artzouchaltzi, A.-M.; Boboridis, K.; Kita, M. Association between vitamin D receptor gene polymorphisms and Graves' disease: A systematic review and meta-analysis. *Endocrine* **2019**, *65*, 244–251. [CrossRef]

89. Inoue, N.; Watanabè, M.; Ishido, N.; Katsumata, Y.; Kagawa, T.; Hidaka, Y.; Iwatani, Y. The functional polymorphisms of VDR, GC and CYP2R1 are involved in the pathogenesis of autoimmune thyroid diseases. *Clin. Exp. Immunol.* **2014**, *178*, 262–269. [CrossRef]

90. Mazokopakis, E.E.; Papadomanolaki, M.G.; Tsekouras, K.C.; Evangelopoulos, A.D.; Kotsiris, D.A.; Tzortzinis, A.A. Is vitamin D related to pathogenesis and treatment of Hashimoto's thyroiditis? *Hell. J. Nucl. Med.* **2015**, *18*, 222–227.

91. Mukhopadhyay, S.; Chaudhary, S.; Dutta, D.; Kumar, M.; Saha, S.; Mondal, S.A.; Kumar, A. Vitamin D supplementation reduces thyroid peroxidase antibody levels in patients with autoimmune thyroid disease: An open-labeled randomized controlled trial. *Indian J. Endocrinol. Metab.* **2016**, *20*, 391–398. [CrossRef]

92. Krysiak, R.; Szkróbka, W.; Okopień, B. The Effect of Vitamin D on Thyroid Autoimmunity in Levothyroxine-Treated Women with Hashimoto's Thyroiditis and Normal Vitamin D Status. *Exp. Clin. Endocrinol. Diabetes* **2017**, *125*, 229–233. [CrossRef]

93. Krysiak, R.; Szkróbka, W.; Okopień, B. The effect of vitamin D and selenomethionine on thyroid antibody titers, hypothalamic-pituitary-thyroid axis activity and thyroid function tests in men with Hashimoto's thyroiditis: A pilot study. *Pharmacol. Rep.* **2019**, *71*, 243–247. [CrossRef]

94. Chahardoli, R.; Saboor-Yaraghi, A.-A.; Amouzegar, A.; Khalili, D.; Vakili, A.Z.; Azizi, F. Can Supplementation with Vitamin D Modify Thyroid Autoantibodies (Anti-TPO Ab, Anti-Tg Ab) and Thyroid Profile (T3, T4, TSH) in Hashimoto's Thyroiditis? A Double Blind, Randomized Clinical Trial. *Horm. Metab. Res.* **2019**, *51*, 296–301. [CrossRef] [PubMed]

95. Krysiak, R.; Kowalcze, K.; Okopień, B. The effect of vitamin D on thyroid autoimmunity in non-lactating women with postpartum thyroiditis. *Eur. J. Clin. Nutr.* **2016**, *70*, 637–639. [CrossRef] [PubMed]

96. Dizdar, O.S.; Şimşek, Y.; Çakır, I.; Yetmis, M.; Başpinar, O.; Gokay, F. Effects of Vitamin D treatment on thyroid autoimmunity. *J. Res. Med. Sci.* **2016**, *21*, 85. [CrossRef] [PubMed]

97. Wang, S.; Wu, Y.; Zuo, Z.; Zhao, Y.; Wang, K. The effect of vitamin D supplementation on thyroid autoantibody levels in the treatment of autoimmune thyroiditis: A systematic review and a meta-analysis. *Endocrine* **2018**, *59*, 499–505. [CrossRef] [PubMed]

98. Koehler, V.F.; Filmann, N.; Mann, W.A. Vitamin D Status and Thyroid Autoantibodies in Autoimmune Thyroiditis. *Horm. Metab. Res.* **2019**, *51*, 792–797. [CrossRef]

99. Krysiak, R.; Kowalcze, K.; Okopień, B. The effect of vitamin D on thyroid autoimmunity in euthyroid men with autoimmune thyroiditis and testosterone deficiency. *Pharmacol. Rep.* **2019**, *71*, 798–803. [CrossRef]

100. Krysiak, R.; Kowalcze, K.; Okopień, B. Selenomethionine potentiates the impact of vitamin D on thyroid autoimmunity in euthyroid women with Hashimoto's thyroiditis and low vitamin D status. *Pharmacol. Rep.* **2019**, *71*, 367–373. [CrossRef]

101. Mirhosseini, N.; Brunel, L.; Muscogiuri, G.; Kimball, S. Physiological serum 25-hydroxyvitamin D concentrations are associated with improved thyroid function-observations from a community-based program. *Endocrine* **2017**, *58*, 563–573. [CrossRef]

102. Cho, Y.Y.; Chung, Y.J. Vitamin D supplementation does not prevent the recurrence of Graves' disease. *Sci. Rep.* **2020**, *10*, 16–17. [CrossRef]

103. Grove-Laugesen, D.; Malmstroem, S.; Ebbehoj, E.; Riis, A.L.; Watt, T.; Hansen, K.W.; Rejnmark, L. Effect of 9 months of vitamin D supplementation on arterial stiffness and blood pressure in Graves' disease: A randomized clinical trial. *Endocrine* **2019**, *66*, 386–397. [CrossRef] [PubMed]

104. Grove-Laugesen, D.; Cramon, P.K.; Malmstroem, S.; Ebbehoj, E.; Watt, T.; Hansen, K.W.; Rejnmark, L. Effects of Supplemental Vitamin D on Muscle Performance and Quality of Life in Graves' Disease: A Randomized Clinical Trial. *Thyroid* **2020**, *30*, 661–671. [CrossRef] [PubMed]

105. Sanders, K.M.; Nicholson, G.C.; Ebeling, P.R. Is High Dose Vitamin D Harmful? *Calcif. Tissue Int.* **2013**, *92*, 191–206. [CrossRef] [PubMed]

106. Durup, D.; Jørgensen, H.L.; Christensen, J.; Schwarz, P.; Heegaard, A.-M.; Lind, B. A Reverse J-Shaped Association of All-Cause Mortality with Serum 25-Hydroxyvitamin D in General Practice: The CopD Study. *J. Clin. Endocrinol. Metab.* **2012**, *97*, 2644–2652. [CrossRef]

107. Institute of Medicine. *Dietary Reference Intakes for Calcium and Vitamin D*; The National Academies Press: Cambridge, MA, USA, 2011.

108. Holick, M.F.; Binkley, N.; Bischoff-Ferrari, H.A.; Gordon, C.M.; Hanley, D.A.; Heaney, R.P.; Murad, M.; Weaver, C.M. Evaluation, Treatment, and Prevention of Vitamin D Deficiency: An Endocrine Society Clinical Practice Guideline. *J. Clin. Endocrinol. Metab.* **2011**, *96*, 1911–1930. [CrossRef]

109. Bizzaro, G.; Shoenfeld, Y. Vitamin D and thyroid autoimmune diseases: The known and the obscure. *Immunol. Res.* **2015**, *61*, 107–109. [CrossRef]

110. Kozai, M.; Yamamoto, H.; Ishiguro, M.; Harada, N.; Masuda, M.; Kagawa, T.; Takei, Y.; Otani, A.; Nakahashi, O.; Ikeda, S.; et al. Thyroid Hormones Decrease Plasma 1α,25-Dihydroxyvitamin D Levels Through Transcriptional Repression of the Renal 25-Hydroxyvitamin D31α-Hydroxylase Gene (CYP27B1). *Endocrinology* **2013**, *154*, 609–622. [CrossRef]

111. Hu, S.; Rayman, M.P. Multiple Nutritional Factors and the Risk of Hashimoto's Thyroiditis. *Thyroid* **2017**, *27*, 597–610. [CrossRef]

112. Botelho, I.M.B.; Neto, A.M.; Silva, C.A.; Tambascia, M.A.; Alegre, S.M.; Zantut-Wittmann, D.E. Vitamin D in Hashimoto's thyroiditis and its relationship with thyroid function and inflammatory status. *Endocr. J.* **2018**, *65*, 1029–1037. [CrossRef]

113. Kawicka, A.; Regulska-Ilow, B. Metabolic disorders and nutritional status in autoimmune thyroid diseases. *Postępy Hig. Med. Dośw.* **2015**, *69*, 80–90. [CrossRef]

114. Walker, M.D.; Zylberberg, H.M.; Green, P.H.R.; Katz, M.S. Endocrine complications of celiac disease: A case report and review of the literature. *Endocr. Res.* **2019**, *44*, 27–45. [CrossRef] [PubMed]

115. Roy, A.; Laszkowska, M.; Sundström, J.; Lebwohl, B.; Green, P.H.; Kämpe, O.; Ludvigsson, J.F. Prevalence of Celiac Disease in Patients with Autoimmune Thyroid Disease: A Meta-Analysis. *Thyroid* **2016**, *26*, 880–890. [CrossRef] [PubMed]

116. Krysiak, R.; Szkróbka, W.; Okopień, B. The Effect of Gluten-Free Diet on Thyroid Autoimmunity in Drug-Naïve Women with Hashimoto's Thyroiditis: A Pilot Study. *Exp. Clin. Endocrinol. Diabetes* **2019**, *127*, 417–422. [CrossRef] [PubMed]

117. Jyotsna, V.P.; Sahoo, A.; Singh, A.K.; Sreenivas, V.; Gupta, N. Bone mineral density in patients of Graves disease pre- & post-treatment in a predominantly vitamin D deficient population. *Indian J. Med. Res.* **2012**, *135*, 36–41. [CrossRef] [PubMed]

118. Proal, A.D.; Albert, P.J.; Marshall, T.G. Dysregulation of the Vitamin D Nuclear Receptor May Contribute to the Higher Prevalence of Some Autoimmune Diseases in Women. *Ann. N. Y. Acad. Sci.* **2009**, *1173*, 252–259. [CrossRef]

119. Waterhouse, J.C.; Perez, T.H.; Albert, P.J. Reversing Bacteria-Induced Vitamin D Receptor Dysfunction Is Key to Autoimmune Disease. *Ann. N. Y. Acad. Sci.* **2009**, *1173*, 757–765. [CrossRef]

Article

Vitamin D Correction Down-Regulates Serum Amyloid P Component Levels in Vitamin D Deficient Arab Adults: A Single-Arm Trial

Osama E. Amer [1], Malak N. K. Khattak [1], Abdullah M. Alnaami [1], Naji J. Aljohani [2] and Nasser M. Al-Daghri [1,*]

[1] Chair for Biomarkers of Chronic Diseases, Biochemistry Department, College of Science, King Saud University, Riyadh 11451, Saudi Arabia; osamaemam@gmail.com (O.E.A.); malaknawaz@yahoo.com (M.N.K.K.); aalnaami@yahoo.com (A.M.A.)
[2] Specialized Diabetes and Endocrine Center, King Fahad Medical City, Riyadh 11525, Saudi Arabia; najij@hotmail.com
* Correspondence: aldaghri2011@gmail.com; Tel.: +966-14675939; Fax: +966-14675931

Received: 23 August 2020; Accepted: 18 September 2020; Published: 21 September 2020

Abstract: Vitamin D (VD) has been observed to have anti-inflammatory properties. However, the effects of VD supplementation on the serum amyloid P component (SAP) has not been established. This study aimed to investigate the effect of VD supplementation on serum SAP levels in Arab adults. A total of 155 VD-deficient adult Saudis (56 males and 99 females) were recruited in this non-randomized, 6-month, single-arm trial. The intervention was as follows; cholecalciferol 50,000 international units (IU) every week for the first 2 months, followed by 50,000 twice a month for the next two months, and for the last two months, 1000 IU daily. Serum 25(OH)D, SAP, C-reactive protein (CRP), lipid profile, and glucose were assessed at baseline and post-intervention. At post-intervention, VD levels were significantly increased, while SAP levels significantly decreased in all study participants. Remarkably, this reduction in SAP was more significant in males than females after stratification. SAP was inversely correlated with VD overall (r = −0.17, $p < 0.05$), and only in males (r = −0.27, $p < 0.05$) after stratification according to sex after 6 months of VD supplementation. Such a relationship was not observed at baseline. VD supplementation can favorably modulate serum SAP concentrations in Arab adults, particularly in males.

Keywords: vitamin D; SAP; amyloidosis; Arab; vitamin D supplementation

---

1. Introduction

Vitamin D (VD) is a fat-soluble secosteroid hormone, having both autocrine and endocrine roles [1]. While the main roles of VD include calcium homeostasis and bone metabolism [2,3], the presence of vitamin D receptors (VDR) in major cell types of the body gives it multiple extra-skeletal functions, one of which is modulation of inflammatory pathways [4,5].

The anti-inflammatory and immune-modulating properties of vitamin D (VD) are well-established [6]. Multiple studies consistently reveal the beneficial effects of VD supplementation in terms of increasing levels of anti-inflammatory markers and decreasing the production of inflammatory cytokines [7–9]. In a recent systematic review and meta-analysis involving 10 clinical trials and 924 participants, Chen and colleagues concluded that supplementation with VD can decrease C-reactive protein (CRP) levels, a well-known acute-phase inflammatory marker predictive of cardiovascular events, by as much as 2.21 mg/L [10]. Other inflammatory markers have been investigated, such as (IL)-10, IL-6 and TNF-α, all of which have been observed to be significantly associated with varying levels of 25(OH)D status [8,11,12].

Another acute-phase inflammation-induced protein is the serum amyloid P component (SAP), not to be confused with serum amyloid protein. Together, SAP and CRP are the short pentraxins chiefly produced by hepatocytes [13]. In humans, SAP contributes to host defense, either via opsonins or through complement activation. In a calcium-dependent way, SAP binds to several lipoprotein ligands, which suggests that this process could have significant inferences in amyloidosis and atherosclerosis in humans. Moreover, many studies support the fact that SAP has a significant role in inflammatory regulation. [14]. Significantly, SAP and CRP share structural characteristics (being organized in five identical subunits arranged in a pentameric radial symmetry) and biological functions, including activation of the complement system and pathogen recognition [13]. In calcium-free conditions, SAP pentamers physically interact with CRP pentamers to form very stable mixed decamers [15], which could have functional consequences on inflammation activation [16]. In a nested case-control proteomic analysis study, sera from 60 obese women with gestational diabetes mellitus (GDM) identified three candidate predictors of GDM: SAP, afamin, and vitronectin [17]. Lastly, for cardiovascular disease (CVD), SAP is considered as a valuable biomarker, as it contributes to CVD pathogenesis through modulating innate immunity and inflammation [18].

We hypothesize that improving VD status can favorably regulate SAP activity. In this single-arm trial, we aim to evaluate for the first time the effects of vitamin D supplementation on serum SAP levels in Saudi adults with VD deficiency.

2. Methods

2.1. Study Design and Participants

In this 6-month, single-arm trial, a total of 250 overweight Saudi adult males and females aged 30–50 years with 25(OH) D deficiency (<50 nmol/L) were selected randomly from the Vitamin D School database of the Chair for Biomarkers of Chronic Diseases (CBCD) in King Saud University (KSU, Riyadh, KSA). In brief, this database was taken from a capital-wide, multi-center observational study done in primary and secondary schools in Riyadh, Saudi Arabia [19,20]. Written informed consent was obtained from all participants before enrolment. A generalized questionnaire was taken from all participants, including demographic information, and present and past medical history. This intervention study was conducted from December 2015 to May 2016 (the cold season in Riyadh). The present study was part of a bigger project registered in the Saudi Clinical Trials Registry (SCTR) (E1-15-1667) Riyadh, Saudi Arabia, and was approved by the Scientific Research Ethics Committee at King Fahd Medical City (16-018), Riyadh, Saudi Arabia. Exclusion criteria were as follows: those with chronic clinical conditions (cancer, cardiovascular diseases (CVD), T2DM, osteopenia/osteoporosis, gastrointestinal disease, liver and renal dysfunction, and thyroid conditions), on VD supplementation or any medication and those with baseline 25(OH)D ≥ 50 nmol/L. Out of 250 enrolled participants, 13 were excluded for having one of the mentioned conditions, and another 27 for having basal 25(OH)D levels ≥50 nmol/L. Overall, 210 participants (75 males and 135 females) were able to complete the study (Figure 1).

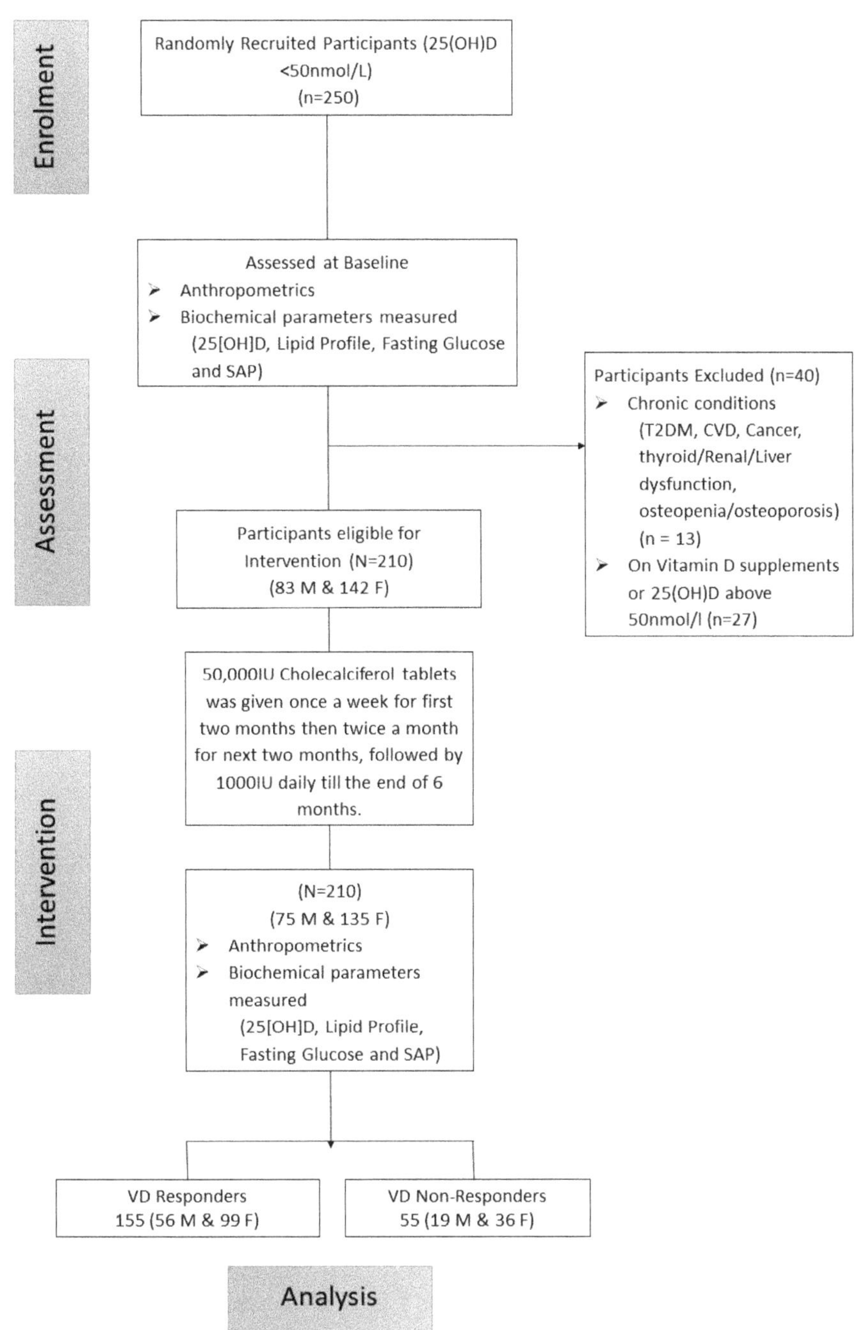

Figure 1. Participant flow chart.

2.2. Anthropometry and Biochemical Assessments

Anthropometrics which were determined included height (rounded off to the nearest 0.5 cm), weight (rounded off to the nearest 0.1 kg), waist and hip circumference (centimeters), and mean blood pressure (systolic and diastolic in mmHg) (average of two readings). Body mass index (BMI) was calculated as weight in kilograms divided by height in square meters. Fasting blood samples were collected and transferred immediately to a non-heparinized tube for centrifugation. Fasting glucose, lipid profile, were measured using a chemical analyzer (Konelab, Espoo, Finland). Serum 25(OH)D was measured by using commercially available kits using Roche Elecsys Modular Analytics Cobas e411 utilizing electrochemiluminescence immunoassay (Roche Diagnostics, Mannheim, Germany). Serum levels of SAP and CRP were measured using ELISA kits (Abcam®, Cambridge, UK and R & D SYSTEMS®, Minneapolis, MN, USA, respectively) following manufacturers' instructions. To minimize inter-assay variability, all samples were analyzed simultaneously and the actual variations were well within the inter- and intra-assay ranges. All measurements were done at baseline and post-intervention.

2.3. Intervention

VD supplementation was given to all participants in the following manner: (1) 50,000 IU cholecalciferol tablets given once a week for the first two months (VitaD50000; Synergy pharma, Dubai, UAE); (2) 50,000 IU cholecalciferol tablets twice a month for the next two months; and (3) 1000 IU daily (VitaD1000; Synergy pharma, Dubai, UAE) in the last 2 months as maintenance. The Short Message Service (SMS) was used to encourage participants to take their recommended doses of VD. For compliance, all participants had to return blister packs to quantify unconsumed tablets every month before a fresh refill was given. Intervention doses were according to the national and regional recommendations on management of vitamin D deficiency [21,22]. For stratification purposes, post-intervention responders were defined as those who achieved 25 (OH)D levels above 50 nmol/L, while non-responders were those who did not achieve 25 (OH)D levels > 50 nmol/L.

2.4. Statistical Analysis

Data were analyzed using the Statistical Package for Social Sciences (SPSS 22.0, SPSS, Inc., Chicago, IL, USA). Continuous data were presented as mean ± standard deviation (SD) for normal variables and non-normal variables were presented in median (first and third) percentiles. All categorical variables were presented as frequency and percentages. The Independent *T*-test and Mann-Whitney U test were used to compare baseline differences between normal and non-normal variables, respectively. Bonferroni correction was done for multiple comparisons at baseline to minimize type 1 error. The paired *T* test and Wilcoxon sign rank test were performed to check the mean and median differences at baseline and after intervention. Pearson's and Spearman's correlation were performed to determine associations of SAP with other parameters. The Bonferroni-adjusted *p*-value for baseline comparisons was $p < 0.0038$. A *p*-value < 0.05 was considered significant for the rest of the analysis.

3. Results

A total of 210 (75 males and 135 females) Saudi adults deficient in vitamin D were included in this 6-month interventional study. Table 1 shows the clinical characteristics of participants before and after intervention for responders and non-responders to VD supplementation. At baseline and using the Bonferroni-corrected *p*-value, responders were significantly older than non-responders ($p = 0.007$). Similarly, WHR measures were significantly higher in responders than non-responders ($p < 0.001$). Baseline BMI, blood pressure, and other parameters were not significantly different between groups.

Table 1. Clinical characteristics of study participants at baseline and after 6-month intervention.

Parameters	Responders				Non-Responders				* p-Value
	Before	After	Change	p-Value	Before	After	Change	p-Value	
N (M/F)	155 (56/99)				55 (19/36)				
Age (year)		39.9 ± 10.6				34.6 ± 11.2			0.007
BMI (kg/m^2)		29.2 ± 4.9				27.1 ± 5.1			0.98
WHR		0.94 ± 0.13				0.87 ± 0.11			<0.001
SBP (mmHg)		124.5 ± 13.6				125.3 ± 15.4			0.73
DBP (mmHg)		77.4 ± 8.9				78.4 ± 9.5			0.50
TC (mmol/L)	5.15 ± 1.2	5.16 ± 1.2	0.02 (−0.16–0.19)	0.86	5.10 ± 0.9	5.51 ±1.2	0.44 (0.13–0.8)	0.006	0.67
HDL-C (mmol/L)	1.04 ± 0.4	1.16 ± 0.4	0.11 (0.03–0.19)	0.007	1.1 ± 0.5	1.3 ± 0.4	0.22 (0.08–0.37)	0.003	0.88
LDL-C (mmol/L)	3.26 ± 0.9	3.15 ± 1.0	−0.11 (−0.26–0.05)	0.17	3.3 ± 0.8	3.4 ± 0.9	0.14 (−0.14–0.41)	0.33	0.79
TG (mmol/L)	1.72 ± 1.1	1.75 ± 1.0	0.03 (−0.13–0.19)	0.70	1.4 ± 0.8	1.7 ± 0.9	0.29 (0.04–0.54)	0.02	0.08
25(OH)D (nmol/L)	31.7 ± 11.7	63.8 ± 19.8	32.1 (29.0–35.2)	<0.001	31.9 ± 15.3	29.1 ± 12.4	−2.8 (−6.2–0.7)	0.11	0.60
Glucose (mmol/L)	5.48 ± 0.9	5.55 ± 0.9	0.07 (−0.10–0.23)	0.43	5.4 ± 0.9	5.5 ± 0.8	0.13 (−0.17–0.44)	0.39	0.93
SAP (mg/L)	44.9 (3.1–84.8)	41.2 (2.5–69.1)	−2.8 (−18.2–1.74)	0.002	40.0 (16.9–57.6)	42.4 (21.7–67.3)	1.42(−5.6–8.4)	0.29	0.44
CRP (μg/mL)	22.8 (4.5–53.5)	13.5 (4.9–34.5)	−8.2 (−16.6–2.24)	0.014	39.9(14.9–75.9)	34.7 (8.0–80.4)	−3.5 (−20.1–6.9)	0.74	0.038

Note: Data presented as mean ± SD for normal and median (first–third) percentiles for non-normal variables. BMI—body mass index; WHR—waist-hip ratio; SBP—systolic blood pressure; DBP—diastolic blood pressure; TC—total cholesterol; TG—triglycerides; * Bonferroni corrected p-value applied to actual p-values; significant at $p < 0.0038$. M: Male; F: Female; SAP: serum amyloid P component; CRP: C-reactive protein.

Post-intervention, 25(OH)D and HDL-cholesterol levels significantly increased after 6 months (p-values < 0.001 and 0.007, respectively) in the responders group. In contrast, SAP levels significantly decreased post-intervention (p = 0.002), as well as CRP levels (p = 0.014). No significant changes were observed in other parameters. Among non-responders, no changes in 25(OH)D levels were observed post-intervention. The same non-significance was observed for SAP and CRP levels. Total cholesterol, HDL-cholesterol, and triglycerides all significantly increased after intervention (p-values = 0.006, 0.003 and 0.02, respectively). For the rest of the other parameters, no significant differences were observed (Table 1).

Table 2 shows the between-group comparisons of both responders and non-responders. Serum 25(OH)D increased over time, and this was clinically significant in favor of the responders, even after adjusting for age and BMI (p < 0.001). A clinically significant decrease in SAP levels was observed over time, again in favor of responders, and this effect remained significant even after adjusting for age and BMI (p = 0.001).

Table 2. Between-group comparisons in 25(OH)D and Serum Amyloid P Component.

Parameters	25(OH) D (nmol/L)		Group Effect	Group Effect (Adjusted)
	Responders	**Non-Responders**		
Baseline	31.7 ± 11.7	31.9 ± 15.3		
6 month	63.8 ± 19.8	29.1 ± 12.4		
Mean Difference	32.1 (29.0–35.2)	−2.8 (−6.2–0.7)	<0.001	<0.001
Change (%)	103%	8.8%		
Time effect	<0.001			
Time effect (adjusted)	<0.001			
Serum Amyloid P Component (mg/L)				
	Responders	**Non-Responders**		
Baseline	44.9 (3.1–84.8)	40.0 (16.9–57.6)		
6 month	41.2 (2.5–69.1)	42.4 (21.7–67.3)		
Mean Difference	−2.8 (−18.2–1.74)	1.42(−5.6–8.4)	0.004	0.001
Change (%)	6.2%	3.6%		
Time effect	0.017			
Time effect (adjusted)	<0.001			

Note: Data presented as mean ± standard deviation, median (first and 75th) percentiles, and mean and median change (95% CI); adjusted for age and BMI; significant at p < 0.05.

Table 3 shows comparisons of responders' characteristics pre- and post-intervention according to sex. Levels of 25(OH)D significantly increased over time in both sexes (p < 0.001). Similarly, HDL was significantly increased in both sexes. SAP was significantly decreased over time in both sexes; remarkably, this reduction in SAP was more significant in males [55.7 (31.2–78.4) vs. 57.3 (27.7–100.9), p = 0.01] than in females [28.9 (1.4–62.4) vs. 38.4 (1.3–74.1), p = 0.046]. Conversely, CRP was significantly reduced post-intervention in females [7.8 (4.4–32.4) vs. 22.2 (3.9–61.6), p = 0.036], but no significant difference was observed in males. No reduction in glucose levels was observed in both sexes; contrary to what was expected, glucose had a significant increase in males [5.87 ± 0.9 vs. 5.61 ± 0.9, p = 0.029], with no significant difference observed in females.

Table 4 shows the bivariate correlation coefficients of SAP with other study parameters in responder participants at baseline, where SAP had a significantly positive relationship with systolic BP (r = 0.20, p < 0.05) and diastolic BP (r = 0.33, p < 0.01) in our study participants overall. This relationship was observed for diastolic BP only in males (r = 0.30, p < 0.05) after stratification according to sex. At baseline, SAP also had a significant inverse correlation with HDL-cholesterol (r = −0.30, p < 0.01). Overall, this clinically significant inverse correlation persisted in females (r = −0.37, p < 0.01) but not in males after stratification according to sex. In addition, SAP had a significantly positive correlation with glucose (r = 0.32, p < 0.05) in males at baseline, as well as with CRP overall and in both sexes (p < 0.001).

Post-intervention, SAP was inversely correlated with VD overall (r = −0.17, $p < 0.05$) and only in males (r = −0.27, $p < 0.05$) after stratification according to sex, whereas such a relationship was not observed at baseline. Triglycerides had a significant positive correlation with SAP only in females post-intervention (r = 0.23, $p < 0.05$) but not in males.

Table 5 shows the delta change correlation analysis between SAP and other parameters. Overall, there was a significant inverse relationship between Δ SAP and Δ HDL (r = −0.30, $p < 0.01$), and it was positively correlated with Δ CRP (r = 0.28; $p < 0.01$) in our study population. After stratification according to sex, Δ SAP was inversely correlated with Δ HDL (r= −0.31; $p < 0.05$) and Δ triglycerides (r = −0.27; $p < 0.05$) only in males.

Table 6 shows the responders' characteristics pre- and post-intervention using the SAP cut-off values, a normal reference interval for serum SAP concentration, for both sexes (males; 32 mg/L and females; 24 mg/L) [23]. Of the participants, 98 (42 males and 56 females) had high values of serum SAP above referenced normal levels. Over time, 25(OH)D significantly increased in both sexes ($p < 0.001$). Remarkably, post-supplementation with VD, the reduction in SAP serum levels was more significant in this sub-group compared to the main group in both sexes (in males, −9.5 (−33.9–7.1), $p = 0.007$ vs. −1.75 (−21.7–7.4), $p = 0.011$; in females, −13.9 (−33.3–2.2), $p < 0.001$ vs. −0.57 (−16.5–1.2), $p = 0.046$).

Table 3. Clinical characteristics of responders at baseline and after 6-month intervention according to sex.

Parameters	Male				Female			
N (M/F)	56				99			
Age (year)	41.9 ± 9.6				37.9 ± 11.0			
BMI (kg/m^2)	28.4 ± 3.8				29.6 ± 5.4			
WHR	0.98 ± 0.10				0.92 ± 0.15			
SBP (mmHg)	130.4 ± 11.5				120.3 ± 13.6			
DBP (mmHg)	81.2 ± 7.1				74.8 ± 9.1			
	Baseline	6-Month	Δ	p-Value	Baseline	6-Month	Δ	p-Value
TC (mmol/L)	5.02 ± 1.2	5.11 ± 1.4	0.09 (−0.19–0.37)	0.52	5.22 ± 1.2	5.20 ± 1.1	−0.03 (−0.24–0.19)	0.79
HDL-C (mmol/L)	0.94 ± 0.3	1.02 ± 0.4	0.08 (0.001–0.16)	0.05	1.11 ± 0.5	1.24 ± 0.4	0.13 (0.01–0.25)	0.03
LDL-C (mmol/L)	3.16 ± 0.8	3.06 ± 1.04	−0.10 (−0.4–0.19)	0.50	3.32 ± 0.9	3.20 ± 0.9	−0.12 (−0.31–0.07)	0.23
TG (mmol/L)	1.89 ± 1.4	1.97 ± 1.2	0.08 (0.001–0.16)	0.54	1.61 ± 0.9	1.62 ± 0.8	0.003 (−0.19–0.20)	0.98
25(OH)D (nmol/L)	34.5 ± 10.3	64.6 ± 17.8	30.1 (24.9–35.3)	<0.001	30.1 ± 12.2	63.3 ± 20.9	33.2 (29.3–37.1)	<0.001
Glucose (mmol/L)	5.61 ± 0.9	5.87 ± 0.9	0.25 (0.03–0.47)	0.029	5.41 ± 1.0	5.38 ± 0.8	−0.03 (−0.26–0.20)	0.80
SAP (mg/L)	57.3 (27.7–100.9)	55.7 (31.2–78.4)	−1.75 (−21.7–7.4)	0.011	38.4 (1.3–74.1)	28.9 (1.4–62.4)	−0.57 (−16.5–1.2)	0.046
CRP (µg/mL)	26.8 (4.7–48.5)	22.1 (5.2–34.8)	−2.42 (−10.9–3.6)	0.20	22.2 (3.9–61.6)	7.8 (4.4–32.4)	−0.10 (−28.9–2.20)	0.036

Note: Data presented as mean ± SD for normal, and median (1st–3rd) percentiles for non-normal variables. BMI—body mass index; WHR—waist-hip ratio; SBP—systolic blood pressure; DBP—diastolic blood pressure; TC—total cholesterol; TG—triglycerides; Significant at $p < 0.05$.

Table 4. Bivariate associations of SAP among responders at baseline and after 6-month intervention.

Parameters	Baseline			6-Month		
	All	Males	Females	All	Males	Females
Age (year)	0.08	−0.04	0.04			
BMI (kg/m^2)	−0.03	−0.05	0.04			
WHR	−0.07	−0.17	−0.18			
Systolic BP	0.20 *	0.19	0.03			
Diastolic BP	0.33 **	0.30 *	0.20			
Total Cholesterol	−0.06	−0.08	−0.03	0.04	0.03	0.10
HDL-C	−0.30 **	0.11	−0.37 **	−0.06	0.22	−0.10
LDL-C	−0.05	−0.18	0.02	0.02	0.14	0.01
Triglycerides	0.10	−0.03	0.14	0.12	−0.19	0.23 *
25(OH)D	−0.04	0.11	−0.16	−0.17 *	−0.27 *	−0.16
Glucose	0.13	0.32 *	0.04	0.12	0.04	0.14
CRP	0.55 **	0.54 **	0.55 **	0.39 **	0.61 **	0.47 **

Note: Data presented as coefficient (R); * denotes significance at 0.05 level; ** denotes significance at 0.01 level.

Table 5. Delta change associations between study parameters among responders.

All Participants								
	Δ SAP	Δ CRP	Δ Cholesterol	Δ HDL	Δ LDL	Δ Triglycerides	Δ VD	Δ Glucose
Δ SAP	1.00							
Δ CRP	0.28 **	1.00						
Δ TC	−0.05	0.26 *	1.00					
Δ HDL-C	−0.30 **	−0.16	0.33 **	1.00				
Δ LDL-C	0.11	0.31 **	0.75 **	−0.01	1.00			
Δ Triglycerides	−0.05	0.05	0.29 **	−0.06	−0.15	1.00		
Δ 25(OH)D	0.01	−0.12	−0.04	0.15	0.00	−0.10	1.00	
Δ Glucose	0.06	0.34 **	0.19 *	−0.09	0.15	0.16	−0.19 *	1.00
Males								
Δ SAP	1.00							
Δ CRP	0.11	1.00						
Δ TC	−0.07	0.17	1.00					
Δ HDL-C	−0.31 *	0.10	0.00	1.00				
Δ LDL-C	0.01	0.31	0.83 **	0.17	1.00			
Δ Triglycerides	−0.27 *	−0.11	0.25	−0.18	−0.24	1.00		
Δ 25(OH)D	−0.02	−0.13	−0.21	0.15	−0.25	0.00	1.00	
Δ Glucose	0.22	0.06	−0.11	−0.04	−0.08	−0.09	−0.05	1.00
Females								
Δ SAP	1.00							
Δ CRP	0.09	1.00						
Δ TC	−0.06	0.08	1.00					
Δ HDL-C	−0.16	−0.13	0.29 **	1.00				
Δ LDL-C	0.02	0.06	0.71 **	−0.04	1.00			
Δ Triglycerides	−0.07	0.26	0.42 **	−0.05	0.10	1.00		
Δ 25(OH)D	0.10	−0.03	0.07	0.00	0.02	−0.02	1.00	
Δ Glucose	−0.12	0.07	0.25 *	−0.17	0.25 *	0.13	−0.14	1.00

Note: Data presented as coefficient (R); * denotes significance at 0.05 level; ** denotes significance at 0.01 level. VD: Vitamin D.

Table 6. Clinical characteristics of responders at baseline and after 6-month intervention using SAP cut-off levels.

Parameters	Male (SAP > 30 mg/L)					Female (SAP > 24 mg/L)				
	Baseline	6-Month	Δ	Effect Size	p-Value	Baseline	6-Month	Δ	Effect Size	p-Value
N (M/F)			42					56		
Age (year)			41.4 ± 8.8					38.8 ± 11.6		
BMI (kg/m^2)			27.7 ± 3.0					29.8 ± 4.4		
WHR			0.98 ± 0.06					0.89 ± 0.10		
SBP (mmHg)			132.0 ± 11.2					121.5 ± 13.4		
DBP (mmHg)			82.6 ± 6.7					76.8 ± 8.4		
T. Chol (mmol/L)	4.9 ± 1.1	5.0 ± 1.4	0.1 (−0.1–0.3)	0.12	0.41	5.2 ± 1.3	5.2 ± 1.2	0.03 (−0.3–3)	0.026	0.85
HDL-C (mmol/L)	0.97 ± 0.3	1.1 ± 0.3	0.1 (0.05–0.2)	0.52	0.002	1.0 ± 0.4	1.2 ± 0.5	0.21 (0.05–0.4)	0.36	0.01
LDL-C (mmol/L)	3.1 ± 0.8	3.0 ± 0.9	−0.08 (−0.3–0.2)	0.10	0.51	3.3 ± 0.9	3.1 ± 1.0	−0.12 (−0.4–0.1)	0.13	0.32
TG (mmol/L)	0.03 ± 0.8	0.15 ± 0.5	0.1 (−0.05–0.3)	0.21	0.16	0.04 ± 1.1	0.21 ± 0.8	0.17 (0.03–0.3)	0.34	0.016
VD (nmol/L)	34.9 ± 9.7	63 ± 14.1	28 (23–33)	1.61	<0.001	28.6 ± 11.4	59.6 ± 20.7	30.9 (26–36)	1.53	<0.001
Glucose (mmol/L)	5.8 ± 0.9	5.9 ± 0.9	0.14 (−0.1–0.4)	0.21	0.25	5.5 ± 1.1	5.4 ± 0.9	−0.06 (−0.4–0.3)	0.06	0.71
SAP (mg/L)	82 (53–109)	65 (44–88)	−9.5 (−34–7)	0.39	0.007	64 (46–104)	58.4 (38–76)	−13.9 (−33–2.2)	0.56	<0.001
CRP (μg/mL)	36.2 (9–49)	27.5 (10–41)	−3.0 (−16–4)	0.32	0.07	27.7 (5–62)	9.1 (5.1–33)	−0.4 (−30.1–3.6)	0.33	0.038

Note: Data presented as mean ± SD for normal and median (first–third) percentiles for non-normal variables. Significant at $p < 0.05$.

4. Discussion

The present interventional study is, to our knowledge, the first to show a clinically significant reduction in serum SAP levels after 6 months of VD correction. Remarkably, the post-intervention reduction in serum SAP levels was even more significant than without applying the cut-off values. At baseline, SAP levels were inversely correlated with cardiometabolic factors, such as BMI and HDL-cholesterol, and positively correlated with blood pressure, with no association between VD and SAP. However, at post-intervention, our results showed a significant inverse correlation between SAP and VD among responders, and this significant correlation persisted in males after stratification for sex.

The link between SAD and VD based on the present results is most likely tied to their associations with cardiometabolic factors. SAP has a key role in innate immunity and cardiometabolism [24]. Furthermore, like VD, it is also directly influenced by calcium [25]. In a calcium-dependent manner, SAP binds to many different lipoprotein ligands, and this can have a significant contribution in the progression of amyloidosis and atherosclerosis [26,27]. In fact, it has been found in the plaques of advanced human atherosclerosis and is proposed to have an active role in atherogenesis [28]. Previous studies indicated a significant increase in SAP levels in the early phase of post-acute myocardial infarction [29]. Furthermore, SAP deficiency prevents the atherosclerotic process [30] and other pathological processes, such as fibrosis, hypercoagulation, and inflammation [24,31]. Lastly, pentraxins including SAP have been demonstrated to be involved in obesity and other states of a chronic low-grade inflammatory [32]. Hence, VD supplementation can reduce the cardiovascular risk associated to overweight and obesity by reducing the pro-inflammatory pentraxin SAP.

Another highlight of the present study is the significant inverse correlation post-supplementation between SAP and VD only among male responders, which highlights sex-specific extra-skeletal properties of VD correction. Previously, we found that VD deficiency and its association with cardio-metabolic risk factors were mostly limited to males, in a study which involved more than 3000 Saudi adolescents and adults. This led us to believe that correction of VD status could prove more beneficial to men than women, at least in terms of extra-skeletal benefits [20]. One explanation that we have also recently documented at the proteomic level is that the conversion of 25(OH)D to its active form, 1,25(OH)D2, is higher in men than women, and this can be linked to the sex hormone metabolism [33].

Lastly, it is worthy to discuss that the primary grouping variable used in the present study to elicit differences between circulating SAP was the participants' response to vitamin D supplementation. Despite monitoring all participants for compliance and adherence, it was anticipated that some will not be able to achieve full vitamin D sufficiency despite large boluses of vitamin D. The failure to achieve full vitamin D correction despite above-average supplementation has been a consistent dilemma in Saudi Arabia and the rest of the region, and this has been fully acknowledged by national and regional experts, prompting vitamin D guidelines unique to the Middle-Eastern region and the Gulf Cooperation Council (GCC) countries in particular (21, 22). A recent genetic study within the Saudi community that could partially explain the non-responsiveness to exogenous vitamin D sources are the variants in vitamin D binding proteins (rs7041 and rs4588), carriers of which are three to 12 times more likely to be non-responders to vitamin D treatment [34].

The authors acknowledge some limitations. First, we used the non-responders as our comparator group, since we wanted to clearly delineate that the modest but significant changes in circulating SAP was associated with acute changes in vitamin D status brought about by a favorable response to vitamin D supplementation. Furthermore, since VD deficiency is very common in Saudi Arabia, the use of a true control group (without supplementation) is inappropriate, given that the inclusion criteria are participants with VD deficiency. Whether the present results will be the same using a control group remains to be investigated. Second, important factors influencing VD status were not measured in the current study, such as sunlight exposure, season, and outdoor physical activity, and as such, essential adjustments were not carried out. Nevertheless, this is the first study of its kind to investigate the effects of VD supplementation on SAP levels.

5. Conclusions

This is the first study to demonstrate the inverse relationship between serum VD and SAP. The present study showed that VD correction can significantly reduce serum SAP concentrations, particularly in male participants. As such, one of the cardiometabolic benefits of VD supplementation is through modulation of serum SAP levels, which can decrease risk for atherosclerosis, plaque formation, and multi-organ fibrosis. Further investigations are needed to determine whether prolonged states of vitamin D sufficiency can reverse atherosclerotic and fibrotic conditions through normalcy of SAP levels.

Author Contributions: N.M.A.-D. and O.E.A. contributed in study conception and design, A.M.A. dealt in recruitment of subjects and procurement of samples, O.E.A. and A.M.A. analyzed samples, M.N.K.K. analyzed data. O.E.A. wrote the manuscript, N.M.A.-D. and N.J.A. edited and reviewed the manuscript. All authors have read and agreed to the published version of the manuscript.

Funding: This study is funded by the Deanship of Scientific Research Chairs, Chair for Biomarkers of Chronic Diseases, Department of Biochemistry College of Science in King Saud University, Riyadh, Saudi Arabia.

Acknowledgments: The authors thank Deanship of Scientific Research, KSU and Chair for Biomarkers of Chronic Diseases (CBCD) for its continuous support.

Conflicts of Interest: The authors declare no conflict of interest.

References

1. Vanchinathan, V.; Lim, H.W. A dermatologist's perspective on vitamin D. *Mayo Clin. Proc.* **2012**, *87*, 372–380. [CrossRef] [PubMed]
2. Lee, J.H.; O'Keefe, J.H.; Bell, D.; Hensrud, D.D.; Holick, M.F. Vitamin D deficiency an important, common, and easily treatable cardiovascular risk factor? *J. Am. Coll. Cardiol.* **2008**, *52*, 1949–1956. [CrossRef]
3. Pfeifer, M.; Begerow, B.; Minne, H.W. Vitamin D and muscle function. *Osteoporos. Int.* **2002**, *13*, 187–194. [CrossRef] [PubMed]
4. Heaney, R.P. Vitamin D in health and disease. *Clin. J. Am. Soc. Nephrol.* **2008**, *3*, 1535–1541. [CrossRef] [PubMed]
5. Cohen-Lahav, M.; Douvdevani, A.; Chaimovitz, C.; Shany, S. The anti-inflammatory activity of 1,25-dihydroxyvitamin D3 in macrophages. *J. Steroid Biochem. Mol. Biol.* **2007**, *103*, 558–562. [CrossRef] [PubMed]
6. Gysemans, C.A.; Cardozo, A.K.; Callewaert, H.; Giulietti, A.; Hulshagen, L.; Bouillon, R.; Eizirik, D.L.; Mathieu, C. 1,25-Dihydroxyvitamin D3 modulates expression of chemokines and cytokines in pancreatic islets: Implications for prevention of diabetes in nonobese diabetic mice. *Endocrinology* **2005**, *146*, 1956–1964. [CrossRef]
7. Hossein-Nezhad, A.; Mirzaei, K.; Keshavarz, S.A.; Ansar, H.; Saboori, S.; Tootee, A. Evidences of dual role of vitamin D through cellular energy homeostasis and inflammation pathway in risk of cancer in obese subjects. *Minerva Med.* **2013**, *104*, 295–307.
8. Calton, E.K.; Keane, K.N.; Newsholme, P.; Soares, M.J. The Impact of Vitamin D Levels on Inflammatory Status: A Systematic Review of Immune Cell Studies. *PLoS ONE* **2015**, *10*, e0141770. [CrossRef]
9. Prietl, B.; Treiber, G.; Pieber, T.R.; Amrein, K. Vitamin D and immune function. *Nutrients* **2013**, *5*, 2502–2521. [CrossRef]
10. Chen, N.; Wan, Z.; Han, S.-F.; Li, B.-Y.; Zhang, Z.-L.; Qin, L.-Q. Effect of vitamin D supplementation on the level of circulating high-sensitivity C-reactive protein: A meta-analysis of randomized controlled trials. *Nutrients* **2014**, *6*, 2206–2216. [CrossRef]
11. Calton, E.K.; Keane, K.N.; Soares, M.J. The potential regulatory role of vitamin D in the bioenergetics of inflammation. *Curr. Opin. Clin. Nutr. Metab. Care* **2015**, *18*, 367–373. [CrossRef] [PubMed]
12. Rodriguez, A.J.; Mousa, A.; Ebeling, P.R.; Scott, D.; de Courten, B. Effects of vitamin D supplementation on inflammatory markers in heart failure: A systematic review and meta-analysis of randomized controlled trials. *Sci. Rep.* **2018**, *8*, 1169. [CrossRef] [PubMed]
13. Christner, R.B.; Mortensen, R.F. Specificity of the binding interaction between human serum amyloid P-component and immobilized human C-reactive protein. *J. Biol. Chem.* **1994**, *269*, 9760–9766. [PubMed]

14. Lu, J.; Marjon, K.D.; Mold, C.; Du Clos, T.W.; Sun, P.D. Pentraxins and Fc receptors. *Immunol. Rev.* **2012**, *250*, 230–238. [CrossRef] [PubMed]

15. Deban, L.; Bottazzi, B.; Garlanda, C.; de la Torre, Y.M.; Mantovani, A. Pentraxins: Multifunctional proteins at the interface of innate immunity and inflammation. *Biofactors* **2009**, *35*, 138–145. [CrossRef] [PubMed]

16. Steel, D.M.; Whitehead, A.S. The major acute phase reactants: C-reactive protein, serum amyloid P component and serum amyloid A protein. *Immunol. Today* **1994**, *15*, 81–88. [CrossRef]

17. Ravnsborg, T.; Svaneklink, S.; Andersen, L.L.T.; Larsen, M.R.; Jensen, D.M.; Overgaard, M. First-trimester proteomic profiling identifies novel predictors of gestational diabetes mellitus. *PLoS ONE* **2019**, *14*, e0214457. [CrossRef]

18. Maekawa, Y.; Nagai, T.; Anzai, A. Pentraxins: CRP and PTX3 and cardiovascular disease. *Inflamm. Allergy Drug Targets* **2011**, *10*, 229–235. [CrossRef]

19. Al-Daghri, N.M.; Al-Attas, O.S.; Wani, K.; Alnaami, A.M.; Sabico, S.; Al-Ajlan, A.; Chrousos, G.P.; Alokail, M.S. Sensitivity of various adiposity indices in identifying cardiometabolic diseases in Arab adults. *Cardiovasc. Diabetol.* **2015**, *14*, 101. [CrossRef]

20. Al-Daghri, N.M.; Al-Othman, A.; Albanyan, A.; Al-Attas, O.S.; Alokail, M.S.; Sabico, S.; Chrousos, G.P. Perceived stress scores among Saudi students entering universities: A prospective study during the first year of university life. *Int. J. Environ Res. Public Health* **2014**, *11*, 3972–3981. [CrossRef]

21. Al-Daghri, N.M.; Al-Saleh, Y.; Aljohani, N.; Sulimani, R.; Al-Othman, A.M.; Alfawaz, H.; Fouda, M.; Al-Amri, F.; Shahrani, A.; Alharbi, M.; et al. Vitamin D status correction in Saudi Arabia: An experts' consensus under the auspices of the European Society for Clinical and Economic Aspects of Osteoporosis, Osteoarthritis, and Musculoskeletal Diseases (ESCEO). *Arch. Osteoporos.* **2017**, *12*, 1. [CrossRef] [PubMed]

22. Al Saleh, Y.; Beshyah, S.A.; Hussein, W.; Almadani, A.; Hassoun, A.; Al Mamari, A.; Ba-Essa, E.; Al-Dhafiri, E.; Hassanein, M.; Fouda, M.A.; et al. Diagnosis and management of vitamin D deficiency in the Gulf Cooperative Council (GCC) countries: An expert consensus summary statement from the GCC vitamin D advisory board. *Arch. Osteoporos.* **2020**, *15*, 35. [CrossRef] [PubMed]

23. Nelson, S.R.; Tennent, G.A.; Sethi, D.; Gower, P.E.; Ballardie, F.W.; Amatayakul-Chantler, S.; Pepys, M.B. Serum amyloid P component in chronic renal failure and dialysis. *Clin. Chim. Acta* **1991**, *200*, 191–199. [CrossRef]

24. Bottazzi, B.; Inforzato, A.; Messa, M.; Barbagallo, M.; Magrini, E.; Garlanda, C.; Mantovani, A. The pentraxins PTX3 and SAP in innate immunity, regulation of inflammation and tissue remodelling. *J. Hepatol.* **2016**, *64*, 1416–1427. [CrossRef] [PubMed]

25. Emsley, J.; White, H.E.; O'hara, B.P.; Oliva, G.; Srinivasan, N.; Tickle, I.J.; Blundell, T.L.; Pepys, M.B.; Wood, S.P. Structure of pentameric human serum amyloid P component. *Nature* **1994**, *367*, 338–345. [CrossRef]

26. Ashton, A.W.; Boehm, M.K.; Gallimore, J.R.; Pepys, M.B.; Perkins, S.J. Pentameric and decameric structures in solution of serum amyloid P component by X-ray and neutron scattering and molecular modelling analyses. *J. Mol. Biol.* **1997**, *272*, 408–422. [CrossRef]

27. Li, X.A.; Yutani, C.; Shimokado, K. Serum amyloid P component associates with high density lipoprotein as well as very low density lipoprotein but not with low density lipoprotein. *Biochem. Biophys. Res. Commun.* **1998**, *244*, 249–252. [CrossRef]

28. Li, X.A.; Hatanaka, K.; Ishibashi-Ueda, H.; Yutani, C.; Yamamoto, A. Characterization of serum amyloid P component from human aortic atherosclerotic lesions. *Arterioscler. Thromb. Vasc. Biol.* **1995**, *15*, 252–257. [CrossRef]

29. Cubedo, J.; Padró, T.; Badimon, L. Coordinated proteomic signature changes in immune response and complement proteins in acute myocardial infarction: The implication of serum amyloid P-component. *Int. J. Cardiol.* **2013**, *168*, 5196–5204. [CrossRef]

30. Zheng, L.; Wu, T.; Zeng, C.; Li, X.; Li, X.; Wen, D.; Ji, T.; Lan, T.; Xing, L.; Li, J.; et al. SAP deficiency mitigated atherosclerotic lesions in ApoE(-/-) mice. *Atherosclerosis* **2016**, *244*, 179–187. [CrossRef]

31. Xi, D.; Luo, T.; Xiong, H.; Liu, J.; Lu, H.; Li, M.; Hou, Y.; Guo, Z. SAP: Structure, function, and its roles in immune-related diseases. *Int. J. Cardiol.* **2015**, *187*, 20–26. [CrossRef] [PubMed]

32. Ogawa, T.; Kawano, Y.; Imamura, T.; Kawakita, K.; Sagara, M.; Matsuo, T.; Kakitsubata, Y.; Ishikawa, T.; Kitamura, K.; Hatakeyama, K.; et al. Reciprocal contribution of pentraxin 3 and C-reactive protein to obesity and metabolic syndrome. *Obesity* **2010**, *18*, 1871–1874. [CrossRef] [PubMed]

33. Al-Daghri, N.M.; Al-Attas, O.S.; Johnston, H.E.; Singhania, A.; Alokail, M.S.; Alkharfy, K.M.; Abd-Alrahman, S.H.; Sabico, S.L.; Roumeliotis, T.I.; Manousopoulou-Garbis, A.; et al. Whole serum 3D LC-nESI-FTMS quantitative proteomics reveals sexual dimorphism in the milieu intérieur of overweight and obese adults. *J. Proteome Res.* **2014**, *13*, 5094–5105. [CrossRef] [PubMed]
34. Al-Daghri, N.M.; Mohammed, A.K.; Bukhari, I.; Rikli, M.; Abdi, S.; Ansari, M.G.A.; Sabico, S.; Hussain, S.D.; Alenad, A.; Al-Saleh, Y.; et al. Efficacy of vitamin D supplementation according to vitamin D-binding protein polymorphisms. *Nutrition* **2019**, *63–64*, 148–154. [CrossRef]

 *nutrients*

Article

Associations of Vitamin D Deficiency, Parathyroid hormone, Calcium, and Phosphorus with Perinatal Adverse Outcomes. A Prospective Cohort Study

Íñigo María Pérez-Castillo [1,†], Tania Rivero-Blanco [1], Ximena Alejandra León-Ríos [1], Manuela Expósito-Ruiz [2], María Setefilla López-Criado [3] and María José Aguilar-Cordero [1,4,*,†]

[1] Andalusian Plan for Research, Development and Innovation, CTS 367, University of Granada, 18001 Granada, Spain; perezcastillo@correo.ugr.es (Í.M.P.-C.); taniarivero89@gmail.com (T.R.-B.); ximenaleonr18@gmail.com (X.A.L.-R.)

[2] Foundation for Biomedical Research in Eastern Andalusia (FIBAO), 18014 Granada, Spain; manuela.exposito.ruiz@juntadeandalucia.es

[3] Obstetrics and Gynecology Service, Virgen de las Nieves University Hospital, 18014 Granada, Spain; mmefilla@gmail.com

[4] Department of Nursing, Faculty of Health Sciences, University of Granada, 18071 Granada, Spain

* Correspondence: mariajaguilar@telefonica.net; Tel.: +34-657-84-17-51

† These authors contributed equally to this work.

Received: 15 September 2020; Accepted: 22 October 2020; Published: 26 October 2020

Abstract: Vitamin D deficiency during pregnancy has been linked to perinatal adverse outcomes. Studies conducted to date have recommended assessing interactions with other vitamin D-related metabolites to clarify this subject. We aimed to evaluate the association of vitamin D deficiency during early pregnancy with preterm birth. Secondary outcomes included low birth weight and small for gestational age. Additionally, we explored the role that parathyroid hormone, calcium and phosphorus could play in the associations. We conducted a prospective cohort study comprising 289 pregnant women in a hospital in Granada, Spain. Participants were followed-up from weeks 10–12 of gestation to postpartum. Serum 25-hydroxyvitamin D, parathyroid hormone, calcium, and phosphorus were measured within the first week after recruitment. Pearson's χ^2 test, Mann–Whitney U test, binary and multivariable logistic regression models were used to explore associations between variables and outcomes. 36.3% of the participants were vitamin D deficient (<20 ng/mL). 25-hydroxyvitamin D concentration was inversely correlated with parathyroid hormone ($\rho = -0.146$, $p = 0.013$). Preterm birth was associated with vitamin D deficiency in the multivariable model, being this association stronger amongst women with parathyroid hormone serum levels above the 80th percentile (adjusted odds ratio (aOR) = 6.587, 95% CI (2.049, 21.176), $p = 0.002$). Calcium and phosphorus were not associated with any studied outcome. Combined measurement of 25-hydroxyvitamin D and parathyroid hormone could be a better estimator of preterm birth than vitamin D in isolation.

Keywords: vitamin D deficiency; perinatal adverse outcomes; 25-hydroxyvitamin D; parathyroid hormone; PTH; calcium; phosphorus; cohort study

---

1. Introduction

Vitamin D deficiency is considered to be a pandemic [1] whose global prevalence varies widely depending on the studied population, dietary intake, ultraviolet-B light exposure, ethnicity, and age, amongst other factors [2]. The severe deficiency of this secosteroid is associated with skeletal disorders as well as other pathologies outside bone metabolism [3]. During pregnancy, vitamin D deficiency has been linked to pregnancy and perinatal adverse outcomes such as pre-eclampsia, gestational diabetes mellitus, preterm birth, and low birth weight [4].

Nutrients **2020**, *12*, 3279; doi:10.3390/nu12113279   www.mdpi.com/journal/nutrients

Preterm birth (PTB) is the leading cause of mortality in children under five years old worldwide, and its global prevalence has been estimated to be 10.6% of all births accounting for 14.84 million newborns in 2014 [5]. PTB is regarded as a syndrome resulting from different mechanisms such as uteroplacental dysfunction, inflammation, and infection, along with other immunological processes [6]. Vitamin D exerts important immunomodulatory effects decreasing levels of IL-1, IL-6 and TNF-α produced by macrophages [7], regulating the activity of lymphocytes B and T [8], and inducing human cathelicidin production [9], thus playing an important role in both innate and adaptive immune responses. However, studies evaluating the association between vitamin D deficiency during pregnancy and prematurity have not reached consensus on their results [10,11].

According to the WHO, low birth weight (LBW) is defined as a birth weight of less than 2500 g [12], whereas small for gestational age (SGA) is defined as weight below 10th percentile for the gestational age and depends on the reference population [13]. Several authors have associated vitamin D deficiency during pregnancy with LBW and SGA [14–17]. Possible mechanisms of action of vitamin D on fetal growth might consist of anti-inflammatory properties, regulation of genes implicated in angiogenesis, promotion of trophoblast invasion and control of fetal glucose availability [13,14,16,18]. Meta-analyses conducted to evaluate the associations between vitamin D deficiency, PTB, LBW or SGA have not yielded strong evidence [19–23].

25-hydroxyvitamin D (which will be referred to as vitamin D throughout the paper) plays a key role in calcium and phosphorus homeostasis and its concentration is regulated by the parathyroid hormone (PTH). However, vitamin D is usually measured in isolation and some authors have highlighted the importance of assessing interactions with vitamin D-related metabolites when evaluating associations with pregnancy and perinatal outcomes [24–26]. In this regard, low maternal calcium concentrations have been associated with LBW [27] and PTB [17,28], but evidence remains unclear. Furthermore, several authors have described the concept of functional vitamin D deficiency characterized by secondary hyperparathyroidism, which refers to elevated levels of PTH in combination with low levels of vitamin D [26,29]. According to this concept, calcium metabolic stress rather than vitamin D insufficiency would be an etiological factor for fetal growth impairment as a consequence of secondary hyperparathyroidism. In line with this idea, a recent study has suggested that the combined measurement of 25-hydroxyvitamin D and PTH during pregnancy could be a better determinant of fetal growth restriction [25]. In this same study, calcium levels were elevated only among pregnant women with high PTH levels in combination with low concentrations of vitamin D [25]. Therefore, calcium measurement proved to be an interesting addition to previous studies [26,29]. Finally, the synthesis of 1,25-dihydroxyvitamin D, which is the most active form of vitamin D, is closely regulated by PTH, calcium and phosphorus [3]. Hence, 25-hydroxyvitamin D deficiency could not be indicative of low levels of 1,25-dihydroxyvitamin D without considering the impact that other metabolites might have on the association [24].

The main purpose of the present research is to study the association between vitamin D deficiency during early pregnancy and PTB. Secondary outcomes of the study consist of evaluating the influence of vitamin D deficiency during pregnancy on the odds of LBW, and SGA, as well as to explore the role that metabolites related to the metabolism of this secosteroid, namely PTH, calcium, and phosphorus, could play in the associations. We hypothesized that vitamin D deficiency defined as maternal 25-hydroxyvitamin D levels below 20 ng/mL is associated with higher odds of preterm birth, low birth weight and small for gestational age, these associations being stronger among women with parathyroid hormone levels above the 80th percentile.

2. Materials and Methods

To achieve the proposed objectives, we conducted a prospective cohort study at the University Hospital Complex "Virgen de las Nieves" of Granada, Spain, a medical center with 2956 deliveries in 2019. Pregnant women were recruited from 2018 to 2019 and followed-up from weeks 10–12 of gestation to one month postpartum. This study was approved by the Ethics Committee of the University of Granada, number 72-2015, and conducted in accordance with the principles of the Declaration of

Helsinki, reviewed in Fortaleza, Brazil, in 2003. Results of the present study are reported following the STROBE statement guidelines for cohort studies [30].

2.1. Participants Data

Women were approached in their first prenatal visit at the obstetrics and gynecology services of the hospital complex. Inclusion criteria included pregnant women older than 16 years old, able to speak Spanish, and capable of signing for informed consent between 10–12 weeks of gestation determined by ultrasonography. Exclusion criteria at enrollment consisted of pregnant women with the intention to give birth in a different hospital. Other exclusion criteria consisted of women undergoing voluntary interruption of pregnancy, miscarriage, stillbirth, and multiple pregnancy. Previous history of pregnancy adverse outcomes was not an exclusion criterion for the present study.

The required sample size for the present study was calculated based on the results obtained in another study conducted by Perez-Ferre et al., who observed a prevalence of preterm birth amongst vitamin D deficient women (<20 ng/mL) of 22.9% and a prevalence of preterm birth amongst vitamin D sufficient women (>20 ng/mL) of 8.25% with a vitamin D sufficiency/deficiency ratio of 0.69 [31]. To achieve a power of 80% to detect differences in the null hypothesis H_0:p1 = p2, using χ^2 test with a confidence level of 95%, we estimated a sample size of 203 participants. Given the prospective design of the study, we estimated 20% of lost to follow-up. Hence, final calculated minimum sample size consisted of 244 participants to be included in the study.

Sociodemographic characteristics of participants were collected at recruitment by researchers from self-report and medical records. Considered variables consisted of maternal age, pre-gestational body mass index (BMI), smoking habit during pregnancy (defined as >1 or 0 cigarettes per day), parity and gravidity, history of previous pregnancy, and perinatal adverse outcomes (LBW, SGA, PTB, pre-eclampsia, gestational diabetes mellitus, miscarriage, and stillbirth), ethnicity and seasonality of sampling. Women with pre-gestational BMI >30 were classified as obese. Data regarding vitamin D supplementation at recruitment was not collected. However, Spain is a country without vitamin D supplementation policy, and vitamin D supplementation among Spanish pregnant women is uncommon in comparison with other European countries [32].

2.2. Clinical and Biochemical Procedures

Fasting maternal blood samples were obtained during the week of enrolment. Sampling was performed by venipuncture in tubes containing anticoagulant (EDTA, Ethylenediaminetetraacetic acid) and were immediately transported to the laboratory for analysis.

25-hydroxyvitamin D and intact-PTH (1–84) were quantified by microparticle chemiluminescence immunoassay (CMIA) using an Alinity I® analyzer (Abbott, Wiesbaden, Germany). Briefly, CMIA analysis is based on the use of paramagnetic microparticles coated with antibodies. Regarding 25-hydroxyvitamin D, it is first separated from the vitamin D-binding protein (DBP) to be mixed with the anti-vitamin D antibody-coated microparticle. The complex is labeled with acridinium afterwards. The reaction conjugate is incubated to be later washed-out, and the correlation between emitted chemiluminescence light measured in relative light units (RLU) and the 25-hydroxyvitamin D or intact-PTH concentration is calculated. According to the manufacturer, the method detection limit for the 25-hydroxyvitamin D assay is 3.5 ng/mL (8.85 nmol/L) and intra-assay coefficient of variation is 3.6% at 39.8 ng/mL (99.4 nmol/L) whilst the quoted PTH assay detection limit is 0.5 pg/mL (0.05 pmol) and intra-assay coefficient of variation is 2.6% at 63.8 pg/mL (6.76 pmol/L).

Calcium and phosphorus were analyzed using an Alinity C® analyzer (Abbott, Wiesbaden, Germany). Calcium was analyzed by arsenazo-III colorimetric assay measuring absorbance at 660 nm whilst phosphorus was analyzed by phosphomolybdate assay measuring absorbance at 340 nm.

Vitamin D deficiency was defined as serum 25-hydroxyvitamin D concentrations <20 ng/mL (50 nmol/L) whilst vitamin D insufficiency was defined as serum 25-hydroxyvitamin D concentrations <30 ng/mL (75 nmol/L). Used cut-off points were based on other studies [22]. Chosen cut-off points differed from those recommended by the American Institute of Medicine [33], however, optimal vitamin D cut-off points during pregnancy remain controversial and consensus on this matter has not been reached to date [34]. The seasonality of sampling was considered as a potential confounder given the existing association between sun exposure and vitamin D concentration [35]. Due to a lack of consensus, we considered elevated PTH levels as concentrations above the 80th percentile in line with another author [26]. Therefore, women with elevated PTH levels were those with PTH concentrations ≥31.9 pg/mL.

Women were followed-up in subsequent prenatal visits, and cases of pre-eclampsia and gestational diabetes mellitus were diagnosed. Values of maternal diastolic and systolic blood pressure, proteinuria, and glucose tolerance test results were collected by researchers during routine controls. Blood pressure was measured using a validated automatic tensiometer and the measurement was repeated within 15 min. De novo systolic blood pressure >140 mm/Hg and diastolic blood pressure >90 mm/Hg measurements were considered as gestational hypertension and women were further evaluated by the obstetrician. Proteinuria was defined as urine protein-to-creatinine ratio above 0.3 mg/mg and was assessed in routine controls after week 20 of gestation. Proteins in urine were quantified using benzethonium chloride turbidimetric method and creatinine was analyzed using alkaline picrate colorimetric assay. An oral glucose tolerance test was performed between weeks 24–28 of gestation. Blood glucose was analyzed using a hexokinase/glyceraldehyde 3-phosphate dehydrogenase activity assay kit.

Cases of pre-eclampsia were defined according to the International Society for the Study of Hypertension in Pregnancy (ISSHP) 2018 classification [36], and GDM cases were defined in line with the American Diabetes Association criteria [37]. Cases of miscarriage and stillbirth, type of delivery, and values of gestational age at delivery and birth weight were documented from medical records. Low birth weight was defined as live birth with less than 2500 g at delivery in accordance with the International Classification of Diseases, 10th Edition [12]. Preterm birth was defined as live birth with less than 37 weeks of gestation [38]. Small for gestational age cases were considered as live births with weight below 10th percentile for the gestational age [13] and were calculated using Spanish reference percentile charts from 2010–2014, based on gender, parity, and type of delivery [39].

2.3. Statistical Analysis

All statistical analyses were performed using the software SPSS version 25 (IBM Corp®, Armonk, NY, USA). Normality of continuous variables was examined using Kolmogorov–Smirnov test. Categorical variables were reported as percentages, and continuous variables were reported as mean ± standard deviation or median and interquartile range based on normality test results. Differences between participants depending on vitamin D cut-off points were analyzed using Pearson's chi-square (χ^2) test for categorical variables and the Mann–Whitney U test for continuous variables. The Spearman correlation test was used to evaluate the strength of association between vitamin D and concentrations of PTH, calcium, and phosphorus. A scatter plot was provided to graphically represent statistically significant correlations. For each outcome, bivariate analysis was performed to evaluate possible confounders based on the literature. Variables with p-values < 0.20 in bivariate analysis were chosen for adjustment in multivariable analysis. This cut-off is supported by the literature [40,41]. Other related variables strongly supported by the scientific literature were also considered for adjustment when applicable. Odds ratios (ORs) with 95% confidence interval (95% CI) were calculated for each chosen outcome and biomarker using bivariate and multivariable logistic regression models. In logistic regression models, parathyroid hormone, calcium and phosphorus were analyzed as continuous variables whilst vitamin D deficiency was a categorical variable (<20 ng/mL/≥20 ng/mL)

Finally, we provided binary logistic regression unadjusted and adjusted models to examine the associations between concentrations of vitamin D <20 ng/mL and <30 ng/mL along with the PTH 80th percentile and the odds of PTB, LBW, SGA in the cohort of study. A sensitivity analysis was conducted to evaluate consistency of the results using the PTH 75th percentile.

3. Results

3.1. Cohort of Study

We approached 500 women for study participation, 380 of whom signed informed consent and were enrolled in the study. After follow-up, a completed dataset from 303 women and their children was available (20.26% lost to follow-up). A final analytical sample of 289 women fulfilled inclusion criteria and was available for the present study Figure 1.

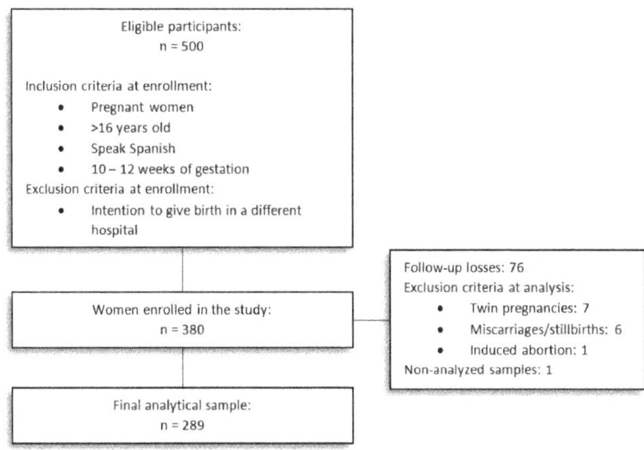

Figure 1. Flow diagram of participants.

3.2. Characteristics of Participants

The sociodemographic characteristics of participants based on vitamin D cut-off points (<20 ng/mL or $\geq$20 ng/mL), are presented in Table 1 and concentrations of calcium, phosphorus and parathyroid hormone are reported in Table 2. Results of the Kolmogorov–Smirnov test showed that maternal age, BMI, calcium, phosphorus, and PTH concentrations were non-normally distributed across vitamin D cut-off points. All expected numbers were higher than five in Pearson's χ^2 test for categorical variables. Vitamin D levels were normally distributed amongst participants. Serum 25-hydroxyvitamin D mean concentration was 22.36 $\pm$ 6.3 ng/mL. Thirty-four women had sufficient levels of vitamin D ($\geq$30 ng/mL) (11.76%), 150 were vitamin D insufficient (20–29.9 ng/mL) (51.9%), and the 105 remaining women suffered vitamin D deficiency (<20 ng/mL) (36.33%). Median maternal age was 33 (29–36) years old, whilst the median pre-pregnancy BMI was 25.1 (21.9–29.3). 52 participants were obese (18%). With respect to the history of previous pregnancy adverse outcomes, 67 women had history of miscarriage or stillbirth; one had history of pre-eclampsia; five had history of gestational diabetes mellitus; and 12 women had history of preterm birth. Regarding ethnicity, three African women were lost to follow-up, and most of the ethnic women approached did not fulfill the inclusion criteria (speak Spanish). Therefore, all women who completed the study were Caucasian. Only obesity (pre-pregnancy BMI $\geq$30), preterm birth and maternal blood parathyroid hormone concentration varied significantly across the chosen vitamin D cut-off points ($p < 0.05$).

Table 1. Characteristics of participants based on 25-hydroxyvitamin D cut-off points.

Variable	All Participants (*n* = 289)	Vitamin D < 20 ng/mL (*n* = 105)	Vitamin D ≥ 20 ng/mL (*n* = 184)	*p*-Value
Age	33 (29–36)	34 (30–35.5)	32 (28–36)	0.358
Seasonality				
Spring	42 (14.5%)	12 (11.4%)	30 (16.3%)	
Summer	21 (7.3%)	5 (4.8%)	16 (8.7%)	0.159
Autumn	216 (74.7%)	82 (78.1%)	134 (72.83%)	
Winter	10 (3.5%)	6 (5.7%)	4 (2.17%)	
Smoking				
Yes	36 (12.5%)	13 (12.4%)	23 (12.5%)	0.976
No	253 (87.5%)	92 (87.6%)	161 (87.5%)	
Obesity				
BMI ≥ 30	52 (18%)	27 (25.7%)	25 (13.59%)	0.01 *
BMI < 30	237 (82%)	78 (74.3%)	159 (86.41%)	
Parity				
Nulliparity	163 (56.4%)	54 (51.4%)	75 (40.76%)	0.198
Multiparity	126 (43.6%)	51 (48.6%)	109 (59.24%)	
Preterm birth	17 (5.9%)	10 (9.5%)	7 (3.8%)	0.047 *
Low birth weight	24 (8.3%)	13 (12.4%)	11 (6%)	0.058
Small for gestational age	27 (9.3%)	14 (13.3%)	13 (7.1%)	0.078

BMI: body mass index. Categorical data are presented as absolute frequency (percentages), and continuous data are presented as median (interquartile range). *p*-values were obtained by Pearson's χ^2 test for categorical variables and the Mann–Whitney U test for continuous variables. * *p*-value < 0.05.

Table 2. Concentrations of parathyroid hormone, calcium and phosphorus based on 25-hydroxyvitamin D cut-off points.

Metabolite	All Participants (*n* = 289)	Vitamin D < 20 ng/mL (*n* = 105)	Vitamin D ≥ 20 ng/mL (*n* = 184)	*p*-Value
Parathyroid hormone	21 (16–29.3) pg/mL	25.50 (16.6–34.6) pg/mL	19.6 (15.8–26.4) pg/mL	0.002 *
Calcium	9.2 (8.9–9.4) mg/dL	9.2 (9–9.4) mg/dL	9.2 (8.9–9.4) mg/dL	0.914
Phosphorus	3.6 (3.4–3.9) mg/dL	3.6 (3.4–3.9) mg/dL	3.6 (3.4–3.9) mg/dL	0.899

Data are presented as median (interquartile range). *p*-values were obtained using the Mann–Whitney U test for continuous variables. * *p*-value < 0.05.

The spearman correlation test showed an inverse association between vitamin D and parathyroid hormone concentrations (ρ = −0.146, *p* = 0.013). This correlation was also evident in the scatter plot in Figure 2. On the other hand, neither calcium nor phosphorus were correlated with vitamin D in the Spearman's test (calcium: ρ = 0.022, *p* = 0.705, phosphorus: ρ = −0.024, *p* = 0.689).

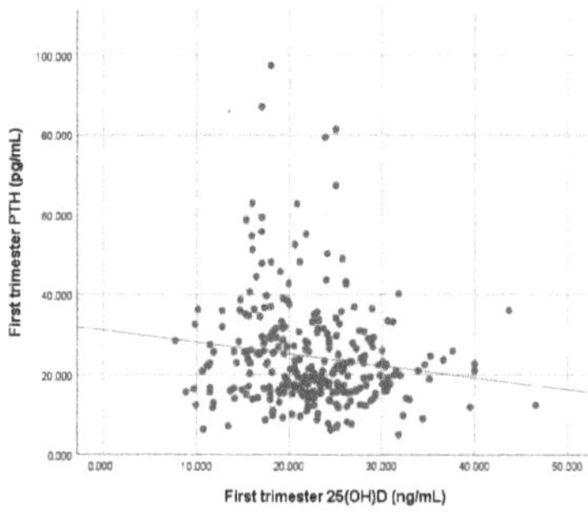

Figure 2. Scatter plot of the correlation between parathyroid hormone and vitamin D. PTH: parathyroid hormone; 25(OH)D: 25-hydroxyvtiamin D.

3.3. Pregnancy and Perinatal Adverse Outcomes

Frequencies of pregnancy and perinatal adverse outcomes observed in the present study compared to estimated global frequencies and estimated frequencies in the USA and Europe are described in Table 3. One pre-eclampsia case was a twin pregnancy, thus being excluded from further analyses. We also excluded type I and pre-gestational type II diabetes cases (four cases) when describing the frequency of gestational diabetes mellitus in the cohort of study.

Table 3. Frequency of pregnancy and perinatal adverse outcomes compared to global and regional frequencies.

Outcome	Frequency	Estimated Global Frequency	Estimated Frequency in the USA	Estimated Frequency in Europe
Preeclampsia	1.7%	4.6% (2010) [42]	3% (2010) [42]	5.3% (2010) [42]
Gestational diabetes mellitus	5.6%	16.9% (2013) [43]	4.6–9.2% (2010) [44]	15.2% (2013) [43]
Cesarean section	21.5%	31% (2011) [45]	31.9% (2018) [46]	25.2% (2010) [47]
Preterm birth	5.9%	10.6% (2014) [5]	10.2% (2018) [46]	8.7% (2014) [5]
Low birthweight	8.3%	14.6% (2015) [48]	8.28% (2018) [46]	6.5% (2015) [48]

With the exemption of LBW, frequencies of adverse outcomes in the cohort of study were lower than average estimated frequencies. Seventeen births were premature (<37 weeks of gestation) (5.9%) and 24 newborns had low birth weight (<2.500 g) (8.3%). When comparing gestational age and birth weight data with the Spanish reference percentile charts [39], we obtained a total of 27 SGA cases in the cohort of study (birth weight < 10th percentile for their gestational age) (9.34%).

3.4. Associations of Vitamin D Deficiency, PTH, Calcium, and Phosphorus with Perinatal Adverse Outcomes

In Table 4, unadjusted and adjusted logistic regression models are presented to describe associations between vitamin D deficiency (<20 ng/mL/<50 nmol/L), parathyroid hormone, calcium, phosphorus continuous concentrations, and perinatal outcomes. Covariables with p-values < 0.20 in bivariate analysis were selected for adjustment in multivariable analysis.

Table 4. Associations between Vitamin D deficiency, parathyroid hormone (PTH), calcium, phosphorus, and adverse perinatal outcomes in the cohort of study.

Outcome		Vitamin D Deficiency	p-Value	Parathyroid Hormone	p-Value	Calcium	p-Value	Phosphorus	p-Value
Preterm Birth	Unadjusted OR	2.662 (0.982–7.217)	0.054	1.030 (1.002–1.058)	0.035 *	2.024 (0.581–7.048)	0.268	1.021 (0.630–1.652)	0.934
	Adjusted OR [1]	3.529 (1.159–10.741)	0.026 *	1.027 (0.997–1.059)	0.083	1.814 (0.513–6.413)	0.355	0.764 (0.240–2.431)	0.648
Low Birth Weight	Unadjusted OR	2.222 (0.958–5.157)	0.063	1.019 (0.993–1.046)	0.156	1.572 (0.566–4.366)	0.386	0.738 (0.282–1.927)	0.535
	Adjusted OR [2]	1.586 (0.586–4.336)	0.361	1.009 (0.977–1.041)	0.597	1.212 (0.355–4.144)	0.758	0.568 (0.189–1.711)	0.315
Small for Gestational Age	Unadjusted OR	2.024 (0.912–4.488)	0.083	0.985 (0.951–1.020)	0.399	1.215 (0.488–3.022)	0.676	0.735 (0.296–1.913)	0.551
	Adjusted OR [3]	1.794 (0.786–4.093)	0.165	0.978 (0.939–1.018)	0.276	1.127 (0.435–2.923)	0.805	0.699 (0.269–1.818)	0.463

Data reported as odds ratios (OR) (95%CI). [1] Adjusted for: history of PTB and pre-eclampsia. [2] Adjusted for: maternal age, smoking habit, pre-eclampsia, and preterm birth. [3] Adjusted for: seasonality, smoking habit, and parity. * p-value < 0.05.

Maternal first-trimester vitamin D deficiency was associated with higher odds of preterm birth in bivariate analysis, but it was not statistically significant (OR = 2.662, 95% CI (0.982, 7.217), p = 0.054). Only after adjusting for history of PTB and cases of pre-eclampsia, did the association become statistically significant (OR = 3.529, 95% CI (1.159, 10.741), p = 0.026). PTH concentration and preterm birth were weakly associated only in bivariate analysis (OR = 1.030, 95% CI (1.002, 1.058), p = 0.035). Regarding birth weight, there was a trend towards higher odds of low birth weight amongst the offspring of vitamin D deficient women. However, this association was not statistically significant neither in bivariate analysis or after adjusting for confounders (OR = 2.222, 95% CI (0.958, 5.157), p = 0.06/aOR = 1.586, 95% CI (0.586, 4.336), p = 0.361). In the same fashion, the relationship between vitamin D deficiency and risk of SGA was not significant neither in crude or adjusted models (OR = 2.024, 95% CI (0.912–4.488), p = 0.083/aOR = 1.794, 95% CI (0.786–4.093), p = 0.165). We did not observe any correlation between calcium and phosphorus concentrations with perinatal outcomes.

In Table 5, we presented the associations between vitamin D deficiency and insufficiency along with the PTH 80th percentile and perinatal outcomes.

Table 5. Associations between combination of maternal serum 25-hydroxyvitamin D and PTH 80th percentile and perinatal adverse outcomes.

	Preterm Birth			Low Birth Weight			Small for GESTATIONAL Age		
	n (%)	OR	aOR [1]	n (%)	OR	aOR [2]	n (%)	OR	aOR [3]
25[OH]D ≥ 20 ng/mL (≥50 nmol/L)									
PTH > 80th	0/26 (0%)	–	–	0/26 0%	–	–	0/26 0%	–	–
PTH ≤ 80th	7/158 (4.4%)	0.561 (0.207–1.517)	0.581 (0.203–1.667)	11/158 (7%)	0.679 (0.294–1.571)	0.899 (0.333–2.432)	13/158 (8.2%)	0.749 (0.339–1.656)	0.857 (0.376–1.954)
25[OH]D < 20 ng/mL (<50 nmol/L)									
PTH > 80th	6/31 (19.4%)	5.389 (1.837–15.812) *	6.223 (1.939–19.970) *	7/31 (19.4%)	4.135 (1.560–10.963) *	2.653 (0.766–9.188)	4/31 (12.9%)	1.514 (0.487–4.705)	1.356 [0.407–4.518]
PTH ≤ 80th	4/74 (5.4%)	0.888 (0.280–2.813)	1.057 (0.313–1.357)	6/74 (8.1%)	0.966 (0.368–2.533)	0.877 (0.268–2.868)	10/74 (13.5%)	1.820 (0.793–4.175)	1.663 (0.705–3.919)
25[OH]D ≥ 30 ng/mL (≥75 nmol/L)									
PTH > 80th	0/4 (0%)	–	–	0/4 0%	–	–	0/4 0%	–	–
PTH ≤ 80th	2/30 (6.7%)	1.162 (0.253–5.346)	1.480 (0.310–7.065)	1/30 (3.3%)	0.354 (0.460–2.718)	0.257 (0.024–2.787)	1/30 (3.3%)	0.309 (0.040–2.363)	0.324 (0.041–2.548)
25[OH]D < 30 ng/mL (<75 nmol/L)									
PTH > 80th	6/53 (11.3%)	2.611 (0.920–7.411)	2.109 (0.673–6.611)	7/53 (13.2%)	1.960 (0.769–4.998)	1.402 (0.442–4.441)	4/53 (7.5%)	0.756 (0.250–2.285)	0.713 (0.226–2.251)
PTH ≤ 80th	9/202 (4.5%)	0.460 (0.172–1.236)	0.492 (0.171–1.419)	16/202 (7.9%)	0.849 (0.349–2.066)	1.188 (0.391–3.615)	22/202 (10.9%)	2.004 (0.733–5.479)	2.202 (0.772–6.181)

Data reported as OR (95%CI). OR: odds ratios. aOR: adjusted odds ratio. [1] Adjusted for pre-eclampsia and history of preterm birth. [2] Adjusted for maternal age, smoking habit, pre-eclampsia, and preterm birth. [3] Adjusted for seasonality, smoking habit, and parity. * p-value < 0.05.

Overall, first-trimester vitamin D insufficiency defined as maternal blood levels of 25-hydroxyvitamin D < 30 ng/mL along with levels of PTH above the 80th percentile correlated with prematurity, but the association was not statistically significant (OR = 2.611, 95% CI (0.92, 7.411), p = 0.071). However, vitamin D deficiency (<20 ng/mL) during the first trimester of gestation was strongly associated with the odds of PTB amongst women with PTH levels above the 80th percentile (OR = 5.389, 95% CI (1.837, 15.812), p = 0.002). Furthermore, this association remained evident after adjusting for preterm birth confounders (aOR = 6.587, 95% CI (2.049, 21.176), p = 0.002). Vitamin D concentrations ≥ 20 ng/mL and PTH levels ≤ 80th percentile did not correlate with PTB (p > 0.05).

Low birth weight was more prevalent amongst women with vitamin D levels <20 ng/mL in combination with PTH levels > 80th percentile (OR = 4.135, 95% CI (1.560, 10.963), p = 0.004). However, this association was rendered statistically non-significant after adjusting for confounders (aOR = 2.653, 95% CI (0.766, 9.188), p = 0.124).

Finally, we did not find any association between SGA and vitamin D deficiency or insufficiency along with the 80th PTH percentile neither in crude nor adjusted models. Sensitivity analyses using the 75th PTH percentile (≥29.25 pg/mL) were performed to evaluate the consistency of the results Table S1. Overall, associations between studied outcomes and combinations of vitamin D deficiency/insufficiency with the 75th PTH percentile were similar to those shown in the main analysis.

4. Discussion

The literature about deficiency of vitamin D and perinatal outcomes is inconsistent, and several authors have suggested that interactions with metabolites linked to the metabolism of vitamin D could play an important role in the associations [24–26]. We conducted a prospective cohort study with 289 pregnant women recruited between weeks 10–12 of gestation in a hospital of Granada, Spain, and associations between 25-hydroxyvitamin D, PTH, calcium, phosphorus, and perinatal adverse outcomes, namely preterm birth, low birth weight and small for gestational age were evaluated. We found a trend towards lower maternal 25-hydroxyvitamin D serum levels in the first trimester of gestation and higher odds of preterm birth. This association was stronger amongst women with elevated levels of PTH (>80th percentile), and it was not attenuated after adjusting for preterm birth confounders. Although a similar association was observed for low birth weight, it was not statistically significant after confounder adjustment. SGA was defined based on weight and weeks of gestation at delivery from Spanish percentile charts [39] did not correlate either with vitamin D or related metabolites.

With the exemption of low birth weight, the prevalence of pregnancy and perinatal adverse outcomes was lower than average estimates in Europe. Preeclampsia is a strong contributor to preterm birth [49]. The small number of preeclampsia cases could partially explain the low preterm birth cases observed in the cohort of study.

4.1. Limitations of the Study

The present study has some limitations to be acknowledged. The prevalence of the main outcome of the study, preterm birth, was more than 30% lower than average estimates in Europe. This could be the cause of the lack of significance observed in the association between vitamin D deficiency and preterm birth and might compromise extrapolation of our results to other populations. Regarding secondary outcomes (SGA and LBW), it is possible that the lack of statistical significance of associations could be consequence of sample size limitations given that they were not included in sample size calculations. We did not use liquid chromatography-tandem mass spectrometry (LC-MS/MS), which is considered the gold-standard method by most authors to analyze 25-hydroxyvitamin D. Due to equipment limitations, we did not directly measure ionized calcium and we could not determine albumin levels thus we were not able to estimate ionized calcium concentration which is the most active form of calcium. Additionally, we could not measure other important bone turnover biomarkers such as alkaline phosphatase which would be of interest when assessing associations between 25-hydroxyvitamin D and PTH. Almost 75% of the samples were obtained during autumn, and all participants were Caucasian. Therefore, it was not possible to adjust the results for ethnicity, and seasonality adjustment could be inaccurate. These are important factors that can potentially influence maternal vitamin D blood levels [22].

4.2. Deficiency of Vitamin D and Preterm Birth

Spain is a Mediterranean country with high levels of sun exposure. Despite this fact, vitamin D deficiency is highly prevalent among Spanish pregnant women [50]. This situation is known as the "Mediterranean paradox," and it has been estimated that 41%–90% of all pregnant women living in Mediterranean countries have vitamin D levels below sufficiency [51]. In line with this data, only 11.76% of study participants had sufficient vitamin D levels (>30 ng/mL) whilst more than one-third of the women had levels below 20 ng/mL, which implies a high prevalence of vitamin D deficiency amongst participants. The observed ratio of vitamin D sufficiency/insufficiency is consistent with the results obtained by Perez-Ferre et al., who conducted a prospective cohort study in 266 pregnant women during weeks 24–28 of gestation in Madrid, Spain, finding a significant association between vitamin D deficiency and preterm birth using the same vitamin D cut-off points, in both unadjusted and adjusted logistic regression models (OR = 3.31, 95% CI (1.52, 7.19), $p = 0.002$/aOR = 3.80, 95% CI (1.32, 10.97), $p = 0.013$) [31]. However, in the present study, we only observed a statistically significant association between vitamin D deficiency (<20 ng/mL) and preterm birth after adjusting for confounders with statistical significance in the univariate model ($p < 0.20$). Differences between both studies could be attributed to our significantly smaller number of PTB cases and different sampling time. Another study conducted in a Spanish cohort of 2382 pregnant women could not find any association between 25-hydroxyvitamin D and perinatal outcomes, including PTB and SGA. However, almost 50% of the participants had sufficient levels of vitamin D, which implies a low rate of vitamin D insufficiency in comparison with average estimates [52].

Using similar study designs, several authors have explored the link between vitamin D deficiency and prematurity in other countries yielding negative results [10,53], whilst other studies have found a positive association [11,54]. Authors of these studies state the necessity of conducting well-designed randomized clinical trials to further clarify this subject. However, meta-analyses of randomized clinical trials have failed to verify an association between vitamin D supplementation and lower odds of preterm birth [4,55]. In this sense, randomized clinical trials conducted to date not only have to face important ethical issues but also lack relevant criteria related to nutrients studies [56]. One important criterium that is usually overlooked is the optimization of the status of associated nutrients in order to ensure the causality of observed associations [57].

4.3. Vitamin D Associated Metabolites and Perinatal Outcomes

Vitamin D regulates calcium and phosphorus homeostasis, and its production is controlled by PTH [58]. Santorelli et al. measured 25-hydroxyvitamin D, PTH, and calcium in a heterogeneous population composed of 1010 pregnant women differentiating between white and Pakistani participants. They observed that higher calcium levels were associated with lower odds of PTB amongst white participants, whilst vitamin D exerted a protective effect on the overall risk of SGA. However, none of the studied metabolites were associated with SGA in white participants [17]. In the present study, we did not observe any significant association between calcium and preterm birth in Caucasian pregnant women. Nonetheless, due to sample limitations, we were not able to examine the impact that ethnicity could have on the analyses.

Other authors have explored the concept of functional vitamin D deficiency in pregnancy as a cause of calcium metabolic stress, which could ultimately lead to perinatal adverse outcomes associated with the deficiency of this secosteroid. This concept has been applied to examine the association between vitamin D deficiency, gestational hypertensive disorders, and fetal growth restriction [26,29]. Scholl et al. observed a higher incidence of SGA cases amongst pregnant women with PTH > 62 pg/mL in combination with 25-hydroxyvitamin D < 20 ng/mL or calcium intakes below 60% of the estimated average requirement (OR = 2.23, 95% CI (1.23, 4.33)) [29]. In the same line, Hemmingway et al. found a 2.38-fold increased risk of SGA amongst pregnant women with serum 25-hydroxyvitamin D levels < 12 ng/mL (<30 nmol/L) in combination with PTH > 80th percentile in the cohort of study (RR = 2.38, 95% CI (1.31, 4.33)). However, this association was not statistically significant after confounder adjustment [26]. More recently, Meng et al. prospectively measured PTH, calcium, and 25-hydroxyvitamin D in 3407 participants in China, finding that maternal 25-hydroxyvitamin D levels <12 ng/mL and <20 ng/mL (<50 nmol/L) along with PTH concentrations >75th percentile were associated with increased risk of SGA and lower mean birth weight compared to vitamin D sufficient women. This association was not attenuated in sensitivity analyses (PTH > 80th percentile) [25]. On the other hand, Tao et al. evaluated the effect of the duration of vitamin D supplementation (400–600 IU/d) on fetal growth, finding a direct association between more prolonged vitamin D supplementation and higher weeks of gestation and weight at delivery independently of calcium and phosphorus concentrations [59]. In the present study, we found a correlation between low birth weight and vitamin D < 20 ng/mL in combination with high levels of PTH (>80th). However, this association was not significant after adjustment for confounders, which implies that gestational age at delivery was the main underlying factor for the association. In the same fashion, the risk of SGA was not correlated with vitamin D or PTH in any subgroup analysis. Nonetheless, we observed that women with PTH levels > 80th percentile and 25-hydroxyvitamin D < 20 ng/mL had more than five times higher odds of PTB compared to the reference group, and this relationship persisted after adjusting for confounders. These results were consistent with those obtained in the sensitivity analysis using the 75th PTH percentile instead Table S1. It is possible that vitamin D deficiency could exert an effect on birth weight by influencing the length of gestation [23]. Finally, neither calcium nor phosphorus concentrations were associated with any studied outcome.

Our results do not support the hypothesis that elevated levels of PTH in combination with vitamin D deficiency are associated with fetal growth restriction. However, reference levels for PTH during pregnancy are not firmly established, and SGA is defined depending on specific reference charts and, thus, results could not be extrapolated to other populations.

5. Conclusions

In the present study, we observed that vitamin D deficiency defined as 25-hydroxyvitamin D concentrations below 20 ng/mL, in combination with parathyroid hormone maternal levels above the 80th percentile during the first trimester of gestation, was a better estimator of preterm birth than the assessment of vitamin D deficiency in isolation. However, we did not observe the same association with low birth weight after controlling for weeks of gestation or small for gestational age. Interventional studies with vitamin D supplementation would benefit from measuring parathyroid hormone in order to demonstrate a potential causal association between deficiency of vitamin D and perinatal adverse outcomes.

Supplementary Materials: The following are available online at http://www.mdpi.com/2072-6643/12/11/3279/s1, Table S1: Associations between combination of maternal serum 25-hydroxyvitamin D and PTH 75th percentile and perinatal adverse outcomes.

Author Contributions: Conceptualization, M.J.A.-C.; formal analysis, Í.M.P.-C.; investigation, M.J.A.-C., Í.M.P.-C., X.A.L.-R., T.R.-B.; writing—original draft preparation, Í.M.P.-C.; writing—review and editing, M.J.A.-C., T.R.-B., X.A.L.-R., M.E.-R., M.S.L.-C.; supervision, M.S.L.-C., M.J.A.-C.; project administration, M.J.A.-C.; funding acquisition, M.J.A.-C. All authors have read and agreed to the published version of the manuscript.

Funding: Research reported in this publication was funded by the Spanish Ministry of Science, Innovation, and Universities (Project FIS-ISCIII, PI17/02305) and co-founded by FEDER, "investing in your future".

Acknowledgments: We thank the participants for their collaboration in the present study, which made possible this research. We also thank the gynecologists and midwives of the hospital complex "Virgen de las Nieves" of Granada for their support in this project. This paper is part of the doctoral thesis of Í.M.P.-C. in the Clinical Medicine and Public Health Doctoral Program of the University of Granada, Spain.

Conflicts of Interest: The authors declare no conflict of interest. The funders had no role in the design of the study; in the collection, analyses, or interpretation of data; in the writing of the manuscript, or in the decision to publish the results.

References

1. Holick, M.F. The Vitamin D Deficiency Pandemic: Approaches for diagnosis, treatment and prevention. *Rev. Endocr. Metab. Disord.* **2017**, *18*, 153–165. [CrossRef] [PubMed]
2. Roth, D.E.; Abrams, S.A.; Aloia, J.; Bergeron, G.; Bourassa, M.W.; Brown, K.H.; Calvo, M.S.; Cashman, K.D.; Combs, G.; De-Regil, L.M.; et al. Global prevalence and disease burden of vitamin D deficiency: A roadmap for action in low- and middle-income countries. *Ann. N. Y. Acad. Sci.* **2018**. [CrossRef] [PubMed]
3. Wacker, M.; Holiack, M.F. Vitamin D-effects on skeletal and extraskeletal health and the need for supplementation. *Nutrients* **2013**, *5*, 111–148. [CrossRef] [PubMed]
4. Pérez-López, F.R.; Pasupuleti, V.; Mezones-Holguin, E.; Benites-Zapata, V.A.; Thota, P.; Deshpande, A.; Hernandez, A.V. Effect of vitamin D supplementation during pregnancy on maternal and neonatal outcomes: A systematic review and meta-analysis of randomized controlled trials. *Fertil. Steril.* **2015**, *103*, 1278–1288. [CrossRef]
5. Chawanpaiboon, S.; Vogel, J.P.; Moller, A.-B.; Lumbiganon, P.; Petzold, M.; Hogan, D.; Landoulsi, S.; Jampathong, N.; Kongwattanakul, K.; Laopaiboon, M.; et al. Global, regional, and national estimates of levels of preterm birth in 2014: A systematic review and modelling analysis. *Lancet Glob. Health* **2019**, *7*, e37–e46. [CrossRef]

6. Goldenberg, R.L.; Culhane, J.F.; Iams, J.D.; Romero, R. Epidemiology and causes of preterm birth. *Lancet* **2008**, *371*, 75–84. [CrossRef]

7. Wei, S.-Q.; Qi, H.-P.; Luo, Z.-C.; Fraser, W.D. Maternal vitamin D status and adverse pregnancy outcomes: A systematic review and meta-analysis. *J. Matern. Fetal. Neonatal Med.* **2013**, *26*, 889–899. [CrossRef]

8. Agarwal, S.; Kovilam, O.; Agrawal, D.K. Vitamin D and its impact on maternal-fetal outcomes in pregnancy: A critical review. *Crit. Rev. Food Sci. Nutr.* **2018**, *58*, 755–769. [CrossRef]

9. Chung, C.; Silwal, P.; Kim, I.; Modlin, R.L.; Jo, E.K. Vitamin D-Cathelicidin axis: At the crossroads between protective immunity and pathological inflammation during infection. *Immune Netw.* **2020**, *20*, 1–26. [CrossRef]

10. Morgan, C.; Dodds, L.; Langille, D.B.; Weiler, H.A.; Armson, B.A.; Forest, J.-C.; Giguere, Y.; Woolcott, C.G. Cord blood vitamin D status and neonatal outcomes in a birth cohort in Quebec, Canada. *Arch. Gynecol. Obstet.* **2016**, *293*, 731–738. [CrossRef]

11. Chen, Y.-H.; Fu, L.; Hao, J.-H.; Wang, H.; Zhang, C.; Tao, F.-B.; Xu, D.-X. Influent factors of gestational vitamin D deficiency and its relation to an increased risk of preterm delivery in Chinese population. *Sci. Rep.* **2018**, *8*, 3608. [CrossRef]

12. World Health Organization. *ICD-10: International Statistical Classification of Diseases and Related Health Problems: Tenth Revision*, 2nd ed.; World Health Organization: Geneva, Switzerland, 2004.

13. WHO Expert Committee on Physical Status. The use and interpretation of anthropometry (1993: Geneva, Switzerland) & World Health Organization. In *Physical Status: The Use of and Interpretation of Anthropometry, Report of A WHO Expert Committee*; World Health Organization: Geneva, Switzerland, 1995; Available online: https://apps.who.int/iris/handle/10665/37003 (accessed on 20 July 2020).

14. Chen, Y.H.; Fu, L.; Hao, J.H.; Yu, Z.; Zhu, P.; Wang, H.; Xu, Y.Y.; Zhang, C.; Tao, F.B.; Xu, D.X. Maternal vitamin D deficiency during pregnancy elevates the risks of small for gestational age and low birth weight infants in Chinese population. *J. Clin. Endocrinol. Metab.* **2015**, *100*, 1912–1919. [CrossRef] [PubMed]

15. Wang, H.; Xiao, Y.; Zhang, L.; Gao, Q. Maternal early pregnancy vitamin D status in relation to low birth weight and small-for-gestational-age offspring. *J. Steroid Biochem. Mol. Biol.* **2018**, *175*, 146–150. [CrossRef]

16. Monier, I.; Baptiste, A.; Tsatsaris, V.; Senat, M.V.; Jani, J.; Jouannic, J.M.; Winer, N.; Elie, C.; Souberbielle, J.C.; Zeitlin, J.; et al. First trimester maternal vitamin D status and risks of preterm birth and small-for-gestational age. *Nutrients* **2019**, *11*, 3042. [CrossRef]

17. Santorelli, G.; Whitelaw, D.; Farrar, D.; West, J.; Lawlor, D.A. Associations of maternal vitamin D, PTH and calcium with hypertensive disorders of pregnancy and associated adverse perinatal outcomes: Findings from the born in Bradford cohort study. *Sci. Rep.* **2019**, *9*, 1–9. [CrossRef]

18. Bodnar, L.M.; Catov, J.M.; Zmuda, J.M.; Cooper, M.E.; Parrott, M.S.; Roberts, J.M.; Marazita, M.L.; Simhan, H.N. Maternal serum 25-hydroxyvitamin D concentrations are associated with small-for-gestational age births in white women. *J. Nutr.* **2010**, *21*, 999–1006. [CrossRef]

19. Qin, L.-L.; Lu, F.-G.; Yang, S.-H.; Xu, H.-L.; Luo, B.-A. Does maternal vitamin D deficiency increase the risk of preterm birth: A meta-analysis of observational studies. *Nutrients* **2016**, *8*, 301. [CrossRef]

20. Zhou, S.-S.; Tao, Y.-H.; Huang, K.; Zhu, B.-B.; Tao, F.-B. Vitamin D and risk of preterm birth: Up-to-date meta-analysis of randomized controlled trials and observational studies. *J. Obstet. Gynaecol. Res.* **2017**, *43*, 247–256. [CrossRef] [PubMed]

21. Chen, Y.; Zhu, B.; Wu, X.; Li, S.; Tao, F. Association between maternal vitamin D deficiency and small for gestational age: Evidence from a meta-analysis of prospective cohort studies. *BMJ Open* **2017**, *7*, e016404. [CrossRef]

22. Aguilar-cordero, M.J.; Lasserrot-cuadrado, A.; Mur-villar, N.; León-ríos, X.A. Vitamin D, preeclampsia and prematurity: A systematic review and meta-analysis of observational and interventional studies. *Midwifery* **2020**, *87*, 102707. [CrossRef] [PubMed]

23. Maugeri, A.; Barchitta, M.; Blanco, I.; Agodi, A. Effects of vitamin D supplementation during pregnancy on birth size: A systematic review and meta-analysis of randomized controlled trials. *Nutrients* **2019**, *11*, 442. [CrossRef] [PubMed]

24. Dalmar, A.; Raff, H.; Chauhan, S.P.; Singh, M.; Siddiqui, D.S. Serum 25-hydroxyvitamin D, calcium, and calcium-regulating hormones in preeclamptics and controls during first day postpartum. *Endocrine* **2015**, *48*, 287–292. [CrossRef]

25. Meng, D.H.; Zhang, Y.; Hu, H.L.; Li, J.J.; Yin, W.J.; Ma, S.S.; Tao, R.X.; Zhu, P. The role of PTH during pregnancy on the relationship between maternal vitamin D deficiency and fetal growth restriction: A prospective birth cohort study. *Br. J. Nutr.* **2020**, *124*, 432–439. [CrossRef] [PubMed]

26. Hemmingway, A.; Kenny, L.C.; Malvisi, L.; Kiely, M.E. Exploring the concept of functional vitamin D deficiency in pregnancy: Impact of the interaction between 25-hydroxyvitamin D and parathyroid hormone on perinatal outcomes. *Am. J. Clin. Nutr.* **2018**, *108*, 821–829. [CrossRef]

27. Buppasiri, P.; Lumbiganon, P.; Thinkhamrop, J.; Ngamjarus, C.; Laopaiboon, M.; Medley, N. Calcium supplementation (other than for preventing or treating hypertension) for improving pregnancy and infant outcomes. *Cochrane Database Syst. Rev.* **2015**, *2*, CD007079. [CrossRef] [PubMed]

28. Hofmeyr, G.J.; Lawrie, T.A.; Atallah, A.N.; Duley, L.; Torloni, M.R. Calcium supplementation during pregnancy for preventing hypertensive disorders and related problems. *Cochrane Database Syst. Rev.* **2014**, *6*, CD001059. [CrossRef]

29. Scholl, T.O.; Chen, X.; Stein, T.P. Maternal calcium metabolic stress and fetal growth. *Am. J. Clin. Nutr.* **2014**, *99*, 918–925. [CrossRef]

30. Von Elm, E.; Altman, D.G.; Egger, M.; Pocock, S.J.; Gøtzsche, P.C.; Vandenbrouckef, J.P. The strengthening the reporting of observational studies in epidemiology (STROBE) statement: Guidelines for reporting observational studies. *Bull. World Health Organ.* **2007**, *85*, 867–872. [CrossRef]

31. Perez-Ferre, N.; Torrejon, M.J.; Fuentes, M.; Fernandez, M.D.; Ramos, A.; Bordiu, E.; Del Valle, L.; Rubio, M.A.; Bedia, A.R.; Montañez, C.; et al. Association of low serum 25-hydroxyvitamin D levels in pregnancy with glucose homeostasis and obstetric and newborn outcomes. *Endocr. Pract.* **2012**, *18*, 676–684. [CrossRef]

32. Oliver, E.M.; Grimshaw, K.E.C.; Schoemaker, A.A.; Keil, T.; McBride, D.; Sprikkelman, A.B.; Ragnarsdottir, H.S.; Trendelenburg, V.; Emmanouil, E.; Reche, M.; et al. Dietary habits and supplement use in relation to national pregnancy recommendations: Data from the EuroPrevall birth cohort. *Matern. Child Health J.* **2014**, *18*, 2408–2425. [CrossRef]

33. IOM (Institute of Medicine). *Dietary Reference Intakes for Calcium and Vitamin D*; The National Academies Press: Washington, DC, USA, 2011.

34. Giustina, A.; Adler, R.A.; Binkley, N.; Bollerslev, J.; Bouillon, R.; Dawson-Hughes, B.; Ebeling, P.R.; Feldman, D.; Formenti, A.M.; Lazaretti-Castro, M.; et al. Consensus Statement from 2nd International Conference on Controversies in Vitamin D. *Rev. Endocr. Metab. Disord.* **2020**, *21*, 89–116. [CrossRef] [PubMed]

35. Karras, S.N.; Fakhoury, H.; Muscogiuri, G.; Grant, W.B.; van den Ouweland, J.M.; Colao, A.M.; Kotsa, K. Maternal vitamin D levels during pregnancy and neonatal health: Evidence to date and clinical implications. *Ther. Adv. Musculoskelet. Dis.* **2016**, *8*, 124–135. [CrossRef] [PubMed]

36. Brown, M.A.; Magee, L.A.; Kenny, L.C.; Karumanchi, S.A.; McCarthy, F.P.; Saito, S.; Hall, D.R.; Warren, C.E.; Adoyi, G.; Ishaku, S. The hypertensive disorders of pregnancy: ISSHP classification, diagnosis & management recommendations for international practice. *Pregnancy Hypertens.* **2018**, *13*, 291–310. [CrossRef] [PubMed]

37. Gavin, J.R.; Alberti, K.G.M.M.; Davidson, M.B.; DeFronzo, R.A.; Drash, A.; Gabbe, S.G.; Genuth, S.; Harris, M.I.; Kahn, R.; Keen, H.; et al. Report of the expert committee on the diagnosis and classification of diabetes mellitus. *Diabetes Care* **2002**, *25*, 5–20. [CrossRef]

38. World Health Organization. WHO: Recommended definitions, terminology and format for statistical tables related to the perinatal period and use of a new certificate for cause of perinatal deaths. *Acta Obstet. Gynecol. Scand.* **1977**, *56*, 247–253.

39. Terán, J.M.; Varea, C.; Bernis, C.; Bogin, B.; González-González, A. Nuevas curvas de peso al nacer por paridad y tipo de parto para la población Española. *Gac. Sanit.* **2017**, *31*, 116–122. [CrossRef] [PubMed]

40. Hosmer, D.W.; Lemeshow, S. *Aplied Logistic Regression*, 2nd ed.; John Wiley & Sons: Hoboken, NJ, USA, 2000.

41. Zhang, Z. Model building strategy for logistic regression: Purposeful selection. *Ann. Transl. Med.* **2016**, *4*, 4–10. [CrossRef]

42. Abalos, E.; Cuesta, C.; Grosso, A.L.; Chou, D.; Say, L. Global and regional estimates of preeclampsia and eclampsia: A systematic review. *Eur. J. Obstet. Gynecol. Reprod. Biol.* **2013**, *170*, 1–7. [CrossRef]

43. Guariguata, L.; Linnenkamp, U.; Beagley, J.; Whiting, D.R.; Cho, N.H. Global estimates of the prevalence of hyperglycaemia in pregnancy. *Diabetes Res. Clin. Pract.* **2014**, *103*, 176–185. [CrossRef]

44. DeSisto, C.L.; Kim, S.Y.; Sharma, A.J. Prevalence estimates of gestational diabetes mellitus in the United States, pregnancy risk assessment monitoring system (PRAMS), 2007–2010. *Prev. Chronic Dis.* **2014**, *11*, 1–9. [CrossRef]

45. Vogel, J.P.; Betrán, A.P.; Vindevoghel, N.; Souza, J.P.; Torloni, M.R.; Zhang, J.; Tunçalp, Ö.; Mori, R.; Morisaki, N.; Ortiz-Panozo, E.; et al. Use of the Robson classification to assess caesarean section trends in 21 countries: A secondary analysis of two WHO multicountry surveys. *Lancet Glob. Health* **2015**, *3*, e260–e270. [CrossRef]

46. Martin, J.A.; Hamilton, B.E.; Osterman, M.J.K.; Driscoll, A.K. Births: Final data for 2018. *Natl. Vital Stat. Rep.* **2019**, *68*, 1980–2018.

47. EURO-PERISTAT Project with SCPE and EUROCAT. European Perinatal Health Report. The Health and Care of Pregnant Women and Babies in Europe in 2010. Available online: www.europeristat.com (accessed on 20 July 2020).

48. Blencowe, H.; Krasevec, J.; de Onis, M.; Black, R.E.; An, X.; Stevens, G.A.; Borghi, E.; Hayashi, C.; Estevez, D.; Cegolon, L.; et al. National, regional, and worldwide estimates of low birthweight in 2015, with trends from 2000: A systematic analysis. *Lancet Glob. Health* **2019**, *7*, e849–e860. [CrossRef]

49. Jeyabalan, A. Epidemiology of preeclampsia: Impact of obesity. *Nutr. Rev.* **2013**, *71*, 18–25. [CrossRef]

50. Rodríguez-Dehli, A.C.; Galán, I.R.; Fernández-Somoano, A.; Navarrete-Muñoz, E.M.; Espada, M.; Vioque, J.; Tardón, A. Prevalencia de deficiencia e insuficiencia de vitamina D y factores asociados en mujeres embarazadas del norte de España. *Nutr. Hosp.* **2015**, *31*, 1633–1640. [CrossRef]

51. Karras, S.; Paschou, S.A.; Kandaraki, E.; Anagnostis, P.; Annweiler, C.; Tarlatzis, B.C.; Hollis, B.W.; Grant, W.B.; Goulis, D.G. Hypovitaminosis D in pregnancy in the mediterranean region: A systematic review. *Eur. J. Clin. Nutr.* **2016**, *70*, 979–986. [CrossRef] [PubMed]

52. Rodriguez, A.; García-Esteban, R.; Basterretxea, M.; Lertxundi, A.; Rodríguez-Bernal, C.; Iñiguez, C.; Rodriguez-Dehli, C.; Tardōn, A.; Espada, M.; Sunyer, J.; et al. Associations of maternal circulating 25-hydroxyvitamin D3 concentration with pregnancy and birth outcomes. *BJOG Int. J. Obstet. Gynaecol.* **2015**, *122*, 1695–1704. [CrossRef]

53. Choi, R.; Kim, S.; Yoo, H.; Cho, Y.Y.; Kim, S.W.; Chung, J.H.; Oh, S.; Lee, S.-Y. High Prevalence of Vitamin D Deficiency in Pregnant Korean Women: The First trimester and the winter season as risk factors for vitamin D deficiency. *Nutrients* **2015**, *7*, 3427–3448. [CrossRef]

54. Miliku, K.; Vinkhuyzen, A.; Blanken, L.M.; McGrath, J.J.; Eyles, D.W.; Burne, T.H.; Hofman, A.; Tiemeier, H.; Steegers, E.A.; Gaillard, R.; et al. Maternal vitamin D concentrations during pregnancy, fetal growth patterns, and risks of adverse birth outcomes. *Am. J. Clin. Nutr.* **2016**, *103*, 1514–1522. [CrossRef]

55. Palacios, C.; Kostiuk, L.K.; Peñas-Rosas, J.P. Vitamin D supplementation for women during pregnancy. *Cochrane Database Syst. Rev.* **2019**, *7*, CD008873. [CrossRef]

56. Wagner, C.L.; Hollis, B.W. The implications of vitamin D Status during pregnancy on mother and her developing child. *Front. Endocrinol. (Lausanne)* **2018**, *9*, 500. [CrossRef] [PubMed]

57. Heaney, R.P. Guidelines for optimizing design and analysis of clinical studies of nutrient effects. *Nutr. Rev.* **2014**, *72*, 48–54. [CrossRef] [PubMed]

58. Barrera, D.; Díaz, L.; Noyola-Martínez, N.; Halhali, A. Vitamin D and inflammatory cytokines in healthy and preeclamptic pregnancies. *Nutrients* **2015**, *7*, 6465–6490. [CrossRef] [PubMed]

59. Tao, R.X.; Meng, D.H.; Li, J.J.; Tong, S.L.; Hao, J.H.; Huang, K.; Tao, F.B.; Zhu, P. Current recommended vitamin D prenatal supplementation and fetal growth: Results from the China-Anhui birth cohort study. *J. Clin. Endocrinol. Metab.* **2018**, *103*, 244–252. [CrossRef] [PubMed]

Publisher's Note: MDPI stays neutral with regard to jurisdictional claims in published maps and institutional affiliations.

Review

Effects of Vitamin D Supplementation on Lipid Profile in Adults with the Metabolic Syndrome: A Systematic Review and Meta-Analysis of Randomized Controlled Trials

Fatme AlAnouti [1,*], Myriam Abboud [1], Dimitrios Papandreou [1], Nadine Mahboub [2,3], Suzan Haidar [2] and Rana Rizk [4]

[1]   Department of Health Sciences, College of Natural and Health Sciences, Zayed University, Abu Dhabi 144534, UAE; myriam.abboud@zu.ac.ae (M.A.); dimitrios.papandreou@zu.ac.ae (D.P.)

[2]   Department of Nutrition and Food Sciences, Faculty of Arts and Sciences, Lebanese International University, Beirut 14404, Lebanon; nadine.baltagi@liu.edu.lb (N.M.); suzan.haidar@liu.edu.lb (S.H.)

[3]   Department of Health Promotion, Faculty of Health, Medicine and Life Sciences, Maastricht University, 6200 MD Maastricht, The Netherlands

[4]   Institut National de Santé Publique, d'Épidémiologie Clinique et de Toxicologie (INSPECT-Lb), Beirut 14404, Lebanon; rana.rizk@inspect-lb.org

*   Correspondence: Fatme.Alanouti@zu.ac.ae

Received: 14 October 2020; Accepted: 29 October 2020; Published: 30 October 2020

Abstract: Background: Metabolic syndrome (MetS) increases the risk of cardiovascular disease, with atherogenic dyslipidemia being a major contributing factor. Methods: A systematic review was conducted following the Preferred Reporting Items for Systematic Reviews and Meta-Analyses (PRISMA) statement to assess whether vitamin D supplementation (VDS) alleviates dyslipidemia in adults with MetS. Scientific databases (PUBMED, MEDLINE, CINAHL, EMBASE, Cochrane Library, ClinicalTrials.gov, International Clinical Trials Registry Platform) and the gray literature were searched for randomized controlled trials of VDS, reporting on blood lipids. A narrative review, meta-analyses, sensitivity analyses, and appraisal of the risk of bias and overall quality of evidence produced were conducted. Results: Seven studies were included, and four were meta-analyzed. The risk of bias was generally low, and the final quality of evidence was low or very low. VDS, whether in high or low dose, significantly increased endline vitamin D blood levels; did not affect total, low-density, high-density cholesterol levels, and novel lipid-related biomarkers; yet, significantly increased triglycerides (TG) levels compared with placebo (MD: 30.67 (95%CI: 4.89–56.45) mg/dL; $p = 0.02$ for low-dose VDS; and MD: 27.33 (95%CI: 2.06–52.59) mg/dL; $p = 0.03$ for high-dose VDS). Pertaining heterogeneity was high ($I^2 = 86\%$; and $I^2 = 51\%$, respectively), and some included studies had significantly higher baseline TG in the intervention arm. The sensitivity analyses revealed robust results. Conclusion: VDS seems not to affect blood lipids in adults with MetS.

Keywords: vitamin D supplementation; metabolic syndrome; dyslipidemia; cholesterol; triglycerides; adult; systematic review; meta-analysis

---

1. Introduction

Metabolic syndrome (MetS) is a conglomeration of cardiometabolic disorders that collectively increases a person's risk for developing type 2 diabetes mellitus (T2DM) and cardiovascular disease (CVD) [1,2]. Over the last two decades, the number of people diagnosed with MetS has increased considerably, encompassing 20% to 25% of the adult population and presenting an enormous public health issue [3,4].

The precise definition of MetS varies slightly between guidelines issued by expert groups including the World Health Organization (WHO); the National Cholesterol Education Program Third Adult Treatment Panel (NCEP ATP III); the International Diabetes Federation (IDF); and the American Heart Association/National Heart, Lung, and Blood Institute [5]. Yet, the core components of this syndrome consist of glucose intolerance, hypertension, dyslipidemia—specifically, reduced high-density lipoprotein cholesterol (HDL-C), elevated triglycerides (TG), and central obesity [6].

Individuals with MetS are at an increased risk for CVD, with atherogenic dyslipidemia (low HDL-C and hypertriglyceridemia) being a major underlying cause for its development [7]. Atherogenic dyslipidemia emerges as the greatest competitor of low-density lipoprotein cholesterol (LDL-C) among lipid risk factors for CVD [8–10]. Achieving a better understanding of this atherogenic dyslipidemia and factors associated with it may provide clues and further insight into possible interventions that may reduce the risk of CVD in this patient population [5,10].

Vitamin D supplementation (VDS) is among those interventions suggested to alleviate atherogenic dyslipidemia in patients with MetS [11]. Vitamin D is a fat-soluble vitamin that has an integral role in skeletal and immune system disorders [12], along with numerous metabolic functions, including glucose homeostasis, insulin regulation of body weight, and a potent modifier of cardiovascular risk [13]. Vitamin D deficiency, or low levels of 25-hydroxyvitamin D, is associated with a higher risk of MetS. Additionally, suboptimal levels of the vitamin may increase the severity of the syndrome [14,15]. Concentrations of 25-hydroxyvitamin D are lower in patients with MetS compared with those without it [16], and the prevalence of MetS is reduced by half if individuals have high 25-hydroxyvitamin D concentrations [17]. Specifically, vitamin D might modulate the atherogenic components of MetS. A significant inverse relationship has been observed between higher levels of serum 25-hydroxyvitamin D and hypertriglyceridemia, in addition to a positive association with HDL-C [18–20]. Nevertheless, some studies report a controversial association between low levels of 25-hydroxyvitamin D and MetS and its individual components [21,22]. Numerous randomized controlled trials (RCTs) have investigated the effect of VDS on dyslipidemia among patients with MetS and found conflicting results [16,23,24]. Therefore, the aim of this systematic review and meta-analysis is to summarize the available evidence of RCTs to establish the impact of VDS on dyslipidemia among adult patients with MetS.

2. Materials and Methods

2.1. Review Design

The review was conducted according to the Preferred Reporting Items for Systematic Reviews and Meta-Analyses (PRISMA) statement [25] and following a predefined protocol that was registered at the OSF registries (DOI: 10.17605/OSF.IO/XBJM8). Ethical approval was not required for the current study.

2.2. Criteria for Study Inclusion

This systematic review included randomized controlled trials (RCTs) conducted on adults with the metabolic syndrome, including an intervention group that received supplementation with vitamin D and a control group, where dyslipidemia was reported as an outcome.

RCTs supplementing vitamin D3 or D2 in any form to the intervention group, and a placebo or a lower dose of vitamin D provided to the control group; investigating at least one of the dyslipidemia components of the metabolic syndrome (Total Cholesterol (TC), LDL-C, HDL-C, or TG) measured in the fasting state; including adult participants, as defined by the investigators—e.g., aged > 18 years at baseline, suffering from the metabolic syndrome (irrespective of the definition adopted)—were included. Only RCTs with a minimum duration of 4 weeks were included to ensure that the intervention had sufficient time to produce an effect. Additionally, RCTs involving a co-intervention were included if both arms of the study received the same co-intervention.

Studies were excluded if they were conducted on healthy participants, or participants with chronic or acute conditions other than the metabolic syndrome, or participants receiving medication known to influence vitamin D metabolism.

2.3. Search Strategy

The search strategy considered two key concepts: (1) vitamin D and (2) metabolic syndrome. For each concept, Medical Subject Headings (MeSH) and keywords were mapped. Search terms included but were not limited to vitamin D, cholecalciferol, ergocalciferol, or calcidol, combined with metabolic syndrome. The following databases were searched: PUBMED, MEDLINE, CINAHL, EMBASE, the Cochrane Library, ClinicalTrials.gov, and the International Clinical Trials Registry Platform (ICTRP) [26,27]. No language restrictions were applied to the search; however, the timeline was limited to studies published after the year 1998—when the first definition of the metabolic syndrome was issued by the World Health Organization [3]—until 31 July 2020. The electronic search strategy was validated by a medical information specialist and is described in the Supplement. Bibliographies of included RCTs and relevant reviews were also hand-searched for eligible studies.

2.4. Study Selection

Two pairs of authors screened titles and/or abstracts retrieved by the search and identified studies that potentially meet the inclusion criteria outlined above. The two pairs then reviewed the full texts of potentially eligible studies independently and in duplicate, and assessed them for eligibility. To ensure the validity of the study selection process, a calibration exercise was first conducted. Disagreements were solved through consensus or with the help of a third reviewer.

2.5. Data Extraction

Two pairs of authors extracted data from eligible studies independently and in duplicate using a data extraction form. A calibration exercise was first conducted to ensure the validity of the data extraction process. For all eligible records, the authors recorded characteristics of the study, details of the population, interventions (type, form, and the dose of vitamin D in experimental groups, comparator, and duration), outcomes assessed, as well the main findings. Serum 25OHD was converted to nmol/L, if it was reported as ng/mL by multiplying by a factor of 2.496. Serum TC, LDL-C, HDL-C, and TG were converted to mmol/, if they were reported as mg/dL, using the respective multiplication factors: 0.0259 for TC, LDL-C, and HDL-C, and 0.0113 for TG.

2.6. Quality Assessment

Two pairs of authors assessed independently and in duplicate the risk of bias of included RCTs following the Cochrane criteria (sequence generation, allocation concealment, blinding of participants and outcome assessors, incomplete outcome data, and selective outcome reporting) [28]. Each potential source of bias was graded as low, high, or unclear risk. Disagreements were solved through consensus or with the help of a third reviewer.

The overall quality of the evidence generated by the meta-analysis was assessed according to the Grading of Recommendations Assessment, Development and Evaluation (GRADE) methodology (high risk of bias, imprecision, indirectness, heterogeneity, and publication bias). The evidence was presented using GRADE Evidence Profiles developed in the GRADEpro software [29].

2.7. Data Synthesis

When a meta-analysis was not possible, a narrative review of the findings was performed. Meta-analyses were conducted when participants, treatments, and the outcomes were similar enough to allow pooling. Standard meta-analyses comparing VDS with no supplementation in patients with metabolic syndrome were performed using RevMan version 5.3 (The Cochrane Collaboration,

The Nordic Cochrane Centre). A fixed-effects model was used when analyzing two studies and a random-effects model when analyzing more than two studies. The results were reported on as a weighted mean difference with 95% confidence intervals. The I^2 statistic was used to assess heterogeneity among different studies. The I^2 metric ranges from 0 to 100%, with higher values indicating greater heterogeneity. In cases of moderate to substantial heterogeneity, with I^2 values greater than 50%, the potential causes were explored and reported on, and relevant sensitivity analyses were conducted.

3. Results

3.1. Search Results

Details of the search process are presented in Figure 1. Seven studies were included in the systematic review. Out of the seven included studies, four yielded data that could be combined in the meta-analysis.

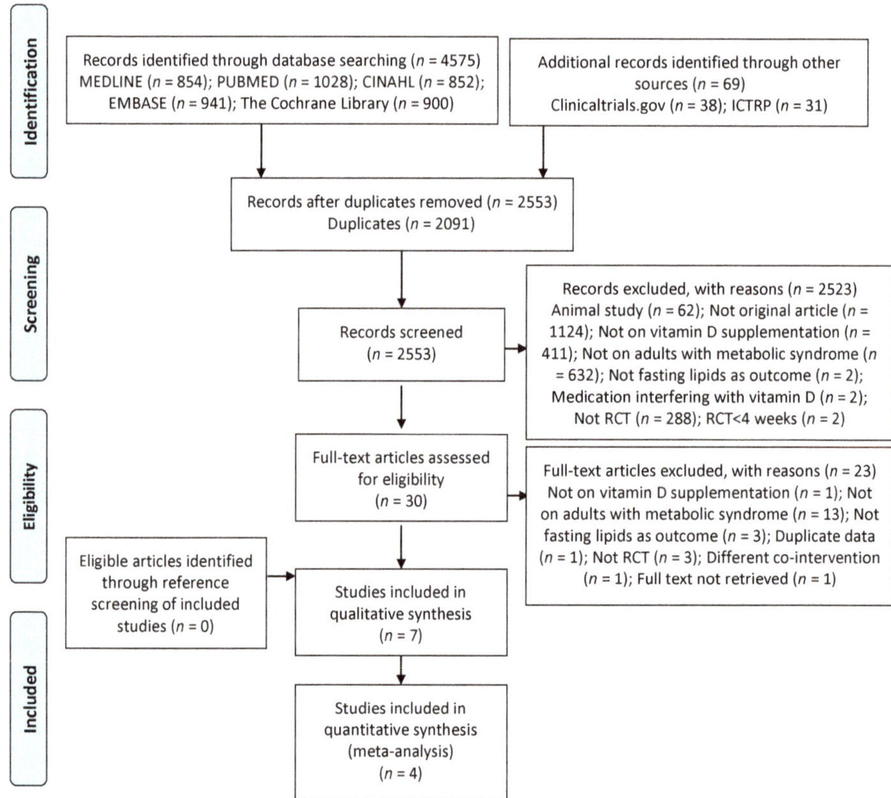

Figure 1. Preferred Reporting Items for Systematic Reviews and Meta-Analyses (PRISMA) diagram of study selection. ICTRP: International Clinical Trials Registry Platform; RCT: Randomized Controlled Trial.

3.2. Characteristics of Included Studies

Characteristics of included studies are given in Table 1. The studies by Makariou [30–32] were conducted on the same sample, but reported on different outcomes in three different manuscripts.

Three of the studies were conducted in Greece [30–32], two in Iran [33,34], one in Thailand [33] and one in China [34]. The number of trial participants varied from 50 to 123, and a mean age ranging between 40 and 65 years. All of the studies were conducted on participants suffering from the MetS, diagnosed either by the NCEP-ATP III [30–34], the IDF [35], or the joint interim statement between several major organizations [36]. The follow-up period varied from 8 weeks [33] to 1 year [34].

In three studies, the intervention consisted of vitamin D3 supplementation with dietary intervention [30–32], one study used vitamin D (without specifying its type) supplementation with physical activity [35], two studies supplemented only with vitamin D3 [34,36], and one study supplemented with vitamin D2 [33]. The average daily dose of VDS ranged from 700 IU [34] to 7142.85 IU [36], whereby four studies were supplemented with 2000 IU per day [30–32,35]. Four RCTs were placebo-controlled [33–36], and in the other three RCTs [30–32], the comparator was dietary intervention according to the NCEP-ATP III guidelines. Only Yin et al. [34] included a co-intervention in the form of calcium supplementation.

As for study outcomes detailed in Table 2, TC, TG, HDL-C, and LDL-C were analyzed in four studies [30,33,35,36]. In addition, Salekzamani et al. [36] assessed TG/HDL-C and LDL-C/HDL-C. Yin et al. [34] analyzed TG, HDL-C, and LDL-C only. Novel lipid-related biomarkers were also assessed in Makariou et al. [30]—i.e., apo A1 and Apo B; Makariou et al. [32]—i.e., oxidized-LDL, oxidized-LDL/LDL, and oxidized LDL/ApoB; Makariou et al. [31]—i.e., sLDL-C and mean LDL size.

3.3. Assessment of Risk of Bias

The assessment of the risk of bias of included studies is presented in Figure 2. The quality of the RCTs design and reporting was low in general and varied across studies. Random allocation of participants was reported in the three studies by Makariou et al. [30–32] and in the study by Salekzamani et al. [36], and was unclear in the other three studies [33–35]. Only Salekzamani et al. [36] gave sufficient detail to ascertain adequate allocation concealment, while this was unclear in the other studies [30–35]. Blinding of participants was impossible in the studies by Makariou et al. [30–32], reflecting a high risk of bias, and was guaranteed only in the study by Salekzamani et al. [36] and Wongwiwatthanananukit et al. [33]. All trials had adequate blinding of outcome assessment, complete outcome data, and low selective reporting bias.

3.4. Results of Included Studies

Table 2 describes the findings from the included studies. All the included studies [30–36] reported a significant increase in vitamin D status in the intervention groups at endline. Regarding end-point values of lipid parameters, Makariou et al. [30], Wongwiwatthanananukit et al. [33], and Yin et al. [34] found no significant differences in TC, TG, LDL-C, and HDL-C between the compared groups. In Farag et al. [35], TG at baseline was significantly higher in the vitamin D group compared with the other groups, and HDL-C was significantly higher in the vitamin D + physical activity group compared with the other groups, which hindered the direct comparison between the three groups at endline. The authors reported that endline TC was significantly lower in the vitamin D group compared with the other groups, LDL-C was significantly lower in the vitamin D group compared with the placebo group, and HDL-C was significantly higher in the vitamin D + physical activity group compared with the other groups. Regarding within-group changes, there was a greater significant decrease in TC and LDL-C in the vitamin D + physical activity group compared with the placebo group; and no other differences in changes in TG and HDL-C between baseline and endline were noted in the three groups [35]. Additionally, in Salekzamani et al. [36], at baseline, TG and TG/HDL-C were significantly higher in the intervention group than the control group. At endline, the authors reported a greater decrease in TG and TG/HDL-C in the vitamin D group compared with the C group, but did not find significant changes in other parameters, namely TC, HDL-C, LDL-C, and LDL-C/HDL-C [36]. Similarly, no significant changes in novel lipid-related biomarkers were noted with VDS in the two studies by Makariou et al. [31,32].

First author, Year	Random Sequence Generation (Selection Bias)	Allocation Concealment (Selection bias)	Blinding of Participants and Personnel (Performance	Blinding of Outcome Assessment (Detection Bias)	Incomplete outcome Data (Attrition Bias)	Selective Reporting (Reporting Bias)	Other Bias
Farag, 2019 [35]	?	?	?	+	+	+	?
Makariou, 2017 [30]	+	?	–	+	+	+	?
Makariou, 2019a [32]	+	?	–	+	+	+	+
Makariou, 2019b [31]	+	?	–	+	+	+	?
Salekzamani [36]	+	+	+	+	+	+	+
Wongwiwatthananukit, 2013 [33]	?	?	+	+	+	+	+
Yin, 2016 [34]	?	?	?	+	+	+	+

+ Low risk of bias   ? Unclear risk of bias   – High risk of bias

Figure 2. Risk of bias of included studies from consensus between a pair of raters.

Table 1. Characteristics of included studies.

First Author, year	Study Design	Geographic Setting/Data Collection Time Period	Study Population	Definition of Metabolic Syndrome	Intervention	Dose, Frequency, Duration	Daily Dose Equivalent	Control	Co-Intervention	Compliance	Drop-Out
Farag, 2019 [35]	Parallel randomized placebo-controlled trial	Halabja (Kurdistan Region of Iraq)/March to May	I1: *n* = 24; I2: *n* = 21; C: *n* = 25 Ethnicity: NR Mean age (SD): I1: 40.54 (5.94); I2: 40.42 (5.89); C: 42.6 (5.62)% Male:I1: 33.3%; I2: 33.3%; C: 52.0%	IDF criteria	I1. Vitamin D without PA I2. Vitamin D + PA	I1: 2000 IU, Daily, 12 weeks I2: 2000 IU, Daily, 12 weeks + 30 min of endurance PA, Daily	2000 IU	Placebo without endurance PA	None	NR	I1: 20% I2: 30% C: 16.66%
Makariou, 2017 [30]	Prospective, randomized, open-label, blinded end-point trial	Greece/March to September	I: *n* = 25; C: *n* = 25 Ethnicity: NR Mean age (SD): I: 52 (9); C: 51 (12)% Male: I: 60%; C: 44%	NCEP-ATP III criteria	Vitamin D3 + dietary intervention according to the NCEP-ATP III guidelines	2000 IU, Daily, 12 weeks	2000 IU	Dietary intervention according to NCEP-ATP III guidelines	None	Compliance with Vitamin D: NR Poor compliance with dietary instructions in both groups	0%
Makariou, 2019a [32]	Prospective, randomized, open-label, blinded end-point trial	Greece/March to September	I: *n* = 25; C: *n* = 25 Ethnicity: NR Mean age (SD): I: 53 (7); C: 52 (15)% Males: I: 60%; C: 40%	NCEP-ATP III criteria	Vitamin D3 + dietary intervention according to NCEP-ATP III guidelines	2000 IU, Daily, 12 weeks	2000 IU	Dietary intervention according to NCEP-ATP III guidelines	None	I: 100%; poor compliance with dietary instructions C: Poor compliance with dietary instructions	0%
Makariou, 2019b [31]	Prospective, randomized, open-label, blinded end-point trial	Greece/March to September	I: *n* = 25; C: *n* = 25 Ethnicity: NR Mean age (SD): I: 53 (7); C: 52 (15)% Males: I: 60%; C: 40%	NCEP-ATP III criteria	Vitamin D3 + dietary intervention according to NCEP-ATP III guidelines	2000 IU, Daily, 12 weeks	2000 IU	Dietary intervention according to NCEP-ATP III guidelines	None	Compliance with Vitamin D: NR Poor compliancewith dietary instructions in both groups	0%

Table 1. *Cont.*

First Author, year	Study Design	Geographic Setting/Data Collection Time Period	Study Population	Definition of Metabolic Syndrome	Intervention	Dose, Frequency, Duration	Daily Dose Equivalent	Control	Co-Intervention	Compliance	Drop-Out
Salekzamani, 2016 [36]	Randomized placebo-controlled, double-blind parallel trial	Tabriz, Iran/October to June	I: $n = 35$; C: $n = 36$ Ethnicity: NR Mean age (SD): 40.49 (5.04)% Males: 49% (data per group: NR)	Criteria of the joint interim statement of the IDF task force on epidemiology and prevention; NHLBI; AHA; World Heart Federation; International Atherosclerosis Society; and International Association for the Study of Obesity	Vitamin D3	50,000 IU, Weekly, 16 weeks	7142.85 IU	Placebo	None	97% in both groups	I: 12.5% C: 10%
Wongwiwatt hananukit, 2013 [33]	Prospective randomized, double-blind, double-dummy, parallel trial	Bangkok, Thailand/January to September	I1: $n = 28$; I2: $n = 28$; C: $n = 28$ Ethnicity: NR Mean age (SD): I1: 62.29 (10.63); I2: 63.61 (13.25); C: 65.07 (11.31)% Male: I1: 53.3%; I2: 50%; C: 50%	NCEP-ATP III criteria	I1: vitamin D2 I2: vitamin D2	I1: 40,000 IU, Weekly, 8 weeks I2: 20,000 IU, Weekly, 8 weeks + 1 placebo capsule, Weekly, 8 weeks	I1: 5714.28 IU I2: 2857.14 IU	Placebo	None	100% in the 3 groups	I1: 6.66% I2: 6.66% C: 6.66%
Yin, 2016 [34]	Randomized placebo-controlled intervention trial	Jinan, North China/ November to February	I: $n = 61$; C: $n = 62$ with vitamin D deficiency (25(OH)D < 50 nmol/L) Ethnicity: Northern Chinese Mean age (SD): 49.5 (8.72)% Male: 54% (data per group: NR)	Updated NCEP-ATP III criteria for Asian Americans	Vitamin D3	700 IU, Daily, 1 year	700 IU	Placebo	600 mg elemental Calcium (CalciumCitrate), Daily	95% in both groups	I: 3.17% C: 1.58%

I: Intervention; C: Control; NR: Not Reported; SD: Standard Deviation; 25(OH)D: 25-Hydroxyvitamin D; NCEP-ATP: National Cholesterol Education Program Adult Treatment Panel III; IDF: International Diabetes Federation; NHLBI: National Heart, Lung, and Blood Institute; AHA: American Heart Association; PA: Physical Activity; IU: International Unit.

Table 2. Outcomes and results of included studies.

First Author, year	Assessment Method: Vitamin D	Assessment Method: Dyslipidemia Outcomes	Baseline 25OHD Level (nmol/L) ng/mL * 2.496 = nmol/L	Endline 25OHD Level (nmol/L) ng/mL * 2.496 = nmol/L	Baseline Dyslipidemia Outcomes HDL-C, LDL-C: mmol/L * 38.67 = mg/dL TG: mmol/L * 88.57 = mg/dL	Endline Dyslipidemia Outcomes HDL-C, LDL-C: mmol/L * 38.67 = mg/dL TG: mmol/L * 88.57 = mg/dL	Conclusion
Farag, 2019 [35]	25(OH)D: measured by immunoassay	TC, TG: measured using enzymatic colorimetric tests HDL-C: measured after precipitation of the apolipoprotein B containing lipoproteins with phosphotungistic acid LDL-C: calculated from serum TC, TG and HDL-C based on relevant formula	Mean (SD) I1: 26.70 (6.98) I2: 25.95 (7.98) C: 30.20 (9.73)	Mean (SD) I1: 57.90 (12.23) I2: 72.38 (13.72) C: 31.44 (9.98)	TC mean (SD) (mg/dL) I1: 173.5 (60.8) I2: 194.7 (32.2) C: 185.9 (39) HDL-C mean (SD) (mg/dL) I1: 34.9 (17.3) I2: 40.9 (14.4) C: 30.04 (8.5) LDL-C mean (SD) (mg/dL) I1: 120.7 (64.4) I2: 149.6 (35.8) C: 150.4 (39.8) TG mean (SD) (mg/dL) I1: 229.3 (113.8) I2: 184.5 (98.5) C: 174.4 (43)	TC mean(SD) (mg/dL) I1: 160.5(33.4) I2: 181.7(31.3) C: 196.8(39.4) HDL-C mean (SD) (mg/dL) I1: 33.7 (10.6) I2: 39 (10) C: 31.8 (7) LDL-C mean (SD) (mg/dL) I1: 107(36.6) I2: 138.3(31.4) C: 158.8(39) TG mean (SD) (mg/dL) I1: 233.8 (97) I2: 178.1 (80.8) C: 158.6 (35.4)	At baseline, TG was significantly higher in the I1 compared with the other groups; and HDL-C was significantly higher in the I2 group compared with the other groups There were NS differences in other study parameters between groups At endline, 25(OH)D was significantly higher in the I1 and I2 group compared with the C group TC was significantly lower in the I1 group compared with the other groups LDL-C was significantly lower in the I1 group compared with the C group HDL-C was significantly higher in I2 compared with the other groups Greater significant decrease in TC and LDL-C in I2 compared with the C group There were NS differences in changes in TG and HDL-C between baseline and endline in the 3 groups
Makariou, 2017 [30]	25(OH)D: measured by enzyme immunoassay	TC, TG, HDL-C: measured enzymatically LDL-C: calculated by the Friedewald equation (when TG < 350 mg/dl) ApoA1, ApoB: measured by immunonephelometry	median (min–max) I: 39.93 (7.48–87.36) C: 24.96 (9.98–117.84)	median (min–max) I: 76.37 (20.96–167.23) C: 32.44 (8.73–92.35)	TC mean (SD) (mg/dL) I: 219 (36) C: 231 (34) HDL-C mean (SD) (mg/dL) I: 48 (10) C: 50 (9) LDL-C mean (SD) (mg/dL) I: 140 (35) C: 147 (26) TG median (min–max) (mg/dL) I: 150 (56–336) C: 146 (84–339) Apo A1 mean (SD) (mg/dL) I: 136 (26) C: 143 (13) Apo B mean (SD) (mg/dL) I: 92 (25) C: 107 (16)	TC mean (SD) (mg/dL) I: 224 (37) C: 2232 (42) HDL-C mean (SD) (mg/dL) I: 49 (9) C: 49 (10) LDL-C mean (SD) (mg/dL) I: 145 (34) C: 152 (37) TG median (min-max) (mg/dL) I: 136 (46–261) C: 131 (73–307) Apo A1, ApoB: NR	At baseline, there were NS differences in study parameters between groups At endline, 25(OH)D was significantly higher in the I group compared with the C group There were NS differences in lipid parameters between groups

Table 2. *Cont.*

First Author, year	Assessment Method: Vitamin D	Assessment Method: Dyslipidemia Outcomes	Baseline 25OHD Level (nmol/L) ng/mL * 2.496 = nmol/L	Endline 25OHD Level (nmol/L) ng/mL * 2.496 = nmol/L	Baseline Dyslipidemia Outcomes HDL-C, LDL-C: mmol/L * 38.67 = mg/dL TG: mmol/L * 88.57 = mg/dL	Endline Dyslipidemia Outcomes HDL-C, LDL-C: mmol/L * 38.67 = mg/dL TG: mmol/L * 88.57 = mg/dL	Conclusion
Makariou, 2019a [32]	25(OH)D: measured by enzyme immunoassay	Oxidized-LDL: measured by a competitive enzyme-linked immunosorbent assay using a specific murine monoclonar antibody Oxidized-LDL/LDL: NR Oxidized-LDL/ApoB: NR	median (95%CI) I: 40.18 (25.70–61.90) C: 24.71 (13.72–39.18)	median (95%CI) I: 76.37 (64.14–103.58) C: 32.94 (19.96–58.65)	Oxidized LDL-C mean (SD) (95%CI) (U/L) I: 70.3 (15.2) (64.8–87.6) C: 67.2 (16.9) (59.2–79.3) Oxidized LDL-C/LDL-C mean (SD) (95%CI) (U/mg) I: 0.05 (0.01) (0.46–0.65) C: 0.06 (0.008) (0.41–0.58) Oxidized LDL-C/ApoB mean(SD) (95%CI) (U/mg) I: 0.08 (0.04) (0.70–1.08) C: 0.07 (0.008) (0.57–0.74)	Oxidized LDL-C mean(SD) (95%CI) (U/L) I: 75.9(21.2) (67.9–89.9) C: 67.3(19.3) (56.9–92.5) Oxidized LDL-C/LDL-C mean(SD) (95%CI) (U/mg) I: 0.05(0.01) (0.46–0.60) C: 0.06(0.02) (0.42–0.78) Oxidized LDL-C/ApoB mean(SD) (95%CI) (U/mg) I: 0.07(0.01) (0.69–0.82) C: 0.08(0.02) (0.66–0.91)	At baseline, Ox-LDL/ApoB (U/mg) was significantly higher in the C group compared with the I group There were NS differences in other study parameters between groups At endline, 25(OH)D was significantly higher in the I group compared with the C group There were NS differences in lipid parameters between groups
Makariou, 2019b [31]	25(OH)D: measured by enzyme immunoassay	sdLDL-C: analyzed electrophoretically sdLDL proportion, mean LDL size: analyzed using the methods of the European Panel On Low-Density Lipoprotein Subclasses	median (min–max) I: 40.18 (8.23–87.60) C: 24.71 (9.98–98.84)	median (min–max) I: 76.37 (20.96–168.72) C: 32.94 (8.73–91.85)	sdLDL median (min–max) (mg/dL) I: 9.0 (0.0–40) C: 7.0 (0.0–22) sdLDL proportion mean (SD) (%) I: 5.7 (5.2) C: 3.8 (2.8) LDL size mean (SD) (nm) I: 264.8 (6.3) C: 266.5 (3.9)	sdLDL median (min–max) (mg/dL) I: 4.0 (0.0–46) C: 5.0 (2.0–25) sdLDL proportion mean (SD) (%) I: 4.5 (4.4) C: 3.3 (2.3) LDL size mean (SD) (nm) I: 266.6 (5.2) C: 267.0 (3.5)	At baseline, there were NS differences in study parameters between groups At endline, 25(OH)D was significantly higher in the I group compared with the C group There were NS difference in lipid parameters between groups

Table 2. *Cont.*

First Author, year	Assessment Method: Vitamin D	Assessment Method: Dyslipidemia Outcomes	Baseline 25OHD Level (nmol/L) ng/mL * 2.496 = nmol/L	Endline 25OHD Level (nmol/L) ng/mL * 2.496 = nmol/L	Baseline Dyslipidemia Outcomes HDL-C, LDL-C: mmol/L * 38.67 = mg/dL TG: mmol/L * 88.57 = mg/dL	Endline Dyslipidemia Outcomes HDL-C, LDL-C: mmol/L * 38.67 = mg/dL TG: mmol/L * 88.57 = mg/dL	Conclusion
Salekzamani, 2016 [36]	25(OH)D: measured by chemiluminescent immunoassay	TG, TC, LDL-C, HDL-C: measured enzymatically TG/HDL-C: NR LDL/HDL-C: NR	Mean (SD) I: 16.45 (15.50) C: 23.47 (21.34)	Mean (SD) I: 78.38 (21.71) C: 21.46 (17.74)	TC mean (SD) (mg/dL) I: 212 (42) C: 200 (39.27) HDL-C mean (SD) (mg/dL) I: 45 (8.08) C: 45 (10.08) LDL-C mean (SD) (mg/dL) I: 114 (33) C: 117 (28) TG mean (SD) (mg/dL) I: 269 (97) C: 185 (61) TG/HDL-C mean (SD) I: 6.05 (2.21) C: 4.22 (1.64) LDL-C/HDL-C mean (SD) I: 2.53 (0.69) C: 2.57 (0.54)	TC mean (SD) (mg/dL) I: 203.21 (34.63) C: 197.14 (33.57) HDL-C mean (SD) (mg/dL) I: 47 ± 6.63 C: 47 ± 8.24 LDL-C mean (SD) (mg/dL) I: 106 ± 25 C: 111 ± 29 TG mean(SD) (mg/dL) I: 242 ± 82 C: 196 ± 72 TG/HDL-C mean (SD) I: 5.20 ± 1.67 C: 4.37 ± 1.99 LDL-C/HDL-C mean (SD) I: 2.27 ± 0.53 C: 2.36 ± 0.64	At baseline, TG and TG/HDL-C were significantly higher in the I group compared with the C group There were NS differences in other study parameters between groups At endline, 25(OH)D significantly increased in the I group and was stable in the C group TG and TG/HDL-C had a greater% change in the I compared with the C group There were NS differences in other lipid parameters between groups
Wongwiwatt hananukit, 2013 [33]	25(OH)D: measured by chemiluminescent immunoassay	TC, TG, HDL-C, LDL-C: NR	Mean (SD) I1: 35.66 (8.36) I2: 37.63 (7.88) C: 40.43 (7.46)	Mean (SD) I1: 75.95 (17.39) I2: 66.89 (15.89) C: 47.39 (16.74)	TC mean (SD) (mg/dL) I1: 180.36 (34.43) I2: 166.89 (20.95) C: 174.29 (38.90) HDL-C mean(SD) (mg/dL) I1: 53.18(12.46) I2: 52.36(11.86) C: 53.43(12.73) LDL-C mean(SD) (mg/dL) I1: 107.00(27.46) I2: 96.68(19.96) C: 102.50(29.51) TG mean(SD) (mg/dL) I1: 139.32(61.26) I2: 132.29(62.36) C: 129.46(59.75)	TC mean (SD) (mg/dL) I1: 182.04 (31.00) I2: 170.54 (39.83) C: 175.96 (39.12) HDL-C mean (SD) (mg/dL) I1: 52.54 (13.49) I2: 50.96 (12.21) C: 53.46 (11.75) LDL-C mean (SD) (mg/dL) I1: 110.54 (27.47) I2: 102.96 (35.09) C: 105.61 (32.31) TG mean (SD) (mg/dL) I1: 144.82 (64.07) I2: 137.79 (53.48) C: 135.75 (71.40)	At baseline, there were NS differences in study parameters between groups At endline, 25(OH)D was significantly higher in the I1 and I2 groups compared with the C group There were NS differences in lipid parameters between groups

Table 2. *Cont.*

First Author, year	Assessment Method: Vitamin D	Assessment Method: Dyslipidemia Outcomes	Baseline 25OHD Level (nmol/L) ng/mL * 2.496 = nmol/L	Endline 25OHD Level (nmol/L) ng/mL * 2.496 = nmol/L	Baseline Dyslipidemia Outcomes HDL-C, LDL-C: mmol/L * 38.67 = mg/dL TG: mmol/L * 88.57 = mg/dL	Endline Dyslipidemia Outcomes HDL-C, LDL-C: mmol/L * 38.67 = mg/dL TG: mmol/L * 88.57 = mg/dL	Conclusion
Yin, 2016 [34]	25(OH)D: measured by double antibody radioimmunoassay	TG, HDL-C: measured by enzymatic colorimetric assay LDL-C: calculated using the Friedwald equation	Mean (SD) I: 36.44 (5.44) C: 35.44 (6.36)	Mean (SD) I: 82.61 (10.90) C: 36.44 (6.98)	HDL-C mean (SD) (mg/dL) I: 41.38 (3.09) C: 38.28 (2.70) LDL-C mean (SD) (mg/dL) I: 126.06 (9.66) C: 123.74 (7.73) TG mean (SD) (mg/dL) I: 295.84 (60.23) C: 280.8 (35.43)	HDL-C mean (SD) (mg/dL) I: 42.15 (2.70) C: 40.22 (2.32) LDL-C mean (SD) (mg/dL) I: 122.97 (9.28) C: 121.42 (8.50) TG mean (SD) (mg/dL) I: 250.66 (36.31) C: 255.1 (19.48)	At baseline, there were NS differences in study parameters between groups At endline, 25(OH)D significantly increased in the I group and was stable in the C group There were NS differences in lipid parameters between groups Similar results were obtained in the obesity and non-obesity subgroups

I: Intervention; C: Control; NR: Not Reported; SD: Standard Deviation; 25(OH)D: 25-Hydroxyvitamin D; min: minimum; max: maximum; TC: Total Cholesterol; TG: Triglycerides; HDL-C: High-Density Lipoprotein Cholesterol; LDL-C: Low-Density Lipoprotein Cholesterol; Apo: Apolipoprotein; sdLDL-C: Small Dense Low-Density Lipoprotein Cholesterol; CI: Confidence Interval; NS: non-significant. *: symbol denoting multiplication.

3.5. Results of the Meta-Analyses

Two of the studies by Makariou et al. [31,32] and that by Yin et al. [34] were not included in the meta-analysis; as Makariou et al. [31,32] solely reported on novel lipid-related biomarkers, namely oxidized LDL-C and small-density LDL-C (sdLDL-C), and the study by Yin et al. [34] was conducted over the period of one year—a duration that is much longer than the other studies. Moreover, in the study by Farag et al. [35], the intervention arm entailing vitamin D + physical activity was excluded from the meta-analysis since the control arm consisted of administration of placebo only, without physical activity. In contrast, the study by Makariou et al. [30] was included in the analysis since both arms entailed a dietary intervention, allowing it to be canceled out.

Based on the administered daily dose equivalent of vitamin D, two sets of meta-analyses were conducted. The first one included the studies by Makariou et al. [30], Farag et al. [35], and the I^2 arm of the study by Wongwiwatthananukit et al. [33]. The analysis consisted of comparing a low dose of VDS versus no supplementation, namely placebo or dietary intervention. The other analysis included the study by Salekzamani et al. [36] and the I1 arm of the study by Wongwiwatthananukit et al. [33] and consisted of comparing a high dose of VDS versus placebo.

Forest plots for the mean difference in LDL-C, HDL-C, TC, and TG for the two sets meta-analyses based on the daily dose equivalent of vitamin D in the intervention arms are presented in Figures 3 and 4, respectively. The first set of meta-analyses revealed no statistically significant difference in LDL-C, HDL-C, and TC between patients receiving low-dose VDS compared with those not receiving it. Furthermore, the meta-analysis revealed a statistically significant increase in TG in the group receiving VDS compared with placebo (mean difference, 30.67 (95% CI, 4.89, 56.45) mg/dL; $p = 0.02$). Yet, the heterogeneity of this analysis was substantially high ($I^2 = 86\%$) (Figure 3). The final quality of evidence of all the meta-analyses was very low (Supplement 2). Supplement 3 presents the results of the sensitivity analyses, which were based on the exclusion of Farag et al. [35], as a source of heterogeneity. The study had inadequate randomization that is reflected in the incomparable baseline TG of the randomized arms. Excluding this study dropped the heterogeneity to none, yet, the sensitivity analyses did not affect the results.

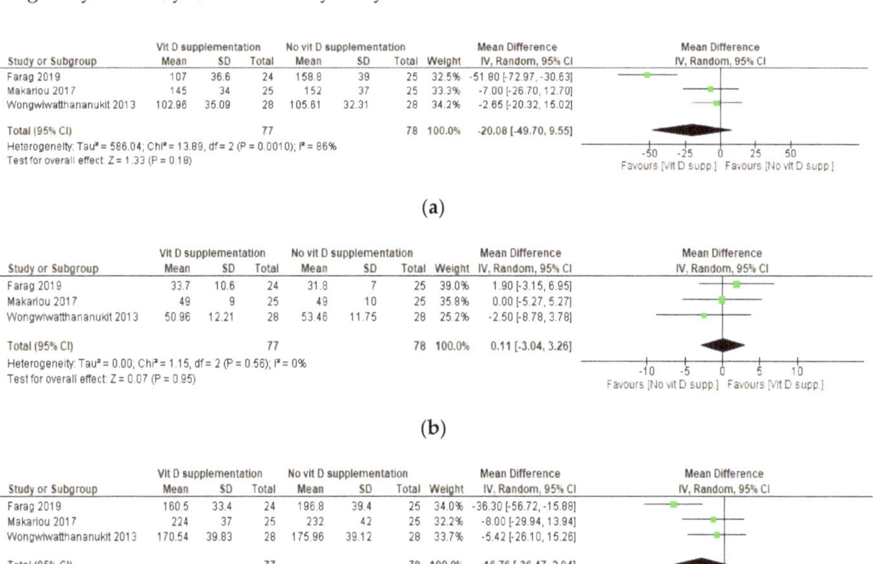

(a)

(b)

(c)

Figure 3. *Cont.*

Study or Subgroup	Vit D supplementation			No vit D supplementation			Weight	Mean Difference IV, Fixed, 95% CI
	Mean	SD	Total	Mean	SD	Total		
Farag 2019	233.8	97	24	158.6	35.4	25	39.1%	75.20 [33.99, 116.41]
Wongwiwatthananukit 2013	137.79	53.48	28	135.75	71.4	28	60.9%	2.04 [-31.00, 35.08]
Total (95% CI)			52			53	100.0%	30.67 [4.89, 56.45]

Heterogeneity: Chi² = 7.37, df = 1 (P = 0.007); I² = 86%
Test for overall effect: Z = 2.33 (P = 0.02)

(**d***)

Figure 3. Meta-analysis of effects of low-dose VDS on LDL-C, HDL-C, TC, and TG. Mean differences for each study are represented by squares, and 95% Confidence Intervals are represented by the lines through the squares. The pooled mean differences are represented by diamonds. Between-study heterogeneity was assessed with the use of the I^2 statistic. VDS: Vitamin D Supplementation; LDL-C: Low-density Lipoprotein Cholesterol; HDL-C: High-density Lipoprotein Cholesterol; TC: Total Cholesterol; TG: Triglycerides. (**a**) Forest plot of mean differences in LDL-C (in mg/dL) between subjects receiving low-dose VDS compared with those not receiving VDS. (**b**) Forest plot of mean differences in HDL-C (in mg/dL) between subjects receiving low-dose VDS compared with those not receiving VDS. (**c**) Forest plot of mean differences in TC (in mg/dL) between subjects receiving low-dose VDS compared with those not receiving VDS. (**d**) Forest plot of mean differences in TG (in mg/dL) between subjects receiving low-dose VDS compared with those not receiving VDS. * The study by Makariou et al. [30] was excluded from the primary analysis since the data are reported as median and range. The median cannot be assumed the same as the mean, and the standard deviations cannot be extrapolated from the range since the sample size is small. In addition, the study explicitly reports on the use of median and range when the distribution is skewed. * In the study by Farag et al. [35], TG at baseline was significantly higher in the intervention group compared with the control group.

Similarly, the meta-analyses revealed no statistically significant difference in LDL-C, HDL-C, and TC between patients receiving high-dose VDS compared with those receiving a placebo. Additionally, the meta-analysis revealed a statistically significant increase in TG in the group receiving VDS compared with placebo (mean difference, 27.33 (95% CI, 2.06, 52.59) mg/dL; *p* = 0.03) (Figure 4). The heterogeneity of this analysis was also high (I^2 = 51%). Similar to the first set of meta-analyses, one of the included studies, namely that by Salekzamani et al. [36], had unequal baseline TG levels between the randomized arms. Excluding this study and conducting a sensitivity analysis was impossible, as this set of meta-analyses included only two studies. The final quality of evidence of these meta-analyses was low (Supplement 2).

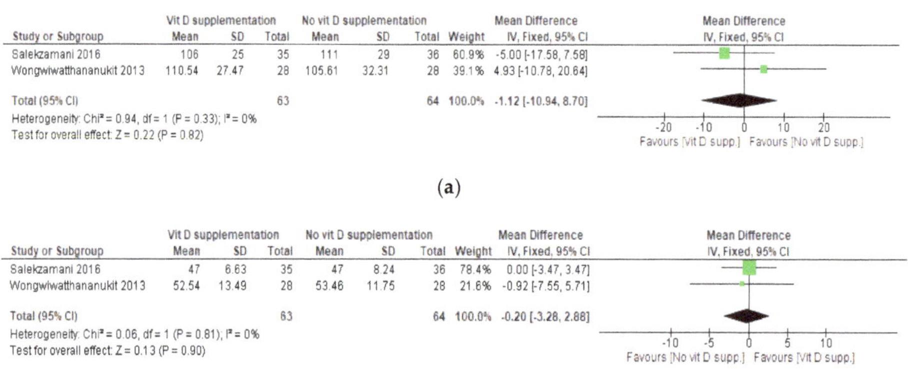

Study or Subgroup	Vit D supplementation			No vit D supplementation			Weight	Mean Difference IV, Fixed, 95% CI
	Mean	SD	Total	Mean	SD	Total		
Salekzamani 2016	106	25	35	111	29	36	60.9%	-5.00 [-17.58, 7.58]
Wongwiwatthananukit 2013	110.54	27.47	28	105.61	32.31	28	39.1%	4.93 [-10.78, 20.64]
Total (95% CI)			63			64	100.0%	-1.12 [-10.94, 8.70]

Heterogeneity: Chi² = 0.94, df = 1 (P = 0.33); I² = 0%
Test for overall effect: Z = 0.22 (P = 0.82)

(**a**)

Study or Subgroup	Vit D supplementation			No vit D supplementation			Weight	Mean Difference IV, Fixed, 95% CI
	Mean	SD	Total	Mean	SD	Total		
Salekzamani 2016	47	6.63	35	47	8.24	36	78.4%	0.00 [-3.47, 3.47]
Wongwiwatthananukit 2013	52.54	13.49	28	53.46	11.75	28	21.6%	-0.92 [-7.55, 5.71]
Total (95% CI)			63			64	100.0%	-0.20 [-3.28, 2.88]

Heterogeneity: Chi² = 0.06, df = 1 (P = 0.81); I² = 0%
Test for overall effect: Z = 0.13 (P = 0.90)

(**b**)

Figure 4. *Cont.*

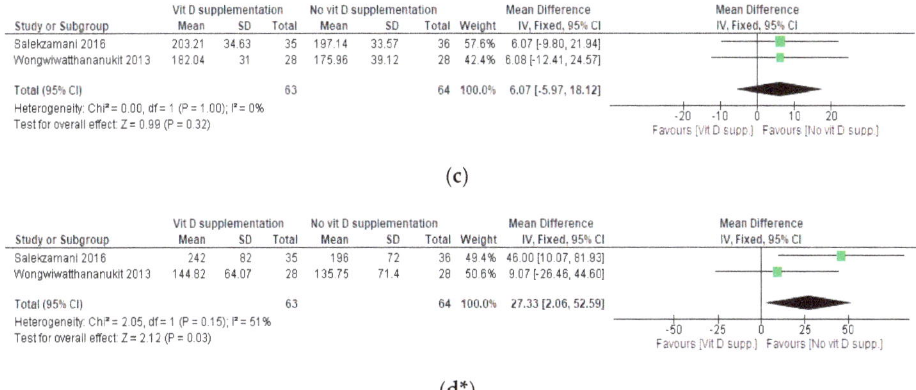

(c)

(d*)

Figure 4. Meta-analysis of effects of high-dose VDS on LDL-C, HDL-C, TC, and TG. Mean differences for each study are represented by squares, and 95% Confidence Intervals are represented by the lines through the squares. The pooled mean differences are represented by diamonds. Between-study heterogeneity was assessed with the use of the I^2 statistic. VDS: Vitamin D Supplementation; LDL-C: Low-density Lipoprotein Cholesterol; HDL-C: High-density Lipoprotein Cholesterol; TC: Total Cholesterol; TG: Triglycerides. (**a**) Forest plot of mean differences in LDL-C (in mg/dL) between subjects receiving a high dose of VDS compared with those not receiving VDS. (**b**) Forest plot of mean differences in HDL-C (in mg/dL) between subjects receiving a high dose of VDS compared with those not receiving VDS. (**c**) Forest plot of mean differences in TC (in mg/dL) between subjects receiving a high dose of VDS compared with those not receiving VDS. (**d**) Forest plot of mean differences in TG (in mg/dL) between subjects receiving a high dose of VDS compared with those not receiving VDS. * The mean differences of the two studies are very different and the heterogeneity is 51%. This might be due to the study by Salekzamani et al. [36] since TG at baseline was significantly higher in the intervention group compared with the control group.

4. Discussion

Vitamin D deficiency is a worldwide public health problem that affects all age groups [37]. It is widespread even in sunny countries [38] and in those that have implemented a rigorous VDS strategy for years [1,2]. The high prevalence of vitamin D deficiency is associated with various factors, including genetics, skin pigmentation, latitude, air pollution, obesity, in addition to behavioral lifestyle factors, such as sun avoidance, reduced outdoor activities, and use of sunscreen [39].

In parallel, MetS has recently surfaced as a major public health problem and a leading risk factor for the progression of T2DM and CVD [8–10]. Specifically, atherogenic dyslipidemia in MetS emerged as a key factor for CVD and a target for future interventions aiming at reducing poor patient outcomes [7,10]. Patients with MetS were reported to have decreased 25-hydroxyvitamin D levels [40]. Accordingly, correcting vitamin D deficiency through VDS was suggested to alleviate MetS, specifically the atherogenic dyslipidemia component of this syndrome [11]. This topic is gaining attention in the research world and is of clinical relevance [41].

To date, the literature presents conflicting results on the effects of VDS on the dyslipidemia component of MetS. Specifically, observational data indicate an inverse association between hypovitaminosis D and dyslipidemia in patients with MetS [16]. However, our findings indicate that correcting suboptimal vitamin D levels through supplementation was not effective in improving dyslipidemia. VDS, whether as D2 or D3, in a high or low dose, for a short or long duration, although significantly increased vitamin D blood levels, did not significantly affect TC, LDL-C, and HDL-C levels, nor the levels of novel lipid-related biomarkers. Furthermore, VDS significantly increased TG levels compared with placebo, although the baseline TG levels of compared arms in two of the included studies [35,36] were not comparable, which limits this finding. Our results are similar to those of other reviews reporting no meaningful changes in blood lipid values secondary to VDS in healthy, obese, or diabetic subjects [42,43].

The direct or indirect mechanisms through which vitamin D influences the lipid profile remain unclear [44]. Since the observational and interventional studies have conflicting evidence, it has been suggested that the association between vitamin D and metabolic disorders may be confounded by obesity rather than being a causal relationship. Obesity reduces the detectable serum levels of 25-hydroxyvitamin D through the sequestration of vitamin D in body fat tissue or decreased skin synthesis of vitamin D due to the limited outdoor activity and sun exposure [40,45]. Moreover, chronic inflammatory processes, which usually present in obese patients, might decrease 25-hydroxyvitamin D levels [46] and simultaneously affect various metabolic parameters. Accordingly, the relationship between vitamin D deficiency and poor metabolic profile may be explained by the fact that both of these factors are prone to cluster in obese subjects. It is thus possible that high vitamin D levels are not the cause of good health, rather its outcome, since healthy people generally stay outdoors longer and have better eating habits [42].

Furthermore, the dose, frequency, and duration of supplementation with vitamin D might also explain the discouraging results of interventional studies. For instance, supplementation for a period of three months may not be long enough to have a significant effect. The concentration of serum vitamin D would need to be in the range 100–150 nmol/L for cardiovascular disease protection [47], whereas the mean endline vitamin D levels in the intervention groups of the studies included in this review fell well below this level. Furthermore, VDS should be administered on a daily basis to ensure stable circulating concentrations for optimal functioning of the endocrine system [48]. Therefore, short treatment durations and bolus doses of some of the included RCTs could explain the null effects. Finally, it is also possible that vitamin D could provide benefits for cardiometabolic health through improvement in markers other than the lipid profile, such as in endothelial function [49], or through its effect on improving serum calcium profile early in the disease course. The latter observation is suggested by RCTs showing improvements in lipid profile in non-lean healthy subjects with low dietary calcium intake following vitamin D and calcium supplementation [50].

It is worthy to note that, to date, there is no consensus on the most suitable approach to correct vitamin D deficiency, and we lack information on the form, dose, frequency, and duration of vitamin D intervention that would be required to improve the metabolic components of MetS.

Multiple determinants may affect vitamin D status including genetic variation which could have a clinically important impact on response to VDS treatment among different individuals with identical doses [51,52]. For a better understanding of the regulation of vitamin D metabolism and its relation to dyslipidemia, variants of several genes including VDR which encodes the vitamin D receptor, DHCR7 which encodes the enzyme 7-dehydrocholesterol reductase, CYP2R1 which encodes the hepatic enzyme 25- hydroxylase, CYP24A1 which encodes 24-hydroxylase, and GC which encodes DBP the transporting protein for vitamin D DBP should all be considered [53]. Moreover, it is possible that single nucleotide polymorphisms (SNPs) in genotypes could modify the optimal vitamin D status required to reduce MetS disease outcomes [54]. Other confounding factors such as seasonal variation (vitamin D levels rise in summer and drop in winter) and geographic latitude have an important impact on vitamin D status and its correlation with health risk assessment [55]. Since the reviewed RCTs have not examined the genetic predisposition, nor the seasonal effect, it could be misleading to firmly conclude that VDS had no impact on dyslipidemia among MetS patients and hence further investigations are still warranted.

5. Strengths and Limitations

To our knowledge, this is the first review to systematically assess the effect of VDS and its effect on dyslipidemia, specifically in adults with MetS. The main strength of our review is that we included only RCTs, which generally had a low risk of bias. Another strength is that we conducted this review according to a predefined protocol, following standard methods for reporting systematic reviews (Moher, 2010), and using a comprehensive and sensitive search strategy with multiple databases and gray literature. We also employed several sensitivity analyses to assess the robustness of our results, whereby in cases of moderate to substantial heterogeneity, we explored and reported on the potential causes. However, our findings are limited by the small number of identified studies, their small sample sizes, and short duration. Furthermore, three of the included studies [32,35,36] started out with significantly higher baseline lipid levels in the intervention group, which limits the results generated by this review and pertaining meta-analyses.

6. Conclusions

Physiological mechanisms throughout epidemiological data suggest a link between vitamin D deficiency and MetS. Yet, we report inconsistent results on the relationship between vitamin D status and dyslipidemia in adults with MetS, mainly pointing towards a lack of effect, despite improvement in vitamin D status. Our results should be interpreted with caution given the limited number of included RCTs, the small sample size, and limited intervention period. It is plausible that potentially the associations between vitamin D and cardiometabolic health are not causal; this was also suggested regarding the link between vitamin D and a wide range of acute and other chronic health disorders [56]. Despite the fact that the positive outcome of VDS for improving dyslipidemia among patients with MetS was weak, this does not eliminate the beneficial effect of vitamin D in this subpopulation of patients as an anti-inflammatory hormone which mediates muscle strength and homeostasis [57]. The use of vitamin D status for clinical implications has been well established for many diseases including CVDs [58,59]. Several mendelian randomization studies have supported the protective role of VDS against some diseases such as MS [60]. Hence, further studies are needed before making any solid conclusions about the vitamin D status for clinical implications for dyslipidemia in the context of MetS. Till then, it remains crucial to achieve vitamin D sufficiency in patients with MetS.

Supplementary Materials: The following are available online at http://www.mdpi.com/2072-6643/12/11/3352/s1, Supplement 1: Search strategy. Supplement 2: GRADEing of the meta-analysis. Supplement 3: Sensitivity analyses.

Author Contributions: F.A. and R.R. were involved in the concept and design. S.H. performed the searches. F.A., M.A., and D.P. conducted the title and abstract screening. S.H., N.M., and R.R. conducted the full text screening and performed the data extraction and quality assessment. All the authors contributed to writing the draft manuscript. All authors have read and agreed to the published version of the manuscript.

Funding: This study was funded by the cluster grant R18030 awarded by Zayed University, United Arab Emirates. The funding body was not involved in the design of the study and collection, analysis, and interpretation of data or in writing the manuscript.

Acknowledgments: We would like to thank Aida Farha for her assistance in developing the search strategy, and Sally Yaacoub and Joanne Khabsa for their assistance in conducting the meta-analysis.

Conflicts of Interest: The authors declare that they have no competing interests.

References

1. Challa, A.S.; Makariou, S.E.; Siomou, E.C. The relation of vitamin D status with metabolic syndrome in childhood and adolescence: An update. *J. Pediatr. Endocrinol. Metab.* **2015**, *28*. [CrossRef] [PubMed]
2. Ostman, C.; Smart, N.A.; Morcos, D.; Duller, A.; Ridley, W.; Jewiss, D. The effect of exercise training on clinical outcomes in patients with the metabolic syndrome: A systematic review and meta-analysis. *Cardiovasc. Diabetol.* **2017**, *16*. [CrossRef] [PubMed]
3. Alberti, K.G.M.M.; Zimmet, P.Z. Definition, diagnosis and classification of diabetes mellitus and its complications. Part 1: Diagnosis and classification of diabetes mellitus. Provisional report of a WHO consultation. *Diabet. Med.* **1998**, *15*, 539–553. [CrossRef]
4. Azhdari, M.; Karandish, M.; Mansoori, A. Metabolic benefits of curcumin supplementation in patients with metabolic syndrome: A systematic review and meta-analysis of randomized controlled trials. *Phytother. Res.* **2019**, *33*, 1289–1301. [CrossRef]

5. Ginsberg, H.N.; MacCallum, P.R. The Obesity, Metabolic Syndrome, and Type 2 Diabetes Mellitus Pandemic: Part I. Increased Cardiovascular Disease Risk and the Importance of Atherogenic Dyslipidemia in Persons With the Metabolic Syndrome and Type 2 Diabetes Mellitus. *J. Cardiometab. Syndr.* **2009**, *4*, 113–119. [CrossRef]

6. Yamaoka, K.; Tango, T. Effects of lifestyle modification on metabolic syndrome: A systematic review and meta-analysis. *BMC Med.* **2012**, *10*, 138. [CrossRef]

7. Ginsberg, H.N.; MacCallum, P.R. The Obesity, Metabolic Syndrome, and Type 2 Diabetes Mellitus Pandemic: II. Therapeutic Management of Atherogenic Dyslipidemia. *J. Clin. Hypertens.* **2009**, *11*, 520–527. [CrossRef]

8. Tenenbaum, A.; Fisman, E.Z.; Motro, M.; Adler, Y. Atherogenic dyslipidemia in metabolic syndrome and type 2 diabetes: Therapeutic options beyond statins. *Cardiovasc. Diabetol.* **2006**, *5*, 20. [CrossRef]

9. Grundy, S.M. Hypertriglyceridemia, atherogenic dyslipidemia, and the metabolic syndrome. *Am. J. Cardiol.* **1998**, *81*, 18B–25B. [CrossRef]

10. Musunuru, K. Atherogenic dyslipidemia: Cardiovascular risk and dietary intervention. *Lipids* **2010**, *45*, 907–914. [CrossRef] [PubMed]

11. Moukayed, M.; Grant, W.B. Linking the metabolic syndrome and obesity with vitamin D status: Risks and opportunities for improving cardiometabolic health and well-being. *Diabetes Metab. Syndr. Obes. Targets Ther.* **2019**, *12*, 1437–1447. [CrossRef] [PubMed]

12. Al-Daghri, N.M.; Al-Attas, O.; Yakout, S.; Aljohani, N.; Al-Fawaz, H.; Alokail, M.S. Dietary products consumption in relation to serum 25-hydroxyvitamin D and selenium level in Saudi children and adults. *Int. J. Clin. Exp. Med.* **2015**, *8*, 1305. [PubMed]

13. Barbalho, S.M.; Tofano, R.J.; de Campos, A.L.; Rodrigues, A.S.; Quesada, K.; Bechara, M.D.; de Alvares Goulart, R.; Oshiiwa, M. Association between vitamin D status and metabolic syndrome risk factors. *Diabetes Metab. Syndr. Clin. Res. Rev.* **2018**, *12*, 501–507. [CrossRef]

14. Hypponen, E.; Boucher, B.J.; Berry, D.J.; Power, C. 25-Hydroxyvitamin D, IGF-1, and Metabolic Syndrome at 45 years of Age: A Cross-Sectional Study in the 1958 British Birth Cohort. *Diabetes* **2008**, *57*, 298–305. [CrossRef] [PubMed]

15. Wimalawansa, S.J. Associations of vitamin D with insulin resistance, obesity, type 2 diabetes, and metabolic syndrome. *J. Steroid Biochem. Mol. Biol.* **2018**, *175*, 177–189. [CrossRef] [PubMed]

16. Prasad, P.; Kochhar, A. Interplay of vitamin D and metabolic syndrome: A review. *Diabetes Metab. Syndr. Clin. Res. Rev.* **2016**, *10*, 105–112. [CrossRef]

17. Parker, J.; Hashmi, O.; Dutton, D.; Mavrodaris, A.; Stranges, S.; Kandala, N.-B.; Clarke, A.; Franco, O.H. Levels of vitamin D and cardiometabolic disorders: Systematic review and meta-analysis. *Maturitas* **2010**, *65*, 225–236. [CrossRef]

18. Ford, E.S.; Ajani, U.A.; McGuire, L.C.; Liu, S. Concentrations of serum vitamin D and the metabolic syndrome among US adults. *Diabetes Care* **2005**, *28*, 1228–1230. [CrossRef]

19. Martini, L.A.; Wood, R.J. Vitamin D status and the metabolic syndrome. *Nutr. Rev.* **2006**, *64*, 479–486. [CrossRef]

20. Lu, L.; Yu, Z.; Pan, A.; Hu, F.B.; Franco, O.H.; Li, H.; Li, X.; Yang, X.; Chen, Y.; Lin, X. Plasma 25-hydroxyvitamin D concentration and metabolic syndrome among middle-aged and elderly Chinese individuals. *Diabetes Care* **2009**, *32*, 1278–1283. [CrossRef]

21. Scragg, R.; Sowers, M.; Bell, C. Serum 25-hydroxyvitamin D, diabetes, and ethnicity in the Third National Health and Nutrition Examination Survey. *Diabetes Care* **2004**, *27*, 2813–2818. [CrossRef] [PubMed]

22. Chiu, K.C.; Chu, A.; Go, V.L.W.; Saad, M.F. Hypovitaminosis D is associated with insulin resistance and β cell dysfunction. *Am. J. Clin. Nutr.* **2004**, *79*, 820–825. [CrossRef] [PubMed]

23. Ponda, M.P.; Dowd, K.; Finkielstein, D.; Holt, P.R.; Breslow, J.L. The short-term effects of vitamin D repletion on cholesterol: A randomized, placebo-controlled trial. *Arterioscler. Thromb. Vasc.* **2012**, *32*, 2510–2515. [CrossRef] [PubMed]

24. Maki, K.C.; Rubin, M.R.; Wong, L.G.; McManus, J.F.; Jensen, C.D.; Lawless, A. Effects of vitamin D supplementation on 25-hydroxyvitamin D, high-density lipoprotein cholesterol, and other cardiovascular disease risk markers in subjects with elevated waist circumference. *Int. J. Food. Sci. Nutr.* **2011**, *62*, 318–327. [CrossRef]

25. Moher, D.; Liberati, A.; Tetzlaff, J.; Altman, D.G. Preferred reporting items for systematic reviews and meta-analyses: The PRISMA statement. *Int. J. Surg.* **2010**, *8*, 336–341. [CrossRef]

26. Available online: http://www.ClinicalTrials.gov (accessed on 31 July 2020).

27. International Clinical Trials Registry Platform (ICTRP). Available online: https://apps.who.int/trialsearch/ (accessed on 31 July 2020).

28. *Cochrane Handbook for Systematic Reviews of Interventions*; Higgins, J.P.; Thomas, J.; Chandler, J.; Cumpston, M.; Li, T.; Page, M.J.; Welch, V.A. (Eds.) John Wiley & Sons: New York, NY, USA, 2019.

29. GRADEpro. Available online: www.gradepro.org (accessed on 20 September 2020).

30. Makariou, S.E.; Elisaf, M.; Challa, A.; Tentolouris, N.; Liberopoulos, E.N. No effect of vitamin D supplementation on cardiovascular risk factors in subjects with metabolic syndrome: A pilot randomised study. *Arch. Med. Sci. Atheroscler. Dis.* **2017**, *2*, 52–60. [CrossRef]

31. Makariou, S.E.; Elisaf, M.; Challa, A.; Tellis, C.C.; Tselepis, A.D.; Liberopoulos, E.N. No effect of vitamin D administration plus dietary intervention on emerging cardiovascular risk factors in patients with metabolic syndrome. *J. Nutr. Intermed. Metab.* **2019**, *16*, 100093. [CrossRef]

32. Makariou, S.E.; Elisaf, M.; Challa, A.; Tellis, C.; Tselepis, A.D.; Liberopoulos, E.N. Effect of combined vitamin D administration plus dietary intervention on oxidative stress markers in patients with metabolic syndrome: A pilot randomized study. *Clin. Nutr. ESPEN* **2019**, *29*, 198–202. [CrossRef]

33. Wongwiwatthananukit, S.; Sansanayudh, N.; Phetkrajaysang, N.; Krittiyanunt, S. Effects of vitamin D 2 supplementation on insulin sensitivity and metabolic parameters in metabolic syndrome patients. *J. Endocrinol Investig.* **2013**, *36*, 558–563.

34. Yin, X.; Yan, L.; Lu, Y.; Jiang, Q.; Pu, Y.; Sun, Q. Correction of hypovitaminosis D does not improve metabolic syndrome risk profile in a Chinese population: A randomized controlled trial for 1 year. *Asia Pac. J. Clin. Nutr.* **2016**, *25*, 71–77.

35. Farag, H.A.M.; Hosseinzadeh-Attar, M.J.; Muhammad, B.A.; Esmaillzadeh, A.; Hamid el Bilbeisi, A. Effects of vitamin D supplementation along with endurance physical activity on lipid profile in metabolic syndrome patients: A randomized controlled trial. *Diabetes Metab. Syndr. Clin. Res. Rev.* **2019**, *13*, 1093–1098. [CrossRef] [PubMed]

36. Salekzamani, S.; Mehralizadeh, H.; Ghezel, A.; Salekzamani, Y.; Jafarabadi, M.A.; Bavil, A.S.; Gargari, B.P. Effect of high-dose vitamin D supplementation on cardiometabolic risk factors in subjects with metabolic syndrome: A randomized controlled double-blind clinical trial. *J. Endocrinol. Investig.* **2016**, *39*, 1303–1313. [CrossRef]

37. Van Schoor, N.M.; Lips, P. Worldwide vitamin D status. *Best Pract. Res. Clin. Endocrinol. Metab.* **2011**, *25*, 671–680. [CrossRef] [PubMed]

38. Mendes, M.M.; Charlton, K.; Thakur, S.; Ribeiro, H.; Lanham-New, S.A. Future perspectives in addressing the global issue of vitamin D deficiency. *Proc. Nutr. Soc.* **2020**, *79*, 246–251. [CrossRef] [PubMed]

39. Grant, W.B.; Bhattoa, H.P.; Pludowski, P. Determinants of Vitamin D Deficiency from Sun Exposure. In *Vitamin D*; Elsevier: Amsterdam, The Netherlands, 2018; pp. 79–90. ISBN 9780128099636.

40. Vimaleswaran, K.S.; Berry, D.J.; Lu, C.; Tikkanen, E.; Pilz, S.; Hiraki, L.T.; Cooper, J.D.; Dastani, Z.; Li, R.; Houston, D.K.; et al. Causal relationship between obesity and vitamin D status: Bi-directional Mendelian randomization analysis of multiple cohorts. *PLoS Med.* **2013**, *10*, e1001383. [CrossRef] [PubMed]

41. Aquino, S.L.S.; Cunha, A.T.O.; Lima, J.G.; Sena-Evangelista, K.C.M.; Oliveira, A.G.; Cobucci, R.N.; Pedrosa, L.F.C. Effects of vitamin D supplementation on fasting glucose, dyslipidemia, blood pressure, and abdominal obesity among patients with metabolic syndrome: A protocol for systematic review and meta-analysis of randomized controlled trials. *Syst. Rev.* **2020**, *9*. [CrossRef] [PubMed]

42. Challoumas, D. Vitamin D supplementation and lipid profile: What does the best available evidence show? *Atherosclerosis* **2014**, *235*, 130–139. [CrossRef]

43. Wang, H.; Xia, N.; Yang, Y.; Peng, D.Q. Influence of vitamin D supplementation on plasma lipid profiles: A meta-analysis of randomized controlled trials. *Lipids Health Dis.* **2012**, *11*, 42. [CrossRef]

44. Jorde, R.; Grimnes, G. Vitamin D and metabolic health with special reference to the effect of vitamin D on serum lipids. *Prog. Lipid Res.* **2011**, *50*, 303–312. [CrossRef]

45. Wortsman, J.; Matsuoka, L.Y.; Chen, T.C.; Lu, Z.; Holick, M.F. Decreased bioavailability of vitamin D in obesity. *Am. J. Clin. Nutr.* **2000**, *72*, 690–693. [CrossRef]

46. Mangin, M.; Sinha, R.; Fincher, K. Inflammation and vitamin D: The infection connection. *Inflamm. Res.* **2014**, *63*, 803–819. [CrossRef] [PubMed]

47. Lugg, S.T.; Howells, P.A.; Thickett, D.R. Optimal Vitamin D Supplementation Levels for Cardiovascular Disease Protection. *Dis. Markers* **2015**, *2015*, 1–10. [CrossRef] [PubMed]

48. Hollis, B.W.; Wagner, C.L. The Role of the Parent Compound Vitamin D with Respect to Metabolism and Function: Why Clinical Dose Intervals Can Affect Clinical Outcomes. *J. Clin. Endocrinol. Metab.* **2013**, *98*, 4619–4628. [CrossRef] [PubMed]

49. Tabrizi, R.; Vakili, S.; Lankarani, K.; Akbari, M.; Jamilian, M.; Mahdizadeh, Z.; Mirhosseini, N.; Asemi, Z. The Effects of Vitamin D Supplementation on Markers Related to Endothelial Function Among Patients with Metabolic Syndrome and Related Disorders: A Systematic Review and Meta-Analysis of Clinical Trials. *Horm. Metab. Res.* **2018**, *50*, 587–596. [CrossRef]

50. Major, G.C.; Alarie, F.; Doré, J.; Phouttama, S.; Tremblay, A. Supplementation with calcium+ vitamin D enhances the beneficial effect of weight loss on plasma lipid and lipoprotein concentrations. *Am. J. Clin. Nutr.* **2007**, *85*, 54–59.

51. Ahn, J.; Yu, K.; Stolzenberg-Solomon, R.; Simon, K.C.; McCullough, M.L.; Gallicchio, L.; Jacobs, E.J.; Ascherio, A.; Helzlsouer, K.; Jacobs, K.B.; et al. Genome-wide association study of circulating vitamin D levels. *Hum. Mol. Genet.* **2010**, *19*, 2739–2745. [CrossRef]

52. Wang, T.J.; Zhang, F.; Richards, J.B.; Kestenbaum, B.; Van Meurs, J.B.; Berry, D.; Kiel, D.P.; Streeten, E.A.; Ohlsson, C.; Koller, D.L.; et al. Common genetic determinants of vitamin D insufficiency: A genome-wide association study. *Lancet* **2010**, *376*, 180–188. [CrossRef]

53. Shea, M.K.; Benjamin, E.J.; Dupuis, J.; Massaro, J.M.; Jacques, P.F.; D'Agostino, R.B., Sr.; Ordovas, J.M.; O'Donnell, C.J.; Dawson-Hughes, B.; Vasan, R.S.; et al. Genetic and non-genetic correlates of vitamins K and D. *Eur. J. Clin. Nutr.* **2009**, *63*, 458–464. [CrossRef]

54. McGrath, J.J.; Saha, S.; Burne, T.H.; Eyles, D.W. A systematic review of the association between common single nucleotide polymorphisms and 25-hydroxyvitamin D concentrations. *J. Steroid Biochem. Mol. Biol.* **2010**, *121*, 471–477. [CrossRef]

55. Rosecrans, R.; Dohnal, J.C. Seasonal vitamin D changes and the impact on health risk assessment. *Clin. Biochem.* **2014**, *47*, 670–672. [CrossRef]

56. Autier, P.; Boniol, M.; Pizot, C.; Mullie, P. Vitamin D status and ill health: A systematic review. *Lancet Diabetes Endocrinol.* **2014**, *2*, 76–89. [CrossRef]

57. Salekzamani, S.; Bavil, A.S.; Mehralizadeh, H.; Jafarabadi, M.A.; Ghezel, A.; Gargari, B.P. The effects of vitamin D supplementation on proatherogenic inflammatory markers and carotid intima media thickness in subjects with metabolic syndrome: A randomized double-blind placebo-controlled clinical trial. *Endocrine* **2017**, *57*, 51–59. [CrossRef]

58. Chamoli, R.; Bahuguna, G.; Painuly, A.M. Miraculous Physiological activities of vitamin D and its role in human health and diseases. *Sci. Cult.* **2019**, *85*, 395–400. [CrossRef]

59. Manson, J.E.; Bassuk, S.S.; Buring, J.E. VITAL Research Group. Principal results of the vitamin D and OmegA-3 TriaL (VITAL) and updated meta-analyses of relevant vitamin D trials. *J. Steroid Biochem. Mol. Biol.* **2020**, *198*, 105522. [CrossRef]

60. Manousaki, D.; Mitchell, R.; Dudding, T.; Haworth, S.; Harroud, A.; Forgetta, V.; Shah, R.L.; Luan, J.A.; Langenberg, C.; Timpson, N.J.; et al. Genome-wide association study for vitamin D levels reveals 69 independent loci. *Am. J. Hum. Genet.* **2020**, *106*, 327–337. [CrossRef]

Publisher's Note: MDPI stays neutral with regard to jurisdictional claims in published maps and institutional affiliations.

Review

Evidence Regarding Vitamin D and Risk of COVID-19 and Its Severity

Joseph Mercola [1,*], William B. Grant [2] and Carol L. Wagner [3]

[1]   Natural Health Partners, LLC, 125 SW 3rd Place, Cape Coral, FL 33991, USA
[2]   Sunlight, Nutrition, and Health Research Center, P.O. Box 641603, San Francisco, CA 94164-1603, USA; wbgrant@infionline.net
[3]   Department of Pediatrics, Shawn Jenkins Children's Hospital, Medical University of South Carolina, 10 McClennan Banks Drive, MSC 915, Charleston, SC 29425, USA; wagnercl@musc.edu
*   Correspondence: dr@mercola.com; Tel.: +1-239-599-9529

Received: 4 October 2020; Accepted: 29 October 2020; Published: 31 October 2020

Abstract: Vitamin D deficiency co-exists in patients with COVID-19. At this time, dark skin color, increased age, the presence of pre-existing illnesses and vitamin D deficiency are features of severe COVID disease. Of these, only vitamin D deficiency is modifiable. Through its interactions with a multitude of cells, vitamin D may have several ways to reduce the risk of acute respiratory tract infections and COVID-19: reducing the survival and replication of viruses, reducing risk of inflammatory cytokine production, increasing angiotensin-converting enzyme 2 concentrations, and maintaining endothelial integrity. Fourteen observational studies offer evidence that serum 25-hydroxyvitamin D concentrations are inversely correlated with the incidence or severity of COVID-19. The evidence to date generally satisfies Hill's criteria for causality in a biological system, namely, strength of association, consistency, temporality, biological gradient, plausibility (e.g., mechanisms), and coherence, although experimental verification is lacking. Thus, the evidence seems strong enough that people and physicians can use or recommend vitamin D supplements to prevent or treat COVID-19 in light of their safety and wide therapeutic window. In view of public health policy, however, results of large-scale vitamin D randomized controlled trials are required and are currently in progress.

Keywords: cathelicidin; COVID-19; endothelial dysfunction; IL-6; immune system; inflammation; MMP-9; SARS-CoV-2; vitamin D; 25-hydroxyvitamin D

---

1. Introduction

Until the 21st century, vitamin D was primarily recognized for its role in regulating calcium and bone health and preventing rickets [1]. In the last 20 years, however, research has shown that vitamin D also profoundly influences immune cells and generally lowers inflammation [2,3]. Vitamin D is a powerful epigenetic regulator, influencing more than 2500 genes [4] and impacting dozens of our most serious health challenges, including cancer [5,6], diabetes mellitus [7], acute respiratory tract infections [8], and autoimmune diseases such as multiple sclerosis [9].

According to the Worldometer website [10], the world had recorded 40,628,492 cases and 1,122,733 deaths from COVID-19 by 19 October 2020.

There are a number of findings regarding COVID-19 that may be related to vitamin D status.

- Seasonal dependence: it began in winter in the northern hemisphere and both case and death rates were lowest in summer, especially in Europe, and rates began increasing again in July, August, or September in various European countries [10]; it is thus generally inversely correlated with solar UVB doses and vitamin D production [11,12].

- African Americans and Hispanics have higher COVID-19 case and death rates than European Americans [13,14], possibly due to darker skin pigmentation and lower 25-hydroxyvitamin D [25(OH)D] concentrations [15].
- Much of the damage from COVID-19 is thought to be related to the "cytokine storm", which is manifested as hyperinflammation and tissue damage [16].
- The body's immune system becomes dysregulated in severe COVID-19 [17].

This narrative review examines the evidence indicating that vitamin D could play important roles in reducing the risk and severity of and death from infections, including COVID-19.

2. Findings Regarding Vitamin D and COVID-19

2.1. Vitamin D Deficiency Increases the Risk and Severity of COVID-19

Mainly owing to the recency and novelty of the SARS-CoV-2 virus, the evidence that vitamin D status affects the risk of COVID-19 comes primarily from observational and ecological studies. Clinical trials involving vitamin D supplementation and incidence of COVID-19 have not been reported to date. Of the 48 clinical trials on vitamin D and COVID-19 listed in the Clinical Trials registry maintained by the U.S. government [18], only four will investigate prevention, and three of those are enrolling health care workers, a group that is highly exposed to COVID-19.

Table 1 lists the findings from observational studies regarding 25(OH)D concentration and COVID-19 as of 15 October 2020, listed in ascending order by date first posted online. The table lists the study parameters and findings as well as the strengths and limitations of the studies. Two of the studies used 25(OH)D concentrations 10–14 years before the COVID-19 incidence data; the others generally used 25(OH)D concentrations at the time of hospital admission. Many of the studies have small numbers of COVID-19 patients. Other than the two studies with long intervals between 25(OH)D concentrations and COVID-19, and one observational study from Austria, the studies found inverse correlations between COVID-19 severity and/or risk of death.

Table 1. Summary of observational study findings regarding COVID-19 and 25(OH)D concentrations posted at pubmed.gov by 27 September 2020.

	Location	Participants	Outcomes vs. 25(OH)D (ng/mL)	Strengths, Limitations	Reference
1	UK	449 C19 patients 348,598 controls from UK Biobank	Incidence for 25(OH)D <10 vs. >10 Univariable OR = 1.37 (1.07–1.76, p = 0.01) Multivariable OR = 0.92 (0.71–1.21, p = 0.56)	Some confounding variables should not be used since they affect 25(OH)D concentrations [19,20] 25(OH)D data were from blood drawn from 2006 to 2010 Participant 25(OH)D concentrations change over time, reducing correlations with disease outcomes [21]	Hastie [22]
2	Switzerland	27 patients PCR+ for SARS-CoV-2; 80 patients PCR– 1377 controls with 25(OH)D measured in same period in 2019	Patients PCR+ had mean 25(OH)D = 11 vs. 25 for patients PCR– (p = 0.004) Controls had 25(OH)D = 25, not significantly different from patients PCR– (p = 0.08)	PCR+ is for antibodies; may not be active COVID-19 Small number of PCR+	D'Avolio [23]

Table 1. *Cont.*

	Location	Participants	Outcomes vs. 25(OH)D (ng/mL)	Strengths, Limitations	Reference
3	UK, Newcastle upon Tyne	92 C19, non-ITU; 42 C19, ITU Patients were supplemented with vitamin D_3 at doses inversely correlated with baseline 25(OH)D concentration	Non-ITU vs. ITU: 25(OH)D 19 ± 15 vs. 13 ± 7 ($p = 0.30$) 25(OH)D <20 vs. >20 ($p = 0.02$) RR for death, 25(OH)D = 0.97 (0.42–2.23, $p = 0.94$)	Lack of correlation of death with baseline 25(OH)D was likely due to graded supplementation with vitamin D	Panagiotou [24]
5	Italy	42 C19 hospitalized patients; mean age 65 ± 13 years, 88 with ARDS	!L6 for 25(OH)D >30: 80 ± 40 pg/L; for 25(OH)D <10, 240 ± 470 pg/L After 10 days, patients with 25(OH)D <10 had a 50% mortality vs. 5% for 25(OH)D <10 ($p = 0.02$)	Patients with 25(OH)D <10 ng/mL had a mean age of 74 ± 11 years vs. 63 ± 15 years for patients with 25(OH)D ≥10 ng/mL	Carpagnano [25]
6	Korea	50 C19 patients with PCR+, 150 controls; mean age = 52 ± 20 years	C19 vs. control: 16 (SD 8) vs. 25 (SD 13) ($p < 0.001$); ≤20, 74% vs. 43% ($p = 0.003$); ≤10, 24% vs. 7% ($p = 0.001$)	Strengths: measured B vitamin, folate, selenium and zinc concentrations as well as 25(OH)D Weaknesses: small number of patients; incomplete analysis of data for C19 outcomes	Im [26]
7	Russia	80 C19 patients with community-acquired pneumonia	Severe: 25(OH)D = 12 ± 6 ng/mL; moderate to severe: 25(OH)D = 19 ± 14 ng/mL Death: 25(OH)D = 11 ± 6 ng/mL; discharged: 18 ± 6 ng/mL Obesity rates: 62% for severe, 15% for discharged, $p < 0.001$	Strengths: studied the effect of obesity Weaknesses: small numbers	Karonova [27]
8	Mexico	172 hospitalized C19 patients	Mean 25(OH)D = 17 ± 7 ng/mL for hospitalized C19 patients Survivors: mean age = 48 ± 13 years; 25(OH)D = 17 ± 7 ng/mL Death: mean age = 65 ± 12 years; 25(OH)D = 14 ± 6 ng/mL (p value for difference in 25(OH)D = 0.0008)	Weaknesses: survivors were much younger than non-survivors Comorbid factors not reported	Tort [28]
9	UK	105 patients with C19 symptoms; 70 C19 PCR+, 35 PCR–; mean age = 80 ± 10 years	PCR+: 25(OH)D = 11 (8–19); PCR–: 25(OH)D = 21 (13–129) ($p = 0.0008$) Comorbid diseases were not significantly correlated with ≤12 vs. >12;	PCR+ is for antibodies; may not be active COVID-19	Baktash [29]

Table 1. *Cont.*

	Location	Participants	Outcomes vs. 25(OH)D (ng/mL)	Strengths, Limitations	Reference
10	UK	656 C19, 203 died from C19; 340,824 controls from UK Biobank	Incidence for 25(OH)D <10 vs. >10 Univariable OR = 1.56 (1.28–1.90, $p < 0.0001$) Multivariable OR = 1.10 (0.88–1.37, $p = 0.40$) Death for 25(OH)D <10 ng/mL vs. >10 ng/mL Univariable OR = 1.61 (1.14–2.27, $p = 0.0007$) Multivariable OR = 1.21 (0.83–1.76, $p = 0.31$)	Same comments as for earlier UK Biobank study	Hastie [30]
11	Germany	185 C19; median age = 60 years	Multivariable HR for death for 25(OH)D <12: IMV/D, 6.1 (2.8–13.4, $p < 0.001$); D, 14.7 (4.2–52.2, $p < 0.001$)	Strengths: HR adjusted for age, gender, and comorbidities Weaknesses: Small number of IMV and deaths	Radujkovic [31]
12	Austria	109 C19 hospitalized patients; mean age = 58 ± 14 years	Mild: 26 ± 12 Moderate: 22 ± 8 Severe: 20 ± 10 ($p = 0.12$) PTH increased significantly with age ($p = 0.001$)	The vitamin D finding may have been limited owing to the high mean 25(OH)D concentrations Mild C19 patients had mean age = 46 ± 16 years; moderate and severe patients has mean age = 60 ± 13 years PTH increases with age [32]	Pizzini [33]
13	Spain	80 emergency department patients with a PCR+ test within the past three months; retrospective study	49 non-severe C19, 25(OH)D = 19 ng/mL; 31 severe C19, 25(OH)D = 13 ng/mL ($p = 0.15$) For patients under 65 years, 30 non-severe C19, 25(OH)D = 22 (11–31) ng/mL; 10 severe C19, 25(OH)D = 11 (9–12) ng/mL ($p = 0.009$) Multivariable OR for severe C19 for 25(OH)D <20 ng/mL = 3.2 (95% CI, 0.9 to 11.4, $p = 0.07$)	Weaknesses: small study; prevalence of advanced chronic kidney disease was higher in severe than non-severe cases (45% vs. 24%, $p = 0.054$)	Macaya [34]
14	China	62 C19 patients, 80 healthy controls	age, 25(OH)D: controls: 43 years, 29 (23–33) ng/mL; mild/moderate C19: 39 (30–49) years, 23 (18–27) ng/mL; severe/critical C19: 65 (54–69) years, 15 (13–20) ng/mL Multivariate OR for severe/critical C19 for 25(OH)D <20 ng/mL = 15 (1.2 to 187, $p = 0.03$)	Strengths: many factors measured Weaknesses: the severe/critical patients were much older than mild/moderate patients and controls	Ye [35]

Abbreviations: ARDS, acute respiratory distress syndrome; C19, COVID-19 patients; D, death; HR, hazard ratio; IMV, invasive mechanical ventilation; ITU, intensive treatment unit; OR, odds ratio; PCR, polymerase chain reaction; PTH, parathyroid hormone; RR, relative risk; SD, standard deviation.

The study from Newcastle upon Tyne, UK, supplemented patients with vitamin D_3 depending on their baseline 25(OH)D concentration [24]. Those with 25(OH)D concentration <5 ng/mL were given

300,000 IU vitamin D_3 followed by 1600 IU/d. Those with 25(OH)D between 5 and 10 ng/mL were given 200,000 IU vitamin D followed by 800 IU/d. Those with 25(OH)D between 10 and 16 ng/mL were given 100,000 IU vitamin D followed by 800 IU/d. Those with 25(OH)D between 16–30 ng/mL were given 800 IU/d, while those with 25(OH)D >30 ng/mL were not given vitamin D. Probably as a result, baseline 25(OH)D concentrations were not associated with mortality ($p = 0.94$).

Table 2 presents data on SARS-CoV-2 positivity for large populations independent of whether the participants had symptomatic COVID-19.

Table 2. Summary of observational study findings regarding SARS-CoV-2 positivity in general populations and 25(OH)D concentrations by date of first publication up to October 15, 2020.

	Location	Participants	Outcomes vs. 25(OH)D (ng/mL)	Strengths, Limitations	Reference
1	Israel	Data from a hospital in Tel Aviv involving patients who had previous 25(OH)D measurements and were tested for SARS-CoV-2 using PCR 782 patients PCR+ 7025 patients PCR–	Univariate: 20–29 vs. >30: OR = 1.59 (1.24–2.02, $p = 0.005$); <20 vs. >30, OR = 1.58 (1.13–2.09, $p = 0.0002$). Multivariate: <30 vs. >30, OR = 1.50 (1.13–1.98, $p = 0.001$)	Strengths: large number of participants. Weakness: PCR+ is not COVID-19.	Merzon [36]
2	US	489 C19 patients, PCR+; mean age = 49 ± 18 years with 25(OH)D concentrations were from preceding 12 months	124 <20 vs. 287 >20, RR = 1.77 (1.12–2.81, $p = 0.02$)	Strengths: this is a retrospective study in which serum 25(OH)D concentrations and vitamin D supplementation history were obtained during the preceding 12 months.	Meltzer [37]
3	US	191,779 patients tested for 25(OH)D and SARS-CoV-2 positivity during the past year by Quest Diagnostics	SARS-CoV-2 positivity for 25(OH)D <20 = 12.5% (95% CI, 12.2–12.8%); positivity for 25(OH)D >55 = 5.9% (95% CI, 5.5–6.4%). For 25(OH)D <20, SARS-CoV-2 positivity rates were: black non-Hispanic, 19%; Hispanic, 16%; white non-Hispanic, 9%	Strengths: large number of participants and is a retrospective study. 25(OH)D concentrations were seasonally adjusted. Weaknesses: SARS-CoV-2 positivity is a precursor to COVID-19, but many with positivity do not develop COVID-19. There may be bias in who was tested since the tests were ordered by physicians.	Kaufman[38]

The study from Israel reported that 25(OH)D concentration inversely correlated with COVID-19 in both univariate and multivariate regression analyses except for multivariate hospitalization of patients [36]. For hospitalization of patients, the only significant factor in the multivariate hospitalization was age 50 years and older, implying that vitamin D status becomes less important with age. Yet, the study from the UK with patients of mean age 80 ± 10 years reported that 25(OH)D concentration was significantly lower for COVID-19 PCR+ patients than COVID-19 PCR– patients [29].

The observational study from the U.S. based on test data from Quest Diagnostics (Secaucus, NJ, USA) [38] is the largest observational study to date, with data for 191,779 patients with a mean age of 50 years (interquartile range, 40–65 years) tested for SARS-CoV-2 between 9 March and 19 June with 25(OH)D tests in the preceding 12 months at Quest Diagnostics. The study reported the following rates of SARS-CoV-2 positivity vs. 25(OH)D concentration: 39,120 patients <20 ng/mL, 12.5% (95% CI, 12.2–12.8%); 27,870 patients, 30–34 ng/mL, 8.1% (7.8–8.4%); 12,321 patients, >55 ng/mL, 5.9% (5.5–6.4%).

The finding that the SARS-CoV-2-positive rate in the U.S. varied from 6.5% for 25(OH)D concentration between 40 and 50 ng/mL to approximately 11.3% for 25(OH)D = 20 ng/mL may be due to the effect of vitamin D in reducing survival and replication of the virus by induction of cathelicidin and defensins as well as by increasing concentrations of free ACE2 [39], thereby preventing SARS-CoV-2 from entering cells via the ACE2 receptor [39]. The regression fit to all the data indicates that SARS-CoV-2 positivity is 40% lower for 25(OH)D >50 ng/mL than for 20 ng/mL, the value recommended by the Institute of Medicine [40,41]. The SARS-CoV-2-adjusted OR (aOR) for northern (>40°) vs. southern (<32°) was 2.66 (95% CI, 2.54–2.79), whereas that for central (32°–40°) vs. southern was 1.22 (1.16–1.38).

Regarding the higher rates in the northern states, a genetic variation was evident in SARS-CoV-2 from the original spike protein amino acid D614 form in China to the D614G mutated form it took in Europe [42]. (The Spike D614GF amino acid change is caused by an A-to-G nucleotide mutation at position 23,403 in the Wuhan reference strain.) The D614G form has greater transmission and was introduced to New York by people returning from Europe. Thus, that genetic change probably accounts for some of the higher SARS-CoV-2 positivity rate in the north. However, the shape of the 25(OH)D positivity rate is similar for all three latitude regions.

As for race/ethnic differences, African Americans have increased rates of social determinants predisposing them to COVID-19, such as lower income, education, and employment as well as higher rates of existing conditions such as diabetes, hypertension, cardiovascular disease, obesity, and lung disease [43]. Those factors may help explain why black people and Hispanic people have 7% and 4% higher SARS-CoV-2 positivity rates, respectively, than white people at 30 ng/mL. Nonetheless, the SARS-CoV-2 positive rate spread was much higher for black and Hispanic people than for white people near 20 ng/mL (18%, 16%, and 9%, respectively) than near 60 ng/mL (11%, 9%, and 5%, respectively), suggesting that vitamin D status plays a role in the increased COVID-19 rate for black and Hispanic people.

It can be argued that the association of low serum 25(OH)D concentrations with various diseases is due to "reverse causation", i.e., that the disease state lowers the concentrations in proportion to the severity of the disease. That argument was made to explain why randomized controlled trials (RCTs) with vitamin D supplementation often fail to support observational studies reporting inverse correlations between 25(OH)D concentration and disease risk [44,45]. There are several counters to that argument.

One is that many vitamin D RCTs did not enroll participants with low 25(OH)D concentrations and did not supplement with sufficient vitamin D to produce a significant change in health outcome. Robert Heaney pointed out that vitamin D RCTs should be guided by serum 25(OH)D concentrations, not vitamin D dose [46] (see also, [47]). In addition, more recent RCTs have found that vitamin D supplementation can reduce risk of some of the non-skeletal health disorders considered by Autier in 2017: cancer incidence and death according to secondary analyses [48], cancer mortality rate [49] and progression from prediabetes to diabetes in the secondary analyses [7].

A second argument is that the 25(OH)D concentrations used in prospective observational studies are obtained from blood drawn prior to the disease outcomes of interest. Only three observational studies listed in Table 1 were prospective studies with less than one year lag between blood draw and COVID-19 or SARS-CoV-2 positivity [36–38].

A counter argument is that there is evidence that an acute-inflammatory disease state lowers 25(OH)D concentrations. A systematic review summarized results from eight studies reported between 1992 and 2013 regarding changes in 25(OH)D concentrations after acute inflammatory insult [50]. Four studies involved surgery. One involving 19 patients undergoing cardiopulmonary bypass reported an 8 ng/mL drop in five minutes with return to near baseline after 24 h [51]. Three involving knee or knee/hip arthroplasty or orthopedic surgery reported two-day decreases of 7 ng/mL [52], 4 ng/mL [53] and 1 ng/mL for males, 3 ng/mL for females [54]. There was no significant change for malarial infection [55] and a one ng/mL decrease for acute pancreatitis [56]. The largest decease was

15 ng/mL after three days for an injection of bisphosphonate [57]. The nearest outcome to COVID-19 was malaria infection, for which no change was found. Thus, from these studies, it is unclear whether acute inflammation not associated with surgery results in reduction in 25(OH)D.

2.2. Vitamin D and Treatment of COVID-10

A study by Ohaegbulam and colleagues involved four COVID-19$^+$ patients in New York [58]. Two, a male aged 41 years and a female aged 57 years, were given five daily 50,000 IU vitamin D_2 doses, whereas another two, males aged 53 and 74 years, were given five daily 1000 IU vitamin D_3 doses. Baseline 25(OH)D concentration was between 17 and 22 ng/mL, whereas achieved 25(OH)D was 40 and 51 ng/mL for patients treated with high-dose vitamin D and 19 and 20 ng/mL for those treated with standard-dose vitamin D.

Biomarkers of inflammation were significantly reduced with high-dose treatment: CRP went from 31 to 2 mg/dL and from 17 to 8 mg/dL, compared with 13 to 22 mg/dL and 21 to 18 mg/dL for low-dose treatment; IL-6 went from 14 and 10 pg/mL to <5 pg/mL for high-dose treatment and from <5 and 6 pg/mL to <5 and 11 pg/mL for low-dose treatment.

The length of stay was 10 days for the high-dose patients and 13 and 14 days for the low-dose patients. The oxygen requirement went from zero and 15 L to zero for the high-dose patients and from 2 and 3 L to 2 and 7 L for the low-dose patients. The strengths of this study include that high-dose vitamin D_3 supplementation was used and that baseline and post-supplementation values for many parameters were measured. The main limitation was that only two patients were supplemented with high-dose vitamin D_3.

The results of pilot RCT of treatment of COVID-19 patients in Spain with calcifediol were announced on August 29 [59]. (Calcifediol [25(OH)D] is often used in Spain. It raises serum 25(OH)D concentration more quickly but does not last as long in the serum as a result of its lower lipophilia [60].) The mean age of the patients was 53 ± 10 years. None of the prognostic factors evaluated except previous high blood pressure [15 (58%) without treatment vs. 11 (24%) with treatment] significantly affected the outcome. In this study, 50 patients were given soft capsules of 0.532 mg of calcifediol on the day of admission, then 0.266 ng on day 3 and 7, and then weekly until discharge or admission to the intensive care unit (ICU). Thus, those in the treatment arm received approximately 130,000 IU of vitamin D during the first week, then approximately 33,000 IU/week thereafter. Serum 25(OH)D concentrations were not measured, but the calcifediol dose in the treatment arm was high enough to raise 25(OH)D concentration by approximately 20 ng/mL.

Forty-nine of the calcifediol-treated patients did not require the ICU, whereas 13 of the 26 not receiving that treatment did require the ICU. In addition, two of the patients admitted to the ICU died. The odds ratio (OR) for ICU in treated vs. control patients was 0.02 (95% CI, 0.002 to 0.17), which increased slightly when adjusted for hypertension and type 2 diabetes mellitus [OR = 0.03 (95% CI, 0.003 to 0.25)]. A meta-analysis of 34 studies found that hypertension was a significant risk factor for several or fatal COVID-19 compared to non-severe/non-fatal COVID-19: OR = 3.2 (95% CI 2.5 to 4.1) [61]. Thus, prevalence of hypertension should have been considered when dividing patients into treatment and control groups. The results of this study cannot be used for policy decisions. The main value of this study is that it is a pilot study for a study involving 1000 COVID-19 patients.

A "quasi-experimental study" of bolus vitamin D supplementation of residents in a French nursing home was conducted preceding and during a COVID-19 outbreak in the nursing home [62]. Residents were normally given a bolus dose of 80,000 IU vitamin D_3 every two to three months. COVID-19 affected many of the residents starting in March 2020.

Fifty seven of the residents, who had received 80,000 IU vitamin D_3 in the preceding month, were included in the "intervention group" while nine who had not were included in the "comparator group". The mean age of the residents was 87 ± 9 years. The mean follow-up time was 36 ± 7 days. Forty-seven (83%) of the intervention group survived compared to only four (44%) of the comparator group ($p = 0.02$). The fully adjusted HR for mortality according to vitamin D supplementation was 0.11 (95% CI, 0.03 to 0.48, $p = 0.003$).

A clinical trial was conducted regarding bolus vitamin D dose (100,000 IU vitamin D_3) supplementation involving 30 older (71 ± 6 years) and ten younger (38 ± 8 years) subjects and ten older controls (71 ± 10 years) [63]. Baseline 25(OH)D was 27 ± 8 ng/mL, rising to 42 ± 9 ng/mL within six days, then falling in a linear fashion to 32 ng/mL after 70 days. Thus, bolus vitamin D_3 supplementation monthly would be appropriate for nursing-home residents.

2.3. Vitamin D Helps Immune Cells Produce Antimicrobial Peptides

Many studies have shown that vitamin D activates immune cells to produce AMPs, which include molecules known as cathelicidins and defensins [64–67]. AMPs have a broad spectrum of activity, not only antimicrobial but also antiviral, and can inactivate the influenza virus [68]. The antiviral effects of AMPs are the result of, among other effects, the destruction of envelope proteins by cathelicidin [69–71]. See Figure 1.

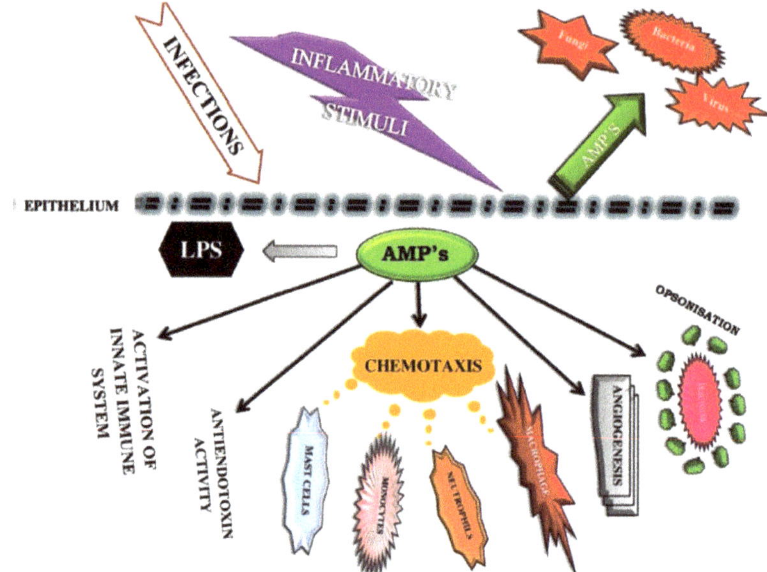

Figure 1. The cascade of events by the innate immune system in response to viral infections. Among the functions of AMPs (antimicrobial peptides) is chemotaxis, the movement of cells in response to a chemical stimulus, here macrophages, mast cells, monocytes, and neutrophils. Other effects include activation of the innate immune system, effects on angiogenesis, antiendotoxin activity, and opsonization (the molecular mechanism whereby pathogenic molecules, microbes, or apoptotic cells (antigenic substances) are connected to antibodies, complement, or other proteins to attach to the cell surface receptors on phagocytes and NK cells). LMS (lipopolysaccharide)

Cathelicidins are a distinct class of proteins present in innate immunity of mammals. In humans, the primary form of cathelicidin is known as LL-37 [72]. LL-37 also blocks viral entry into the cell similarly to what is seen with other antimicrobial peptides [73].

2.4. Vitamin D Reduces Inflammatory Cytokine Production

Elevated inflammation is an important risk factor for COVID-19 [16]. For example, much of the pathogenesis surrounding COVID-19 infection involves microvascular injury induced by hypercytokinemia, namely, by an important inflammatory cytokine—interleukin 6 (IL-6) [74,75]. Thus, it is useful to examine the role of vitamin D in reducing inflammation.

A number of reviews have suggested that one of the hallmarks of COVID-19 severity is the presence of a "cytokine storm" [76–79]. The "cytokine storm" is defined as the state of out-of-control release of a variety of inflammatory cytokines [79]. Observational studies, however, have found that cytokine concentrations are elevated in COVID-19 patients compared to controls, but not as high as in some other diseases.

A study in the Netherlands compared cytokine levels in critically ill patients [80]. The study involved 46 COVID-19 patients, 51 with septic shock with acute respiratory tract syndrome (ARDS), 15 with septic shock without ARDS, 30 with out-of-hospital cardiac arrest (OHCA), and 62 with trauma. Levels of (TNF_α) for COVID-19 patients were lower than for septic shock patients but higher than for OHCA or trauma patients. Levels of IL-6 and IL-8 for COVID-19 patients were lower than for septic shock patients but comparable with those for OHCA and trauma patients.

A recent review examined whether IL-6 concentrations might affect the outcome of COVID-19 [75]. The evidence presented included age-stratified IL-6 concentrations from a healthy Italian population were highly correlated with age-stratified Italian COVID-19 deaths, which in turn were highly correlated with age-stratified COVID-19 death rates in the UK. The researchers also cited trials of vitamin D supplementation and its effect on IL-6 concentrations, of which eight of 11 showed a significant lowering of IL-6. People for whom a significant lowering was not found were healthy older adults, asthma patients, and prediabetic adults. That reviewshowed how IL-6 increases the severity of COVID-19 by upregulating angiotensin-converting enzyme 2 (ACE2) receptors and induction of macrophage cathepsin L. Cathepsin L mediates the cleavage of the S1 subunit of the coronavirus surface spike glycoprotein. That cleavage is necessary for coronavirus entry into human host cells, virus–host cell endosome membrane fusion, and viral RNA release for the next round of replication [81].

A study from Ireland investigated cytokine concentrations of healthy controls, stable COVID-19 patients, ICU COVID-19 patients, and ICU community-acquired pneumonia patients [75]. ICU-COVID-19 patients had the highest concentrations of IL-1β, IL-6, IL-6 to IL-10 ratio, and tumor necrosis factor receptor superfamily member 1A (TNFR1). Stable COVID-19 patients had concentrations that were between the levels noted for healthy controls and those of ICU COVID-19 patients for all of the cytokines. ICU-community-acquired pneumonia patients had inflammatory cytokine concentrations between stable and ICU COVID-19 patients but higher IL-10 concentrations.

A study of COVID-19 hyperinflammation (COV-HI) was conducted on 269 polymerase chain reaction (PCR)-confirmed COVID-19 patients admitted to two UK hospitals in March [82]. Hyperinflammation was defined as CRP concentration greater than 150 mg/L or doubling within 24 h from greater than 50 mg/L, or a ferritin concentration greater than 1500 µg/L. Ninety (33%) of the patients met the criteria for COV-HI at admission. Forty percent of COV-HI patients died compared to 26% of the non-COV-HI patients. Meeting the COV-HI criteria was significantly associated with risk of next-day escalation of respiratory support or death (hazard ratio = 2.24 (95% CI, 1.62 to 2.87)).

Another study developed a more extensive set of criteria for COV-HI [83]. The criteria included elevated temperature, macrophage activation (elevated ferritin), haemotological dysfunction related to neutrophils and lymphocytes, coagulopathy (elevated D-dimer), hepatic injury (elevated lactate dehydrogenase or aspartate aminotransferase concentration), and cytokinaemia (elevated IL-6, triglyceride, or CRP concentration). It is not clear whether vitamin D supplementation could affect any of these factors other than cytokinaemia.

Other papers have noted that concentrations of many cytokines are elevated in COVID-19 patients [75,84,85].

There are several reasons why the cytokine storm is associated with severe COVID-19 and death [86,87]. As outlined in the review by Hojyo [86], the hypothesis that the main cause of death of COVID-19 is ARDS with cytokine storms can be explained by at least two reasons. One is intravascular coagulation as an important cause of multiorgan injury, which is mainly mediated by inflammatory cytokines such as IL-6 [88]. The other is that the SARS-CoV-2 virus affects endothelial cells, causing further cell death, which leads to vascular leakage and induces a cytopathic effect on airway epithelial cells [89].

2.5. Type II Pneumocytes and Surfactants in the Lungs

The type II pneumocytes in the lung are the primary target for coronaviruses because the ACE2 receptors to which the virus binds are highly expressed on those cells. One problem with COVID-19 is that it impairs the function of type II pneumocytes, which then decreases the surfactant concentration in the alveolar–air interface [90]. That is important because surfactant prevents the collapse of the alveoli.

Surfactant allows alveoli to stay open and compliant during both inhalation and exhalation. During inhalation, alveoli may collapse if they do not contain surfactant. If they collapse, gas exchange across the alveoli wall cannot occur. Without surfactant, each breath taken is like blowing up a collapsed balloon and then letting the air out of that balloon (lungs) and then doing it all over again with the next breath cycle. Simply put, having enough surfactant is necessary for alveoli to stay open and gas exchange to occur. Another aspect of surfactant is its protein A (SP-A), which binds to influenza A viruses via its sialic acid residues and thereby neutralizes the virus [91]. Surfactant protein D clears influenza A from the lungs of mice [92]. There is some evidence that $1\alpha,25(OH)_2D$ increases surfactant production [93]. Such activity can be generalized to other viruses.

2.6. Vitamin D, Angiotensin II, and ACE2 Receptors

Angiotensin-converting enzyme (ACE) is part of the renin–angiotensin system (RAS), which controls blood pressure by regulating the volume of bodily fluids. Angiotensin-converting enzyme 1 (ACE1) converts the hormone angiotensin I to the active vasoconstrictor angiotensin II [94]. Angiotensin II is a natural peptide hormone best known for increasing blood pressure through stimulating aldosterone [95] ACE2 normally consumes angiotensin I, thereby lowering its concentrations. However, SARS-CoV-2 infection downregulates ACE2, leading to excessive accumulation of angiotensin II.

Cell cultures of human alveolar type II cells with vitamin D have shown that the SARS-CoV-2 virus interacts with the ACE2 receptor expressed on the surface of lung epithelial cells. Once the virus binds to the ACE2 receptor, it reduces its activity and, in turn, promotes ACE1 activity, forming more angiotensin II, which increases the severity of COVID-19 [96,97]. That effect may also be related to the vitamin D binding protein [98].

The vitamin D metabolite calcitriol increases expression of ACE2 in the lungs of experimental animals [99]. (Calcitriol has also been found to increase ACE2 protein expression in rat microglia BV2 cells [100].) The additional ACE2 expressed as a consequence of vitamin D supplementation might reduce lung injury [101] because it can promote binding of the virus to the pulmonary epithelium. As mentioned, calcitriol also induces α-1-antitrypsin synthesis, which is vital for lung integrity and repair, by CD4[+] T cells, which is required for the increased production of anti-inflammatory IL-10. Calcitriol should not be used to treat COVID-19 given the risk of hypercalcemia; however, vitamin D supplementation increases calcitriol concentrations [102] through its regulated conversion in the proximal tubules of the kidney and in extrarenal cells at the nuclear membrane.

High concentrations of angiotensin II may cause ARDS or cardiopulmonary injury. Renin, by contrast, is a proteolytic enzyme and a positive regulator of angiotensin II. Vitamin D is a potent inhibitor of renin. Vitamin D supplementation prevents angiotensin II accumulation and decreases proinflammatory activity of angiotensin II by suppressing the release of renin in patients infected with COVID, thus reducing the risk of ARDS, myocarditis, or cardiac injury [103].

Although vitamin D increases expression of ACE2, which promotes the binding of the virus, it prevents the constriction response of the lung blood vessel in COVID-19, as illustrated in Figure 2 [104] (permission to reuse granted by copyright holder). ARDS is also due to a variety of mechanisms, including cytokine storm, neutrophil activation, and increased (micro)coagulation, and it is likely that vitamin D supplementation would counter those mechanisms [105]. ARDS is responsible for approximately 70% of fatal COVID-19 cases [106].

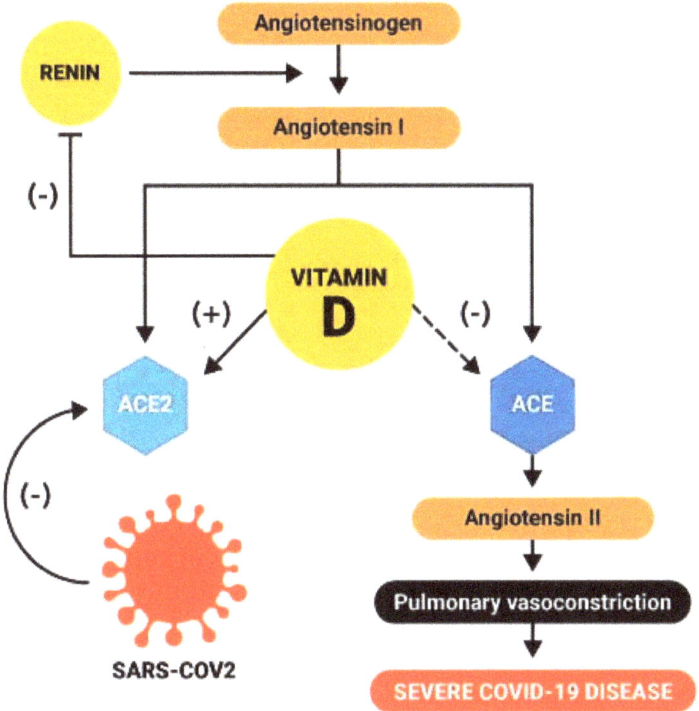

The role of vitamin D in COVID-19. SARS-CoV2 binds to the ACE2 of alveolar cells and disturbs the ratio of ACE2/ACE activity. It increases ACE activity and, in turn, results in more angiotensin II formation causing pulmonary vasoconstriction to precipitate severe COVID-19. Vitamin D induces ACE2 expression, which limits the formation of angiotensin II via ACE and reduces lung injury. Besides, vitamin D supplementation may have a protective role against COVID-19. *(Dashed line indicates indirect effect)*

Figure 2. The role of vitamin D regarding ACE in response to SARS-CoV-2. ACE: angiotensin-converting enzyme.

2.7. Reduces Risk of Endothelial Dysfunction

Jun Zhang and colleagues outlined how endothelial dysfunction (ED) contributes to COVID-19-associated vascular inflammation and coagulopathy, two hallmarks of severe COVID-19 [107]. Four stages of ED were identified that contribute to inflammation and coagulopathy. Stage 1 is Type I endothelial cell (EC) activation after infection by SARS-CoV-2 entering through the ACE2 receptor. That results in the loss of anticoagulant molecules. Stage 2 is Type II EC activation which leads to the de novo synthesis of procoagulant molecules. Stage 3 is endothelial apoptosis involving endothelial detachment and denudation of basement membrane. Stage 4 is endothelial necrosis.

A number of papers have discussed how vitamin D can reduce risk of ED. In a second review, Zhang and colleagues notes that vitamin D likely protects against ED by reducing oxidative stress and NF-κB activation [108,109]. A recent review outlined how vitamin D maintains endothelial function by reducing the production of reactive oxygen species as well as reducing proinflammatory mediators such as TNF-α and IL-6 and suppressing the NF-κB pathway [110]. A laboratory study involving mice as well as type II alveolar epithelial cells found that vitamin D attenuated lung injury by stimulating type II alveolar epithelial cell proliferation and migration, reducing epithelial cell apoptosis and inhibiting TGF-β-induced epithelial mesenchymal transition [111].

2.8. Matrix Metalloproteinase 9

Matrix metalloproteinase-9 (MMP-9) is a member of the family of proteases that degrade extracellular matrix remodeling proteins. MMP-9 has been widely studied in acute lung injury and acute lung disease [112]. A study in Norway investigated correlations between respiratory failure and MMP-9 in 21 COVID-19 patients with respiratory failure in comparison with seven COVID-19 patients without respiratory failure [113].

Respiratory failure was defined as arterial partial pressure of oxygen to fraction of inspired oxygen ratio (P/F ratio) <40 kPa (300 mmHg), corresponding to the threshold in ARDS. The researchers found a significant inverse correlation of the P/F ratio with respect to the $\log_{10}$ (MMP-9) as well as significantly higher MMP-9 concentrations for P/F below the threshold than above it. In a study of 171 healthy British Bangladeshi adults, vitamin D status was the sole determinant of circulating MMP-9 (inversely) and an independent determinant of CRP (inversely) [114].

A search of pubmed.gov for articles regarding vitamin D, MMP-9 and infections did not find any related to viral infections, but did find some regarding bacterial infections. A laboratory study in the UK found that *Mycobacterium tuberculosis* induced the production of MMP-1, MMP-7, and MMP-10 [115]. MMP-9 gene expression, secretion and activity were significantly inhibited by $1\alpha,25(OH)_2D_3$ irrespective of infection.

2.9. RAS-Mediated Bradykinin Storm

Several recent publications looked at the role of bradykinin (BK) in the progression of COVID-19. Jacobson used the Summit supercomputer at Oak Ridge National Lab in Tennessee is the second fastest supercomputer in the world and in the summer of 2020 analyzed data on more than 40,000 genes from 17,000 samples from COVID-19 patients [116]. The analysis revealed that SARS-CoV-2 actively upregulates ACE2 receptors in places where they're typically expressed at low levels, including the lungs.

Additionally, an imbalance in RAS was also found, represented by decreased ACE in combination with increases in ACE2, renin, angiotensin, key RAS receptors, kininogen and many kallikrein enzymes that activate it, and both BK receptors, which produces a BK storm [116]. Since BK dilates blood vessels and increases permeability, excessive BK leads to fluid to soft tissue fluid accumulation. This leads to several adverse effects seen in COVID-19 patients, including on the heart, vascular system, pulmonary system, brain, and muscles [116]. The authors suggested that vitamin D could reduce the risk of the BK storm through several mechanisms including regulation of RAS.

Renin is the enzyme that catalyzes the first step in the activation pathway of angiotensinogen by cleaving angiotensinogen to form angiotensin I, which is then converted to angiotensin II by angiotensin I converting enzyme. In the COVID samples Jacobson analyzed renin levels were increased 380-fold compared to controls. Vitamin D is a negative endocrine RAS modulator and inhibits renin expression and generation [40] and it appears likely that vitamin D deficiency amelioration would limit the COVID-19 BK storm. However, further investigation is needed to evaluate the role of vitamin D in this context

2.10. Summary: How Vitamin D Might Reduce Risk, Severity, and Death from COVID-19

Many reviews consider the ways in which vitamin D reduces the risk of viral infections [8,15,76,117–125]. Vitamin D probably reduces the risk of viral respiratory infections because it influences several immune pathways [126].

Vitamin D appears to decrease the risk of respiratory tract infections, including COVID-19, through six potential mechanisms:

- Inactivates some viruses by stimulating antiviral mechanisms such as antimicrobial peptides, as discussed in Section 2.3.
- Reduces proinflammatory cytokines through modulating the immune system, as discussed in Section 2.4.
- Increases ACE2 concentrations and reduces risk of death from ensuing ARDS, as discussed in Section 2.5.
- Reduces risk of endothelial dysfunction, as discussed in Section 2.7.
- Reduces MMP-9 concentrations, as discussed in Section 2.8.
- Reduces risk of the bradykinin storm, as discussed in Section 2.9.

However, much further research is required to confirm the mechanisms by which vitamin D reduces the risk and severity of COVID-19.

2.11. Vitamin D Seasonality and COVID-19

Since epidemic influenza rates are higher in winter than in summer [127], it was expected that COVID-19 would also exhibit a seasonal dependence. Two recent papers provide evidence on monthly and seasonal variation of viral infections. One in 2019 performed a systematic analysis of global patterns in monthly activity of influenza virus, respiratory syncytial virus, parainfluenza virus and metapneumovirus [128].

The second one, published in 2020, did the same for the global seasonality of human seasonal coronaviruses [129]. For nearly all of these viruses, infection rates in northern mid and high latitudes are highest from November through March. They examined correlations of meteorological conditions with coronavirus infections, finding the highest correlation with low temperature combined with high relative humidity. In winter, high relative humidity is associated with low absolute humidity. Low absolute humidity was found as an important factor for transmission of epidemic influenza [130].

A recent analysis of influenza seasonality in northern Europe found that low temperature was the most important factor facilitating transmission, followed by solar UV radiation and low humidity [131]. That paper also noted that high humidity favors transmission in tropical and subtropical zones, in accordance with the findings by Li et al. [129]. According to data posted at WorldoMeter [10], COVID-19 case rates in Northern Europe peaked in spring, were very low in summer, then started rising in July (e.g., Spain), August (e.g., Italy) or September (e.g., the UK).

At higher latitudes in the southern hemisphere, COVID-19 rates were very low through April, then started to rise in June and continued rising into October as in Argentina. On the other hand, in the tropical South American countries such as Brazil, COVID-19 rates started rising in April, peaking around early August then declined, in general agreement with other coronavirus infections [129]. Of course, a number of factors help determine the case rate including the extent to which social distancing and mask wearing are practiced, when school attendance begins, and solar and meteorological factors. However, mortality rates were only high in the spring. Most likely the low mortality rates in September were due to the COVID-19 rates being highest for those aged 20 to 29 years [132]. Yet, with time, COVID-19 rates will increase among the elderly as well.

2.12. Racial/Ethnic Disparities

As mentioned in the introduction, African American and Hispanic individuals have higher COVID-19 case and death rates than European Americans [13,14], possibly due to darker skin pigmentation and lower 25(OH)D concentrations [15]. Confounding these findings, however, is that both African Americans and Hispanics are also at greater risk of COVID-19 due to other factors such as working and living in close proximity to many people and having higher rates of hypertension and other chronic diseases such as type II diabetes, often associated with COVID-19 [133].

The findings regarding SARS-CoV-2 positivity by race/ethnicity from the Quest Diagnostics data set are useful regarding racial/ethnic variations in risk of COVID-19 [38]. Mean serum 25(OH)D concentrations for different racial/ethnic groups in the U.S. can be used to estimate the effect of vitamin D status on the risk of COVID-19. Figure 2 shows that Black non-Hispanics with 25(OH)D ≤20 ng/mL had a 19% SARS-CoV-2 positivity, Hispanics with 25(OH)D concentration = 21 ng/mL had 15% positivity, while white non-Hispanics with 25(OH)D concentrations near 26 ng/mL had a positivity near 8%. If black non-Hispanics had a mean 25(OH)D concentration near 26 ng/mL, it is projected that they would have a positivity of approximately 17%.

Thus, the contribution of vitamin D status to positivity higher than for white non-Hispanics is 2%(19%–8%) ~20%, while that for Hispanics is 2%(15%–8%) to ~30%. Thus, while disparities in vitamin D status do not explain much of the ethnic/racial differences in SARS-CoV-2 positivity, if black non-Hispanics were to raise their mean serum 25(OH)D concentration to 50 ng/mL, they could lower risk by approximately 40%, Hispanics by ~50%, and white non-Hispanics by ~25%. A recent letter suggested that African Americans have a high risk of severe disease and mortality by SARS-CoV-2 due to vitamin D deficiency [134]. The mechanism proposed was reduced ACE2 due to vitamin D deficiency.

An analysis of physician deaths in the UK showed that 18 of 19 doctors and dentists who died by 22 April 2020, were of black, Asian, and mixed ethnicity [135]. Presumably, they were not of low socioeconomic status and had similar contact with patients as their white counterparts. They could have had low vitamin D status due to darker skin and/or vegetarian or vegan diets. In England, a study involving white residents reported that vegans and vegetarians have 25(OH)D concentrations as much as 8 ng/mL lower than those of meat eaters [136].

2.13. Vitamin D Reduces Risk of COVID-19 in a Causal Manner

Hill's criteria for causality offer a scientific approach to determine causal relationships in biological systems [137]. The important criteria for vitamin D include temporality, strength of association, dose–response relationship, consistency of findings, plausibility (e.g., mechanisms), accounting for alternate explanations, experiment (e.g., randomized controlled trial), and coherence with known facts.

Annweiler and colleagues evaluated the evidence that vitamin D reduces the risk and severity of COVID-19 in a causal manner [138]. An updated summary of the evidence is presented in Table 3. Most of the criteria are satisfied. A number of mechanisms have been identified or proposed regarding how vitamin D could reduce risk of COVID-19. Further experimental verification is warranted for some of them.

Table 3. Hill's criteria for causality applied to vitamin D and COVID-19.

Criterion	Evidence	Reference
Strength of association	A retrospective study in Chicago found a 77% increased risk of COVID-19 for 25(OH)D <20 ng/mL vs. >20 ng/mL	[37]
Consistency	Thirteen of 16 observational studies of COVID-19 or SARS-CoV-2 positivity reported inverse correlations with respect to 25(OH)D concentration. Two studies that did not find an inverse association used 25(OH)D values from more than a decade prior to COVID-19 and in the multivariable analysis used some confounding factors that affect 25(OH)D	Tables 1 and 2

Table 3. *Cont.*

Criterion	Evidence	Reference
Temporality	Four retrospective studies found inverse correlations between serum 25(OH)D and incidence of COVID-19 or SARS-CoV-2 positivity	[34,36–38]
Biological gradient	The large observational study of SARS-CoV-2 positivity found a large decrease as serum 25(OH)D increased from <20 to 50 ng/mL	[38]
Plausibility	Mechanisms have been proposed to explain how vitamin D reduces risk of SARS-CoV-2 infection and COVID-19	Discussed in this review
Coherence with known facts	Serum 25(OH)D concentrations are inversely correlated with risk and outcome of many diseases, also supported by RCTs in several cases	[5,7,8,44,139]
Experiment	Two intervention studies provide weak experimental support. Many RCTs are either planned or in progress to evaluate the role of vitamin D supplementation on COVID-19 risk and outcomes [18]	[58,59]
Analogy	Vitamin D supplementation reduces risk of some acute respiratory tract infections	[8]
Account for confounding factors	Univariate or multivariate regression analyses with confounding factors	[29,31,36,37]

The pilot calcifediol treatment RCT conducted in Spain was of low quality due to the low number of participants and failure to measure 25(OH)D concentrations [59]. While the meta-analysis of acute respiratory tract infections found a significant reduction with respect to vitamin D supplementation in RCTs [8], vitamin D supplementation does not reduce risk of all respiratory tract infections, e.g., pneumonia in infancy and early childhood [140].

Hill stated: "None of my nine viewpoints can bring indisputable evidence for or against the cause-and-effect hypothesis and none can be required as a *sine qua non*. What they can do, with greater or less strength, is to help us to make up our minds on the fundamental question—is there any other way of explaining the set of facts before us, is there any other answer equally, or more, likely than cause and effect?" p. 36 in [137].

Evidence-based medicine (EBM) has generally come to mean a heavy reliance on RCTs. Yet, that was only one type of evidence proposed by Sackett, the father of EBM. The practice of evidence-based medicine means integrating individual clinical expertise with the best available external clinical evidence from systematic research. By best available external clinical evidence, we mean clinically relevant research, often from the basic sciences of medicine, but especially from patient-centered clinical research into the accuracy and precision of diagnostic tests (including the clinical examination), the power of prognostic markers, and the efficacy and safety of therapeutic, rehabilitative, and preventive regimens [141].

Indeed, several reviews of EBM have discussed the relative roles of RCTs and observational studies. A review from 2004 compared results from RCTs and observational studies for four health outcomes, reporting that if a reasonable number of each type of study was available, the results were very similar [142]. A review from 2010 proposed a hierarchy with meta-analysis on top, followed by systematic review, RCT, and so on [143].

A review tabulated the ways both RCTs and their meta-analyses could have biased results, either in domains or in design [144]. One design bias is the wrong dose, often a problem with vitamin D

RCTs in that vitamin D doses have generally been 1000 IU/d or less until recently. Another problem is enrolling participants with relatively high 25(OH)D concentrations and giving doses too low to be effective [46]. Finally, a review published in 2017 compared RCTs with "real-world studies" (observational studies) [145]. Among other strengths, observational studies generally include more diverse and larger populations than RCTs.

Regarding the comparison of findings for vitamin D from observational studies and RCTs, they are in general agreement—though with some caveats. RCTs support the role of vitamin D supplementation in reducing the risk of acute respiratory tract infections (ARTIs) [8]. However, an RCT reporting that vitamin D supplementation reduced risk of influenza type A for schoolchildren showed no reduction for influenza type B [146]. Vitamin D_3 supplementation (14,000 IU/wk) did not result in a lower risk of tuberculosis infection, tuberculosis disease, or ARTIs than placebo among vitamin D-deficient schoolchildren in Mongolia [147]. Thus, vitamin D supplementation does not reduce risk of all types of respiratory tract infections in all places.

2.14. Other Nutrients That May Augment the Effectiveness of Vitamin D Supplementation

Magnesium supplementation is recommended when taking vitamin D supplements. Magnesium facilitates vitamin D-related processes. All the enzymes that metabolize vitamin D seem to require magnesium, which acts as a cofactor in the enzymatic reactions in the liver and kidneys [148]. The dose of magnesium should be in the range of 250–500 mg/d. Magnesium activates more than 600 enzymes and influences extracellular calcium concentrations [149]. It is essential for the stability of cell function, RNA and DNA synthesis, and cell repair, as well as maintaining the cell's antioxidant status. Magnesium is an important cofactor for activating a wide range of transporters and enzymes [150,151], many of which involve vitamin D metabolism.

Although vitamin D is likely to be the most important nutrient to optimize COVID-19 prevention, other nutrients, such as magnesium, vitamin K_2 and other micronutrients, are also known to impact the immune system and infection risk [152–154].

3. Conclusions

As discussed here, there is emerging evidence that higher serum 25(OH)D concentrations are associated with the reduced risk and severity of COVID-19. It might do so through a variety of mechanisms, such as maintaining intact epithelial layers, reducing the survival and replication of viruses, reducing the production of pro-inflammatory cytokines, and increasing ACE2 concentrations. More research is required to evaluate the mechanisms whereby vitamin D might reduce the risk of COVID-19.

The strongest evidence to date comes from 14 observational studies that report inverse correlations between serum 25(OH)D concentrations and SARS-CoV-2 positivity and/or COVID-19 incidence, severity and/or death. Hill's criteria for causality in a biological system are largely satisfied for vitamin D in reducing risk of COVID-19, with the exception of successful large-scale vitamin D supplementation RCTs demonstrating significantly reduced risk of or improved outcome for COVID-19. Such RCTs are now under way [18,155]. Until then, individuals and physicians can use vitamin D supplementation as they wish, but public health policies likely will not include vitamin D to reduce risk or death from COVID-19 until large-scale RCTs are reported demonstrating significant reductions in COVID-19 incidence, severity, and/or death from vitamin D supplementation.

Author Contributions: Conceptualization, J.M.; methodology, J.M. and W.B.G.; writing—original draft preparation, J.M.; writing—review and editing, J.M., W.B.G., and C.L.W.; visualization, J.M. All authors have read and agreed to the published version of the manuscript.

Funding: This research received no external funding.

Conflicts of Interest: J.M. sells vitamin D and other supplements; W.B.G. receives funding from Bio-Tech Pharmacal, Inc. (Fayetteville, AR, USA). C.L.W. has no conflicts of interest to declare.

References

1. Chibuzor, M.T.; Graham-Kalio, D.; Osaji, J.O.; Meremikwu, M.M. Vitamin D, calcium or a combination of vitamin D and calcium for the treatment of nutritional rickets in children. *Cochrane Database Syst. Rev.* **2020**, *4*, CD012581. [CrossRef] [PubMed]

2. Agrawal, D.K.; Yin, K. Vitamin D and inflammatory diseases. *J. Inflamm. Res.* **2014**, *7*, 69–87. [CrossRef] [PubMed]

3. Panfili, F.M.; Roversi, M.; D'Argenio, P.; Rossi, P.; Cappa, M.; Fintini, D. Possible role of vitamin D in Covid-19 infection in pediatric population. *J. Endocrinol. Investig.* **2020**, 1–9. [CrossRef]

4. Carlberg, C. Vitamin D Signaling in the Context of Innate Immunity: Focus on Human Monocytes. *Front. Immunol.* **2019**, *10*, 2211. [CrossRef] [PubMed]

5. Manson, J.E.; Cook, N.R.; Lee, I.M.; Christen, W.; Bassuk, S.S.; Mora, S.; Gibson, H.; Gordon, D.; Copeland, T.; D'Agostino, D.; et al. Vitamin d supplements and prevention of cancer and cardiovascular disease. *N. Engl. J. Med.* **2019**, *380*, 33–44. [CrossRef] [PubMed]

6. Grant, W.B.; Al Anouti, F.; Moukayed, M. Targeted 25-hydroxyvitamin D concentration measurements and vitamin D3 supplementation can have important patient and public health benefits. *Eur. J. Clin. Nutr.* **2020**, *74*, 366–376. [CrossRef]

7. Pittas, A.G.; Dawson-Hughes, B.; Sheehan, P.; Ware, J.H.; Knowler, W.C.; Aroda, V.R.; Brodsky, I.; Ceglia, L.; Chadha, C.; Chatterjee, R.; et al. Vitamin D Supplementation and Prevention of Type 2 Diabetes. *N. Engl. J. Med.* **2019**, *381*, 520–530. [CrossRef]

8. Martineau, A.R.; Jolliffe, D.A.; Greenberg, L.; Aloia, J.F.; Bergman, P.; Dubnov-Raz, G.; Esposito, S.; Ganmaa, D.; Ginde, A.A.; Goodall, E.C.; et al. Vitamin D supplementation to prevent acute respiratory infections: Individual participant data meta-analysis. *Health Technol. Assess.* **2019**, *23*, 1–44. [CrossRef]

9. Hayes, C.E.; Ntambi, J.M. Multiple Sclerosis: Lipids, Lymphocytes, and Vitamin D. *Immunometabolism* **2020**, *2*, 2. [CrossRef]

10. Covid-19 Coronavirus Pandemic. Available online: https://www.worldometers.info/coronavirus/ (accessed on 27 June 2020).

11. Mitri, J.; Muraru, M.D.; Pittas, A.G. Vitamin D and type 2 diabetes: A systematic review. *Eur. J. Clin. Nutr.* **2011**, *65*, 1005–1015. [CrossRef]

12. Kroll, M.H.; Bi, C.; Garber, C.C.; Kaufman, H.W.; Liu, D.; Caston-Balderrama, A.; Zhang, K.; Clarke, N.; Xie, M.; Reitz, R.E.; et al. Temporal Relationship between Vitamin D Status and Parathyroid Hormone in the United States. *PLoS ONE* **2015**, *10*, e0118108. [CrossRef] [PubMed]

13. Yancy, C.W. COVID-19 and African Americans. *JAMA* **2020**, *323*, 1891. [CrossRef] [PubMed]

14. Yehia, B.R.; Winegar, A.; Fogel, R.; Fakih, M.; Ottenbacher, A.; Jesser, C.; Bufalino, A.; Huang, R.-H.; Cacchione, J. Association of Race With Mortality Among Patients Hospitalized With Coronavirus Disease 2019 (COVID-19) at 92 US Hospitals. *JAMA Netw. Open* **2020**, *3*, e2018039. [CrossRef] [PubMed]

15. Ginde, A.A.; Liu, M.C.; Camargo, C.A. Demographic Differences and Trends of Vitamin D Insufficiency in the US Population, 1988-2004. *Arch. Intern. Med.* **2009**, *169*, 626–632. [CrossRef] [PubMed]

16. Caricchio, R.; Gallucci, M.; Dass, C.; Zhang, X.; Gallucci, S.; Fleece, D.; Bromberg, M.; Criner, G.J. Preliminary predictive criteria for COVID-19 cytokine storm. *Ann. Rheum. Dis.* **2020**. [CrossRef]

17. Qin, C.; Zhou, L.; Hu, Z.; Zhang, S.; Yang, S.; Tao, Y.; Xie, C.; Ma, K.; Shang, K.; Wang, W.; et al. Dysregulation of Immune Response in Patients with Coronavirus 2019 (COVID-19) in Wuhan, China. *Clin. Infect. Dis.* **2020**, *71*, 762–768. [CrossRef]

18. ClinicalTrials.gov. Studies for Vitamin D, Covid19. Available online: https://clinicaltrials.gov/ct2/results?cond=COVID19&term=vitamin+D&cntry=&state=&city=&dist= (accessed on 29 June 2020).

19. VanderWeele, T.J. Principles of confounder selection. *Eur. J. Epidemiol.* **2019**, *34*, 211–219. [CrossRef]

20. Grant, W.B.; McDonnell, S.L. Statistical error in Vitamin d concentrations and covid-19 infection in UK Biobank. *Diabetes Metab. Syndr. Clin. Res. Rev.* **2020**, *14*, 893–894. [CrossRef]

21. Roy, A.S.; Matson, M.; Herlekar, R. Response to 'vitamin d concentrations and covid-19 infection in uk biobank'. *Diabetes Metab. Syndr.* **2020**, *14*, 777. [CrossRef]

22. Hastie, C.E.; Mackay, D.F.; Ho, F.; Celis-Morales, C.A.; Katikireddi, S.V.; Niedzwiedz, C.L.; Jani, B.D.; Welsh, P.; Mair, F.S.; Gray, S.R.; et al. Vitamin D concentrations and COVID-19 infection in UK Biobank. *Diabetes Metab. Syndr. Clin. Res. Rev.* **2020**, *14*, 561–565. [CrossRef]

23. D'Avolio, A.; Avataneo, V.; Manca, A.; Cusato, J.; De Nicolo, A.; Lucchini, R.; Keller, F.; Cantu, M. 25-hydroxyvitamin d concentrations are lower in patients with positive pcr for sars-cov-2. *Nutrients* **2020**, *12*, 1359. [CrossRef] [PubMed]

24. Panagiotou, G.; Tee, S.A.; Ihsan, Y.; Athar, W.; Marchitelli, G.; Kelly, D.; Boot, C.S.; Stock, N.; Macfarlane, J.; Martineau, A.R.; et al. Low serum 25-hydroxyvitamin D (25[OH]D) levels in patients hospitalised with COVID-19 are associated with greater disease severity: Results of a local audit of practice. *Clin. Endocrinol. (Oxf.)* **2020**. [CrossRef]

25. Carpagnano, G.E.; Di Lecce, V.; Quaranta, V.N.; Zito, A.; Buonamico, E.; Capozza, E.; Palumbo, A.; Di Gioia, G.; Valerio, V.N.; Resta, O. Vitamin D deficiency as a predictor of poor prognosis in patients with acute respiratory failure due to COVID-19. *J. Endocrinol. Investig.* **2020**, 1–7. [CrossRef]

26. Im, J.H.; Je, Y.S.; Baek, J.; Chung, M.H.; Kwon, H.Y.; Lee, J.S. Nutritional status of patients with coronavirus disease 2019 (covid-19). *Int. J. Infect. Dis.* **2020**, *100*, 390–393. [CrossRef] [PubMed]

27. Karonova, T.L.; Andreeva, A.T.; Vashukova, M.A. Serum 25(Oh)D Level in Patients with Covid-19. *J. Infectol.* **2020**, *12*, 21–27. (In Russian) [CrossRef]

28. Pérez, R.A.R.; Nieto, A.V.P.; Martínez-Cuazitl, A.; Mercado, E.A.M.; Tort, A.R. La deficiencia de vitamina D es un factor de riesgo de mortalidad en pacientes con COVID-19. *Rev. Sanid. Mil.* **2020**, *74*, 106–113. [CrossRef]

29. Baktash, V.; Hosack, T.; Patel, N.; Shah, S.; Kandiah, P.; Abbeele, K.V.D.; Mandal, A.K.J.; Missouris, C.G. Vitamin D status and outcomes for hospitalised older patients with COVID-19 2020. *Postgrad. Med. J.* **2020**. [CrossRef]

30. Hastie, C.E.; Pell, J.P.; Sattar, N. Vitamin D and COVID-19 infection and mortality in UK Biobank. *Eur. J. Nutr.* **2020**, 1–4. [CrossRef] [PubMed]

31. Radujkovic, A.; Hippchen, T.; Tiwari-Heckler, S.; Dreher, S.; Boxberger, M.; Merle, U. Vitamin D Deficiency and Outcome of COVID-19 Patients. *Nutrients* **2020**, *12*, 2757. [CrossRef]

32. Valcour, A.; Blocki, F.; Hawkins, D.M.; Rao, S.D. Effects of Age and Serum 25-OH-Vitamin D on Serum Parathyroid Hormone Levels. *J. Clin. Endocrinol. Metab.* **2012**, *97*, 3989–3995. [CrossRef]

33. Pizzini, A.; Aichner, M.; Sahanic, S.; Böhm, A.; Egger, A.; Hoermann, G.; Kurz, K.; Widmann, G.; Bellmann-Weiler, R.; Weiss, G.; et al. Impact of Vitamin D Deficiency on COVID-19—A Prospective Analysis from the CovILD Registry. *Nutrients* **2020**, *12*, 2775. [CrossRef] [PubMed]

34. Macaya, F.; Paeres, C.E.; Valls, A.; Fernández-Ortiz, A.; Del Castillo, J.G.; Martín-Sánchez, J.; Runkle, I.; Herrera, M.; Ángel, R. Interaction between age and vitamin D deficiency in severe COVID-19 infection. *Nutrición Hospitalaria* **2020**. [CrossRef]

35. Ye, K.; Tang, F.; Liao, X.; Shaw, B.A.; Deng, M.; Huang, G.; Qin, Z.; Peng, X.; Xiao, H.; Chen, C.; et al. Does Serum Vitamin D Level Affect COVID-19 Infection and Its Severity?-A Case-Control Study. *J. Am. Coll. Nutr.* **2020**, 1–8. [CrossRef] [PubMed]

36. Merzon, E.; Tworowski, D.; Gorohovski, A.; Vinker, S.; Cohen, A.G.; Green, I.; Morgenstern, M.F. Low plasma 25(OH) vitamin D level is associated with increased risk of COVID-19 infection: An Israeli population-based study. *FEBS J.* **2020**. [CrossRef]

37. Meltzer, D.O.; Best, T.J.; Zhang, H.; Vokes, T.; Arora, V.; Solway, J. Association of Vitamin D Status and Other Clinical Characteristics with COVID-19 Test Results. *JAMA Netw. Open* **2020**, *3*, e2019722. [CrossRef]

38. Kaufman, H.W.; Niles, J.K.; Kroll, M.H.; Bi, C.; Holick, M.F. SARS-CoV-2 positivity rates associated with circulating 25-hydroxyvitamin D levels. *PLoS ONE* **2020**, *15*, e0239252. [CrossRef]

39. Mahdavi, A.M. A brief review of interplay between vitamin D and angiotensin-converting enzyme 2: Implications for a potential treatment for COVID-19. *Rev. Med. Virol.* **2020**, *30*, 2119. [CrossRef] [PubMed]

40. Ross, A.C.; Manson, J.E.; Abrams, S.A.; Aloia, J.F.; Brannon, P.M.; Clinton, S.K.; Durazo-Arvizu, R.A.; Gallagher, J.C.; Gallo, R.L.; Jones, G.; et al. The 2011 dietary reference intakes for calcium and vitamin d: What dietetics practitioners need to know. *J. Am. Diet. Assoc.* **2011**, *111*, 524–527. [CrossRef]

41. Ross, A.C.; Manson, J.E.; Abrams, S.A.; Aloia, J.F.; Brannon, P.M.; Clinton, S.K.; Durazo-Arvizu, R.A.; Gallagher, J.C.; Gallo, R.L.; Jones, G.; et al. The 2011 Report on Dietary Reference Intakes for Calcium and Vitamin D from the Institute of Medicine: What Clinicians Need to Know. *J. Clin. Endocrinol. Metab.* **2011**, *96*, 53–58. [CrossRef]

42. Korber, B.; Fischer, W.M.; Gnanakaran, S.; Yoon, H.; Theiler, J.; Abfalterer, W.; Hengartner, N.; Giorgi, E.E.; Bhattacharya, T.; Foley, B.; et al. Tracking changes in sars-cov-2 spike: Evidence that d614g increases infectivity of the covid-19 virus. *Cell* **2020**, *182*, 812–827.e19. [CrossRef]

43. Strickland, O.L.; Powell-Young, Y.; Reyes-Miranda, C.; Alzaghari, O.; Giger, J.N. African-americans have a higher propensity for death from covid-19: Rationale and causation. *J. Natl. Black Nurses Assoc.* **2020**, *31*, 1–12. [PubMed]

44. Autier, P.; Boniol, M.; Pizot, C.; Mullie, P. Vitamin D status and ill health: A systematic review. *Lancet Diabetes Endocrinol.* **2014**, *2*, 76–89. [CrossRef]

45. Autier, P.; Mullie, P.; Macacu, A.; Dragomir, M.; Boniol, M.; Coppens, K.; Pizot, C.; Boniol, M. Effect of vitamin D supplementation on non-skeletal disorders: A systematic review of meta-analyses and randomised trials. *Lancet Diabetes Endocrinol.* **2017**, *5*, 986–1004. [CrossRef]

46. Heaney, R.P. Guidelines for optimizing design and analysis of clinical studies of nutrient effects. *Nutr. Rev.* **2013**, *72*, 48–54. [CrossRef]

47. Grant, W.B.; Boucher, B.J.; Bhattoa, H.P.; Lahore, H. Why vitamin D clinical trials should be based on 25-hydroxyvitamin D concentrations. *J. Steroid Biochem. Mol. Biol.* **2018**, *177*, 266–269. [CrossRef]

48. Orkaby, A.R.; Djousse, L.; Manson, J.E. Vitamin D supplements and prevention of cardiovascular disease. *Curr. Opin. Cardiol.* **2019**, *34*, 700–705. [CrossRef] [PubMed]

49. Zhang, X.; Niu, W. Meta-analysis of randomized controlled trials on vitamin D supplement and cancer incidence and mortality. *Biosci. Rep.* **2019**, *39*, 39. [CrossRef]

50. Silva, M.C.; Furlanetto, T.W. Does serum 25-hydroxyvitamin D decrease during acute-phase response? A systematic review. *Nutr. Res.* **2015**, *35*, 91–96. [CrossRef] [PubMed]

51. Krishnan, A.; Ochola, J.; Mundy, J.; Jones, M.; Kruger, P.; Duncan, E.; Venkatesh, B. Acute fluid shifts influence the assessment of serum vitamin D status in critically ill patients. *Crit. Care* **2010**, *14*, R216. [CrossRef]

52. Reid, D.; Toole, B.J.; Knox, S.; Talwar, D.; Harten, J.; O'Reilly, D.S.J.; Blackwell, S.; Kinsella, J.; McMillan, D.C.; Wallace, A.M.; et al. The relation between acute changes in the systemic inflammatory response and plasma 25-hydroxyvitamin D concentrations after elective knee arthroplasty. *Am. J. Clin. Nutr.* **2011**, *93*, 1006–1011. [CrossRef]

53. Waldron, J.L.; Ashby, H.; Cornes, M.; Bechervaise, J.; Razavi, C.; Thomas, O.L.; Chugh, S.; Deshpande, S.; Ford, C.; Gama, R. Vitamin D: A negative acute phase reactant. *J. Clin. Pathol.* **2013**, *66*, 620–622. [CrossRef]

54. Krishnan, A.V.; Trump, N.L.; Johnson, C.S.; Feldman, D. The role of vitamin D in cancer prevention and treatment. *Endocrinol. Metab. Clin. N. Am.* **2010**, *39*, 401–418. [CrossRef]

55. Newens, K.; Filteau, S.; Tomkins, A. Plasma 25-hydroxyvitamin D does not vary over the course of a malarial infection. *Trans. R. Soc. Trop. Med. Hyg.* **2006**, *100*, 41–44. [CrossRef]

56. Bang, U.C.; Novovic, S.; Andersen, A.M.; Fenger, M.; Hansen, M.B.; Jensen, J.-E.B. Variations in Serum 25-Hydroxyvitamin D during Acute Pancreatitis: An Exploratory Longitudinal Study. *Endocr. Res.* **2011**, *36*, 135–141. [CrossRef] [PubMed]

57. Bertoldo, F.; Pancheri, S.; Zenari, S.; Boldini, S.; Giovanazzi, B.; Zanatta, M.; Valenti, M.T.; Carbonare, L.D.; Cascio, V.L. Serum 25-hydroxyvitamin D levels modulate the acute-phase response associated with the first nitrogen-containing bisphosphonate infusion. *J. Bone Miner. Res.* **2010**, *25*, 447–454. [CrossRef] [PubMed]

58. Ohaegbulam, K.C.; Swalih, M.; Patel, P.; Smith, M.A.; Perrin, R. Vitamin D Supplementation in COVID-19 Patients. *Am. J. Ther.* **2020**, *27*, e485–e490. [CrossRef] [PubMed]

59. Castillo, M.E.; Costa, L.M.E.; Barrios, J.M.V.; Díaz, J.F.A.; Miranda, J.L.; Bouillon, R.; Gomez, J.M.Q. Effect of calcifediol treatment and best available therapy versus best available therapy on intensive care unit admission and mortality among patients hospitalized for COVID-19: A pilot randomized clinical study. *J. Steroid Biochem. Mol. Biol.* **2020**, *203*, 105751. [CrossRef]

60. Henríquez, M.S.; Romero, M.J.G.D.T. Cholecalciferol or Calcifediol in the Management of Vitamin D Deficiency. *Nutrients* **2020**, *12*, 12. [CrossRef]

61. Zhou, Y.; Yang, Q.; Chi, J.; Dong, B.; Lv, W.; Shen, L.; Wang, Y. Comorbidities and the risk of severe or fatal outcomes associated with coronavirus disease 2019: A systematic review and meta-analysis. *Int. J. Infect. Dis.* **2020**, *99*, 47–56. [CrossRef]

62. Annweiler, C.; Hanotte, B.; De L'Eprevier, C.G.; Sabatier, J.-M.; Lafaie, L.; Célarier, T. Vitamin D and survival in COVID-19 patients: A quasi-experimental study. *J. Steroid Biochem. Mol. Biol.* **2020**, 105771. [CrossRef]

63. Ilahi, M.; Armas, L.A.G.; Heaney, R.P. Pharmacokinetics of a single, large dose of cholecalciferol. *Am. J. Clin. Nutr.* **2008**, *87*, 688–691. [CrossRef] [PubMed]

64. Beard, J.A.; Bearden, A.; Striker, R. Vitamin D and the anti-viral state. *J. Clin. Virol.* **2011**, *50*, 194–200. [CrossRef]

65. Shin, D.-M.; Jo, E.-K. Antimicrobial Peptides in Innate Immunity against Mycobacteria. *Immune Netw.* **2011**, *11*, 245–252. [CrossRef]

66. Dimitrov, V.; White, J.H. Species-specific regulation of innate immunity by vitamin D signaling. *J. Steroid Biochem. Mol. Biol.* **2016**, *164*, 246–253. [CrossRef] [PubMed]

67. Martineau, A.R.; Jolliffe, D.A.; Demaret, J. Vitamin D and Tuberculosis. *Vitamin D* **2018**, *2*, 915–935.

68. Cannell, J.J.; Vieth, R.; Umhau, J.C.; Holick, M.F.; Grant, W.B.; Madronich, S.; Garland, C.F.; Giovannucci, E. Epidemic influenza and vitamin D. *Epidemiol. Infect.* **2006**, *134*, 1129–1140. [CrossRef] [PubMed]

69. Barlow, P.G.; Findlay, E.G.; Currie, S.M.; Davidson, D.J. Antiviral potential of cathelicidins. *Futur. Microbiol.* **2014**, *9*, 55–73. [CrossRef]

70. Crane-Godreau, M.A.; Clem, K.J.; Payne, P.; Fiering, S. Vitamin D Deficiency and Air Pollution Exacerbate COVID-19 through Suppression of Antiviral Peptide LL37. *Front. Public Health* **2020**, *8*, 232. [CrossRef]

71. Kara, M.; Ekiz, T.; Ricci, V.; Kara, Ö.; Chang, K.-V.; Özçakar, L. 'Scientific Strabismus' or two related pandemics: Coronavirus disease and vitamin D deficiency. *Br. J. Nutr.* **2020**, *124*, 736–741. [CrossRef]

72. Dürr, U.H.; Sudheendra, U.; Ramamoorthy, A. LL-37, the only human member of the cathelicidin family of antimicrobial peptides. *Biochim. Biophys. Acta (BBA) Biomembr.* **2006**, *1758*, 1408–1425. [CrossRef]

73. Leikina, E.; Delanoe-Ayari, H.; Melikov, K.; Cho, M.-S.; Chen, A.; Waring, A.J.; Wang, W.; Xie, Y.; Loo, J.A.; I Lehrer, R.; et al. Carbohydrate-binding molecules inhibit viral fusion and entry by crosslinking membrane glycoproteins. *Nat. Immunol.* **2005**, *6*, 995–1001. [CrossRef]

74. Morris, G.; Bortolasci, C.C.; Puri, B.K.; Olive, L.; Marx, W.; O'Neil, A.; Athan, E.; Carvalho, A.F.; Maes, M.; Walder, K.; et al. The pathophysiology of sars-cov-2: A suggested model and therapeutic approach. *Life Sci.* **2020**, *258*, 118166. [CrossRef] [PubMed]

75. McElvaney, O.J.; McEvoy, N.L.; McElvaney, O.F.; Carroll, T.P.; Murphy, M.P.; Dunlea, D.M.; Choileáin, O.N.; Clarke, J.; O'Connor, E.; Hogan, G.; et al. Characterization of the Inflammatory Response to Severe COVID-19 Illness. *Am. J. Respir. Crit. Care Med.* **2020**, *202*, 812–821. [CrossRef] [PubMed]

76. Grant, W.B.; Lahore, H.; McDonnell, S.L.; Baggerly, C.A.; French, C.B.; Aliano, J.L.; Bhattoa, H.P. Evidence that Vitamin D Supplementation Could Reduce Risk of Influenza and COVID-19 Infections and Deaths. *Nutrients* **2020**, *12*, 988. [CrossRef]

77. Bhaskar, S.; Sinha, A.; Banach, M.; Mittoo, S.; Weissert, R.; Kass, J.S.; Rajagopal, S.; Pai, A.R.; Kutty, S. Cytokine Storm in COVID-19—Immunopathological Mechanisms, Clinical Considerations, and Therapeutic Approaches: The REPROGRAM Consortium Position Paper. *Front. Immunol.* **2020**, *11*, 1648. [CrossRef] [PubMed]

78. Fara, A.; Mitrev, Z.; Mitrev, Z.; Assas, B.M. Cytokine storm and COVID-19: A chronicle of pro-inflammatory cytokines. *Open Biol.* **2020**, *10*, 200160. [CrossRef] [PubMed]

79. Mahmudpour, M.; Roozbeh, J.; Keshavarz, M.; Farrokhi, S.; Nabipour, I. COVID-19 cytokine storm: The anger of inflammation. *Cytokine* **2020**, *133*, 155151. [CrossRef]

80. Kox, M.; Waalders, N.J.B.; Kooistra, E.J.; Gerretsen, J.; Pickkers, P. Cytokine Levels in Critically Ill Patients With COVID-19 and Other Conditions. *JAMA* **2020**, *324*, 1565. [CrossRef]

81. Liu, T.; Luo, S.; Libby, P.; Shi, G.-P. Cathepsin L-selective inhibitors: A potentially promising treatment for COVID-19 patients. *Pharmacol. Ther.* **2020**, *213*, 107587. [CrossRef]

82. Manson, J.J.; Crooks, C.; Naja, M.; Ledlie, A.; Goulden, B.; Liddle, T.; Khan, E.; Mehta, P.; Martin-Gutierrez, L.; Waddington, E.K.; et al. COVID-19-associated hyperinflammation and escalation of patient care: A retrospective longitudinal cohort study. *Lancet Rheumatol.* **2020**, *2*, e594–e602. [CrossRef]

83. Webb, B.J.; Peltan, I.D.; Jensen, P.; Hoda, D.; Hunter, B.; Silver, A. Clinical criteria for covid-19-associated hyperinflammaotry syndrome: A cohort study. *Lancet Rheumatol.* **2020**. [CrossRef]

84. Lin, S.-H.; Zhao, Y.-S.; Zhou, D.-X.; Zhou, F.-C.; Xu, F. Coronavirus disease 2019 (COVID-19): Cytokine storms, hyper-inflammatory phenotypes, and acute respiratory distress syndrome. *Genes Dis.* **2020**. [CrossRef]

85. Meftahi, G.H.; Jangravi, Z.; Sahraei, H.; Bahari, Z. The possible pathophysiology mechanism of cytokine storm in elderly adults with COVID-19 infection: The contribution of "inflame-aging". *Inflamm. Res.* **2020**, *69*, 825–839. [CrossRef] [PubMed]

86. Hojyo, S.; Uchida, M.; Tanaka, K.; Hasebe, R.; Tanaka, Y.; Murakami, M.; Hirano, T. How COVID-19 induces cytokine storm with high mortality. *Inflamm. Regen.* **2020**, *40*, 1–7. [CrossRef] [PubMed]

87. Iwasaki, M.; Saito, J.; Zhao, H.; Sakamoto, A.; Hirota, K.; Ma, D. Inflammation triggered by sars-cov-2 and ace2 augment drives multiple organ failure of severe covid-19: Molecular mechanisms and implications. *Inflammation* **2020**. [CrossRef]

88. Merad, M.; Martin, J.C. Pathological inflammation in patients with COVID-19: A key role for monocytes and macrophages. *Nat. Rev. Immunol.* **2020**, *20*, 355–362. [CrossRef]

89. Park, W.B.; Kwon, N.J.; Choi, S.J.; Kang, C.K.; Choe, P.G.; Kim, J.Y.; Yun, J.; Lee, G.W.; Seong, M.W.; Kim, N.J.; et al. Virus isolation from the first patient with sars-cov-2 in korea. *J. Korean Med. Sci.* **2020**, *35*, e84. [CrossRef]

90. Bombardini, T.; Picano, E. Angiotensin-Converting Enzyme 2 as the Molecular Bridge between Epidemiologic and Clinical Features of COVID-19. *Can. J. Cardiol.* **2020**, *36*, 784.e1–784.e2. [CrossRef]

91. Benne, C.A.; Kraaijeveld, C.A.; Van Strijp, J.A.G.; Brouwer, E.; Harmsen, M.; Verhoef, J.; Van Golde, L.M.G.; Van Iwaarden, J.F. Interactions of Surfactant Protein a with Influenza A Viruses: Binding and Neutralization. *J. Infect. Dis.* **1995**, *171*, 335–341. [CrossRef]

92. Levine, A.M.; Whitsett, J.A.; Hartshorn, K.L.; Crouch, E.C.; Korfhagen, T.R. Surfactant Protein D Enhances Clearance of Influenza A Virus from the Lung In Vivo. *J. Immunol.* **2001**, *167*, 5868–5873. [CrossRef]

93. Phokela, S.S.; Peleg, S.; Moya, F.R.; Alcorn, J.L. Regulation of human pulmonary surfactant protein gene expression by 1α,25-dihydroxyvitamin D3. *Am. J. Physiol. Cell. Mol. Physiol.* **2005**, *289*, L617–L626. [CrossRef]

94. Gemmati, D.; Bramanti, B.; Serino, M.L.; Secchiero, P.; Zauli, G.; Tisato, V. COVID-19 and Individual Genetic Susceptibility/Receptivity: Role of ACE1/ACE2 Genes, Immunity, Inflammation and Coagulation. Might the Double X-Chromosome in Females Be Protective against SARS-CoV-2 Compared to the Single X-Chromosome in Males? *Int. J. Mol. Sci.* **2020**, *21*, 3474. [CrossRef] [PubMed]

95. Li, Y.C.; Qiao, G.; Uskokovic, M.; Xiang, W.; Zheng, W.; Kong, J. Vitamin D: A negative endocrine regulator of the renin–angiotensin system and blood pressure. *J. Steroid Biochem. Mol. Biol.* **2004**, 387–392. [CrossRef] [PubMed]

96. Rolf, J.D. Clinical characteristics of Covid-19 in China. *N. Engl. J. Med.* **2020**, *382*, 1860. [CrossRef] [PubMed]

97. Bavishi, C.; Maddox, T.M.; Messerli, F.H. Coronavirus Disease 2019 (COVID-19) Infection and Renin Angiotensin System Blockers. *JAMA Cardiol.* **2020**, *5*, 745. [CrossRef] [PubMed]

98. Speeckaert, M.M.; Delanghe, J.R. Association between low vitamin D and COVID-19: Don't forget the vitamin D binding protein. *Aging Clin. Exp. Res.* **2020**, *32*, 1207–1208. [CrossRef]

99. Xu, J.; Yang, J.; Chen, J.; Luo, Q.; Zhang, Q.; Zhang, H. Vitamin D alleviates lipopolysaccharide-induced acute lung injury via regulation of the renin-angiotensin system. *Mol. Med. Rep.* **2017**, *16*, 7432–7438. [CrossRef]

100. Cui, C.; Xu, P.; Li, G.; Qiao, Y.; Han, W.; Geng, C.; Liao, D.; Yang, M.; Chen, D.; Jiang, P. Vitamin D receptor activation regulates microglia polarization and oxidative stress in spontaneously hypertensive rats and angiotensin II-exposed microglial cells: Role of renin-angiotensin system. *Redox Biol.* **2019**, *26*, 101295. [CrossRef]

101. Aygun, H. Vitamin D can prevent COVID-19 infection-induced multiple organ damage. *Naunyn-Schmiedeberg's Arch. Pharmacol.* **2020**, *393*, 1157–1160. [CrossRef]

102. Zittermann, A.; Ernst, J.B.; Birschmann, I.; Dittrich, M. Effect of Vitamin D or Activated Vitamin D on Circulating 1,25-Dihydroxyvitamin D Concentrations: A Systematic Review and Metaanalysis of Randomized Controlled Trials. *Clin. Chem.* **2015**, *61*, 1484–1494. [CrossRef]

103. Hanff, T.C.; O'Harhay, M.; Brown, T.S.; Cohen, J.B.; Mohareb, A.M. Is There an Association Between COVID-19 Mortality and the Renin-Angiotensin System? A Call for Epidemiologic Investigations. *Clin. Infect. Dis.* **2020**, *71*, 870–874. [CrossRef] [PubMed]

104. Kumar, D.; Gupta, P.; Banerjee, D. Letter: Does vitamin d have a potential role against Covid-19? *Aliment. Pharmacol. Ther.* **2020**, *52*, 409–411. [CrossRef] [PubMed]

105. Quesada-Gomez, J.M.; Castillo, M.E.; Bouillon, R. Vitamin d receptor stimulation to reduce acute respiratory distress syndrome (ards) in patients with Coronavirus sars-cov-2 infections: Revised ms sbmb 2020_166. *J. Steroid Biochem. Mol. Biol.* **2020**, *202*, 105719. [CrossRef]

106. Tay, M.Z.; Poh, C.M.; Rénia, L.; Macary, P.A.; Ng, L.F.P. The trinity of COVID-19: Immunity, inflammation and intervention. *Nat. Rev. Immunol.* **2020**, *20*, 363–374. [CrossRef] [PubMed]

107. Zhang, J.; Tecson, K.M.; McCullough, P.A. Endothelial dysfunction contributes to COVID-19-associated vascular inflammation and coagulopathy. *Rev. Cardiovasc. Med.* **2020**, *21*, 315–319. [CrossRef]

108. Zhang, J.; McCullough, P.A.; Tecson, K.M. Vitamin D deficiency in association with endothelial dysfunction: Implications for patients withCOVID-19. *Rev. Cardiovasc. Med.* **2020**, *21*, 339–344. [CrossRef]

109. Kanikarla-Marie, P.; Jain, S.K. 1,25(OH) 2 D 3 inhibits oxidative stress and monocyte adhesion by mediating the upregulation of GCLC and GSH in endothelial cells treated with acetoacetate (ketosis). *J. Steroid Biochem. Mol. Biol.* **2016**, *159*, 94–101. [CrossRef]

110. Kim, D.-H.; Meza, C.A.; Clarke, H.; Kim, J.-S.; Hickner, R.C. Vitamin D and Endothelial Function. *Nutrients* **2020**, *12*, 575. [CrossRef]

111. Zheng, S.; Yang, J.; Hu, X.; Li, M.; Wang, Q.; Dancer, R.C.A.; Parekh, D.; Gao-Smith, F.; Thickett, D.R.; Jin, S. Vitamin d attenuates lung injury via stimulating epithelial repair, reducing epithelial cell apoptosis and inhibits tgf-beta induced epithelial to mesenchymal transition. *Biochem. Pharmacol.* **2020**, *177*, 113955. [CrossRef]

112. Davey, A.; McAuley, D.F.; O'Kane, C. Matrix metalloproteinases in acute lung injury: Mediators of injury and drivers of repair. *Eur. Respir. J.* **2011**, *38*, 959–970. [CrossRef]

113. Ueland, T.; Holter, J.; Holten, A.; Müller, K.; Lind, A.; Ke, M.; Dudman, S.; Aukrust, P.; Dyrhol-Riise, A.; Heggelund, L.; et al. Distinct and early increase in circulating MMP-9 in COVID-19 patients with respiratory failure. *J. Infect.* **2020**, *81*, e41–e43. [CrossRef] [PubMed]

114. Timms, P.; Mannan, N.; Hitman, G.; Noonan, K.; Mills, P.; Syndercombe-Court, D.; Aganna, E.; Price, C.; Boucher, B.J. Circulating MMP9, vitamin D and variation in the TIMP-1 response with VDR genotype: Mechanisms for inflammatory damage in chronic disorders? *QJM Int. J. Med.* **2002**, *95*, 787–796. [CrossRef] [PubMed]

115. Coussens, A.K.; Timms, P.M.; Boucher, B.J.; Venton, T.R.; Ashcroft, A.T.; Skolimowska, K.H.; Newton, S.M.; Wilkinson, K.A.; Davidson, R.N.; Griffiths, C.J.; et al. 1α,25-dihydroxyvitamin D3inhibits matrix metalloproteinases induced byMycobacterium tuberculosisinfection. *Immunology* **2009**, *127*, 539–548. [CrossRef] [PubMed]

116. Garvin, M.R.; Alvarez, C.; Miller, J.I.; Prates, E.T.; Walker, A.M.; Amos, B.K.; Mast, E.A.; Justice, A.; Aronow, B.; Jacobson, D. A mechanistic model and therapeutic interventions for COVID-19 involving a RAS-mediated bradykinin storm. *eLife* **2020**, *9*, 9. [CrossRef]

117. Gombart, A.F.; Pierre, A.; Maggini, S. A Review of Micronutrients and the Immune System–Working in Harmony to Reduce the Risk of Infection. *Nutrients* **2020**, *12*, 236. [CrossRef]

118. Rondanelli, M.; Miccono, A.; Lamburghini, S.; Avanzato, I.; Riva, A.; Allegrini, P.; Faliva, M.A.; Peroni, G.; Nichetti, M.; Perna, S. Self-Care for Common Colds: The Pivotal Role of Vitamin D, Vitamin C, Zinc, and Echinacea in Three Main Immune Interactive Clusters (Physical Barriers, Innate and Adaptive Immunity) Involved during an Episode of Common Colds—Practical Advice on Dosages and on the Time to Take These Nutrients/Botanicals in order to Prevent or Treat Common Colds. *Evid.-Based Complement. Altern. Med.* **2018**, *2018*, 1–36. [CrossRef]

119. Lang, P.O.; Aspinall, R. Vitamin D Status and the Host Resistance to Infections: What It Is Currently (Not) Understood. *Clin. Ther.* **2017**, *39*, 930–945. [CrossRef]

120. Abhimanyu, A.; Coussens, A.K. The role of UV radiation and vitamin D in the seasonality and outcomes of infectious disease. *Photochem. Photobiol. Sci.* **2017**, *16*, 314–338. [CrossRef]

121. Jolliffe, D.A.; Griffiths, C.J.; Martineau, A.R. Vitamin D in the prevention of acute respiratory infection: Systematic review of clinical studies. *J. Steroid Biochem. Mol. Biol.* **2013**, *136*, 321–329. [CrossRef]

122. Goodall, E.C.; Granados, A.; Luinstra, K.; Pullenayegum, E.M.; Coleman, B.L.; Loeb, M.; Smieja, M. Vitamin D3and gargling for the prevention of upper respiratory tract infections: A randomized controlled trial. *BMC Infect. Dis.* **2014**, *14*, 273. [CrossRef]

123. Laaksi, I.; Ruohola, J.-P.; Tuohimaa, P.; Auvinen, A.; Haataja, R.; Pihlajamäki, H.; Ylikomi, T. An association of serum vitamin D concentrations < 40 nmol/L with acute respiratory tract infection in young Finnish men. *Am. J. Clin. Nutr.* **2007**, *86*, 714–717. [CrossRef] [PubMed]

124. Cannell, J.J.; Vieth, R.; Willett, W.; Zasloff, M.; Hathcock, J.N.; White, J.H.; Tanumihardjo, S.A.; Larson-Meyer, D.E.; Bischoff-Ferrari, H.A.; Lamberg-Allardt, C.J.; et al. Cod Liver Oil, Vitamin A Toxicity, Frequent Respiratory Infections, and the Vitamin D Deficiency Epidemic. *Ann. Otol. Rhinol. Laryngol.* **2008**, *117*, 864–870. [CrossRef]

125. Belančić, A.; Kresović, A.; Rački, V. Potential pathophysiological mechanisms leading to increased COVID-19 susceptibility and severity in obesity. *Obes. Med.* **2020**, *19*, 100259. [CrossRef] [PubMed]

126. Parlak, E.; Ertürk, A.; Çağ, Y.; Sebin, E.; Gümüşdere, M. The effect of inflammatory cytokines and the level of vitamin D on prognosis in Crimean-Congo hemorrhagic fever. *Int. J. Clin. Exp. Med.* **2015**, *8*, 18302–18310. [PubMed]

127. Hope-Simpson, R.E. The role of season in the epidemiology of influenza. *J. Hyg.* **1981**, *86*, 35–47. [CrossRef] [PubMed]

128. Li, Y.; Reeves, R.M.; Wang, X.; Bassat, Q.; Brooks, W.A.; Cohen, C.; Moore, D.P.; Nunes, M.; Rath, B.; Campbell, H.; et al. Global patterns in monthly activity of influenza virus, respiratory syncytial virus, parainfluenza virus, and metapneumovirus: A systematic analysis. *Lancet Glob. Health* **2019**, *7*, e1031–e1045. [CrossRef]

129. Li, Y.; Wang, X.; Nair, H. Global Seasonality of Human Seasonal Coronaviruses: A Clue for Postpandemic Circulating Season of Severe Acute Respiratory Syndrome Coronavirus 2? *J. Infect. Dis.* **2020**, *222*, 1090–1097. [CrossRef]

130. Shaman, J.; Pitzer, V.E.; Viboud, C.; Grenfell, B.T.; Lipsitch, M. Absolute humidity and the seasonal onset of influenza in the continental united states. *PLoS Biol.* **2010**, *8*, e1000316. [CrossRef]

131. Ianevski, A.; Zusinaite, E.; Shtaida, N.; Kallio-Kokko, H.; Valkonen, M.; Kantele, A.; Telling, K.; Lutsar, I.; Letjuka, P.; Metelitsa, N.; et al. Low Temperature and Low UV Indexes Correlated with Peaks of Influenza Virus Activity in Northern Europe during 2010–2018. *Viruses* **2019**, *11*, 207. [CrossRef]

132. France, S.P. Heatmap: Covid-19 Incidence per 100,000 Inhabitants by Age Group. Available online: https://guillaumepressiat.shinyapps.io/covid-si-dep/?reg=11%7c93%7c32 (accessed on 2 October 2020).

133. Phillips, N.; Park, I.-W.; Robinson, J.R.; Jones, H.P. The Perfect Storm: COVID-19 Health Disparities in US Blacks. *J. Racial Ethn. Health Disparities* **2020**, 1–8. [CrossRef]

134. Martin Gimenez, V.M.; Inserra, F.; Ferder, L.; Garcia, J.; Manucha, W. Vitamin d deficiency in african americans is associated with a high risk of severe disease and mortality by Sars-Cov-2. *J. Hum. Hypertens.* **2020**. [CrossRef] [PubMed]

135. Cook, T.; Kursumovie, E.; Lennane, S. Exclusive: Deaths of NHS staff from covid-19 analysed. *Health Serv. J.* **2020**, 7027471.

136. Crowe, F.L.; Steur, M.; Allen, E.N.; Appleby, P.N.; Travis, R.C.; Key, T.J. Plasma concentrations of 25-hydroxyvitamin D in meat eaters, fish eaters, vegetarians and vegans: Results from the EPIC–Oxford study. *Public Health Nutr.* **2011**, *14*, 340–346. [CrossRef] [PubMed]

137. Hill, A.B. The Environment and Disease: Association or Causation? *Proc. R. Soc. Med.* **1965**, *58*, 295–300. [CrossRef]

138. Annweiler, C.; Cao, Z.; Sabatier, J.-M. Point of view: Should COVID-19 patients be supplemented with vitamin D? *Maturitas* **2020**, *140*, 24–26. [CrossRef] [PubMed]

139. Rejnmark, L.; Bislev, L.S.; Cashman, K.D.; Eiríksdottir, G.; Gaksch, M.; Gruebler, M.; Grimnes, G.; Gudnason, V.; Lips, P.; Pilz, S.; et al. Non-skeletal health effects of vitamin D supplementation: A systematic review on findings from meta-analyses summarizing trial data. *PLoS ONE* **2017**, *12*, e0180512. [CrossRef]

140. Zisi, D.; Challa, A.; Makis, A. The association between vitamin D status and infectious diseases of the respiratory system in infancy and childhood. *Hormones* **2019**, *18*, 353–363. [CrossRef]

141. Sackett, D.L. Evidence-based medicine. *Semin. Perinatol.* **1997**, *21*, 3–5. [CrossRef]

142. Concato, J. Observational Versus Experimental Studies: What's the Evidence for a Hierarchy? *NeuroRX* **2004**, *1*, 341–347. [CrossRef]

143. Haidich, A.B. Meta-analysis in medical research. *Hippokratia* **2010**, *14*, 29–37.

144. Garattini, S.; Jakobsen, J.C.; Wetterslev, J.; Bertele, V.; Banzi, R.; Rath, A.; Neugebauer, E.A.M.E.; Laville, M.; Masson, Y.; Hivert, V.; et al. Evidence-based clinical practice: Overview of threats to the validity of evidence and how to minimise them. *Eur. J. Intern. Med.* **2016**, *32*, 13–21. [CrossRef] [PubMed]

145. Murthi, P.; Davies-Tuck, M.; Lappas, M.; Singh, H.; Mockler, J.; Rahman, R.; Lim, R.; Leaw, B.; Doery, J.; Wallace, E.M.; et al. Maternal 25-hydroxyvitamin D is inversely correlated with foetal serotonin. *Clin. Endocrinol.* **2016**, *86*, 401–409. [CrossRef] [PubMed]

146. Urashima, M.; Segawa, T.; Okazaki, M.; Kurihara, M.; Wada, Y.; Ida, H. Randomized trial of vitamin D supplementation to prevent seasonal influenza A in schoolchildren. *Am. J. Clin. Nutr.* **2010**, *91*, 1255–1260. [CrossRef] [PubMed]

147. Ganmaa, D.; Uyanga, B.; Zhou, X.; Gantsetseg, G.; Delgerekh, B.; Enkhmaa, D.; Khulan, D.; Ariunzaya, S.; Sumiya, E.; Bolortuya, B.; et al. Vitamin D Supplements for Prevention of Tuberculosis Infection and Disease. *N. Engl. J. Med.* **2020**, *383*, 359–368. [CrossRef] [PubMed]

148. Uwitonze, A.M.; Razzaque, M.S. Role of Magnesium in Vitamin D Activation and Function. *J. Am. Osteopat. Assoc.* **2018**, *118*, 181–189. [CrossRef]

149. Caspi, R.; Altman, T.; Dreher, K.; Fulcher, C.A.; Subhraveti, P.; Keseler, I.M.; Kothari, A.; Krummenacker, M.; Latendresse, M.; Mueller, L.A.; et al. The metacyc database of metabolic pathways and enzymes and the biocyc collection of pathway/genome databases. *Nucleic Acids Res.* **2012**, *40*, D742–D753. [CrossRef] [PubMed]

150. Swaminathan, R. Magnesium Metabolism and its Disorders. *Clin. Biochem. Rev.* **2003**, *24*, 47–66.

151. Noronha, L.J.; Matuschak, G.M. Magnesium in critical illness: Metabolism, assessment, and treatment. *Intensiv. Care Med.* **2002**, *28*, 667–679. [CrossRef]

152. Iddir, M.; Brito, A.; Dingeo, G.; Del Campo, S.S.F.; Samouda, H.; La Frano, M.R.; Bohn, T. Strengthening the Immune System and Reducing Inflammation and Oxidative Stress through Diet and Nutrition: Considerations during the COVID-19 Crisis. *Nutrients* **2020**, *12*, 1562. [CrossRef]

153. Rusciano, D.; Bagnoli, P.; Galeazzi, R. The fight against covid-19: The role of drugs and food supplements. *J. Pharm. Pharm. Res.* **2020**, *3*, 1–15.

154. Junaid, K.; Ejaz, H.; Abdalla, A.E.; Abosalif, K.O.A.; Ullah, M.I.; Yasmeen, H.; Younas, S.; Hamam, S.S.M.; Rehman, A. Effective immune functions of micronutrients against Sars-Cov-2. *Nutrients* **2020**, *12*, 2992. [CrossRef] [PubMed]

155. Lahore, H. Covid-19 Intervention Trial Summary. Available online: https://vitamindwiki.com/tiki-index.php?page_id=11728 (accessed on 26 July 2020).

Publisher's Note: MDPI stays neutral with regard to jurisdictional claims in published maps and institutional affiliations.

Article

Vitamin D Supplementation Associated to Better Survival in Hospitalized Frail Elderly COVID-19 Patients: The GERIA-COVID Quasi-Experimental Study

Gaëlle Annweiler [1,2], Mathieu Corvaisier [3,4], Jennifer Gautier [3], Vincent Dubée [1,5,6], Erick Legrand [1,7], Guillaume Sacco [3,8] and Cédric Annweiler [1,3,8,9,10,*]
on behalf of the GERIA-COVID study group

[1] School of Medicine, Health Faculty, University of Angers, 49045 Angers, France; gaelle.annweiler@gmail.com (G.A.); Vincent.Dubee@chu-angers.fr (V.D.); ErLegrand@chu-angers.fr (E.L.)

[2] Department of Medicine, Clinique de l'Anjou, 49044 Angers, France

[3] Department of Geriatric Medicine, Research Center on Autonomy and Longevity, University Hospital, 49933 Angers, France; Mathieu.Corvaisier@chu-angers.fr (M.C.); JeGautier@chu-angers.fr (J.G.); yogisacco@gmail.com (G.S.)

[4] Department of Pharmacy, Angers University Hospital, 49933 Angers, France

[5] Nantes-Angers Cancer and Immunology Research Center (CRCINA), Inserm, University of Angers, 49000 Angers, France

[6] Department of Infectious and Tropical Diseases, Angers University Hospital, 49933 Angers, France

[7] Department of Rheumatology, Angers University Hospital, 49933 Angers, France

[8] EA4638, Laboratory of Psychology of the Pays de la Loire, University of Angers, 49045 Angers, France

[9] Gérontopôle of Pays de la Loire, 44000 Nantes, France

[10] Robarts Research Institute, Department of Medical Biophysics, Schulich School of Medicine and Dentistry, the University of Western Ontario, London, ON N6A 5K8, Canada

* Correspondence: Cedric.Annweiler@chu-angers.fr; Tel.: +332-4135-4725; Fax: +332-4135-4894

Received: 23 September 2020; Accepted: 28 October 2020; Published: 2 November 2020

Abstract: Background. The objective of this quasi-experimental study was to determine whether bolus vitamin D supplementation taken either regularly over the preceding year or after the diagnosis of COVID-19 was effective in improving survival among hospitalized frail elderly COVID-19 patients. Methods. Seventy-seven patients consecutively hospitalized for COVID-19 in a geriatric unit were included. Intervention groups were participants regularly supplemented with vitamin D over the preceding year (Group 1), and those supplemented with vitamin D after COVID-19 diagnosis (Group 2). The comparator group involved participants having received no vitamin D supplements (Group 3). Outcomes were 14-day mortality and highest (worst) score on the ordinal scale for clinical improvement (OSCI) measured during COVID-19 acute phase. Potential confounders were age, gender, functional abilities, undernutrition, cancer, hypertension, cardiomyopathy, glycated hemoglobin, number of acute health issues at admission, hospital use of antibiotics, corticosteroids, and pharmacological treatments of respiratory disorders. Results. The three groups ($n = 77$; mean $\pm$ SD, 88 $\pm$ 5 years; 49% women) were similar at baseline (except for woman proportion, $p = 0.02$), as were the treatments used for COVID-19. In Group 1 ($n = 29$), 93.1% of COVID-19 participants survived at day 14, compared to 81.2% survivors in Group 2 ($n = 16$) ($p = 0.33$) and 68.7% survivors in Group 3 ($n = 32$) ($p = 0.02$). While considering Group 3 as reference (hazard ratio (HR) = 1), the fully-adjusted HR for 14-day mortality was HR = 0.07 ($p = 0.017$) for Group 1 and HR = 0.37 ($p = 0.28$) for Group 2. Group 1 had longer survival time than Group 3 (log-rank $p = 0.015$), although there was no difference between Groups 2 and 3 (log-rank $p = 0.32$). Group 1, but not Group 2 ($p = 0.40$), was associated with lower risk of OSCI score ≥ 5 compared to Group 3 (odds ratio = 0.08, $p = 0.03$). Conclusions. Regular bolus vitamin D supplementation was associated with less severe COVID-19 and better survival in frail elderly.

Keywords: COVID-19; SARS-CoV-2; vitamin D; therapeutics; quasi-experimental study; older adults

---

1. Introduction

Since December 2019, the COVID-19 caused by SARS-CoV-2 is spreading worldwide, affecting millions of people and leaving hundreds of thousands dead, mostly in older adults. With the lack of effective therapy, chemoprevention, and vaccination [1], focusing on the immediate repurposing of existing drugs gives hope of curbing the pandemic. Importantly, a recent unbiased genomics-guided tracing of the SARS-CoV-2 targets in human cells identified vitamin D among the three top-scoring molecules manifesting potential infection mitigation patterns through their effects on gene expression [2]. In particular, by activating or repressing several genes in the promoter region of which it binds to the vitamin D response element, [3] vitamin D may theoretically prevent or improve COVID-19 adverse outcomes by regulating i) the renin–angiotensin system (RAS), ii) the innate and adaptive cellular immunity, iii) the physical barriers, and iv) the host frailty and comorbidities [4,5]. Consistently, epidemiology shows that hypovitaminosis D is more common from October to March at northern latitudes above 20 degrees, [3] which corresponds to the latitudes with the highest lethality rates of COVID-19 during the first months of winter 2020 [1]. In line with this, significant inverse associations were found in European countries between serum 25-hydroxyvitamin D (25(OH)D) concentration and the number of COVID-19 cases, as well as with COVID-19 mortality [6]. This suggests that increasing serum 25(OH)D concentration may improve the prognosis of COVID-19. However, no large well-designed randomized controlled trial (RCT) has tested the effect of vitamin D supplements on COVID-19 outcomes yet. We had the opportunity to examine the association between the use of bolus vitamin D supplements and COVID-19 outcomes in a sample of hospitalized frail elderly patients infected with SARS-CoV-2. The main objective of this hospital-based quasi-experimental study was to determine whether bolus vitamin D supplementation taken either regularly during the preceding year or after the diagnosis of COVID-19 was effective in improving survival among frail elderly COVID-19 patients. The secondary objective was to determine whether this intervention was also effective in limiting the clinical severity of the infection.

2. Materials and Methods

2.1. Study Population

The study consisted in a quasi-experimental study conducted in one geriatric acute care unit dedicated to COVID-19 patients. Data of the GERIA-COVID study were retrospectively collected from patients' records. The inclusion criteria were as follows: (1) patients hospitalized in the geriatric acute care unit of Angers University Hospital, France, in March–May 2020; (2) no objection from the patient and/or relatives to the use of anonymized clinical and biological data for research purpose. The inclusion criteria for the present analysis were as follows: (1) COVID-19 diagnosed with RT-PCR and/or chest CT-scan; (2) data available on the treatments received, including vitamin D supplementation, since the diagnosis of COVID-19 and over the preceding year at least; (3) data available on the vital status 14 days after the diagnosis of COVID-19. Seventy-seven patients were consecutively diagnosed with COVID-19 during the study period in the unit. All of them were recruited in the GERIA-COVID study. They all met the other inclusion criteria and were included in the present analysis.

2.2. Intervention: Vitamin D Supplementation

The regular intake of bolus vitamin D supplements over the preceding year was systematically noted from the primary care physicians' prescriptions and sought by questioning the patients and their relatives.

"Group 1" was defined as all COVID-19 patients who had received oral boluses of vitamin D supplements over the preceding year. Bolus included the doses of 50,000 IU vitamin D3 per month, or the doses of 80,000 IU or 100,000 IU vitamin D3 every 2–3 months. None received D2 or intramuscular supplements, and no patient in Group 1 received additional supplements following the diagnosis of COVID-19.

"Group 2" was defined as the COVID-19 patients usually not supplemented with vitamin D, but who received an oral supplement of 80,000 IU vitamin D3 within a few hours of the diagnosis of COVID-19.

Finally, "Group 3" was defined as the Comparator group, i.e., all COVID-19 patients who had received no vitamin D supplements, neither over the preceding year nor after the diagnosis of COVID-19; the absence of vitamin D treatment being mostly explained by the patients' refusal to be supplemented, since vitamin D supplementation is recommended with no biological testing in all patients over 65 years of age in France [3].

2.3. Primary Outcome: 14-Day COVID-19 Mortality

The primary outcome was the 14-day mortality. Follow-up started from the day of COVID-19 diagnosis for each patient and continued for 14 days or until death when applicable.

2.4. Secondary Outcome: Ordinal Scale for Clinical Improvement (OSCI) Score for COVID-19 in Acute Phase

The secondary outcome was the score on the 9-point World Health Organization's ordinal scale for clinical improvement (OSCI) for COVID-19 [7]. The OSCI distinguishes between several levels of COVID-19 clinical severity according to the outcomes and dedicated treatments required, with a score ranging from 0 (no clinical or virological sign of infection) to 8 (death). The score was determined by the geriatrician of the hospital unit on admission, then revised regularly according to the clinical course of the patients. The highest score during hospitalization was used for the present analysis, corresponding to the most severe acute phase of COVID-19 for each patient. A score of 3 corresponds to a degree of severity requiring hospitalization (i.e., all GERIA-COVID participants had an OSCI score ≥3 here), a score of 5 corresponds to the introduction of non-invasive ventilation, and a score of 6 to intubation and invasive ventilation [7]. Severe COVID-19 was defined here as a score of 5 or more.

2.5. Covariables

Potential confounders were age, gender, functional abilities, severe undernutrition, history of cancer, hypertension, cardiomyopathy, glycated hemoglobin, number of acute health issues at admission, hospital use of antibiotics, systemic corticosteroids, and pharmacological treatments of respiratory disorders. Functional abilities prior to COVID-19 were measured from 1 to 6 (best) with the iso-resources groups (GIR) [8]. Serum albumin concentration, C-reactive protein (CRP), and glycated hemoglobin were measured at hospital admission. Severe undernutrition was defined as albumin <30 g/L. Acute health issues were defined as diseases with sudden onset and rapid progression, whatever their nature or site [9]. History of hematological and solid cancers, hypertension, and cardiomyopathy were noted from the medical register, and by interviewing patients, their relatives, and family physicians. The use of systemic corticosteroids and/or antibiotics (i.e., quinolones, beta-lactams, sulfonamides, macrolides, lincosamides, aminoglycosides, among others), and/or pharmacological treatments of respiratory disorders (i.e., beta2-adrenergic agonists, inhaled corticosteroids, antihistamines, among others) were noted from prescriptions during hospitalization.

2.6. Statistical Analysis

The participants' characteristics were summarized using means and standard deviations (SD) or frequencies and percentages, as appropriate. As the number of observations was higher than 40, comparisons were not affected by the shape of the error distribution and no transformation was applied [10]. Firstly, comparisons between participants separated into three groups according to the intervention (i.e., regular supplementation versus supplementation initiated after COVID-19 diagnosis versus no supplementation) were performed using analysis of variance (ANOVA) or Mann–Whitney–U and Kruskal–Wallis tests for quantitative variables as appropriate, and using Chi-square test or Fisher exact test for qualitative variables as appropriate. To address the issue of multiple comparisons, analyses were completed by a post hoc Fisher's least significant difference (LSD) test. Secondly, a fully-adjusted Cox regression was used to examine the associations of 14-day mortality (dependent variable) with vitamin D supplementation and covariables (independent variables). The model produces a survival function that provides the probability of death at a given time for the characteristics supplied for the independent variables. Third, the elapsed time to death was studied by survival curves computed according to the Kaplan–Meier method and compared by log-rank test. Finally, a multiple logistic regression was used to examine the association of vitamin D supplementation (independent variable) with severe COVID-19 defined as an OSCI score ≥ 5 (dependent variable), while adjusting for potential confounders. p-values <0.05 were considered significant. All statistics were performed using SPSS (v23.0, IBM Corporation, Chicago, IL, USA) and SAS (v9.4, Sas Institute Inc, Cary, NC, USA).

2.7. Ethics

The study was conducted in accordance with the ethical standards set forth in the Helsinki Declaration (1983). No participant or relatives objected to the use of anonymized clinical and biological data for research purposes. Ethics approval was obtained from the Ethics Board of the University Hospital of Angers, France (2020/100). The study protocol was also declared to the National Commission for Information Technology and civil Liberties (CNIL; ar20-0087v0) and registered on ClinicalTrials.gov (NCT04560608).

3. Results

Seventy-seven participants (mean $\pm$ SD age 88 ± 5 years, range 78–100 years; 49.4% women) were included in this quasi-experimental study. Seventeen participants experienced severe COVID-19, and 62 participants survived COVID-19 at day 14, while 15 died.

Table 1 indicates the characteristics of participants separated into Group 1 who regularly received vitamin D supplements over the preceding year ($n = 29$), Group 2 who received vitamin D supplements after the diagnosis of COVID-19 ($n = 16$), and Group 3 who had not received vitamin D supplements ($n = 32$). The three groups were similar at baseline with no significant difference regarding the age ($p = 0.22$), the functional abilities ($p = 0.36$), the history of various comorbidities, the number of acute health issues at hospital admission ($p = 0.22$), and the use of treatments dedicated to COVID-19 (Table 1). At hospital admission, all participants had an OSCI score for COVID-19 of 3 or more. Only the proportion of women differed between groups ($p = 0.02$). At the end of the study, the proportion of participants experiencing severe COVID-19 was lower in Group 1 (10.3%) compared to Group 3 (31.3%, $p = 0.047$), just like the 14-day mortality (6.9% in Group 1 versus 31.3% in Group 3, $p = 0.02$). In contrast, participants in Group 2 did not experience less severe COVID-19 ($p = 0.75$) and less mortality ($p = 0.50$) than participants in Group 3 (Table 1). Similarly, there were no outcome differences between Groups 1 and 2 ($p = 0.23$ for the onset of severe COVID-19, and $p = 0.33$ for 14-day mortality).

Table 1. Characteristics and comparisons of participants with COVID-19 according to the study groups ($n = 77$).

		Study Groups			p-Value *			
	All COVID-19 Participants ($n = 77$)	Group 1 Regular Vitamin D Supplementation ($n = 29$)	Group 2 Vitamin D Supplementation After COVID-19 Diagnosis ($n = 16$)	Group 3 Non-Supplemented Comparator Group ($n = 32$)	Overall ($n = 77$)	Group 1 Versus Group 3 ($n = 61$)	Group 2 Versus Group 3 ($n = 48$)	Group 1 Versus Group 2 ($n = 45$)
Demographical data								
Age (years), med (IQR)	88 (85–92)	88 (87–93)	85 (84–89)	88 (84–92)	0.22	0.98	0.12	0.10
Female gender	38 (49.4)	20 (69.0)	5(31.3)	13 (40.6)	**0.02**	**0.027**	0.52	**0.015**
GIR score (/6), med (IQR)	4 (2–4)	4 (3–4)	4 (3–5)	4 (2–5)	0.36	0.63	0.34	0.13
Comorbidities								
Severe undernutrition [†]	21 (27.3)	9 (31.0)	3 (18.8)	9 (28.1)	0.67	0.80	0.73	0.49
Hematological and solid cancers	27 (35.1)	10 (34.5)	4 (25.0)	13 (40.6)	0.56	0.62	0.29	0.74
Hypertension	49 (63.6)	18 (62.1)	10 (62.5)	21 (65.6)	0.95	0.77	0.83	0.98
Cardiomyopathy	42 (54.5)	13 (44.8)	11 (68.8)	18 (56.3)	0.30	0.37	0.40	0.12
Glycated hemoglobin (%), med (IQR)	6.2 (5.8–6.7)	6.0 (5.5–6.6)	6.4 (6.0–8.2)	6.2 (5.9–6.7)	0.16	0.19	0.34	0.08
Hospitalization								
Number of acute health issues at hospital admission, med (IQR)	3.0 (2.0–4.0)	3.0 (2.0–4.0)	3.5 (2.0–5.0)	2.5 (1.0–4.0)	0.22	0.18	0.14	0.62
CRP at admission (mg/L), med (IQR)	59.5 (19.5–135.0)	44.0 (19.0–110.0)	69.0 (15.5–140.0)	59.0 (29.0–166.0)	0.47	0.21	0.67	0.63
Use of antibiotics [‡]	59 (76.6)	23 (79.3)	14 (87.5)	22 (68.8)	0.32	0.349	0.29	0.69
Use of systemic corticosteroids	13 (16.9)	6 (20.7)	2 (12.5)	5 (15.6)	0.79	0.607	1.00	0.69
Use of pharmacological treatments of respiratory disorders [‖]	10 (13.0)	1 (3.5)	2 (12.5)	7 (21.9)	0.10	0.055	0.70	0.29
COVID-19 outcomes								
Severe COVID-19 [§]	17 (22.1)	3 (10.3)	4 (25.0)	10 (31.3)	0.14	**0.047**	0.75	0.23
14-day mortality	15 (19.5)	2 (6.9)	3 (18.8)	10 (31.3)	0.06	**0.017**	0.50	0.33

Data presented as n (%) where applicable; COVID-19: Coronavirus Disease 2019; CRP: C-reactive protein; GIR: Iso Resource Groups; IQR: interquartile range; OSCI: Ordinal Scale for Clinical Improvement of the World Health Organization; *: between-group comparisons based on Chi-square test (or Fisher exact test where applicable) and ANOVA (or Mann–Whitney–U or Kruskal–Wallis test where applicable); [†]: serum albumin concentration <30 g/L; [‡]: quinolones, beta-lactams, sulfonamides, macrolides, lincosamides, aminoglycosides, among others; [‖]: beta2-adrenergic agonists, inhaled corticosteroids, antihistamines, among others; [§]: defined as an OSCI score for COVID-19 in acute phase ≥5.

Figure 1 shows a statistically significant and clinically relevant inverse association between regular vitamin D supplementation and 14-day mortality. While considering Group 3 as the reference (hazard ratio (HR) = 1), the HR for mortality in Group 1 was 0.19 (95% confidence interval (95% CI): 0.04; 0.85) (*p* = 0.03) in the unadjusted model, HR = 0.18 (95% CI: 0.04; 0.85) (*p* = 0.03) after partial adjustment for age, gender and GIR score, and HR = 0.07 (95% CI: 0.01; 0.61) (*p* = 0.017) after full adjustment for all potential confounders. In contrast, being supplemented with vitamin D after the diagnosis of COVID-19 (Group 2) was not associated with lower mortality risk (HR = 0.37 (95% CI): 0.06; 2.21), *p* = 0.28). The history of hematological and solid cancers was associated with greater mortality risk (HR = 5.56, *p* = 0.01). Using the season of COVID-19 diagnosis as an additional potential confounder did not affect the results (data not shown). Consistently, Kaplan–Meier distributions showed in Figure 2 that COVID-19 participants in Group 3 had shorter survival time than those in Group 1 (log-rank mboxemphp = 0.015), although there was no difference between Groups 2 and 3 (log-rank *p* = 0.32) and between Groups 1 and 2 (log-rank *p* = 0.22).

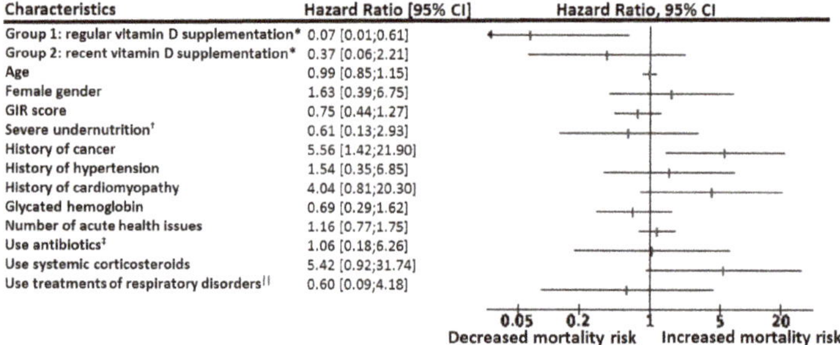

Figure 1. Hazard ratio for 14-day mortality according to vitamin D interventions among participants with COVID-19, adjusted for potential confounders (*n* = 77). CI: confidence interval; COVID-19: coronavirus disease 2019; GIR: Iso Resource Groups; HR: hazard ratio; OSCI: World Health Organization's Ordinal Scale for Clinical Improvement; *: while using Group 3 (no vitamin D supplementation) as a reference (HR = 1); †: serum albumin concentration <30 g/L; ‡: quinolones, beta-lactams, sulfonamides, macrolides, lincosamides, aminoglycosides, among others; ‖: beta2-adrenergic agonists, inhaled corticosteroids, antihistamines, among others.

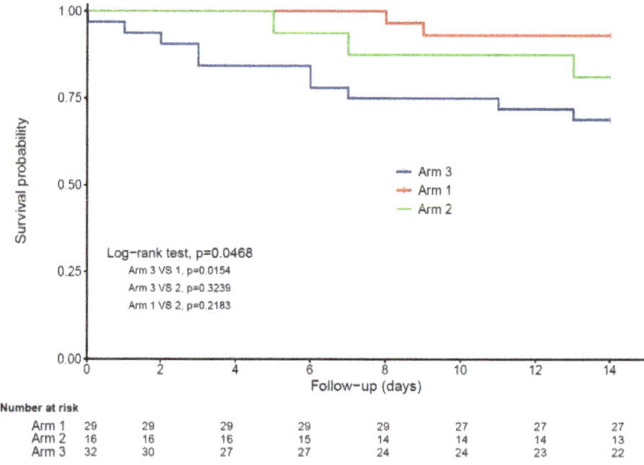

Figure 2. Kaplan–Meier estimates of the cumulative probability of COVID-19 participants' survival according to vitamin D interventions (*n* = 77). Arm 1: regular vitamin D supplementation; Arm 2: vitamin D supplementation initiated after COVID-19 diagnosis; Arm 3: no vitamin D supplementation.

Finally, the multiple logistic regression model in Table 2 revealed that regular vitamin D supplementation (Group 1) was associated with a lower proportion of participants with severe COVID-19 in acute phase (odds ratio (OR) = 0.08 (95% CI): 0.01; 0.81), p = 0.033) compared to Group 3 without vitamin D supplementation. In contrast, Group 2 was not associated with any beneficial effect compared to Group 3 (OR = 0.46 (95% CI): 0.07; 2.85), p = 0.40).

Table 2. Multiple logistic regressions showing the association between vitamin D interventions (independent variable) and the risk of severe COVID-19 * (dependent variable), adjusted for participants' characteristics (n = 77).

	Severe COVID-19 *		
	OR	(95% CI)	p-Value
Interventions			
Group 1: regular vitamin D supplementation	0.08	(0.01; 0.81)	**0.033**
Group 2: vitamin D supplementation initiated after COVID-19 diagnosis	0.46	(0.07; 2.85)	0.40
Group 3: no vitamin D supplementation	1		
Age	1.05	(0.88; 1.25)	0.61
Female gender	1.43	(1.29; 7.13)	0.66
GIR score	0.76	(0.44; 1.33)	0.33
Severe undernutrition [†]	0.42	(0.07; 2.48)	0.34
History of cancer	7.30	(1.37; 38.8)	**0.02**
History of hypertension	0.51	(0.11; 2.33)	0.39
History of cardiomyopathy	10.01	(1.44; 69.88)	**0.02**
Glycated hemoglobin [‡]	0.96	(0.56; 1.63)	0.87
Number of acute health issues at hospital admission	1.19	(0.76; 1.88)	0.45
Use of antibiotics [‖]	1.12	(0.18; 6.85)	0.91
Use of systemic corticosteroids	2.53	(0.34; 17.00)	0.34
Use of pharmacological treatments of respiratory disorders [§]	0.26	(0.02; 2.86)	0.27

CI: confidence interval; COVID-19: coronavirus disease 2019; GIR: Iso Resource Groups; OR: odds ratio; OSCI: World Health Organization's Ordinal Scale for Clinical Improvement; *: defined as OSCI score for COVID-19 in acute phase ≥5; [†]: serum albumin concentration <30 g/L; [‡]: 6 missing data; [‖]: quinolones, beta-lactams, sulfonamides, macrolides, lincosamides, aminoglycosides, among others; §: beta2-adrenergic agonists, inhaled corticosteroids, antihistamines, among others.

4. Discussion

The main finding of this quasi-experimental study is that, irrespective of all measured potential confounders, regular bolus vitamin D3 supplementation was associated with less severe COVID-19 and better survival rate in hospitalized frail elderly. Being supplemented with 80,000 IU vitamin D3 after the diagnosis of COVID-19 was not associated with improved COVID-19 outcomes. These novel findings provide a scientific basis for vitamin D replacement trials attempting to improve COVID-19 prognosis.

To our knowledge, we provide here the first quasi-experimental data comparing the effects of chronic and recent vitamin D supplementations on survival in COVID-19 patients. Growing evidence supports a link between vitamin D and COVID-19. The first reports indicated that adults with hypovitaminosis D were at greater risk of being infected with SARS-CoV-2 (relative risk 1.77 with p < 0.02), [11] and that cases with COVID-19 had lower 25(OH)D concentrations compared to controls without COVID-19 (respectively, 11.1 ng/mL versus 24.6 ng/mL, p = 0.004) [12]. Similarly, significant inverse correlations were found in 20 European countries between the mean serum 25(OH)D concentrations and the number of COVID-19 cases, as well as with mortality [6]. The severity of hypovitaminosis D appears to relate to the prognosis of COVID-19, since COVID-19 cases with hypovitaminosis D were more prone to experience severe COVID-19 (relative risk 1.59 with p = 0.02 if vitamin D insufficiency <30ng/mL) [13]. Finally, hypovitaminosis D was found to be associated with greater COVID-19 mortality risk (incident relative risk 1.56 with p < 0.001 if vitamin D deficiency; p = 0.404 after adjustment) [14]. These results support that enhancing serum 25(OH)D concentration may improve the prognosis of COVID-19, as demonstrated by a pilot controlled trial reporting that the administration of calcifediol versus no calcifediol reduced the need for ICU treatment in

76 hospitalized participants with COVID-19 also receiving best available therapy (mean age, 53 years; 40.8% women) [15]. Following these preliminary findings, larger interventional studies dedicated to COVID-19 with groups properly matched are warranted for investigating the role of vitamin D supplementation on COVID-19 outcomes. Interestingly, previous meta-analyses found that high-dose prophylactic vitamin D supplementation was able to reduce the risk of respiratory tract infections [16]. Based on this observation, we and others are conducting an RCT, the COVIT-TRIAL study, designed to test the effect of high-dose versus standard-dose vitamin D3 on 14-day mortality in COVID-19 older patients (https://clinicaltrials.gov/ct2/show/NCT04344041). While waiting for the recruitment of this RCT to be completed, the findings of the present quasi-experimental study strongly suggest benefits of regular vitamin D3 supplementation on COVID-19 outcomes and survival, which reinforces the recommendations of some scientific societies to supplement all elderly people with vitamin D, in order to improve COVID-19 mortality [17,18]. Additionally, our results support the observation that a single standard dose of 80,000 IU vitamin D3 initiated after the diagnosis of COVID-19 brings no significant benefit on COVID-19 outcomes, which justifies using low-dose vitamin D supplements as a comparator in the COVIT-TRIAL study to determine the effect of higher-dose vitamin D supplements on the prognosis of COVID-19.

How vitamin D supplementation improves COVID-19 outcomes and survival is not fully elucidated. Four mechanisms are likely: regulation of i) the RAS, ii) the innate and adaptive cellular immunity, iii) the physical barriers, and iv) the host frailty and comorbidities [3–5]. First, vitamin D reduces pulmonary permeability in animal models of acute respiratory distress syndrome (ARDS) by modulating the activity of RAS and the expression of the angiotensin-2 converting enzyme (ACE2) [19]. This action is crucial since SARS-CoV-2 reportedly uses ACE2 as a receptor to infect host cells [20] and downregulates ACE2 expression [21]. ACE2 is expressed in many organs, including the endothelium and the pulmonary alveolar epithelial cells, where it has protective effects against inflammation [22]. During COVID-19, downregulation of ACE2 results in an inflammatory chain reaction, the cytokine storm, complicated by ARDS [23]. In contrast, a study in rats with chemically-induced ARDS showed that the administration of vitamin D increased the levels of ACE2 mRNA and proteins [24]. Rats supplemented with vitamin D had milder ARDS symptoms and moderate lung damage compared to controls. Second, many studies have described the antiviral effects of vitamin D, which works either by induction of antimicrobial peptides with direct antiviral activity against enveloped and non-enveloped viruses, or by immunomodulatory and anti-inflammatory effects [25]. These are potentially important during COVID-19 to limit the cytokine storm. Vitamin D can prevent ARDS [26] by reducing the production of pro-inflammatory Th1 cytokines, such as TNFα and interferon γ [26]. It also increases the expression of anti-inflammatory cytokines by macrophages [25]. Third, vitamin D stabilizes physical barriers [4]. These barriers are made up of closely linked cells to prevent outside agents (such as viruses) from reaching tissues susceptible to viral infection. Although viruses alter the integrity of the cell junction, vitamin D contributes to the maintenance of functional tight junctions via E-cadherin [4]. Fourth, the literature over the past decade on the non-bone effects of vitamin D has repeatedly reported that hypovitaminosis D is accompanied by various comorbidities including diabetes mellitus, hypertension, chronic cardiovascular and respiratory diseases, and cancers [3], all conditions that are associated with an increased risk of COVID-19 worsening and death [1]. Prolonged hypovitaminosis D may thus be considered as a factor of poor prognosis of COVID-19, potentiating the risk of cardiorespiratory severity in frail older adults infected with SARS-CoV-2.

All these actions of vitamin D may explain the protective effect of regular long-term vitamin D supplementation, the latter providing the body with a desirable vitamin D environment allowing the various beneficial effects to be expressed and potentiated in the protection against COVID-19. On the contrary, we assume that vitamin D supplementation initiated after the diagnosis of COVID-19 was started too late for the effects of vitamin D to be effective against the infection. It is also possible that the single dose of 80,000 IU was too low to generate protective effects in a very short time, a hypothesis tested in the COVIT-TRIAL RCT.

We also noted, in the present study, a 14-day mortality rate of 31.3% among frail older adults not supplemented with vitamin D (Table 1). This result is consistent with previous literature that points out a special vulnerability of frail older adults. Mortality is less than 1.1% in patients aged <50 years and it increases exponentially after that age up to around 30% [27], especially in frail older adults who have the highest proportion of severe cases of COVID-19 and fatal outcomes [28]. Thus, this result validates the consistency of our cohort and of our main results, notably the protective effect of the regular intake of vitamin D supplements on COVID-19 outcomes.

The strengths of the present study include (i) the originality of the research question on an emerging infection for which there is no scientifically validated treatment [1], (ii) the follow-up and the detailed description of the participants' characteristics allowing the use of multivariate Cox models to measure adjusted longitudinal associations according to three vitamin D regimens, and (iii) the standardized collection of data from a single research center.

Regardless of that, a number of limitations also existed. First, the study participants were restricted to a limited number of hospitalized frail elderly patients who might be unrepresentative of all older adults. It is also possible that the limited sample size in each group had resulted in a lack of power with increased beta risk. Second, although we were able to control for the important characteristics that could modify the association, residual potential confounders might still be present such as the serum concentration of 25(OH)D at baseline—a low level classically ensuring the efficacy of the supplementation [29], or the OSCI score on admission. The OSCI score was collected here in the most acute phase of COVID-19 as it was reported that COVID-19 can get worse between 7–10 days due to the cytokine storm regardless of the initial disease severity [30]. Third, the quasi-experimental design of our study is less robust than an RCT. Participants in the Comparator group did not receive vitamin D placebo. Moreover, there was no randomization. It is plausible that the participants who regularly received vitamin D supplementation (Group 1) were treated better by their family physicians than the others, thereby exhibiting more stable chronic diseases such as cardiovascular comorbidities. It is also plausible that patients or relatives refused taking vitamin D supplementation in Group 3, because the conditions of patients were too severe for them to take the supplements. It should yet be noted that the history did not differ between the 3 groups and that their demographical and health characteristics were similar at baseline, except for the proportion of women (who are likely to suffer from osteoporosis and may have received corresponding treatment that includes vitamin D). While gender is a recognized prognostic factor for COVID-19 [30], the effect of vitamin D supplementation on COVID-19 outcomes persisted after adjustment for all studied confounders including the gender, which allows interpreting the severity and survival differences as being explained by the interventions based on vitamin D supplementation.

5. Conclusions

In conclusion, we were able to report, among hospitalized frail elderly patients with COVID-19, that regular bolus vitamin D3 supplementation was associated with less severe COVID-19 and better survival rate. Vitamin D3 supplementation may represent an effective, accessible, and well-tolerated adjuvant treatment for COVID-19, the incidence of which increases dramatically and for which there are currently no validated treatments. Further large prospective, preferentially interventional studies are needed to confirm whether supplementing older adults regularly with vitamin D3 prevents COVID-19 onset and/or improves COVID-19 outcomes; and whether higher-dose bolus of vitamin D3 given after the diagnosis of COVID-19 is able to improve its prognosis.

Author Contributions: C.A. has full access to all of the data in the study, takes responsibility for the data, the analyses and interpretation, and has the right to publish any and all data, separate and apart from the attitudes of the sponsors. All authors have read and approved the manuscript. Study concept and design: C.A. Acquisition of data: G.A., M.C., J.G., G.S. and C.A. Analysis and interpretation of data: G.A., J.G., V.D., E.L. and C.A. Drafting of the manuscript: G.A. and C.A. Critical revision of the manuscript for important intellectual content: M.C., J.G., V.D., E.L. and G.S. Obtained funding: Not applicable. Statistical expertise: J.G. Administrative, technical, or material support: C.A. Study supervision: C.A.

Funding: G.S. is supported by a postdoctoral grant from the Research Center on Autonomy and Longevity, University Hospital of Angers, France (2019–2020). The sponsor had no role in the design and conduct of the study, in the collection, management, analysis, and interpretation of the data, or in the preparation, review, or approval of the manuscript.

Acknowledgments: The authors wish to thank the GERIA-COVID study group. GERIA-COVID study group: Cédric Annweiler, Marine Asfar, Melinda Beaudenon, Jean Barré, Antoine Brangier, Mathieu Corvaisier, Guillaume Duval, Jennifer Gautier, Mialy Guenet, Jocelyne Loison, Frédéric Noublanche, Marie Otekpo, Hélène Rivière, Guillaume Sacco and Romain Simon. The authors have listed everyone who contributed significantly to the work. Permission has been obtained from all persons named in the Acknowledgments section. There was no compensation for this contribution.

Conflicts of Interest: C.A. serves as an editor for Nutrients. All authors declare that they do not have any other financial and personal conflicts of interest with this manuscript.

References

1. Ahn, D.-G.; Shin, H.-J.; Kim, M.-H.; Lee, S.; Kim, H.-S.; Myoung, J.; Kim, B.-T.; Kim, S.-J. Current Status of Epidemiology, Diagnosis, Therapeutics, and Vaccines for Novel Coronavirus Disease 2019 (COVID-19). *J. Microbiol. Biotechnol.* **2020**, *30*, 313–324. [CrossRef] [PubMed]

2. Glinsky, G.V. Tripartite Combination of Candidate Pandemic Mitigation Agents: Vitamin D, Quercetin, and Estradiol Manifest Properties of Medicinal Agents for Targeted Mitigation of the COVID-19 Pandemic Defined by Genomics-Guided Tracing of SARS-CoV-2 Targets in Human Cells. *Biomed.* **2020**, *8*, 129. [CrossRef]

3. Benhamou, C.-L.; Souberbielle, J.-C.; Cortret, B.; Fardellone, P.; Gauvain, J.-B.; Thomas, T. Vitamin D in adults: GRIO guidelines. *La Presse Médicale.* **2011**, *40*, 673–682.

4. Grant, W.B.; Lahore, H.; McDonnell, S.L.; Baggerly, C.A.; French, C.B.; Aliano, J.L.; Bhattoa, H.P. Evidence that Vitamin D Supplementation Could Reduce Risk of Influenza and COVID-19 Infections and Deaths. *Nutrienrs* **2020**, *12*, 988. [CrossRef] [PubMed]

5. Annweiler, C.; Cao, Z.; Sabatier, J.-M. Point of view: Should COVID-19 patients be supplemented with vitamin D? *Maturitas* **2020**, *140*, 24–26. [CrossRef]

6. Ilie, P.C.; Stefanescu, S.; Smith, L. The role of vitamin D in the prevention of coronavirus disease 2019 infection and mortality. *Aging Clin. Exp. Res.* **2020**, *32*, 1195–1198. [CrossRef] [PubMed]

7. WHO. Coronavirus Disease (COVID-2019) R&D. WHO. Available online: https://www.who.int/teams/blueprint/covid-19 (accessed on 20 September 2020).

8. Vetel, J.M.; Leroux, R.; Ducoudray, J.M. [AGGIR. Practical use. Geriatric Autonomy Group Resources Needs]. *Soins Gérontologie* **1998**, 23–27.

9. Del Galy, A.S.; Bertrand, M.; Bigot, F.; Abraham, P.; Thomlinson, R.; Paccalin, M.; Beauchet, O.; Annweiler, C.; Del Galy, A.S.; Bigot, F.; et al. Vitamin D Insufficiency and Acute Care in Geriatric Inpatients. *J. Am. Geriatr. Soc.* **2009**, *57*, 1721–1723. [CrossRef] [PubMed]

10. Rochon, J.; Gondan, M.; Kieser, M. To test or not to test: Preliminary assessment of normality when comparing two independent samples. *BMC Med. Res. Methodol.* **2012**, *12*, 81. [CrossRef] [PubMed]

11. Meltzer, D.O.; Best, T.J.; Zhang, H.; Vokes, T.; Arora, V.; Solway, J. Association of Vitamin D Status and Other Clinical Characteristics With COVID-19 Test Results. *JAMA Netw. Open* **2020**, *3*, e2019722. [CrossRef]

12. D'Avolio, A.; Avataneo, V.; Manca, A.; Cusato, J.; De Nicolò, A.; Lucchini, R.; Keller, F.; Cantù, M. 25-Hydroxyvitamin D Concentrations Are Lower in Patients with Positive PCR for SARS-CoV-2. *Nutrients* **2020**, *12*, 1359. [CrossRef]

13. Maghbooli, Z.; Sahraian, M.A.; Ebrahimi, M.; Pazoki, M.; Kafan, S.; Tabriz, H.M.; Hadadi, A.; Montazeri, M.; Nasiri, M.; Shirvani, A.; et al. Vitamin D sufficiency, a serum 25-hydroxyvitamin D at least 30 ng/mL reduced risk for adverse clinical outcomes in patients with COVID-19 infection. *PLoS ONE* **2020**, *15*, e0239799. [CrossRef]

14. Hastie, C.E.; Mackay, D.F.; Ho, F.; Celis-Morales, C.A.; Katikireddi, S.V.; Niedzwiedz, C.L.; Jani, B.D.; Welsh, P.; Mair, F.S.; Gray, S.R.; et al. Vitamin D concentrations and COVID-19 infection in UK Biobank. *Diabetes Metab. Syndr. Clin. Res. Rev.* **2020**, *14*, 561–565. [CrossRef] [PubMed]

15. Castillo, M.E.; Costa, L.M.E.; Barrios, J.M.V.; Díaz, J.F.A.; Miranda, J.L.; Bouillon, R.; Gomez, J.M.Q. Effect of calcifediol treatment and best available therapy versus best available therapy on intensive care unit admission and mortality among patients hospitalized for COVID-19: A pilot randomized clinical study. *J. Steroid Biochem. Mol. Biol.* **2020**, *203*, 105751. [CrossRef]

16. Martineau, A.R.; Jolliffe, D.A.; Hooper, R.L.; Greenberg, L.; Aloia, J.F.; Bergman, P.; Dubnov-Raz, G.; Esposito, S.; Ganmaa, D.; Ginde, A.A.; et al. Vitamin D supplementation to prevent acute respiratory tract infections: Systematic review and meta-analysis of individual participant data. *BMJ* **2017**, *356*, i6583. [CrossRef]

17. Chhetri, J.K.; International Association for Gerontology and Geriatrics-Asia/Oceania Region; Chan, P.; Arai, H.; Park, S.C.; Gunaratne, P.S.; Setiati, S.; Assantachai, P. Prevention of COVID-19 in Older Adults: A Brief Guidance from the International Association for Gerontology and Geriatrics (IAGG) Asia/Oceania Region. *J. Nutr. Heal. Aging* **2020**, *24*, 471–472. [CrossRef]

18. French National Academy of Medicine. Vitamin D and Covid-19—Press Release. Available online: http://www.academie-medecine.fr/ wp-content/uploads/2020/05/20.5.22-Vitamine-D-et-coronavirus-ENG.pdf. (accessed on 20 September 2020).

19. Kong, J.; Zhu, X.; Shi, Y.; Liu, T.; Chen, Y.; Bhan, I.; Zhao, Q.; Thadhani, R.; Li, Y.C. VDR Attenuates Acute Lung Injury by Blocking Ang-2-Tie-2 Pathway and Renin-Angiotensin System. *Mol. Endocrinol.* **2013**, *27*, 2116–2125. [CrossRef]

20. Hoffmann, M.; Kleine-Weber, H.; Schroeder, S.; Krüger, N.; Herrler, T.; Erichsen, S.; Schiergens, T.S.; Herrler, G.; Wu, N.-H.; Nitsche, A.; et al. SARS-CoV-2 Cell Entry Depends on ACE2 and TMPRSS2 and Is Blocked by a Clinically Proven Protease Inhibitor. *Cell* **2020**, *181*, 271–280.e8. [CrossRef]

21. Dijkman, R.; Jebbink, M.F.; Deijs, M.; Milewska, A.; Pyrc, K.; Buelow, E.; Van Der Bijl, A.; Van Der Hoek, L. Replication-Dependent Downregulation of Cellular Angiotensin-Converting Enzyme 2 Protein Expression by Human Coronavirus NL63. *J. Gen. Virol.* **2012**, *93*, 1924–1929. [CrossRef] [PubMed]

22. Annweiler, C.; Cao, Z.; Wu, Y.; Faucon, E.; Mouhat, S.; Kovacic, H.; Sabatier, J.-M. Counter-regulatory 'Renin-Angiotensin' System-based Candidate Drugs to Treat COVID-19 Diseases in SARS-CoV-2-infected patients. *Infect. Disord. Drug Targets* **2020**, *20*, 1–2. [CrossRef]

23. Ji, X.; Zhang, C.; Zhai, Y.; Zhang, Z.; Zhang, C.; Xue, Y.; Tan, G.; Niu, G. TWIRLS, an Automated Topic-Wise Inference Method Based on Massive Literature, Suggests a Possible Mechanism via ACE2 for the Pathological Changes in the Human Host after Coronavirus Infection. *bioRxiv* **2020**, preprint. [CrossRef]

24. Yang, J.; Zhang, H.; Xu, J. Effect of Vitamin D on ACE2 and Vitamin D receptor expression in rats with LPS-induced acute lung injury. *Chin. J. Emerg. Med.* **2016**, *25*, 1284–1289.

25. Fabbri, A.; Infante, M.; Ricordi, C. Editorial—Vitamin D status: A key modulator of innate immunity and natural defense from acute viral respiratory infections. *Eur. Rev. Med. Pharmacol. Sci.* **2020**, *24*, 4048–4052.

26. Dancer, R.C.A.; Parekh, D.; Lax, S.; D'Souza, V.; Zheng, S.; Bassford, C.R.; Park, D.; Bartis, D.G.; Mahida, R.; Turner, A.M.; et al. Vitamin D deficiency contributes directly to the acute respiratory distress syndrome (ARDS). *Thorax* **2015**, *70*, 617–624. [CrossRef]

27. Bonanad, C.; García-Blas, S.; Tarazona-Santabalbina, F.; Sanchis, J.; Bertomeu-González, V.; Fácila, L.; Ariza, A.; Núñez, J.; Cordero, A. The Effect of Age on Mortality in Patients With COVID-19: A Meta-Analysis With 611,583 Subjects. *J. Am. Med. Dir. Assoc.* **2020**, *21*, 915–918. [CrossRef]

28. Hewitt, J.; Carter, B.; Vilches-Moraga, A.; Quinn, T.J.; Braude, P.; Verduri, A.; Pearce, L.; Stechman, M.; Short, R.; Price, A.; et al. The effect of frailty on survival in patients with COVID-19 (COPE): A multicentre, European, observational cohort study. *Lancet Public Health* **2020**, *5*, e444–e451. [CrossRef]

29. Heaney, R.P. Guidelines for optimizing design and analysis of clinical studies of nutrient effects. *Nutr. Rev.* **2014**, *72*, 48–54. [CrossRef] [PubMed]

30. Huang, C.; Wang, Y.; Li, X.; Ren, L.; Zhao, J.; Hu, Y.; Zhang, L.; Fan, G.; Xu, J.; Gu, X.; et al. Clinical features of patients infected with 2019 novel coronavirus in Wuhan, China. *Lancet* **2020**, *395*, 497–506. [CrossRef]

Publisher's Note: MDPI stays neutral with regard to jurisdictional claims in published maps and institutional affiliations.

 *nutrients*

Article

Improving Glucose Homeostasis after Parathyroidectomy for Normocalcemic Primary Hyperparathyroidism with Co-Existing Prediabetes

Spyridon Karras [1,*], Cedric Annweiler [2,3], Dimitris Kiortsis [4], Ioannis Koutelidakis [5] and Kalliopi Kotsa [1]

[1] Division of Endocrinology and Metabolism, First Department of Internal Medicine, Medical School, Aristotle University of Thessaloniki, AHEPA University Hospital, 54621 Thessaloniki, Greece; kalm@yahoo.gr

[2] Division of Geriatric Medicine, Department of Neuroscience, Angers University Hospital, 49035 Angers, France; ce.annweiler@chu.angers.fr

[3] Robarts Research Institute, Department of Medical Biophysics, Schulich School of Medicine and Dentistry, The University of Western Ontario, London, ON N6A 3K7, Canada

[4] Department of Nuclear Medicine, University of Ioannina, 45110 Ioannina, Greece; dkiorts@uoi.gr

[5] Second Department of Surgery, Gennimatas General Hospital, Aristotle University of Thessaloniki, 54124 Thessaloniki, Greece; Iokoutel@auth.gr

* Correspondence: karraspiros@yahoo.gr; Tel.: +30-2310324863

Received: 21 October 2020; Accepted: 12 November 2020; Published: 16 November 2020

Abstract: We have previously described increased fasting plasma glucose levels in patients with normocalcemic primary hyperparathyroidism (NPHPT) and co-existing prediabetes, compared to prediabetes per se. This study evaluated the effect of parathyroidectomy (PTx) (Group A), versus conservative follow-up (Group B), in a small cohort of patients with co-existing NPHPT and prediabetes. Sixteen patients were categorized in each group. Glycemic parameters (levels of fasting glucose (fGlu), glycosylated hemoglobin (HbA1c), and fasting insulin (fIns)), the homeostasis model assessment for estimating insulin secretion (HOMA-B) and resistance (HOMA-IR), and a 75-g oral glucose tolerance test were evaluated at baseline and after 32 weeks for both groups. Measurements at baseline were not significantly different between Groups A and B, respectively: fGlu (119.4 ± 2.8 vs. 118.2 ± 1.8 mg/dL, p = 0.451), HbA$_{1c}$ (5.84 ± 0.3 %vs. 5.86 ± 0.4%, p = 0.411), HOMA-IR (3.1 ± 1.2 vs. 2.9 ± 0.2, p = 0.213), HOMA-B (112.9 ± 31.8 vs. 116.9 ± 21.0%, p = 0.312), fIns (11.0 ± 2.3 vs. 12.8 ± 1.4 μIU/mL, p = 0.731), and 2-h post-load glucose concentrations (163.2 ± 3.2 vs. 167.2 ± 3.2 mg/dL, p = 0.371). fGlu levels demonstrated a positive correlation with PTH concentrations for both groups (Group A, rho = 0.374, p = 0.005, and Group B, rho = 0.359, p = 0.008). At the end of follow-up, Group A demonstrated significant improvements after PTx compared to the baseline: fGlu ((119.4 ± 2.8 vs. 111.2 ± 1.9 mg/dL, p = 0.021) (−8.2 ± 0.6 mg/dL)), and 2-h post-load glucose concentrations ((163.2 ± 3.2 vs. 144.4 ± 3.2 mg/dL, p = 0.041), (−18.8 ± 0.3 mg/dL)). For Group B, results demonstrated non-significant differences: fGlu ((118.2 ± 1.8 vs. 117.6 ± 2.3 mg/dL, p = 0.031), (−0.6 ± 0.2 mg/dL)), and 2-h post-load glucose concentrations ((167.2 ± 2.7 vs. 176.2 ± 3.2 mg/dL, p = 0.781), (+9.0 ± 0.8 mg/dL)). We conclude that PTx for individuals with NPHPT and prediabetes may improve their glucose homeostasis when compared with conservative follow-up, after 8 months of follow-up.

Keywords: normocalcemic primary hyperparathyroidism; parathyroidectomy; prediabetes; fasting glucose

1. Introduction

Primary hyperparathyroidism (PHPT) is biochemically confirmed by hypercalcemia and inappropriately increased concentrations of parathyroid hormone (PTH) [1]. Apart from its well-documented musculoskeletal effects, PHPT has been associated with an increased prevalence of metabolic clinical conditions including disorders of glucose homeostasis [2–5]. However, thus far the potential metabolic benefits of parathyroidectomy (PTx) have not been established in these clinical conditions [3–8].

In vivo studies demonstrated that PTH administration has been associated with a reduction in insulin-stimulated glucose uptake, and a decrease in glucose transporter-4 and insulin receptor substrate-1 protein expression [2–4]. PTH also has been reported to down-regulate insulin intracellular signaling, resulting in an increase in peripheral insulin resistance [4–6], and is also inversely correlated with insulin sensitivity [5–7]. Epidemiological studies have also indicated that chronic inappropriate PTH secretion is a predisposing factor for the impairment of glucose homeostasis [7,8].

Normocalcemic primary hyperparathyroidism (NPHPT) is recognized as a new subclinical entity in the field of parathyroid disorders, and is characterized by increased serum PTH concentrations where serum calcium levels are within normal values, after the exclusion of other causes of high PTH [9,10]. In the field of metabolic complications, available results are conflicting regarding the effects of NPHPT as a potential risk factor for the development of cardiovascular and metabolic complications, although it has been associated with increased fasting glucose levels in patients with type 2 diabetes [11–14].

We have also reported that vitamin D deficiency, in combination with increased PTH, is associated with higher fasting glucose profiles in elderly individuals with prediabetes [15] and patients with co-existing NPHPT and prediabetes, compared to individuals with prediabetes per se [16]. However, the effect of a surgical intervention, compared to conservative follow-up in these populations, remains obscure.

This study evaluated the effect of parathyroidectomy (PTx) (Group A), compared to conservative follow-up (Group B), in a small cohort of individuals with co-existing NPHPT and prediabetes.

2. Materials and Methods

2.1. Study Population

The inclusion period was from December 2016 to March 2020. A total of 32 patients with NPHPT and prediabetes were initially included. NPHPT was defined as elevated serum PTH concentration (>65 pg/mL) and normal corrected serum Ca concentration [1,10]. Corrected Ca and PTH concentrations were measured on at least two occasions during a 3–6-month period, to confirm the persistence of the hyperparathyroid state. Prediabetes was diagnosed using the American-Diabetes Association (ADA) criteria, either as impaired fasting-plasma glucose (IFG) (fasting glucose (fGlu): 101–125 mg/dL) or impaired glucose tolerance (IGT) (2-h plasma glucose in the 75-g oral glucose tolerance (OGTT): 140–199 mg/dL), or as HbA1c values between 5.7% (39 mmol/mol) and 6.4% (46 mmol/mol) [17]. Exclusion criteria for both groups were as follows: (a) patients with pre-existing diabetes; (b) patients on any medication that could affect glucose metabolism and PTH dynamics (e.g., loop diuretics, lithium, denosumab); (c) previous medical thyroidectomy/parathyroidectomy; (d) conditions affecting vitamin D metabolism, such as malabsorption syndromes and chronic renal failure (stage 3–5); (e) patients with a body mass index (BMI) of >35 kg/m^2; (f) hypercalciuria (>4 mg/kg body weight/day); (g) patients with familial hypocalciuric hypercalcemia (detected by 24 h urine Ca excretion and calcium-creatinine clearance formula) [1,18].

All participants underwent a physical examination and anthropometric measurements including height, body weight (BW), body mass index (BMI), waist circumference (WC), body fat (BF) mass and percentage, and lean body mass (LBM). Height was measured to the nearest 0.1 cm (cm) with a Holtain wall stadiometer. Waist circumference (WC) was measured midway between the lowest rib and the iliac crest by using an anthropometric tape. BMI was calculated as the ratio of weight in kilograms divided by the height in the meters squared (kg/m2). Body fat (BF) mass and percentage, muscle mass (MM) and lean body mass (LBM) were measured using bioelectrical impedance analysis (BIA) (SC-330S, Tanita Corporation, Tokyo, Japan).

Three months prior to baseline evaluation, all participants were supplemented with a daily vitamin D regimen (ranging from 1.200 to 4.000 IU) according to their latest available 25-hydroxy-vitamin D (25(OH)D) concentrations. This was in order to achieve 25(OH)D concentrations $\geq$ 30 ng/mL, according to international criteria [1,9,10], to avoid the effects of vitamin D deficiency on PTH values [18]. All participants underwent methoxy isobutyl isonitrile (MIBI) scintigraphy and parathyroid gland ultrasonography prior to inclusion. Imaging findings from parathyroid scintigraphy and ultrasonography were positive for the existence of a single parathyroid adenoma in 23 participants, whereas in four participants, findings were negative in ultrasonography and positive in scintigraphy. Finally, in the five individuals where no evidence of a parathyroid lesion was found, a conservative follow-up was suggested. According to a recent re-evaluation of these cases, two of these patients were confirmed as being positive for an adenoma (MIBI scintigraphy).

PTx was suggested to 21 participants at baseline, according to internationally adopted criteria for the management of PHPT [1] or patient preference. A total of five participants to whom PTx was suggested preferred conservative follow-up. Conservative follow-up consisted of a regular (3 month) biochemical evaluation of calcium homeostasis (total calcium (Ca), phosphorus (P), albumin, PTH, 25-hydroxy-vitamin D (25(OH)D), and 24 h urine calcium) for the assessment of hypercalcemia or hypercalciuria. Subsequently, participants were included in two groups, A ($n = 16$) and B ($n = 16$), following PTx or conservative approach, respectively. Anthropometric evaluation was repeated 32 weeks after PTx or conservative approach, for both groups. For the entire study period, participants in both groups followed regular dietetic guidance focused on prediabetes management according ADA recommendations [17]. A similar isocaloric diet was implemented for both groups, aiming at maintaining initial body weight. Participants were contacted via telephone and/or e-mail twice during the intervention to confirm their adherence to diets and to resolve any potential issues.

During the entire study period, participants were advised to maintain a stable level of physical activity, namely 150 min per week of moderate-intensity aerobic exercise, according to the ADA recommendations [17]. We did not use a specific method (e.g., a wearable device) to monitor physical activity. However, participants were strongly advised to conform to the recommendations, and the importance of such compliance was repeatedly emphasized by the research team. For the majority of participants, the suggested level of physical activity was slightly more intense than their habitual exercise status.

All participants were supplemented with a daily vitamin D regimen (1200 to 4000 IU daily) until the end of the study, with target 25(OH)D levels of $\geq$30 ng/mL [18]. Written informed consent was obtained from all participants. The study was conducted according to the Declaration of Helsinki for research involving humans. An ethical approval by the scientific committee was obtained by the Institution involved (Cyan Cross Hospital Ethics Committee).

2.2. Laboratory Evaluation

Laboratory evaluation was conducted both at baseline and after 32 weeks after PTx or conservative approach, for both groups. Blood samples were drawn after a 12-h overnight fast and stored at −20 °C prior to analysis. Fasting samples for total Ca, P, albumin, PTH, 25(OH)D, and glycemic parameters (fGlu, HbA1c, fasting insulin (fIns)) were obtained. Homeostasis model assessment (HOMA) was used for estimating insulin secretion (HOMA-B) and resistance (HOMA-IR) according to previously reported formulas [19]. Subsequently, all participants from both groups underwent a 75-g OGTT to evaluate 2-h post-load glucose response. Glucose and insulin measurements were performed using the Cobas INTEGRA clinical chemistry system (D-68298; Roche® Diagnostics, Mannheim, Germany). Reference ranges were reported previously [16]. Corrected Ca was calculated according to the type Ca (mg/dL) + 0.8X (4 (mg/dL)—albumin (mg/dL)). Vitamin D status was assessed through the measurement of serum 25(OH)D, by competitive electrochemiluminescent immunoassay (Roche® Modular E170). PTH was measured using the electrochemiluminescence immunoassay ECLIA (Roche® Diagnostics GmbA,

Mannheim, Germany). Coefficient variations and reference ranges have been reported previously for both assays [16].

2.3. Statistical Analysis

Continuous data are presented as mean ± standard deviation (SD). Comparisons among groups were performed using a Student's *t*-test for unpaired data or using a Mann–Whitney U test. Proportions were compared with a Fisher exact test. The Pearson correlation coefficient (Pearson's r) was also used for the examination of correlations of normally distributed variables, and Spearman's rank correlation coefficient (rho) for non-normally distributed variables. A *p*-value < 0.05 was considered statistically significant. Two-way ANOVA analysis was conducted for comparisons between groups and within the same group.

Statistical analysis was performed using SPSS 13.0 software. We used the best available data of a suggested background prevalence of NPHPT in women of reproductive age, of 0.06% [12].

The required sample size was calculated using the STATA corporation statistical platform and was based on detecting a change in prevalence from 0.06% in the general population to 1% or more in our population. One per cent prevalence for NPHPT was chosen, as it is generally regarded as the lowest value that may stimulate a change in clinical practice. A one-tailed test with a power of 80% (i.e., β = 0.2) and significance level (α) of 5% showed a minimum requirement of 22 NPHPT patients with binomial 95% confidence intervals around the 1% prevalence of 0.23–3.19%. Descriptive statistics were calculated in Excel® (V14.6, 2010, Microsoft®, Washington, USA).

3. Results

3.1. Baseline

Demographic, anthropometric, and biochemical data are presented in Table 1. Individuals in both groups did not differ with respect to age, female to male ratio, BMI, waist circumference, body fat, and lean body mass.

Table 1. Comparative baseline demographic and anthropometric characteristics between groups.

Parameter	Group A	Group B	*p*-Value *
Participants; Women (*n* (%))	16; 12 (75%)	16; 11 (68.75%)	0.091
Age (years)	58.9 ± 1.0	56.2 ± 3.2	0.391
Weight (kg)	77.2 ± 18.8	77.6 ± 17.1	0.420
BMI (kg/m^2)	28.1 ± 0.7	28.2 ± 1.3	0.814
Waist circumference (cm)	94± 1.9	96.1 ± 3.7	0.543
Body fat (%)	33.6 ± 7.6	35.4 ± 9.1	0.126
Lean body mass (kg)	50.7 ± 12.1	47.5 ± 9.9	0.283

Data are presented as mean ± standard deviation. *: Mann–Whitney test. Abbreviations: PTH: parathyroid hormone; BMI: body mass index; WC: waist circumference; 25(OH)D: 25-hydroxy-vitamin D.

Both groups manifested a similar profile for calciotropic hormones (Table 1).

With respect to glucose homeostasis markers at baseline, fGlu (119.4 ± 2.8 vs. 118.2 ± 1.8 mg/dL, *p* = 0.451), HbA$_{1c}$ (5.84 ± 0.0 vs. 5.86 ± 0.0%, *p* = 0.415), HOMA-IR (3.1 ± 1.2 vs. 2.9 ± 0.2, *p* = 0.211), HOMA-B (112.9 ± 31.8 vs. 116.9 ± 21.0%, *p* = 0.314), fIns (11.0 ± 2.3 vs. 12.8 ± 1.4 µIU/mL, *p* = 0.731), and 2-h post-load glucose concentrations (163.2 ± 3.2 vs. 167.2 ± 3.2 mg/dL, *p* = 0.371), were not different in the two groups at baseline (Table 2).

When calciotropic hormones were evaluated for interactions with glucose homeostasis, fGlu was positively associated with PTH concentrations (Group A, rho = 0.374, *p* = 0.005 and Group B, rho = 0.359, *p* = 0.008).

Table 2. Comparative anthropometric and biochemical data throughout the study.

	Baseline	Week 32	*p*-Value for Trend within Groups *	*p*-Value for Group × Time Interaction *
Weight (kg)				
Group A	77.2 ± 18.8	77.8 ± 18.1	0.714	0.420
Group B	77.6 ± 17.1	78.0 ± 16.8		
BMI (kg/m^2)				
Group A	28.1 ± 0.7	28.4 ± 0.6	0.811	0.652
Group B	28.2 ± 1.3	28.8 ± 1.9		
Waist circumference (cm)				
Group A	94.0± 1.9	96.4 ± 1.2	0.514	0.541
Group B	96.1 ± 3.7	97.1 ± 3.1		
Body fat (%)				
Group A	33.6 ± 7.6	34.7 ± 14.8	0.651	0.134
Group B	35.4 ± 9.1	32.2 ± 7.3		
Lean body mass (kg)				
Group A	50.7 ± 12.1	49.1 ± 15.6	0.783	0.178
Group B	47.5 ± 9.9	48.5 ± 10.4		
PTH (pg/mL)				
Group A	94.2 ± 2.4	44.2 ± 1.4 [t,a]	< 0.01	< 0.01
Group B	96.2 ± 3.2	86.2 ± 2.2		
25-hydroxy-vitamin D (ng/mL)				
Group A	36.3 ± 2.1	32.3 ± 3.1	0.145	0.383
Group B	33.2 ± 1.3	31.2 ± 1.9		
Serum corrected calcium (mg/dL)				
Group A	9.9 ± 0.0	9.1 ± 0.0 [t,a]	0.031	0.045
Group B	9.8 ± 0.1	9.7 ± 0.2		
Serum phosphorus (mg/dL)				
Group A	3.5 ± 0.0	3.9 ± 0.1 [t,a]	0.011	0.031
Group B	3.4 ± 0.1	3.6 ± 0.1		
Fasting glucose (mg/dL)				
Group A	119.4 ± 2.8	111.2 ± 1.9 [t,a]	0.021	0.020
Group B	118.2 ± 1.8	117.6 ± 2.3		
Fasting insulin (µIU/mL)				
Group A	11.0 ± 2.3	10.8 ± 1.1	0.601	0.731
Group B	12.8 ± 1.4	13.1 ± 1.8		
HOMA-IR				
Group A	3.1 ± 1.2	3.0 ± 1.1	0.631	0.213
Group B	2.9 ± 0.2	3.4 ± 1.1		
HOMA-B (%)				
Group A	112.9 ± 31.8	114.1 ± 11.0	0.619	0.312
Group B	116.9 ± 21.0	114.2 ± 19.0		
HbA1c (%)				
Group A	5.84 ± 0.0	5.81 ± 0.0	0.411	0.511
Group B	5.86 ± 0.0	5.88 ± 0.0		
2-h post-load glucose (mg/dL)				
Group A	163.2 ± 3.2	144.4 ± 3.2 [t,a]	0.041	< 0.010
Group B	167.2 ± 2.7	176.2 ± 3.2		

Abbreviations: PTH: parathyroid hormone; BMI: body mass index; WC: waist circumference; 25(OH)D: 25-hydroxy-vitamin D; HbA1c: glycated hemoglobin A1c; HOMA-IR: homeostatic model assessment for insulin resistance; HOMA-B: homeostatic model assessment for beta-cell function. Data are presented as mean ± standard deviation. *: 2-way analysis of variance (ANOVA). [a]: compared to baseline (comparisons within the same group). [t]: compared to B group (comparisons between groups at the same time point).

3.2. Following PTx or Conservative Approach

Anthropometric and biochemical features of calcium homeostasis for both groups are presented in Table 2. Individuals in both groups did not differ with respect to BMI (28.4 ± 0.6 vs. 28.8 ± 1.9 kg/m^2, $p = 0.652$) and waist circumference (96.4 ± 1.2 vs. 97.1 ± 3.1 cm, $p = 0.541$).

At the end of follow-up, Group A demonstrated significant improvements in fGlu after PTx compared to baseline ((119.4 ± 2.8 vs. 111.2 ± 1.9 mg/dL, $p = 0.021$), (−8.2 ± 0.6 mg/dL)), as well as in 2-h post-load glucose concentrations ((163.2 ± 3.2 vs. 144.4 ± 3.2 mg/dL, $p = 0.041$), (−18.8 ± 0.3 mg/dL)). Group B demonstrated non-significant differences in fGlu ((118.2 ± 1.8 vs. 117.6 ± 2.3 mg/dL, $p = 0.031$), (−0.6 ± 0.2 mg/dL)) and 2-h post-load glucose concentrations ((167.2 ± 2.7 vs. 176.2 ± 3.2 mg/dL, $p = 0.781$), (+9.0 ± 0.8 mg/dL)) (Table 2).

As expected, calciotropic homeostasis was significantly improved in Group A compared to Group B, for PTH (44.2 ± 1.4 vs. 86.2 ± 2.2 pg/mL, $p < 0.01$), corrected Ca (9.1 ± 0.0 vs. 9.7 ± 0.2 mg/dL, $p = 0.044$), and P (3.9 ± 0.1 vs. 3.6 ± 0.1 mg/dL, $p = 0.031$), whereas 25(OH)D serum concentrations were similar for the two groups (32.3 ± 3.1 vs. 31.2 ± 1.9 ng/mL, $p = 0.383$) (Table 2). In addition, Group A demonstrated significant improvements vs. Group B in fGlu levels ((111.2 ± 1.9 vs. 117.6 ± 2.3 mg/dL, $p = 0.02$), (−6.4 ± 0.7 mg/dL)) and 2-h post-load glucose concentrations ((144.2 ± 3.2 vs. 176.2± 3.2 mg/dL, $p < 0.01$), (−32 ± 0.4 mg/dL)), respectively (Table 2).

4. Discussion

The purpose of this study was to investigate potential effects of PTx on glucose profiles of patients with co-existing prediabetes and NPHPT, versus conservative treatment. Our results indicated an improvement of glucose homeostasis following PTx, compared to the non-interventional strategy of conservative follow-up. This is the first report identifying an early (32 weeks post-surgery) beneficial glycemic effect of a surgical intervention (PTx) where there is the co-existence of both subclinical entities. The most plausible explanation for this early effect could result from the robust reduction of PTH concentrations after PTx, possibly explaining the improvement of glycemic parameters observed in our analysis.

Previous in-vitro studies indicated that increased concentrations of PTH have an adverse effect on mitochondrial oxidative phosphorylation, as well as adenosine triphosphate (ATP) islet content [20]. The increased milieu of calcium in the cytosol, as a result of the reduction of Ca exocytosis, has been associated with deterioration of glucose-stimulated insulin secretion, particularly in the preclinical or prediabetic state [21].

Of major interest are in vivo animal models which have indicated that PTH administration promotes a plethora of biological phenomena, including a decrease in insulin-stimulated glucose uptake, protein kinase B (AKT) activity (phosphorylated AKT/total AKT protein expression), and a reduction in glucose transporter-4 and insulin receptor substrate-1 protein expression. PTH also induced an increase of insulin receptor substrate-1 (IRS-1) phosphorylation on serine 307, which has been reported to down-regulate insulin intracellular signaling, resulting in an increase in peripheral insulin resistance. These results indicate that chronic PTH over-secretion is implicated in the development of dyshomeostasis on β-cells function, as well as development of insulin resistance in adipose tissue [20]. However, previous clinical human studies have not consistently reported a deterioration of peripheral insulin action [8–10]. In this study, no differences in HOMA-IR or HOMA-B indices between groups were evident, probably as a result of the small sample size.

In this regard, PTH has also been shown to inversely correlate with insulin sensitivity index (ISI), and has also been proven to be an independent negative determinant of ISI, with a decrease of 0.247 μmol/L/m^2/min/pmol/L in ISI for each pg/mL increase in plasma PTH levels [22]. These reports were also confirmed by large-scale observational and epidemiological data, implying that elevated PTH concentrations are positively associated with abnormal glucose metabolism [23,24]. However, these results have not been confirmed by available intervention trials. We recently demonstrated a significant increase of glucose-stimulated glucagon-like peptide 1 secretion after successful PTx for asymptomatic primary hyperparathyroidism [25], indicating a favorable profile in incretion secretion physiology after surgical intervention.

Hagström et al. [12] evaluated the incidence of metabolic diseases (including diabetes mellitus) in 30 subjects with NPHPT (treated either with PTx or conservatively) during a five-year follow-up period, compared with age-matched controls. In their study [12], PTx had no effect on fGlu and HbA1c values. In this study, we also did not observe differences in HbA1c, probably due to the shorter follow-up period, which could outline a potential improvement in this setting. On the other hand, the discrepancy in fGlu concentrations was likely the result of differences in dietary practices and physical exercise between the two studies. In our study, a detailed dietary and exercise regimen was suggested to the participants, whereas in the Hagström et al. [12] study no such supervision was reported. Recently, a case–control study investigating the effect of PTx on cardiovascular risk factors in patients with normocalcemic and hypercalcemic PHPT [26] showed that after PTx improvements in blood pressure, serum total cholesterol, HOMA-IR, and cardiovascular risk scores were reported for both groups.

Our results suggest that in patients with NPHPT the reduction in PTH secretion after PTx results in significant improvement in the context of prediabetes only 32 weeks after surgery, indicating that surgical management of NPHPT might be a rational approach, particularly in cases where a deterioration of glucose homeostasis is evident during conservative follow-up. Although available guidelines [1,2,18] recommend that PTx is suggested primarily for musculoskeletal complications, our results indicate that PTx for metabolic reasons may be a future option.

A relative strength of our study is that we used strict criteria for diagnosing NPHPT and prediabetes, prior to patient inclusion in this analysis [18]. We have also excluded patients with other causes of increased PTH concentrations.

Main limitations include the lack of inclusion of a control group of participants with normal glucose homeostasis and other components of the metabolic syndrome (to draw more complete conclusions about the association between insulin resistance and dysfunction of parathyroid glands), the small sample size, and the use of HOMA indexes to evaluate insulin resistance and beta-cell function, instead of using the gold standard of clamping [27]. Moreover, we were not able to incorporate additional markers for evaluating insulin resistance and insulin secretion, such as the Matsuda index and the insulin secretion-sensitivity index-2. Finally, the follow-up period after PTx was relatively short in order to identify additional beneficial effects on glycemic homeostasis.

Measurements of additional components of the metabolic syndrome were absent, limiting the generalizability of these findings regarding the effects of PTx in the cardiometabolic profile of this cohort, as well as the association of insulin resistance with NPHPT. In this setting, more appropriately designed future studies could improve and elucidate the pathophysiological interplay between these preclinical entities.

In conclusion, these results indicate that PTx in individuals with NPHPT and prediabetes may improve glucose homeostasis compared to conservative follow-up, after 8 months of follow-up. Similar studies are required in larger population groups to confirm these results in the co-existence of these subclinical entities.

Author Contributions: S.K., C.A. and I.K. conceptualized and conducted the study. S.K. wrote and drafted the manuscript. D.K. and K.K., contributed to statistical analysis and final drafting of the manuscript. Final manuscript was approved by all authors. All authors have read and agreed to the published version of the manuscript.

Funding: This research received no external funding.

Conflicts of Interest: The authors declare no conflict of interest.

References

1. Silva, B.C.; Cusano, N.E.; Bilezikian, J.P. Primary hyperparathyroidism. *Best Pract. Res. Clin. Endocrinol. Metab.* **2018**, *32*, 593–607. [CrossRef]
2. Cardenas, M.G.; Vigil, K.J.; Talpos, G.B.; Lee, M.W.; Peterson, E.; Rao, D.S. Prevalence of Type 2 Diabetes Mellitus in Patients with Primary Hyperparathyroidism. *Endocr. Pract.* **2008**, *14*, 69–75. [CrossRef]
3. Kim, H.; Kalkhoff, R.K.; Costrini, N.V.; Cerletty, J.M.; Jacobson, M. Plasma insulin disturbances in primary hyperparathyroidism. *J. Clin. Investig.* **1971**, *50*, 2596–2605. [CrossRef]
4. Kautzky-Willer, A.; Pacini, G.; Niederle, B.; Schernthaner, G.; Prager, R. Insulin secretion, insulin sensitivity and hepatic insulin extraction in primary hyperparathyroidism before and after surgery. *Clin. Endocrinol.* **1992**, *37*, 147–155. [CrossRef] [PubMed]
5. Prager, R.; Schernthaner, G.; Niederle, B.; Roka, R. Evaluation of glucose tolerance, insulin secretion, and insulin action in patients with primary hyperparathyroidism before and after surgery. *Calcif. Tissue Int.* **1990**, *46*, 1–4. [CrossRef] [PubMed]
6. Ljunghall, S.; Palmer, M.; Akerstrom, G.; Wide, L. Diabetes mellitus, glucose tolerance and insulin response to glucose in patients with primary hyperparathyroidism before and after parathyroidectomy. *Eur. J. Clin. Investig.* **1983**, *13*, 373–377. [CrossRef] [PubMed]
7. Koubaity, O.; Mandry, D.; Nguyen-Thi, P.L.; Bihain, F.; Nomine-Criqui, C.; Demarquet, L.; Croise-Laurent, V.; Brunaud, L. Coronary artery disease is more severe in patients with primary hyperparathyroidism. *Surgery* **2020**, *167*, 149–154. [CrossRef] [PubMed]
8. Procopio, M.; Borretta, G. Derangement of glucose metabolism in hyperparathyroidism. *J. Endocrinol. Investig.* **2003**, *26*, 1136–1142. [CrossRef] [PubMed]
9. Lowe, H.; McMahon, D.J.; Rubin, M.R.; Bilezikian, J.P.; Silverberg, S.J. Normocalcemic primary hyperparathyroidism: Further characterization of a new clinical phenotype. *J. Clin. Endocrinol. Metab.* **2007**, *92*, 3001–3005. [CrossRef] [PubMed]
10. Dawood, N.B.; Yan, K.L.; Shieh, A.; Livhits, M.J.; Yeh, M.W.; Leung, A.M. Normocalcaemic primary hyperparathyroidism: An update on diagnostic and management challenges. *Clin. Endocrinol. (Oxf.)* **2020**. [CrossRef]
11. Cakir, I.; Unluhizarci, K.; Tanriverdi, F.; Elbuken, G.; Karaca, Z.; Kelestimur, F. Investigation of insulin resistance in patients with normocalcemic primary hyperparathyroidism. *Endocrine* **2012**, *42*, 419–422. [CrossRef]
12. Hagström, E.; Lundgren, E.; Rastad, J.; Hellman, P. Metabolic abnormalities in patients with normocalcemic hyperparathyroidism detected at a population-based screening. *Eur. J. Endocrinol.* **2006**, *155*, 33–39. [CrossRef] [PubMed]
13. Temizkan, S.; Kocak, O.; Aydin, K.; Ozderya, A.; Arslan, G.; Yucel, N.; Sargin, M. Normocalcemic hyperparathyroidism and insulin resistance. *Endocr. Pract.* **2015**, *21*, 23–29. [CrossRef] [PubMed]
14. Yener Ozturk, F.; Erol, S.; Canat, M.M.; Karatas, S.; Kuzu, I.; Dogan Cakir, S.; Altuntas, Y. Patients with normocalcemic primary hyperparathyroidism may have similar metabolic profile as hypercalcemic patients. *Endocr. J.* **2016**, *63*, 111–118. [CrossRef] [PubMed]
15. Karras, S.N.; Anagnostis, P.; Antonopoulou, V.; Tsekmekidou, X.; Koufakis, T.; Goulis, D.G.; Zebekakis, P.; Kotsa, K. The combined effect of vitamin D and parathyroid hormone concentrations on glucose homeostasis in older patients with prediabetes: A cross-sectional study. *Diabetes Vasc. Dis Res.* **2018**, *15*, 150–153. [CrossRef]
16. Karras, S.N.; Koufakis, T.; Tsekmekidou, X.; Antonopoulou, V.; Zebekakis, P.; Kotsa, K. Increased parathyroid hormone is associated with higher fasting glucose in individuals with normocalcemic primary hyperparathyroidism and prediabetes: A pilot study. *Diabetes Res. Clin. Pract.* **2020**, *160*, 107985. [CrossRef] [PubMed]
17. American Diabetes Association. 2. Classification and Diagnosis of Diabetes: Standards of Medical Care in Diabetes—2018. *Diabetes Care* **2018**, *41* (Suppl. S1), S13–S27. [CrossRef]
18. Bilezikian, J.P. Primary hyperparathyroidism. *J. Clin. Endocrinol. Metab.* **2018**, *103*, 3993–4004. [CrossRef]
19. Matthews, D.R.; Hosker, J.P.; Rudenski, A.S.; Naylor, B.A.; Treacher, D.F.; Turner, R.C. Homeostasis model assessment: Insulin resistance and beta-cell function from fasting plasma glucose and insulin concentrations in man. *Diabetologia* **1985**, *28*, 412–419. [CrossRef]

20. Perna, A.F.; Fadda, G.Z.; Zhou, X.J.; Massry, S.G. Mechanisms of impaired insulin secretion after chronic excess of parathyroid hormone. *Am. J. Physiol.* **1990**, *259*, F210–F216. [CrossRef]

21. Massry, S.G. Sequence of Cellular Events in Pancreatic Islets Leading to Impaired Insulin Secretion in Chronic Kidney Disease. *J. Ren. Nutr.* **2011**, *21*, 92–99. [CrossRef] [PubMed]

22. Chiu, K.C.; Chuang, L.M.; Lee, N.P.; Ryu, J.M.; Mc, G.u.l.l.a.m.J.L.; Tsai, G.P.; Saad, M.F. Insulin sensitivity is inversely correlated with plasma intact parathyroid hormone level. *Metabolism* **2000**, *49*, 1501–1505. [CrossRef] [PubMed]

23. Kumar, S.; Olukoga, A.O.; Gordon, C.; Mawer, E.B.; France, M.; Hosker, J.P.; Davies, M.; Boulton, A.J. Impaired glucose tolerance and insulin insensitivity in primary hyperparathyroidism. *Clin. Endocrinol. (Oxf.)* **1994**, *40*, 47–53. [CrossRef]

24. Wareham, N.J.; Byrne, C.D.; Carr, C.; Day, N.E.; Boucher, B.J.; Hales, C.N. Glucose intolerance is associated with altered calcium homeostasis: A possible link between increased serum calcium concentration and cardiovascular disease mortality. *Metabolism* **1997**, *46*, 1171–1177. [CrossRef]

25. Antonopoulou, V.; Karras, S.N.; Koufakis, T.; Yavropoulou, M.; Katsiki, N.; Gerou, S.; Papavramidis, T.; Kotsa, K. Rising Glucagon-Like Peptide 1 Concentrations After Parathyroidectomy in Patients With Primary Hyperparathyroidism. *J. Surg. Res.* **2019**, *245*, 22–30. [CrossRef]

26. Beysel, S.; Caliskan, M.; Kizilgul, M.; Apaydin, M.; Kan, S.; Ozbek, M.; Cakal, E. Parathyroidectomy improves cardiovascular risk factors in normocalcemic and hypercalcemic primary hyperparathyroidism. *BMC Cardiovasc. Disord.* **2019**, *19*, 106. [CrossRef]

27. Katsuki, A.; Sumida, Y.; Gabazza, E.C.; Murashima, S.; Furuta, M.; Araki-Sasaki, R.; Hori, Y.; Yano, Y.; Adachi, Y. Homeostasis model assessment is a reliable indicator of insulin resistance during follow-up of patients with type 2 diabetes. *Diabetes Care* **2001**, *24*, 362–365. [CrossRef]

Publisher's Note: MDPI stays neutral with regard to jurisdictional claims in published maps and institutional affiliations.

Article

Ultra-Marathon-Induced Increase in Serum Levels of Vitamin D Metabolites: A Double-Blind Randomized Controlled Trial

Jan Mieszkowski [1], Błażej Stankiewicz [2], Andrzej Kochanowicz [1], Bartłomiej Niespodziński [2], Tomasz Kowalik [2], Michał A. Żmijewski [3], Konrad Kowalski [4], Rafał Rola [4,5], Tomasz Bieńkowski [4] and Jędrzej Antosiewicz [6,*]

[1] Department of Gymnastics and Dance, Gdansk University of Physical Education and Sport, 80-336 Gdansk, Poland; mieszkowskijan@gmail.com (J.M.); andrzejkochanowicz@o2.pl (A.K.)

[2] Institute of Physical Education, Kazimierz Wielki University, 85-064 Bydgoszcz, Poland; blazej1975@interia.pl (B.S.); bartlomiej.niespodzinski@ukw.edu.pl (B.N.); tomasz.kowalik@ukw.edu.pl (T.K.)

[3] Department of Histology, Medical University of Gdańsk, 80-211 Gdansk, Poland; michal.zmijewski@gumed.edu.pl

[4] Masdiag Sp. z o.o. Company, 01-882 Warsaw, Poland; konrad.kowalski@masdiag.pl (K.K.); r.rola@doktorant.umk.pl (R.R.); tomasz.bienkowski@masdiag.pl (T.B.)

[5] Chair of Environmental Chemistry and Bioanalytics, Faculty of Chemistry, Nicolaus Copernicus University in Toruń, 87-100 Toruń, Poland

[6] Department of Bioenergetics and Physiology of Exercise, Medical University of Gdansk, 80-210 Gdansk, Poland

* Correspondence: jant@gumed.edu.pl; Tel.: +48-583491456

Received: 20 October 2020; Accepted: 23 November 2020; Published: 25 November 2020

Abstract: Purpose: While an increasing number of studies demonstrate the importance of vitamin D for athletic performance, the effects of any type of exercise on vitamin D metabolism are poorly characterized. We aimed to identify the responses of some vitamin D metabolites to ultra-marathon runs. Methods: A repeated-measures design was implemented, in which 27 amateur runners were assigned into two groups: those who received a single dose of vitamin D_3 (150,000 IU) 24 h before the start of the marathon ($n = 13$) and those ($n = 14$) who received a placebo. Blood samples were collected 24 h before, immediately after, and 24 h after the run. Results: In both groups of runners, serum $25(OH)D_3$, $24,25(OH)_2D_3$, and $3\text{-}epi\text{-}25(OH)D_3$ levels significantly increased by 83%, 63%, and 182% after the ultra-marathon, respectively. The increase was most pronounced in the vitamin D group. Body mass and fat mass significantly decreased after the run in both groups. Conclusions: Ultra-marathon induces the mobilization of vitamin D into the blood. Furthermore, the $24,25(OH)_2D_3$ and $3\text{-}epi\text{-}25(OH)D_3$ increases imply that the exercise stimulates vitamin D metabolism.

Keywords: endurance exercise; $3\text{-}epi\text{-}25(OH)D_3$; $24,25(OH)_2D_3$; $25(OH)D_3$

---

1. Introduction

Vitamin D plays a crucial role in the regulation of multiple physiological processes. Its activity is mainly ascribed to the active form, $1,25(OH)_2D_3$, which acts via a specific vitamin D receptor (VDR). VDR is a transcriptional factor that regulates the expression of approximately 1000 genes. VDR is present in almost all human tissues [1,2]. Consistently, vitamin D deficiency has been associated with multiple morbidities, such as cancer, diabetes, multiple sclerosis, cardiovascular diseases, and others [3–6]. Therefore, it is recognized that vitamin D status is an important risk factor for several diseases of civilization. Moreover, more and more athletes also show a low vitamin D status, which may negatively impact the health, performance, and training efficiency of athletes [7,8].

Vitamin D is produced in the skin in response to ultraviolet (sunlight) exposure. Subsequently, it is hydroxylated at positions 25 and 1 to gain full hormonal activity. On the other hand, 25-OH vitamin D [25(OH)D$_3$] is a good marker of vitamin D status. The kidney, brain, bone, skin, prostate, and white blood cells can convert 25(OH)D$_3$ to its active form [1,25(OH)$_2$D$_3$]. It can be anticipated that low serum levels of 25(OH)D$_3$ will limit the synthesis of the active form in all these tissues. Serum 25(OH)D$_3$ levels are mainly determined by exposure to sunlight and vitamin D supplementation. In addition, higher fat tissue content is associated with lower serum 25(OH)D$_3$ levels, possibly because of its ability to store vitamin D [9,10]. On the other hand, higher physical activity is associated with better vitamin D status, even though many athletes are vitamin D-deficient [11]. Vitamin D supplementation is the easiest way to correct its deficiency and single high doses have been demonstrated to be effective in short periods [12].

Among many factors, exercise-induced release of vitamin D from adipose tissue has been postulated as an important mechanism that leads to increased vitamin D levels in the blood [13]. However, the effect of exercise on vitamin D status is not fully understood. Of note, vitamin D metabolism is involved in the formation of other metabolites, such as 3-*epi*-25(OH)D$_3$ and 24,25(OH)$_2$D$_3$. These metabolites are not considered to be physiologically active. However, based on recent studies, they play an important role in the regulation of general metabolism. For example, 3-*epi*-25(OH)D$_3$ levels are associated with an improved cardiovascular risk profile, and 3-*epi*-1,25(OH)$_2$D$_3$, derived from 3-*epi*-25(OH)D$_3$, effectively reduces blood parathormone without inducing changes in the plasma calcium levels [14,15]. In addition, 24,25(OH)$_2$D$_3$, considered to be an inactive form of vitamin D, protects cells from 1,25(OH)$_2$D$_3$ toxicity and modulates the antioxidant potential by binding catalase [16]. Furthermore, studies involving animal models demonstrated that 24,25(OH)$_2$D$_3$ plays an important role in normal bone integrity, function, and healing [17,18]. Interestingly, high 24,25(OH)$_2$D$_3$ levels are associated with a reduced disability status in multiple sclerosis patients [19].

Collectively, the above observations indicate that vitamin D metabolites have important biological functions, which are far from being completely understood. Therefore, it is crucial to study the effects of exercise on vitamin D metabolism. Exercise stimulates the release of several hundreds of proteins (myokines) into circulation from the skeletal muscle while also stimulating the liberation of bioactive proteins (exerkines) from other tissues [20]. An example of such an exerkine is fibroblast growth factor 23, whose concentration increases after exercise [21]. This exerkine is responsible for the regulation of plasma phosphate levels and modifies vitamin D metabolism by inhibiting the formation of 1,25(OH)$_2$D$_3$ [22].

Ultra-marathon is a type of exercise that pushes the boundaries of human performance. This extreme type of physical activity, involving continuous running over a distance well above the 42 km of a regular marathon run, is associated with enormous energy expenditure. Until now, knowledge about vitamin D metabolism associated with strenuous exercise like ultra-marathon has been limited [23]. Running and walking for extreme durations, so-called ultra-marathons, have become increasingly popular in the last years throughout the world, particularly in the USA, Europe, Japan, and South Africa [24]. Hence, it seems important to evaluate the effect of this type of prolonged exercise on the physiological responses of the human body, both under conditions of supplementation and where supplementation is not provided. Especially vitamin D supplementation could prevent exercise-induced inflammation processes and other adverse body reactions. We proposed that ultra-marathon, which alters the production of hundreds of exerkines [25] and has the potential to reduce the amount of adipose tissue [26], influences vitamin D metabolism.

Here, we performed a double-blind randomized controlled trial to determine the impact of extreme endurance exercise on vitamin D metabolites in relation to vitamin D supplementation. We found that ultra-marathon induced a significant increase in metabolites of vitamin D$_3$ which do not possess classical metabolic effects of the active form of vitamin D$_3$.

2. Materials and Methods

2.1. Experimental Overview

The study was designed as a double-blind randomized controlled trial with parallel groups. Participants were randomly assigned to two groups: the supplementation group and the control group. The supplementation protocol involved a single high dose of vitamin D3 before the start of the ultra-marathon. During the initial visit, data on the subject's age, body composition, and height were collected. All runners were examined by a professional physician. A sample of venous blood was obtained before the ultra-marathon start and immediately after and 24 h after the run to evaluate the vitamin D metabolites. Additionally, before starting the actual experiment, to evaluate body responses to a high dose of vitamin D, profiles of vitamin D analogues were assessed. All laboratory analyses were performed at the Gdansk University of Physical Education (Gdansk, Poland).

2.2. Participants

Twenty-seven male ultra-marathon runners taking part in the Lower Silesian Mountain Runs Festival Ultra-Marathon Race participated in the study. The runners were randomly assigned to the experimental (supplemented, UM-S; $n = 13$) and control (placebo, UM-C; $n = 14$) groups. The characteristics of the groups are shown in Table 1. The participants were physically active ultra-marathon amateur runners. None of the runners had any history of known diseases or reported any intake of medication due to illnesses in six months before the experiment. All runners had previous ultra-marathon race experience (not less than two). During all testing periods and 1 week before testing, the participants refrained from alcohol, caffeine, guarana, theine, tea, chocolate, and any other substance intake which may potentially influence exercise performance. Furthermore, the participants were asked to adopt a similar eating pattern on the days of measurements, based on a randomized diet for their age group and physical intensity. The study protocol was accepted by the Bioethics Committee for Clinical Research at of the Collegium Medicum University of Nicolaus Copernicus (decision number KB-124/2017) and conducted according to the Declaration of Helsinki. The study was registered as clinical trial: NCT03417700. Written informed consent was obtained from all study participants, who were also informed about the possibility of withdrawal of consent at any time and for any reason. Prior to participation, subjects were informed about the study procedures but not about the rationale and study aim, so as to keep them naive about the potential effect of supplementation.

Table 1. Characteristics of the participants ($n = 27$).

Variable	UM-S ($n = 13$)		UM-C ($n = 14$)		Effect Size (η^2)
	Mean ± SD	(95% CI)	Mean ± SD	(95% CI)	
Age (years)	42.00 ± 8.44	(36.00–47.00)	40.00 ± 8.11	(36.00–45.00)	0.01
Body mass	74.29 ± 7.51	(70.12–78.45)	78.64 ± 10.66	(72.20–85.09)	0.05
Body height (cm)	174.80 ± 3.80 *	(172.54–177.45)	181.30 ± 5.43	(178.43–183.56)	0.34
Body mass index	23.92 ± 2.42	(21.10–25.65)	24.18 ± 1.83	(22.95–25.41)	0.01
Fat mass (%)	12.58 ± 3.25	10.25–14.90	12.43 ± 4.65	9.31–15.56	0.02

Note: UM-S, runners given vitamin D3; UM-C, runners without vitamin D supplementation (control group). Significant difference at * $p \leq 0.05$.

2.3. Pilot Study

Although vitamin D metabolism is well documented, to date, changes in serum levels of vitamin D metabolites after administration of a high dose of vitamin D have not been evaluated. To evaluate body responses to a high dose of vitamin D, profiles of vitamin D analogues were assessed. For this purpose, four physically active non-ultra-marathon runners (volunteers) took two doses of vitamin D3

(100,000 or 200,000 IU) 28 day apart. The blood samples were taken at selected time points (days 0 to 45) and profiles of vitamin D analogues were analyzed as described in Section 2.6. The following were analyzed: $25(OH)D_3$, 3-*epi*-$25(OH)D_3$, and $24,25(OH)_2D_3$, and the ratios of the last two compounds to $25(OH)D_3$.

2.4. Ultra-Marathon Run

One day after the first blood sample collection, physician examination, and supplementation, all runners (UM-C and UM-S groups) participated in the Lower Silesian Mountain Run Festival (19 July 2018). The start and finish points were in the town of Lądek Zdrój (Lower Silesian Voivodeship, Poland).

The running festival took place in the Kłodzko Land (latitude of 50° N) and consisted of seven mountain trails, with a maximum course length of 240 km, a maximum altitude of approximately 1425 m a.s.l., and a minimum altitude of approximately 261 m a.s.l. The entire altitude range was approximately 1164 m, and the total ascent and descent was 7670 m. (Figure 1). The run started at 18:00 h and the temperature during the run varied from 18 °C at the start point to 4 °C on the top of the Śnieżnik Mountain. Most of the time, the sky was overcast.

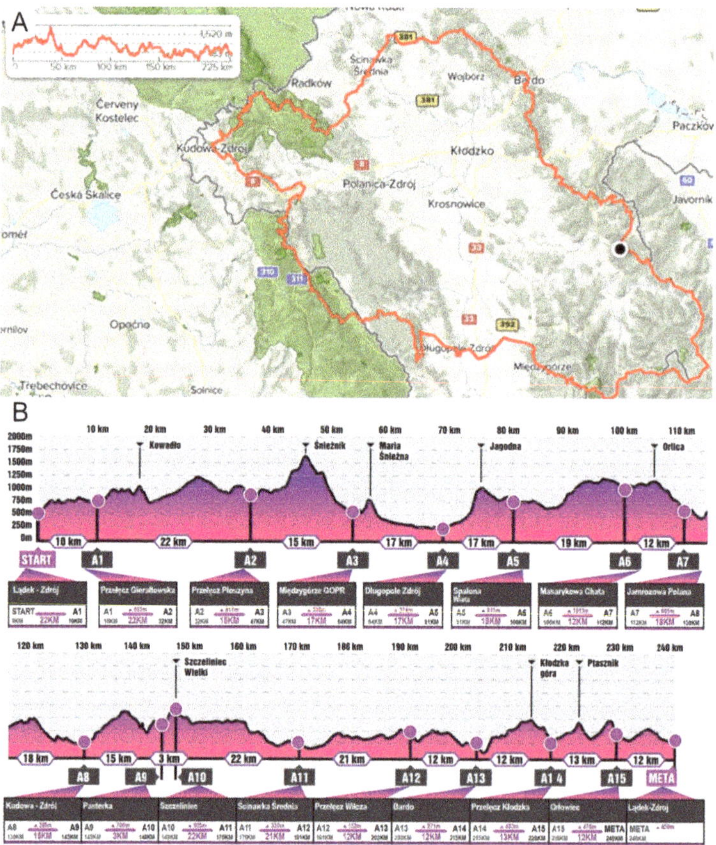

Figure 1. Ultra-marathon track characteristics of the Lower Silesian Mountain Run Festival 2018, Lądek Zdrój. The entire track (**A**) and select track parts and distances (**B**) are shown (Mountain Marathons Foundation).

2.5. Vitamin D Supplementation

Each participant from the experimental group received a single high dose of vitamin D3 (150,000 IU) 24 h before the start of the ultra-marathon. The decision was based on the pilot study, which showed the highest concentration of vitamin D metabolites within 24–48 h. The control group received a placebo solution with the taste (anise), color, and consistency matching those of the vitamin D solution (pure vegetable oil solution).

The participants and researchers had no knowledge of the groups and differences in the supplementation procedures.

2.6. Sample Collection and Measurements of Vitamin D Metabolite Levels

Blood (9 mL) was collected three times: 24 h before and after the race and immediately after the run (up to 5 min after the run). Venous blood samples were collected into Sarstedt S-Monovette tubes (S-Monovette® Sarstedt AG&Co, Nümbrecht, Germany) without anticoagulant for serum separation (with a coagulation accelerator). The serum was separated using standard laboratory procedures, aliquoted, and frozen at $-80\ °C$ until further analysis. Sample preparation was based on serum protein precipitation and derivatization. Quantitative analysis was performed using liquid chromatography coupled with tandem mass spectrometry (QTRAP®4500 (Sciex) coupled with ExionLC HPLC system) with minor changes according to previously published method [27]. Serum samples were analyzed in the positive ion mode, using electrospray ionization. The raw data were collected using Analyst® software, while to process and quantify the collected data MultiQuant® was used. Various reagents were used in the sample preparation procedure. Furthermore, 4-(4′-Dimethylaminophenyl)-1,2,4-triazoline-3,5-dione (DAPTAD) was used as a derivatization agent. It was synthesized by Masdiag Laboratory (Warsaw, Poland). Additionally, solvents such as water, ethyl acetate (POCh S.A., Gliwice, Poland), and methanol (Honeywell, Sigma-Aldrich, Gillingham, Dorset, UK) were used.

Mobile phases were prepared using acetonitrile (ACN) (Honeywell, Sigma-Aldrich, Gillingham, Dorset, UK), water (POCh S.A., Gliwice, Poland), and formic acid (FA) (Merck KGaA, Darmstadt, Germany). All solvents were of LC-MS grade.

The following were determined: $25(OH)D_3$, $24,25(OH)_2D_3$, $3\text{-}epi\text{-}25(OH)D_3$, and $25(OH)D_2$ levels and the ratios of $25(OH)D_3$ to $24,25(OH)_2D_3$ and $25(OH)D_3$ to $3\text{-}epi\text{-}25(OH)D_3$. The concentrations of vitamin D metabolites were corrected to change in plasma volume [28,29].

2.7. Statistical Analysis

Descriptive statistics included mean ± standard deviation (SD) for all measured variables. A two-way ANOVA with repeated measures (2 × 3) was performed to investigate the impact of ultra-marathon running (marathon: 24 h before, immediately after, and 24 h after the run) on vitamin D metabolites and physical characteristics in relation to vitamin D supplementation (group: UM-S, UM-C). In case of a significant interaction, Tukey's post-hoc test was performed to assess differences in specific subgroups. In addition, the effect size was determined by eta-squared statistics (η^2). Values equal to or more than 0.01, 0.06, and 0.14 indicated a small, moderate, and large effect, respectively. All calculations and graphics were conducted using Statistica 12 software (StatSoft, Tulsa, OK, USA). Differences were considered statistically significant when $p \leq 0.05$.

3. Results

In the control study, four volunteers who were not participating in any sport activities were given a single high dose of vitamin D on days 0 and 28, with two doses tested. Regardless of the dose, the $25(OH)D_3$ and $3\text{-}epi\text{-}25(OH)D_3$ levels increased within the first 48–72 h, following which the concentration gradually decreased. A similar relationship was observed for the $24,25(OH)_2D_3$ metabolite, with the highest concentration noted between days 4 and 7 after administration. Its levels remained relatively constant once the peak was reached. Interestingly, the (slow) increase in the 25(OH)D$_3$ levels began 12 h after vitamin D administration, while the 24,25(OH)$_2$D$_3$ levels started to increase 24 h after vitamin D administration (Figure 2).

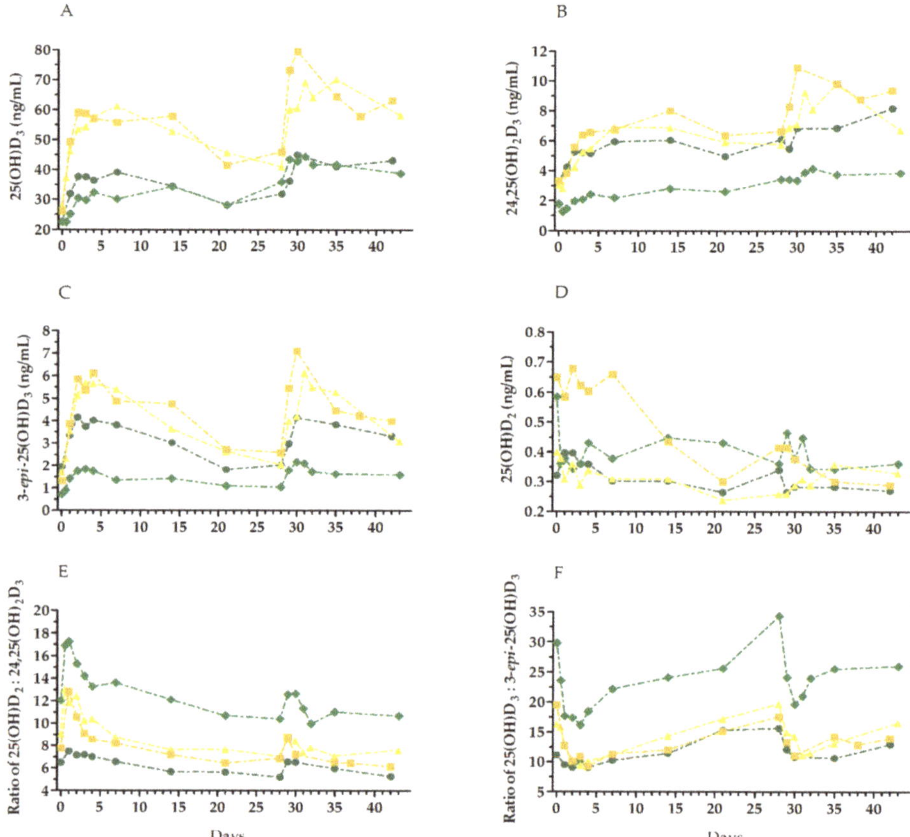

Figure 2. Changes in vitamin D metabolite levels in the serum, and metabolite ratios over time, in four healthy volunteers after receiving two doses of vitamin D$_3$. The doses were given on days 0 and 28. Notation: green lines and symbols—participants administered with 100,000 IU; yellow lines and symbols—participants administered with 200,000 IU. The following were determined: 25(OH)D$_3$ levels (**A**); 24,25(OH)$_2$D$_3$ levels (**B**); f25(OH)D$_3$ levels (**C**); 25(OH)D$_2$ (**D**); ratio of 25(OH)D$_3$ to 24,25(OH)$_2$D$_3$ (**E**); ratio of 25(OH)D$_3$ to 3-*epi*-25(OH)D$_3$ (**F**).

Changes in the serum levels of vitamin D after the ultra-marathon are presented in Figure 3. The two-way ANOVA revealed a significant ultra-marathon effect in all analyzed variables on serum of vitamin D$_3$ levels (Table 2). Regardless of vitamin D$_3$ supplementation, a significant increase in the levels of 25(OH)D$_3$ (82.9%), 24.25(OH)$_2$D$_3$ (63.3%), and 3-*epi*-25(OH)D$_3$ (182.6%), 25(OH)D$_2$ (17.7%) and the ratio of 25(OH)D$_3$ to 24,25(OH)$_2$D$_3$ (45.8%) was observed, as well as a significant decrease in the ratio of 25(OH)D$_3$ to 3-*epi*-25(OH)D$_3$ (18.4%) immediately after the ultra-marathon. A significant increase in the levels of 24,25(OH)$_2$D$_3$, 3-*epi*-25(OH)D$_3$, and 25(OH)D$_2$ and the ratios of 25(OH)D$_3$ to 24,25 (OH)$_2$D$_3$ and 25(OH)D$_3$ to 3-*epi*-25(OH)D$_3$ was also apparent 24 h after the ultra-marathon (Table 2). The two-way ANOVA also revealed a significant group effect in 25(OH)D$_3$, 3-*epi*-25(OH)D$_3$, and 25(OH)D$_3$: 24,25(OH)$_2$D$_3$ ratio (Table 2). The UM-S group showed a 30.1% higher concentration of 25(OH)D$_3$, a 61.8% higher concentration of 3-*epi*-25(OH)D$_3$, and a 24.7% higher ratio of 25(OH)D$_3$ to 24,25(OH)$_2$D$_3$ in comparison with the UM-C group.

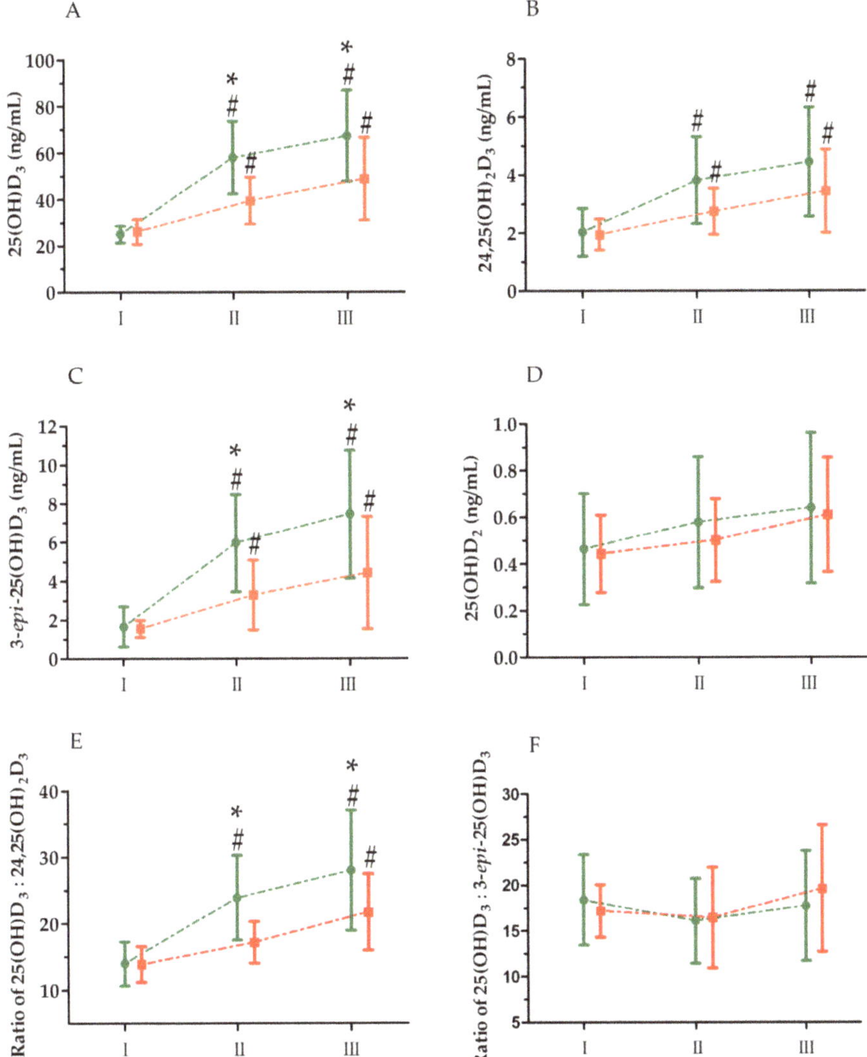

Figure 3. An effect of ultra-marathon on the serum levels of vitamin D metabolites. Data for runners supplemented with vitamin D$_3$ (green symbols and lines) and runners without supplementation (red symbols and lines) are shown. (**A**) 25(OH)D$_3$ levels; (**B**) 24,25(OH)$_2$D$_3$ levels; (**C**) 3-*epi*-25(OH)D$_3$ levels; (**D**) 25(OH)D$_2$ levels; (**E**) 25(OH)D$_3$: 24,25(OH)$_2$D$_3$ ratio; (**F**) 25(OH)D$_3$: 3-*epi*-25(OH)D$_3$ ratio. Time points: I, 24 h before the ultra-marathon; II, immediately after the ultra-marathon; III, 24 h after the ultra-marathon. The values are presented as mean ± SD. * significant difference vs. group without supplementation at $p < 0.05$; # significant difference vs. I time point at $p < 0.05$.

Table 2. Two-way (2 groups × 3 repeated measurements) ANOVA of the serum levels of vitamin D_3 induced by ultra-marathon run.

Variable	Effect	F	df	p	Effect Size (η^2)	Post-Hoc Outcome
$25(OH)D_3$	GR	6.59	1, 30	0.01 *	0.19	S > C I < II, III S-I < S-II, S-III C-I < C-II, C-III S-II > C-II S-III > C-III
	UM	67.00	2, 60	0.01 **	0.70	
	GR × UM	7.43	2, 60	0.01 **	0.20	
$24,25(OH)_2D_3$	GR	2.91	1, 30	0.09	0.08	I < II < III S-I < S-II, S-III C-I < C-II, C-III
	UM	49.90	2, 60	0.01 **	0.62	
	GR × UM	3.72	2, 60	0.02 *	0.11	
$3\text{-}epi\text{-}25(OH)D_3$	GR	7.84	1, 30	0.01 **	0.21	S > C I < II < III S-I < S-II, S-III C-I < C-II, C-III S-II > C-II S-III > C-III
	UM	58.32	2, 60	0.01 **	0.66	
	GR × UM	7.66	2, 60	0.01 **	0.20	
$25(OH)D_2$	GR	0.26	1, 30	0.61	0.01	I < II < III
	UM	34.06	2, 60	0.01 **	0.53	
	GR × UM	1.05	2, 60	0.35	0.03	
Ratio $25(OH)D_3$: $24,25(OH)_2D_3$	GR	6.79	1, 30	0.01 *	0.18	S > C I < II < III S-I < S-II, S-III C-I, C-II < C-III S-II > C-II S-III > C-III
	UM	70.82	2, 60	0.01 **	0.70	
	GR × UM	8.06	2, 60	0.01 **	0.21	
Ratio $25(OH)D_3$: $3\text{-}epi\text{-}25(OH)D_3$	GR	0.06	1, 30	0.81	0.01	II < III
	UM	47.38	2, 60	0.04 *	0.10	
	GR × UM	1.34	2, 60	0.26	0.04	

Note: GR, group; UM, ultra-marathon; S, runners supplemented with vitamin D_3; C, runners without supplementation (control group); I—24 h before the ultra-marathon; II—immediately after the ultra-marathon; III—24 h after the ultra-marathon. Significant difference at * $p \leq 0.05$, ** $p \leq 0.01$.

Furthermore, the analysis of variance of the $25(OH)D_3$ and $3\text{-}epi\text{-}25(OH)D_3$ levels and the ratio of $25(OH)D_3$ to $3\text{-}epi\text{-}25(OH)D_3$ also showed a significant interaction of the group and ultra-marathon factor. An interaction analysis of the 'group' and 'ultra-marathon' factors indicated that the $25(OH)D_3$ and $3\text{-}epi\text{-}25(OH)D_3$ levels and the ratio of $25(OH)D_3$ to $24,25(OH)_2D_3$ immediately after and 24 h after the ultra-marathon were significantly higher in the group supplemented with vitamin D_3 than in the control group (Figure 3).

4. Discussion

The main goal of the current study was to define whether vitamin D supplementation (150,000 IU) affects vitamin D metabolism after an acute exercise, such as an ultra-marathon. We demonstrated that serum $25(OH)D_3$, $24,25(OH)_2D_3$, and $3\text{-}epi\text{-}25(OH)D_3$ levels significantly increased after the ultra-marathon. We also showed that in volunteers supplemented with large dose of vitamin D, similar changes were observed after the run; however, the time of changes differed. In addition, the control experiment demonstrated that application of single high dose of vitamin D is effective in correcting vitamin D deficiency, as reported before [12]. This observation indicates that exercise in addition to supplementation modifies vitamin D metabolism.

According to several reports, vitamin D is stored in adipose tissue and, therefore, increased lipolysis observed during exercise, which leads to vitamin D release into the blood [30,31]. In fact, 30 min of cycling increases $25(OH)D_3$ levels by approximately 20 nmol/L, while 5 weeks of progressive endurance exercise increases $25(OH)D$ levels by 2 nmol/L [13,32]. We confirmed, here, these observations, as we noted almost a 20 nmol/L increase in $25(OH)D$ levels in the control group runners and an even higher increase in the vitamin D-supplemented runners. However, contrary to the findings of Sun et al. [32], in the current study, 25(OH)D levels did not return to baseline values after 24 h but remained elevated. Detailed analysis of individual responses revealed that in 3 of 14 runners from the control group, 25(OH)D$_3$ levels did not increase and even slightly decreased. On the other hand, 25(OH)D$_3$ levels did not change after the run in two runners from the vitamin D-supplemented group. This could be explained by recent studies concerning the responders and non-responders to vitamin D supplementation; however, the detailed molecular mechanism of such variable responsiveness remains to be determined [33]. It cannot be excluded that vitamin D released from adipose tissue is reabsorbed by adipose tissue, and the reabsorption is affected by exercise to a lesser degree than the release. This would explain an earlier observation of a rapid return of 25(OH)D$_3$ levels after exercise to the initial value [13]. The upper body subcutaneous adipose depot is a more important source of plasma fatty acids during exercise than visceral adipose tissue [34]. In addition, visceral adipose tissue accumulates more vitamin D than subcutaneous adipose tissue [35]. Hence, subtle differences in visceral adipose tissue may significantly affect serum levels of vitamin D and vitamin D metabolism during and after exercise. In addition, the adipose tissue content of vitamin D may be significantly different between individuals (4 to 500 ng/g) [36], which could also partially explain the different responses to the ultra-marathon.

In one study involving team-sport athletes, 12-week vitamin D supplementation resulted in increased serum 25(OH)D$_3$ levels; however, a significant increase in 24,25(OH)$_2$D$_3$ levels was observed only after 70,000 IU of vitamin D$_3$ was administered per week, while half of this dose had no effect [37]. Certainly, the observed effects were related to both, supplementation and training. Here, we observed a significant increase in 24,25(OH)$_2$D$_3$ levels in both supplemented and non-supplemented runners. This indicates an increased hydroxylation of vitamin D in C-24 position as a result of extreme exercise, such as an ultra-marathon. In addition, no direct correlation between an increase in 25(OH)D$_3$ and 24,25(OH)$_2$D$_3$ levels was apparent in the above study [37]. That is probably associated with the activation of 24-hydroxylase after a certain cellular 25(OH)D$_3$ threshold is exceeded. In the current study, 25(OH)D$_3$ levels after the ultra-marathon were approximately 40 and 35 ng/mL in runners with and without supplementation, respectively. These levels were much lower than those observed in athletes supplemented with a lower dose of vitamin D (approximately 60 ng/mL) in whom changes in 24,25(OH)$_2$D$_3$ levels were not observed [37]. Furthermore, the ratio of 25(OH)D$_3$ to 24,25(OH)$_2$D$_3$ also significantly increased after the ultra-marathon, which indicated a lack of direct dependency between these two compounds, implying that the exercise stimulates the synthesis of 24,25(OH)$_2$D$_3$. Conversely, in the control subjects, the ratio of 25(OH)D$_3$ to 24,25(OH)$_2$D$_3$ initially rapidly increased as a result of a faster increase in the 25(OH)D$_3$ levels than that of the 24,25(OH)$_2$D$_3$ levels. Over time, the ratio decreased, which was associated with a further production of 24,25(OH)$_2$D$_3$.

Exercise induces the release of several myokines and exerkines [20,38], and some of these molecules possibly stimulate C-24 hydroxylation. However, this requires further study. The 24,25(OH)$_2$D$_3$ metabolite is considered to be an inactive form of vitamin D. Nevertheless, according to recent studies, this metabolite has many biological functions, including protection against 1,25(OH)$_2$D$_3$ toxicity, reduction in inflammation, stimulation of bone healing, and some others [17,39]. Hence, an increase in its levels during exercise could have important implications for the body that should be investigated in future studies.

Another metabolite, whose concentration increased after the ultra-marathon, was 3-*epi*-25(OH)D$_3$. C-3 epimerization is a common metabolic pathway of major metabolites of vitamin D$_3$. 25(OH)D$_3$ undergoes epimerization and 3-*epi*-25(OH)D$_3$ is the most prevalent form [40]. The biological function of 3-*epi*-25-(OH)D$_3$ is not well understood. Its concentration has been reported to be in the range of 0–9.01 ng/mL [41]. In the current study, 3-*epi*-25-(OH)D$_3$ was detectable before and after the ultra-marathon in all runners. Interestingly, 3-*epi*-25(OH)D$_3$ can be converted to 3-*epi*-1,25-(OH)D$_3$, which participates in the suppression of parathormone secretion without inducing hypercalcemia and induces surfactant phospholipid synthesis in pulmonary cells [15,42]. To the best of our knowledge, this is the first paper reporting an exercise-induced increase in 3-*epi*-25(OH)D$_3$ levels. The physiological meaning of this changes remains to be determined. Further, the relative contribution of 3-*epi*-25(OH)D$_3$ to serum vitamin D does not correlate with 25(OH)D$_3$ levels in individuals with hypervitaminosis D [43]. Conversely, here, we showed a decrease in the ratio of 25(OH)D$_3$ to 3-*epi*-25(OH)D$_3$ after the run or supplementation, indicating that increased serum 25(OH)D$_3$ levels lead to a rise in C-3 epimerization. Hence, epimerization may be the first line of defense of the body against high levels of 25(OH)D$_3$, since epimeric forms of 1,25(OH)$_2$D$_3$ are considered to be less biologically active than native forms.

The consumption of vitamins during an ultra-marathon is a common nutritional habit and, along with the growing interest in this type of physical activity, has been studied [44,45]. It should be highlighted that intake of typical antioxidants, such as vitamins C and E, as anti-inflammatory and antioxidative factors in endurance training could even blunt training adaptations and attenuate some of the cellular adaptations in skeletal muscle [46]. Furthermore consumption of those vitamins did not affect physiological aspects related to sport performance and did not improve sport results [47]. That is why there is a need for searching for proper supplementation methods, where vitamin D supplementation can be beneficial for ultra-marathon runners' health.

In conclusion, this is the first report demonstrating that endurance exercise significantly increases serum levels of 24,25(OH)$_2$D$_3$, 3-*epi*-25(OH)D$_3$, and 25(OH)D$_3$, possibly by liberating vitamin D from adipose tissue and stimulating its metabolism. These observations imply that formation of vitamin D metabolites can, on the one hand, protect from vitamin D toxicity, and on the other hand, they can exert some other biological functions, e.g., anti- inflammatory and antioxidative. All of these data indicate that these changes in vitamin D metabolism are a physiological response to endurance exercise. The changes are affected by vitamin D status; thus, one can predict that low adipose or skeletal muscle vitamin stores may negatively influence physiological response to exercise. However, more work is needed to explore the role of vitamin D metabolites in physiological response to exercise.

Author Contributions: Conceptualization, J.M., M.A.Ż. and J.A.; data curation, J.M.; formal analysis, J.M., A.K., B.N., K.K., R.R. and T.B.; investigation, J.M., B.S., T.K. and K.K.; methodology, J.M., B.S., B.N., K.K., R.R. and T.B.; project administration, A.K.; supervision, J.M.; validation, M.A.Ż.; writing—original draft, J.M., B.S., K.K., R.R. and J.A.; writing—review and editing, J.M., A.K., B.N., T.K., M.A.Ż., T.B. and J.A. All authors have read and agreed to the published version of the manuscript.

Funding: This research was funded by the Knowledge Grant of the Polish Ministry of Science and Higher Education (number NRSA 406754).

Acknowledgments: We gratefully acknowledge all the participants involved in the experiment. We also thank the Mountain Marathons Foundation (Kudowa-Zdrój), Piotr Hercog, and Przemysław Ząbecki. The authors would like to thank Joanna Mackie for language assistant.

Conflicts of Interest: The authors declare no conflict of interest. The funders had no role in the design of the study; in the collection, analyses, or interpretation of data; in the writing of the manuscript, or in the decision to publish the results.

References

1. Wang, Y.; Zhu, J.; DeLuca, H.F. Where is the vitamin D receptor? *Arch. Biochem. Biophys.* **2012**, *523*, 123–133. [CrossRef] [PubMed]
2. Wierzbicka, J.; Piotrowska, A.; Żmijewski, M.A. The renaissance of vitamin D. *Acta Biochim. Pol.* **2014**, *61*, 679–686. [CrossRef] [PubMed]
3. Wierzbicka, E.; Szalecki, M.; Pludowski, P.; Jaworski, M.; Brzozowska, A. Vitamin D status, body composition and glycemic control in Polish adolescents with type 1 diabetes. *Minerva Endocrinol.* **2016**, *41*, 445–455. [PubMed]
4. Strange, R.C.; Shipman, K.E.; Ramachandran, S. Metabolic syndrome: A review of the role of vitamin D in mediating susceptibility and outcome. *World J. Diabetes* **2015**, *6*, 896–911. [CrossRef] [PubMed]
5. Prasad, P.; Kochhar, A. Interplay of vitamin D and metabolic syndrome: A review. *Diabetes Metab. Syndr.* **2016**, *10*, 105–112. [CrossRef]

6. Pludowski, P.; Jaworski, M.; Niemirska, A.; Litwin, M.; Szalecki, M.; Karczmarewicz, E.; Michalkiewicz, J. Vitamin D status, body composition and hypertensive target organ damage in primary hypertension. *J. Steroid Biochem. Mol. Biol.* **2014**, *144*, 180–184. [CrossRef]

7. De la Puente Yague, M.; Collado Yurrita, L.; Ciudad Cabanas, M.J.; Cuadrado Cenzual, M.A. Role of Vitamin D in Athletes and Their Performance: Current Concepts and New Trends. *Nutrients* **2020**, *12*, 579. [CrossRef]

8. Ogan, D.; Pritchett, K. Vitamin D and the athlete: Risks, recommendations, and benefits. *Nutrients* **2013**, *5*, 1856–1868. [CrossRef]

9. Janssen, H.C.; Emmelot-Vonk, M.H.; Verhaar, H.J.; van der Schouw, Y.T. Determinants of vitamin D status in healthy men and women aged 40–80 years. *Maturitas* **2013**, *74*, 79–83. [CrossRef]

10. Theuri, G.; Kiplamai, F. Association between vitamin D levels and central adiposity in an eastern Africa outpatient clinical population. *Dermatoendocrinol* **2013**, *5*, 218–221. [CrossRef]

11. Bescos Garcia, R.; Rodriguez Guisado, F.A. Low levels of vitamin D in professional basketball players after wintertime: Relationship with dietary intake of vitamin D and calcium. *Nutr. Hosp.* **2011**, *26*, 945–951. [CrossRef] [PubMed]

12. Mentaverri, R.; Souberbielle, J.C.; Brami, G.; Daniel, C.; Fardellone, P. Pharmacokinetics of a New Pharmaceutical Form of Vitamin D3 100,000 IU in Soft Capsule. *Nutrients* **2019**, *11*, 703. [CrossRef] [PubMed]

13. Sun, X.; Cao, Z.B.; Taniguchi, H.; Tanisawa, K.; Higuchi, M. Effect of an Acute Bout of Endurance Exercise on Serum 25(OH)D Concentrations in Young Adults. *J. Clin. Endocrinol. Metab.* **2017**, *102*, 3937–3944. [CrossRef] [PubMed]

14. Lutsey, P.L.; Eckfeldt, J.H.; Ogagarue, E.R.; Folsom, A.R.; Michos, E.D.; Gross, M. The 25-hydroxyvitamin D3 C-3 epimer: Distribution, correlates, and reclassification of 25-hydroxyvitamin D status in the population-based Atherosclerosis Risk in Communities Study (ARIC). *Clin. Chim. Acta* **2015**, *442*, 75–81. [CrossRef] [PubMed]

15. Brown, A.J.; Ritter, C.S.; Weiskopf, A.S.; Vouros, P.; Sasso, G.J.; Uskokovic, M.R.; Wang, G.; Reddy, G.S. Isolation and identification of 1alpha-hydroxy-3-epi-vitamin D3, a potent suppressor of parathyroid hormone secretion. *J. Cell. Biochem.* **2005**, *96*, 569–578. [CrossRef] [PubMed]

16. Nemere, I.; Wilson, C.; Jensen, W.; Steinbeck, M.; Rohe, B.; Farach-Carson, M.C. Mechanism of 24,25-dihydroxyvitamin D3-mediated inhibition of rapid, 1,25-dihydroxyvitamin D3-induced responses: Role of reactive oxygen species. *J. Cell. Biochem.* **2006**, *99*, 1572–1581. [CrossRef]

17. Seo, E.G.; Einhorn, T.A.; Norman, A.W. 24R,25-dihydroxyvitamin D3: An essential vitamin D3 metabolite for both normal bone integrity and healing of tibial fracture in chicks. *Endocrinology* **1997**, *138*, 3864–3872. [CrossRef]

18. Seo, E.G.; Norman, A.W. Three-fold induction of renal 25-hydroxyvitamin D3-24-hydroxylase activity and increased serum 24,25-dihydroxyvitamin D3 levels are correlated with the healing process after chick tibial fracture. *J. Bone Miner. Res.* **1997**, *12*, 598–606. [CrossRef]

19. Weinstock-Guttman, B.; Zivadinov, R.; Qu, J.; Cookfair, D.; Duan, X.; Bang, E.; Bergsland, N.; Hussein, S.; Cherneva, M.; Willis, L.; et al. Vitamin D metabolites are associated with clinical and MRI outcomes in multiple sclerosis patients. *J. Neurol. Neurosurg. Psychiatry* **2011**, *82*, 189–195. [CrossRef]

20. Safdar, A.; Tarnopolsky, M.A. Exosomes as Mediators of the Systemic Adaptations to Endurance Exercise. *Cold Spring Harb. Perspect. Med.* **2018**, *8*. [CrossRef]

21. Li, D.J.; Fu, H.; Zhao, T.; Ni, M.; Shen, F.M. Exercise-stimulated FGF23 promotes exercise performance via controlling the excess reactive oxygen species production and enhancing mitochondrial function in skeletal muscle. *Metabolism* **2016**, *65*, 747–756. [CrossRef] [PubMed]

22. Shimada, T.; Hasegawa, H.; Yamazaki, Y.; Muto, T.; Hino, R.; Takeuchi, Y.; Fujita, T.; Nakahara, K.; Fukumoto, S.; Yamashita, T. FGF-23 is a potent regulator of vitamin D metabolism and phosphate homeostasis. *J. Bone Miner. Res.* **2004**, *19*, 429–435. [CrossRef]

23. Kasprowicz, K.; Ratkowski, W.; Wolyniec, W.; Kaczmarczyk, M.; Witek, K.; Zmijewski, P.; Renke, M.; Jastrzebski, Z.; Rosemann, T.; Nikolaidis, P.T.; et al. The Effect of Vitamin D3 Supplementation on Hepcidin, Iron, and IL-6 Responses after a 100 km Ultra-Marathon. *Int. J. Environ. Res. Public Health* **2020**, *17*, 2962. [CrossRef] [PubMed]

24. Hoffman, M.D.; Ong, J.C.; Wang, G. Historical analysis of participation in 161 km ultramarathons in North America. *Int. J. Hist. Sport* **2010**, *27*, 1877–1891. [CrossRef] [PubMed]

25. Knechtle, B.; Nikolaidis, P.T. Physiology and Pathophysiology in Ultra-Marathon Running. *Front. Physiol.* **2018**, *9*, 634. [CrossRef]

26. Tiller, N.B.; Roberts, J.D.; Beasley, L.; Chapman, S.; Pinto, J.M.; Smith, L.; Wiffin, M.; Russell, M.; Sparks, S.A.; Duckworth, L.; et al. International Society of Sports Nutrition Position Stand: Nutritional considerations for single-stage ultra-marathon training and racing. *J. Int. Soc. Sports Nutr.* **2019**, *16*, 50. [CrossRef]

27. Rola, R.; Kowalski, K.; Bienkowski, T.; Studzinska, S. Improved sample preparation method for fast LC-MS/MS analysis of vitamin D metabolites in serum. *J. Pharm. Biomed. Anal.* **2020**, *190*, 113529. [CrossRef]

28. Van Beaumont, W. Evaluation of hemoconcentration from hematocrit measurements. *J. Appl. Physiol.* **1972**, *32*, 712–713. [CrossRef]

29. Sherk, V.D.; Chrisman, C.; Smith, J.; Young, K.C.; Singh, H.; Bemben, M.G.; Bemben, D.A. Acute bone marker responses to whole-body vibration and resistance exercise in young women. *J. Clin. Densitom.* **2013**, *16*, 104–109. [CrossRef]

30. Blum, M.; Dolnikowski, G.; Seyoum, E.; Harris, S.S.; Booth, S.L.; Peterson, J.; Saltzman, E.; Dawson-Hughes, B. Vitamin D(3) in fat tissue. *Endocrine* **2008**, *33*, 90–94. [CrossRef]

31. Blum, M.; Dallal, G.E.; Dawson-Hughes, B. Body size and serum 25 hydroxy vitamin D response to oral supplements in healthy older adults. *J. Am. Coll. Nutr.* **2008**, *27*, 274–279. [CrossRef] [PubMed]

32. Sun, X.; Cao, Z.B.; Tanisawa, K.; Taniguchi, H.; Kubo, T.; Higuchi, M. Effects of chronic endurance exercise training on serum 25(OH)D concentrations in elderly Japanese men. *Endocrine* **2018**, *59*, 330–337. [CrossRef] [PubMed]

33. Seuter, S.; Virtanen, J.K.; Nurmi, T.; Pihlajamäki, J.; Mursu, J.; Voutilainen, S.; Tuomainen, T.P.; Neme, A.; Carlberg, C. Molecular evaluation of vitamin D responsiveness of healthy young adults. *J. Steroid Biochem. Mol. Biol.* **2017**, *174*, 314–321. [CrossRef]

34. Votruba, S.B.; Jensen, M.D. Regional fat deposition as a factor in FFA metabolism. *Annu. Rev. Nutr.* **2007**, *27*, 149–163. [CrossRef] [PubMed]

35. Carrelli, A.; Bucovsky, M.; Horst, R.; Cremers, S.; Zhang, C.; Bessler, M.; Schrope, B.; Evanko, J.; Blanco, J.; Silverberg, S.J.; et al. Vitamin D Storage in Adipose Tissue of Obese and Normal Weight Women. *J. Bone Miner. Res.* **2017**, *32*, 237–242. [CrossRef] [PubMed]

36. Didriksen, A.; Burild, A.; Jakobsen, J.; Fuskevag, O.M.; Jorde, R. Vitamin D3 increases in abdominal subcutaneous fat tissue after supplementation with vitamin D3. *Eur. J. Endocrinol.* **2015**, *172*, 235–241. [CrossRef] [PubMed]

37. Owens, D.J.; Tang, J.C.; Bradley, W.J.; Sparks, A.S.; Fraser, W.D.; Morton, J.P.; Close, G.L. Efficacy of High-Dose Vitamin D Supplements for Elite Athletes. *Med. Sci. Sports Exerc.* **2017**, *49*, 349–356. [CrossRef]

38. Pedersen, B.K. Physical activity and muscle-brain crosstalk. *Nat. Rev. Endocrinol.* **2019**, *15*, 383–392. [CrossRef]

39. Chen, Y.; Kong, J.; Sun, T.; Li, G.; Szeto, F.L.; Liu, W.; Deb, D.K.; Wang, Y.; Zhao, Q.; Thadhani, R.; et al. 1,25-Dihydroxyvitamin D(3) suppresses inflammation-induced expression of plasminogen activator inhibitor-1 by blocking nuclear factor-kappaB activation. *Arch. Biochem. Biophys.* **2011**, *507*, 241–247. [CrossRef]

40. Singh, R.J.; Taylor, R.L.; Reddy, G.S.; Grebe, S.K. C-3 epimers can account for a significant proportion of total circulating 25-hydroxyvitamin D in infants, complicating accurate measurement and interpretation of vitamin D status. *J. Clin. Endocrinol. Metab.* **2006**, *91*, 3055–3061. [CrossRef]

41. Bailey, D.; Veljkovic, K.; Yazdanpanah, M.; Adeli, K. Analytical measurement and clinical relevance of vitamin D(3) C3-epimer. *Clin. Biochem.* **2013**, *46*, 190–196. [CrossRef] [PubMed]

42. Rehan, V.K.; Torday, J.S.; Peleg, S.; Gennaro, L.; Vouros, P.; Padbury, J.; Rao, D.S.; Reddy, G.S. 1Alpha,25-dihydroxy-3-epi-vitamin D3, a natural metabolite of 1alpha,25-dihydroxy vitamin D3: Production and biological activity studies in pulmonary alveolar type II cells. *Mol. Genet. Metab.* **2002**, *76*, 46–56. [CrossRef]

43. Granado-Lorencio, F.; Blanco-Navarro, I.; Perez-Sacristan, B.; Donoso-Navarro, E.; Silvestre-Mardomingo, R. Serum levels of 3-epi-25-OH-D3 during hypervitaminosis D in clinical practice. *J. Clin. Endocrinol. Metab.* **2012**, *97*, E2266–E2270. [CrossRef] [PubMed]

44. Nikolaidis, P.T.; Veniamakis, E.; Rosemann, T.; Knechtle, B. Nutrition in Ultra-Endurance: State of the Art. *Nutrients* **2018**, *10*, 1995. [CrossRef]

45. Knez, W.L.; Peake, J.M. The prevalence of vitamin supplementation in ultraendurance triathletes. *Int. J. Sport Nutr. Exerc. Metab.* **2010**, *20*, 507–514. [CrossRef] [PubMed]

46. Costa, R.J.S.; Knechtle, B.; Tarnopolsky, M.; Hoffman, M.D. Nutrition for Ultramarathon Running: Trail, Track, and Road. *Int. J. Sport Nutr. Exerc. Metab.* **2019**, *29*, 130–140. [CrossRef] [PubMed]

47. Taghiyar, M.; Ghiasvand, R.; Askari, G.; Feizi, A.; Hariri, M.; Mashhadi, N.S.; Darvishi, L. The effect of vitamins C and e supplementation on muscle damage, performance, and body composition in athlete women: A clinical trial. *Int. J. Prev. Med.* **2013**, *4*, S24–S30.

Publisher's Note: MDPI stays neutral with regard to jurisdictional claims in published maps and institutional affiliations.

Review

The Benefits of Vitamin D Supplementation for Athletes: Better Performance and Reduced Risk of COVID-19

William B. Grant [1,*], **Henry Lahore** [2] **and Michelle S. Rockwell** [3,4]

[1] Sunlight, Nutrition, and Health Research Center, P.O. Box 641603, San Francisco, CA 94164-1603, USA
[2] VitaminDWiki, 2289 Highland Loop, Port Townsend, WA 98368, USA; hlahore@gmail.com
[3] Department of Human Nutrition, Foods, and Exercise, Virginia Tech, Blacksburg, VA 24061, USA; msrock@vt.edu
[4] Center for Transformative Research on Health Behaviors, Fralin Biomedical Research Institute at Virginia Tech Carilion, Roanoke, VA 24016, USA
* Correspondence: wbgrant@infionline.net; Tel.: +1-415-409-1980

Received: 14 November 2020; Accepted: 2 December 2020; Published: 4 December 2020

Abstract: The COVID-19 pandemic is having major economic and personal consequences for collegiate and professional sports. Sporting events have been canceled or postponed, and even when baseball and basketball seasons resumed in the United States recently, no fans were in attendance. As play resumed, several players developed COVID-19, disrupting some of the schedules. A hypothesis now under scientific consideration is that taking vitamin supplements to raise serum 25-hydroxyvitamin D [25(OH)D] concentrations could quickly reduce the risk and/or severity of COVID-19. Several mechanisms have been identified through which vitamin D could reduce the risks of infection and severity, death, and long-haul effects of COVID-19: (1) inducing production of cathelicidin and defensins to reduce the survival and replication of the SARS-CoV-2 virus; (2) reducing inflammation and the production of proinflammatory cytokines and risk of the "cytokine storm" that damages the epithelial layer of the lungs, heart, vascular system, and other organs; and (3) increasing production of angiotensin-converting enzyme 2, thus limiting the amount of angiotensin II available to the virus to cause damage. Clinical trials have confirmed that vitamin D supplementation reduces risk of acute respiratory tract infections, and approximately 30 observational studies have shown that incidence, severity, and death from COVID-19 are inversely correlated with serum 25(OH)D concentrations. Vitamin D supplementation is already familiar to many athletes and sports teams because it improves athletic performance and increases playing longevity. Thus, athletes should consider vitamin D supplementation to serve as an additional means by which to reduce risk of COVID-19 and its consequences.

Keywords: athletic performance; COVID-19; acute respiratory tract infections; immunity; team sports; vitamin D; 25-hydroxyvitamin D

1. Introduction

The COVID-19 pandemic has had and continues to have a major impact on life and economics globally. Among people affected are athletes and sports teams at all levels of play. The Summer Olympics originally scheduled for summer 2020 in Japan are now postponed until summer 2021. Professional sports organizations have experienced delays, interruptions, and cancelations despite practicing significant measures designed to reduce COVID-19 risk. Collegiate sports seasons have been delayed, canceled, or dramatically altered by the pandemic [1–3]. Some programs have even been eliminated in light of growing financial challenges [4]. High school sports, amateur races, and recreational sports have been put on hold. A Google Forms survey of 692 elite and semi-elite South African athletes conducted in April found that in response to COVID-19 reductions in sports events, many of the athletes consumed excessive amounts of carbohydrates, felt depressed, and required motivation to keep active. [5]. Thus, additional methods to reduce risk of COVID-19 for athletes would be useful, especially if they might also improve athletic performance.

This narrative review outlines the use of vitamin D supplementation to raise serum 25-hydroxyvitamin D [25(OH)D] concentrations to optimal values, which may be at least 40 ng/mL for sports (e.g., [6]). The benefits of vitamin D for athletic performance and general well-being are similarly reviewed.

2. Results

2.1. Introduction to COVID-19

COVID-19 is caused by the body's dysregulated immune response to the SARS-CoV-2 virus [7]. (macrophage activation, associated with the "cytokine storm," promotes the dysregulation of innate immunity [8].) The virus enters largely through the lungs. The virus can enter cells by attaching to the angiotensin-converting enzyme 2 (ACE2) receptor. SARS-CoV-2's binding to ACE2 makes more angiotensin II available to cause damage [9]. The infection also increases inflammation by ramping up production of both proinflammatory and anti-inflammatory cytokines, which can result in a cytokine storm [10]. By increasing inflammation, the cytokine storm injures the epithelial layer of the lungs—which can lead to pneumonia, acute respiratory distress syndrome (ARDS), and sepsis [7]—and later, other internal organs, which can lead to permanent damage. The T-helper 1 (Th1) and macrophage-based proinflammatory cytokines are interleukin 1β (IL-1β), IL-2, IL-6, tumor necrosis factor α, and IL-17 [7,10,11]. Approximately 70% of fatal COVID-19 cases are due to ARDS, whereas sepsis accounts for approximately 28% [7].

COVID-19 can progress through various stages [12]. The first stage is generally limited to upper respiratory tract infection accompanied by fever, fatigue, and muscle ache, whereas nausea and diarrhea are infrequent symptoms at onset [13]. The second stage is pneumonia (infection of the lower respiratory tract) with or without dyspnea (labored breathing). The third stage is complications, which could include ARDS, sepsis, cardiac and kidney injury, and secondary infection [14]. The fourth stage is death or healing. Death is unlikely for athletes because the main risk factors for death are older age [15], various chronic diseases, and elevated systemic inflammation [16].

A rapidly increasing body of research reports the benefits of vitamin D in reducing risk and severity of SARS-CoV-2 infection and COVID-19. A recent review summarized the findings as of mid-October 2020 [17]. By that time, at least 14 observational studies and a few intervention studies as well as several mechanisms related to vitamin D had been published.

2.2. Observational Studies of 25(OH)D and COVID-19

More than 15 observational studies of COVID-19 incidence, severity, and/or death with respect to serum 25(OH)D concentrations have been published in peer-reviewed journals. The findings in those studies are tabulated in a companion paper [17]. Although three studies reported no beneficial effect related to 25(OH)D, the others reported an inverse correlation between 25(OH)D concentrations and severity of COVID-19. Two of those studies that showed no benefit used 25(OH)D concentration values from blood drawn more than a decade before the incidence of COVID-19 and in the multivariable analysis included factors that affect 25(OH)D concentrations. In summary, mean 25(OH)D concentrations <15 ng/mL are generally associated with greater severity and risk of death for COVID-19 patients, whereas mean 25(OH)D concentrations for less severe but still hospitalized COVID-19 patients ranged from 17 to perhaps 30 ng/mL. Thus, the 10 observational studies suggest that 25(OH)D concentrations <30 ng/mL are associated with increased risk of COVID-19 infection but that the risk with respect to higher concentrations cannot be ruled out. Thus, it would be prudent to assume that higher values, such as between 40 and 60 ng/mL, might be the more appropriate range.

An observational study conducted in Chicago included 489 COVID-19 patients with a mean age of 49 ± 18 years, 75% of whom were women. Those patients presented at University of Chicago Medicine between March 3 and April 10, 2020, and had 25(OH)D concentrations measured within the past year [18]. Patients were deemed vitamin D deficient if their most recent serum 25(OH)D concentrations within 1 year before their first COVID-19 tests were <20 ng/mL for 25(OH)D or <18 pg/mL for 1,25(OH)$_2$D. Patients were deemed not deficient if their most recent concentrations were ≥20 ng/mL for 25(OH)D or ≥18 pg/mL for 1,25(OH)$_2$D. The relative risk of COVID-19 was 1.77 (95% confidence interval [CI], 1.12 to 2.81; p = 0.02) for deficient vs. non-deficient vitamin D status. The relative risk for COVID-19 for non-white vs. white race was 2.54 (95% CI, 1.26 to 5.12; p = 0.009).

Observational studies by themselves are not considered reliable indicators of causal relationships because confounding factors may play important roles. For 25(OH)D, sun exposure and diet [19] are two important contributing factors other than vitamin D supplementation, and they may have effects independent of vitamin D such as destroying viruses [20]. In addition, the disease state may affect 25(OH)D concentrations [21,22]. That concern is particularly important for people with chronic diseases. Thus, randomized controlled trials (RCTs) are considered the best way to determine causality.

Observational studies have also offered insight into who is at greater risk of developing COVID-19 and the severity of the disease. In the United States, African Americans and Hispanics have had much higher rates of infection and death than European Americans. [23–25]. In addition, people who are elderly, who are obese, and/or who have chronic conditions are at greater risk [25]. Diabetes is also an important risk factor [26]. Although African Americans and Hispanics have lower 25(OH)D concentrations than European Americans [27], many other factors help explain the incidence–severity relationships such as prevalence of other diseases and working and living in close contact with many people. A recent review outlined the evidence regarding elevated chronic disease rates for African Americans, including cardiovascular disease, diabetes, hypertension, and pulmonary disease [23]. A recent publication outlined the reasons why vitamin D deficiency in African Americans contributes to their increased risk of COVID-19 [28]. In June 2020, another publication noted the increase in COVID-19 death rates of dark-skinned Americans [29].

Seropositivity to SARS-CoV-2 is a precursor to COVID-19, which can develop after a dysregulated immune response to the virus. The correlation between SARS-CoV-2 positivity with respect to deseasonalized serum 25(OH)D concentrations measured within the past year for more than 190,000 patients by Quest Diagnostics was reported recently [30]. Non-Hispanic black people had approximately double the SARS-CoV-2 seropositivity of non-Hispanic white people over the 25(OH)D concentration range from <20 to >60 ng/mL, whereas Hispanic people had seropositivity rates approximately 60% higher than those of non-Hispanic white people. On the basis of the dependence of seropositivity on race and serum 25(OH)D concentration, researchers estimated that mean population serum 25(OH)D concentrations explained 20% of SARS-CoV-2 seropositivity among non-Hispanic black people and 30% of SARS-CoV-2 seropositivity among Hispanic people [17].

2.3. Treating COVID-19 with Vitamin D

The results of the first vitamin D RCT to treat COVID-19 patients were reported in late August 2020 [31]. The mean age of patients was 53 ± 11 years, and 54% of treated patients were males. Fifty were randomized to be treated with calcifediol [25(OH)D] in addition to the standard of care, whereas 26 were treated only with the standard of care. The calcifediol treatment was 0.532 mg on the day of admission and then 0.266 mg on days 3 and 7 and then weekly until discharge or admission to the intensive care unit (ICU). The conversion from calcifediol to cholecalciferol (vitamin D$_3$) was given as 3.2 times the molecular weight of each; therefore, 0.532 mg of calcifediol is approximately 68,000 IU of vitamin D$_3$. Calcifediol has an advantage over vitamin D$_3$ in not having to go through the liver to be processed. However, as reported in the New York study, large doses of vitamin D were effective in treating COVID-19 patients [32]. Whereas only one treated patient had to enter the ICU, 13 of those given only the standard of care treatment had to do so. The univariate risk odds ratio for ICU for patients with calcifediol treatment was 0.02 (95% CI, 0.002 to 0.17).

A second vitamin D RCT to treat COVID-19 patients was reported from India [33]. COVID-19 patients admitted to a tertiary care hospital in north India were invited to the study. The criteria for participation included being mildly symptomatic or asymptomatic with or without comorbidities, that serum 25(OH)D was <20 ng/mL, and that participants were able to take oral vitamin D supplementation (e.g., not requiring invasive ventilation or with significant comorbidities). Forty patients were enrolled: 16 were randomized to receive 60,000 IU/day of vitamin D_3 for 7 days, whereas 24 served as controls. Members of the treatment group who did not achieve a 25(OH)D concentration >50 ng/mL in the 7 days were supplemented with 60,000 IU/day for another 7 days. The mean age was ~50 years (range, 36 to 51 years). Mean 25(OH)D was 9 ng/mL (range, 7 to 13 ng/mL) in the treatment group and 19 ng/mL (range, 8 to 13 ng/mL) in the control group. Serum 25(OH)D increased by 42 ng/mL (range, 39 to 49 ng/mL) in the treatment group and 5 ng/mL (range, 0 to 12 ng/mL) in the control group. Fibrinogen decreased from 4.1 g/L (range, 3.7 to 5.1 g/L) to 3.2 g/L (range, 1.7 to 4.1 g/L) in the treatment group but was essentially unchanged in the control group: 3.7 g/L (range, 3.4 to 4.3 g/L) vs. 3.7 g/L (range, 2.4 to 4.3 g/L) (p = 0.001). As a result, 10 (63%) participants in the intervention group and five (22%) participants in the control arm (p < 0.02) became SARS-CoV-2 RNA negative.

A recent "quasi-experimental" study of high-dose vitamin D supplementation in a French nursing home shows the benefit of maintaining high 25(OH)D concentrations [34]. Sixty-three of 96 elderly residents developed COVID-19. The residents had been receiving single oral doses of 80,000 IU of vitamin D_3 every 2–3 months. During 36 ± 17 days of follow up, 83% (57) residents who had received vitamin D within 1 month before to 1 week after diagnosis of COVID-19 compared to 44% of the nine who did not. The fully adjusted hazard ratio for survival with respect to vitamin D was 0.11 (95% CI, 0.03 to 0.48; p = 0.003). Those authors reported similar results for 77 consecutive COVID-19 patients in a geriatric hospital [35]. Of course, many athletes are larger than nursing home residents and so should take higher daily average vitamin D supplements. As of 9 November 2020, 30 observational studies report that COVID-19 or SARS-CoV-2 positivity was associated with lower serum 25(OH)D concentration [36]. In addition, two small-scale RCTs with vitamin D supplementation have been reported and at least 33 clinical trials have been registered [37].

2.4. Mechanisms of Vitamin D against SARS-CoV-2 and COVID-19

Vitamin D has several main mechanisms by which it reduces risks of COVID-19 [17,38]. One is through mounting a defense against the virus, in part through induction of cathelicidin (LL-37) and defensins. LL-37 acts at several steps in viral infection and is effective against both enveloped and non-enveloped viruses [39]. LL-37 also affects regulatory T cells. In one study, higher levels of LL-37 in serum corresponded to lower expression of IL-17 in the tonsils and to lower levels of its transcription factor, RORC2, both of which are necessary for the development of Th17 cells [40], FOXP3 (a transcription factor involved in inducing regulatory T cells) also was expressed at lower levels [41]. Several papers suggested that IL-17 was involved in the pathology of COVID-19, including risk of thrombosis [42] and ARDS [43]. A 2016 article reported that athletes who took 5000 IU/day of vitamin D_3 for 14 weeks increased mean 25(OH)D from 22 to 50 ng/mL, resulting in a 15% increase in the concentration of cathelicidin in plasma [44].

A second mechanism is to regulate the production of cytokines, generally upregulating anti-inflammatory cytokines such as IL-10, and downregulating proinflammatory cytokines such as IL-6 [45]. Such regulation can reduce risk of the cytokine storm. An ecological study reported that influenza case-fatality rates in the United States during the 1918–1919 influenza pandemic were significantly lower in southwestern communities than in northeastern communities [46].

The mechanism proposed was vitamin D production from solar ultraviolet-B (UVB) exposure through reducing the cytokine storm.

A third mechanism is through increasing concentration of ACE2. That higher concentration counters the effect of SARS-CoV-2's binding to the enzyme ACE2, making more angiotensin II available to cause damage [9]. In addition, increasing ACE2 may shift the balance within the renin–angiotensin–aldosterone system toward the favorable ACE2–Ang-(1-7)–MasR pathway [47]. Thus, vitamin D inhibits mediators of the renin–angiotensin–aldosterone system—present in nearly all cells of the human body—and by inhibiting ACE activity and increasing ACE2, it lowers angiotensin II levels [9] A recent article reported that ACE2 concentrations are inversely correlated with damage to heart and lung tissues [48].

The apparent role of angiotensin II in modulating or suppressing B-cell response may also become of great interest for a better understanding of the pathophysiology of coronavirus infections [49]. As discussed in a recent review, vitamin D affects B-cell activation [8], as discussed in an earlier paper [50].

In general, innate immune responses (Toll-like receptors, type I interferons, macrophages, and dendritic cells) represent the initial host defense against invading pathogens. The innate immune system inhibits virus replication, promotes virus clearance, induces tissue repair, and triggers a prolonged adaptive immune response (T cells produce proinflammatory cytokines via the NF-κB and mitogen-activated protein kinase signaling pathways) against the viruses. Pulmonary and systemic inflammatory responses associated with coronaviruses are usually triggered by the innate immune system when it recognizes the viruses [51,52].

2.5. COVID-19 and Athletes

Of particular concern to athletes is that COVID-19 can cause both short-term and permanent damage to many organs. Damage has been noted in the lungs [53], respiration regulation mechanisms [54], and cardiovascular system [55]. Other organs also are damaged [10]. Organ damage would reduce athletic performance. A recent review concluded that physical function and fitness are impaired following SARS-CoV-2 infection, and impairments can last for a year or more [56]. Thus, it is imperative that athletes try to reduce risk of COVID-19; supplementing with vitamin D appears to be an effective and efficient way to do so if high enough 25(OH)D concentrations are achieved.

Athletes who recover from COVID-19 may have lingering damage or other health concerns such as chronic fatigue, which could be considered a fifth stage of the disease. Lung damage is one concern [57]. More importantly, heart damage also is a concern. Damage to the heart from the cytokine storm can include decrements in heart function as well as myocarditis, acute coronary syndromes, heart failure, arrhythmias, and venous thromboembolism [58]. The clinical syndromes include acute myocardial injury, myocarditis, acute coronary syndromes, heart failure, arrhythmias, and venous thromboembolism [59]. Adverse effects can also befall the musculoskeletal, hematologic, and gastrointestinal systems [60]. Thus, athletes who have had COVID-19 should be monitored by physicians before returning to practice and competition [58].

Another concern is that because physical activity is curtailed during and shortly after COVID-19, maintenance of key physical qualities, such as game-specific contact skills and decision-making ability, are challenged, affecting performance and risk of injury on resumption of training and competition. However, strategies exist that can dramatically mitigate potential losses [61].

Several publications offer recommendations for competitive athletes returning to sports. An infographic has been prepared for graduated return-to-play guidance after COVID-19 infection [62]. Another publication proposed an algorithm for return to sports [63]. Another gave ideas to consider when fans are permitted to attend events [64]. The Australian Institute of Sport presented a framework for rebooting sport in a COVID-19 environment [65], as did the Royal Spanish Football Federation [66]. Unfortunately, those publications do not address the long-haul problem of chronic fatigue, which exercise exacerbates. Inflammatory myocarditis also has been suggested [67].

2.6. Other Micronutrients

Several other micronutrients have been studied regarding their impact on COVID-19 incidence and treatment, including vitamin A (retinol), vitamin C (ascorbic acid), magnesium, selenium, and zinc. Several general reviews discussed the role of micronutrients in improving the immune response to viral infections, including COVID-19 [68–73].

2.7. African Americans

In the United States, many collegiate and professional athletes are of African American or Hispanic race or ethnicity. As a result of dark skin pigmentation, they generally have lower 25(OH)D concentrations than European Americans. For the period 2009–2010, mean 25(OH)D concentrations determined from the National Health and Nutrition Examination Survey (NHANES) dataset were as follows: non-Hispanic black, 18 ± 2 ng/mL; Mexican American, 22 ± 1 ng/mL; non-Hispanic white, 30 ± 1 ng/mL [74]. For the period 2011–2014, the prevalence of 25(OH)D concentration <12 ng/mL in the United States for different ethnicities was as follows: non-Hispanic black, 18%; Hispanic, 6%; non-Hispanic white, 2% [75]. African Americans have stronger bones than European Americans as a result of excreting calcium at lower rates, most likely as an adaptation to life in the hot, dry environment of Africa [76]. Because the classical benefit of vitamin D relates to regulation of calcium and phosphorus absorption and metabolism, many people think that African Americans do not need to increase serum 25(OH)D concentrations. However, it is now realized that for non-skeletal effects, the role of vitamin D is essentially independent of race or ethnicity (Ames, Grant, Willett, in preparation).

African Americans and Hispanic Americans have much higher case and mortality rates of COVID-19 than European Americans or Asian Americans [77]. One reason is that they have higher rates of chronic diseases that are risk factors for COVID-19, including cardiovascular disease, diabetes mellitus, hypertension, and pulmonary disease [23]. Those diseases are associated with chronic systemic inflammation, and COVID-19 might be likely to induce production of enough additional cytokines to result in a cytokine storm [78]. Another reason for the higher disease prevalence in African Americans is their overrepresentation in high-risk broad occupational categories, such as health occupations, as well as working in low-income occupations that put them at greater risk of exposure to COVID-19 than other workers [79]. A third reason is that they have lower 25(OH)D concentrations. For the period 2001–2004, white males aged 20–39 years had mean 25(OH)D concentrations of 26 ng/mL, black men had 15 ng/mL, and Mexican American men had 22 ng/mL [27]. In the UK, "deceased doctors of Black, Asian and minority ethnic (BAME) comprised 94% of the total deaths figures in the UK, notwithstanding that they represent 44% of the workforce. The trend was similar among nurses; 71% of COVID-19 fatalities were in the BAME group, although they account for 20% of the workforce." [80].

2.8. Athletic Performance

For some time, sports teams have been aware of the benefits of vitamin D supplementation to improve athletic performance. A 2009 review by Cannell and colleagues increased the interest of vitamin D among athletes [81]. It reviewed the evidence that many athletes have vitamin D deficiency, that Russian and German investigators showed improved athletic performance though UVB irradiation starting in the 1930s, that athletic performance improves after solar or artificial UVB irradiation, and that vitamin D has been shown to improve athletic performance. Interestingly, after publication of that review, the Chicago Blackhawks ice hockey team was supplemented with 5000 IU/day of vitamin D$_3$ and improved from near the bottom rank in 2009 to win the Stanley Cup in 2010 [82]. Now many sports teams have their players supplement with vitamin D [83,84].

A 2020 review by de la Puente Yague and colleagues outlined the important benefits of vitamin D for athletes [85]. Table 1 presents selected findings related to those benefits.

Table 1. Benefits of higher vitamin D status for athletes.

Benefit	Population	Intervention	Results	Reference
Muscle strength	163 healthy athletes	Vitamin D_3 supplementation (5000 IU/day) in RCTs, meta-analysis	No significant effect *	[86]
Muscle strength	22 adult male white national-level judoka athletes	Bolus dose of 150,000 IU vitamin D_3	25(OH)D concentration increased from 13 to 17 ng/mL and muscle strength by 13% in 8 days	[87]
Muscle strength and power	25 Polish elite judoists	Observational study, 25(OH)D ranged from 8 to 30 ng/mL	Left hand grip, total work during extension of the right and left lower limb, and muscle power increased by 20–30% (r = 0.22 to 0.32)	[88]
Muscle repair	14 recreationally active adults	Intense leg exercise of one leg	Serum 25(OH)D concentrations inversely predicted ($p < 0.05$) muscular weakness (i.e., control leg vs. exercise leg peak isometric force) immediately and days (i.e., 48 and 72 h)	[89]
Muscle repair	30 reportedly healthy and modestly active adult males (31 ± 5 years, 31 ± 8 ng/mL)	Randomly assigned to 4000 IU/day of vitamin D_3 or placebo for 28 days and then subjected to a one-legged exercise routine	Supplemental vitamin D increased serum 25(OH)D concentrations ($p < 0.05$; ≈70%) and enhanced recovery in peak isometric force after the damaging event ($p < 0.05$; ≈8% at 24 h). Supplemental vitamin D attenuated ($p < 0.05$) immediate and delayed (48, 72, or 168 h) increase in circulating biomarkers representative of muscle damage (ALT or AST) without ameliorating muscle soreness ($p > 0.05$)	[90]
Muscle repair	20 males with baseline 25(OH)D = 18 ± 10 ng/mL	Participants performed knee-damaging exercise, supplements with 4000 IU/day of vitamin D_3 or placebo	Supplemental vitamin D_3 increased serum 25(OH)D and improved recovery of peak torque at 48 h and 7 days postexercise	[91]
Stress fractures	118 NCAA Division I athletes in South Carolina	Vitamin D supplementation in winter	Stress fracture rate dropped from 7.5% to 1.7% ($p = 0.009$)	[92]
Lung function	28 active college-age males, Gdansk, Poland	6000 IU/day of vitamin D_3 for 8 weeks or placebo, January to March; mean 25(OH)D increased from 20 to 60 ng/mL	Significant improvements in maximal aerobic and anaerobic power; VO_{2max} test, maximal lung minute ventilation (VE_{max} mL·min^{-1}), maximal breath frequency (BF_{max} 1·min^{-1}) improved significantly	[93]
Immune function	225 endurance athletes in winter, UK	Observational study	A significantly higher proportion of subjects presented with symptoms of URTI in the vitamin D-deficient status group (initial plasma 25(OH)D < 12 ng/mL) than in the optimal vitamin D group (>48 ng/mL); total number of URTI symptom days and median symptom-severity score in vitamin D-deficient group was significantly higher	[6]

Table 1. *Cont.*

Benefit	Population	Intervention	Results	Reference
Heart size	521 male national-level athletes in Qatari; 244 lightly exercising controls	Observational study with respect to serum 25(OH)D concentration	Severely 25(OH)D-deficient athletes (25(OH)D < 10 ng/mL) present significantly smaller cardiac structural parameters than insufficient and sufficient athletes; athletes had larger cardiac parameters than controls	[94]
Traumatic brain injury	Three patients, aged 17, 23, and 31 years	Case series treated with vitamin D, progesterone, omega-3 fatty acids, and glutamine	Reversed coma and improved clinical outcomes	[95]
Stress fractures	600 navy servicewomen diagnosed with stress fracture of the tibia or fibula and 600 controls matched by race, length of service, day of blood draw	Observational study with respect to serum 25(OH)D concentration	OR for stress fracture for high 25(OH)D quintile (mean = 50 ng/mL) vs. low quintile (mean = 14 ng/mL) = 0.51 (95% CI, 0.34 to 0.76, $p < 0.01$)	[96]

25(OH)D, 25-hydroxyvitamin D; 95% CI, 95% confidence interval; ALT, alanine aminotransferase; AST, aspartate aminotransferase; BF_{max}, maximal breath frequency; r, regression coefficient; OR, odds ratio; RCT, randomized controlled trial; URTI, upper respiratory tract infection; VE, maximum minute ventilation; VO_{2max}, maximal oxygen uptake. * Trial lasted only 4 weeks and got half the athletes to levels >31 ng—enough for strength to start to increase.

A review of the effects of vitamin D on muscles noted that vitamin D increases the number of type II, or fast-twitch, muscle cells, including type IIA fibers, which are associated with muscular high-power output [97].

A recent study confirmed the benefit of vitamin D supplementation in maintaining optimal serum 25(OH)D concentrations after the summer season. A 12 week intervention study was conducted in which 19 college swimmers in Virginia were given 5000 IU/day of vitamin D_3 or placebo from August to November [98]. Those in the treatment arm increased mean 25(OH)D concentration from 47 to 61 ng/mL, whereas those in the control arm had 25(OH)D decrease from 44 to 33 ng/mL. Fat-free mass increased in the treatment arm but not in the control arm. Those in the treatment arm performed better on dead-lift and vertical-jump tests than participants in the control arm.

Researchers at Virginia Tech sent questionnaires to all NCAA Division I head athletic trainers to learn about 25(OH)D testing, vitamin D supplementation, and vitamin D-related protocols and procedures [84]. Responses were received from 249 trainers (72% response rate). The 139 programs with a full-time registered dietitian or nutritionist were more likely to have a protocol in place ($p < 0.05$). A range of 25(OH)D concentration targets resulted: 20–30 ng/mL, 3%; 30–40 ng/mL, 6%; 40–50 ng/mL, 27%; >50 ng/mL, 13%; unsure, 51%. Programs that participated in the Football Bowl Subdivision were more likely to have 25(OH)D concentrations measured.

2.9. Observational Studies of 25(OH)D Concentrations in Athletes

Dietary sources of vitamin D such as eggs, fish, and meat do not supply enough vitamin D to affect either athletic performance or risk of COVID-19. An analysis of dietary intake for U.S. professional football players indicated that 24 defensive players were obtaining 180 ± 100 IU/day of vitamin D_3 from dietary intake, whereas 20 offensive players were obtaining 150 ± 90 IU/day [99].

An analysis was reported for 25(OH)D concentrations in 2011 for 80 members of one U.S. football team [100]. The mean age was 27 ± 4 years and the mean 25(OH)D concentration was 27 ± 12 ng/mL. Sixty-seven players were black and 13 were white or Polynesian. Twenty-one (31%) black players had 25(OH)D <20 ng/mL, whereas no white or Polynesian players did. However, only 15 (22%) black players had 25(OH)D >32 ng/mL vs. 10 (77%) white or Polynesian players.

An analysis was reported of 25(OH)D concentrations for 33 professional football players in the National Football League ca. 2014 [101]. By race, black players had mean 25(OH)D of 27 ± 9 ng/mL, white players had 48 ± 14 ng/mL, and players of other races had 23 ± 5 ng/mL.

An observational study in Poland looked at changes of biomarkers of iron, inflammation, and vitamin D during an 8 month competitive season [102]. Among the participants, 14 players had an average of 20 ± 5 years of training plus competition; 10 non-athletes served as controls. A measure of inflammation, IL-6, increased by 77% (95% CI, 35% to 131%) between athletes and controls. Serum 25(OH)D concentrations decreased by 12% (95% CI, 20% to 3%) between athletes and controls. Systemic inflammation is an important hallmark of chronic disease [103], so taking steps to slow the increase in systemic inflammation with playing time and age, such as through vitamin D supplementation, could reduce risk of chronic diseases later in life.

A study of 25(OH)D concentrations for 105 professional ice hockey players from three teams in Canada and the United States was conducted in September 2015 [104]. The results showed 13% with insufficient 25(OH)D (<32 ng/mL), 22% with sufficient 25(OH)D (≥32 to 39.9 ng/mL), and 65% with ideal 25(OH)D concentration (≥40 ng/mL). Evidently the 2009 publication by Cannell and colleagues [81] in this journal had a lasting impact on the sport. Interestingly, the authors noted that vitamin D-sufficient players were nearly 3 years older than those who were vitamin D insufficient. The researchers suggested that the players' higher vitamin D status enabled them to have a longer playing career.

A review of 25(OH)D concentrations, fractures, and rates of being drafted into the National Basketball Association (NBA) in round 1 or 2 was conducted for 279 athletes participating in the 2009–2013 NBA Combine [105]. The number of players in each vitamin D category were as follows: deficiency [25(OH)D = 20 ng/mL], 32%; insufficiency (20–30 ng/mL), 41%; sufficiency (>30 ng/mL), 27%. Approximately 55% of players had sustained at least one fracture, but rates were independent of 25(OH)D concentration. The rate of being drafted into the NBA increased with increasing vitamin D status: 70% for deficient, 82% for insufficient, and 85% for sufficient ($p = 0.007$).

A vitamin D supplementation study was conducted on 10 male and 10 female collegiate basketball players [106]. Five with mean baseline 25(OH)D concentration of 36 ± 6 ng/mL took 5000 IU/day of vitamin D_3, whereas 13—11 of whom were African American, with mean baseline of 23 ± 3 ng/mL—took 10,000 IU/day. Five months later in postseason, those taking 5000 IU/day lost 4 ± 4 ng/mL, whereas those taking 10,000 IU/day gained 14 ± 11 ng/mL.

High vitamin D intake and high 25(OH)D concentrations have few adverse effects—hypercalcemia being the most severe. The symptoms of hypercalcemia may include neuropsychiatric manifestations, such as difficulty in concentration, confusion, apathy, drowsiness, depression, psychosis, and in extreme cases, a stupor and coma [107]. Only a few of the symptoms would be present in mild hypercalcemia. However, hypercalcemia seldom has serious long-term consequences if corrected, as shown in a case in which a health adviser recovered from hypercalcemia after taking 1 million IU/day of vitamin D_3 for a month, during which his 25(OH)D concentration reached 900 ng/mL [108]. Once his 25(OH)D concentration dropped below 400 ng/mL, his hypercalcemia vanished. A study in Minnesota involving 20,308 total 25(OH)D concentration measurements over a 10 year period reported only one case of clinical toxicity associated with hypercalcemia; the concentration was 364 ng/mL [109]. One effect of high-dose vitamin D supplementation is increased absorption of calcium from the gastrointestinal tract [110]. Calcium supplementation has been linked to increased risk of myocardial infarction [111]. Thus, it is recommended that calcium supplementation be reduced when taking high-dose vitamin D.

In a meta-analysis of 48 studies with 19,833 participants in vitamin D RCTs, kidney stones were reported in only nine trials, with a tendency for fewer subjects reporting stones in the vitamin D arm than in the placebo arm (risk ratio [RR] = 0.66; 95% CI, 0.41 to 1.09; $p = 0.10$). In 37 studies, hypercalcemia was shown with increased risk shown for the vitamin D group (RR = 1.54; 95% CI, 1.09 to 2.18; $p = 0.01$). Similar increased risk of hypercalciuria was shown in 14 studies for the vitamin

D group (RR = 1.64; 95% CI, 1.06 to 2.53; p = 0.03). [112]. However, one study used 100,000 IU/day, two studies used vitamin D_2, and 11 studies included calcium. Eleven of the vitamin D_2 or calcium studies had findings that supported the placebo to cause hypercalcemia, whereas only three had findings that supported vitamin D_2 supplementation. If those 14 studies, representing 63% of the data, are omitted from the analysis, it appears very likely that the risk of hypercalcemia due to vitamin D supplementation would not be significant.

A later meta-analysis was reported by the same team, this time including studies with >2800 IU/day of vitamin D_2 or D_3 for a year or longer, involving 15 studies with 3150 participants [113]. "Long-term high-dose vitamin D supplementation did not increase total adverse events compared to placebo in 1731 participants from 10 studies (RR = 1.05; 95% CI = 0.88, 1.24; p = 0.61), nor kidney stones in 1336 participants from 5 studies (RR = 1.26; 95% CI = 0.35, 4.58; p = 0.72). However, there was a trend for vitamin D to increase risk of hypercalcemia in 2598 participants from 10 studies (RR = 1.93; 95% CI = 1.00, 3.73; p = 0.05); while its effect on hypercalciuria in only 276 participants from 3 studies was inconclusive (RR = 1.93; 95% CI = 0.83, 4.46; p = 0.12)." However, if one study that involved vitamin D supplementation not appropriate for athletes—100,000 IU of vitamin D_3 per day—is omitted from the meta-analysis, the risk ratio would not have been significant.

By contrast, a psychiatric hospital in Ohio found no relationship between vitamin D and hypercalcemia: "During this time, we have admitted over 4700 patients, the vast majority of whom agreed to supplementation with either 5000 or 10,000 IUs/day. Due to disease concerns, a few agreed to larger amounts, ranging from 20,000 to 50,000 IUs/day. There have been no cases of vitamin D_3 induced hypercalcemia or any adverse events attributable to vitamin D_3 supplementation in any patient" [114].

2.10. Other Health Benefits of Vitamin D

For people likely to be athletes from their teenage years to their mid-30s, several health outcomes that may be affected by vitamin D status are of interest. Table 2 lists some of those outcomes along with the evidence for beneficial effects of vitamin D.

Table 2. Evidence for beneficial effects of vitamin D for selected outcomes.

Outcome	Population	Intervention	Results	Reference
Progression to diabetes mellitus	2423 participants with prediabetes	Randomized to receive 4000 IU/day of vitamin D_3 or placebo, 24 month duration	Various groups had reduced progression to diabetes mellitus in the secondary analyses	[115]
Progression to diabetes mellitus	2423 participants with prediabetes	Randomized to receive 4000 IU/day of vitamin D_3 or placebo, 24 month duration	Hazard ratio for diabetes for an increase of 10 ng/mL in intratrial 25(OH)D level was 0.75 (95% CI, 0.68 to 0.82) among those assigned to vitamin D and 0.90 (95% CI, 0.80 to 1.02) among those assigned to placebo	[116]
Acute respiratory tract infection	25 eligible RCTs; IPD obtained for 10,933 of 11,321 participants	Vitamin D supplementation	Protective effects seen in individuals receiving daily or weekly vitamin D without additional bolus doses (aOR = 0.81; 95% CI, 0.72 to 0.91)	[117]
Pregnancy and birth outcomes	Pregnant women	Review	Having 25(OH)D concentration >40 ng/mL during pregnancy has many important benefits for both mother and fetus/infant	[118]

95% CI, 95% confidence interval; aOR, adjusted odds ratio; IPD, individual participant data; RCT, randomized controlled trial.

3. Discussion

Debate is ongoing regarding the advisability of measuring serum 25(OH)D concentrations. The benefits include that such measurements can help guide vitamin D supplementation doses [119]. Many factors affect the relationship between vitamin D dose and serum 25(OH)D concentration, including body mass, genetics related to absorption of vitamin D from the gastrointestinal tract, conversion from vitamin D to 25(OH)D, and baseline 25(OH)D concentrations. On the negative side is the cost and time required. In the past few years, mail-in blood spot tests have been developed that are inexpensive, convenient, and accurate [120].

Government agencies and disease organizations offer many recommendations regarding vitamin D supplementation and 25(OH)D concentrations. Two better known ones are from the U.S. Institute of Medicine [121] and the U.S. Endocrine Society [122]. The Institute of Medicine recommendations were based on requirements for bone health, recommending 600 IU/day up to age 70 years and 800 IU/day for people older than 70 years, with 20 ng/mL considered an adequate concentration. The Endocrine Society recommendation was for patients, advising 1000–2000 up to 4000 IU/day of vitamin D supplementation, with 30 ng/mL considered sufficient. The consensus statement from a vitamin D conference held in Warsaw, Poland, in 2017 stated: "The bone-centric guidelines recommend a target 25(OH)D concentration of 20ng/mL (50nmol/L), and age-dependent daily vitamin D doses of 400-800IU. The guidelines focused on pleiotropic effects of vitamin D recommend a target 25(OH)D concentration of 30 ng/mL (75 nmol/L), and age-, body weight-, disease-status, and ethnicity dependent vitamin D doses ranging between 400 and 2000 IU/day." [123]. However, another analysis "estimated, for example, that doses of 1885, 2802 and 6235 IU per day are required for normal weight, overweight and obese individuals respectively to achieve natural 25(OH)D concentrations (defined as 23 to 68 ng/mL)." [124].

Most of the action of vitamin D is due to the hormonal metabolite, 1,25(OH)$_2$D, entering vitamin D receptors attached to chromosomes, thereby affecting gene expression. A study was conducted involving "30 healthy adults randomized to receive 600, 4,000 or 10,000 IU/day of vitamin D$_3$ for 6 months. Circulating parathyroid hormone (PTH), 25(OH)D, calcium and peripheral white blood cells broad gene expression were evaluated. We observed a dose-dependent increase in 25(OH)D concentrations, decreased PTH and no change in serum calcium. A plateau in PTH concentrations was achieved at 16 weeks in the 4000 and 10,000 IU/day groups. There was a dose-dependent 25(OH)D alteration in broad gene expression with 162, 320 and 1289 genes up- or down-regulated in their white blood cells, respectively." [125]. That finding offers additional justification for 10,000 IU/day of vitamin D$_3$ for athletes.

A review from Italy took a more cautionary view of vitamin D supplementation by athletes. The review noted that vitamin D can confer several benefits, including reduced risk of cancer, better brain health, improved immune system and reduced inflammation, and better muscle function by decreasing oxidative stress and supporting mitochondrial function. However, those authors also noted that some athletes take high doses to improve performance but run the risk of vitamin D toxicity manifested as hypercalcemia [126]. Also mentioned was the increased risk of prostate and pancreatic cancer at high levels of 1α,25(OH)$_2$D. Findings of observational studies do indicate that mild prostate cancer incidence rates increase with increasing 25(OH)D concentrations [127]. However, prostate cancer mortality rates decrease with increasing 25(OH)D concentrations [128]. What appears to explain that dichotomy is that the classic role of vitamin D is to regulate absorption of calcium and phosphorus from the gastrointestinal tract and that calcium and phosphorus concentrations are associated with prostate cancer risk [129]. High calcium intake is a risk factor for aggressive prostate cancer for African Americans but not European Americans [130].

Of course, athletes also require other nutrients for optimal health and performance. A 2018 review discusses which nutrients might need to be supplemented for athletes [131].

Vitamin D comes in two forms, cholecalciferol (vitamin D$_3$) and ergocalciferol (vitamin D$_2$). Cholecalciferol is made by animals, whereas ergocalciferol is made by fungi, including yeast. In general, cholecalciferol is considered better than ergocalciferol, in part because it raises serum 25(OH)D concentration for longer and in part because it is more likely to produce beneficial health outcomes. A meta-analysis of 52 trials with a total of 75,454 participants reported that all-cause mortality rates in trials with vitamin D_3 were significantly lower (RR = 0.95; 95% CI, 0.90 to 1.00), whereas those with vitamin D_2 had an increased risk (RR = 1.03; 95% CI, 0.98 to 1.09). A systematic review of vitamin D supplementation regarding muscle strength in athletes indicated that vitamin D_3 had a positive impact on muscle strength [132].

Diet also affects serum 25(OH)D concentrations. A study conducted in England showed that meat eaters and fish eaters had much higher 25(OH)D than vegetarians and vegans: mean 25(OH)D concentrations were 30, 29, 26, and 22 ng/mL, respectively [19]. Animal products such as eggs, fish, and meat can have vitamin D as both vitamin D_3 and 25(OH)D_3 [133]. However, because food frequency tables generally do not include the contribution of vitamin D from 25(OH)D, it is generally overlooked in dietary intake studies.

4. Conclusions

Athletes and people associated with them could benefit from better athletic performance, better health, and reduced risk for COVID-19 by maintaining serum 25(OH)D concentrations above 40 ng/mL. To achieve that concentration could take supplementation of vitamin D_3 at perhaps 4000–10,000 IU/day depending on body size, skin pigmentation, and other personal factors. The 10,000 IU/day dosing level will yield a good serum concentration of vitamin D in several months. If a high concentration is desired sooner for sports performance or to avoid COVID-19, a person should consider starting with a bolus dose.

Vitamin D supplementation can be useful in reducing risk of COVID-19 and its severity, but it should not be the only measure employed. Athletes should also follow official guidelines such as regarding wearing masks, social distancing, and periodic testing.

Author Contributions: Conceptualization, W.B.G. and H.L.; methodology, W.B.G. and H.L.; writing (original draft), W.B.G.; writing (review and editing), W.B.G., H.L., and M.S.R. All authors have read and agreed to the published version of the manuscript.

Funding: This research received no external funding.

Conflicts of Interest: W.B.G. and H.L. receive funding from Bio-Tech Pharmacal, Inc. (Fayetteville, AR). M.S.R. has no conflict of interest to declare.

References

1. ESPN. The Coronavirus and College Sports: NCAA Reopening Plans, Latest News, Program Cuts, More. Available online: https://www.espn.com/college-football/story/_/id/29036650/the-coronavirus-college-sports-ncaa-reopening-plans-latest-news-program-cuts-more (accessed on 12 November 2020).

2. Evans, A.B.; Blackwell, J.; Dolan, P.; Fahlen, J.; Hoekman, R.; Lenneis, V. Sport in the face of the COVID-19 pandemic: Towards an agenda for research in the sociology of sport. *Eur. J. Sport Soc.* **2020**, *17*, 85–95. [CrossRef]

3. Garcia-Garcia, B.; James, M.; Koller, D.; Lindholm, J.; Mavromati, D.; Parrish, R.; Rodenberg, R. The impact of Covid-19 on sports: A mid-way assessment. *Int. Sports Law J.* **2020**, *20*, 115–119. [CrossRef]

4. Swanson, R.; Smith, A.B. COVID-19 and the cutting of college athletic teams. *Sport Soc.* **2020**, *23*, 1724–1735. [CrossRef]

5. Pillay, L.; Janse van Rensburg, D.C.C.; Jansen van Rensburg, A.; Ramagole, D.A.; Holtzhausen, L.; Dijkstra, H.P.; Cronje, T. Nowhere to hide: The significant impact of coronavirus disease 2019 (COVID-19) measures on elite and semi-elite South African athletes. *J. Sci. Med. Sport* **2020**, *23*, 670–679. [CrossRef] [PubMed]

6. He, C.S.; Handzlik, M.; Fraser, W.D.; Muhamad, A.; Preston, H.; Richardson, A.; Gleeson, M. Influence of vitamin D status on respiratory infection incidence and immune function during 4 months of winter training in endurance sport athletes. *Exerc. Immunol. Rev.* **2013**, *19*, 86–101. [PubMed]

7. Tay, M.Z.; Poh, C.M.; Renia, L.; MacAry, P.A.; Ng, L.F.P. The trinity of COVID-19: Immunity, inflammation and intervention. *Nat. Rev. Immunol.* **2020**, *20*, 363–374. [CrossRef] [PubMed]

8. Malaguarnera, L. Vitamin D3 as Potential Treatment Adjuncts for COVID-19. *Nutrients* **2020**, *12*, 3512. [CrossRef] [PubMed]

9. Mansur, J.L.; Tajer, C.; Mariani, J.; Inserra, F.; Ferder, L.; Manucha, W. Vitamin D high doses supplementation could represent a promising alternative to prevent or treat COVID-19 infection. *Clin. Investig. Arterioscler.* **2020**, *32*, 267–277. [CrossRef] [PubMed]

10. Garg, S.; Garg, M.; Prabhakar, N.; Malhotra, P.; Agarwal, R. Unraveling the mystery of Covid-19 cytokine storm: From skin to organ systems. *Dermatol. Ther.* **2020**. [CrossRef] [PubMed]

11. Tang, Y.; Liu, J.; Zhang, D.; Xu, Z.; Ji, J.; Wen, C. Cytokine Storm in COVID-19: The Current Evidence and Treatment Strategies. *Front. Immunol.* **2020**, *11*, 1708. [CrossRef]

12. Matricardi, P.M.; Dal Negro, R.W.; Nisini, R. The first, holistic immunological model of COVID-19: Implications for prevention, diagnosis, and public health measures. *Pediatr. Allergy Immunol.* **2020**, *31*, 454–470. [CrossRef] [PubMed]

13. Guan, W.J.; Ni, Z.Y.; Hu, Y.; Liang, W.H.; Ou, C.Q.; He, J.X.; Liu, L.; Shan, H.; Lei, C.L.; Hui, D.S.C.; et al. Clinical Characteristics of Coronavirus Disease 2019 in China. *N. Engl. J. Med.* **2020**, *382*, 1708–1720. [CrossRef] [PubMed]

14. Zhou, F.; Yu, T.; Du, R.; Fan, G.; Liu, Y.; Liu, Z.; Xiang, J.; Wang, Y.; Song, B.; Gu, X.; et al. Clinical course and risk factors for mortality of adult inpatients with COVID-19 in Wuhan, China: A retrospective cohort study. *Lancet* **2020**, *395*, 1054–1062. [CrossRef]

15. Richardson, S.; Hirsch, J.S.; Narasimhan, M.; Crawford, J.M.; McGinn, T.; Davidson, K.W.; Barnaby, D.P.; Becker, L.B.; Chelico, J.D.; Cohen, S.L.; et al. Presenting Characteristics, Comorbidities, and Outcomes Among 5700 Patients Hospitalized With COVID-19 in the New York City Area. *JAMA* **2020**, *323*, 2052–2059. [CrossRef] [PubMed]

16. Lippi, G.; Mattiuzzi, C.; Sanchis-Gomar, F.; Henry, B.M. Clinical and demographic characteristics of patients dying from COVID-19 in Italy vs China. *J. Med. Virol.* **2020**, *92*, 1759–1760. [CrossRef]

17. Mercola, J.; Grant, W.B.; Wagner, C.L. Evidence Regarding Vitamin D and Risk of COVID-19 and its Severity. *Nutrients* **2020**, *12*, 3361. [CrossRef]

18. Meltzer, D.O.; Best, T.J.; Zhang, H.; Vokes, T.; Arora, V.; Solway, J. Association of Vitamin D Status and Other Clinical Characteristics with COVID-19 Test Results. *JAMA Netw. Open* **2020**, *3*, e2019722. [CrossRef]

19. Crowe, F.L.; Steur, M.; Allen, N.E.; Appleby, P.N.; Travis, R.C.; Key, T.J. Plasma concentrations of 25-hydroxyvitamin D in meat eaters, fish eaters, vegetarians and vegans: Results from the EPIC-Oxford study. *Public Health Nutr.* **2011**, *14*, 340–346. [CrossRef]

20. Ianevski, A.; Zusinaite, E.; Shtaida, N.; Kallio-Kokko, H.; Valkonen, M.; Kantele, A.; Telling, K.; Lutsar, I.; Letjuka, P.; Metelitsa, N.; et al. Low Temperature and Low UV Indexes Correlated with Peaks of Influenza Virus Activity in Northern Europe during 2010(–)2018. *Viruses* **2019**, *11*, 207. [CrossRef]

21. Autier, P.; Boniol, M.; Pizot, C.; Mullie, P. Vitamin D status and ill health: A systematic review. *Lancet Diabetes Endocrinol.* **2014**, *2*, 76–89. [CrossRef]

22. Autier, P.; Mullie, P.; Macacu, A.; Dragomir, M.; Boniol, M.; Coppens, K.; Pizot, C.; Boniol, M. Effect of vitamin D supplementation on non-skeletal disorders: A systematic review of meta-analyses and randomised trials. *Lancet Diabetes Endocrinol.* **2017**, *5*, 986–1004. [CrossRef]

23. Alcendor, D.J. Racial Disparities-Associated COVID-19 Mortality among Minority Populations in the US. *J. Clin. Med.* **2020**, *9*, 2442. [CrossRef] [PubMed]

24. Rozenfeld, Y.; Beam, J.; Maier, H.; Haggerson, W.; Boudreau, K.; Carlson, J.; Medows, R. A model of disparities: Risk factors associated with COVID-19 infection. *Int. J. Equity Health* **2020**, *19*, 126. [CrossRef]

25. Ebinger, J.E.; Achamallah, N.; Ji, H.; Claggett, B.L.; Sun, N.; Botting, P.; Nguyen, T.T.; Luong, E.; Kim, E.H.; Park, E.; et al. Pre-existing traits associated with Covid-19 illness severity. *PLoS ONE* **2020**, *15*, e0236240. [CrossRef] [PubMed]

26. Azar, W.S.; Njeim, R.; Fares, A.H.; Azar, N.S.; Azar, S.T.; El Sayed, M.; Eid, A.A. COVID-19 and diabetes mellitus: How one pandemic worsens the other. *Rev. Endocr. Metab. Disord.* **2020**. [CrossRef]

27. Ginde, A.A.; Liu, M.C.; Camargo, C.A., Jr. Demographic differences and trends of vitamin D insufficiency in the US population, 1988–2004. *Arch. Intern. Med.* **2009**, *169*, 626–632. [CrossRef]

28. Martin Gimenez, V.M.; Inserra, F.; Ferder, L.; Garcia, J.; Manucha, W. Vitamin D deficiency in African Americans is associated with a high risk of severe disease and mortality by SARS-CoV-2. *J. Hum. Hypertens.* **2020**. [CrossRef]

29. Flagg, A.; Sharma, D.; Fenn, L.; Stobbe, M. COVID-19's Toll on People of Color Is Worse than We Knew. Available online: https://www.themarshallproject.org/2020/08/21/covid-19-s-toll-on-people-of-color-is-worse-than-we-knew (accessed on 1 September 2020).

30. Kaufman, H.W.; Niles, J.K.; Kroll, M.H.; Bi, C.; Holick, M.F. SARS-CoV-2 positivity rates associated with circulating 25-hydroxyvitamin D levels. *PLoS ONE* **2020**, *15*, e0239252. [CrossRef]

31. Entrenas Castillo, M.; Entrenas Costa, L.M.; Vaquero Barrios, J.M.; Alcala Diaz, J.F.; Miranda, J.L.; Bouillon, R.; Quesada Gomez, J.M. Effect of Calcifediol Treatment and best Available Therapy versus best Available Therapy on Intensive Care Unit Admission and Mortality Among Patients Hospitalized for COVID-19: A Pilot Randomized Clinical study. *J. Steroid Biochem. Mol. Biol.* **2020**, *203*, 105751. [CrossRef]

32. Ohaegbulam, K.C.; Swalih, M.; Patel, P.; Smith, M.A.; Perrin, R. Vitamin D Supplementation in COVID-19 Patients: A Clinical Case Series. *Am. J. Ther.* **2020**, *27*, e485–e490. [CrossRef]

33. Rastogi, A.; Bhansali, A.; Khare, N.; Suri, V.; Yaddanapudi, N.; Sachdeva, N.; Puri, G.D.; Malhotra, P. Short term, high-dose vitamin D supplementation for COVID-19 disease: A randomised, placebo-controlled, study (SHADE study). *Postgrad. Med. J.* **2020**. [CrossRef] [PubMed]

34. Annweiler, C.; Hanotte, B.; Grandin de l'Eprevier, C.; Sabatier, J.M.; Lafaie, L.; Celarier, T. Vitamin D and survival in COVID-19 patients: A quasi-experimental study. *J. Steroid Biochem. Mol. Biol.* **2020**, *204*, 105771. [CrossRef] [PubMed]

35. Annweiler, G.; Corvaisier, M.; Gautier, J.; Dubee, V.; Legrand, E.; Sacco, G.; Annweiler, C. Vitamin D Supplementation Associated to Better Survival in Hospitalized Frail Elderly COVID-19 Patients: The GERIA-COVID Quasi-Experimental Study. *Nutrients* **2020**, *12*, 3377. [CrossRef]

36. Lahore, H. COVID-19 Treated by Vitamin D—Studies, Reports, Videos. Available online: https://vitamindwiki.com/tiki-index.php?page_id=11728 (accessed on 1 September 2020).

37. ClinicalTrials.gov. Studies for Vitamin D, COVID19. Available online: https://clinicaltrials.gov/ct2/results?cond=COVID19&term=vitamin+D&cntry=&state=&city=&dist= (accessed on 29 June 2020).

38. Grant, W.B.; Lahore, H.; McDonnell, S.L.; Baggerly, C.A.; French, C.B.; Aliano, J.L.; Bhattoa, H.P. Evidence that Vitamin D Supplementation Could Reduce Risk of Influenza and COVID-19 Infections and Deaths. *Nutrients* **2020**, *12*, 988. [CrossRef] [PubMed]

39. Brice, D.C.; Diamond, G. Antiviral Activities of Human Host Defense Peptides. *Curr. Med. Chem.* **2020**, *27*, 1420–1443. [CrossRef]

40. Soyer, O.U.; Akdis, M.; Ring, J.; Behrendt, H.; Crameri, R.; Lauener, R.; Akdis, C.A. Mechanisms of peripheral tolerance to allergens. *Allergy* **2013**, *68*, 161–170. [CrossRef]

41. Elenius, V.; Palomares, O.; Waris, M.; Turunen, R.; Puhakka, T.; Ruckert, B.; Vuorinen, T.; Allander, T.; Vahlberg, T.; Akdis, M.; et al. The relationship of serum vitamins A, D, E and LL-37 levels with allergic status, tonsillar virus detection and immune response. *PLoS ONE* **2017**, *12*, e0172350. [CrossRef]

42. Raucci, F.; Mansour, A.A.; Casillo, G.M.; Saviano, A.; Caso, F.; Scarpa, R.; Mascolo, N.; Iqbal, A.J.; Maione, F. Interleukin-17A (IL-17A), a key molecule of innate and adaptive immunity, and its potential involvement in COVID-19-related thrombotic and vascular mechanisms. *Autoimmun. Rev.* **2020**, *19*, 102572. [CrossRef]

43. Pacha, O.; Sallman, M.A.; Evans, S.E. COVID-19: A case for inhibiting IL-17? *Nat. Rev. Immunol.* **2020**, *20*, 345–346. [CrossRef]

44. He, C.S.; Fraser, W.D.; Tang, J.; Brown, K.; Renwick, S.; Rudland-Thomas, J.; Teah, J.; Tanqueray, E.; Gleeson, M. The effect of 14 weeks of vitamin D3 supplementation on antimicrobial peptides and proteins in athletes. *J. Sports Sci.* **2016**, *34*, 67–74. [CrossRef]

45. Bilezikian, J.P.; Bikle, D.; Hewison, M.; Lazaretti-Castro, M.; Formenti, A.M.; Gupta, A.; Madhavan, M.V.; Nair, N.; Babalyan, V.; Hutchings, N.; et al. Mechanisms in Endocrinology: Vitamin D and COVID-19. *Eur. J. Endocrinol.* **2020**, *183*, R133–R147. [CrossRef] [PubMed]

46. Grant, W.B.; Giovannucci, E. The possible roles of solar ultraviolet-B radiation and vitamin D in reducing case-fatality rates from the 1918-1919 influenza pandemic in the United States. *Dermatoendocrinology* **2009**, *1*, 215–219. [CrossRef] [PubMed]

47. Akhtar, S.; Benter, I.F.; Danjuma, M.I.; Doi, S.A.R.; Hasan, S.S.; Habib, A.M. Pharmacotherapy in COVID-19 patients: A review of ACE2-raising drugs and their clinical safety. *J. Drug Target.* **2020**, *28*, 683–699. [CrossRef] [PubMed]

48. Aygun, H. Vitamin D can prevent COVID-19 infection-induced multiple organ damage. *Naunyn Schmiedebergs Arch. Pharmacol.* **2020**, *393*, 1157–1160. [CrossRef] [PubMed]

49. Mangge, H.; Pruller, F.; Schnedl, W.; Renner, W.; Almer, G. Beyond Macrophages and T Cells: B Cells and Immunoglobulins Determine the Fate of the Atherosclerotic Plaque. *Int. J. Mol. Sci.* **2020**, *21*, 4082. [CrossRef]

50. Baeke, F.; Takiishi, T.; Korf, H.; Gysemans, C.; Mathieu, C. Vitamin D: Modulator of the immune system. *Curr. Opin. Pharmacol.* **2010**, *10*, 482–496. [CrossRef]

51. Li, G.; Fan, Y.; Lai, Y.; Han, T.; Li, Z.; Zhou, P.; Pan, P.; Wang, W.; Hu, D.; Liu, X.; et al. Coronavirus infections and immune responses. *J. Med. Virol.* **2020**, *92*, 424–432. [CrossRef]

52. Santos, R.N.D.; Maeda, S.S.; Jardim, J.R.; Lazaretti-Castro, M. Reasons to avoid vitamin D deficiency during COVID-19 pandemic. *Arch. Endocrinol. Metab.* **2020**, *64*, 498–506.

53. Nowakowski, A.C.H. Brave New Lungs: Aging in the Shadow of COVID-19. *J. Gerontol. B Psychol. Sci. Soc. Sci.* **2020**. [CrossRef]

54. Baig, A.M. Computing the Effects of SARS-CoV-2 on Respiration Regulatory Mechanisms in COVID-19. *ACS Chem. Neurosci.* **2020**, *1*, 2416–2421. [CrossRef]

55. Zhu, H.; Rhee, J.W.; Cheng, P.; Waliany, S.; Chang, A.; Witteles, R.M.; Maecker, H.; Davis, M.M.; Nguyen, P.K.; Wu, S.M. Cardiovascular Complications in Patients with COVID-19: Consequences of Viral Toxicities and Host Immune Response. *Curr. Cardiol. Rep.* **2020**, *22*, 32. [CrossRef] [PubMed]

56. Rooney, S.; Webster, A.; Paul, L. Systematic Review of Changes and Recovery in Physical Function and Fitness after Severe Acute Respiratory Syndrome-Related Coronavirus Infection: Implications for COVID-19 Rehabilitation. *Phys. Ther.* **2020**, *100*, 1717–1729. [CrossRef] [PubMed]

57. Hull, J.H.; Loosemore, M.; Schwellnus, M. Respiratory health in athletes: Facing the COVID-19 challenge. *Lancet Respir. Med.* **2020**, *8*, 557–558. [CrossRef]

58. Baggish, A.; Drezner, J.A.; Kim, J.; Martinez, M.; Prutkin, J.M. Resurgence of sport in the wake of COVID-19: Cardiac considerations in competitive athletes. *Br. J. Sports Med.* **2020**, *54*, 1130–1131. [CrossRef] [PubMed]

59. Goha, A.; Mezue, K.; Edwards, P.; Nunura, F.; Baugh, D.; Madu, E. COVID-19 and the heart: An update for clinicians. *Clin. Cardiol.* **2020**, *43*, 1216–1222. [CrossRef]

60. Metzl, J.D.; McElheny, K.; Robinson, J.N.; Scott, D.A.; Sutton, K.M.; Toresdahl, B.G. Considerations for Return to Exercise Following Mild-to-Moderate COVID-19 in the Recreational Athlete. *HSS J.* **2020**, 1–6. [CrossRef]

61. Stokes, K.A.; Jones, B.; Bennett, M.; Close, G.L.; Gill, N.; Hull, J.H.; Kasper, A.M.; Kemp, S.P.T.; Mellalieu, S.D.; Peirce, N.; et al. Returning to Play after Prolonged Training Restrictions in Professional Collision Sports. *Int. J. Sports Med.* **2020**. [CrossRef]

62. Graham, N.S.N.; Junghans, C.; Downes, R.; Sendall, C.; Lai, H.; McKirdy, A.; Elliott, P.; Howard, R.; Wingfield, D.; Priestman, M.; et al. SARS-CoV-2 infection, clinical features and outcome of COVID-19 in United Kingdom nursing homes. *J. Infect.* **2020**, *81*, 411–419. [CrossRef]

63. Oikonomou, E.; Papanikolaou, A.; Anastasakis, A.; Bournousouzis, E.; Georgakopoulos, C.; Goudevenos, J.; Ioakeimidis, N.; Kanakakis, J.; Lazaros, G.; Papatheodorou, S.; et al. Proposed algorithm for return to sports in competitive athletes who have suffered COVID-19. *Hell. J. Cardiol.* **2020**. [CrossRef]

64. Carmody, S.; Murray, A.; Borodina, M.; Gouttebarge, V.; Massey, A. When can professional sport recommence safely during the COVID-19 pandemic? Risk assessment and factors to consider. *Br. J. Sports Med.* **2020**, *54*, 946–948. [CrossRef]

65. Hughes, D.; Saw, R.; Perera, N.K.P.; Mooney, M.; Wallett, A.; Cooke, J.; Coatsworth, N.; Broderick, C. The Australian Institute of Sport framework for rebooting sport in a COVID-19 environment. *J. Sci. Med. Sport* **2020**, *23*, 639–663. [CrossRef] [PubMed]

66. Herrero-Gonzalez, H.; Martin-Acero, R.; Del Coso, J.; Lalin-Novoa, C.; Pol, R.; Martin-Escudero, P.; De la Torre, A.I.; Hughes, C.; Mohr, M.; Biosca, F.; et al. Position statement of the Royal Spanish Football Federation for the resumption of football activities after the COVID-19 pandemic (June 2020). *Br. J. Sports Med.* **2020**, *54*, 1133–1134. [CrossRef] [PubMed]

67. Udelson, J.E.; Curtis, M.A.; Rowin, E.J. Return to Play for Athletes After Coronavirus Disease 2019 Infection-Making High-Stakes Recommendations as Data Evolve. *JAMA Cardiol.* **2020**. [CrossRef] [PubMed]

68. Calder, P.C.; Carr, A.C.; Gombart, A.F.; Eggersdorfer, M. Optimal Nutritional Status for a Well-Functioning Immune System Is an Important Factor to Protect against Viral Infections. *Nutrients* **2020**, *12*, 1181. [CrossRef]
69. BourBour, F.; Mirzaei Dahka, S.; Gholamalizadeh, M.; Akbari, M.E.; Shadnoush, M.; Haghighi, M.; Taghvaye-Masoumi, H.; Ashoori, N.; Doaei, S. Nutrients in prevention, treatment, and management of viral infections; special focus on Coronavirus. *Arch. Physiol. Biochem.* **2020**, 1–10. [CrossRef]
70. Ferreira, A.O.; Poloninin, H.C.; Dijkers, E.C.F. Postulated Adjuvant Therapeutic Strategies for COVID-19. *J. Pers. Med.* **2020**, *10*, 80. [CrossRef]
71. Fiorino, S.; Zippi, M.; Gallo, C.; Sifo, D.; Sabbatani, S.; Manfredi, R.; Rasciti, E.; Rasciti, L.; Giampieri, E.; Corazza, I.; et al. The Rationale for a Multi-Step Therapeutic Approach Based on Antivirals, Drugs, and Nutrients with Immunomodulatory Activity in Patients with Coronavirus-Sars2-Induced Disease of Different Severity. *Br. J. Nutr.* **2020**, 1–37. [CrossRef]
72. Jayawardena, R.; Sooriyaarachchi, P.; Chourdakis, M.; Jeewandara, C.; Ranasinghe, P. Enhancing immunity in viral infections, with special emphasis on COVID-19: A review. *Diabetes Metab. Syndr.* **2020**, *14*, 367–382. [CrossRef]
73. Shakoor, H.; Feehan, J.; Al Dhaheri, A.S.; Ali, H.I.; Platat, C.; Ismail, L.C.; Apostolopoulos, V.; Stojanovska, L. Immune-boosting role of vitamins D, C, E, zinc, selenium and omega-3 fatty acids: Could they help against COVID-19? *Maturitas* **2021**, *143*, 1–9. [CrossRef]
74. Schleicher, R.L.; Sternberg, M.R.; Lacher, D.A.; Sempos, C.T.; Looker, A.C.; Durazo-Arvizu, R.A.; Yetley, E.A.; Chaudhary-Webb, M.; Maw, K.L.; Pfeiffer, C.M.; et al. The vitamin D status of the US population from 1988 to 2010 using standardized serum concentrations of 25-hydroxyvitamin D shows recent modest increases. *Am. J. Clin. Nutr.* **2016**, *104*, 454–461. [CrossRef]
75. Herrick, K.A.; Storandt, R.J.; Afful, J.; Pfeiffer, C.M.; Schleicher, R.L.; Gahche, J.J.; Potischman, N. Vitamin D status in the United States, 2011–2014. *Am. J. Clin. Nutr.* **2019**, *110*, 150–157. [CrossRef] [PubMed]
76. Redmond, J.; Palla, L.; Yan, L.; Jarjou, L.M.; Prentice, A.; Schoenmakers, I. Ethnic differences in urinary calcium and phosphate excretion between Gambian and British older adults. *Osteoporos. Int.* **2015**, *26*, 1125–1135. [CrossRef] [PubMed]
77. Holmes, L., Jr.; Enwere, M.; Williams, J.; Ogundele, B.; Chavan, P.; Piccoli, T.; Chinacherem, C.; Comeaux, C.; Pelaez, L.; Okundaye, O.; et al. Black-White Risk Differentials in COVID-19 (SARS-COV2) Transmission, Mortality and Case Fatality in the United States: Translational Epidemiologic Perspective and Challenges. *Int. J. Environ. Res. Public Health* **2020**, *17*, 4322. [CrossRef]
78. Vepa, A.; Bae, J.P.; Ahmed, F.; Pareek, M.; Khunti, K. COVID-19 and ethnicity: A novel pathophysiological role for inflammation. *Diabetes Metab. Syndr.* **2020**, *14*, 1043–1051. [CrossRef] [PubMed]
79. St-Denis, X. Sociodemographic Determinants of Occupational Risks of Exposure to COVID-19 in Canada. *Can. Rev. Sociol.* **2020**, *57*, 399–452. [CrossRef] [PubMed]
80. Barsoum, Z. Coronavirus (COVID-19) Pandemic and Health Workers of an Ethnic Group-A Slant on a Shocking Report. *SN Compr. Clin. Med.* **2020**. [CrossRef]
81. Cannell, J.J.; Hollis, B.W.; Sorenson, M.B.; Taft, T.N.; Anderson, J.J. Athletic performance and vitamin D. *Med. Sci. Sports Exerc.* **2009**, *41*, 1102–1110. [CrossRef]
82. Council, V.D. The Chicago Blackhawks are the First Vitamin D Team in Modern Professional Sports History. Available online: https://www.prnewswire.com/news-releases/the-chicago-blackhawks-are-the-first-vitamin-d-team-in-modern-professional-sports-history-95041864.html (accessed on 14 August 2020).
83. Owens, D.J.; Allison, R.; Close, G.L. Vitamin D and the Athlete: Current Perspectives and New Challenges. *Sports Med.* **2018**, *48*, 3–16. [CrossRef]
84. Rockwell, M.; Hulver, M.; Eugene, E. Vitamin D Practice Patterns in National Collegiate Athletic Association Division I Collegiate Athletics Programs. *J. Athl. Train.* **2020**, *55*, 65–70. [CrossRef]
85. de la Puente Yague, M.; Collado Yurrita, L.; Ciudad Cabanas, M.J.; Cuadrado Cenzual, M.A. Role of Vitamin D in Athletes and Their Performance: Current Concepts and New Trends. *Nutrients* **2020**, *12*, 579. [CrossRef]
86. Han, Q.; Li, X.; Tan, Q.; Shao, J.; Yi, M. Effects of vitamin D3 supplementation on serum 25(OH)D concentration and strength in athletes: A systematic review and meta-analysis of randomized controlled trials. *J. Int. Soc. Sports Nutr.* **2019**, *16*, 55. [CrossRef] [PubMed]
87. Wyon, M.A.; Wolman, R.; Nevill, A.M.; Cloak, R.; Metsios, G.S.; Gould, D.; Ingham, A.; Koutedakis, Y. Acute Effects of Vitamin D3 Supplementation on Muscle Strength in Judoka Athletes: A Randomized Placebo-Controlled, Double-Blind Trial. *Clin. J. Sport Med.* **2016**, *26*, 279–284. [CrossRef] [PubMed]

88. Ksiazek, A.; Dziubek, W.; Pietraszewska, J.; Slowinska-Lisowska, M. Relationship between 25(OH)D levels and athletic performance in elite Polish judoists. *Biol. Sport* **2018**, *35*, 191–196. [PubMed]

89. Barker, T.; Henriksen, V.T.; Martins, T.B.; Hill, H.R.; Kjeldsberg, C.R.; Schneider, E.D.; Dixon, B.M.; Weaver, L.K. Higher serum 25-hydroxyvitamin D concentrations associate with a faster recovery of skeletal muscle strength after muscular injury. *Nutrients* **2013**, *5*, 1253–1275. [CrossRef]

90. Barker, T.; Schneider, E.D.; Dixon, B.M.; Henriksen, V.T.; Weaver, L.K. Supplemental vitamin D enhances the recovery in peak isometric force shortly after intense exercise. *Nutr. Metab.* **2013**, *10*, 69. [CrossRef]

91. Owens, D.J.; Sharples, A.P.; Polydorou, I.; Alwan, N.; Donovan, T.; Tang, J.; Fraser, W.D.; Cooper, R.G.; Morton, J.P.; Stewart, C.; et al. A systems-based investigation into vitamin D and skeletal muscle repair, regeneration, and hypertrophy. *Am. J. Physiol. Endocrinol. Metab.* **2015**, *309*, E1019–E1031. [CrossRef]

92. Williams, K.; Askew, C.; Mazoue, C.; Guy, J.; Torres-McGehee, T.M.; Jackson Iii, J.B. Vitamin D3 Supplementation and Stress Fractures in High-Risk Collegiate Athletes—A Pilot Study. *Orthop. Res. Rev.* **2020**, *12*, 9–17. [CrossRef]

93. Kujach, S.; Lyzwinski, D.; Chroboczek, M.; Bialowas, D.; Antosiewicz, J.; Laskowski, R. The Effect of Vitamin D3 Supplementation on Physical Capacity among Active College-Aged Males. *Nutrients* **2020**, *12*, 1936. [CrossRef]

94. Allison, R.J.; Close, G.L.; Farooq, A.; Riding, N.R.; Salah, O.; Hamilton, B.; Wilson, M.G. Severely vitamin D-deficient athletes present smaller hearts than sufficient athletes. *Eur. J. Prev. Cardiol.* **2015**, *22*, 535–542. [CrossRef]

95. Matthews, L.R.; Danner, O.K.; Ahmed, Y.A.; Dennis-Griggs, D.M.; Fredeick, A.; Clark, C.; Moore, R.; DuMornay, W.; Childs, E.W.; Wilson, K.L. Combination therapy with vitamin D3, progesterone, omega3 fatty acids and glutamine reverses coma and improves clinical outcomes in patients with severe traumatic brain injuries: A case serie. *Int. J. Case Rep. Images* **2013**, *4*, 143–148. [CrossRef]

96. Burgi, A.A.; Gorham, E.D.; Garland, C.F.; Mohr, S.B.; Garland, F.C.; Zeng, K.; Thompson, K.; Lappe, J.M. High serum 25-hydroxyvitamin D is associated with a low incidence of stress fractures. *J. Bone Miner. Res.* **2011**, *26*, 2371–2377. [CrossRef] [PubMed]

97. Koundourakis, N.E.; Avgoustinaki, P.D.; Malliaraki, N.; Margioris, A.N. Muscular effects of vitamin D in young athletes and non-athletes and in the elderly. *Hormones* **2016**, *15*, 471–488. [CrossRef] [PubMed]

98. Rockwell, M.S.; Frisard, M.I.; Rankin, J.W.; Zabinsky, J.S.; McMillan, R.P.; You, W.; Davy, K.P.; Hulver, M.W. Effects of Seasonal Vitamin D3 Supplementation on Strength, Power, and Body Composition in College Swimmers. *Int. J. Sport Nutr. Exerc. Metab.* **2020**. [CrossRef]

99. Smarkusz, J.; Zapolska, J.; Witczak-Sawczuk, K.; Ostrowska, L. Characteristics of a diet and supplementation of American football team players: Following a fashionable trend or a balanced diet? *Roczniki Państwowego Zakładu Higieny* **2019**, *70*, 49–57. [CrossRef] [PubMed]

100. Maroon, J.C.; Mathyssek, C.M.; Bost, J.W.; Amos, A.; Winkelman, R.; Yates, A.P.; Duca, M.A.; Norwig, J.A. Vitamin D profile in National Football League players. *Am. J. Sports Med.* **2015**, *43*, 1241–1245. [CrossRef]

101. Blue, M.N.; Trexler, E.T.; Hirsch, K.R.; Smith-Ryan, A.E. A profile of body composition, omega-3 and vitamin D in National Football League players. *J. Sports Med. Phys. Fit.* **2019**, *59*, 87–93. [CrossRef]

102. Dzedzej, A.; Ignatiuk, W.; Jaworska, J.; Grzywacz, T.; Lipinska, P.; Antosiewicz, J.; Korek, A.; Ziemann, E. The effect of the competitive season in professional basketball on inflammation and iron metabolism. *Biol. Sport* **2016**, *33*, 223–229. [CrossRef]

103. Kruit, A.; Zanen, P. The association between vitamin D and C-reactive protein levels in patients with inflammatory and non-inflammatory diseases. *Clin. Biochem.* **2016**, *49*, 534–537. [CrossRef]

104. Mehran, N.; Schulz, B.M.; Neri, B.R.; Robertson, W.J.; Limpisvasti, O. Prevalence of Vitamin D Insufficiency in Professional Hockey Players. *Orthop. J. Sports Med.* **2016**, *4*, 2325967116677512. [CrossRef]

105. Grieshober, J.A.; Mehran, N.; Photopolous, C.; Fishman, M.; Lombardo, S.J.; Kharrazi, F.D. Vitamin D Insufficiency Among Professional Basketball Players: A Relationship to Fracture Risk and Athletic Performance. *Orthop. J. Sports Med.* **2018**, *6*, 2325967118774329. [CrossRef]

106. Sekel, N.M.; Gallo, S.; Fields, J.; Jagim, A.R.; Wagner, T.; Jones, M.T. The Effects of Cholecalciferol Supplementation on Vitamin D Status Among a Diverse Population of Collegiate Basketball Athletes: A Quasi-Experimental Trial. *Nutrients* **2020**, *12*, 370. [CrossRef]

107. Marcinowska-Suchowierska, E.; Kupisz-Urbanska, M.; Lukaszkiewicz, J.; Pludowski, P.; Jones, G. Vitamin D Toxicity-A Clinical Perspective. *Front. Endocrinol.* **2018**, *9*, 550. [CrossRef]

108. Araki, T.; Holick, M.F.; Alfonso, B.D.; Charlap, E.; Romero, C.M.; Rizk, D.; Newman, L.G. Vitamin D intoxication with severe hypercalcemia due to manufacturing and labeling errors of two dietary supplements made in the United States. *J. Clin. Endocrinol. Metab.* **2011**, *96*, 3603–3608. [CrossRef]

109. Dudenkov, D.V.; Yawn, B.P.; Oberhelman, S.S.; Fischer, P.R.; Singh, R.J.; Cha, S.S.; Maxson, J.A.; Quigg, S.M.; Thacher, T.D. Changing Incidence of Serum 25-Hydroxyvitamin D Values Above 50 ng/mL: A 10-Year Population-Based Study. *Mayo Clin. Proc.* **2015**, *90*, 577–586. [CrossRef]

110. von Restorff, C.; Bischoff-Ferrari, H.A.; Theiler, R. High-dose oral vitamin D3 supplementation in rheumatology patients with severe vitamin D3 deficiency. *Bone* **2009**, *45*, 747–749. [CrossRef]

111. Li, K.; Kaaks, R.; Linseisen, J.; Rohrmann, S. Associations of dietary calcium intake and calcium supplementation with myocardial infarction and stroke risk and overall cardiovascular mortality in the Heidelberg cohort of the European Prospective Investigation into Cancer and Nutrition study (EPIC-Heidelberg). *Heart* **2012**, *98*, 920–925.

112. Malihi, Z.; Wu, Z.; Stewart, A.W.; Lawes, C.M.; Scragg, R. Hypercalcemia, hypercalciuria, and kidney stones in long-term studies of vitamin D supplementation: A systematic review and meta-analysis. *Am. J. Clin. Nutr.* **2016**, *104*, 1039–1051. [CrossRef]

113. Malihi, Z.; Wu, Z.; Lawes, C.M.M.; Scragg, R. Adverse events from large dose vitamin D supplementation taken for one year or longer. *J. Steroid Biochem. Mol. Biol.* **2019**, *188*, 29–37. [CrossRef]

114. McCullough, P.J.; Lehrer, D.S.; Amend, J. Daily oral dosing of vitamin D3 using 5000 TO 50,000 international units a day in long-term hospitalized patients: Insights from a seven year experience. *J. Steroid Biochem. Mol. Biol.* **2019**, *189*, 228–239. [CrossRef]

115. Pittas, A.G.; Dawson-Hughes, B.; Sheehan, P.; Ware, J.H.; Knowler, W.C.; Aroda, V.R.; Brodsky, I.; Ceglia, L.; Chadha, C.; Chatterjee, R.; et al. Vitamin D Supplementation and Prevention of Type 2 Diabetes. *N. Engl. J. Med.* **2019**, *381*, 520–530. [CrossRef]

116. Dawson-Hughes, B.; Staten, M.A.; Knowler, W.C.; Nelson, J.; Vickery, E.M.; LeBlanc, E.S.; Neff, L.M.; Park, J.; Pittas, A.G.; D2d Research Group. Intratrial Exposure to Vitamin D and New-Onset Diabetes Among Adults With Prediabetes: A Secondary Analysis From the Vitamin D and Type 2 Diabetes (D2d) Study. *Diabetes Care* **2020**, *43*, 2916–2922. [CrossRef]

117. Martineau, A.R.; Jolliffe, D.A.; Greenberg, L.; Aloia, J.F.; Bergman, P.; Dubnov-Raz, G.; Esposito, S.; Ganmaa, D.; Ginde, A.A.; Goodall, E.C.; et al. Vitamin D supplementation to prevent acute respiratory infections: Individual participant data meta-analysis. *Health Technol. Assess.* **2019**, *23*, 1–44. [CrossRef]

118. Wagner, C.L.; Hollis, B.W. The Implications of Vitamin D Status During Pregnancy on Mother and her Developing Child. *Front. Endocrinol.* **2018**, *9*, 500. [CrossRef]

119. Grant, W.B.; Al Anouti, F.; Moukayed, M. Targeted 25-hydroxyvitamin D concentration measurements and vitamin D3 supplementation can have important patient and public health benefits. *Eur. J. Clin. Nutr.* **2020**, *74*, 366–376. [CrossRef]

120. Makowski, A.J.; Rathmacher, J.A.; Horst, R.L.; Sempos, C.T. Simplified 25-Hydroxyvitamin D Standardization and Optimization in Dried Blood Spots by LC-MS/MS. *J. AOAC Int.* **2017**, *100*, 1328–1336. [CrossRef]

121. Ross, A.C.; Manson, J.E.; Abrams, S.A.; Aloia, J.F.; Brannon, P.M.; Clinton, S.K.; Durazo-Arvizu, R.A.; Gallagher, J.C.; Gallo, R.L.; Jones, G.; et al. The 2011 report on dietary reference intakes for calcium and vitamin D from the Institute of Medicine: What clinicians need to know. *J. Clin. Endocrinol. Metab.* **2011**, *96*, 53–58. [CrossRef]

122. Holick, M.F.; Binkley, N.C.; Bischoff-Ferrari, H.A.; Gordon, C.M.; Hanley, D.A.; Heaney, R.P.; Murad, M.H.; Weaver, C.M.; Endocrine, S. Evaluation, treatment, and prevention of vitamin D deficiency: An Endocrine Society clinical practice guideline. *J. Clin. Endocrinol. Metab.* **2011**, *96*, 1911–1930. [CrossRef]

123. Pludowski, P.; Holick, M.F.; Grant, W.B.; Konstantynowicz, J.; Mascarenhas, M.R.; Haq, A.; Povoroznyuk, V.; Balatska, N.; Barbosa, A.P.; Karonova, T.; et al. Vitamin D supplementation guidelines. *J. Steroid Biochem. Mol. Biol.* **2018**, *175*, 125–135. [CrossRef]

124. Veugelers, P.J.; Pham, T.M.; Ekwaru, J.P. Optimal Vitamin D Supplementation Doses that Minimize the Risk for Both Low and High Serum 25-Hydroxyvitamin D Concentrations in the General Population. *Nutrients* **2015**, *7*, 10189–10208. [CrossRef]

125. Shirvani, A.; Kalajian, T.A.; Song, A.; Holick, M.F. Disassociation of Vitamin D's Calcemic Activity and Non-calcemic Genomic Activity and Individual Responsiveness: A Randomized Controlled Double-Blind Clinical Trial. *Sci. Rep.* **2019**, *9*, 17685. [CrossRef]

126. Di Luigi, L.; Antinozzi, C.; Piantanida, E.; Sgro, P. Vitamin D, sport and health: A still unresolved clinical issue. *J. Endocrinol. Investig.* **2020**, *43*, 1689–1702. [CrossRef]
127. Gao, J.; Wei, W.; Wang, G.; Zhou, H.; Fu, Y.; Liu, N. Circulating vitamin D concentration and risk of prostate cancer: A dose-response meta-analysis of prospective studies. *Ther. Clin. Risk Manag.* **2018**, *14*, 95–104. [CrossRef]
128. Song, Z.Y.; Yao, Q.; Zhuo, Z.; Ma, Z.; Chen, G. Circulating vitamin D level and mortality in prostate cancer patients: A dose-response meta-analysis. *Endocr. Connect.* **2018**, *7*, R294–R303. [CrossRef]
129. Wilson, K.M.; Shui, I.M.; Mucci, L.A.; Giovannucci, E. Calcium and phosphorus intake and prostate cancer risk: A 24-y follow-up study. *Am. J. Clin. Nutr.* **2015**, *101*, 173–183. [CrossRef]
130. Batai, K.; Murphy, A.B.; Ruden, M.; Newsome, J.; Shah, E.; Dixon, M.A.; Jacobs, E.T.; Hollowell, C.M.; Ahaghotu, C.; Kittles, R.A. Race and BMI modify associations of calcium and vitamin D intake with prostate cancer. *BMC Cancer* **2017**, *17*, 64. [CrossRef]
131. Larson-Meyer, D.E.; Woolf, K.; Burke, L. Assessment of Nutrient Status in Athletes and the Need for Supplementation. *Int. J. Sport Nutr. Exerc. Metab.* **2018**, *28*, 139–158. [CrossRef]
132. Chiang, C.M.; Ismaeel, A.; Griffis, R.B.; Weems, S. Effects of Vitamin D Supplementation on Muscle Strength in Athletes: A Systematic Review. *J. Strength Cond. Res.* **2017**, *31*, 566–574. [CrossRef]
133. Liu, J.; Arcot, J.; Cunningham, J.; Greenfield, H.; Hsu, J.; Padula, D.; Strobel, N.; Fraser, D.R. New data for vitamin D in Australian foods of animal origin: Impact on estimates of national adult vitamin D intakes in 1995 and 2011–13. *Asia Pac. J. Clin. Nutr.* **2015**, *24*, 464–471.

Publisher's Note: MDPI stays neutral with regard to jurisdictional claims in published maps and institutional affiliations.

Article

Exploring the Impact of Individual UVB Radiation Levels on Serum 25-Hydroxyvitamin D in Women Living in High Versus Low Latitudes: A Cross-Sectional Analysis from the D-SOL Study

Marcela M. Mendes [1,2,*], Kathryn H. Hart [1], Susan A. Lanham-New [1] and Patrícia B. Botelho [2]

1   Department of Nutritional Sciences, Faculty of Health and Medical Sciences, University of Surrey, Guildford GU2 7XH, UK; k.hart@surrey.ac.uk (K.H.H.); s.lanham-new@surrey.ac.uk (S.A.L.-N.)
2   Department of Nutrition, Faculty of Health Sciences, University of Brasília, Brasília 70910-900, Brazil; patriciaborges.nutri@gmail.com
*   Correspondence: m.moraesmendes@surrey.ac.uk; Tel.: +44-014-8368-9222

Received: 30 October 2020; Accepted: 7 December 2020; Published: 11 December 2020

Abstract: Vitamin D can be synthesized in the skin via sunlight exposure as well as ingested through diet. Vitamin D deficiency is currently a major global public health issue, with increasing prevalence in both low and high latitude locations. This cross-sectional analysis aimed to compare the intensity of individual Ultraviolet B radiation levels between women of the same ethnicity living in England and Brazil, respectively; and to investigate the association with circulating 25(OH)D concentrations. We analysed data from 135 Brazilian women (England, $n = 56$, 51° N; Brazil, $n = 79$, 16° S) recruited for the D-SOL study (Interaction between Vitamin **D** Supplementation and **S**unlight Exposure in Women Living in **O**pposite **L**atitudes). Serum 25(OH)D concentrations were analysed by high performance liquid chromatography tandem mass spectrometry (HPLC-MS), individual UVB radiation via UVB dosimeter badges and dietary intake via 4-day diet diaries. Anthropometric, skin phototype, sociodemographic and lifestyle patterns were also assessed. Mean serum 25(OH)D concentration of England residents was significantly lower than Brazil residents. Daily individual UVB radiation level showed a strong significant positive correlation with serum 25(OH)D concentrations. The required UVB radiation to achieve 75 nmol/L was 2.2 SED and 38.8% of the total variance in 25(OH)D concentrations was explained uniquely by daily individual UVB radiation, after controlling for the influence of age and body mass index. Thus, these results highlight the strong positive association between serum 25(OH)D concentrations and individual UVB radiation and the influence of different individual characteristics and behaviours. Collectively, these factors contribute to meaningful, country-specific, public health strategies and policies for the efficient prevention and treatment of vitamin D inadequacy.

Keywords: vitamin D; sunlight exposure; latitude; vitamin D intake

---

1. Introduction

Prolonged and severe vitamin D deficiency can lead to rickets in children and osteomalacia/osteoporosis in adults [1–3]. Vitamin D is naturally present in very few foods and in small quantities. The main source of Vitamin D is considered to be casual exposure of the skin to the UVB portion of sunlight (290–350 nm), which converts the molecule 7-dehydrocholesterol naturally present in the epidermis, to pre-vitamin D [2–5].

Sunlight will need to travel through the atmosphere before reaching the skin to produce vitamin D. The availability of UVB radiation is determined mainly by the solar zenith angle, which is influenced by latitude, season and time of day. Therefore, the solar irradiation on the surface of the earth, and concomitantly the local UVB radiation, will depend on the verticalization of the solar zenith angle in each particular latitude [3,6]. In other words, populations living within the tropics, namely low latitude locations (e.g., South America), are exposed to substantially higher levels of solar UVB radiation throughout the year than those living in high latitude locations (i.e., Europe) [6]. For instance, the annual total horizontal irradiation in Surrey, UK (latitude 51° N) is approximately 1000 kWh/m^2 [7] in comparison to the annual average of 5500 kWh/m^2 in Goiás, Brazil (16° S) [8]. Nevertheless, recent reports of an increasing prevalence of low vitamin D concentrations in both low and high latitude locations show that vitamin D deficiency is rapidly becoming, if not already, a major global public health issue [9–13].

Advisory agencies (Government led or otherwise) have consistently highlighted the challenges in establishing reference values for adequate vitamin D recommendations, particularly due to the individual variation as well as the influence of external environmental factors [14–16]. The inappropriateness of direct comparison of data from studies conducted in different locations is mainly due to significant variations in results between different laboratories, different latitudes and different populations/ethnic groups—and therefore influencing factors, adding greatly to the difficulty in finding a global consensus. Moreover, very few studies to date have investigated the relationship between actual individual exposure to UVB radiation and serum 25(OH)D concentrations [6,17,18], with most data derived from in vitro or animal studies. Another important limitation to the current recommendations regarding optimal dietary intakes and sunlight exposure to maintain adequate levels, is that they are generally based on studies of Caucasian populations in high latitude countries, with limited robust data for other ethnicity and/or different geographical locations. Consequently, there is a substantial lack of evidence on the effect of individual sunlight exposure in low latitude countries and in their native non-Caucasian populations.

In order to develop effective vitamin D guidelines, we need to fully understand the actual impact of sunlight on 25(OH)D concentrations, based not only on UVB radiation availability (latitude) but also individual UVB radiation exposure, as well as the relative contribution of key influential factors, such as dietary intake, adiposity, skin pigmentation and lifestyle. Therefore, the aim of this cross-sectional analysis was to compare the intensity of individual UVB radiation levels between Brazilian adult women living in England and Brazilian adult women living in Brazil, and to investigate the association with circulating 25(OH)D concentrations.

2. Materials and Methods

This is a cross-sectional analysis of data collected by the D-SOL study (Interaction between vitamin D supplementation and Sunlight exposure in women living in Opposite Latitudes; registered at clinicaltrials.gov as NCT03318029). The study was approved by the University of Surrey (UEC/2016/009/FHMS) and Federal University of Goiás Ethics Committees and by the Brazilian National Ethics Committee (CONEP) (CAAE 62149516.9.0000.5083, CEP-UFG n° 2013222; CONPEP n° 1972029; respectively). All participants at commencement of the study provided written informed consent.

2.1. Study Location

The D-SOL study was conducted at the University of Surrey, Guildford, Surrey (51° N), England, from December 2016 to March 2017 (high latitude group) and at the University of Goiás, Goiás (16° S), Brazil, from June to September 2017 (low latitude group). Surrey has a temperate climate (throughout the year a maximum of ~25 °C and minimum of ~0 °C) with a summer mean temperature of 22 °C and winter mean temperature of 6 °C, whereas Goiás has a typical tropical savannah climate (throughout the year a maximum of ~36 °C and minimum of ~15 °C), with a summer mean temperature of 26 °C and winter mean temperature of 24 °C. The UV index never exceeds 8 in the UK (peaking towards the end of June), and in clear contrast, the minimum in Brazil is 8 in winter, reaching up to 14 during hotter months.

2.2. Study Design

Participants were selected if they were female of Brazilian nationality (born in Brazil and having at least one parent born in Brazil), aged 20–59 years. Exclusion criteria included: currently receiving treatment for medical conditions that are likely to affect vitamin D metabolism (osteoporosis therapy, anti-estrogens treatment, antiepileptic drugs, cancer treatment); taking supplements containing vitamin D (if the prospective participants agreed to stop vitamin D supplementation to join the study, a wash-out period of 8 weeks prior to commencing the trial was accepted), being pregnant or planning a pregnancy during the study period, being post-menopausal (defined as permanent cessation of menstruation) and living in the UK for less than 3 months at the commencement of the study (for England participants only).

Participants in England were recruited through advertisements distributed locally in Surrey and London. Brazilian institutions in the UK, such as the Brazilian Embassy in London and the Brazilian Researchers Association (ABEP-UK), agreed to circulate a recruitment letter to their contact lists. Participants in Brazil were also recruited locally form the residents of the city of Goiânia, Goiás. For both trials, recruitment via social media and online platforms was also used.

2.3. Anthropometric Measures

For measurement of weight participants were asked to remove shoes, socks and heavy coats before stepping on the scale (England: Tanita Body Composition Analyser MC-180MA, Tanita Coopertatives, Tokyo, Japan; Brazil: standard weighing scale, Balmak®). Waist circumference was measured with a non-extendable standard measure tape, at the narrowest point of the torso, to the nearest 0.1 cm. If this point could not be estimated, the level of the umbilicus was used as the reference point. Body Mass Index (BMI) was classified according to the World Health Organization as underweight (<18.5 kg/m^2), normal weight (18.5–24.9 kg/m^2), overweight (25.0–29.9 kg/m^2) or obese (>30 kg/m^2) [19].

2.4. Lifestyle and Sun Exposure

A lifestyle questionnaire, adapted for each country, was administered to assess for cultural and general lifestyle aspects. Dietary intake of participants was determined by 4 consecutive days of estimated diet diaries, commencing on a Sunday to ensure weekdays and weekend days were represented. Participants were trained by the research team on how to correctly complete the diary.

Participants were instructed to maintain their habitual sunlight exposure and sun protective measures (if part of their habitual routine) as well as their usual dietary intake for the duration of the study and were requested to report any significant changes to their habits or normal routine.

To determine individual exposure to ambient UVB radiation, participants were asked to wear individual polysulphone film badge dosimeters (provided by the University of Manchester, UK) on their outer clothing. Participants were instructed to wear their dosimeters around the upper shoulder/chest region from sunrise to sunset for a full week (7 consecutive days), starting the day after blood samples were taken. All dosimeters, for both the England and Brazil trials, were read at the University of Surrey, prior to and after use, with a Cecil Aquarius CE7200 Double Beam Spectrophotometer (which has a CV < 1%) at 330 nm, to detect change in absorbency [6]. The amount of UVB captured by each dosimeter badge was then translated to a standard erythematous dose (SED): 1 SED is equal to 100 J m^{-2} of erythemal (sun burning) UVB radiation. A measure of 3 SED is roughly equivalent to one minimal erythema dose [MED] in unacclimatized, sensitive white skin. An exposure of 5–8 SED will result in moderate sunburn and 10 SED or more can result in a painful, blistering sunburn, in unacclimatized, sensitive white skin [18].

2.5. Skin Pigmentation

Race and skin type were both self-reported via the lifestyle questionnaire. Race categories were based on the Brazilian ethno-race national demographic spectrum [20], which includes: White, Black,

Brown/Mixed ("Pardo" in Portuguese), Indigenous and "Yellow" (Asian-descendent). Participants were asked to indicate which category they most identified with. Skin type was based on the Fitzpatrick validated classification for skin photo-types, which classifies the skin according to the ability of each skin type to tan under sun exposure and its sensitivity and tendency to turn red under solar radiation [21]. Participants were asked to choose one category only that best represented the effect of sunlight exposure on their skins. For the purpose of this study, Fitzpatrick's photo-types were combined into skin type categories as follow: Type I (Always burns; Never tan; Very sensitive to the Sun) and II (Always burning; Very little tan; Sun sensitive) were classified as 'white'; Type III (Burns moderately; Bronze moderately; Normal sensitivity to the Sun) and IV (Burns a little; Always tan; Normal sensitivity to the Sun) as 'brown'; Type V (Rarely burns; Always tan; Not sensitive to the sun) and VI (Never burn; Totally pigmented; Insensitive to the Sun) as 'black'.

2.6. Laboratory Analysis

An overnight fasted (8 h) blood sample was collected by venipuncture by trained phlebotomists in both centres. For serum, the collected blood samples were left to clot for 1 h at room temperature followed by centrifugation at 3000 g for 10 min at 4 °C (England trial: Sigma 3–16 K Centrifuge, SciQuip, Shropshire, UK; Brazil trial: Eppendorf™ 5702R Centrifuge, UK). Processed serum samples were distributed into aliquots and stored at −80 °C at the University of Surrey, prior to analysis. Samples collected in Brazil followed the exact same procedures and were temporarily stored at −80 °C at the University of Goiás. The samples were then sent by air to the UK to be stored at the University of Surrey, prior to analysis.

All samples, from both countries, were analysed for 25(OH)D, serum calcium and serum albumin at Imperial College London. Serum 25(OH)D concentrations were determined by HPLC-MS/MS method on a Waters Acuity TQD using a PFP column following supported liquid extraction (SLE). Laboratory intra- and interassay CVs were 5.6% and 7.8%, respectively. Serum calcium was measured by using an endpoint spectrophotometric reaction based on the o-cresolphthalein complexone methodology, and serum albumin was measured by using an endpoint spectrophotometric reaction based on the bromocresol green solution dye binding methodology. Serum calcium concentrations were adjusted for albumin concentrations.

Due to the lack of global consensus as to the definition of vitamin D status, for the purpose of this study vitamin D deficiency was defined as 25(OH)D concentrations below 25 nmol/L, as suggested by the UK Scientific Advisory Committee on Nutrition [14]; insufficiency as concentrations between 25–49.9 nmol/L and adequacy between 50–74.9 nmol/L, as recommended by the US Institute of Medicine [15]; and optimal levels as concentrations above 75 nmol/L, as proposed by the US Endocrine Society [16].

2.7. Statistical Analysis

Statistical analysis of the data was undertaken using SPSS software for Windows (version 26.0, 2019; IBM Corp, Armonk, NY, USA). Data were tested for normality using the Kolmogorov-Smirnov test. Non-normally distributed variables were log transformed and reported in the original scale. Non-parametric tests were used when log transforming did not normalize the data. For categorical variables, frequency and percentage were reported. The distribution of skin type and BMI classification were compared between countries using chi squared tests.

Mean circulating serum 25(OH)D concentrations were compared between different aspects of lifestyle, individual characteristics and individual UVB radiation levels using independent t-tests, or Mann-Whitney U tests for non-normally distributed data; or one-way ANOVA with post-hoc Tukey tests, or Kruskal-Wallis for non-normally distributed data. Standard linear regression models were run to investigate the predictive ability of individual daily sunlight exposure on circulating serum 25(OH)D concentrations.

A p value of <0.05 was considered significant.

3. Results

Of the 335 participants enrolled for the D-SOL study after the screening process, 135 participants were included in this cross-sectional analysis (n = 56 in England and 79 in Brazil). Reasons for exclusion at screening are detailed in Figure 1. In the Brazil cohort, one participant did not have valid laboratory results and was therefore excluded from the database as previously described in Mendes et al. (2020) [22]. In the England cohort, five participants were post-menopausal. There were no differences between analyses including or excluding the five post-menopausal women, and therefore, only analyses including these participants are reported here.

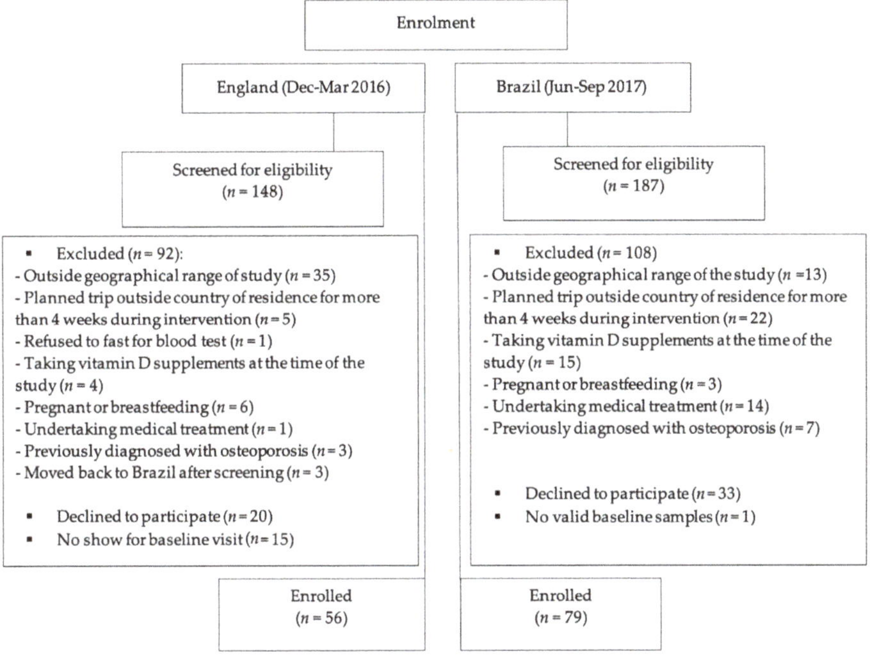

Figure 1. Flow diagram of participant enrolment.

3.1. Participant Characteristics

Participant characteristics, specifically age, weight, BMI, vitamin D and calcium intake and 25(OH)D, PTH and calcium serum concentrations, have been previously published in Mendes et al., 2019 [22]. Brazilian women living in England were older, heavier and had a greater waist circumference than those living in Brazil (p < 0.01). There were no significant differences between Brazilian women living in England and in Brazil for BMI classification distributions although, in line with the weight data, the mean BMI was significantly greater for Brazilian women residing in England.

Figure 2 shows the difference in serum 25(OH)D concentrations between the two countries, with concentrations ranging from 5.0 to 73.5 nmol/L for participants living in England and from 36.2 to 148.6 nmol/L for those living in Brazil. Mean serum 25(OH)D concentration of England residents was significantly lower than Brazil residents (36.0 ± 14.9 nmol/L and 75.0 ± 22.1 nmol/L, respectively p < 0.001), as previously published [22]. The statistical significance remained after controlling for daily UVB radiation level, age, BMI and waist circumference (ANCOVA, p < 0.001). Only participants living in Brazil had serum 25(OH)D concentrations above 100 nmol/L (n = 3), of which two women had concentrations above 130 nmol/L (134.9 and 148.6 nmol/L). There were no significant differences in serum albumin-corrected calcium concentrations between England and Brazil residents

(2.3 ± 0.07 and 2.2 ± 0.06 mmol/L, respectively, p = 0.066), with all participants having concentrations within the normal range of 2.1–2.6 mmol/L.

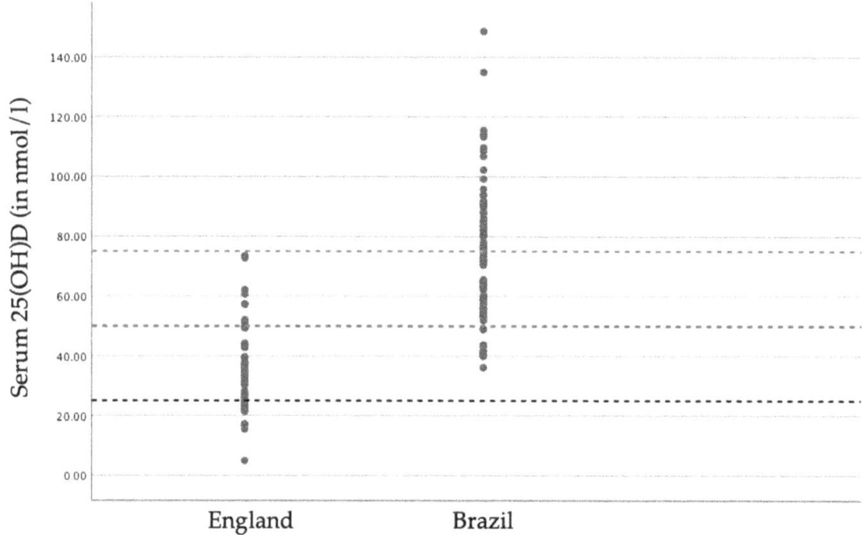

Figure 2. Serum 25(OH)D concentrations for women living in England (n = 56) and women living in Brazil (n = 79). Hashed lines represent thresholds of 25 (deficiency), 50 (insufficiency) and 75 (optimal) nmol/L.

The proportions of women with deficient and insufficient levels were significantly higher in those living in England than in Brazil (p < 0.001). Amongst women living in England 25% had serum 25(OH)D concentrations below 25 nmol/L, while there were no participants with concentrations below this threshold living in Brazil. There were no participants living in England with concentrations above 75 nmol/L, while half (50.6%) of the participants in Brazil presented levels above this threshold. The majority (82.1%) of women living in England in comparison to 11.4% of those living in Brazil, had concentrations below the insufficiency cut off value of 50 nmol/L.

Overall (n = 135), serum 25(OH)D concentration was negatively correlated with age (r = −0.282, p = 0.001) and waist circumference (r = −0.361, p < 0.001), and there was a trend for a negative association with BMI (r = −0.169, p = 0.052) (Table 1). Within each country, no correlations were found between serum 25(OH)D concentration and age nor anthropometry (BMI or waist circumference) (all p > 0.406). Overall, women younger than 30 years of age had significantly higher mean 25(OH)D concentrations than those aged 30–44 years (64.7 ± 27.4 and 51.6 ± 27.2 nmol/L, p = 0.027); however, no significant differences amongst age groups were observed within each country (Table 1).

The proportion of white women in the England cohort was significantly higher (p = 0.012) while in the Brazil cohort there was an even distribution between white and brown (51.9% and 44.3%, respectively). With respect to skin type classification, the majority of participants classified themselves as type III and IV (63% overall, 66.1% of England and 60.8% of Brazil residents) (Table 1). There were no significant differences in mean serum 25(OH)D concentrations between ethno-race or skin type (all p > 0.081). Amongst England residents, those living in Southern England for more than 2 years had significantly lower 25(OH)D concentrations than those that had recently moved to the UK (less than a year) (p = 0.039) (Table 2).

Table 1. Socio-demographic characteristics of adult Brazilian women overall and by country of residence and associations with serum 25(OH)D concentrations ($n = 135$) [1].

					All		England		Brazil	
	All	England ($n = 56$)	Brazil ($n = 79$)	p	25(OH)D in nmol/L	p^2	25(OH)D in nmol/L	p^2	25(OH)D in nmol/L	p^2
Age (years)	31.39 ± 8.7	35.55 ± 9.0	28.43 ± 7.1	<0.001 [2]						
<30	71 (53%)	18 (32%)	53 (67%)		64.7 ± 27.4 [a]	0.030	13.2 ± 3.1	0.483	74.1 ± 24.4	0.761
30–44	48 (35%)	29 (52%)	19 (24%)		51.6 ± 27.2 [a]		15.7 ± 2.9		78.2 ± 18.0	
>44	16 (12%)	9 (16%)	7 (9%)		54.8 ± 21.1		14.9 ± 4.9		72.5 ± 12.9	
Ethno-race [n (%)] [R1]										
White	85 (63%)	44 (78.6%)	41 (51.9%)	0.012 [3]	55.6 ± 25.8	0.207	37.4 ± 15.9	0.529	75.2 ± 19.3	0.422
Black	3 (2.2%)	1 (1.8%)	2 (2.5%)		43.1 ± 15.3		25.7		51.8 ± 4.1	
Brown	45 (33.3%)	10 (17.9%)	35 (44.3)		66.3 ± 29.5		30.9 ± 8.2		76.4 ± 25.2	
Yellow	1 (0.7%)	0	1 (1.3%)		59.5		N/A		59.5	
Indigenous	1 (0.7%)	1 (1.8%)	0		42.9		42.9		N/A	
Skin type [n (%)] [R2]										
Type I and II	42 (31.1%)	17 (30.4%)	25 (31.6%)	0.589 [3]	61.0 ± 24.9	0.705	37.2 ± 12.5	0.927	77.3 ± 16.6	0.812
Type III and IV	85 (63%)	37 (66.1%)	48 (60.8%)		57.4 ± 28.9		35.7 ± 16.0		74.9 ± 25.5	
Type V and VI	8 (5.9%)	2 (3.6)	6 (7.6)		63.0 ± 21.0		33.9 ± 13.9		72.7 ± 11.4	
Years in England [n (%)]							51.0 ± 14.9 [a]	0.039		
<1 year	N/A	6 (10.73%)	N/A	N/A			37.4 ± 11.3			
1–2 years		10 (17.85%)					33.5 ± 14.6 [a]			
>2 years		40 (71.42%)								

[1] Values are mean ± Standard deviation or n (%). Statistical analysis: [2] One-way ANOVA with post-hoc Tukey's test. Values in same column with same superscript letters are significantly different ($p < 0.05$). [3] Pearson Chi Squares. References: [R1] Official ethno-racial categories for the Brazilian population by the Brazilian Institute of Geography and Statistics (Brown: mixed; Yellow: Japanese-descendant; Indigenous: native Indian) [20] [R2] The Fitzpatrick Classification Scale for Skin Types [21]. N/A: Not Applicable.

Table 2. Lifestyle characteristics of adult Brazilian women overall and by country of residence ($n = 135$) [1].

	All	England($n = 56$)	Brazil($n = 79$)	p^2	All		England		Brazil	
					25(OH)D in nmol/L	p^3	25(OH)D in nmol/L	p^3	25(OH)D in nmol/L	p^3
Alcohol Consumption										
No	58 (43%)	34 (60.7%)	36 (45.6%)	0.467	57.57 ± 25.6	0.674	37.9 ± 16.9	0.253	76.9 ± 24.0	0.396
Yes	77 (57%)	22 (39.3%)	43 (54.4%)		59.7 ± 28.7		33.2 ± 11.1		72.6 ± 19.6	
Smoke										
No	126 (93.3%)	5 (8.9%)	75 (94.9%)	0.375	60.2 ± 22.6	0.861	43.2 ± 10.1	0.266	81.6 ± 11.6	0.545
Yes	9 (6.7%)	51 (91.1%)	4 (5.1%)		58.6 ± 27.8		35 ± 15.2		74.6 ± 22.6	
Milk consumption/day										
Never	39 (28.7%)	13 (23.2%)	28 (35.5%)	0.001	65.9 ± 30.11	0.009	37.4 ± 16.7	0.928	78.6 ± 25.7	0.613
Less than 1 mug	31 (23%)	17 (30.4%)	14 (17.7%)		51.84 ± 23.5		37.1 ± 16.2		68.9 ± 18.9	
1 mug (280 mL)	38 (28.1)	9 (16.1%)	29 (36.7%)		65.0 ± 26.5		33.5 ± 11.7		74.7 ± 21.8	
2 mugs (560 mL)	21 (15.6%)	13 (21.3%)	8 (10.1%)		49.3 ± 24.7		34.7 ± 15.7		73.1 ± 16.7	
More than 2 mugs	4 (3%)	4 (7.1%)	0		31.3 ± 6.74		31.3 ± 6.7		N/A	
Egg consumption/week										
Never	4 (2.9%)	1 (1.8%)	3 (3.8%)	0.203	62.5 ± 15.7	0.002	52.1	0.091	67.8 ± 18.2	0.095
Less than once	21 (15.6%)	9 (16.1%)	12 (15.2%)		51.4 ± 23.3 [a]		30.2 ± 14.3		66.8 ± 16.1	
Once a week	29 (21.5%)	17 (30.4%)	12 (15.2%)		47.1 ± 23.1 [b]		30.6 ± 9.1		70.4 ± 15.0	
2–5 times	61 (45.2%)	24 (42.9%)	37 (46.8%)		60.3 ± 26.6		39.5 ± 15.7		73.9 ± 23.4	
>5 times	20 (14.8%)	5 (8.9%)	15 (19%)		77.5 ± 30.5 [a,b]		44.9 ± 21.5		88.3 ± 25.1	
Oily fish consumption/week										
Never	31 (22.9%)	10 (17.9%)	20 (25.3%)	0.005	56.1 ± 23.3	0.047	31.9 ± 12.1	0.034	68.1 ± 17.4	0.215
Less than once	58 (43%)	17 (30.4%)	41 (51.9%)		64.6 ± 30.5		31.6 ± 16.2 [a]		78.4 ± 23.4	
Once	33 (24.4%)	21 (37.5%)	12 (15.2%)		51.2 ± 24.2		37.13 ± 8.7		75.8 ± 22.9	
2–5 times	12 (8.9%)	8 (14.3%)	4 (5.1%)		52.7 ± 20.8		48.4 ± 21.2 [a]		61.1 ± 20.0	
>5 times	1 (0.7%)	0	1 (1.3%)		109.7		N/A		109.7	
Liver consumption/week										
Never	100 (77%)	45(81.8%)	55 (70%)	0.520	59.1 ± 28.1	0.645	36.9 ± 15.5	0.594	76.9 ± 22.8	0.483
Less than once	28 (20.7%)	9 (16%)	19 (24%)		55.9 ± 25.5		30.3 ± 10.1		68.9 ± 21.2	
Once	2 (1.5%)	1 (2%)	1 (3%)		55.8 ± 35.1		31.0		80.7	
2–5 times	1 (0.7%)	0	1 (3%)		84.7		N/A		84.7	
>5 times	0	0	0		n/a		n/a		n/a	
Supplement use (within the last year)										
None	86 (70%)	32 (57.2%)	54 (68.4%)	0.002	61.4 ± 28.3	0.189	36.3 ± 17.2	0.866	75.5 ± 23.1	0.382
Vitamin D	15 (11.1%)	3 (5.4%)	12 (15.2%)		64.5 ± 22.1		42.9 ± 7.2		68.9 ± 21.3	
Fish/fish liver oil	3 (2.2%)	2 (3.6%)	1 (1.3%)		45.4 ± 30.5		28.2 ± 9.0		80.0	
Fish/liver oil with vit. D	14 (10.4%)	5 (8.9%)	9 (11.4%)		55.2 ± 22.6		30.4 ± 5.3		69.0 ± 15.1	
Multivitamin with vit. D	13 (9.6%)	12 (21.4%)	1 (1.3%)		41.9 ± 25.3		35.9 ± 14.0		113.3	
Calcium with vit. D	4 (3%)	2 (3.6%)	2 (2.5%)		64.2 ± 34.3		37.7 ± 2.1		90.8 ± 26.72	

[1] Values are n (%). [2] Statistical analysis: Chi Squares. [3] One-way ANOVA with post-hoc Tukey's test or t-test. Values in same column with same superscript letters are significantly different ($p < 0.05$). N/A: not applicable.

3.2. Dietary Habits

Participant data by country for vitamin D intake have been previously published [22]. Overall (n = 119), mean habitual vitamin D dietary intake was 2.45 ± 1.91 µg/day. Vitamin D and calcium intakes were significantly higher in England residents compared to Brazil residents ($p < 0.001$ and $p = 0.003$, respectively). In total ($n = 119$), 21.8% had dietary vitamin D intakes below 1 µg/day and 100% had intakes below the IOM Recommended Daily Allowance (RDA) of 15 µg/day [23].

Those consuming eggs more than five times per week had significantly higher 25(OH)D concentrations (77.5 ± 30.5 nmol/L) than those consuming eggs once a week or less (47.1 ± 23.1 nmol/L, $p = 0.001$ and 51.4 ± 23.3 nmol/L, $p = 0.014$, respectively). There was also a difference in mean 25(OH)D concentration according to frequency of milk consumption, although post-hoc tests did not identify which groups differed significantly ($p = 0.009$). The same was observed for oily fish consumption within the overall sample, and a clearer difference was observed within England participants, with those consuming fish 2–5 times per week having significantly higher concentrations (48.4 ± 21.2 nmol/L) compared to those consuming it less than once a week (31.6 ± 16.2 nmol/L, $p = 0.034$). There were no significant differences according to frequency of liver consumption or level of supplement intake within the last year (Table 2).

3.3. Sun Exposure Behaviour

The patterns of all sunlight-related behaviour reported by the life-style questionnaire were significantly different between women living in England and women living in Brazil ($p \leq 0.05$), except for sunscreen use in general and sun protection factor (SPF) during holidays ($p > 0.05$) (Table 3).

Table 3. Sun-exposure behaviours of adult Brazilian women overall and by country of residence ($n = 135$) [1].

	All	**England ($n = 56$)**	**Brazil ($n = 79$)**	p [2]
Body parts exposed				
Face only	1 (0.7%)	1 (1.8%)	0	<0.001
Hands and face	48 (35.6%)	33 (58.9%)	15 (19%)	
Hands/face + arms and/or legs	71 (52.6%)	15 (26.8%)	56 (70.9%)	
Hands/face + arms/legs + torso	15 (11.1%)	7 (12.5%)	8 (10.1%)	
Sunscreen use				
No	41 (30.4%)	14 (25%)	27 (34.2%)	0.253
Yes	94 (69.6%)	42 (75%)	52 (65.8%)	
SPF at home [§]				
15	11 (11.7%)	10 (23.8%)	1 (2.27%)	0.003
20	14 (14.8%)	12 (28.5%)	1 (2.27%)	
30	30 (31.9%)	6 (14.2%)	23 (54.5%)	
40 or over	32 (34.0%)	14 (32.5%)	18 (40.9%)	
Missing	7 (7.4%)	0	7	
SPF on holidays [§]				
15	8 (8.4%)	7 (16.7%)	1 (0.19%)	0.06
20	3 (3.2)	2 (4.7%)	1 (0.19%)	
30	34 (35.8%)	17 (40.6%)	17 (32.6%)	
40 or over	39 (41%)	15 (35.8%)	24 (46.1%)	
Missing	11 (11.6%)	1 (2.1%)	9 (21.4%)	
Natural sunbathing habit				
No	92 (68.1%)	33 (58.9%)	59 (74.7%)	0.05
Yes	43 (31.9%)	23 (41.1%)	20 (25.3%)	
Artificial sunbed use				
No	131 (97%)	52 (92.9%)	79 (100%)	0.036
Yes	3 (2.2%)	3 (5.4%)	0	
Missing	1 (0.8%)	1 (1.7%)	0	

[1] Values are n (%). [2] Statistical analysis: Pearson Chi Squares. [§] Amongst participants who said "Yes" to previous item. *SPF*: Sun Protection Factor.

Overall and within just the Brazil participants, there were significant differences in serum 25(OH)D concentrations according to the number of body parts exposed (both $p < 0.04$). Amongst Brazil participants, those reporting the usual exposure of hands and face + arms and/or legs had significantly higher 25(OH)D concentrations than those exposing their hands and face only (78.5 ± 21.8 and 62.2 ± 20.0 nmol/L, respectively; p = 0.029).

In total, around 70% if participants reported habitual use of sunscreen, although for non-holiday use those living in England were most likely to report SPF of 15 or 20 (52.3%) compared to 95.4% of those living in Brazil who reported using SPF of 30, 40 or over (p = 0.003). England residents were more likely than Brazil to report natural sunbathing (41.1% and 25.3%, respectively, p = 0.05), whilst only 3 women in England, and none in Brazil, reported having ever used an artificial sunbed. Self-reported sunbathers were most likely to report higher SPF use on holiday (60% reported using 30, 40 or more compared to only 9% using SPF 15 or 20). There were no significant differences in 25(OH)D concentrations between sunscreen users and non-users. However, overall, those reporting using a SPF 15 sunscreen during holidays had significantly lower levels than those reporting the use of SPF ≥ 40 (34.1 ± 16.6 and 62.0 ± 29.1, respectively; p = 0.034).

3.4. Individual UVB Radiation Levels

The average daily individual UVB radiation levels at the beginning of winter are shown in Figure 3, labelled by country of residence. Values are expressed in units of standard erythema dose (SED) [18]. Individual UVB radiation levels differed significantly between the two countries with concentrations ranging from 0.0031 to 0.0984 SED for England residents and from 0.3283 to 12.0393 SED for Brazil residents (mean values 0.035 ± 0.026 and 1.75 ± 2.32 SED, respectively p < 0.001). All England dwelling participants recorded daily exposure levels of less than 1 SED compared to around half of those living in Brazil (53.6%).

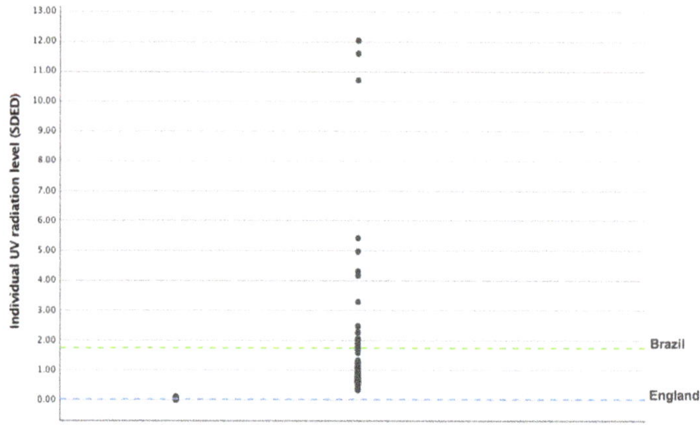

Figure 3. Daily individual Ultraviolet B radiation levels for women living in England (n = 46) and women living in Brazil (n = 69). Hashed lines represent mean daily individual UVB radiation level for women living in England (measured between October to March) and Brazil (measured between June to September).

Overall, vitamin D status was associated with individual UVB radiation (Table 4), with women presenting serum 25(OH)D concentrations above 75 nmol/L having significantly higher mean UVB radiation (2.26 ± 3.04 SED) than those with deficient (0.02 ± 0.01 SED), insufficient (0.25 ± 0.43 SED) or suboptimal (0.98 ± 1.00 SED) status (p < 0.001). Within those living in Brazil there was also a significant association (p < 0.040), with those in the higher vitamin D status groups presenting higher UVB radiation. There were no differences in women living in England, in which a mean SED below 0.04 was observed for all vitamin D status groups.

Overall, daily individual UVB radiation level showed a strong significant positive correlation with serum 25(OH)D concentrations (n = 112, r = 0.673, p < 0.001; with n = 3 outliers removed from analysis due to daily SED > 10) (Figure 4) and remained statistically significant after controlling for vitamin D intake (dietary intake), age and BMI ($r = 0.669$, $p < 0.001$). In this linear model (Figure 4), a daily exposure of 0.07 SED and 2.2 SED predicted a serum 25(OH)D concentration of 50 nmol/L and 75 nmol/L, respectively.

Table 4. Association between Vitamin D status and mean individual UVB radiation level [1].

Vitamin D status		All			England			Brazil	
	n	Mean ± SD	p^2	n	Mean ± SD	p^2	n	Mean ± SD	p^2
<25 nmol/L	12	0.02 ± 0.01 [a]	<0.001	12	0.02 ± 0.018	0.666	0	N/A	0.040
25–49.9 nmol/L	34	0.25 ± 0.43 [b]		26	0.038 ± 0.028		8	0.94 ± 0.42	
50–74.9 nmol/L	33	0.98± 1.00 [c]		8	0.037 ± 0.027		25	1.28 ± 0.97	
>75 nmol/L	36	2.26 ± 3.04 [a,b,c]		0	N/A		36	2.26 ± 3.04	

[1] Values: mean ± SD. [2] Statistical analysis: one-way ANOVA with post-hoc Tukey's test. Values in same column with same superscript letters are significantly different ([a] $p = 0.002$; [b] $p < 0.001$; [c] $p = 0.020$). N/A: not applicable.

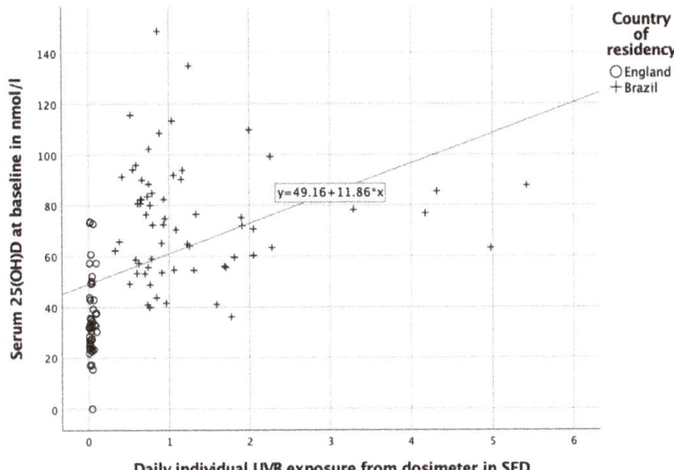

Figure 4. Relationship between baseline serum 25(OH)D concentration and baseline individual daily sunlight exposure level in participants with daily individual UVB exposure levels below 10 SED ($n = 112$).

3.5. Prediction of Circulating 25(OH)D Concentrations: Mathematical Modeling

Preliminary analyses ensured no violation of normality, linearity, generalizability (sample size), multicollinearity and homoscedasticity. Due to the differences in mean age and anthropometric measures and the significant correlations with serum 25(OH)D, a hierarchical multiple regression analysis was used to investigate the ability of daily individual UVB radiation levels (SED) to predict 25(OH)D concentrations (nmol/L), after controlling for the influence of age, weight and BMI.

Age and BMI were entered at Step 1, explaining 7.7% of the variance in 25(OH)D concentrations. After entry of daily individual sunlight exposure level at step 2, the total variance explained by the model as a whole was 46.5%. The added UVB radiation measure explained an additional 38.8% of the variance 25(OH)D concentrations, after controlling for the influence of age and BMI, ($F_{(3, 111)} = 32.16$, $p < 0.001$). In the final model, only UVB radiation made a unique statistically significant contribution to the prediction of 25(OH)D concentrations. According to the slope coefficient for daily individual UVB radiation levels, 25(OH)D concentration increased by 20.2 nmol/L for each extra SED of UVB radiation, regardless of age and BMI ($p < 0.001$).

4. Discussion

Individual daily UVB radiation levels were strongly and positively correlated with serum vitamin D concentrations. Moreover, 38.8% of the total variance in 25(OH)D concentrations was explained uniquely by daily individual UVB radiation, after controlling for the influence of age and BMI. Mean serum 25(OH)D concentration of Brazilian women living in England was significantly lower than those living in Brazil. If the threshold of 75 nmol/L recommended by the Endocrine Society [16] is applied, suboptimal status was universal amongst England residents and affected half (49.3%) of the women living in Brazil.

Although significant differences in serum vitamin D concentrations between the two countries were expected, it is no less remarkable that concentrations in ethnically identical adult women ranged from 5.0 to 73.5 nmol/L for participants living in England and from 36.2 to 148.6 nmol/L for those living in Brazil. Similar serum concentrations to those observed in Brazil have been previously reported in Africa, ranging from 70 to 170 nmol/L [12], and amongst high UVB-exposure individuals (tanners, surfers, and outdoor workers, ≤162 nmol/L) [12].

The prevalence of deficiency found in this study for women living in England (25%) was similar to the yearly average of 21.7% deficiency reported for UK Caucasian adult women [14]. Brazilian women who had been living in England for more than 2 years had lower 25(OH)D concentrations than their Brazilian peers with shorter residency, indicating a worsening of status over time. Amongst the women in this study living in Brazil, there were no records of vitamin D deficiency and just less than half of the sample had insufficient or sub optimal levels. Studies conducted previously in Brazil have observed a high prevalence of insufficiency (ranging from 28–38%) and sub-optimal status (ranging from 43–81%) [24–27]. To our knowledge, our study was the first to investigate vitamin D status in the State of Goiás, located in mid-west Brazil, and the higher prevalence reported previously (in samples from southern or northern Brazil) may reflect regional and cultural differences, i.e., variations in season patterns and climate, and cultural habits with regards to sun exposure behaviour and food intake.

It was estimated in this cohort, that each extra SED of UVB radiation (a safe limit of daily UVB radiation level enough to produce vitamin D for skin types I-IV [28]), would increase 25(OH)D concentration by 20.2 nmol/L, independent of age and BMI. Furthermore, a daily individual UVB radiation of around 2 SED would be required to maintain serum 25(OH)D at 75 nmol/L. It is surprising that such a high proportion of participants in Brazil presented relatively low individual daily exposure levels (53.6% < 1SED), considering the high minimum winter UVB index in Brazil of 8. However, this may reflect known influential factors such as behaviour towards sunlight exposure, pollution, weather variations, sun avoidance due to concerns regarding skin damage and time spent indoors (i.e., work, physical activity and commuting inside vehicles) [29].

The importance of direct skin exposure was also shown, since those women living in Brazil who exposed more body parts had a better Vitamin D status. In fact, a recent study conducted in Ireland, with 5138 community-dwelling participants aged >60 years and blood sampling throughout the year, showed that individuals who avoided sun exposure were at higher risk of deficiency (<40 nmol/L), whilst those who reported enjoying sun exposure tended to be vitamin D sufficient (≥50 nmol/L). Moreover, UVB dose and sunshine enjoyment seemed to improve prediction of vitamin D deficiency in individuals not taking supplements [30]. Such observations build on the affirmation that individual UVB radiation level is better determined by individual behaviour towards sunlight rather than estimated local UVB radiation availability and so recommendations need to help individuals to optimize their own sun exposure to balance safety and vitamin D production.

Additionally, although there were no differences in mean 25(OH)D concentrations between sunscreen users and non-users, those reporting using a SPF 15 sunscreen during holidays had significantly lower Vitamin D levels than those reporting the use of SPF of ≥40. This suggests that higher factor use may be a marker for higher overall exposure and greater likelihood of sunbathing. In fact, it was observed in this study that self-reported sunbathers were more likely to use higher factor sunscreen on holiday, supporting this behavioural association. A similar study conducted in Australia did not observed any statistically significant association between the 25(OH)D concentrations and the use of sunscreens, but also reported that participants who used sunscreen presented some of the highest 25(OH)D concentrations. The authors hypothesized that the use of sunscreens is likely to be an indicator of increased sun exposure in general [10]. Such observations reinforce the importance of considering habitual behaviour towards sunlight and how it can affect 25(OH)D cutaneous production in the skin when determining vitamin D and sunlight exposure recommendations for different populations. From a holistic point of view, taking high factor sunscreen alone as an indicator of lower UVB radiation reaching the skin, for instance, could potentially underestimate individual exposure UVB radiation level if higher SPF is also a marker for greater length of time spent in the sun.

The influence of skin pigmentation has also been well observed in studies showing poorer vitamin D status in dark-skinned compared to lighter-skinned individuals, with higher amounts of UVB required in pigmented skins to achieve adequate 25(OH)D concentrations. In the present study, no associations were found between 25(OH)D concentrations and self-declared ethno-race or skin type. The reason for this is likely due to the potential inconsistency in self-declared ethno-race and skin colour in the Brazilian population because of subjective definitions and cultural influence on ethnic identification. This was indeed observed in this study where 63% of participants identified themselves as white and 33.3% as brown while, conversely, 31.1% classified themselves as type I and II (white) and 63% as type III and IV (light and moderate brown). This inconsistency suggests that simple classifications of ethnicity and skin colour to investigate the effect of skin pigmentation on vitamin D status might not be appropriate for some populations or countries. Other more objective methods such as measures of melanin density via spectral reflectance of the skin [10] or classification by a trained researcher based on observed skin type characteristics may be better options in certain populations.

There are still very few studies that have investigated the effects of UVB exposure in vivo in South America, and specifically in Brazil, which has led to formulation of recommendations based on data derived from studies conducted mainly in the USA and Europe, where the UVB availability and the sun exposure habits are considerably different from those observed in low latitude countries. The strength of this cross-sectional analysis is the directly comparable data on serum 25(OH)D concentrations and daily individual UVB radiation measurement that represents personal and habitual solar radiation in real-life scenarios. This study addresses key knowledge gaps with two parallel studies, using identical methodologies to examine same sex and ethnicity individuals (minimizing confounding due to cultural habits and skin pigmentation), directly comparing individual UVB radiation levels and habitual behaviour towards sunlight between high and low latitudes. Further strengths of the present study include serum 25(OH)D measurement via liquid chromatography–mass spectrometry, the gold-standard method for assessing vitamin D status and data collection during the same season in both countries. Furthermore, to date, this is also the first study to measure habitual UBV radiation levels using a personal UVB dosimeter in Brazil and the first study to show the strong correlation of individual levels with vitamin D serum concentrations in the Brazilian population.

It is important to note that these findings may not be generalizable to other groups such as men, children, adolescents and pregnant or older women and other ethnic groups with different characteristics, habits or culture. Participants in this study had a generally healthy BMI and therefore, findings may not reflect populations with overweight or obesity due to the known influence of adiposity on vitamin D status. The study was conducted during wintertime in England in order to achieve a minimal sun exposure so this cohort would represent the minimal habitual UVB exposure in comparison to a high (but not extreme) exposure in Brazil. Additionally, the study was conducted in Brazil also during wintertime so that we could minimize the extreme radiation and boosted sunlight exposure habits during summer. Alternatively, the same study could have been done during summertime in both latitudes in order to compare differences in the influence of sun exposure when sunlight availability is at its highest.

5. Conclusions

The prevalence of vitamin D deficiency and insufficiency was extremely high in adult Brazilian women residing in southern England. This study has highlighted the strong positive association between vitamin D serum 25(OH)D concentrations and individual UVB radiation. Given the perhaps previously underappreciated variation in individual UVB radiation levels this study highlights the importance of measuring rather than assuming individual exposure. Vitamin D deficiency or inadequacy, strongly associated with low individual UVB radiation levels, could put these women at a greater risk of poor bone health at the end of winter, particularly in England. Further work should focus on extending the sample to include a wider demographic range, including more overweight individuals, males, and other ethnic and age groups in different latitudes.

Author Contributions: Conceptualization, M.M.M., K.H.H., S.A.L.-N. and P.B.B.; methodology M.M.M., K.H.H., S.A.L.-N. and P.B.B.; formal analysis, M.M.M.; investigation, M.M.M.; resources, M.M.M., K.H.H., S.A.L.-N. and P.B.B.; data curation, M.M.M.; writing—original draft preparation, M.M.M.; writing—review and editing, M.M.M., K.H.H., S.A.L.-N. and P.B.B; supervision, K.H.H., S.A.L.-N. and P.B.B.; project administration, M.M.M., K.H.H., S.A.L.-N. and P.B.B.; funding acquisition M.M.M., K.H.H., S.A.L.-N. and P.B.B. All authors have read and agreed to the published version of the manuscript.

Funding: This research was funded by the Brazilian National Council for Scientific and Technological Development, grant number 235046/2014.

Acknowledgments: The authors would like to acknowledge the collaboration of the Faculty of Nutrition, Federal University of Goiás and its collaborators.

Conflicts of Interest: S.A.L.-N. is Research Director of D3Tex Ltd., which holds the UK and Gulf Corporation Council (GCC) Patent for the use of UVB transparent material for vitamin D deficiency prevention. The company had no role in the design of the study, in the collection, analyses, or interpretation of data, in the writing of the manuscript or in the decision to publish the results.

References

1. Lips, P.; Hosking, D.; Lippuner, K.; Norquist, J.M.; Wehren, L.; Maalouf, G.; Ragi-Eis, S.; Chandler, J. The prevalence of vitamin D inadequacy amongst women with osteoporosis: An international epidemiological investigation. *J. Intern. Med.* **2006**, *260*, 245–254. [CrossRef] [PubMed]
2. Adams, J.S.; Hewison, M. Update in Vitamin D. *J. Clin. Endocrinol. Metab.* **2010**, *95*, 471–478. [CrossRef] [PubMed]
3. DeLuca, H.F. Overview of general physiologic features and functions of vitamin D. *Am. J. Clin. Nutr.* **2004**, *80*, 1689S–1696S. [CrossRef] [PubMed]
4. Raiten, D.J.; Picciano, M.F. Vitamin D and health in the 21st century: Bone and beyond. Executive summary. *Am. J. Clin. Nutr.* **2004**, *80*, 1673S–1677S. [CrossRef] [PubMed]
5. Reid, I.R.; Bolland, M.J.; Grey, A. Effects of vitamin D supplements on bone mineral density: A systematic review and meta-analysis. *Lancet* **2014**, *383*, 146–155. [CrossRef]
6. Webb, A.R.; Kift, R.; Durkin, M.T.; O'Brien, S.; Vail, A.; Berry, J.L.; Rhodes, L.E. The role of sunlight exposure in determining the vitamin D status of the U.K. white adult population. *Br. J. Dermatol.* **2010**, *163*, 1050–1055. [CrossRef]
7. global Solar Atlas. Available online: https://apps.who.int/iris/handle/10665/42459 (accessed on 10 December 2020).
8. Centro de Referência para Energia Solar e Eólica Sérgio de Salvo. Available online: http://cresesb.cepel.br/ (accessed on 10 December 2020).
9. Cashman, K.D.; Dowling, K.G.; Škrabáková, Z.; Gonzalez-Gross, M.; Valtueña, J.; De Henauw, S.; Moreno, L.; Damsgaard, C.T.; Michaelsen, K.F.; Mølgaard, C.; et al. Vitamin D deficiency in Europe: Pandemic? *Am. J. Clin. Nutr.* **2016**, *103*, 1033–1044. [CrossRef]
10. Kimlin, M.G.; Harrison, S.L.; Nowak, M.; Moore, M.; Brodie, A.; Lang, C. Does a high UV environment ensure adequate Vitamin D status? *J. Photochem. Photobiol. B Biol.* **2007**, *89*, 139–147. [CrossRef]
11. Unger, M.D.; Cuppari, L.; Titan, S.M.; Magalhães, M.C.T.; Sassaki, A.L.; dos Reis, L.M. Vitamin D status in a sunny country: Where has the sun gone? *Clin. Nutr.* **2010**, *29*, 784–788. [CrossRef]

12. Bouillon, R.; Van Schoor, N.M.; Gielen, E.; Boonen, S.; Mathieu, C.; Vanderschueren, D.; Lips, P. Optimal Vitamin D Status: A Critical Analysis on the Basis of Evidence-Based Medicine. *J. Clin. Endocrinol. Metab.* **2013**, *98*, E1283–E1304. [CrossRef]

13. Prentice, A. Vitamin D deficiency: A global perspective. *Nutr. Rev.* **2008**, *66*, S153–S164. [CrossRef] [PubMed]

14. Scientific Advisory Committee on Nutrition. Vitamin D and Health. Available online: https://assets.publishing.service.gov.uk/government/uploads/system/uploads/attachment_data/file/537616/SACN_Vitamin_D_and_Health_report.pdf (accessed on 10 December 2020).

15. Institute of Medicine. *Dietary Reference Intakes for Calcium and Vitamin D*; Institute of Medicine: Washington, DC, USA, 2011.

16. Holick, M.F.; Binkley, N.; Bischoff-Ferrari, H.A.; Gordon, C.M.; Hanley, D.A.; Heaney, R.P.; Murad, M.H.; Weaver, C.M. Evaluation, Treatment, and Prevention of Vitamin D Deficiency: An Endocrine Society Clinical Practice Guideline. *J. Clin. Endocrinol. Metab.* **2011**, *96*, 1911–1930. [CrossRef] [PubMed]

17. Engelsen, O. The Relationship between Ultraviolet Radiation Exposure and Vitamin D Status. *Nutrients* **2010**, *2*, 482–495. [CrossRef] [PubMed]

18. Shahriari, M.; Kerr, P.E.; Slade, K.; Grant-Kels, J.E. Vitamin D and the skin. *Clin. Dermatol.* **2010**, *28*, 663–668. [CrossRef]

19. World Health Organization. *Body Mass Index Classification. Global Database BMI*; World Health Organization: Geneva, Switzerland, 2004.

20. Instituto Brasileiro de Geografia e Estatística. Censo Demográfico 2010. In *Características da População e dos Domicílios: Resultados do Universo*; IBGE: Rio de Janeiro, Brazil, 2011.

21. Fitzpatrick, T.B. The validity and practicality of sun-reactive skin types I through VI. *Arch Dermatol.* **1988**, *124*, 869–871. [CrossRef]

22. Mendes, M.M.; Hart, K.; Lanham-New, S.; Botelho, P.B. Suppression of Parathyroid Hormone as a Proxy for Optimal Vitamin D Status: Further Analysis of Two Parallel Studies in Opposite Latitudes. *Nutrients* **2020**, *12*, 942. [CrossRef]

23. Ross, A.C.; Taylor, C.L.; Yaktine, A.L.; Del Valle, H.B. (Eds.) *Institute of Medicine Committee to Review Dietary Reference Intakes for Vitamin D Calcium*; The National Academies Collection: Reports Funded by National Institutes of Health), Dietary Reference Intakes for Calcium and Vitamin D; National Academies Press (US) National Academy of Sciences: Washington, DC, USA, 2011.

24. Lopes, V.M.; Lopes, J.R.C.; Brasileiro, J.P.B.; De Oliveira, I.; Lacerda, R.P.; Andrade, M.R.D.; Tierno, N.I.Z.; De Souza, R.C.C.; Da Motta, L.A.C.R. Highly prevalence of vitamin D deficiency among Brazilian women of reproductive age. *Arch. Endocrinol. Metab.* **2017**, *61*, 21–27. [CrossRef]

25. Gouveia, N.; Corrallo, F.P.; de Leon, A.C.P.; Junger, W.; de Freitas, C.U. Air Pollution and Hospitalizations in the Largest Brazilian Metropolis. Available online: https://www.scielo.br/scielo.php?script=sci_arttext&pid=S0034-89102017000100299 (accessed on 10 December 2020).

26. Bandeira, F.; Griz, L.; Freese, E.; Lima, D.C.; Thé, A.C.; Diniz, E.T.; Marques, T.F.; Lucena, C.S. Vitamin D deficiency and its relationship with bone mineral density among postmenopausal women living in the tropics. *Arq. Bras. Endocrinol. Metabol.* **2010**, *54*, 227–232. [CrossRef]

27. Eloi, M.; Horvath, D.V.; Szejnfeld, V.L.; Ortega, J.C.; Rocha, D.A.C.; Szejnfeld, J.; Castro, C.H.D.M. Vitamin D deficiency and seasonal variation over the years in São Paulo, Brazil. *Osteoporos. Int.* **2016**, *27*, 3449–3456. [CrossRef]

28. World Health Organization. Global Solar UV Index: A Practical Guide. Available online: https://globalsolaratlas.info/download/world (accessed on 10 December 2020).

29. Mendes, M.M.; Darling, A.L.; Hart, K.; Morse, S.; Murphy, R.J.; Lanham-New, S.A. Impact of high latitude, urban living and ethnicity on 25-hydroxyvitamin D status: A need for multidisciplinary action? *J. Steroid Biochem. Mol. Biol.* **2019**, *188*, 95–102. [CrossRef]

30. O'Sullivan, F.; Laird, E.; Kelly, D.; Van Geffen, J.; Van Weele, M.; McNulty, H.; Hoey, L.; Healy, M.; McCarroll, K.; Cunningham, C.; et al. Ambient UVB Dose and Sun Enjoyment Are Important Predictors of Vitamin D Status in an Older Population. *J. Nutr.* **2017**, *147*, 858–868. [CrossRef] [PubMed]

Nutrients **2020**, *12*, 3805

Review

The Association between Vitamin D Status and Autism Spectrum Disorder (ASD): A Systematic Review and Meta-Analysis

Zuqun Wang [1,2], **Rui Ding** [1,2] **and Juan Wang** [1,2,*]

[1] Department of Biomedical Informatics, School of Basic Medical Sciences, Peking University, Beijing 100191, China; wzq_quince@bjmu.edu.cn (Z.W.); dingrui@hsc.pku.edu.cn (R.D.)

[2] Autism Research Center, Peking University Health Science Center, Beijing 100191, China

* Correspondence: wjuan@hsc.pku.edu.cn; Tel.: +86-1380-119-7157

Citation: Wang, Z.; Ding, R.; Wang, J. The Association between Vitamin D Status and Autism Spectrum Disorder (ASD): A Systematic Review and Meta-Analysis. *Nutrients* **2021**, *13*, 86. https://doi.org/10.3390/nu13010086

Received: 19 November 2020
Accepted: 25 December 2020
Published: 29 December 2020

Abstract: The association between vitamin D status and autism spectrum disorder (ASD) is well-investigated but remains to be elucidated. We quantitatively combined relevant studies to estimate whether vitamin D status was related to ASD in this work. PubMed, EMBASE, Web of Science, and the Cochrane Library were searched to include eligible studies. A random-effects model was applied to pool overall estimates of vitamin D concentration or odds ratio (OR) for ASD. In total, 34 publications involving 20,580 participants were identified in this present study. Meta-analysis of 24 case–control studies demonstrated that children and adolescents with ASD had significantly lower vitamin D concentration than that of the control group (mean difference (MD): -7.46 ng/mL, 95% confidence interval (CI): -10.26; -4.66 ng/mL, $p < 0.0001$, $I^2 = 98\%$). Quantitative integration of 10 case–control studies reporting OR revealed that lower vitamin D was associated with higher risk of ASD (OR: 5.23, 95% CI: 3.13; 8.73, $p < 0.0001$, $I^2 = 78.2\%$). Analysis of 15 case–control studies barring data from previous meta-analysis reached a similar result with that of the meta-analysis of 24 case–control studies (MD: -6.2, 95% CI: -9.62; -2.78, $p = 0.0004$, $I^2 = 96.8\%$), which confirmed the association. Furthermore, meta-analysis of maternal and neonatal vitamin D showed a trend of decreased early-life vitamin D concentration in the ASD group (MD: -3.15, 95% CI: -6.57; 0.26, $p = 0.07$, $I^2 = 99\%$). Meta-analysis of prospective studies suggested that children with reduced maternal or neonatal vitamin D had 54% higher likelihood of developing ASD (OR: 1.54, 95% CI: 1.12; 2.10, $p = 0.0071$, $I^2 = 81.2\%$). These analyses indicated that vitamin D status was related to the risk of ASD. The detection and appropriate intervention of vitamin D deficiency in ASD patients and pregnant and lactating women have clinical and public significance.

Keywords: autism spectrum disorder; vitamin D; maternal and neonatal vitamin D; systematic review; meta-analysis

1. Introduction

Autism spectrum disorder (ASD) is a group of neurodevelopmental disorders characterized by impaired social interaction and communication, repetitive and stereotyped behaviors, and restricted interests [1]. Autism is becoming increasingly common. Recently, the Centers for Disease Control and Prevention reported that the prevalence of autism among 8-year-old children in the United States in 2016 was 1/54, with a 4.3:1 ratio of males to females [2]. Autistic individuals manifest problematic behaviors, such as attacking, self-injury, resistance to orders, and failure of normal conversation, and are usually comorbid with social-anxiety, attention-deficit/hyperactivity, sleep-wake disorders, and obsessive-compulsive disorders. As such, it is difficult for them to obtain the same education levels as their neurotypical peers, find full-time jobs, or live independently [1,3]. Fortunately, evidence indicates that appropriate and early intervention could help autistic individuals improve their symptoms and life quality [4]. Notably, research shows that nutritional and dietary intervention is an effective way to improve nutritional status, non-verbal IQ, and autism symptoms [5]. Therefore, it is essential to identify physiological dysfunction and abnormal nutritional status in autistic individuals and then take corresponding interventions.

Autism is a multifactorial disorder resulting from an interaction of genetic and environmental factors. Hundreds of autism risk genes and various environmental factors have been discovered [6,7]. Possible environmental factors include folic acid deficiency, neonatal hypoxia, maternal obesity, and gestational diabetes mellitus [7]. Recently, emerging evidence suggests that vitamin D deficiency might be an unfavorable factor of autism [8]. Vitamin D is a steroid hormone; it is primarily synthesized in the skin under UV-B light, and a small amount is derived from dietary intake [9]. Although the underlying mechanisms between vitamin D and autism are unclear, there is some biological evidence indicating the potential link. Vitamin D-metabolizing enzymes and vitamin D receptors are widely expressed in immune cells, the placenta, and the developing and adult brain; high levels of vitamin D surface receptor (protein disulfide isomerase family A members 3, PDIA3) are found in the cortex and hippocampus, which suggests the association between vitamin D, and brain development and function [10–12]. Indeed, vitamin D has important effects on brain development and function, including neuronal differentiation, proliferation and apoptosis, regulating synaptic plasticity, the ontogeny of the dopaminergic system, immunomodulation, and reducing oxidative burden [10]. In addition, vitamin D plays an important role in the regulation of gene expression. One study showed that 223 ASD risk genes in the SFARI database were vitamin D3-sensitive genes, which meant that these ASD related genes might be regulated by vitamin D [13].

Besides the plausible biological explanations, some epidemiological studies also reached related conclusions. A large number of case–control studies investigating the vitamin D status of children and adolescents with ASD from different countries and races showed that autistic children and adolescents had lower vitamin D status [14–31], but seven studies reached the opposite conclusions [32–37]. Moreover, several prospective studies investigated the role of maternal and neonatal vitamin D deficiency in autism onset. A nested case–control study from a Swedish population cohort suggested that neonatal vitamin D was slightly associated with a later risk of ASD, but maternal vitamin D was not related to ASD [38]. In contrast, a cohort study based on a Netherlands birth cohort revealed that neonatal vitamin D was not associated with ASD, but pregnant women with deficient vitamin D concentration at mid-gestation had a more than twofold chance to give birth to autistic infants [39]. Another nested case–control study from China supported the point that reduced neonatal vitamin D was associated with a higher risk of ASD [40], but studies from Canada and the United States did not support this point [41–43].

A meta-analysis on 11 case–control studies found that, compared with healthy children, children with ASD had 8.63 ng/mL lower 25(OH)D concentration overall [44]. However, as mentioned above, case–control studies measuring vitamin D levels of children and adolescents with ASD have constantly been emerging in recent years, with some contradictions in the results. In addition, several prospective studies have been performed on early-life vitamin D levels before ASD diagnosis, which provide evidence on whether autism onset is associated with reduced early-life vitamin D, also with inconsistent results. Individual vitamin D level is affected by many factors, including sunlight, diet, ethnicity, genetic polymorphism, and physiological conditions. Researchers could not completely control these factors between ASD and control groups, and the etiology and clinical manifestations of ASD are of high heterogeneity. Meta-analysis can merge the results of multiple studies and increase the sample size to reach a consistent conclusion and increase the credibility of that conclusion. Therefore, we conducted an updated meta-analysis of the case–control studies and meta-analysis of prospective studies to investigate links between vitamin D and ASD and to explore the potential source of heterogeneity between studies.

2. Materials and Methods

This systematic review and meta-analysis were conducted and reported in accordance with the Preferred Reporting Items for Systematic Reviews and Meta-Analyses (PRISMA) guidelines [45] and Meta-Analyses of Observational Studies in Epidemiology (MOOSE) guidelines [46]. The protocol was registered in the International Prospective Register of Systematic Reviews (PROSPERO) with a registration number of CRD42020161819 at www.crd.york.ac.uk/PROSPERO.

2.1. Search Strategy and Study Selection

We performed a systematic literature search in PubMed, EMBASE, Web of Science, and the Cochrane Library from database inception to 27 November 2019 to identify studies on vitamin D and the risk of autism. The search terms we used were MeSH phrases combined with text words relating to autism ("autism" OR "autistic" OR "ASD") and vitamin D ("vitamin D" OR "1,25 dihydroxyvitamin d3" OR "d3,1,25 dihydroxyvitamin" OR "25 hydroxyvitamin d2" OR "25 hydroxyvitamin d3" OR "25(OH)D OR 1 alpha, 25 dihydroxy 20 epi vitamin d" OR "1,25 dihydroxy 20 epi vitamin d3"). No restrictions were applied for the languages, date, and location of the studies. Besides database searching, we manually checked the reference lists of the identified studies and relevant reviews. Two reviewers (WZQ and DR) independently checked the titles and abstracts of each paper to filter irrelevant papers and then read the full texts of the remaining studies to identify studies that met the eligibility criteria. Any disagreement was resolved by discussion.

2.2. Eligibility Criteria

Eligibility criteria were set according to PICOS approach: the participants (P), the interventions or exposure (I), the comparison (C), the outcome (O), the study design (S), as follows. Each letter in PICOS means a component: Participants: children or adolescents aged less than 18, pregnant women, and neonates. Intervention/exposure: insufficient or deficient vitamin D level in peripheral blood. Comparison: sufficient vitamin D level in peripheral blood. Outcome: autism spectrum disorder. Study design: case–control, cohort, and nested case–control studies. Duplicated studies with the same data were excluded. No limits were applied for the form of vitamin D. Studies had to report the mean and standard deviation of vitamin D concentration or odds ratio (OR)/relative risk (RR) for ASD incidence. Studies were excluded if participants were reported to be comorbid with any other disease that could affect vitamin D levels, such as epilepsy and ADHD.

2.3. Data Extraction

Data were independently extracted from eligible studies by two reviewers (WZQ and DR), including the first author's name, year, country, study design, sample size, participants' age, gender ratio, diagnostic criteria, sample for detecting, vitamin D measurement method, mean ± SD of vitamin D concentrations, *p*-value compared to controls, adjusted variable or confounding variable, and OR/RR (95% confidence interval (CI)) for ASD incidence.

2.4. Study Quality Assessment

For eligible studies, we used the Newcastle–Ottawa Scale (NOS) [47] to assess whether they had the general characteristics of an observational study. This scale comprised three aspects: study-participant selection, 0–4; the comparability of study participants, 0–3; the exposure or outcome of studies, 0–3, ranging from 0 to 9, among which 0–6 was regarded as low-quality, and 7–9 was high-quality. Two reviewers (WZQ and DR) independently evaluated the eligible studies. Any discrepancies were resolved by discussion.

2.5. Statistical Analysis

For the continuous variable, mean ± SD of vitamin D concentration was obtained to calculate the overall effect size and 95% confidence interval (95% CI). Mean difference (MD) was used to describe the difference of mean concentration between ASD and control groups in each study. If the concentration of vitamin D was present in nmol/L, we converted it into ng/mL, according to the formula, 1 ng/mL = 2.5 nmol/L. Meta-analysis of all eligible case–control studies was performed. In order to check the robustness of the results, we conducted an additional meta-analysis, barring data from previous meta-analyses [44]. In addition, prospective studies that reported mean $\pm$ SD were combined to examine if there was any difference in maternal and neonatal vitamin D between the ASD and control groups.

For the categorical variable, we obtained OR/RR from eligible studies to calculate the pooled OR/RR and 95% CI. Some case–control studies did not report vitamin D concentration but provided OR and 95% CI for the possibility of ASD exposed to vitamin D insufficiency or deficiency. Therefore, a meta-analysis on case–control studies with OR was performed. Prospective studies included nested case–control and cohort studies, providing either OR or RR. The prevalence of ASD was less than 10%, so we assumed that OR was approximatively equal to the RR and conducted meta-analysis from prospective studies.

Considering the anticipated large heterogeneity, we used the DerSimonian–Laird random-effects model for all meta-analyses. If there was no or low heterogeneity, the fixed-effects model was used. Cochran's Q test and I^2 statistic were used to measure heterogeneity. I^2 referred to the percentage of heterogeneity, and $I^2 \geq 50\%$ indicated greater heterogeneity. Subgroup and meta-regression analyses were created to explore sources of heterogeneity. Subgroup analysis was based on the study population, measurement method, latitude, location, sample size, age, number of adjusted variables, and study quality. To eliminate the influence of individual studies, especially small sample and low-quality studies, leave-one-out sensitivity analysis was conducted. The funnel plot was used to evaluate publication bias, and Egger's linear regression was conducted to check the symmetry of the funnel plot. A forest plot was created to visualize the overall effect size and 95% CI of the studies. All statistical analyses were performed in R (version 3.6.2), including the meta (version 4.11-0) and metaphor (version 2.1-0) packages. Two-tailed $p < 0.05$ was statistically significant.

3. Results

Figure 1 shows the process of identifying the eligible publications. A total of 1088 articles were retrieved through the databases: 221 records in PubMed, 439 records in EMBASE, 380 records in Web of Science, and 48 records in the Cochrane Library. In total, 16 articles were manually searched in the reference lists of relevant publications. After removing duplicates, 598 articles remained to be screened by title and abstract. Of these, 67 articles were left for full-text reading. Lastly, 34 eligible articles (a total of 20,580 participants) were included in the meta-analysis. Of these, 26 case–control studies [14–32,34–37,48–50] (1792 ASDs, 1969 controls) reported the blood vitamin D concentration of children and adolescents; three case–control studies [42,43,51], and two nested case–control studies [38,40] (2687 ASDs, 3574 controls) examined the neonatal vitamin D concentration of participants; one case–control study [52] and one nested case–control study [38] (517 ASDs, 642 controls) assessed maternal vitamin D concentration of the ASD and control groups; two cohort studies [39,41] (5442 neonates, 3957 pregnant women) investigated the OR/RR for ASD incidence after being exposed to early-life vitamin D deficiency or insufficiency. The participants of two articles included not only neonates but also pregnant women, so there were 36 total studies from 34 articles.

The detailed study characteristics of each eligible study are demonstrated in Tables S1–S3. In general, these studies were published from 2010 to 2019 and involved participants from Asia ($n = 18$), America ($n = 6$), Europe ($n = 5$), and Africa ($n = 5$). The ASD diagnostic criteria used in studies were DSM-IV, DSM-IV-TR, ADOS, ADIR, DSM-V, ICD-9, ICD-10, ICD-F84.0, or a combination of the above. All eligible studies measured a total of 25(OH)D2 and 25(OH)D3 or 25(OH)D3 from serum, plasma, or dried blood spot as the biomarker of vitamin D; 25(OH)D3 was considered approximately equal to the total of 25(OH)D2 and 25(OH)D3, so the form of vitamin D in each study was not distinguished. The quality scores of the included studies are shown in Table S4, ranging from 4 to 9, of which, 9 studies were evaluated as low-quality, 25 studies were high-quality, and the median NOS score of all studies was 7.

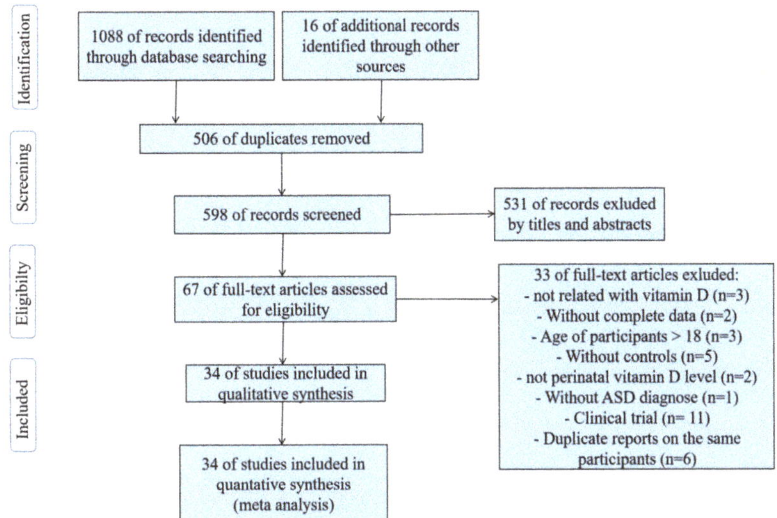

Figure 1. Flow chart of identification of eligible studies.

3.1. Meta-Analysis of Case–Control Studies Involving Children and Adolescents

A total of 24 case–control studies were included in the meta-analysis, providing mean $\pm$ SD vitamin D concentration in children and adolescents with and without ASD, of which two samples were plasma and 22 were serum. Meta-analysis showed that vitamin D concentration of the ASD group was 7.46 ng/mL lower than that of the control group (95% CI: -10.26; -4.66 ng/mL, $p < 0.0001$; Figure 2, Table S5) using a random effects model, with a large heterogeneity ($I^2 = 98\%$, $p < 0.01$). In subgroup analysis, vitamin D measured by ELISA (MD: -10.19 ng/mL, 95% CI: -17.53; -2.86 ng/mL, $p = 0.006$) and radioimmunoassay (MD: -4.33, 95% CI: -6.81; -1.85, $p = 0.0006$) in the ASD group was significantly reduced compared to that of the control group, with slightly decreased heterogeneity between studies, while statistical significance disappeared in studies measured by HPLC (MD: -9.13, 95% CI: -19.33; 1.06; 1.06, $p = 0.079$) and LC–MS/MS (MD: -4.32, 95% CI: -15.20; 6.56, $p = 0.436$). With regard to latitude, subgroup analysis with a latitude below 30 (MD: -13.3, 95% CI: -20.83; -5.76, $p = 0.0005$) and between 30 and 40 (MD: -3.81, 95% CI: -5.83; -1.79, $p = 0.0002$) illustrated that vitamin D concentration was significantly lower in the ASD group than that in the control group. However, when latitude was beyond 40, there was no significant difference between the two groups. Studies performed in different areas showed quite different results. Subjects with ASD in Africa had largely reduced vitamin D concentration compared with that of the control group (MD: -15.56, 95% CI: -24.77; -6.35, $p = 0.0009$). In Asia, the difference in vitamin D concentration was also significant between the ASD and control groups (MD: -6.2, 95% CI: -9.15; -3.25, $p < 0.0001$). However, in Europe, subgroup analysis demonstrated that the ASD group had higher but nonsignificant vitamin D concentration than that of the control group (MD: 3.03, 95% CI: -6.78; 12.83, $p = 0.545$). In America, vitamin D levels did not differ between subjects with and without ASD (MD: -6.01, 95% CI: -13.42; 1.39, $p = 0.11$). More details about subgroup analysis are shown in Table S5. Univariate meta-regression analysis indicated that latitude ($p = 0.0107$) was associated with a mean difference of vitamin D concentration between the two groups, accounting for 8.08% of heterogeneity.

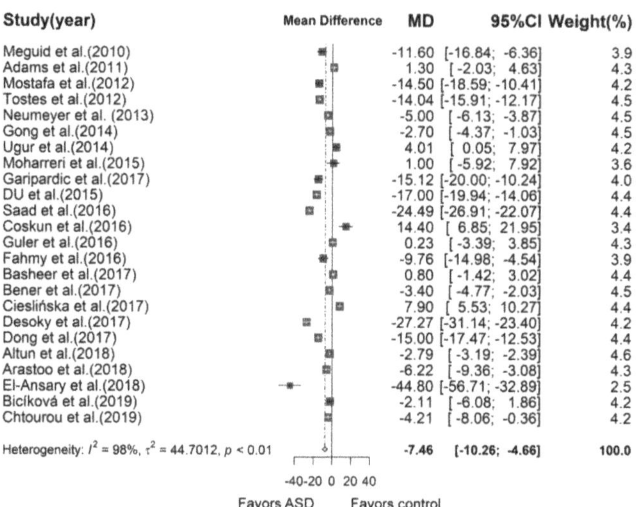

Figure 2. Forest plot of meta-analysis of 24 case control studies based on vitamin D concentration, showing that children and adolescents with autism spectrum disorder (ASD) have an average of 7.46 ng/mL lower vitamin D concentration than that of the controls, with 98% heterogeneity. MD, mean difference; CI, confidence interval.

Since 10 case–control studies reported OR, meta-analysis based on OR was conducted. Results indicated that reduced vitamin D status was significantly associated with increased risk of ASD (OR: 5.23, 95% CI: 3.13; 8.73, $p < 0.0001$, Figure 3, Table S6). However, there was high heterogeneity between studies ($I^2 = 78.2\%$, $p < 0.0001$). The criteria for vitamin D deficiency or insufficiency were inconsistent. Included studies regarded 20 or 30 ng/mL as the cutoff of vitamin D insufficiency or deficiency. In the subgroup analysis, the association was significant when cutoff was 30 ng/mL (OR: 6.13, 95% CI: 3.39; 11.09, $p < 0.0001$, $I^2 = 83.7\%$), but association was nonsignificant when cutoff was 20 ng/mL (OR: 2.83, 95% CI: 0.91; 8.72, $p = 0.07$, $I^2 = 40.2\%$). On the basis of latitude, assessment methods, age, study quality, and number of adjusted variables, all subgroups demonstrated a significant association between reduced vitamin D status and increased risk of ASD (Table S6).

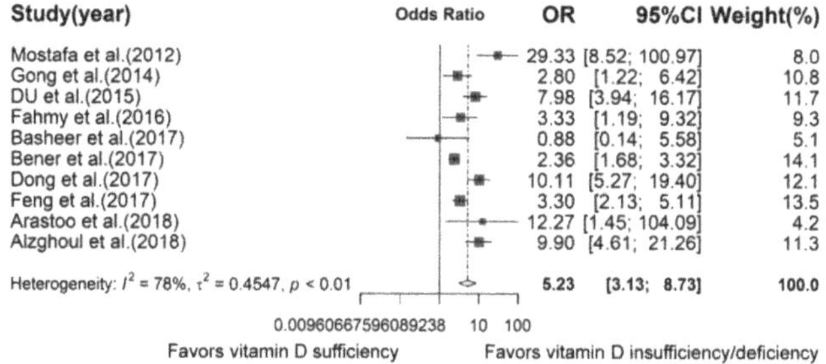

Figure 3. Forest plot of meta-analysis of 10 case–control studies based on odds ratio (OR), demonstrating that children with vitamin D insufficiency and deficiency are about 5.23 times more likely to develop ASD than vitamin D sufficient children are, with 78% heterogeneity. CI, confidence interval.

3.2. Meta-Analysis of Case–Control Studies Barring Data from Previous Meta-Analysis

Previous meta-analysis integrated case–control studies measured vitamin D concentration in ASD and control groups before May 2015. In order to evaluate the robustness of the association between ASD and vitamin D status, we carried out meta-analysis of 15 studies excluding studies before May 2015. Results showed that children and adolescents with

ASD had 6.2 ng/mL lower vitamin D concentration than that of the control group (95% CI: -9.62; -2.78, $p = 0.0004$, $I^2 = 96.8\%$; Figure 4, Table S7), which was similar to results of the previous meta-analysis [44]. Significant difference in vitamin D status between the two groups was observed in several subgroups: latitude between 30 and 40, mean age of participants >5, high study quality, adjusted variable = 2 (Table S7). Univariate meta-regression analysis suggested that age ($p = 0.0486$) had a slightly significant effect on the mean difference of vitamin D concentration between the two groups.

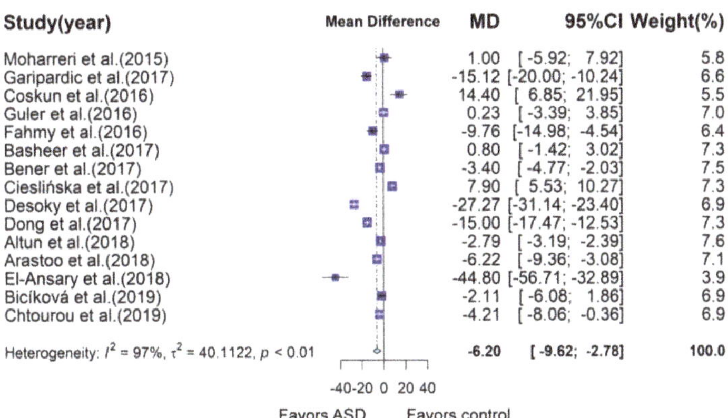

Figure 4. Forest plot of meta-analysis of 15 case–control studies based on vitamin D concentration, showing that children and adolescents with ASD have an average of 6.20 ng/mL lower vitamin D concentration than that of the control in studies conducted after May 2015, with 97% heterogeneity. MD, mean difference; CI, confidence interval.

3.3. Meta-Analysis of Prospective Studies about Neonates and Pregnant Women

Meta-analysis of maternal and neonatal vitamin D concentration indicated that there was a trend of lower vitamin D concentration in subjects with ASD (MD: -3.15, 95% CI: -6.57; 0.26, $p = 0.07$, $I^2 = 99\%$; Figure 5, Table S8). Subgroup analysis showed that maternal and neonatal vitamin D concentration in the ASD group was 3.04 and 3.28 ng/mL lower than that of the control group (95% CI: -6.86; 0.77, $p = 0.102$, $I^2 = 93.3\%$ and 95% CI: -8.47; 1.91, $p = 0.228$, $I^2 = 99.3\%$), respectively, but no statistical significance was observed.

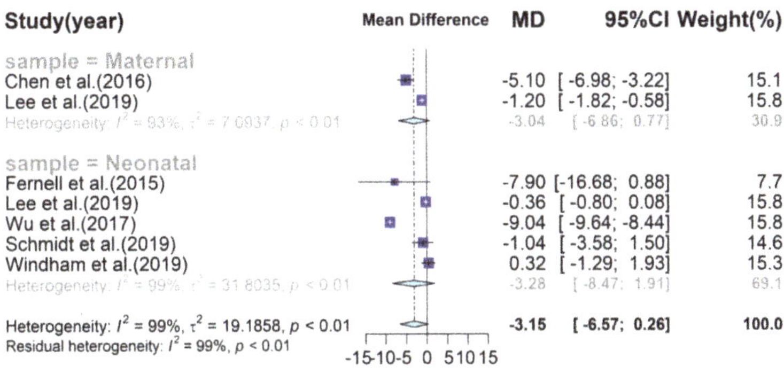

Figure 5. Forest plot of meta-analysis of seven studies based on vitamin D concentration. Overall, children with ASD tend to have 3.15 ng/mL lower neonatal or maternal vitamin D concentration than that of children without ASD, with 97% heterogeneity. MD, mean difference; CI, confidence interval.

Nested case–control and cohort studies were also summarized to a pooled OR of 1.54 (95% CI: 1.12; 2.10; Figure 6, Table S9), suggesting that the lower level of maternal and neonatal vitamin D caused a 54% higher risk of later ASD onset. Subgroup analysis demonstrated that decreased maternal vitamin D concentration contributed to the development of ASD (OR: 2.72, 95% CI: 1.61; 4.59, p = 0.0002, I^2 = 44.7%) but neonatal vitamin D concentration was not found to be significantly related to the risk of ASD (OR: 1.2, 95% CI: 0.94; 1.54, p = 0.15, I^2 = 65.3%).

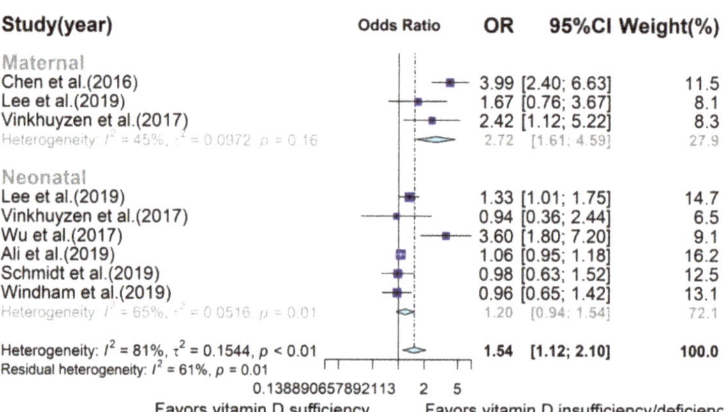

Figure 6. Forest plot of meta-analysis of nine prospective studies based on OR. Overall, children with lower maternal or neonatal vitamin D levels have a 54% higher chance to develop ASD, with 81% heterogeneity. OR, odds ratio; CI, confidence interval.

3.4. Sensitivity Analysis and Publication Bias

Leave-one-out sensitivity analysis was applied to check if there was any individual study affecting the overall results. For three separate meta-analyses of case–control studies, we did not find any outlier that significantly influenced the results. Intriguingly, with respect to meta-analysis on maternal and neonatal vitamin D concentrations, sensitivity analysis suggested that the elimination of Wu et al. (2017) [40] (MD: −1.43, 95% CI: −2.63; −0.24, p = 0.0189, I^2 = 83.3%) or Windham et al. (2019) [43] (MD: −3.79, 95% CI: −7.58; −0.002, p = 0.0499, I^2 = 99.1%), respectively, led to a significant difference between the ASD and control groups. Sensitivity analysis of the meta-analysis of prospective studies did not significantly alter the summarized results. The funnel plots shown in Figures S1–S5 indicate that all meta-analyses have no publication bias, and p > 0.05 in both the Egger's test and Begg's test.

4. Discussion

The present meta-analysis confirmed that children and adolescents with ASD have significantly lower vitamin D concentration than that of healthy children and adolescents, which was consistent with previous meta-analysis [44]. Both meta-analyses of the ORs in 10 case–control studies and of vitamin D concentrations in 15 case–control studies conducted after May 2015 yielded the same findings, which increased the credibility of the results. Furthermore, overall estimates of vitamin D concentrations in prospective studies indicated that early-life vitamin D levels of both maternal and neonatal vitamin D tended to be lower in subjects later diagnosed with ASD. Meta-analysis of ORs in prospective studies showed that decreased early-life vitamin D led to a 54% higher risk of later diagnosed ASD. In subgroup analysis, maternal vitamin D was shown to be associated with ASD, but neonatal vitamin D was not.

There are several probable reasons for the phenomenon that children and adolescents with ASD have lower vitamin D concentration than that of healthy controls. First, the lifestyle habits of ASD children are different from healthy children. Compared with healthy children, ASD children are pickier eaters, eating limited kinds of foods and consuming less vitamin D [53]. Moreover, one study showed that ASD children spent less time on outdoor activities than healthy controls did in the second year of life, which suggested that autistic children were less exposed to solar UV-B, indicating that they received less vitamin D from cutaneous synthesis [54]. These factors may be partly responsible for the lower vitamin D status in ASD children. Second, vitamin D levels may be related to genetic factors. Vitamin D metabolic and vitamin D receptor gene variants that were shown to be associated with ASD risk might influence vitamin D status [55–57]. The use of drugs like antiepileptic drugs might also cause vitamin D loss [58].

In the subgroup analysis, we found that case–control studies from different areas yielded different estimates. Studies from African countries were estimated to show maximal vitamin D concentration difference between ASD and control groups, followed by Asian countries; estimates from American and European countries were not significant. This may be due to different health care services and awareness of autism management. Furthermore, latitude was associated with mean difference according to the meta-regression, although latitude can only account for 8% heterogeneity. Subgroup analysis stratified by latitude showed that participants with and without ASD from lower-latitude areas presented larger vitamin D concentration differences. Studies from the latitude and area subgroups were little overlapped. Data from these meta-analyses showed the mean vitamin D level of children with and without ASD from low-latitude areas was higher than that from high and medium latitude areas, respectively, which suggested that latitude was an important factor influencing vitamin D level. However, the difference in vitamin D concentration between children with and without ASD in low-latitude areas became larger; that is to say, increasing light exposure in autistic children did not cause an equal increase in vitamin D as in the healthy controls. Therefore, we hypothesized that children with ASD might show a weak ability of cutaneous synthesis of vitamin D, which remains to be investigated.

HPLC and LC–MS/MS are gold standards for measuring vitamin D concentration, but the vitamin D level between ASD and control groups was not significantly different in the subgroup analysis of these two methods. Considering the special status of HPLC and LC–MS/MS, we combined these two subgroups for additional meta-analysis and found that vitamin D concentration in the ASD group was lower than that in the control group (MD: −6.72, 95% CI: −13.99; 0.55, $p = 0.07$, $I^2 = 97.8\%$), but it was still not statistically significant. These two subgroups consisted of six studies, of which the results of three studies were significant ($p < 0.05$) and the other three were not ($p > 0.05$). The average vitamin D concentration of the ASD groups in the six studies was in the range of 10–30 ng/mL (Table S1), which meant that children with ASD were with low vitamin D status. With regard to the control groups, average vitamin D concentration in three studies with significant results was above 30 ng/mL (Table S1), but in the three studies with nonsignificant results, it was in the range of 10–30 ng/mL (Table S1), which was comparable to the ASD group. Vitamin D deficiency or insufficiency are global problems. Thus, we speculated that the reason why there was no significant difference in vitamin D concentration between the ASD and control groups was because the control children were also in low vitamin D states.

Vitamin D status after ASD diagnosis is affected by lifestyle characteristics, such as diet and outdoor activities. Fortunately, early-life vitamin D status is not affected by children's lifestyle factors. Vitamin D can be transferred to the fetus through the placenta to support fetal development, so a fetus' vitamin D status depends on maternal vitamin D concentration. However, nearly one-half of pregnant women lack vitamin D [59]. The neonatal period is also a sensitive period of neurodevelopment, and vitamin D plays a pivotal role in this period. Vitamin D deficiency during neurodevelopmental periods could result in brain-structure alterations and behavioral problems [60,61]. Thus, we combined maternal and neonatal vitamin D studies to determine whether their insufficiency or deficiency was associated with ASD diagnosis. Our results showed that early-life vitamin

D deficiency led to a slight increase in ASD risk. However, there were some points about cutoffs to distinguishing vitamin D status. First, cutoffs in the studies were different; some chose 20 ng/mL as the criterion [62], while others chose the lowest quantile as vitamin D deficiency. Second, neonatal vitamin D was lower than maternal vitamin D, but researchers used the same standard to define maternal and neonatal vitamin D deficiency. Different cutoffs might exert an influence on the pooled estimates. On the other hand, vitamin D was measured at a single time and no other time-point in all studies, which could not represent the average level of vitamin D concentration during developmental stages. At this point, we can only speculate whether early-life vitamin D is linked to ASD risk. Future large-scale birth cohort studies and well-designed randomized controlled-trial studies about vitamin D supplementation effects of ASD are needed to confirm this association.

Several underlying mechanisms may explain the association between vitamin D deficiency and ASD. A large number of studies showed that vitamin D significantly contributes to neurodevelopment, playing important roles in neurogenesis, cell proliferation, differentiation, apoptosis, and neurotransmitter metabolism [63]. Vitamin D also has anti-inflammatory and antioxidant properties; for instance, vitamin D supplementation decreased serum interleukin 10 and 12 concentration and increased total antioxidant capacity [64]. In addition, one study showed that vitamin D deficiency induced increased reactive oxygen species (ROS) in the periphery and brain and caused excitatory and inhibitory neurotransmitter imbalance in animal models [65]. Autism is regarded as a condition that affects the brain, and increasing evidence has indicated that oxidative stress and inflammation are involved in the pathogenesis of autism, which may be related to vitamin D deficiency.

4.1. Implication

Our findings indicated that vitamin D level in children and adolescents with autism is significantly lower than that in healthy controls, which has clinical implications. Considering the importance of vitamin D and the high prevalence of vitamin D deficiency, regular screening of vitamin D levels in autistic individuals and necessary intervention are recommended. Furthermore, pregnant and lactating women consume more vitamin D than usual and are generally deficient in vitamin D [66]; maternal and neonatal vitamin D status may be associated with subsequent diagnosis of ASD. Vitamin D status should be included in routine screening during pregnancy and lactation in order to provide appropriate clinical intervention.

4.2. Limitations

Several limitations in this study should be considered. First, the causal relationship between vitamin D and autism could not be confirmed. Case–control studies of vitamin D levels in autistic individuals cannot provide evidence of causation. In the prospective studies, researchers only measured vitamin D levels at one point, which may have failed to reflect vitamin D status across developmental stages. Second, large heterogeneity was observed across studies, which may have resulted from differences in demographic characteristics, measurement methods of vitamin D, seasons, and adjusted variables. Although a random-effects model was used, a substantial amount of heterogeneity remained. Therefore, the effect sizes of meta-analysis should be interpreted with caution. However, a significant relationship between vitamin D status and autism persisted in most subgroups stratified by multiple study characteristics. Third, different cutoffs were chosen in the prospective studies to define vitamin D deficiency, under which circumstances, the same vitamin D level may have belonged to different categories. In addition, in the same study, researchers applied the same kind of standard to determine maternal and neonatal vitamin D status, which led to bias because maternal vitamin D concentration is much higher than neonatal vitamin D concentration is.

5. Conclusions

Our findings suggest that vitamin D status has an association with autism. Some caution should be taken when results are interpreted because of the substantial heterogeneity between studies and unconfirmed causality. Given the high prevalence of vitamin D deficiency in children and adolescents with autism and pregnant and lactating women, screening and appropriate interventions for vitamin D may have significant effects on autism prevention and treatment. Further studies are needed to investigate the causal relationship between vitamin D and autism and elucidate its mechanism.

Supplementary Materials: The following are available online at https://www.mdpi.com/2072-6643/13/1/86/s1, Table S1: Characteristics of case control studies; Table S2: Characteristics of studies involved in neonatal vitamin D; Table S3: Characteristics of studies involved in maternal vitamin D; Table S4: Newcastle–Ottawa quality scale for case control and cohort study assessment; Table S5: Subgroup analysis of case–control studies; Table S6: Subgroup analysis of case–control studies reporting OR; Table S7: Subgroup analysis of case–control studies barring data from previous meta-analysis; Table S8: Subgroup analysis of studies measuring maternal and neonatal vitamin D concentration; Table S9: Subgroup analysis of prospective studies reporting OR; Figure S1: Funnel plot and Egger and Begg test of meta-analysis of 24 case–control studies; Figure S2: Funnel plot and Egger and Begg test of meta-analysis of 10 case control studies; Figure S3: Funnel plot and Egger and Begg test of meta-analysis of 15 case control studies; Figure S4: Funnel plot and Egger and Begg test of meta-analysis of seven studies reporting maternal and neonatal vitamin D concentration; Figure S5: Funnel plot and Egger and Begg test of meta-analysis of nine prospective studies reporting OR.

Author Contributions: Conceptualization, J.W.; methodology, J.W.; software, Z.W.; validation, J.W., Z.W. and R.D.; formal analysis, Z.W.; investigation, Z.W. and R.D.; resources, Z.W. and R.D.; data curation, Z.W., writing—original draft preparation, Z.W.; writing—review and editing, J.W. and Z.W.; visualization, Z.W. and R.D.; supervision, J.W., project administration, Z.W. All authors have read and agreed to the published version of the manuscript.

Funding: This study was supported by a Peking University grant [48014Y0114 to J.W.].

Institutional Review Board Statement: Not applicable.

Informed Consent Statement: Not applicable.

Data Availability Statement: All data obtained from published papers. Source PubMed, EMBASE, Web of Science and the Cochrane Library.

Conflicts of Interest: The authors declare no conflict of interest.

References

1. Lord, C.; Elsabbagh, M.; Baird, G.; Veenstra-Vanderweele, J. Autism spectrum disorder. *Lancet* **2018**, *392*, 508–520. [CrossRef]
2. Maenner, M.J.; Shaw, K.A.; Baio, J.; Washington, A.; Patrick, M.; DiRienzo, M.; Christensen, D.L.; Wiggins, L.D.; Pettygrove, S.; Andrews, J.G.; et al. Prevalence of Autism Spectrum Disorder Among Children Aged 8 Years—Autism and Developmental Disabilities Monitoring Network, 11 Sites, United States. *MMWR Surveill. Summ.* **2020**, *69*, 1–12. [CrossRef] [PubMed]
3. Lai, M.; Kassee, C.; Besney, R.; Bonato, S.; Hull, L.; Mandy, W.; Szatmari, P.; Ameis, S.H. Prevalence of co-occurring mental health diagnoses in the autism population: A systematic review and meta-analysis. *Lancet Psychiatry* **2019**, *6*, 819–829. [CrossRef]
4. Landa, R.J. Efficacy of early interventions for infants and young children with, and at risk for, autism spectrum disorders. *Int. Rev. Psychiatry* **2018**, *30*, 25–39. [CrossRef] [PubMed]
5. Adams, J.; Audhya, T.; Matthews, J.S.; Li, K.; Naviaux, J.C.; Naviaux, R.K.; Adams, R.L.; Coleman, D.M.; Quig, D.W.; Geis, E.; et al. Comprehensive Nutritional and Dietary Intervention for Autism Spectrum Disorder—A Randomized, Controlled 12-Month Trial. *Nutrients* **2018**, *10*, 369. [CrossRef] [PubMed]
6. Linhong, Z.; Lin, X.; Wang, M.; Hu, Y.; Xue, K.; Gu, S.; Lv, L.; Huang, S.; Xie, W. Potential role of genomic imprinted genes and brain developmental related genes in autism. *BMC Med. Genom.* **2020**, *13*, 54. [CrossRef]
7. Lord, C.; Brugha, T.S.; Charman, T.; Cusack, J.; Dumas, G.; Frazier, T.; Jones, E.J.H.; Jones, R.M.; Pickles, A.; State, M.W.; et al. Autism spectrum disorder. *Nat. Rev. Dis. Prim.* **2020**, *6*, 1–23. [CrossRef]
8. Mazahery, H.; Camargo, C.A.; Conlon, C.; Beck, K.L.; Kruger, M.C.; Von Hurst, P. Vitamin D and Autism Spectrum Disorder: A Literature Review. *Nutrients* **2016**, *8*, 236. [CrossRef]
9. Saraff, V.; Shaw, N. Sunshine and vitamin D. *Arch. Dis. Child.* **2016**, *101*, 190–192. [CrossRef]
10. Bivona, G.; Gambino, C.M.; Iacolino, G.; Ciaccio, M. Vitamin D and the nervous system. *Neurol. Res.* **2019**, *41*, 827–835. [CrossRef]

11. Landel, V.; Stephan, D.; Cui, X.; Eyles, D.; Féron, F. Differential expression of vitamin D-associated enzymes and receptors in brain cell subtypes. *J. Steroid Biochem. Mol. Biol.* **2018**, *177*, 129–134. [CrossRef] [PubMed]

12. Wang, Y.; Zhu, J.; DeLuca, H.F. Where is the vitamin D receptor? *Arch. Biochem. Biophys.* **2012**, *523*, 123–133. [CrossRef] [PubMed]

13. Trifonova, E.A.; Klimenko, A.I.; Mustafin, Z.S.; Lashin, S.A.; Kochetov, A.V. The mTOR Signaling Pathway Activity and Vitamin D Availability Control the Expression of Most Autism Predisposition Genes. *Int. J. Mol. Sci.* **2019**, *20*, 6332. [CrossRef] [PubMed]

14. Meguid, N.A.; Hashish, A.F.; Anwar, M.; Sidhom, G. Reduced Serum Levels of 25-Hydroxy and 1,25-Dihydroxy Vitamin D in Egyptian Children with Autism. *J. Altern. Complement. Med.* **2010**, *16*, 641–645. [CrossRef] [PubMed]

15. Mostafa, G.A.; Al-Ayadhi, L.Y. Reduced serum concentrations of 25-hydroxy vitamin D in children with autism: Relation to autoimmunity. *J. Neuroinflammation* **2012**, *9*, 201. [CrossRef] [PubMed]

16. Tostes, M.H.F.D.S.; Polonini, H.C.; Gattaz, W.F.; Raposo, N.R.B.; Baptista, E.B. Low serum levels of 25-hydroxyvitamin D (25-OHD) in children with autism. *Trends Psychiatry Psychother.* **2012**, *34*, 161–163. [CrossRef]

17. Neumeyer, A.M.; Gates, A.; Ferrone, C.; Lee, H.; Misra, M. Bone Density in Peripubertal Boys with Autism Spectrum Disorders. *J. Autism Dev. Disord.* **2013**, *43*, 1623–1629. [CrossRef]

18. Gong, Z.-L.; Luo, C.-M.; Wang, L.; Shen, L.; Wei, F.; Tong, R.-J.; Liu, Y. Serum 25-hydroxyvitamin D levels in Chinese children with autism spectrum disorders. *Neuroreport* **2013**, *25*, 23–27. [CrossRef]

19. Garipardic, M.; Doğan, M.; Bala, K.A.; Mutluer, T.; Kaba, S.; Aslan, O.; Üstyol, L. Association of Attention Deficit Hyperactivity Disorder and Autism Spectrum Disorders with Mean Platelet Volume and Vitamin D. *Med. Sci. Monit.* **2017**, *23*, 1378–1384. [CrossRef]

20. Du, L.; Shan, L.; Wang, B.; Feng, J.Y.; Xu, Z.D.; Jia, F.Y. Serum levels of 25-hydroxyvitamin D in children with autism spectrum disorders. *Chin. J. Contemp. Pediatr.* **2015**, *17*, 68–71. [CrossRef]

21. Saad, K.; Abdel-Rahman, A.A.; Elserogy, Y.M.; Al-Atram, A.A.; Cannell, J.J.; Bjørklund, G.; Abdel-Reheim, M.K.; Othman, H.A.K.; El-Houfey, A.A.; El-Aziz, N.H.R.A.; et al. Vitamin D status in autism spectrum disorders and the efficacy of vitamin D supplementation in autistic children. *Nutr. Neurosci.* **2015**, *19*, 346–351. [CrossRef] [PubMed]

22. Fahmy, S.F.; Sabri, N.A.; El Hamamsy, M.H.; El Sawi, M.; Zaki, O.K. Vitamin D intake and sun exposure in autistic children. *Int. J. Pharm. Sci. Res.* **2016**, *7*, 1043–1049. [CrossRef]

23. Bener, A.; Khattab, A.O.; Bhugra, D.; Hoffmann, G.F. Iron and vitamin D levels among autism spectrum disorders children. *Ann. Afr. Med.* **2017**, *16*, 186–191. [CrossRef] [PubMed]

24. Desoky, T.; Hassan, M.H.; Fayed, H.M.; Sakhr, H.M. Biochemical assessments of thyroid profile, serum 25-hydroxycholecalciferol and cluster of differentiation 5 expression levels among children with autism. *Neuropsychiatr. Dis. Treat.* **2017**, *13*, 2397–2403. [CrossRef] [PubMed]

25. Dong, H.Y.; Wang, B.; Li, H.H.; Shan, L.; Jia, F.Y. Correlation between serum 25-hydroxyvitamin D level and core symptoms of autism spectrum disorder in children. *Chin. J. Pediatr.* **2017**, *55*, 916–919.

26. Feng, J.; Shan, L.; Du, L.; Wang, B.; Li, H.; Wang, W.; Wang, T.; Dong, H.; Yue, X.; Xu, Z.; et al. Clinical improvement following vitamin D3 supplementation in Autism Spectrum Disorder. *Nutr. Neurosci.* **2016**, *20*, 284–290. [CrossRef]

27. Altun, H.; Kurutaş, E.B.; Şahin, N.; Güngör, O.; Fındıklı, E. The Levels of Vitamin D, Vitamin D Receptor, Homocysteine and Complex B Vitamin in Children with Autism Spectrum Disorders. *Clin. Psychopharmacol. Neurosci.* **2018**, *16*, 383–390. [CrossRef]

28. Arastoo, A.A.; Khojastehkia, H.; Rahimi, Z.; Khafaie, M.A.; Hosseini, S.A.; Mansouri, M.T.; Yosefyshad, S.; Abshirini, M.; Karimimalekabadi, N.; Cheraghi, M.; et al. Evaluation of serum 25-Hydroxy vitamin D levels in children with autism Spectrum disorder. *Ital. J. Pediatr.* **2018**, *44*, 150. [CrossRef]

29. El-Ansary, A.; Cannell, J.J.; Bjørklund, G.; Bhat, R.S.; Al Dbass, A.M.; Alfawaz, H.A.; Chirumbolo, S.; Al-Ayadhi, L. In the search for reliable biomarkers for the early diagnosis of autism spectrum disorder: The role of vitamin D. *Metab. Brain Dis.* **2018**, *33*, 917–931. [CrossRef]

30. Bičíková, M.; Máčová, L.; Ostatníková, D.; Hanzlíková, L. Vitamin D in autistic children and healthy controls. *Physiol. Res.* **2019**, *68*, 317–320. [CrossRef]

31. Chtourou, M.; Naifar, M.; Grayaa, S.; Hajkacem, I.; Ben Touhemi, D.; Ayadi, F.; Moalla, Y. Vitamin d status in TUNISIAN children with autism spectrum disorders. *Clin. Chim. Acta* **2019**, *493*, S619–S620. [CrossRef]

32. Adams, J.B.; Audhya, T.; McDonough-Means, S.; Rubin, R.A.; Quig, D.W.; Geis, E.; Gehn, E.; Loresto, M.; Mitchell, J.; Atwood, S.; et al. Nutritional and metabolic status of children with autism vs. neurotypical children, and the association with autism severity. *Nutr. Metab.* **2011**, *8*, 34. [CrossRef]

33. Uğur, Ç.; Gürkan, C.K. Serum vitamin D and folate levels in children with autism spectrum disorders. *Res. Autism Spectr. Disord.* **2014**, *8*, 1641–1647. [CrossRef]

34. Hashemzadeh, M.; Moharreri, F.; SOltanifar, A. Comparative study of vitamin D levels in children with autism spectrum disorder and normal children: A case-control study. *J. Fundam. Ment. Health* **2015**, *17*, 197–201.

35. Guler, S.; Yesil, G.; Ozdil, M.; Ekici, B.; Onal, H. Sleep disturbances and serum vitamin D levels in children with autism spectrum disorder. *Int. J. Clin. Exp. Med.* **2016**, *9*, 14691–14697.

36. Basheer, S.; Natarajan, A.; Van Amelsvoort, T.; Venkataswamy, M.M.; Ravi, V.; Srinath, S.; Girimaji, S.C.; Christopher, R. Vitamin D status of children with Autism Spectrum Disorder: Case-control study from India. *Asian J. Psychiatry* **2017**, *30*, 200–201. [CrossRef]

37. Cieślińska, A.; Kostyra, E.; Chwała, B.; Moszyńska, M.; Fiedorowicz, E.; Teodorowicz, M.; Savelkoul, H.F. Vitamin D Receptor Gene Polymorphisms Associated with Childhood Autism. *Brain Sci.* **2017**, *7*, 115. [CrossRef]

38. Lee, B.K.; Eyles, D.W.; Magnusson, C.; Newschaffer, C.J.; McGrath, J.; Kvaskoff, D.; Ko, P.; Dalman, C.; Karlsson, H.; Gardner, R.M. Developmental vitamin D and autism spectrum disorders: Findings from the Stockholm Youth Cohort. *Mol. Psychiatry* **2019**, 1–11. [CrossRef]
39. Vinkhuyzen, A.; Eyles, D.W.; Burne, T.H.J.; Blanken, L.M.; Kruithof, C.J.; Verhulst, F.; Jaddoe, V.W.; Tiemeier, H.; McGrath, J. Gestational vitamin D deficiency and autism-related traits: The Generation R Study. *Mol. Psychiatry* **2018**, *23*, 240–246. [CrossRef]
40. Erratum Re: Relationship Between Neonatal Vitamin D at Birth and Risk of Autism Spectrum Disorders: The NBSIB Study. *J. Bone Miner. Res.* **2018**, *33*, 550. [CrossRef]
41. Ali, Y.; Anderson, L.N.; Smile, S.; Chen, Y.; Borkhoff, C.M.; Koroshegyi, C.; Lebovic, G.; Parkin, P.C.; Birken, C.S.; Szatmari, P.; et al. Prospective cohort study of vitamin D and autism spectrum disorder diagnoses in early childhood. *Autism* **2018**, *23*, 584–593. [CrossRef]
42. Schmidt, R.J.; Niu, Q.; Eyles, D.W.; Hansen, R.L.; Iosif, A. Neonatal vitamin D status in relation to autism spectrum disorder and developmental delay in the CHARGE case–Control study. *Autism Res.* **2019**, *12*, 976–988. [CrossRef] [PubMed]
43. Windham, G.; Pearl, M.; Anderson, M.C.; Poon, V.; Eyles, D.; Jones, K.L.; Lyall, K.; Kharrazi, M.; Croen, L.A. Newborn vitamin D levels in relation to autism spectrum disorders and intellectual disability: A case–control study in California. *Autism Res.* **2019**, *12*, 989–998. [CrossRef] [PubMed]
44. Wang, T.; Shan, L.; Du, L.; Feng, J.; Xu, Z.; Staal, W.G.; Yue, X.-J. Serum concentration of 25-hydroxyvitamin D in autism spectrum disorder: A systematic review and meta-analysis. *Eur. Child Adolesc. Psychiatry* **2015**, *25*, 341–350. [CrossRef]
45. Moher, D.; Liberati, A.; Tetzlaff, J.; Altman, D.G.; The PRISMA Group. Preferred reporting items for systematic reviews and meta-analyses: The PRISMA statement. *PLoS Med.* **2009**, *6*, e1000097. [CrossRef]
46. Stroup, D.F.; Berlin, J.A.; Morton, S.C.; Olkin, I.; Williamson, G.D.; Rennie, D.; Moher, D.; Becker, B.J.; Sipe, T.A.; Thacker, S.B.; et al. Meta-analysis of Observational Studies in EpidemiologyA Proposal for Reporting. *JAMA* **2000**, *283*, 2008–2012. [CrossRef]
47. Stang, A. Critical evaluation of the Newcastle-Ottawa scale for the assessment of the quality of nonrandomized studies in meta-analyses. *Eur. J. Epidemiol.* **2010**, *25*, 603–605. [CrossRef]
48. Ugur, C.; Gurkan, K. Vitamin D levels in children with autism spectrum disorders. *Klinik Psikofarmakoloji Bulteni* **2014**, *24*, S117–S118.
49. Coşkun, S.; Şimşek, Ş.; Camkurt, M.A.; Çim, A.; Çelik, S.B. Association of polymorphisms in the vitamin D receptor gene and serum 25-hydroxyvitamin D levels in children with autism spectrum disorder. *Gene* **2016**, *588*, 109–114. [CrossRef]
50. Alzghoul, L.; Al-Eitan, L.N.; Aladawi, M.; Odeh, M.; Abu Hantash, O. The Association Between Serum Vitamin D3 Levels and Autism Among Jordanian Boys. *J. Autism Dev. Disord.* **2019**, *50*, 3149–3154. [CrossRef]
51. Fernell, E.; Bejerot, S.; Westerlund, J.; Miniscalco, C.; Simila, H.; Eyles, D.W.; Gillberg, C.; Humble, M.B. Autism spectrum disorder and low vitamin D at birth: A sibling control study. *Mol. Autism* **2015**, *6*, 1–9. [CrossRef]
52. Chen, J.; Xin, K.; Wei, J.; Zhang, K.; Xiao, H. Lower maternal serum 25(OH) D in first trimester associated with higher autism risk in Chinese offspring. *J. Psychosom. Res.* **2016**, *89*, 98–101. [CrossRef] [PubMed]
53. Figuerola, P.E.; Canals, J.; Fernández-Cao, J.C.; Val, V.A. Differences in food consumption and nutritional intake between children with autism spectrum disorders and typically developing children: A meta-analysis. *Autism* **2018**, *23*, 1079–1095. [CrossRef] [PubMed]
54. Liu, D.; Zhan, J.-Y.; Shao, J. Environmental risk factors for autism spectrum disorders in children. *Chin. J. Contemp. Pediatr.* **2015**, *17*, 1147–1153.
55. Schmidt, R.J.; Hansen, R.L.; Hartiala, J.; Allayee, H.; Sconberg, J.L.; Schmidt, L.C.; Volk, H.E.; Tassone, F. Selected vitamin D metabolic gene variants and risk for autism spectrum disorder in the CHARGE Study. *Early Hum. Dev.* **2015**, *91*, 483–489. [CrossRef]
56. Bahrami, A.; Sadeghnia, H.R.; Tabatabaeizadeh, S.-A.; Bahrami-Taghanaki, H.; Behboodi, N.; Esmaeili, H.; Ferns, G.A.; Ghayour-Mobarhan, M.; Avan, A. Genetic and epigenetic factors influencing vitamin D status. *J. Cell. Physiol.* **2018**, *233*, 4033–4043. [CrossRef]
57. Ramos-Lopez, E.; Brück, P.; Jansen, T.; Herwig, J.; Badenhoop, K. CYP2R1 (vitamin D 25-hydroxylase) gene is associated with susceptibility to type 1 diabetes and vitamin D levels in Germans. *Diabetes/Metabol. Res. Rev.* **2007**, *23*, 631–636. [CrossRef]
58. Yıldız, E.P.; Poyrazoglu, Ş.; Bektas, G.; Kardelen, A.D.; Aydinli, N. Potential risk factors for vitamin D levels in medium- and long-term use of antiepileptic drugs in childhood. *Acta Neurol. Belg.* **2017**, *5*, 111–453. [CrossRef]
59. Woon, F.C.; Chin, Y.S.; Ismail, I.H.; Batterham, M.; Latiff, A.H.A.; Gan, W.Y.; Appannah, G.; Hussien, S.H.M.; Edi, M.; Tan, M.L.; et al. Vitamin D deficiency during pregnancy and its associated factors among third trimester Malaysian pregnant women. *PLoS ONE* **2019**, *14*, e0216439. [CrossRef]
60. López-Vicente, M.; Sunyer, J.; Lertxundi, N.; González, L.; Rodríguez-Dehli, C.; Sáenz-Torre, M.E.; Vrijheid, M.; Tardón, A.; Llop, S.; Torrent, M.; et al. Maternal circulating Vitamin D3 levels during pregnancy and behaviour across childhood. *Sci. Rep.* **2019**, *9*, 14792. [CrossRef]
61. Berg, A.O.; Jørgensen, K.N.; Nerhus, M.; Athanasiu, L.; Popejoy, A.B.; Bettella, F.; Norbom, L.C.B.; Gurholt, T.P.; Dahl, S.R.; Andreassen, O.A.; et al. Vitamin D levels, brain volume, and genetic architecture in patients with psychosis. *PLoS ONE* **2018**, *13*, e0200250. [CrossRef] [PubMed]
62. Holick, M.F. Vitamin D Deficiency. *N. Engl. J. Med.* **2007**, *357*, 266–281. [CrossRef] [PubMed]

63. Lisi, G.; Ribolsi, M.; Siracusano, A.; Niolu, C. Maternal Vitamin D and its Role in Determining Fetal Origins of Mental Health. *Curr. Pharm. Des.* **2020**, *26*, 2497–2509. [CrossRef] [PubMed]
64. Heidari, H.; Amani, R.; Feizi, A.; Askari, G.; Kohan, S.; Tavasoli, P. Vitamin D Supplementation for Premenstrual Syndrome-Related inflammation and antioxidant markers in students with vitamin D deficient: A randomized clinical trial. *Sci. Rep.* **2019**, *9*, 14939. [CrossRef] [PubMed]
65. Kasatkina, L.A.; Tarasenko, A.S.; Krupko, O.A.; Kuchmerovska, T.M.; Lisakovska, O.O.; Trikash, I. Vitamin D deficiency induces the excitation/inhibition brain imbalance and the proinflammatory shift. *Int. J. Biochem. Cell Biol.* **2020**, *119*, 105665. [CrossRef] [PubMed]
66. Li, H.; Ma, J.; Huang, R.; Wen, Y.; Liu, G.; Xuan, M.; Yang, L.; Yang, J.; Song, L. Prevalence of vitamin D deficiency in the pregnant women: An observational study in Shanghai, China. *Arch. Public Health* **2020**, *78*, 31. [CrossRef]

Review

The Health Effects of Vitamin D and Probiotic Co-Supplementation: A Systematic Review of Randomized Controlled Trials

Myriam Abboud [1,*], Rana Rizk [2], Fatme AlAnouti [1], Dimitrios Papandreou [1], Suzan Haidar [3] and Nadine Mahboub [3,4]

[1] Department of Health, College of Natural and Health Sciences, Zayed University, Dubai 19282, UAE; fatme.alanouti@zu.ac.ae (F.A.); dimitrios.papandreou@zu.ac.ae (D.P.)

[2] Institut National de Santé Publique, d'Épidémiologie Clinique et de Toxicologie (INSPECT-Lb), Beirut, Lebanon; rana.rizk@inspect-lb.org

[3] Department of Nutrition and Food Sciences, Faculty of Arts and Sciences, Lebanese International University, Beirut 657314, Lebanon; suzan.haidar@liu.edu.lb (S.H.); nadine.baltagi@liu.edu.lb (N.M.)

[4] Department of Health Promotion, Faculty of Health, Medicine and Life Sciences, Maastricht University, 6229 GT Maastricht, The Netherlands

* Correspondence: myriam.abboud@zu.ac.ae

Citation: Abboud, M.; Rizk, R.; AlAnouti, F.; Papandreou, D.; Haidar, S.; Mahboub, N. The Health Effects of Vitamin D and Probiotic Co-Supplementation: A Systematic Review of Randomized Controlled Trials. *Nutrients* **2021**, *13*, 111. https://doi.org/10.3390/nu13010111

Received: 24 November 2020
Accepted: 24 December 2020
Published: 30 December 2020

Publisher's Note: MDPI stays neutral with regard to jurisdictional clai-ms in published maps and institutio-nal affiliations.

Abstract: Evidence of synergic health effects of co-supplementation with vitamin D and probiotics is emerging. Following the Preferred Reporting Items for Systematic Reviews and Meta-Analyses PRISMA statement, scientific databases and the grey literature were searched, and a narrative review and risk of bias assessment were conducted. Seven randomized controlled trials were included, which had low risk of bias. Six studies were double-blind, and once single-blind, extended over 6–12 weeks, and included 50–105 participants. Conditions explored included schizophrenia, gestational diabetes, type 2 diabetes and coronary heart disease, polycystic ovarian syndrome, osteopenia, irritable bowel syndrome (IBS), and infantile colic. Supplementation frequency was daily or bi-monthly, with mainly vitamin D3, and *Lactobacillus*, *Bifidobacterium*, and *Streptococcus*. Comparators were placebo, vitamin D, lower vitamin D dose, and probiotics and lower vitamin D dose. The co-supplementation yielded greater health benefits than its comparators did in all studies except in one assessing IBS. Beneficial effects included decreased disease severity, improved mental health, metabolic parameters, mainly insulin sensitivity, dyslipidemia, inflammation, and antioxidative capacity, and lower use of healthcare. Co-supplementation of vitamin D and probiotics generated greater health benefits than its comparators did. More studies in other diseases and various populations are needed to confirm these findings and to elucidate the optimal form, composition, and frequency of this co-supplementation.

Keywords: vitamin D; probiotic; supplementation; adults; randomized controlled trial; systematic review

1. Introduction

The gut microbiota refers to the assemblage of microorganisms, including bacteria, viruses, and fungi, located in the gastrointestinal (GI) tract [1]. There has been increasing emphasis on the role of the microbiota in physiology, suggesting that it can be considered as another human organ [2]. Furthermore, emerging evidence suggests that this invisible organ is a key driver of human health and disease. Gut microbiota plays a critical role in maintaining metabolic and immune health, synthesis of vitamins, obtaining inaccessible nutrients from the diet, renewal of epithelial cells, fat storage, maintaining intestinal barrier integrity, and brain development [3,4]. Dysbiosis, or alteration in the gut microbiota composition, is a crucial risk factor for the development of several disorders such as inflammatory bowel disease, obesity, diabetes, asthma, and allergies [5,6]. The gut microbiota composition is affected by intrinsic and extrinsic factors like genetics, age, dietary changes, in addition to physiological and psychological stress [2,7].

Specifically, vitamin D and the vitamin D receptor (VDR) were shown to modulate the gut microbiota [8]. Increased VDR expression may decrease microbial dysbiosis, enhance barrier function, increase the expression of antimicrobial peptides, decrease pro-inflammatory cytokines, and increase the commensal production of short-chain fatty acids [2,8]. Likewise, probiotics, which are ingestible nonpathogenic living microorganisms, were also shown to improve the balance of intestinal microbiota by regulating microbial components and metabolites [9]. Probiotics simulate the immune system, balance commensal and pathogenic bacteria, and reestablish homeostasis. They protect barrier integrity, alter toxic compounds, and host products. Thus, they ameliorate inflammation and prevent and repair cell damage [9].

Vitamin D deficiency and defects in VDR signaling have been related to several metabolic, cardiovascular, neurodevelopmental and cancer diseases [10,11]. Yet, interventional studies have conflicting evidence on the effect of vitamin D supplementation in their treatment [12–16]. Similarly, human probiotic supplementation studies generated conflicting evidence regarding the effectiveness of probiotics in the treatment of several health conditions such as allergies, GI disorders, metabolic syndrome, and obesity [17–20].

Recently, a promising evidence of synergic effects of combined supplementation with vitamin D and probiotics in modulating the gut microbiota and metabolome, in addition to fostering healthy microbe–host interactions, is emerging [9,21,22]. This co-supplementation holds a preventive and therapeutic potential with crucial clinical implications. Biologically plausible mechanisms support this interplay. Probiotics were shown to increase vitamin D intestinal absorption, and increase VDR protein expression and transcriptional activity [9]. Likewise, VDR status seems to be crucial in regulating the mechanisms of action of probiotics and modulating their anti-inflammatory, immunomodulatory and anti-infective benefits, suggesting a two-sided pathway [6,8].

The aim of this systematic review is to investigate the literature and summarize the available evidence of randomized controlled trials (RCTs) on the various health effects of a combined supplementation of vitamin D and probiotics among children and adults.

2. Materials and Methods

2.1. Review Design

The reporting of this systematic review followed the Preferred Reporting Items for Systematic Reviews and Meta-Analyses (PRISMA) statement [23]. A predefined protocol for this systematic review was registered at the OSF registries.

2.2. Criteria for Study Inclusion

Randomized controlled trials (RCTs) conducted on adults or children, healthy or with disease other than those known to influence vitamin D metabolism, and including an intervention group that received a co-supplementation of vitamin D and probiotics, and a control group of placebo, or a lower dose of vitamin D or probiotics, or a different form of vitamin D, or different strains of probiotics, were included in this systematic review. RCTs with a duration of a minimum of 1 month were included; this duration was deemed sufficient for the intervention to produce an effect. Additionally, RCTs involving other co-interventions were included, only if both arms received the same co-intervention.

Studies were excluded if they were non-randomized, uncontrolled, involving participants taking medication known to influence vitamin D metabolism or with conditions affecting vitamin D metabolism such as chronic kidney disease, chronic liver disease, or malabsorption states, or entailing a supplementation with either vitamin D or probiotics.

2.3. Search Strategy

The systematic search included Medical Subject Headings (MeSH) and keywords for three concepts: (1) vitamin D, (2) probiotics, and (3) randomized controlled trial, and was conducted in PubMed, MEDLINE, CINAHL, EMBASE, the Cochrane Library, ClinicalTrials.gov, and the International Clinical Trials Registry Platform (ICTRP), from inception until 4 November 2020, without language restrictions. The electronic search strategy, detailed in the Tables S1 and S2, was validated by a medical information specialist. Reference lists of included RCTs and relevant reviews were also hand-searched for eligible studies.

2.4. Study Selection

The titles and/or abstracts retrieved by the search were screened by two pairs of authors, and the full text of all relevant papers was assessed for eligibility independently and in duplicate. A calibration exercise was conducted before study selection to ensure the validity of the process. Inconsistencies were discussed amongst reviewers, and unresolved discrepancies were settled by a third reviewer.

2.5. Data Extraction

Data from the selected articles were extracted by two pairs of authors using a data extraction form. Changes from baseline for the intervention were compared with the control in all the parameters analyzed. A calibration exercise was first conducted. Disagreements were resolved through discussion or with the help of a third reviewer.

2.6. Quality Assessment

The risk of bias for the included studies was assessed using the Cochrane criteria (sequence generation, allocation concealment, blinding of participants and outcome assessors, incomplete outcome data, and selective outcome reporting) [24], whereby each potential source of bias was graded as low, high, or unclear risk. The process was carried out by two pairs of authors independently and in duplicate. They underwent a calibration exercise before performing the assessment of risk of bias. Conflicts were resolved through discussion amongst the pair of reviewers or through consultation with a third reviewer.

2.7. Data Synthesis

A narrative review of the findings was performed and is included in Table S2.

3. Results

3.1. Search Results

Study selection process is detailed in Figure 1, whereby seven studies meeting the inclusion criteria were included in the systematic review.

3.2. Characteristics of Included Studies

Characteristics of included RCTs are detailed in Table 1. The studies were published between 2015 [25,26] and 2019 [27,28]. Five studies were conducted in Iran [27–31], one in Italy [25], and one in the United Kingdom [26]. All the studies were randomized double-blind [26–31], except for Savino et al. [25], which was single-blind. The duration of the studies ranged between 6 [29,31] to 12 weeks [25–28,30]. The number of participants ranged between 50 [31] and 105 [25]. The studies were conducted on infants [25], pregnant women [29], and other adults with diseases [26–28,30,31]. Health conditions that were studied included schizophrenia [27], gestational diabetes mellitus (GDM) [29], type 2 diabetes mellitus (T2DM) and coronary heart disease (CHD) [30], polycystic ovarian syndrome (PCOS) [28], osteopenia [31], irritable bowel syndrome [26], and infantile colic [25].

In the studies by Ghaderi et al. [27], Ostadmohammadi et al. [28], Raygan et al. [30] interventions consisted of a co-supplementation with vitamin D and probiotics, and the control group received placebo only [27,28,30]. In the study by Tazzyman et al. [26], the intervention group received a co-supplementation with vitamin D and probiotics, one of the control groups received a placebo, and the other one received placebo and vitamin D [26]. In Savino et al. [25], the intervention group received vitamin D and probiotics, but the control group received vitamin D only. In the study by Jafarnejad et al. [31], the intervention group received probiotics, yet vitamin D was supplemented in all groups.

This co-intervention rendered the comparison between the intervention group receiving probiotics and vitamin D and the control group receiving placebo and a similar dose of vitamin D. Additionally, in the study by Jamilian et al. [29] the intervention consisted of a co-supplementation with vitamin D and probiotics; one of the control groups received probiotics, and the other one received placebo. Yet, in this study [29], all the groups also received a lower dose of vitamin D. This co-intervention rendered the comparison between the intervention group receiving probiotics and a high dose of vitamin D, the first control group receiving probiotics and a lower dose of vitamin D, and the second control group receiving placebo and a lower dose of vitamin D [29].

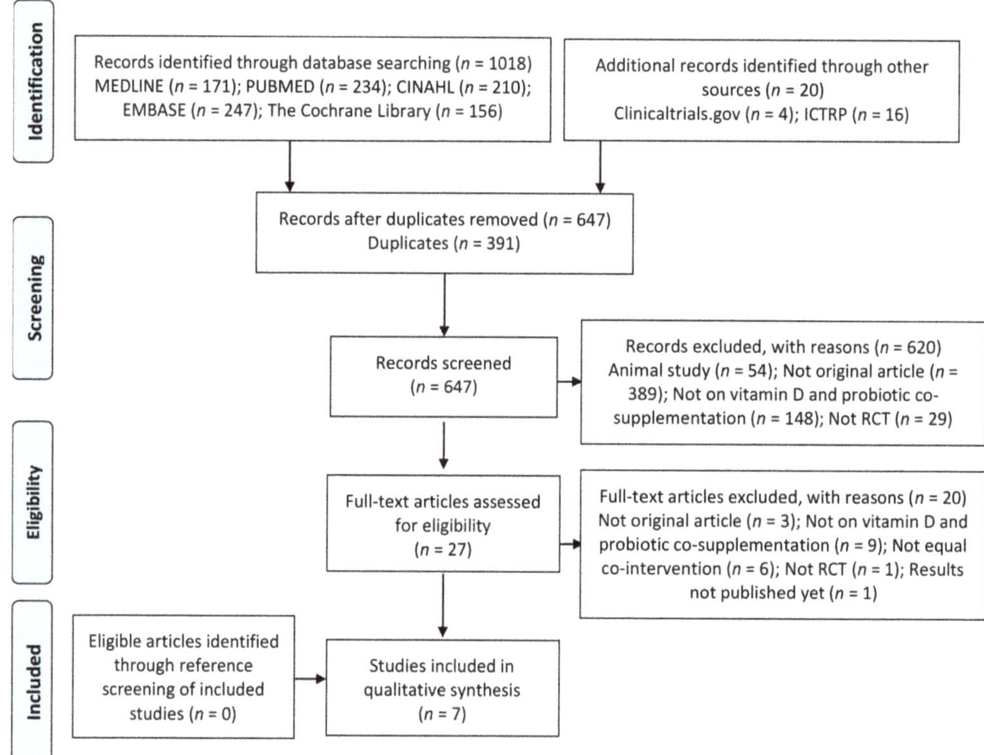

Figure 1. Preferred Reporting Items for Systematic Reviews and Meta-Analyses (PRISMA) Diagram of Study Selection.

Table 1. Characteristics of included studies.

First Author, Year, Country	Study Design	Duration	Study Population	Intervention	Control	Co-Intervention	Compliance/Drop-out
Ghaderi, 2019, Iran [27]	Randomized, double-blind, placebo-controlled trial	12 weeks	n = 60, aged 25–65, 93.33% men, diagnosed with schizophrenia using DSM-IV-TR criteria with disease duration $\geq$2 years, PANSS score $\geq$55, treated with chlorpromazine (300–1000 mg/day, except clozapine) and anticholinergic agents (Trihexyphenidyl, 4–8 mg/day) during the last 6 months	Vitamin D3 and probiotic supplement: - Vitamin D3: 50,000 IU every 2 weeks; DDE = 3571.4 IU - Probiotics: 8×10^9 CFU/day containing *Lactobacillus acidophilus*, *Bifidobacterium bifidum*, *Lactobacillus reuteri*, and *Lactobacillus fermentum* (each 2×10^9 CFU/day)	Placebo similar shape and packaging	None	Compliance: >90%Drop out: I: 13.33% C: 13.33% (Intention-to-treat analysis)
Jafarnejad, 2017, Iran [31]	Randomized, double-blind, placebo-controlled clinical trial	6 weeks	n = 50, age 50–72 years, women with mild bone loss (osteopenia) diagnosed based on the World Health Organization criteria (T-score between −1.0 and −2.5)	Probiotic supplement: *Lactobacillus casei* 1.3×10^{10} CFU, *Bifidobacterium longum* 5×10^{10} CFU, *Lactobacillus acidophilus* 1.5×10^{10} CFU, *Lactobacillus rhamnosus* 3.5×10^9 CFU, *Lactobacillus bulgaricus* 2.5×10^8 CFU, *Bifidobacterium breve* 1×10^{10} CFU, and *Streptococcus thermophilus* 1.5×10^8 CFU/500 mg	Placebo similar in shape, size, odor, color and packaging	Vitamin D (200 IU daily) and Calcium (500 mg daily)	Compliance 100% Drop out: I: 20% C: 16%
Jamilian, 2018, Iran [29]	Randomized, double-blind, placebo-controlled clinical trial	6 weeks	n = 87, women with GDM diagnosed by a "one-step" 2-h 75-g oral glucose tolerance test based on the ADA guidelines	Vitamin D and probiotic supplement: - Vitamin D: 50,000 IU every 2 weeks; DDE = 3571.4 IU - Probiotics: 8×10^9 CFU/g probiotic containing *Lactobacillus acidophilus*, *Bifidobacterium bifidum*, *L. reuteri*, and *Lactobacillus fermentum* (each 2×10^9 CFU/g)	C1: 8×10^9 CFU/day of probiotic supplements C2: Placebo Similar in appearance, color, shape, size, odor, taste and packaging	Vitamin D3: 1000 IU and Vitamin B9: 400 mg, daily from the beginning of pregnancy, and Ferrous sulfate: 60 mg, daily from the secondtrimester	Compliance: 100% Drop out: I: 0% C1: 6.66% C2: 10%
Ostadmohammadi, 2019, Iran [28]	Randomized, double-blind, placebo-controlled clinical trial	12 weeks	n = 60, aged 18–40 years, women with PCOS, diagnosed based on the Rotterdam criteria with BMI: 17–34 kg/m^2 and insulin resistance: 1.4–4	Vitamin D and probiotic supplement: - Vitamin D: 50,000 IU every 2 weeks; DDE = 3571.4 IU - Probiotics: 8×10^9 CFU/day containing *Lactobacillus acidophilus*, *Bifidobacterium bifidum*, *Lactobacillus reuteri* and *Lactobacillus fermentum* (each 2×10^9 CFU/g)	Placebo similar in appearance, color, shape, size, odor, taste and packaging	None	Compliance 100%; No drop out
Raygan, 2018, Iran [30]	Randomized, double-blind, placebo-controlled clinical trial	12 weeks	n = 60, age 45–85 years, 50% men, with T2DM diagnosed based on the criteria of the ADA and with CHD diagnosed as per the AHA with 2- and 3-vessel CHD	Vitamin D3 and probiotic supplement: - Vitamin D3: 50,000 IU every 2 weeks; DDE = 3571.4 IU - Probiotics: 8×10^9 CFU/g - containing *Lactobacillus acidophilus*, *Bifidobacterium bifidum*, *Lactobacillus reuteri*, and *Lactobacillus fermentum* (each 2×10^9 CFU/g)	Placebo similar in appearance, color, shape, size, odor, taste and packaging	None	Compliance > 90% Drop out: I: 13.33% C: 13.33% (Intention-to-treat analysis)

Table 1. *Cont.*

First Author, Year, Country	Study Design	Duration	Study Population	Intervention	Control	Co-Intervention	Compliance/Drop-out
Savino, 2015, Italy [25]	Single-blind, randomized controlled, parallel-group trial	12 weeks	$n = 105$, newborns aged less than 10 days of life, 48.5% boys, with gestational age between 37 and 42 weeks, birth weight from 2500 to 4300 g, and normal physical examination	Vitamin D and probiotic supplement: - Vitamin D3: 400 IU daily - Probiotics: *Lactobacillus reuteri* DSM 17938 (10^8 CFU)	Vitamin D (400 IU daily)	None	No infants lost to follow-ups
Tazzyman, 2015, United Kingdom [26]	Double-blind, randomized, three-arm parallel design trial	12 weeks	$n = 51$, 7.8% men, with previous clinical diagnosis of IBS and met the Rome III criteria and stratified according to vitamin D status at baseline (deficient: 25(OH)D <20 ng/mL; repleted: 25(OH)D >20 ng/mL)	Vitamin D3 and probiotic supplement: - Vitamin D3: sublingual liquid spray, 3000 IU daily - Probiotics: *Lactobacillus acidophilus*, CUL60 (NCIMB 30157), CUL21 (NCIMB 30156), *Bifidobacterium bifidum* CUL20 (NCIMB 30153) and *Bifidobacterium animalis subsp. lactis* CUL34 (NCIMB 30172) 2.5×10^{10} CFU per capsule	C1: Double placebo C2: Placebo and Vitamin D3 (400 IU daily) Similar in form, containing identical buffers	None	Compliance: 98% Drop out: 0%

25(OH)D: 25-hydroxyvitamin D; ADA: American Diabetes Association; AHA: American Heart Association; BMI: Body Mass Index; C: Control; CFU: Colony Forming Units; CHD: Coronary Heart Disease; DDE: Daily Dose Equivalent; DSM-IV-TR: Diagnostic and Statistical Manual of Mental Disorders, Fourth Edition, Text Revision; GDM: Gestational Diabetes Mellitus; I: Intervention; IBS: Irritable Bowel Syndrome; IU: International Unit; PANSS: The Positive and Negative Syndrome Scale; PCOS: Polycystic Ovary Syndrome; T2DM: Type 2 Diabetes Mellitus; TDD: Total Daily Dose.

The frequency of supplement administration ranged between daily [25,26,31] and bi-monthly [27–30]. Probiotic supplementation was given in the form of a capsule in all studies [25–31], whereas supplementation of vitamin D was either in the form of a capsule [26–31] or sublingual liquid spray [25]. The form of vitamin D supplemented was not specified in the studies by Jamilian et al. [29], Ostadmohammadi et al. [28], and Jafarnejad et al. [31], and studies by Ghaderi et al. [27], Raygan et al. [30], Tazzyman et al. [26], and Savino et al. [25] used vitamin D3, and the daily dose equivalent ranged from 200 International Units (IU) [31] to 4571.4 IU [29]. Probiotic strains that were investigated included *Lactobacillus* in all the studies [25–31], *Bifidobacterium* in all the studies [26–31] except for the one by Savino et al. [25], and *Streptococcus* only in Jafarnejad et al. [31]. The supplemented doses greatly varied across studies, and in the majority of the studies, it consisted of 8×10^9 Colony Forming Units (CFU) per day.

There was a high rate of compliance in all studies [25–31], and the drop-out rate ranged from 0% [25,26,28,29] to 20% [31], and was almost equal between the compared groups in all studies [25–28,30,31], except in Jamilian et al. [29].

3.3. Assessment of Risk of Bias

Risk of bias assessment of included RCTs is available in Table 2. In general, the quality of the RCTs design and reporting was high. In all studies [25–31], random allocation of participants was adequate, and allocation was concealed. Blinding of participants and personnel was reported in all of the included studies [26–31], except in the one by Saviano et al. [25], where both patients and physicians, except outcome assessors, were aware of their allocation. All studies reported complete outcome data [25,26,28,29,31], except for the studies conducted by Ghaderi et al. [27] and Raygan et al. [30] who did not mention how missing data were dealt with. Finally, in all studies [25–31], all pre-specified outcomes were reported on.

Table 2. Risk of bias of included studies from consensus between a pair of raters.

First Author, YEAR	Random Sequence Generation (Selection Bias)	Allocation Concealment (Selection Bias)	Blinding of Participants and Personnel (Performance Bias)	Blinding of Outcome Assessment (Detection Bias)	Incomplete Outcome Data (Attrition Bias)	Selective Reporting (Reporting Bias)	Other Bias
Ghaderi, 2019 [27]	Low	Low	Low	Low	High	Low	Low
Jafarnejad, 2017 [31]	Low	Low	Low	Low	Low	Low	Low
Jamilian, 2018 [29]	Low	Low	Low	Low	Low	Low	Low
Ostadmohammadi, 2019 [28]	Low	Low	Low	Low	Low	Low	Low
Raygan, 2018 [30]	Low	Low	Low	Low	High	Low	Low
Savino, 2015 [25]	Low	Low	High	Low	Low	Low	Low
Tazzyman, 2015 [26]	Low	Low	Low	Low	Low	Low	Low

● Low risk of bias ? Unclear risk of bias ● High risk of bias.

3.4. Results of Included Studies

The outcomes assessed and the findings of included RCTs are presented in Table 3. In Ghaderi et al. [27], Ostadmohammadi et al. [28], Raygan et al. [30], and Savino et al. [25], co-supplementation with probiotics and vitamin D yielded greater health benefits than either placebo [27,28,30] or vitamin D on its own [25]. Specifically, in Ghaderi et al. [27], the co-supplementation, compared with placebo, had a favorable effect on schizophrenia symptoms severity, as well as other metabolic outcomes, mainly insulin sensitivity, inflammation, and antioxidative capacity. In Ostadmohammadi et al. [28], vitamin D and probiotic co-supplementation in women with PCOS, compared with placebo, had beneficial effects on mental health parameters, namely depression, anxiety and stress, as well as hormonal, inflammatory, and antioxidative parameters, and on the symptoms of PCOS, specifically, hirsutism. However, the co-supplementation was not associated with improvements in sex hormone-binding globulin, nor with other symptoms of PCOS, namely acne and alopecia, nor were there improvement in sleep quality [28]. In Raygan et al. [30], combined supplementation with vitamin D and probiotics for people with T2DM and CHD, compared with placebo, improved anxiety and depression, insulin sensitivity, inflammatory markers, antioxidative capacity and dyslipidemia, specifically high-density lipoprotein-cholesterol. However, this intervention did not result in a better control of fasting glucose, other markers of dyslipidemia, specifically triglycerides, very low and low lipoprotein-cholesterol, nor with blood pressures [30]. In the study by Savino et al. [25], compared with vitamin D supplementation alone, vitamin D and probiotic co-supplementation to newborns was associated with a reduction of more than two pediatric consultations and phone calls regarding infantile colic over a 12-week period. The co-supplementation was also associated with a lower use of pain-relieving agents and of infant formula [25].

In the study by Jamilian et al. [29], all women with GDM in all groups were being supplemented with 1000 IU (low dose) vitamin D. The group supplemented with probiotics and high dose vitamin D, compared with placebo and low dose vitamin D, showed greater improvement in glucose control, insulin sensitivity, dyslipidemia, inflammatory markers, and antioxidative capacity [29]. Additionally, upon birth, newborns of mothers in this arm had lower incidence of both hyperbilirubinemia and hospitalization [29]. Moreover, the group supplemented with probiotics and high dose vitamin D, compared with probiotics and low dose vitamin D, exhibited a greater improvement in dyslipidemia, inflammation and antioxidative capacity [29]. Furthermore, newborns had better health outcomes [29]. Similarly, in the study by Jafarnejad et al. [31], all groups received 200 IU of vitamin D, and the group receiving probiotics had improvement in osteopenia markers (bone resorption and turnover), namely, bone-specific alkaline phosphatase, collagen type 1 cross-linked C-telopeptide, tumor necrosis factor α, and parathyroid hormone, but did not show an improvement in bone mineral density nor other serum indicators of osteopenia [31], compared with the group receiving placebo and vitamin D.

The only study where the co-supplementation was not found to be more effective than its comparators was the one conducted by Tazzyman et al. [26], where no significant difference in the symptoms of irritable bowel syndrome (IBS) was evident, between co-supplementation with probiotics and vitamin D, compared with vitamin D alone, or with placebo.

Table 3. Outcomes and results of included studies.

First Author, Year, Country	Outcome Measures	Results	Conclusion
Ghaderi, 2019, Iran [1] [27]	BMI: weight in kg divided by height in meters squared (height and weight measured withoutshoes and in light clothing by a trained staff) Serum 25-hydroxyvitamin D: ELISA kit Severity of psychiatric symptoms: PANSS Domains of cognitive function: BPRS scores TAC: method of ferric reduction antioxidant power developed by Benzie and Strain GSH: Beutler method MDA: Thiobarbituric acid reactive substances spectrophotometric Test Serum hs-CRP: ELISA kit NO: Griess Method Serum insulin: ELISA kit HOMA-IR and QUICKI: calculated using standard formula FPG and lipid profiles: Enzymatic kits	At baseline and end line: No significant difference between-groups, in height, age, weight, BMI and METs At baseline: Significant difference between-groups for positive PANSS score, BPRS, GSH and plasma NO At end line: In the I group compared with the C group: Significant greater decrease in MDA (-0.3 ± 0.9 vs. $+0.2 \pm 0.4$ μmol/L), serum hs-CRP (-2.3 ± 3.0 vs. -0.3 ± 0.8 mg/L), FPG (-7.0 ± 9.9 vs. -0.2 ± 9.9 mg/dL), serum insulin (-2.7 ± 2.3 vs. $+0.4 \pm 2.0$ μIU/mL), HOMA-IR (-0.8 ± 0.7 vs. $+0.1 \pm 0.7$), TG (-7.8 ± 25.2 vs. $+10.1 \pm 30.8$ mg/dL), TC (-4.9 ± 15.0 vs. $+5.9 \pm 19.5$ mg/dL), and TC/HDL-C (-0.1 ± 0.6 vs. $+0.3 \pm 0.8$) Significant greater increase in 25-hydroxyvitamin D ($+9.1 \pm 4.1$ vs. $+0.2 \pm 0.4$ ng/mL), general PANSS score (-3.1 ± 4.7 vs. $+0.3 \pm 3.9$), total PANSS score (-7.4 ± 8.7 vs. -1.9 ± 7.5), plasma TAC ($+51.1 \pm 129.7$ vs. -20.7 ± 53.3 mmol/L), QUICKI ($+0.02 \pm 0.01$ vs. $+0.0003 \pm 0.01$) No significant difference in the change of BPRS score and other metabolic profilesIn the analysis adjusting for baseline values of biochemical parameters, age and BMI, and controlling for potential confounders: The difference in changes in TC/HDL between the two groups became non-significant The difference in changes in negative PANSS score, BPRS and plasma GSH became statistically significant Other metabolic profiles did not change statically	Probiotic and vitamin D co-supplementation for 12 weeks to patients with chronic schizophrenia had beneficial effects on the general and total PANSS scores, as well as their metabolic profiles, compared with placebo
Jafarnejad, 2017, Iran [31]	Nutrient intake: 3-day dietary recall (2 weekdays and one weekend day), through monthly interview throughout the study period; nutrient analysis: by Nutritionist IV software modified for Iranian foods Physical activity: daily physical activity questionnaires validated by Kelishady et al. and calculated as metabolic equivalents/day Body weight: measured wearing light clothes without shoes using digital scales with 100-g precision Height: measured using a stadiometer with 0.5-cm precision in a normal standing position without shoes. BMI: weight in kilograms divided by height in meters squared BMD: dual energy X-ray absorptiometry Bone and pro-inflammatory biomarkers (TNF-α and IL-1b), Total serum levels of BALP, Osteocalcin, CTX, Vitamin D, RANKL, Osteoprotegrin, Serum TNF-α and IL-1b, Serum PTH, Urinary deoxypyridinoline: ELISA kits Serum calcium, phosphorus, magnesium, albumin, creatinine, alkaline phosphatase, and urinary amounts of calcium, phosphorus, magnesium, and creatinine: Pars Azmoon kits	At baseline: No significant differences between-groups At end line: Significant between-group differences in BALP (U/L) (I: 19.65 ± 1.66 at baseline and 16.53 ± 0.90 at end line vs. C: 17.81 ± 1.35 at baseline and 18.63 ± 1.29 at end line); CTX (ng/mL) (I: 0.41 ± 0.02 at baseline and 0.35 ± 0.02 at end line vs. C: 0.45 ± 0.02 at baseline and 0.42 ± 0.02 at end line); TNF-α (pg/mL) (I: 4.24 ± 0.5 at baseline and 3.73 ± 0.43 at end line vs. 3.83 ± 0.47 at baseline and 4.32 ± 0.5 at end line); PTH (pg/mL) (I: 31.92 ± 1.39 at baseline and 29.05 ± 1.53 at end line vs. C: 30.65 ± 1.44 at baseline and 32.81 ± 1.72 at end line) No significant between-group difference in Spinal BMD, Total hip BMD, RANKL, osteoprotegrin, RANKL/ osteoprotegrin ratio, deoxypyridinoline, osteocalcin, IL-1, Vitamin D, serum calcium, 24-h urinary Calcium, Serum phosphorus, 24-h urinary phosphorus, Serum magnesium, 24-h urinary magnesium, Serum creatinine, 24-h urinary creatinine, ALP, Albumin	Supplementation with probiotics, vitamin D and calcium for 6 weeks to postmenopausal osteopenic women showed a possible role in suppressing bone resorption and bone turnover, but did not affect bone density and other serum indicators compared with placebo, vitamin D and calcium

Table 3. *Cont.*

First Author, Year, Country	Outcome Measures	Results	Conclusion
Jamilian, 2018, Iran [2,3] [29]	BMI: weight in kg divided by height in meters squared (height and weight measured withoutshoes and in light clothing by a trained staff) Polyhydramnios: sonographic estimation method at post-intervention and defined as an AFI in excess of 25 cm Preterm delivery: defined as delivery occurred at <37 weeks of pregnancy Newborn's macrosomia: defined as birth weight of >4000 g. 2.5 Serum 25-hydroxyvitamin D: ELISA kit Serum insulin: ELISA kit HOMA-IR and QUICKI: calculated according to the standard formula FPG, serum TG, VLDL-C, TC, LDL-C and HDL-C: enzymatic kits Serum hs-CRP: ELISA kit Plasma NO: Griess method TAC: method of ferric reducing antioxidant power developed by Benzie and Strain GSH: Beutler method MDA: Thiobarbituric acid reactive substances spectrophotometric Test Newborns' hyperbilirubinemia: when the total serum bilirubin levels were at ≥15 mg/dL (257 mmol/L) among infants 25–48 h old, 18 mg/dL (308 mmol/L) in infants 49–72 h old, and 20 mg/dL (342 mmol/L) in infants >72 h old	At baseline and end line: No significant difference between-groups, in age, height, weight, BMI, METs and intakes of macro- and micronutrients At end line: In the I group compared with the C1 group Significant greater decrease in TG (β −15.82 mg/dL), VLDL-C (β −3.16 mg/dL) and hs-CRP (β −0.32 mg/L) Significant greater increase in serum 25-hydroxyvitamin D (β 16.16 ng/mL), TAC (β 63.26 mmol/L) and GSH (β 53.61 mmol/L)Lower incidence of hyperbilirubinemiain newborns (10.0% vs. 13.8%) Lower incidence of newborns' hospitalization (10.0% vs. 10.3%) No significant changes in other pregnancy outcomes In the I group compared with C2 group: Significant greater decrease in FPG (β −10.99 mg/dL), serum insulin (β −1.95 mIU/mL), HOMA-IR (β −0.76; 95%), TG (β −37.56 mg/dL), VLDL-C (β −7.51 mg/dL), HDL/TC B: (−0.52), hs-CRP (β −1.80 mg/L) and MDA (β −0.43 mmol/L) Significant greater increase in 25-hydroxyvitamin D (β 18.21 ng/mL), QUICKI (β 0.01) HDL-C (β 4.09 mg/dL) and TAC (β 97.77 mmol/L) No significant changes in other metabolic parameters Lower incidence of hyperbilirubinemia in newborns (10.0% vs. 35.7%) Lower incidence of newborns' hospitalization (10.0% vs. 32.1%) No significant changes in other pregnancy outcomes In the C1 group compared with the C2 group Significant greater decrease in FPG (β −8.60 mg/dL), Insulin (β −1.34 µIU/mL), HOMA-IR (β −0.54), TG (β −21.73 mg/dL), VLDL-C (β −4.34 mg/dL) and hs-CRP (β −1.36 mg/L), and MDA (β −0.50 µmol/L) Significant greater increase in serum 25-hydroxyvitamin D (β 2.05 ng/mL)	High dose of vitamin D and probiotic co-supplementation for 6 weeks to women with GDM had beneficial effects on metabolic status and newborns' outcomes compared with placebo and low dose of vitamin D or probiotic supplementation and a low dose of vitamin D
Ostadmohammadi, 2019, Iran [2,3] [28]	Hirsutism: mFG scoring system Mental health: BDI, GHQ-28 and DASS Quality of sleep: PSQI Serum 25-hydroxyvitamin D: ELISA kit Serum total testosterone and SHBG: ELISA kits hs-CRP: ELISA kit Plasma NO: Griess method TAC: Benzie and Strain method GSH: Beutler method MDA: Thiobarbituric acid reactive substances spectrophotometric Test	At baseline: No significant difference between-groups for mean age, height and dietary macro- and micro-nutrient intakes. At end line: In the I group compared with the C group: Significant greater decrease in BDI (β −0.58), GHQ (β − 0.93), DASS (β − 0.90), total testosterone (β − 0.19 ng/mL), hirsutism (β − 0.95), hs-CRP (β − 0.67 mg/L) and MDA (β − 0.25 µmol/L) Significant greater increase in TAC (β 82.81 mmol/L) and GSH (β 40.42 µmol/L) No significant effect on serum SHBG and plasma NO levels, acne, alopecia and PSQI	Vitamin D and probiotic co-supplementation for 12 weeks to women with PCOS had beneficial effects on mental health parameters, but did not affect serum SHBG, plasma NO levels, acne, alopecia and PSQI, compared with placebo

Table 3. *Cont.*

First Author, Year, Country	Outcome Measures	Results	Conclusion
Raygan, 2018, Iran 1 [30]	Serum 25-hydroxyvitamin D: ELISA FPG and lipid profiles: Enzymatic kit Insulin: ELISA kit HOMA-IR and QUICKI: standard formula Hs-CRP: ELISA kit Plasma TAC: Benzie and Strain method GSH: Beutler and Gelbart method MDA: spectrophotometric test NO: Griess method SBP and DBP: sphygmomanometer (Not detailed) Mental health: BDI, BAI, GHQ-28	At baseline and end line: No significant differences between-groups in mean age, height, weight, BMI and METs and macro and micronutrient intakes At end line: In the I group compared with the C group: Significant greater decrease in BDI (-2.8 ± 3.8 vs. -0.9 ± 2.1), BAI (-2.1 ± 2.3 vs. -0.8 ± 1.4) and GHQ scores (-3.9 ± 4.1 vs. -1.1 ± 3.4), Insulin (μIU/mL) (-2.8 ± 3.8 vs. $+0.2 \pm 4.9$), HOMA-IR (-1.0 ± 1.6 vs. -0.1 ± 1.5), and hs-CRP (ng/mL) (-950.0 ± 1811.2 vs. $+260.5 \pm 2298.2$) Significant greater increase in 25-hydroxyvitamin D (ng/mL) ($+11.8 \pm 5.9$ vs. $+0.1 \pm 1.4$), QUICKI ($+0.03 \pm 0.04$ vs. -0.001 ± 0.01), serum HDL-cholesterol (mg/dL) ($+2.3 \pm 3.5$ vs. -0.5 ± 3.8), plasma NO (μmol/L) ($+1.7 \pm 4.0$ vs. -1.4 ± 6.7) and plasma TAC (mmol/L) ($+12.6 \pm 41.6$ vs. -116.9 ± 324.2) No significant different changes in FPG, Triglycerides, VLDL-Cholesterol, LDL-Cholesterol, GSH, MDA, SBP and DBP	Vitamin D and probiotic co-supplementation for 12 weeks to diabetic people with CHD had beneficial effects on mental health, glycemic control, HDL-cholesterol levels, hs-CRP, NO and TAC, but did not affect other metabolic profiles and blood pressures, compared with placebo
Savino, 2015, Italy [25]	Administration of pain-relieving agents (cimetropium bromide at least three times per week or simethicone at least five times per week): daily reporting by parents % of infants switching from exclusive breastfeeding to partial or exclusive formula feeding: not detailed Number of phone-calls and visits due to infantile colic: noted by the pediatrician.	In the I group compared with the C group: - Significantly lower use of pain-relieving agents: Cimetropium bromide: RR: 0.04 (95%CI: 0.01–0.31); Simethicone: RR: 0.24 (95%CI: 0.14–0.41) - Significantly lower use of infant formula: RR: 0.37 (95%CI: 0.17–0.80) - Significantly lower number of calls to the pediatrician: 5.04 ± 2.64 vs. 8.40 ± 3.58 - Significantly lower number of visits in the pediatric ambulatory: 2.66 ± 1.77 vs. 4.98 ± 1.89	Vitamin D and probiotic co-supplementation for 12 weeks to newborns was associated with a reduction of pediatric consultations for infantile colic, use of pain-relieving agents and of infant formula, compared with vitamin D supplementation
Tazzyman, 2015, United Kingdom [26]	Serum 25(OH)D: Cobas e411 automated immunoassay Dietary intake: Food frequency questionnaire analyzed using FETA open source software IBS symptom: questionnaire assessing abdominal pain (pain severity and number of days with pain), bloating, bowel habits (minimum and maximum bowel movement per day and satisfaction with bowel habit) and quality of life	At baseline: No significant differences between-groups At end line: In the I and C2 groups compared with the C1 group: - Significantly higher 25OHD (ng/mL) (37.2 ± 9.3 and 37.1 ± 11.7 vs. 25.3 ± 8.0) No significant between-group differences for any symptom tested, and total symptom severity (same results obtained for participants who were 25(OH)D-deficient at baseline)	Vitamin D and probiotic co-supplementation had no significant effect on the symptoms of IBS, compared with vitamin D alone, or placebo

25(OH)D: 25-hydroxyvitamin D; AFI: Amniotic Fluid Index; BAI: Beck Anxiety Inventory; BALP: Bone-Specific Alkaline Phosphatase; BDI: Beck Depression Inventory; BMD: Bone Mineral Density; BMI: Body Mass Index; BPRS: Brief Psychiatric Rating Scale; C: Control; CHD: Coronary Heart Disease; CI: Confidence Interval; CXT: Collagen Type 1 Cross-Linked C-Telopeptide; DASS: Depression Anxiety and Stress Scale; DBP: Diastolic Blood Pressure; ELISA: Enzyme-Linked Immunosorbent Assay; FBG: Fasting plasma glucose; GDM: Gestational Diabetes Mellitus; GHQ-28: General Health Questionnaire-28; GSH: Total Glutathione; HDL-C: High-Density Lipoprotein Cholesterol; HOMA-IR: Homeostasis Model of Assessment-Insulin Resistance; hs-CRP: High-Sensitivity C-Reactive Protein; I: Intervention; IBS: Irritable Bowel Syndrome; IL: Interleukin; LDL-C: Low-Density Lipoprotein Cholesterol; MDA: Malondialdehyde; MET: Metabolic Equivalent; mFG: modified Ferriman-Gallwey; NO: Nitric oxide; PCOS: Polycystic Ovary Syndrome; PSQI: Pittsburgh Sleep Quality Index; PTH; Parathyroid Hormone; QUICKI: Quantitative Insulin Sensitivity Check Index; RANKL: Serum Total Receptor Activator of Nuclear Factor-kB Ligand; RR: Relative Risk; SBP: Systolic Blood Pressure; SHGB: Sex Hormone-Binding Globulin; T2DM: Type 2 Diabetes Mellitus; TAC: Total Antioxidant Capacity; TC: Total cholesterol; TG: TNF: Tumor Necrosis Factor; Triglycerides; VLDL-C: Very Low-Density Lipoprotein Cholesterol. [1] Significance obtained for the time × group interaction, computed by analysis of the one-way repeated measures ANOVA. [2] Outcome measures refer to the change in values of measures of interest between baseline and end line in each group. [3] β: difference in the mean outcomes measures between treatment groups, and significance obtained from multiple regression model (adjusted for baseline values of each variable).

4. Discussion

So far, probiotic or vitamin D trials have shown major inconsistency in preventive or therapeutic effects on various health outcomes. The emergence of promising experimental studies on the interplay between vitamin D/VDR and probiotics in modulating the gut microbiota and influencing health and disease has led to several clinical trials of a combined supplementation in human subjects. Our exhaustive search identified seven eligible studies, which were included in our review. Our results show that a combined supplementation with vitamin D and probiotics was mostly more beneficial than placebo, vitamin D or probiotics alone in improving health outcomes in various populations, and suggest a dose-dependent effect.

Vitamin D deficiency had long been seen as a concern in metabolic and inflammatory disorders [32–34]. In the included studies, the majority of inflammatory markers improved with the co-supplementation. It is now evident that VDR expression regulates responses to inflammation through numerous mechanisms, such as inhibiting the nuclear factor-kappa B (NF-kB) pathway and activating autophagy [6]. VDR has an essential role for innate immune cells in intestinal inflammation, whereby the deletion of VDR in macrophages and granulocytes significantly increases the expression of pro-inflammatory cytokines in the colon [35]. In contrast, VDR signaling stimulates anti-inflammatory cytokine secretion [36]. Being a transcription factor, VDR can regulate the expression and signaling of target genes involved in intestinal inflammation and dysbiosis, such as *Atg16l1* [6]. A genome-wide association study of the gut microbiota showed that VDR gene variation in humans influences the intestinal microbiota [37]. Genetic variation at the VDR locus significantly influences microbial co-metabolism and the gut–liver axis [37]. Another study in VDR knockout mice found that the lack of VDR in the intestine leads to dysbiosis, with profound alterations in the gut microbiome profile characterized by an increased abundance of *Bacteroidaceae* [38]. However, to date, the mechanisms behind the change of human VDR protein after using vitamin D supplementation and its role in regulating the gut microbiome in health and inflammation are not entirely known [6]. In parallel, the anti-inflammatory markers and properties of probiotics are reliant on VDR expression [39]. There are data showing that probiotic treatment enhances VDR expression and activity in the host. In a probiotic mono-associated pig model, treatment with *Lactobacillus plantarum* in cultured intestinal epithelial cells resulted in an increase in VDR expression and cathelicidin mRNA [39]. Other data show that probiotics did not inhibit inflammation in mice lacking VDR [39]. Future research is needed to enhance our understanding of the complex interplay of nuclear receptors and probiotics, specifically VDR's contribution to probiotic-induced anti-inflammation and its potential role in inflammatory conditions such as inflammatory bowel diseases [39].

Besides, our review documented improvement in insulin sensitivity, anti-oxidative patterns, and dyslipidemia markers with co-supplementation of vitamin D and probiotics. The same positive direction was also highlighted elsewhere [6,8]. Previous research documented a functional link existing between probiotic metabolism and nuclear receptors involved in regulating insulin sensitivity [22]. In a mice model of genetic dyslipidemia and intestinal inflammation, supplementation with a mixture of probiotic strains, including *Streptococcus thermophiles, Bifidobacterium breve, Bifidobacterium lactis, Lactobacillus acidophilus, Lactobacillus plantarum, Lactobacillus paracasei*, and *Lactobacillus helveticus* modified the nuclear receptors' expression including VDR, and caused their direct transactivation, leading to reversing insulin resistance in liver and fat tissues and protecting against steatohepatitis and atherosclerosis [40]. Yet, these results although emanating from high-quality studies, are far from being conclusive, and future trials are needed before we can confidently establish the effectiveness and superiority of this co-supplementation.

More human experimental studies are needed to fully elucidate the interplay between nuclear receptors and probiotics in metabolic diseases. Shaping our understanding of this unexplored path might pave the way for multi-target preventive and therapeutic strategies, especially in situations where dietary and lifestyle changes have failed [22].

Additionally, improvement in mental health has been reported in this review. Vitamin D is involved in numerous brain processes including neuroimmunomodulation, neuroprotection, as well as brain development; all of which suggests a link between vitamin D and mental health [41,42]. Vitamin D may positively affect mental health through up-regulating tyrosine hydroxylase gene expression and increasing bioavailability of key neurotransmitters, such as norepinephrine and dopamine [43]. In parallel, mechanisms through which gut bacteria can affect mental status include microflora biosynthesis and the regulation of neurotransmitters, including serotonin [44] and gamma aminobutyric acid (GABA) [45]. Existing evidence also pinpoint an association between mood disorders and gut microbiota, and specify a role of the gut–brain axis in the physiopathology of clinical depression [46]. It is highly plausible that the synergism in vitamin D and probiotics' anti-inflammatory, antioxidant, and immunomodulatory effects might augment their impact on mental health. This is yet to be confirmed by future interventional human studies.

The only study in this review that reported null results with the co-supplementation was a trial by Tazzyman et al. [26] which did not show any improvement in the symptoms of patient with IBS whose vitamin D was repleted. This study had a limited sample size (underpowered trial), and a limited duration of follow-up. Additionally, in that study, the group receiving placebo showed an improvement in vitamin D levels, which might be due to seasonal differences in sun exposure, and a placebo effect was observed on symptom scores. The authors speculated that increased sunlight exposure had increased vitamin D levels which in turn improved IBS symptoms. All of these limitations may have prevented the authors from detecting a significant difference in symptom scores between the placebo and supplemented groups. Additionally, individuals might need higher doses of vitamin D plus probiotic supplementation for a longer period of time to provide appropriate circulating levels for improving symptoms.

Understanding the mechanisms of the interplay between vitamin D and probiotics in modulating the gut microbiota and regulating host responses, and exploring the effectiveness of this form of supplementation in high-quality human studies are crucial before applying it to prevent and manage disease. Studies included in this review had revealed thoroughly the superiority of co-supplementing with vitamin D and probiotics. Vitamin D has shown benefits in cellular restoration and reducing inflammation. The latter has been implicated in the pathophysiology of an unlimited number of conditions and diseases. VDR expression and transcriptional activity can be a research focus for future genetic studies. In parallel, data about probiotics and their role in optimizing microbiota and absorption pathways would be very useful not only for vitamin D but for many other nutrients or enzymes involved to boost immunity and host response.

5. Strengths and Limitations

To our knowledge, this is the first review to systematically compile human interventional evidence on the effectiveness of a combined supplementation of vitamin D and probiotics. Our review has numerous strengths [47]. It was conducted following standard methods for reporting systematic reviews [23], and according to a pre-defined protocol, which was published a priori. To increase the comprehensiveness of our search, we searched multiple scientific databases and the grey literature, and did not limit our search to any publication language or time. All the steps of study inclusion, data extraction and quality assessment were conducted in duplicate. We only included RCTs, and assessed their risk of bias using a validated tool; and, in general, the included studies were of high quality. However, included trials were limited in number, and conditions assessed. They were also limited by the small sample size, and short duration of follow-up. Moreover, only two studies [25,26] provided details regarding the strain of bacteria in the used probiotics. None of the studies provided analyses of the gut microbiota, disabling us from establishing whether the co-supplementation changed the composition of the microbiota, or ascertaining whether the observed changes were due to changes in the gut microbiota. Furthermore, we could not pool the studies in a meta-analysis due to the heterogeneity in the populations, conditions assessed, outcomes, doses and forms of vitamin D supplemented, and doses and strains of probiotics supplemented.

6. Conclusions

A combined supplementation with vitamin D and probiotics seems to play a role on the physiological and psychological attributes of the human body, and represents a novel insight in the management of chronic diseases. The findings of this systematic review suggest a superiority of vitamin D and probiotics supplementation over placebo, vitamin D or probiotics alone, and propose a dose-dependent effect. However, solid conclusions cannot be drawn at this level, and these findings remain certainly not robust enough and should be interpreted with caution. Future high-quality studies in other disease areas and various populations are needed to confirm these findings and to inform on the form, composition, and frequency of this co-supplementation for optimal outcomes.

Supplementary Materials: The following are available online at https://www.mdpi.com/2072-6643/13/1/111/s1, Table S1: Characteristics of included studies, Table S2: Outcomes and results of included studies.

Author Contributions: Conceptualization M.A. and R.R.; design M.A. and R.R.; methodology M.A. and R.R.; S.H. performed the searches; project administration F.A. and D.P.; funding acquisition S.H., R.R. and M.A. writing-review and editing S.H., R.R. and M.A.; writing-original draft N.M. All authors have read and agreed to the published version of the manuscript.

Funding: College of Natural and Health Sciences, Zayed University, Dubai, United Arab Emirates. The funding body will not be involved in the design of the study and collection, analysis, and interpretation of data or in writing the manuscript. Cluster grant R18030.

Institutional Review Board Statement: Not applicable to this review.

Informed Consent Statement: Not applicable to this review.

Acknowledgments: We would like to thank Aida Farha for her assistance in developing the search strategy.

Conflicts of Interest: The authors declare that they have no competing interests.

References

1. Quigley, E.M.M. Microbiota-Brain-Gut Axis and Neurodegenerative Diseases. *Curr. Neurol. Neurosci. Rep.* **2017**, *17*, 94. [CrossRef]
2. Ogbu, D.; Xia, E.; Sun, J. Gut instincts: Vitamin D/vitamin D receptor and microbiome in neurodevelopment disorders. *Open Biol.* **2020**, *10*, 200063. [CrossRef]
3. Sun, J. Dietary Vitamin D, Vitamin D Receptor, and Microbiome. *Curr. Opin. Clin. Nutr. Metab. Care* **2018**, *21*, 471–474. [CrossRef] [PubMed]
4. Strandwitz, P. Neurotransmitter modulation by the gut microbiota. *Brain Res.* **2018**, *1693*, 128–133. [CrossRef] [PubMed]
5. Butel, M.-J.; Waligora-Dupriet, A.-J.; Wydau-Dematteis, S. The developing gut microbiota and its consequences for health. *J. Dev. Orig. Health Dis.* **2018**, *9*, 590–597. [CrossRef] [PubMed]
6. Bakke, D.; Chatterjee, I.; Agrawal, A.; Dai, Y.; Sun, J. Regulation of Microbiota by Vitamin D Receptor: A Nuclear Weapon in Metabolic Diseases. *Nucl. Recept. Res.* **2018**, *5*. [CrossRef] [PubMed]
7. Hawrelak, J.A.; Myers, S.P. The Causes of Intestinal Dysbiosis: A Review. *Altern. Med. Rev.* **2004**, *9*, 180–197.
8. Battistini, C.; Nassani, N.; Saad, S.M.; Sun, J. Probiotics, Vitamin D, and Vitamin D Receptor in Health and Disease. In *Lactic Acid Bacteria*; Cavalcanti de Albuquerque, M.A., de Moreno de LeBlanc, A., LeBlanc, J.G., Bedani, R., Eds.; CRC Press: Boca Raton, FL, USA, 2020; pp. 93–105. ISBN 978-0-429-42259-1.
9. Shang, M.; Sun, J. Vitamin D/VDR, Probiotics, and Gastrointestinal Diseases. *Curr. Med. Chem.* **2017**, *24*, 876–887. [CrossRef]
10. Holick, M.F.; Chen, T.C. Vitamin D deficiency: A worldwide problem with health consequences. *Am. J. Clin. Nutr.* **2008**, *87*, 1080S–1086S. [CrossRef]
11. Trehan, N.; Afonso, L.; Levine, D.L.; Levy, P.D. Vitamin D Deficiency, Supplementation, and Cardiovascular Health. *Crit. Pathw. Cardiol.* **2017**, *16*, 109–118. [CrossRef]
12. AlAnouti, F.; Abboud, M.; Papandreou, D.; Mahboub, N.; Haidar, S.; Rizk, R. Effects of Vitamin D Supplementation on Lipid Profile in Adults with the Metabolic Syndrome: A Systematic Review and Meta-Analysis of Randomized Controlled Trials. *Nutrients* **2020**, *12*, 3352. [CrossRef] [PubMed]
13. Abboud, M. Vitamin D Supplementation and Blood Pressure in Children and Adolescents: A Systematic Review and Meta-Analysis. *Nutrients* **2020**, *12*, 1163. [CrossRef] [PubMed]

14. Spedding, S. Vitamin D and Depression: A Systematic Review and Meta-Analysis Comparing Studies with and without Biological Flaws. *Nutrients* **2014**, *6*, 1501–1518. [CrossRef] [PubMed]
15. Zmijewski, M.A. Vitamin D and human health. *Int. J. Mol. Sci.* **2019**, *20*, 145. [CrossRef]
16. Batacchi, Z.; Robinson-Cohen, C.; Hoofnagle, A.N.; Isakova, T.; Kestenbaum, B.; Martin, K.J.; Wolf, M.S.; De Boer, I.H. Effects of vitamin D2 supplementation on vitamin D3 metabolism in health and CKD. *Clin. J. Am. Soc. Nephrol.* **2017**, *7*, 1498–1506. [CrossRef]
17. Sivamaruthi, B.S.; Kesika, P.; Suganthy, N.; Chaiyasut, C. A Review on Role of Microbiome in Obesity and Antiobesity Properties of Probiotic Supplements. *BioMed Res. Int.* **2019**, 1–20. [CrossRef]
18. Isolauri, E. Probiotics in the Development and Treatment of Allergic Disease. *Gastroenterol. Clin.* **2012**, *41*, 747–762. [CrossRef]
19. Tomaro-Duchesneau, C.; Saha, S.; Malhotra, M.; Jones, M.L.; Labbé, A.; Rodes, L.; Kahouli, I.; Prakash, S. Effect of orally administered L. fermentum NCIMB 5221 on markers of metabolic syndrome: An in vivo analysis using ZDF rats. *Appl. Microbiol. Biotechnol.* **2014**, *98*, 115–126. [CrossRef]
20. Varankovich, N.V.; Nickerson, M.T.; Korber, D.R. Probiotic-based strategies for therapeutic and prophylactic use against multiple gastrointestinal diseases. *Front. Microbiol.* **2015**, *6*. [CrossRef]
21. Mohammadi-Sartang, M.; Bellissimo, N.; Mazloomi, S.M.; Fararouie, M.; Bedeltavana, A.; Famouri, M.; Mazloom, Z. The effect of daily fortified yogurt consumption on weight loss in adults with metabolic syndrome: A 10-week randomized controlled trial. *Nutr. Metab. Cardiovasc. Dis.* **2018**, *28*, 565–574. [CrossRef]
22. Shang, M.; Sun, J. Vitamin D, VDR, and probiotics in health and disease: A mini review. *CAB Rev.* **2017**, *24*, 876–887.
23. Moher, D.; Liberati, A.; Tetzlaff, J.; Altman, D.G. Preferred Reporting Items for Systematic Reviews and Meta-Analyses: The PRISMA Statement. *PLoS Med.* **2009**, *6*, e1000097. [CrossRef] [PubMed]
24. Higgins, J.P.; Thomas, J.; Chandler, J.; Cumpston, M.; Li, T.; Page, M.J.; Welch, V.A. (Eds.) *Cochrane Handbook for Systematic Reviews of Interventions*; John Wiley & Sons: New York, NY, USA, 2019.
25. Savino, F.; Ceratto, S.; Poggi, E.; Cartosio, M.E.; Cordero di Montezemolo, L.; Giannattasio, A. Preventive effects of oral probiotic on infantile colic: A prospective, randomised, blinded, controlled trial using *Lactobacillus reuteri* DSM 17938. *Benef. Microbes* **2015**, *6*, 245–251. [CrossRef] [PubMed]
26. Tazzyman, S.; Richards, N.; Trueman, A.R.; Evans, A.L.; Grant, V.A.; Garaiova, I.; Plummer, S.F.; Williams, E.A.; Corfe, B.M. Vitamin D associates with improved quality of life in participants with irritable bowel syndrome: Outcomes from a pilot trial. *BMJ Open Gastroenterol.* **2015**, *2*, e000052. [CrossRef] [PubMed]
27. Ghaderi, A.; Banafshe, H.R.; Mirhosseini, N.; Moradi, M.; Karimi, M.-A.; Mehrzad, F.; Bahmani, F.; Asemi, Z. Clinical and metabolic response to vitamin D plus probiotic in schizophrenia patients. *BMC Psychiatry* **2019**, *19*, 77. [CrossRef] [PubMed]
28. Ostadmohammadi, V. Vitamin D and probiotic co-supplementation affects mental health, hormonal, inflammatory and oxidative stress parameters in women with polycystic ovary syndrome. *J. Ovarian. Res.* **2019**, *12*. [CrossRef]
29. Jamilian, M.; Amirani, E.; Asemi, Z. The effects of vitamin D and probiotic co-supplementation on glucose homeostasis, inflammation, oxidative stress and pregnancy outcomes in gestational diabetes: A randomized, double-blind, placebo-controlled trial. *Clin. Nutr.* **2019**, *38*, 2098–2105. [CrossRef]
30. Raygan, F. The effects of vitamin D and probiotic co-supplementation on mental health parameters and metabolic status in type 2 diabetic patients with coronary heart disease: A randomized, double-blind, placebo-controlled trial. *Prog. Neuropsychopharmacol.* **2018**, *8*, 50–55. [CrossRef]
31. Jafarnejad, S.; Djafarian, K.; Fazeli, M.R.; Yekaninejad, M.S.; Rostamian, A.; Keshavarz, S.A. Effects of a Multispecies Probiotic Supplement on Bone Health in Osteopenic Postmenopausal Women: A Randomized, Double-blind, Controlled Trial. *J. Am. Coll. Nutr.* **2017**, *36*, 497–506. [CrossRef]
32. Autier, P.; Boniol, M.; Pizot, C.; Mullie, P. Vitamin D status and ill health: A systematic review. *Lancet Diabetes Endocrinol.* **2014**, *2*, 76–89. [CrossRef]
33. Schippa, S.; Conte, M.P. Dysbiotic Events in Gut Microbiota: Impact on Human Health. *Nutrients* **2014**, *6*, 5786–5805. [CrossRef] [PubMed]
34. Zeng, M.Y.; Inohara, N.; Nuñez, G. Mechanisms of inflammation-driven bacterial dysbiosis in the gut. *Mucosal Immunol.* **2017**, *10*, 18–26. [CrossRef] [PubMed]
35. Leyssens, C.; Verlinden, L.; De Hertogh, G.; Kato, S.; Gysemans, C.; Mathieu, C.; Carmeliet, G.; Verstuyf, A. Impact on Experimental Colitis of Vitamin D Receptor Deletion in Intestinal Epithelial or Myeloid Cells. *Endocrinology* **2017**, *158*, 2354–2366. [CrossRef] [PubMed]
36. Barragan, M.; Good, M.; Kolls, J. Regulation of Dendritic Cell Function by Vitamin D. *Nutrients* **2015**, *7*, 8127–8151. [CrossRef] [PubMed]
37. Wang, J.; Thingholm, L.B.; Skiecevičienė, J.; Rausch, P.; Kummen, M.; Hov, J.R.; Degenhardt, F.; Heinsen, F.-A.; Rühlemann, M.C.; Szymczak, S.; et al. Genome-wide association analysis identifies variation in vitamin D receptor and other host factors influencing the gut microbiota. *Nat. Genet.* **2016**, *48*, 1396–1406. [CrossRef] [PubMed]
38. Wu, S.; Zhang, Y.; Lu, R.; Xia, Y.; Zhou, D.; Petrof, E.O.; Claud, E.C.; Chen, D.; Chang, E.B.; Carmeliet, G.; et al. Intestinal epithelial vitamin D receptor deletion leads to defective autophagy in colitis. *Gut* **2015**, *64*, 1082–1094. [CrossRef]
39. Yoon, S.S.; Sun, J. Probiotics, Nuclear Receptor Signaling, and Anti-Inflammatory Pathways. *Gastroenterol. Res. Pract.* **2011**, *2011*, 1–16. [CrossRef]

40. Mencarelli, A.; Cipriani, S.; Renga, B.; Bruno, A.; D'Amore, C.; Distrutti, E.; Fiorucci, S. VSL#3 Resets Insulin Signaling and Protects against NASH and Atherosclerosis in a Model of Genetic Dyslipidemia and Intestinal Inflammation. *PLoS ONE* **2012**, *7*, e45425. [CrossRef]
41. Fernandes de Abreu, D.A.; Eyles, D.; Féron, F. Vitamin D, a neuro-immunomodulator: Implications for neurodegenerative and autoimmune diseases. *Psychoneuroendocrinology* **2009**, *34*, S265–S277. [CrossRef]
42. Bertone-Johnson, E.R. Vitamin D and the occurrence of depression: Causal association or circumstantial evidence? *Nutr. Rev.* **2009**, *67*, 481–492. [CrossRef]
43. Humble, M.B. Vitamin D, light and mental health. *J. Photochem. Photobiol. B* **2010**, *101*, 142–149. [CrossRef] [PubMed]
44. Yano, J.M.; Yu, K.; Donaldson, G.P.; Shastri, G.G.; Ann, P.; Ma, L.; Nagler, C.R.; Ismagilov, R.F.; Mazmanian, S.K.; Hsiao, E.Y. Indigenous Bacteria from the Gut Microbiota Regulate Host Serotonin Biosynthesis. *Cell* **2015**, *161*, 264–276. [CrossRef] [PubMed]
45. Barrett, E.; Ross, R.P.; O'Toole, P.W.; Fitzgerald, G.F.; Stanton, C. γ-Aminobutyric acid production by culturable bacteria from the human intestine. *J. Appl. Microbiol.* **2012**, *113*, 411–417. [CrossRef]
46. Schmidt, C. Mental health: Thinking from the gut. *Nature* **2015**, *25*, S12–S15. [CrossRef] [PubMed]
47. Shea, B.J.; Reeves, B.C.; Wells, G.; Thuku, M.; Hamel, C.; Moran, J.; Moher, D.; Tugwell, P.; Welch, V.; Kristjansson, E.; et al. AMSTAR 2: A critical appraisal tool for systematic reviews that include randomised or non-randomised studies of healthcare interventions, or both. *BMJ* **2017**, j4008. [CrossRef]

nutrients

Review

Oral Vitamin D Therapy in Patients with Psoriasis

Ana Maria Alexandra Stanescu [1], Anca Angela Simionescu [2,*] and Camelia Cristina Diaconu [3]

[1] Department of Family Medicine, "Carol Davila" University of Medicine and Pharmacy, 050474 Bucharest, Romania; alexandrazotta@yahoo.com

[2] Department of Obstetrics and Gynecology, Filantropia Clinical Hospital, "Carol Davila" University of Medicine and Pharmacy, 050474 Bucharest, Romania

[3] Department of Internal Medicine, Clinical Emergency Hospital of Bucharest, "Carol Davila" University of Medicine and Pharmacy, 050474 Bucharest, Romania; drcameliadiaconu@gmail.com

*   Correspondence: anca.simionescu@umfcd.ro; Tel.: +40-213-188-930

Abstract: Vitamin D treatment is effective when applied topically to the skin for plaque-type psoriasis. Oral vitamin D supplementation might be effective as an adjuvant treatment option in psoriasis. This umbrella review aimed to highlight the current knowledge regarding the use of oral vitamin D for treatment of patients with psoriasis. We performed a literature search and identified 107 eligible full-text articles that were relevant to the research interest. Among these, 10 review articles were selected, and data were extracted. A data synthesis showed that only a few studies monitored oral vitamin D efficacy in patients with psoriasis. No studies investigated the optimal dose of systemic vitamin D in psoriasis. However, most studies did not observe side effects for doses within a relatively narrow range (0.25 to 2 μg/day). These results suggest that more large-scale studies are needed to determine the efficacy, optimal dose, and adverse effects of vitamin D administration in patients with psoriasis.

Keywords: psoriasis; oral vitamin D; treatment

Citation: Stanescu, A.M.A.; Simionescu, A.A.; Diaconu, C.C. Oral Vitamin D Therapy in Patients with Psoriasis. *Nutrients* **2021**, *13*, 163. https://doi.org/10.3390/nu13010163

Received: 10 December 2020
Accepted: 4 January 2021
Published: 6 January 2021

Publisher's Note: MDPI stays neutral with regard to jurisdictional clai-ms in published maps and institutio-nal affiliations.

1. Introduction

Vitamin D is an essential nutrient in humans; it is produced by the body through exposure to the sun (the primary source of vitamin D), or more precisely, to mild ultraviolet B (UVB) light. Other sources of vitamin D include food and dietary supplements [1]. In 1928, the chemist and medical doctor Adolf Otto Reinhold Windaus was awarded the Nobel Prize for chemistry for the discovery of vitamin D [1–3]. Chemically, vitamin D_2 was first characterized in 1932, and vitamin D_3 was characterized in 1936. Currently, vitamin D is known as a hormone that regulates calcium-phosphorus homeostasis and protects the integrity of the skeletal system [4]. Vitamin D levels are influenced by many factors, including the season, period of sun exposure, time of the day, latitude, use of sunscreen, clothing, skin color, body weight, and medical conditions [5,6].

When epidermal cells are exposed to UVB, 7-dehydrocholesterol can be transformed into pre-vitamin D, which isomerizes to vitamin D_3 [7]. Next, vitamin D_3 undergoes 25-hydroxylation, through an enzymatic conversion in the liver, to form 25(OH) vitamin D (calcidiol), the primary circulating form of vitamin D. The plasma half-life of 25(OH) vitamin D is 2–3 weeks. Calcidiol is converted in the kidneys by 1-alpha-hydroxylation to the most active form, 1,25(OH)2D (calcitriol), which has a plasma half-life of 4–6 h [8]. This entire process is modulated by parathyroid hormone, hypophosphatemia, growth hormone, and other mediators.

Psoriasis is a chronic autoimmune skin disease with a strong genetic predisposition, characterized by sustained inflammation and followed by uncontrolled proliferation of ker-atinocytes and dysfunctional differentiation [9]. The first-line therapy for mild-to-moderate psoriasis is topical administration of corticosteroids and vitamin D analogues [10,11]. Keratinocytes and lymphocytes that infiltrate the lesions express the vitamin D receptor, which explains the effectiveness of this therapy in psoriasis [12].

The pathogenesis of psoriasis is not fully elucidated. The development of psoriasis plaques is mediated by Th1 cells and connected to keratinocyte hyperproliferation. This connection could explain the efficacy of immunosuppressive and antiproliferative vitamin D-like compounds, such as calcipotriol, in psoriasis [13]. Ligands for vitamin D receptor inhibit the expression of pro-inflammatory cytokines produced by T lymphocytes (i.e., IL-2, IFN-γ, IL-6, and IL-8) [14]. Thus, the biological activity of vitamin D_3 analogues leads to suppression of the T cell-mediated immune response. Moreover, dendritic antigen-presenting cells are modulated by 1a,25(OH)2D3 and its analogues, which inhibit the differentiation, maturation, activation, and survival of these cells [15]. Given current knowledge, it is reasonable to assume that epidermal production of vitamin D could be at least partially affected in skin psoriatic lesions, which may contribute to worsening symptoms.

Current knowledge, which holds that vitamin D treatment applied to the skin is effective, has given rise to the possibility that oral vitamin D supplementation might be an effective adjuvant treatment option in psoriasis. Due to the controversial and understudied nature of this topic, this umbrella review aimed to summarize current evidence, with an emphasis on clinical outcomes, on oral vitamin D treatment in patients with psoriasis. The need for this umbrella review derives from the controversies on this subject and the lack of systematic investigations.

2. Materials and Methods

2.1. Search Strategy

Our review strategy was based on the Preferred Reporting Items for Systematic reviews and Meta-Analyses (PRISMA) checklist [16]. We performed a literature search in August 2020 in PubMed and Scopus. The search included the period 2010–2020, and we used the following search terms: "oral vitamin D" AND "psoriasis" AND "treatment" [all text].

2.2. Inclusion and Exclusion Criteria

We scanned the full text of each identified article for relevance to the research interest. All articles written in the English language that addressed oral vitamin D and its analogue treatment in patients with psoriasis and a posttreatment score evaluation (PASI score—Psoriasis Area Severity Index and patient global assessment) were included. Based on the umbrella review typology, we only selected review-type articles, including clinical cases. We excluded articles that did not have a main focus on oral vitamin D administration in psoriasis as a monotherapy, those that only mentioned a phrase regarding this type of administration, studies having less than 2 patients included or psoriasis-associated diseases, studies that compared vitamin D effects and corticosteroids, and reports from meeting abstracts. We did not apply restrictions on the age of inclusion or the type or severity of psoriasis.

2.3. Data Extraction

The data were extracted and summarized in a table (Table 1). The characteristics of individual studies included in the review articles were the number of patients, type of study, and study location.

Table 1. The characteristics of the original studies included in the analyzed reviews.

Authors and Year	Type of Study	Number of Patients	Study Location	Reviews Including the Original Study from the First Column
Morimoto et al., 1986 [17]	Open-design study	21	Japan	Kamangar et al., 2013 [18] Lourenceti et al., 2018 [19] Soleymani et al., 2015 [20] Millsop et al., 2014 [21] Bouillon et al., 2018 [22]
Takamoto et al., 1986 [23]	Descriptive study	7	Japan	Kamangar et al., 2013 [18] Lourenceti et al., 2018 [19]
Smith et al., 1988 [24]	Descriptive study	14	USA	Kamangar et al., 2013 [18] Lourenceti et al., 2018 [19] Millsop et al., 2014 [21] Bouillon et al., 2018 [22] Hambly et al., 2017 [25]
Holland et al., 1989 [26]	Descriptive study	15	UK	Hambly et al., 2017 [25]
Huckins et al., 1990 [27]	Open-label trial	6	USA	Kamangar et al., 1990 [18] Lourenceti et al., 2018 [19]
Siddiqui et al., 1990 [28]	Prospective randomized double-blind control study	41	Saudi Arabia	Millsop et al., 2014 [21] Zuccotti et al., 2018 [29]
Lugo-Somolinos et al., 1990 [30]	Descriptive study	10	Puerto Rico	Hambly et al., 2017 [25]
El-Alzhari et al., 1993 [31]	Descriptive study	8	USA	Lourenceti et al., 2018 [19] Millsop et al., 2014 [21]
Perez et al., 1996 [32]	Open trial	85	USA	Kamangar et al., 2013 [18] Lourenceti et al., 2018 [19] Soleymani et al., 2015 [20] Millsop et al., 2014 [21] Barrea et al., 2017 [33] Bouillon et al., 2018 [22] Hambly et al., 2017 [25]
Gaal et al., 2009 [34]	Case-control	10	USA	Kamangar et al., 2013 [18] Zuccotti et al., 2018 [29]
Finamor et al., 2013 [35]	Open-label clinical trial	9	Hungary	Lourenceti et al., 2018 [19] Millsop et al., 2014 [21] Umar et al., 2018 [36] Hambly et al., 2017 [25]
Hata et al., 2014 [37]	Randomized placebo-controlled	16	Brazil	Hambly et al., 2017 [25]
Jarret et al., 2018 [38]	Randomized double blind, placebo-controlled study	65	USA	Zuccotti et al., 2018 [29]
Ingram et al., 2018 [39]	Randomized double blind, placebo-controlled study	101	New Zealand	
Disphanurat et al., 2019 [40]	Randomized double blind, placebo-controlled study	45	Thailand	Marino et al., 2019 [41]

3. Results

We followed the PRISMA principles in developing this review (Figure 1). In total, after searching for keywords, we identified 395 records. Duplicates were removed, and after applying the other search criteria, we screened 107 eligible full-text articles. According to the established criteria, 10 review articles were included in the final analysis.

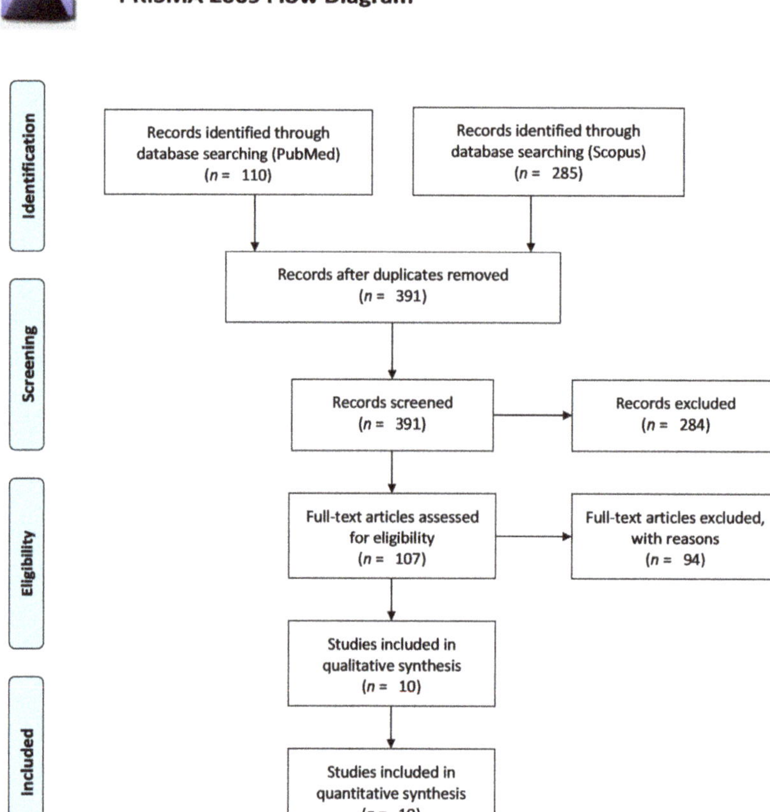

Figure 1. PRISMA 2009 Flow Diagram. From Moher, D.; Liberati, A.; Tetzlaff, J.; Altman, D.G. The Prisma Group (2009). Preferred reporting items for systematic reviews and meta-analyses: The PRISMA statement. PLoS Med 6(7): e1000097. https://doi.org/10.1371/journal.pmed.1000097. For more information, visit www.prisma-statement.org [16].

A relatively small number of studies have investigated the effectiveness of oral vitamin D in patients with psoriasis. Accordingly, we identified a fairly small number of systemic reviews and meta-analyses. Some reviews discussed transient oral administration of vitamin D in other contexts or as a subset of cutaneous vitamin D therapy. The characteristics of the included original studies are shown in Table 1.

To our knowledge, the first case of psoriasis treated with 1-alpha hydroxyvitamin D_3 for osteoporosis was reported in 1985, and the treatment resulted in psoriasis remission [29]. This case led to further research on the effects of systemic vitamin D administration on psoriasis. In 2013, Kamangar et al. studied oral vitamin D in patients with psoriasis and in patients with psoriatic arthritis. In most cases, psoriasis improved visibly after treatment with 0.25 µg to 1 µg/day of 1,25-(OH)2D3, with no adverse effects. The authors concluded that oral vitamin D was a safe and effective therapeutic option for treating psoriasis vulgaris [18]. Treatment effectiveness after oral administration of vitamin D_3 and D_2 in patients with psoriasis, based on the original studies, is shown in Tables 2 and 3.

Table 2. Treatment effectiveness after oral administration of vitamin D$_3$ in patients with psoriasis.

Individual Studies, Year	Dose	Duration of Administration	Efficacy	Type/Severity of Psoriasis	Effectiveness	Treatment Side Effects
Morimoto et al., 1986 [17]	1.0 μg/day 1α-(OH)D3 (40 IU/day)	6 months	2.7 +/− 0.6 months	Psoriasis vulgaris	More than moderate improvement (+2) in 76% of patients	No
	0.5 μg/day 1,25-(OH)2-D3 (20 IU/day)	6 months	3 months	Psoriasis vulgaris	Moderate improvement (+2) in 25% of patients	No
Takamoto et al., 1986 [23]	1.0 μg/day 1α-(OH)D3 (40 IU/day)	12 months	more than 8 months	Psoriasis vulgaris	— Complete remission and marked improvement (+3 up to +4) in 28.57% — Minimal improvement (+1) in 15% of patients	No
Smith et al., 1988 [24]	0.25 μg (10 IU) once or twice/day increased by 0.25 to 0.5 μg/day every 2 weeks to a maximum of 2.0 μg (80 IU)/day 1,25-(OH)2-D3	2 months	less than 2 months	moderate to severe psoriasis	— 50% of patients +4 — 21.43% of patients +2/+3 — 21.43% of patients +1 — 7.14% of patients 0	No
Holland et al., 1989 [26]	1.0 μg/day 1α-(OH)D3 (40 IU)	6 months	6–8 weeks	Plaque psoriasis	46.67% of patients had complete resolution of lesions (+4), 2 within 6 weeks and the rest after 4–6 months of therapy.	No
Huckins et al., 1990 [27]	1.0 μg/day 0.5 μg/day increased by 0.25 μg/day every 2 weeks to a maximum of 2.0 μg (80 IU)/day 1,25-(OH)2-D3	6 months	2–3 months	Psoriatic arthritis	— 44.44% of patients marked improvement (+3) — 22.22% of patients presented worsening of their psoriasis during the trial	hypercalciuria in 20% of patients
Siddiqui et al., 1990 [28]	1 μg/day alpha-calcidol	12 weeks	Not specified	Psoriasis vulgaris	45% of patients showed slight improvement (+1).	
Lugo-Somolinos et al., 1990 [30]	0.5 μg/day 1α,25-(OH)2 -D3 (20 IU)		after 3 months	Moderate to severe psoriasis	40% of patients showed moderate improvement.	No
El-Alzhari et al., 1993 [31]	0.5 μg/day increased by 0.5 μg biweekly to a maximal dosage of 2.0 μg daily. 1,25-(OH)2-D3	6 months	2 months	Psoriasis vulgaris moderate to severe	— 12.5% of patients marked improvement (+3) — 12.5% of patients had moderate improvement (+2) — 75% of patients had mild improvement or no improvement (0 to +1)	No
Perez et al., 1996 [32]	0.5 μg/day increments of 0.5 μg every 2 weeks 1,25-(OH)2-D3	6 months–3 years	6 months	Psoriasisvulgaris	Global severity score for the patients' lesions had a mean value of 7.7 ± 1.2; the mean global severity score significantly decreased to 3.2 ± 1.9. The mean baseline PASI score was 18.4 ± 1.0; at 6 and 36 months of treatment the mean PASI score was reduced to 9.7 ± 0.8 and 7.0 ± 1.3, respectively.	No
Gaal et al., 2009 [34]	0.25 μg twice daily 1α-(OH)D3	6 months	Not specified	Psoriatic arthritis	PASI scores were 12.8 +/−14.3 vs. 11.9 +/− 14.4. on average.	No
Finamor et al., 2013 [35]	35,000 IU per day vit. D$_3$	6 months	Not specified	Psoriasis vulgaris moderate to severe	The clinical condition of all patients significantly improved (+3 to +4).	-
Hata et al., 2014 [37]	4000 IU/day vit. D$_3$	6 months	Not specified	Mild psoriasis	No change in PASI score (0)	No
Jarret et al., 2018 [38]	100,000 IU/month (3300 IU/day) vit. D$_3$	4 years	Not specified	Mild psoriasis	The trial results do not support the use of monthly vitamin D3 supplementation (100,000 IU per month) as a treatment for mild psoriasis in patients over 50 years old.	
Ingram et al., 2018 [39]	200,000 IU at baseline, then 100,000 IU/month vit. D$_3$	11 months	6 months	Chronic psoriasis	No benefit	Not specified

Legend: PASI = psoriasis area severity index score; RCT = randomized clinical trial. 250 μg = 10,000 IU. The degree of improvement of psoriasis lesions was scored by the authors using a 5-point scale: 0, no effect; + 1, minimal improvement up to 25% improved; +2, moderate improvement, 26% to 50% improved; +3, marked improvement, 51% to 75% improved; +4, >75% improved to clear lesions; by PASI score; or by Global Severity Score.

Table 3. Treatment effectiveness of oral vitamin D$_2$ administered in patients with psoriasis.

Individual Studies/ Year	Dose	Period of Administration	Efficacy Observed	Type/Severity of Psoriasis	Effectiveness	Treatment Side Effects
Disphanurat et al., 2019 [40]	20,000 IU/every 2 weeks vit. D$_2$	6 months	3–6 months	Chronic plaque-type psoriasis—mild psoriasis	PASI score decreased at 3 and 6 months, moderate improvement	No

The patients were monitored clinically in the included studies, with one of the most commonly used scores being the PASI score, which takes into account the overall severity score and the percentage of body surface area affected by psoriasis. The PASI score has been used to monitor the effectiveness of antipsoriatic medication since 1978 [42].

Table 4 details the scores and clinical modalities used to determine the clinical efficacy of orally administered vitamin D in psoriasis.

Table 4. Psoriasis outcome measures used for treatment effectiveness.

Authors	Evaluation
Morimoto et al. [17]	Clinical photographs taken at every examination Clinical score: complete remission (+4), marked improvement (+3), moderate improvement (+2), slight improvement (+1), no change (o), deterioration (−1).
Smith et al. [24]	Clinical examination Clinical score: no change (0), minimal improvement up to 25% improved (+1), 26% to 50% improved (+2), 51% to 75% improved (+3), >75% improved to clear (+4).
Takamoto et al. [23]	Clinical examination: complete remission (4) (complete flattering of plaques including borders, percentage of area improved: 95% or more); marked improvement (3) (nearly complete flattering of all plaques still palpable, area improved: 50–90%); definite improvement (2) (partial flattering of plaque, less scaling and less erythema, area improved: 20–50%), minimal improvement (1) (slightly less scaling and less erythema, area improved: 5–20%); no change (0); aggravation (−1) by the percentage of skin involvement was improved.
Huckins et al. [27]	Clinical photographs taken at every examination Clinical score of erythema: deterioration (−1), no change (0), mild improvement (1), moderate improvement (2), marked improvement (3)
Gaal et al. [34]	— PASI score
Perez et al. [32]	Clinical photographs taken at every examination PASI score, global severity score Global Improvement Scale: deterioration (−1), no change (0), mild improvement (1), moderate improvement (2), excellent improvement (3)
El-Azhary et al. [31]	Clinical evaluation of the percentage of body surface involved Grading the erythema, scale, and thickness of the lesions as worsening (−1), no improvement (0), mild improvement (+1), moderate improvement (+2), marked improvement (+3).
Finamor et al. [35]	— PASI score
Siddiqui et al. [28]	PASI score Worsening PASI score (−1), no improvement (0), slight improvement (+1), moderate improvement (+2), marked improvement (+3).
Holland et al. [26]	— Clinical photographs taken — Clinical criteria
Hata et al. [37]	PASI score Punch biopsies of psoriatic skin lesion and uninvolved skin
Jarret et al. [38]	— PASI score — Physician's Global Assessment Score — Dermatology Life Quality Index — Psoriasis Disability Index
Ingram et al. [39]	— PASI score
Disphanurat et al. [40]	— PASI score

Lourencetti and Morgado de Abreu analyzed 10 clinical studies published between 1986 and 2013 from the perspective of vitamin D administration in patients with several forms of psoriasis of varying degrees of severity. The dose ranged from 0.25 to 4 µg/day. These authors observed predominantly good efficacy and tolerance, with side effects noted only at high doses. They concluded that this therapeutic alternative was safe and effective for treating psoriasis [19]. In the context of psoriasis, Soleymani et al. also addressed some concerns about oral vitamin D effects on calcium absorption in the gut, and subsequent systemic calcium homeostasis [20].

The diagnostic marker used for vitamin D deficiency is serum 25(OH)D, its cut-off level varying over the years. The normal serum 25(OH)D levels are estimated to extend from about 25 to 225 nmol/L (10 to 90 ng/mL) and there seems to be a correlation between the low-level of 25(OH)D and the risk of chronic diseases. UVA/UVB phototherapy significantly increased the 25(OH)D serum level in patients with psoriasis and atopic dermatitis and reduced serum parathormone concentrations. There is no study demonstrating the correlation between serum 25(OH)D levels and severity of psoriasis [43,44]

Dietary calcium absorption enhancement could be avoided by taking vitamin D orally in the evening [32,45]. Serum vitamin D levels in patients with psoriasis were correlated with seasonal variations and disease severity [46]. A linear correlation could not be demonstrated, but numerous studies have shown low serum vitamin D levels in patients with psoriasis [20]. There is limited data on the dose-dependence of vitamin D deficiency in the pathogenesis of psoriasis and on the role of vitamin D deficiency in the therapeutic response. Vitamin D 1,25(OH) may act in psoriasis as an inhibitor of T-cell proliferation and Th1 development. Vitamin D 1,25(OH) modulates antigen-presenting cell function; induces hyporesponsiveness to antigens; inhibits the production of IL2, IL-17, IL-8, and interferon-gamma; increases the production of IL-10; and increases regulatory T cells [47]. A study using high doses of vitamin D_3 (more than 60,000 IU) reported the resolution of anti-TNFα-induced psoriasiform lesions in a patient with rheumatoid arthritis and vitamin D deficiency [48].

The doses of vitamin D administered in the reviewed studies were mostly empirical; high doses of D_3 were used after the year 2014. The changes in serum concentrations of vitamin D metabolite 25(OH)D were used to monitor the side effects and were not related to the degree of improvement or worsening in psoriasis lesions. A vitamin D_2 dose higher than 40,000 IU was associated with hypercalcemia toxicity [40].

Millsop et al. analyzed six prospective trials on oral vitamin D treatment for psoriasis. In addition to describing the overall results, they pointed out that the possible side effects of oral vitamin D supplementation included hypercalcemia, hypercalciuria, and kidney stones, and long-term vitamin D overdoses could lead to bone demineralization [21]. Some studies reported increases in blood calcium and vitamin D levels or an increase in urinary calcium after starting oral supplementation, but no patient experienced adverse clinical side effects [31].

Zuccotti et al. addressed nutritional strategies for psoriasis. They also discussed oral vitamin D administration in psoriasis; although the patients did not show significant improvements, the authors concluded that vitamin D supplementation might aid in preventing psoriasis-related comorbidities. The proposed mechanism was that vitamin D might represent a key modulator of immune and inflammatory pathways. They hypothesized that, in psoriasis, an interruption of the immunological homeostasis and a reduction of the inflammation process might be due to low vitamin D levels, which can reduce the number of circulating regulatory T cells [29].

Barrea et al. addressed several aspects of the role of vitamin D in psoriasis, including oral vitamin D supplementation. They suggested that intakes of oral vitamin D up to 10,000 IU daily were not associated with harmful effects; this dose was comparable to the maximum cutaneous vitamin D production, and no study has reported vitamin D intoxication from cutaneous synthesis alone. Although the doses and durations of vitamin D administration were not mentioned, they highlighted results from two studies: One found a clinical improvement of the PASI score in 88% of the patients, and the other reported moderate or better improvements in 25–50% of patients with psoriasis [33].

Another study that did not highlight the dosage or duration of vitamin D administration suggested that the results were somewhat contradictory, concluding that the data were insufficient to determine the effectiveness of oral vitamin D administration in psoriasis [49].

Marino et al. mentioned a single study that compared the effects of 60,000 IU oral vitamin D in 45 patients vs. a placebo for six months. The results showed an increase in serum vitamin D and reductions in the PASI [41].

Bouillon et al. referred to a study that did not find any association between vitamin D supplementation and the induction of psoriasis in over 70,000 women [22,50]. In contrast, Hambly et al. reviewed several studies that administered systemic vitamin D to patients with psoriasis. Improvements were reported in many cases, and no adverse effects were reported. However, they concluded that further studies are needed [25].

Analyzing the dose-dependence relationship for the outcomes of using oral vitamin D in psoriasis, we noticed several differences and ambiguities in what could influence this relationship. Starting with 1986 and until 2013, the doses administered had a uniform character, between 0.25 µg/day and 2.0 µg/day (10–80 IU/day), very low compared to the doses of vitamin D used at the current time, even in other diseases. The outcome of the administered doses could be influenced by several factors not sufficiently documented, for example, the degree of sun exposure, which is quite challenging to monitor, considering that sun exposure of the whole body at a peak time for 1–2 h causes up to 20,000 IU vitamin D_3 to enter the circulation [51]. Other variables are represented by the patient's weight, skin tone, the circulating serum level of vitamin D, and the vitamin D deposits. The number of patients enrolled in existing studies is small, and studies are still very few. Given all this, it is not easy to achieve a dose-dependence relationship for the outcomes. More well-documented studies are needed.

From another perspective, namely, that of vitamin D toxicity, the reviewed studies showed no signs of toxicity in the patients followed, most likely due to the low doses used. McCullough et al. showed remarkable clinical benefit at doses ranging from 25,000 IU/day to 60,000 IU/day in psoriasis, cancer, and asthma, without the development of toxicity or hypercalcemia [52]. In another publication, the same authors argued that the administration of 10,000 IU/day to 25,000 IU/day of oral vitamin D is safe for the population [53].

Vitamin D is biologically inactive and treatment with vitamin D refers to its active metabolites: cholecalciferol (vitamin D_3) and ergocalciferol (vitamin D_2). Vitamin D_3 is more frequently administered than calcitriol or alpha-calcidol, since it is safer and less expensive. Keratinocytes and immune lymphocyte T cells express vitamin D receptor (VDR) and contain enzymes able to convert active metabolites of vitamin D, 25(OH)D-calcidiol to active 1,25(OH)2D-calcitriol. Alterations in calcitriol levels and polymorphisms of the VDR gene have been shown to be associated with several malignant and autoimmune diseases, including psoriasis vulgaris [52,53].

Since the body has been shown to make up to 10,000 to 25,000 IU of vitamin D_3 a day in response to adequate ultraviolet-B (UVB) exposure, it could be presumed that taking daily supplements of vitamin D_3 in doses up to this amount may prevent or treat chronic diseases associated with vitamin D deficiency. Vitamin D level as a risk factor and also as a treatment option is studied in cancer, cardiovascular diseases, osteoporosis, autoimmune diseases, influenza, type 2 diabetes mellitus, Alzheimer disease, and depression in the postpartum and non-postpartum periods [54–61]. Vitamin D_3 exerts significant control over normal cellular metabolism via plasma membranes ion channels and via VDR genes located near autoimmune and cancer-associated genes [53].

Compared to existing studies regarding the administration of vitamin D in psoriasis, vitamin D administration in cancer has been much more studied. Several studies have looked at the effectiveness of various doses, various frequencies of administration, and types of vitamin D such as cholecalciferol: 400–4800 IU/day, 20,000 IU/week, 30,000–100,000 IU/month, 120,000 IU every two months, 100,000 IU every three months,

100,000 IU every four months, or 500,000 IU once/year; ergocalciferol: 1000 IU daily; calcitriol 0.25–0.50 µg daily or 0.25 µg twice daily; alfacalcidol: 1.0 µg daily [62,63]. One very recent study evaluated vitamin D supplementation, which has been associated with a reduced mortality in patients with psoriasis [64]. We want to draw attention to a broad plan for the administration of vitamin D that has not yet been studied to treat psoriasis.

4. Conclusions

Although vitamin D has been used successfully for many years as a topical therapy in the fight against psoriasis, only recently have studies examined systemic vitamin D administration in psoriasis. We examined the pros and cons of this treatment, with the aim of determining whether systemic vitamin D would be a feasible therapeutic option for these patients. Among the existing reviews, very few were systematic in design. Indeed, from 1985 to the present, only a few studies have monitored the effectiveness of oral vitamin D in patients with psoriasis; consequently, the reviews were insufficient and inconclusive. Most studies did not observe side effects for doses within a relatively narrow range (0.25 to 2 µg/day). No evidence has been reported about the efficacy of the highest doses of systemic vitamin D in psoriasis. However, most studies did not observe side effects. Based on these results, we can conclude that more large-scale studies are needed to determine the efficacy, optimal dosing, and adverse effects of vitamin D administration in patients with psoriasis.

Author Contributions: Conceptualization, A.M.A.S., A.A.S., and C.C.D.; methodology, A.M.A.S. and A.A.S.; software, A.M.A.S., A.A.S., and C.C.D. validation, A.M.A.S., A.A.S., and C.C.D.; formal analysis, A.M.A.S. and A.A.S.; investigation A.M.A.S. and A.A.S.; resources, A.M.A.S., A.A.S., and C.C.D.; data curation, A.M.A.S. and A.A.S.; writing—original draft preparation, A.M.A.S.; writing—review and editing, A.M.A.S., A.A.S., and C.C.D.; visualization, A.M.A.S., A.A.S., and C.C.D.; supervision, A.M.A.S., A.A.S., and C.C.D.; project administration, A.M.A.S. All authors have read and agreed to the published version of the manuscript.

Funding: This research received no external funding.

Conflicts of Interest: The authors declare no conflict of interest.

References

1. Martens, P.J.; Gysemans, C.; Verstuyf, A.; Mathieu, A.C. Vitamin D's effect on immune function. *Nutrients* **2020**, *12*, 1248. [CrossRef]
2. Windaus, A.; Linsert, O.; Luttringhaus, A.; Weidlich, G. Über das krystallisierte Vitamin D2. *Ann. Chem. Liebigs* **1932**, *492*, 226–241. [CrossRef]
3. Windaus, A.; Schenck, F.; von Werder, F. Über das antirachitisch wirksame bestrahlungs-produkt aus 7-dehydrocholesterin. *Z Physiol. Chem. Hoppe Seylers* **1936**, *241*, 100–103. [CrossRef]
4. Charoenngam, N.; Holick, M.F. Immunologic effects of vitamin D on human health and disease. *Nutrients* **2020**, *12*, 2097. [CrossRef] [PubMed]
5. Brandão-Lima, P.N.; Santos, B.D.C.; Aguilera, C.M.; Freire, A.R.S.; Martins-Filho, P.R.S.; Pires, L.V. Vitamin D food fortification and nutritional status in children: A systematic review of randomized controlled trials. *Nutrients* **2019**, *11*, 2766. [CrossRef] [PubMed]
6. Holick, M.F. Vitamin D deficiency. *N. Engl. J. Med.* **2007**, *357*, 266–281. [CrossRef]
7. Misra, M.; Pacaud, D.; Petryk, A.; Collett-Solberg, P.F.; Kappy, M. Drug and therapeutics committee of the Lawson Wilkins pediatric endocrine society. Vitamin D deficiency in children and its management: Review of current knowledge and recommendations. *Pediatrics* **2008**, *122*, 398–417. [CrossRef]
8. Chang, S.W.; Lee, H.C. Vitamin u8h brD and health—The missing vitamin in humans. *Pediatrics Neonatol.* **2019**, *60*, 237–244. [CrossRef]
9. Rendon, A.; Schäkel, K. Psoriasis pathogenesis and treatment. *Int. J. Mol. Sci.* **2019**, *20*, 1475. [CrossRef]
10. Slominski, A.; Kim, T.K.; Zmijewski, M.A.; Janjetovic, Z.; Li, W.; Chen, J.; Kusniatsova, E.I.; Semak, I.; Postlethwaite, A.; Miller, D.D.; et al. Novel vitamin D photoproducts and their precursors in the skin. *Dermatoendocrinology* **2013**, *5*, 7–19. [CrossRef]
11. Stanescu, A.M.A.; Grajdeanu, I.V.; Iancu, M.A.; Pantea Stoian, A.; Bratu, O.G.; Socea, B.; Socea, L.I.; Diaconu, C.C. Correlation of oral vitamin D administration with the severity of psoriasis and the presence of metabolic syndrome. *Rev. Chim.* **2018**, *69*, 1668–1672. [CrossRef]
12. Burfield, L.; Burden, A.D. Psoriasis. *J. R. Coll. Physicians Edinb.* **2013**, *43*, 334–338. [CrossRef] [PubMed]

13. Brown, A.; Slatopolsky, E. Vitamin D analogs: Therapeutic applications and mechanisms for selectivity. *Mol. Asp. Med.* **2008**, *29*, 433–452. [CrossRef] [PubMed]

14. Piotrowska, A.; Wierzbicka, J.; Żmijewski, M.A. Vitamin D in the skin physiology and pathology. *Acta Biochim. Pol.* **2016**, *63*, 17–29. [CrossRef] [PubMed]

15. Penna, G.; Adorini, L. 1 Alpha,25-dihydroxyvitamin D3 inhibits differentiation, maturation, activation, and survival of dendritic cells leading to impaired alloreactive T cell activation. *J. Immunol.* **2000**, *164*, 2405–2411. [CrossRef] [PubMed]

16. Moher, D.; Liberati, A.; Tetzlaff, J.; Altman, D.G. The PRISMA group. Preferred reporting items for systematic reviews and meta-analyses: The PRISMA statement. *PLoS Med.* **2009**, *6*, e1000097. [CrossRef] [PubMed]

17. Morimoto, S.; Yoshikawa, K.; Kozyka, T.; Kitano, Y.; Imanaka, S.; Fukuo, K.; Koh, E.; Kumahara, Y. An open study of vitamin D3 treatment in psoriasis vulgaris. *Br. J. Dermatol.* **1986**, *115*, 421–429. [CrossRef]

18. Kamangar, F.; Koo, J.; Heller, M.; Lee, E.; Bhutani, T. Oral vitamin D, still a viable treatment option for psoriasis. *J. Dermatolog. Treat.* **2013**, *24*, 261–267. [CrossRef]

19. Lourencetti, M.; Abreu, M.M. Use of active metabolites of vitamin D orally for the treatment of psoriasis. *Rev. Assoc. Med. Bras.* **2018**, *64*, 643–648. [CrossRef]

20. Soleymani, T.; Hung, T.; Soung, J. The role of vitamin D in psoriasis: A review. *Int. J. Dermatol.* **2015**, *54*, 383–392. [CrossRef]

21. Millsop, J.W.; Bhatia, B.K.; Debbaneh, M.; Koo, J.; Liao, W. Diet and psoriasis, part III: Role of nutritional supplements. *J. Am. Acad. Dermatol.* **2014**, *71*, 561–569. [CrossRef]

22. Bouillon, R.; Marcocci, C.; Carmeliet, G.; Bikle, D.; White, J.H.; Dawson-Hughes, B.; Lips, P.; Munns, C.F.; Lazaretti-Castro, M.; Giustina, A.; et al. Skeletal and extraskeletal actions of vitamin D: Current evidence and outstanding questions. *Endocr. Rev.* **2019**, *40*, 1109–1151. [CrossRef] [PubMed]

23. Takamoto, S.; Onishi, T.; Morimoto, S.; Imanaka, S.; Yukawa, S.; Kozuka, T.; Kitano, Y.; Seino, Y.; Kumahara, Y. Effect of 1 alpha-hydroxycholecalciferol on psoriasis vulgaris: A pilot study. *Calcif. Tissue Int.* **1986**, *39*, 360–364. [CrossRef]

24. Smith, E.L.; Pincus, S.H.; Donovan, L.; Holick, M.F. A novel approach for the evaluation and treatment of psoriasis. Oral or topical use of 1,25-dihydroxyvitamin D3 can be a safe and effective therapy for psoriasis. *J. Am. Acad. Dermatol.* **1988**, *19*, 516–528. [CrossRef]

25. Hambly, R.; Kirby, B. The relevance of serum vitamin D in psoriasis: A review. *Arch. Dermatol. Res.* **2017**, *309*, 499–517. [CrossRef] [PubMed]

26. Holland, D.B.; Wood, E.J.; Roberts, S.G.; West, M.R.; Cunliffe, W.J. Epidermal keratin levels during oral 1-alpha-hydroxyvitamin D3 treatment for psoriasis. *Skin Pharmacol.* **1989**, *2*, 68–76. [CrossRef]

27. Huckins, D.; Felson, D.T.; Holick, M. Treatment of psoriatic arthritis with oral 1,25-dihydroxyvitamin D3: A pilot study. *Arthritis Rheum.* **1990**, *33*, 1723–172715. [CrossRef]

28. Siddiqui, M.A.; Al-Khawajah, M.M. Vitamin D3 and psoriasis: A randomized double-blind placebo-controlled study. *J. Dermatol. Treat.* **1990**, *1*, 243–245. [CrossRef]

29. Zuccotti, E.; Oliveri, M.; Girometta, C.; Ratto, D.; Di Iorio, C.; Occhinegro, A.; Rossi, P. Nutritional strategies for psoriasis: Current scientific evidence in clinical trials. *Eur. Rev. Med. Pharmacol. Sci.* **2018**, *22*, 8537–8551. [CrossRef]

30. Lugo-Somolinos, A.; Sanchez, J.L.; Haddock, L. Efficacy of 1, alpha 25-dihydroxyvitamin D (Calcitriol) in the treatment of psoriasis vulgaris: An open study. *Bol. Asoc. Med. P R* **1990**, *82*, 450–453.

31. el-Azhary, R.A.; Peters, M.S.; Pittelkow, M.R.; Kao, P.C.; Muller, S.A. Efficacy of vitamin D3 derivatives in the treatment of psoriasis vulgaris: A preliminary report. *Mayo Clin. Proc.* **1993**, *68*, 835–841. [CrossRef]

32. Perez, A.; Raab, R.; Chen, T.C.; Turner, A.; Holick, M.F. Safety and efficacy of oral calcitriol (1,25-dihydroxyvitamin D3) for the treatment of psoriasis. *Br. J. Dermatol.* **1996**, *134*, 1070–1078. [CrossRef]

33. Barrea, L.; Savanelli, M.C.; Di Somma, C.; Napolitano, M.; Megna, M.; Colao, A.; Savastano, S. Vitamin D and its role in psoriasis: An overview of the dermatologist and nutritionist. *Rev. Endocr. Metab. Disord.* **2017**, *18*, 195–205. [CrossRef]

34. Gaál, J.; Lakos, G.; Szodoray, P.; Kiss, J.; Horváth, I.; Horkay, E.; Nagy, G.; Szegedi, A. Immunological and clinical effects of alphacalcidol in patients with psoriatic arthropathy: Results of an open, follow-up pilot study. *Acta Derm. Venereol.* **2009**, *89*, 140–144.

35. Finamor, D.C.; Sinigaglia-Coimbra, R.; Neves, L.C.; Gutierrez, M.; Silva, J.J.; Torres, L.D.; Surano, F.; Neto, D.J.; Novo, N.F.; Juliano, Y.; et al. A pilot study assessing the effect of prolonged administration of high daily doses of vitamin D on the clinical course of vitiligo and psoriasis. *Dermatoendocrinology* **2013**, *5*, 222–234. [CrossRef]

36. Umar, M.; Sastry, K.S.; Al Ali, F.; Al-Khulaifi, M.; Wang, E.; Chouchane, A. I: Vitamin D and the pathophysiology of inflammatory skin diseases. *Skin Pharmacol. Physiol.* **2018**, *31*, 74–86. [CrossRef]

37. Hata, T.; Audish, D.; Kotol, P.; Coda, A.; Kabigting, F.; Miller, J.; Alexandrescu, D.; Boguniewicz, M.; Taylor, P.; Aertker, L.; et al. A randomized controlled double-blind investigation of the effects of vitamin D dietary supplementation in subjects with atopic dermatitis. *J. Eur. Acad. Dermatol. Venereol.* **2014**, *28*, 781–789. [CrossRef]

38. Jarrett, P.; Camargo, C.A., Jr.; Coomarasamy, C.; Scragg, R. A randomized, double-blind, placebo-controlled trial of the effect of monthly vitamin D supplementation in mild psoriasis. *J. Dermatolog. Treat.* **2018**, *29*, 324–328. [CrossRef]

39. Ingram, M.A.; Jones, M.B.; Stonehouse, W.; Jarrett, P.; Scragg, R.; Mugridge, O.; von Hurst, P.R. Oral vitamin D3 supplementation for chronic plaque psoriasis: A randomized, double-blind, placebo-controlled trial. *J. Dermatolog. Treat.* **2018**, *29*, 648–657. [CrossRef]

40. Disphanurat, W.; Viarasilpa, W.; Chakkavittumrong, P.; Pongcharoen, P. The clinical effect of oral vitamin D2 supplementation on psoriasis: A double-blind, randomized, placebo-controlled study. *Dermatol. Res. Pract.* **2019**, *2019*, 5237642. [CrossRef]
41. Marino, R.; Misra, M. Extra-sekeletal effects of vitamin D. *Nutrients* **2019**, *11*, 1460. [CrossRef]
42. Fredriksson, T.; Pettersson, U. Severe psoriasis—Oral therapy with a new retinoid. *Dermatologica* **1978**, *157*, 238–244. [CrossRef]
43. Heaney, R.P. Guidelines for optimizing design and analysis of clinical studies of nutrient effects. *Nutr. Rev.* **2014**, *72*, 48–54. [CrossRef]
44. Le, P.; Tu, J.; Gebauer, K.; Brown, S. Serum 25-hydroxyvitamin D increases with UVB and UVA/UVB phototherapy (broadband UVB, narrowband UVB (NBUVB) and heliotherapy) in patients with psoriasis and atopic dermatitis. *Australas J. Dermatol.* **2016**, *57*, 115–121. [CrossRef]
45. Tremezaygues, L.; Reichrath, J. Vitamin D analogs in the treatment of psoriasis: Where are we standing and where will we be going? *Dermatoendocrinology* **2011**, *3*, 180–186. [CrossRef]
46. Orgaz-Molina, J.; Buenda-Eisman, A.; Arrabal-Polo, M.A.; Ruiz, J.C.; Arias-Santiago, S. Deficiency of serum concentration of 25-hydroxyvitamin D in psoriatic patients: A case-control study. *J. Am. Acad. Dermatol.* **2012**, *67*, 931–938. [CrossRef]
47. Adams, J.S.; Hewison, M. Update in vitamin D. *J. Clin. Endocrinol. Metab.* **2010**, *95*, 471–478. [CrossRef]
48. Werner de Castro, G.R.; Neves, F.S.; Pereira, I.A.; Fialho, S.C.; Ribeiro, G.; Zimmermann, A.F. Resolution of adalimumab-induced psoriasis after vitamin D deficiency treatment. *Rheumatol. Int.* **2011**, *32*, 1313–1316. [CrossRef]
49. Thompson, K.G.; Kim, N. Dietary supplements in dermatology: A review of the evidence for zinc, biotin, vitamin D, nicotinamide, and Polypodium. *J. Am. Acad. Dermatol.* **2020**. [CrossRef]
50. Merola, J.F.; Han, J.; Li, T.; Qureshi, A.A. No association between vitamin D intake and incident psoriasis among US women. *Arch. Dermatol. Res.* **2014**, *306*, 305–307. [CrossRef]
51. Nair, R.; Maseeh, A. Vitamin D: The "sunshine" vitamin. *J. Pharmacol. Pharmacother.* **2012**, *3*, 118–126. [CrossRef] [PubMed]
52. McCullough, P.J.; Lehrer, D.S.; Amend, J.J. Daily oral dosing of vitamin D3 using 5000 TO 50,000 international units a day in long-term hospitalized patients: Insights from a seven year experience. *J. Steroid Biochem. Mol. Biol.* **2019**, *189*, 228–239. [CrossRef] [PubMed]
53. McCullough, P.; Amend, J.J. Results of daily oral dosing with up to 60,000 international units (iu) of vitamin D3 for 2 to 6 years in 3 adult males. *J. Steroid Biochem. Mol. Biol.* **2017**, *173*, 308–312. [CrossRef] [PubMed]
54. Garland, C.F.; French, C.B.; Baggerly, L.L.; Heaney, R.P. Vitamin D supplement doses and serum 25-hydroxyvitamin D in the range associated with cancer prevention. *Anticancer Res.* **2011**, *31*, 617–622.
55. Dawson-Hughes, B.; Staten, M.A.; Knowler, W.C.; Nelson, J.; Vickery, E.M.; LeBlanc, E.S.; Neff, L.M.; Park, J.; Pittas, A.G.; D2d Research Group. Intratrial exposure to vitamin D and new-onset diabetes among adults with prediabetes: A secondary analysis from the vitamin D and type 2 diabetes (D2d) study. *Diabetes Care* **2020**, *43*, 2916–2922. [CrossRef]
56. Danik, J.S.; Manson, J.E. Vitamin d and cardiovascular disease. *Curr. Treat. Options Cardiovasc. Med.* **2012**, *14*, 414–424. [CrossRef]
57. Chai, B.; Gao, F.; Wu, R.; Dong, T.; Gu, C.; Lin, Q.; Zhang, Y. Vitamin D deficiency as a risk factor for dementia and Alzheimer's disease: An updated meta-analysis. *BMC Neurol.* **2019**, *19*, 284. [CrossRef]
58. Aghajafari, F.; Letourneau, N.; Mahinpey, N.; Cosic, N.; Giesbrecht, G. Vitamin D deficiency and antenatal and postpartum depression: A systematic review. *Nutrients* **2018**, *10*, 478. [CrossRef]
59. Cuomo, A.; Giordano, N.; Goracci, A.; Fagiolini, A. Depression and vitamin D deficiency: Causality, assessment, and clinical practice implications. *Neuropsyvchiatry* **2017**, *7*, 606–614. [CrossRef]
60. Gruber-Bzura, B.M. Vitamin D and influenza-prevention or therapy? *Int. J. Mol. Sci.* **2018**, *19*, 2419. [CrossRef]
61. American Geriatrics Society Workgroup on Vitamin D Supplementation for Older Adult. Recommendations abstracted from the American geriatrics consensus statement on Vitamin D for prevention of falls and their consequences. *J. Am. Geriatr. Soc.* **2014**, *62*, 147–152. [CrossRef] [PubMed]
62. Goulão, B.; Stewart, F.; Ford, J.A.; MacLennan, G.; Avenell, J.A. Cancer and vitamin D supplementation: A systematic review and meta-analysis. *Am. J. Clin. Nutr.* **2018**, *107*, 652–663. [CrossRef]
63. Bittenbring, J.T.; Neumann, F.; Altmann, B.; Achenbach, M.; Reichrath, J.; Ziepert, M.; Geisel, J.; Regitz, E.; Held, G.; Pfreundschuh, M. Vitamin D deficiency impairs rituximab-mediated cellular cytotoxicity and outcome of patients with diffuse large B-cell lymphoma treated with but not without rituximab. *J. Clin. Oncol.* **2014**, *32*, 3242–3248. [CrossRef] [PubMed]
64. Fu, H.; Tang, Z.; Wang, Y.; Ding, X.; Rinaldi, G.; Rahmani, J.; Xing, F. Relationship between vitamin D level and mortality in adults with psoriasis: A retrospective cohort study of NHANES data. *Clin. Ther.* **2020**. [CrossRef]

Commentary

Vitamin D Supplementation to Prevent COVID-19 Infections and Deaths—Accumulating Evidence from Epidemiological and Intervention Studies Calls for Immediate Action

Hermann Brenner [1,2]

1   German Cancer Research Center, Division of Clinical Epidemiology and Aging Research, 69120 Heidelberg, Germany; h.brenner@dkfz.de; Tel.: +49-6221-42-1300
2   Network Aging Research, Heidelberg University, 69115 Heidelberg, Germany

Abstract: The COVID-19 pandemic poses an unprecedented threat to human health, health care systems, public life, and economy around the globe. The repertoire of effective therapies for severe courses of the disease has remained limited. A large proportion of the world population suffers from vitamin D insufficiency or deficiency, with prevalence being particularly high among the COVID-19 high-risk populations. Vitamin D supplementation has been suggested as a potential option to prevent COVID-19 infections, severe courses, and deaths from the disease, but is not widely practiced. This article provides an up-to-date summary of recent epidemiological and intervention studies on a possible role of vitamin D supplementation for preventing severe COVID-19 cases and deaths. Despite limitations and remaining uncertainties, accumulating evidence strongly supports widespread vitamin D supplementation, in particular of high-risk populations, as well as high-dose supplementation of those infected. Given the dynamics of the COVID-19 pandemic, the benefit–risk ratio of such supplementation calls for immediate action even before results of ongoing large-scale randomized trials become available.

Keywords: COVID-19; mortality; prevention; supplementation; vitamin D

Citation: Brenner, H. Vitamin D Supplementation to Prevent COVID-19 Infections and Deaths—Accumulating Evidence from Epidemiological and Intervention Studies Calls for Immediate Action. *Nutrients* **2021**, *13*, 411. https://doi.org/10.3390/nu13020411

Academic Editor: Spyridon Karras
Received: 28 December 2020
Accepted: 24 January 2021
Published: 28 January 2021

Publisher's Note: MDPI stays neutral with regard to jurisdictional claims in published maps and institutional affiliations.

1. Introduction

The COVID-19 pandemic poses an unprecedented threat to human health, health care systems, public life, and economy around the globe. More than 1.7 million people died from SARS-CoV-2 infections by December 2020 [1], with daily numbers of deaths rapidly re-increasing in fall of 2020 in many countries on the northern hemisphere. Options to effectively treat severe cases remain very limited, while intensive care needs stretch or overstretch available capacities in many countries. There are hopes to cope with the pandemic with newly developed vaccines, but it will take many more months before vaccines with proven efficacy and safety will be globally available.

Readily available measures to limit the toll of the pandemic are thus of paramount importance. Based on a comprehensive review of the evidence on the role of vitamin D in preventing the toll of respiratory infections from the pre-COVID-19 era, Grant et al. recommended vitamin D supplementation for both preventing and treating COVID-19 infections already at the onset of the pandemic [2]. However, such supplementation has not been widely implemented. This article provides an up-to-date summary of epidemiological and intervention studies conducted during the COVID-19 pandemic on a possible role of vitamin D supplementation for preventing COVID-19 cases and deaths, by either preventing COVID-19 infections, preventing severe course of the disease and deaths among those infected, or both. Finally, an outlook will be given on ongoing trials, and public health and clinical implications of current evidence are discussed.

2. Potential Role of Vitamin D Supplementation for Preventing COVID-19 Infection

In a study of 349,598 participants of the U.K. Biobank participants with known baseline vitamin D levels, 449 confirmed COVID-19 infections occurred from 16 March 2020 to 14 April 2020 [3]. Vitamin D levels were inversely related to the risk of COVID-19 infections in univariate analysis, but this association did not persist after adjustment for covariates in multivariable analyses. By contrast, very strong associations were observed between ethnicity and risk of infection, which were slightly attenuated, but persisted after adjustment for vitamin D status, with adjusted odds ratios of 5.3 and 2.65 for blacks and South Asians versus whites, respectively. However, in this study, vitamin D levels were measured in 2006–2010, i.e., 10–14 years prior to the COVID-19 pandemic, and may have been poor indicators of vitamin D status in 2020. In fact, ethnicity, which is strongly and stably associated with vitamin D status, may have been a much better proxy for vitamin D status in 2020, which may explain the observed patterns.

In a study among 7807 members of a large health maintenance organization in Israel tested for COVID-19 from 1 February to 30 April 2020 and at least one preceding vitamin D measurement, the adjusted odds ratios for low vitamin D levels with COVID-19 positivity and hospitalization due to COVID-19 infection were 1.45 ($p < 0.001$) and 1.95 ($p = 0.061$), respectively [4]. However, in this study, the threshold for low vitamin level was rather high (plasma 25-hydroxy-vitamin D [25(OH)D] < 30 ng/mL), and the low vitamin D group included 85% of the population. As much stronger health effects of insufficient vitamin D levels are commonly seen at levels of vitamin D insufficiency (plasma 25(OH)D < 20 ng/mL) or deficiency (plasma 25(OH)D < 12.5 ng/mL) [5], it would be of utmost interest to complement these analyses by a thorough dose–response analysis.

Such dose–response relationships were evaluated in a cohort of >190,000 patients from the United States in whom results of SARS-CoV-2 results performed mid-March through mid-June 2020 were linked to 25(OH)D results from the preceding 12 months [6]. In this cohort, a clear inverse relationship between circulating 25(OH)D levels and SARS-CoV-2 positivity was observed. The SARS-CoV-2 positivity rate was higher in the 39,190 patients with "deficient" 25(OH)D values (<20 ng/mL) (12.5%, 95% confidence interval (C.I.) 12.2–12.8%) than in the 27,870 patients with "adequate" values (30–34 ng/mL) (8.1%, 95% C.I. 7.8–8.4%) and the 12,321 patients with values ≥55 ng/mL (5.9%, 95% C.I. 5.5–6.4%). Those who had a circulating level of 25(OH)D < 20 ng/mL had a 54% higher positivity rate compared with those who had a blood level of 30–34 ng/mL in multivariable analysis. The risk of SARS-CoV-2 positivity continued to decline until the serum levels reached 55 ng/mL. The relationship persisted across latitudes, races/ethnicities, both sexes, and age ranges.

Inverse associations between vitamin D levels and COVID-19 infection were furthermore consistently reported from a case-control study of 201 hospitalized patients and 201 matched controls from Iran [7] and in a cross-sectional study among 392 healthcare workers from the United Kingdom [8]. Finally, negative correlations between mean levels of vitamin D and COVID-19 infection and mortality rates were also reported from an ecological study including 20 European countries [9].

3. Potential Role of Vitamin D Supplementation for Preventing Severe Course of Disease and Death from COVID-19 Infection

A recent clinic-based cohort study among 185 patients diagnosed with and treated for COVID-19 at a University Hospital in Germany showed more than 80% lower risk of invasive mechanical ventilation or death (primary endpoint) and more than 90% lower mortality among patients with sufficient vitamin D levels compared with patients with vitamin D deficiency even after multivariable adjustment for age, gender, and comorbidities [10], suggesting that close to 90% of deaths in this cohort were statistically associated with vitamin D insufficiency [11]. Increased mortality was likewise seen for those with vitamin D levels below the median in a cohort of 30 patients admitted to an intensive care unit in Greece (28-day mortality 5/15 versus 0/15, $p = 0.01$) [12].

In a recent "quasi-experimental" study from France [13], risks of severe course of the disease and of dying within 14 days of admission to a geriatric hospital unit were more than 90% lower among patients who were regularly supplemented with vitamin D over the preceding year ($n = 29$) compared with patients with no vitamin D supplementation ($n = 32$). Despite the limited numbers of cases, strong associations were statistically significant and persisted after multivariable adjustment for confounders. Intermediate and statistically non-significant results were seen for the small group ($n = 16$) of patients who received vitamin D supplementation after COVID-19 diagnosis. In another quasi-experimental study from the same group [14], 82.5% of 57 nursing home residents who received bolus vitamin D3 supplementation either in the week following the suspicion or diagnosis of COVID-19 or during the previous month survived during a mean follow-up of 36 days compared with only 4 out of 9 residents (44%) without such therapy. Despite the small case numbers, the association was highly statistically significant, with an adjusted hazard ratio (95% CI) of 0.11 (0.03–0.48). Vitamin D supplementation was likewise associated with significantly reduced mortality during the COVID-19 pandemic in a cohort of 157 residents of an Italian nursing home [15]. The so far largest "quasi-experimental study" was most recently reported from United Kingdom [16]. A total of 986 participants admitted with COVID-19 to three hospitals were studied, of whom 151 (16.0%) received booster therapy with vitamin D (in its "parent" form, cholecalciferol; approximately 280,000 IU in a time period of up to 7 weeks). In the primary cohort of 444 patients from one hospital, cholecalciferol booster therapy was associated with strongly reduced risk of COVID-19 mortality, with an odds ratio (95% CI) of 0.13 (0.05–0.35, $p < 0.001$) after adjustment for multiple potential confounders. This finding was replicated in a validation cohort of 541 patients from two other hospitals (odds ratio 0.38, 95% CI 0.17–0.84, $p = 0.018$).

However, despite major efforts to control for confounding, such observational or quasi experimental studies remain prone to residual confounding by uncontrolled or imperfectly measured covariates. It is well established that serum 25OHD is a negative acute phase reactant, and associations may in part reflect reverse causality. Such factors could lead to overestimation of the beneficial effects of having adequate vitamin D levels or of vitamin D supplementation. In the "quasi-experimental studies", confounding by indication, i.e., selective supplementation of vitamin D among those with lowest baseline vitamin D status or in highest need of supplementation, could also lead to underestimation of supplementation effects. On the other hand, if supplementation initiated after the start of mortality follow-up (e.g., several days after diagnosis or hospitalization) is considered as intervention, "immortal time bias" may lead to overestimation of beneficial intervention effects unless appropriate precautions are taken in the analysis, as those with a delayed start of the intervention would necessarily have survived (had been "immortal") up to such initiation [17]. The final answer as to a causal role of vitamin D supplementation will thus have to come from randomized controlled trials.

A first pilot study of such a randomized trial has been reported from Cordoba, Spain, in which 76 consecutive patients hospitalized with COVID-19, a clinical picture of acute respiratory infection, confirmed by a radiographic pattern of viral pneumonia and by a positive SARS-CoV-2 PCR with CURB65 severity scale (recommending hospital admission in case of total score >1), were enrolled [18]. All hospitalized patients received as best available therapy the same standard care (per hospital protocol) including a combination of hydroxychloroquine and azithromycin. Eligible patients were randomly allocated at a 2:1 ratio on the day of admission to take oral calcifediol (25-hydroxyvitamin D 3, 0.532 mg = 21.280 IU), or not. Patients in the calcifediol treatment group continued with oral calcifediol (0.266 mg = 10.640 IU) on days 3 and 7, and then weekly until discharge or intensive care unit (ICU) admission. Outcomes of effectiveness included the rate of ICU admission and deaths. Only 1 of 50 patients treated with calcifediol (2%), but 13 of 26 untreated patients (50%), required ICU admission, resulting in a multivariate adjusted odds ratio (95% CI) of ICU admission of 0.03 (0.003–0.25). Despite such adjustment, concerns have been expressed with respect to imperfect blinding, as well as uneven distribution of and imperfect control for potential confounders. However, a comprehensive mathematical reanalysis by an independent group concluded that the randomization, large effect size, and high statistical significance address many of these concerns [19]. In particular, it showed that random assignment of patients to treatment and control groups was highly unlikely to distribute comorbidities or other prognostic indicators sufficiently unevenly to account for the large effect size, and that imperfect blinding would need to have had an implausibly large effect to account for the reported results. The authors concluded that the trial provided sufficient evidence to warrant immediate, well-designed pivotal clinical trials of early calcifediol administration in a broader cohort of inpatients and outpatients with COVID-19.

In a more recently published randomized placebo-controlled trial from India [20], high-dose oral cholecalciferol supplementation (60,000 IU per day for at least 7 days in asymptomatic or mildly symptomatic SARS-CoV-2 RNA positive vitamin D deficient (25(OH)D < 20 ng/mL) individuals very effectively overcame vitamin D deficiency ($p < 0.001$), enhanced SARS-CoV-2 viral clearance ($p = 0.018$), and decreased fibrinogen levels ($p = 0.007$). No significant differences were seen for other inflammatory markers. In a multicenter, double-blind, randomized, placebo-controlled trial conducted in two centers (a quaternary hospital and a field hospital) in Sao Paulo, Brazil, involving 240 hospitalized patients with severe COVID-19 (116 with vitamin D deficiency), a single dose of 200,000 IU of vitamin D3 supplementation was likewise safe and effective in increasing 25-hydroxyvitamin D levels, but did not significantly reduce hospital length of stay or any other clinically-relevant outcomes compared with placebo [21]. It has been suggested that oral supplementation with vitamin D3 (a slower-acting treatment than oral supplementation with calcifediol) of this mostly obese population of patients (mean body mass index (BMI) 31.6 kg/m^2) may have been provided too late to significantly affect clinically relevant outcomes (randomization occurred on average 10 days after onset of symptoms, with 90% of patients requiring supplemental oxygen at baseline) [19].

4. Ongoing Trials

Timely conduction, completion, and publication of further well-designed studies including, but not restricted to large-scale randomized clinical trials is paramount for more fully exploring and defining the role of vitamin D supplementation in preventing occurrence and severe course of COVID-19 infections [22]. Several large-scale trials are currently under way, with the main results expected at some time in 2021. Key characteristics of some of the major trials are outlined below.

CORONAVIT, an open-label, phase 3, randomised clinical trial conducted in the United Kingdom, investigates whether implementation of a test-and-treat approach to correction of sub-optimal vitamin D status results in reduced risk and/or severity of COVID-19 and other acute respiratory infections [23]. The trial started on 27 October 2020 and is designed to recruit 6200 U.K. residents ≥16 years. Participants in the intervention group with 25(OH)D level < 30 ng/mL are offered a daily dose of 800 IU or 3200 IU cholecalciferol, while the control group receives standard of care (national recommendation of 400 IU/day vitamin D). The primary endpoint is the proportion of participants experiencing at least one doctor-diagnosed or laboratory-confirmed acute respiratory infection of any cause over 6 months. The secondary endpoints include multiple COVID-19-specific endpoints, such as proportions of participants developing antigen test-positive COVID-19, seroconverting to SARS-CoV-2, requiring hospitalization due to COVID-19, hospitalised for COVID-19 requiring ventilatory support, and dying of COVID-19, along with other or more generic endpoints, such as proportions of participants who experience influenza requiring hospitalization, dying of influenza, dying of any acute respiratory infection, and dying of any cause.

In the COVIDIOL trial in Cordoba, Spain, the above described pilot study among 76 participants [18] is followed by a trial involving 1008 patients aged 18–90 years diagnosed with COVID-19 and radiological image compatible with inflammatory pleuropulmonary exudate [24]. The intervention group receives the best available treatment plus oral calcefediol (0.532 mg = 21,280 IU on day 1, 0.266 mg = 10.640 IU on days 3, 7, 14, 21 and 28), while the control group receives the best available treatment only. The primary endpoints are ICU admission and deaths within 28 days. The secondary endpoints include, among others, time from onset of symptoms to discharge of patients in conventional hospitalization, time until admission to ICU with mechanical ventilation, and time until mechanical ventilation is removed.

In the CoVitTrial, a multicenter randomized trial conducted in France, 260 high-risk patients aged 65 year or older diagnosed with COVID-19 infections within the preceding 3 days and seen in hospitalization or consultation or in nursing home are recruited [25]. The trial compares the offer of a single high oral dose of vitamin D3 (400,000 IU) with the offer of a single low oral dose of vitamin D3 (50,000 IU). The primary endpoint is death from any cause during the 14 days following the inclusion and intervention. The secondary endpoints include, among others, death from any cause during 28 days and clinical evolution during 14 days and 28 days based on the change of the World Health Organization (WHO) Ordinal Scale for Clinical Improvement for COVID-19.

In a pragmatic, cluster randomized, double-blinded trial in the United States (The vitamin D for COVID-19 (VIVID) trial, n = 2700 in total nationwide), 1500 newly diagnosed individuals with COVID-19, together with up to one close household contact each (~1200 contacts), are recruited nation-wide via social media; community advocacy groups and equity initiatives; and flyers and electronic communications distributed in healthcare centers, low income residential housing organizations, COVID-19 testing centers, and other avenues. Participants are randomized to either vitamin D3 (9600 IU/day loading dose on days 1 and 2, then 3200 IU/day) or placebo in a 1:1 ratio and a household cluster design [26]. The study duration is 4 weeks. The primary outcome for newly diagnosed individuals is the occurrence of hospitalization and/or mortality. Key secondary outcomes include symptom severity scores among cases and changes in the infection (seroconversion) status for their close household contacts. Changes in vitamin D 25(OH)D levels will be assessed and their relation to study outcomes will be explored.

5. Public Health and Clinical Implications

Vitamin D insufficiency and deficiency are widespread in many countries, in particular among the elderly, calling for public health action even before the COVID-19 pandemic [27]. Randomized controlled trials (RCTs) have shown efficacy and safety of vitamin D supplementation in preventing various adverse health outcomes, such as hip fractures, acute respiratory infections, or deaths from cancer [28–30]. Widespread vitamin D supplementation, at least for the elderly and the high risk groups, thus seems to be prudent even in the absence of the COVID-19 pandemic and is recommended or practiced to some extent in a few countries [31]. Nevertheless, vitamin D levels have remained inadequate in most countries [27], with prevalence of vitamin D deficiency remaining highest among nursing home residents [32], the group at highest risk for COVID-19 infection and death. The ongoing COVID-19 pandemic, globally accounting for more than 10 million new cases and 200,000 deaths per month in the second half of 2020, calls for immediate efforts to enhance vitamin D status of populations at risk and of those infected with COVID-19 even before results of the ongoing large trials become available, which will not be before spring to summer of 2021. Besides the recent epidemiological evidence outlined above, a major protective impact of vitamin D supplementation on risk and course of COVID 19 infections is strongly supported by long known and well established molecular mechanisms of vitamin D, such as its immunomodulatory effects, as outlined in detail elsewhere [2,33–37]. Vitamin D supplementation could thus be a most cost-effective, readily available tool that could potentially prevent millions of COVID-19 infections and tens if not hundreds of thousands COVID-19 deaths, and at the same time, prevent overstretching of health care systems, beyond its established beneficial effects on other health outcomes.

Obviously, vitamin D supplementation should complement, not replace established and other efforts to cope with the COVID-19 pandemic, such as social distancing, wearing of masks, and hygiene measures with which it shares protective effects not only against COVID-19 infections, but also other infections, such as other acute respiratory infections including influenza. Although there is hope that widespread vaccination will finally end or at least widely control the current COVID-19 pandemic, its long-term safety and effectiveness are yet to be demonstrated. In the meantime, but also in the long run, vitamin D supplementation, for which safety and effectiveness with respect to acute respiratory infections has long been established, and which is a very low-cost measure, should be widely applied. Even a minor effect on protection from infection that might turn the COVID-19 effective reproduction number from slightly above one (as estimated for many countries shortly before or during lockdown measures of varying intensity during most of the second half of 2020) to slightly below one could make the difference between further exponential growth or regression of the pandemic. In the absence of specific contraindications, supplementation with safe, but sufficient doses (e.g., ranging from 800 to 4000 IU/day for older adults depending on individual factors, such as age and sex [38,39], body mass index, or comorbidity) should thus be strongly promoted for the population at large and the high-risk population in particular, not only to those with already manifest COVID-19 infection. Despite remaining uncertainties with respect to optimal dosing, evidence from vitamin D trials with other endpoints suggests supplementation should preferably be done on a regular basis rather than by occasional high-dose bolus therapy. For patients with manifest COVID-19 infection, initiation of high-dose supplementation as early as possible after diagnosis should be strongly considered whenever there are no specific contraindications against such treatment. At the very least, such strategies would help to reduce the burden of established adverse consequences of widespread vitamin D insufficiency and deficiency, which would be a great achievement by itself. In the best case, they might add to this the even greater achievement of curbing the ongoing COVID-19 pandemic with all its adverse consequences even prior to and beyond widespread availability of vaccination. Immediate action is warranted.

Funding: There was no extramural funding for this commentary.

Institutional Review Board Statement: Not applicable.

Informed Consent Statement: Not applicable.

Data Availability Statement: Not applicable.

Conflicts of Interest: The author declares no competing financial interest.

References

1. Johns Hopkins University Corona Virus Resource Center. Available online: https://coronavirus.jhu.edu/ (accessed on 23 December 2020).
2. Grant, W.B.; Lahore, H.; McDonnell, S.L.; Baggerly, C.A.; French, C.B.; Aliano, J.L.; Bhattoa, H.P. Evidence that vitamin D supplementation could reduce risk of influenza and COVID-19 infections and deaths. *Nutrients* **2020**, *12*, 988. [CrossRef] [PubMed]
3. Hastie, C.E.; Mackay, D.F.; Ho, F.; Celis-Morales, C.A.; Katikireddi, S.V.; Niedzwiedz, C.L.; Jani, B.D.; Welsh, P.; Mair, F.S.; Gray, S.R.; et al. Vitamin D concentrations and COVID-19 infection in UK Biobank. *Diabetes Metab. Syndr.* **2020**, *14*, 561–565. [CrossRef] [PubMed]
4. Merzon, E.; Tworowski, D.; Gorohovski, A.; Vinker, S.; Golan Cohen, A.; Green, I.; Frenkel-Morgenstern, M. Low plasma 25(OH) vitamin D level is associated with increased risk of COVID-19 infection: An Israeli population-based study. *FEBS J.* **2020**, *287*, 3693–3702. [CrossRef] [PubMed]
5. Brenner, H.; Jansen, L.; Saum, K.U.; Holleczek, B.; Schöttker, B. Vitamin D Supplementation Trials Aimed at Reducing Mortality Have Much Higher Power When Focusing on People with Low Serum 25-Hydroxyvitamin D Concentrations. *J. Nutr.* **2017**, *147*, 1325–1333. [CrossRef] [PubMed]
6. Kaufman, H.W.; Niles, J.K.; Kroll, M.H.; Bi, C.; Holick, M.F. SARS-CoV-2 positivity rates associated with circulating 25-hydroxyvitamin D levels. *PLoS ONE* **2020**, *15*, e0239252. [CrossRef]

7. Abdollahi, A.; Sarvestani, H.K.; Rafat, Z.; Ghaderkhani, S.; Mahmoudi-Aliabadi, M.; Jafarzadeh, B.; Mehrtash, V. The Association Between the Level of Serum 25(OH) Vitamin D, Obesity, and underlying Diseases with the risk of Developing COVID-19 Infection: A case-control study of hospitalized patients in Tehran, Iran. *J. Med. Virol.* **2020**. Epub ahead of print. [CrossRef]
8. Faniyi, A.A.; Lugg, S.T.; Faustini, S.E.; Webster, C.; Duffy, J.E.; Hewison, M.; Shields, A.; Nightingale, P.; Richter, A.G.; Thickett, D.R. Vitamin D status and seroconversion for COVID-19 in UK healthcare workers. *Eur. Respir. J.* **2020**, 2004234, Epub ahead of print.
9. Ilie, P.C.; Stefanescu, S.; Smith, L. The role of vitamin D in the prevention of coronavirus disease 2019 infection and mortality. *Aging Clin. Exp. Res.* **2020**, *32*, 1195–1198. [CrossRef]
10. Radujkovic, A.; Hippchen, T.; Tiwari-Heckler, S.; Dreher, S.; Boxberger, M.; Merle, U. Vitamin D Deficiency and Outcome of COVID-19 Patients. *Nutrients* **2020**, *12*, 2757. [CrossRef]
11. Brenner, H.; Schöttker, B. Vitamin D insufficiency may account for almost nine of ten COVID-19 deaths: Time to act. *Nutrients* **2020**, *12*, E3642. [CrossRef]
12. Vassiliou, A.G.; Jahaj, E.; Pratikaki, M.; Orfanos, S.E.; Dimopoulou, I.; Kotanidou, A. Low 25-Hydroxyvitamin D Levels on Admission to the Intensive Care Unit May Predispose COVID-19 Pneumonia Patients to a Higher 28-Day Mortality Risk: A Pilot Study on a Greek ICU Cohort. *Nutrients* **2020**, *12*, E3773. [CrossRef] [PubMed]
13. Annweiler, G.; Corvaisier, M.; Gautier, J.; Dubée, V.; Legrand, E.; Sacco, G.; Annweiler, C. Vitamin D Supplementation Associated to Better Survival in Hospitalized Frail Elderly COVID-19 Patients: The GERIA-COVID Quasi-Experimental Study. *Nutrients* **2020**, *12*, E3377. [CrossRef] [PubMed]
14. Annweiler, C.; Hanotte, B.; Grandin de l'Eprevier, C.; Sabatier, J.M.; Lafaie, L.; Célarier, T. Vitamin D and survival in COVID-19 patients: A quasi-experimental study. *J. Steroid Biochem. Mol. Biol.* **2020**, *204*, 105771. [CrossRef] [PubMed]
15. Cangiano, B.; Fatti, L.M.; Danesi, L.; Gazzano, G.; Croci, M.; Vitale, G.; Gilardini, L.; Bonadonna, S.; Chiodini, I.; Caparello, C.F.; et al. Mortality in an Italian nursing home during COVID-19 pandemic: Correlation with gender, age, ADL, vitamin D supplementation, and limitations of the diagnostic tests. *Aging* **2020**, *12*, 24522–24534.
16. Ling, S.F.; Broad, E.; Murphy, R.; Pappachan, J.M.; Pardesi-Newton, S.; Kong, M.-F.; Jude, E.B. High-Dose Cholecalciferol Booster Therapy is Associated with a Reduced Risk of Mortality in Patients with COVID-19: A Cross-Sectional Multi-Centre Observational Study. *Nutrients* **2020**, *12*, 3799. [CrossRef]
17. Suissa, S. Immortal time bias in pharmaco-epidemiology. *Am. J. Epidemiol.* **2008**, *167*, 492–499. [CrossRef]
18. Castillo, M.E.; Costa, L.M.E.; Barrios, J.M.V.; Díaz, J.F.A.; Miranda, J.L.; Bouillon, R.; Gomez, J.M.Q. Effect of calcifediol treatment and best available therapy versus best available therapy on intensive care unit admission and mortality among patients hospitalized for COVID-19: A pilot randomized clinical study. *J. Steroid Biochem. Mol. Biol.* **2020**, *203*, 105751. [CrossRef]
19. Jungreis, I.; Kellis, M. Mathematical analysis of Córdoba calcifediol trial suggests strong role for Vitamin D in reducing ICU admissions of hospitalized COVID-19 patients. *MedRxiv* **2020**. [CrossRef]
20. Rastogi, A.; Bhansali, A.; Khare, N.; Suri, V.; Yaddanapudi, N.; Sachdeva, N.; Puri, G.D.; Malhotra, P. Short term, high-dose vitamin D supplementation for COVID-19 disease: A randomised, placebo-controlled, study (SHADE study). *Postgrad Med. J.* **2020**. Epub ahead of print. [CrossRef] [PubMed]
21. Murai, I.H.; Fernandes, A.L.; Sales, L.P.; Pinto, A.J.; Goessler, K.F.; Duran, C.S.C.; Silva, C.S.C.; Franco, A.S.; Macedo, M.B.; Pereira, R.M.; et al. Effect of Vitamin D$_3$ Supplementation vs. Placebo on Hospital Length of Stay in Patients with Severe COVID-19: A Multicenter, Double-blind, Randomized Controlled Trial. *medRxiv* **2020**. [CrossRef]
22. Camargo, C.A., Jr.; Martineau, A.R. Vitamin D to prevent COVID-19: Recommendations for the design of clinical trials. *FEBS J.* **2020**, *287*, 3689–3692. [CrossRef] [PubMed]
23. Trial of Vitamin D to Reduce Risk and Severity of COVID-19 and Other Acute Respiratory Infections (CORONAVIT). Available online: https://clinicaltrials.gov/ct2/show/NCT04579640 (accessed on 23 December 2020).
24. Prevention and Treatment With Calcifediol of COVID-19 Induced Acute Respiratory Syndrome (COVIDIOL). Available online: https://clinicaltrials.gov/ct2/show/NCT04366908 (accessed on 23 December 2020).
25. COvid-19 and Vitamin D Supplementation: A Multicenter Randomized Controlled Trial of High Dose Versus Standard Dose Vitamin D3 in High-risk COVID-19 Patients (CoVitTrial). Available online: https://clinicaltrials.gov/ct2/show/NCT04344041 (accessed on 23 December 2020).
26. Wang, R.; DeGruttola, V.; Lei, Q.; Mayer, K.H.; Redline, S.; Hazra, A.; Mora, S.; Willett, W.C.; Ganmaa, D.; Manson, J.E. The vitamin D for COVID-19 (VIVID) trial: A pragmatic cluster-randomized design. *Contemp. Clin. Trials* **2021**, *100*, 106176. [CrossRef] [PubMed]
27. Cashman, K.D.; Dowling, K.G.; Škrabáková, Z.; Gonzalez-Gross, M.; Valtueña, J.; De Henauw, S.; Moreon, L.; Damsgaard, C.T.; Michaelsen, K.F.; Kiely, M.; et al. Vitamin D deficiency in Europe: Pandemic? *Am. J. Clin. Nutr.* **2016**, *103*, 1033–1044. [CrossRef] [PubMed]
28. Weaver, C.M.; Alexander, D.D.; Boushey, C.J.; Dawson-Hughes, B.; Lappe, J.M.; LeBoff, M.S.; Liu, S.; Looker, A.C.; Wallace, T.C.; Wang, D.D.; et al. Calcium plus vitamin D supplementation and risk of fractures: An updated meta-analysis from the National Osteoporosis Foundation. *Osteoporos. Int.* **2016**, *27*, 367–376. [CrossRef] [PubMed]
29. Martineau, A.R.; Jolliffe, D.A.; Hooper, R.L.; Greenberg, L.; Aloia, J.F.; Bergman, P.; Dubnov-Raz, G.; Esposito, S.; Ganmaa, D.; Ginde, A.A.; et al. Vitamin D supplementation to prevent acute respiratory tract infections: Systematic review and meta-analysis of individual participant data. *BMJ* **2017**, *356*, i6583. [CrossRef] [PubMed]

30. Keum, N.; Lee, D.H.; Greenwood, D.C.; Manson, J.E.; Giovannucci, E. Vitamin D supplementation and total cancer incidence and mortality: A meta-analysis of randomized controlled trials. *Ann. Oncol.* **2019**, *30*, 733–743. [CrossRef]

31. Rizzoli, R.; Boonen, S.; Brandi, M.L.; Bruyère, O.; Cooper, C.; Kanis, J.A.; Kaufman, J.-M.; Ringe, J.D.; Weryha, G.; Reginster, J.-Y.; et al. Vitamin D supplementation in elderly or postmenopausal women: A 2013 update of the 2008 recommendations from the European Society for Clinical and Economic Aspects of Osteoporosis and Osteoarthritis (ESCEO). *Curr. Med. Res. Opin.* **2013**, *29*, 305–313. [CrossRef]

32. Griffin, T.P.; Wall, D.; Blake, L.; Griffin, D.G.; Robinson, S.M.; Bell, M.; Mulkerrin, E.C.; O'Shea, P.M. Vitamin D Status of Adults in the Community, in Outpatient Clinics, in Hospital, and in Nursing Homes in the West of Ireland. *J. Gerontol. A Biol. Sci. Med. Sci.* **2020**, *75*, 2418–2425. [CrossRef] [PubMed]

33. Brenner, H.; Holleczek, B.; Schöttker, B. Vitamin D Insufficiency and Deficiency and Mortality from Respiratory Diseases in a Cohort of Older Adults: Potential for Limiting the Death Toll during and beyond the COVID-19 Pandemic? *Nutrients* **2020**, *12*, 2488. [CrossRef]

34. Cantorna, M.T. Mechanisms underlying the e_ect of vitamin D on the immune system. *Proc. Nutr. Soc.* **2010**, *69*, 286–289. [CrossRef]

35. Lemire, J.M.; Adams, J.S.; Kermani-Arab, V.; Bakke, A.C.; Sakai, R.; Jordan, S.C. 1,25-Dihydroxyvitamin D3 suppresses human T helper/inducer lymphocyte activity in vitro. *J. Immunol.* **1985**, *134*, 3032–3035. [PubMed]

36. Cantorna, M.T.; Snyder, L.; Lin, Y.D.; Yang, L. Vitamin D and 1,25(OH)2D regulation of T cells. *Nutrients* **2015**, *7*, 3011–3021. [CrossRef] [PubMed]

37. Jeffery, L.E.; Burke, F.; Mura, M.; Zheng, Y.; Qureshi, O.S.; Hewison, M.; Walker, L.S.K.; Lammas, D.A.; Raza, K.; Sansom, D.M. 1,25-Dihydroxyvitamin D3 and IL-2 combine to inhibit T cell production of inflammatory cytokines and promote development of regulatory T cells expressing CTLA-4 and FoxP3. *J. Immunol.* **2009**, *183*, 5458–5467. [CrossRef] [PubMed]

38. Spanier, J.A.; Nashold, F.E.; Mayne, C.G.; Nelson, C.D.; Hayes, C.E. Vitamin D and estrogen synergy in Vdr-expressing CD4(+) T cells is essential to induce Helios(+)FoxP3(+) T cells and prevent autoimmune demyelinating disease. *J. Neuroimmunol.* **2015**, *286*, 48–58. [CrossRef]

39. Pagano, M.T.; Peruzzu, D.; Ruggieri, A.; Ortona, E.; Gagliardi, M.C. Vitamin D and Sex Differences in COVID-19. *Front. Endocrinol.* **2020**, *11*, 567824. [CrossRef]

Article

Diverse Effects of Combinations of Maternal-Neonatal VDR Polymorphisms and 25-Hydroxyvitamin D Concentrations on Neonatal Birth Anthropometry: Functional Phenocopy Variability Dependence, Highlights the Need for Targeted Maternal 25-Hydroxyvitamin D Cut-Offs during Pregnancy

Spyridon N. Karras [1,*], Erdinç Dursun [2,3], Merve Alaylıoğlu [2], Duygu Gezen-Ak [2], Cedric Annweiler [4,5], Dimitrios Skoutas [6], Dimosthenis Evangelidis [7] and Dimitrios Kiortsis [8]

[1] National Scholarship Foundation, 55535 Thessaloniki, Greece
[2] Brain and Neurodegenerative Disorders Research Laboratories, Department of Medical Biology, Cerrahpasa Faculty of Medicine, Istanbul University-Cerrahpasa, 34381 Istanbul, Turkey; erdincdu@gmail.com (E.D.); mariabioanalysis@yahoo.gr (M.A.); duygugezenak@gmail.com (D.G.-A.)
[3] Department of Neuroscience, Institute of Neurological Sciences, Istanbul University-Cerrahpasa, 34381 Istanbul, Turkey
[4] Division of Geriatric Medicine, Department of Neuroscience, Angers University Hospital, 49035 Angers, France; ceannweiler@chu-angers.fr
[5] Robarts Research Institute, Department of Medical Biophysics, Schulich School of Medicine and Dentistry, The University of Western Ontario, London, ON N6A 3K7, Canada
[6] Sarafianos General Private Hospital, 54631 Thessaloniki, Greece; skoutasd@otenet.gr
[7] Epsom and St Helier University Hospital NHS, London SM5 1AA, UK; dimos@doctors.org.uk
[8] Department of Nuclear Medicine, University of Ioannina, 45110 Ioannina, Greece; dkiorts@uoi.gr
* Correspondence: karraspiros@yahoo.gr; Tel.: +30-2310324863

Citation: Karras, S.N.; Dursun, E.; Alaylıoğlu, M.; Gezen-Ak, D.; Annweiler, C.; Skoutas, D.; Evangelidis, D.; Kiortsis, D. Diverse Effects of Combinations of Maternal-Neonatal VDR Polymorphisms and 25-Hydroxyvitamin D Concentrations on Neonatal Birth Anthropometry: Functional Phenocopy Variability Dependence, Highlights the Need for Targeted Maternal 25-Hydroxyvitamin D Cut-Offs during Pregnancy. *Nutrients* **2021**, *13*, 443. https://doi.org/10.3390/nu13020443

Academic Editors: Spyridon N. Karras and Pawel Pludowski

Received: 1 January 2021
Accepted: 26 January 2021
Published: 29 January 2021

Publisher's Note: MDPI stays neutral with regard to jurisdictional claims in published maps and institutional affiliations.

Abstract: Vitamin D receptor (VDR) polymorphisms have been associated with a plethora of adverse pregnancy and offspring outcomes. The aim of this study was to evaluate the combined effect of maternal and neonatal VDR polymorphisms (ApaI, TaqI, BsmI, FokI, Tru9I) and different maternal and neonatal 25(OH)D cut-offs on neonatal birth anthropometry. This cross-sectional study included data and samples from a cohort of mother–child pairs at birth. A detailed neonatal anthropometry analysis at birth was also conducted. Different 25(OH)D cut-offs for neonates and mothers were included, according to their vitamin D status at birth: for neonates, cut-offs of [25(OH)D $\leq$ 25 and > 25 nmol/L] and [25(OH)D $\leq$ 50 nmol/L] were adopted, whereas for mothers, a 25(OH)D cut-off of [25(OH)D $\leq$ 50 and > 50 nmol/L)] was investigated. Following this classification, maternal and neonatal VDR polymorphisms were evaluated to investigate the potential different effects of different neonatal and maternal 25(OH)D cut-offs on neonatal birth anthropometry. A total of 69 maternal-neonatal dyads were included in final analysis. Weight, neck rump length, chest circumference, abdominal circumference, abdominal circumference (iliac), high thigh circumference, middle thigh circumference, lower arm radial circumference, and lower leg calf circumference of neonates who had the TAQl SNP TT genotype and maternal 25(OH)D < 50 nmol/L were significantly higher than that of neonates who had the Tt or tt genotypes ($p = 0.001$, Hg = 1.341, $p = 0.036$, Hg = 0.976, $p = 0.004$, Hg = 1.381, $p = 0.001$, Hg = 1.554, $p = 0.001$, Hg = 1.351, $p = 0.028$, Hg = 0.918, $p = 0.008$, Hg = 1.090, $p = 0.002$, Hg = 1.217, and $p = 0.020$, Hg = 1.263, respectively). Skin fold high anterior was significantly lower in neonates who had the BSMI SNP BB genotype compared to that of neonates with Bb or bb genotypes ($p = 0.041$, Hg = 0.950), whereas neck rump length was significantly higher in neonates who had the FOKI SNP FF genotype compared to that of neonates who had Ff or ff genotypes ($p = 0.042$, Hg = 1.228). Regarding neonatal VDR polymorphisms and cut-offs, the abdominal circumference (cm) of neonates who had the TAQI SNP TT genotype and 25(OH)D < 25 nmol/L were significantly higher than that of neonates who had the Tt or tt genotypes ($p = 0.038$, Hg = 1.138). In conclusion, these results indicate that the maternal TAQI VDR polymorphism significantly affected neonatal birth anthropometry when maternal 25(OH) concentrations were <50 nmol/L, but not for a higher

---

cut-off of >50 nmol/L, whereas this effect is minimally evident in the presence of neonatal TAQI polymorphism with neonatal 25(OH)D values <25 nmol/L. The implication of these findings could be incorporated in daily clinical practice by targeting a maternal 25(OH)D cut-off >50 nmol/L, which could be protective against any effect of genetic VDR variance polymorphism on birth anthropometry.

Keywords: vitamin D; pregnancy; neonatal health; polymorphism; birth anthropometry

1. Introduction

International nutritional recommendations during pregnancy comprise a fundamental guide for the optimal fetal development. Although there is almost universal agreement regarding macronutrient and folic requirements as well as the monitoring of weight gain during pregnancy, there are significant controversies regarding supplementation with specific biological molecules including vitamins [1].

A plethora of observational trials indicate maternal hypovitaminosis D during pregnancy as a significant risk factor for the development of adverse pregnancy outcomes and impairment of future offspring metabolic health [2]. Despite the wide inconsistency in available randomized trials, it is considered that vitamin D supplementation during pregnancy might reduce the risk of low birth weight, gestational diabetes, and preeclampsia [2,3]. In this regard, a sufficient maternal vitamin D profile during pregnancy is a critical component for the development of optimal neonatal vitamin D status at birth and during early infancy, since maternal 25-hydroxy-vitamin D [25(OH)D] comprises the main pool of vitamin D for the fetus [3] and serum fetal (cord blood) 25(OH)D concentrations correlate strongly with maternal 25(OH)D concentrations [4,5]. There is wide controversy, about the definition of maternal vitamin D deficiency during pregnancy worldwide, especially regarding the optimal thresholds of maternal 25(OH)D concentrations ($\geq$50 nmol/L vs. $\geq$75 nmol/L) [6,7]. On the other hand, different criteria are used to define the optimal neonatal vitamin D status (sufficiency >50 nmol/L, insufficiency 30–50 nmol/L, deficiency <30 nmol/L) [8].

In addition to this ongoing scientific argument, several parameters that could affect the daily clinical interpretation of available results, beyond maternal and neonatal 25(OH)D cut-offs including ultraviolet B (UVB) variations, country-specific dietary patterns, and public health policies are largely misinterpreted or ignored [8,9]. The potential influence of the specific genetic background of each individual for decreasing pregnancy complications and optimizing neonatal health has also been the objective of several previous observational studies, which mainly focused on the specific sequence variants of the vitamin D receptor (VDR). Specific maternal VDR polymorphisms have been associated with adverse pregnancy and offspring outcomes [10–12] and could demonstrate either detrimental or protective effects [10–13] on the development of maternal and neonatal hypovitaminosis D and other outcomes.

The extent of the potential effects of specific neonatal VDR polymorphisms related to neonatal 25(OH)D by adopting international cut-offs has not been fully elucidated so far [2,7]. By taking into account that maternal and neonatal vitamin D status at birth are dynamic parameters, in order to understand the mechanistic basis by which a polymorphism is associated with a particular pregnancy or offspring outcome, it is necessary to know whether that polymorphism is functional in different states of vitamin D equilibrium. In this regard, specific VDR polymorphisms have been associated with a decrease in birth weight and neonatal skin folds at birth [14] and increased risk for small for gestational age births in black and white women [15,16]. In addition, placental genetic variations in vitamin D metabolism were also associated through a sex-specific pattern with birth weight [17]. A recent meta-analysis of available results reported that birth weight and other anthropometric neonatal outcomes are affected by specific patterns of VDR polymorphisms [18]. Moreover, maternal genetic variations in GC, the gene encoding vitamin-D binding protein, have also been reported to affect the relationships between the maternal and cord-blood concentrations of 25(OH)D and birth weight [19]. However, robust evidence of such an association is currently unavailable, given that various studies have presented significant heterogeneity in terms of maternal and neonatal criteria for vitamin D status, study design, sample size, and racial descent of participants.

It has been hypothesized [20] that specific sequence VDR variants exert variable degrees of functionality associated with a specific neonatal outcome including birth height, weight, and additional birth anthropometry parameters, according to different cut-offs and available maternal and neonatal VDR polymorphisms. Moreover, a combined clinical (in terms of different maternal/neonatal 25(OH) D cut-offs) and VDR genotype association study focusing on a specific outcome could be mechanistically proven to be more productive than a study of individual polymorphisms or genome-wide associations of polymorphisms of unknown function. The aim of this study was to evaluate the combined effect of maternal and neonatal VDR polymorphisms (ApaI, TaqI, BsmI, FokI, Tru9I) and different maternal and neonatal 25(OH) D cut-offs on neonatal birth anthropometry.

2. Methods

2.1. Inclusion and Exclusion Criteria

This study included data and samples from a cohort of mother–child pairs at birth that have been previously described [3]. Pregnant women on regular follow-up were recruited from the Maternity Unit of the 1st Department of Obstetrics and Gynecology, Aristotle University, Thessaloniki, Greece (latitude 40°N). All women were fair skinned. The inclusion criterion was full-term pregnancy (gestational week 37–42). Maternal exclusion criteria were primary hyperparathyroidism, secondary osteoporosis, heavy alcohol use ($\geq$7 alcohol units per week or $\geq$6 units at any time during pregnancy), hyperthyroidism, nephritic syndrome, inflammatory bowel disease, rheumatoid arthritis, osteomalacia, obesity [body mass index (BMI) > 30 kg/m^2], gestational diabetes, preexisting diabetes mellitus, and use of medications affecting calcium (Ca) or vitamin D status (e.g., corticosteroids) including vitamin D supplements. Neonatal exclusion criteria were being small-for-gestational age (SGA) and presence of severe congenital anomalies. Informed consent was obtained from all mothers. The study was conducted from January 2018 to September 2018. The protocol received approval from the Bioethics Committee of the Aristotle University of Thessaloniki, Greece (approval number 1/19-12-2011).

2.2. Biochemical and Hormonal Assays

Blood samples were obtained from mothers by antecubital venipuncture 30–60 min before delivery. Umbilical cord blood was collected immediately after clamping from the umbilical vein. Serum and umbilical cord specimens were stored at -20 °C prior to analysis for the following parameters: Ca, phosphorus (P), parathyroid hormone (PTH), and 25(OH)D. Serum Ca and P determinations were performed using the Cobas INTEGRA clinical chemistry system (D-68298; Roche Diagnostics, Mannheim, Germany). The inter- and intra-assay coefficients of variation (CVs) were 1.0% and 3.5% for Ca, and 1.3% and 2.5% for P, respectively. PTH determinations were performed using the electro-chemiluminescence immunoassay ECLIA (Roche Diagnostics GmbA, Mannheim, Germany). Reference range for PTH was 15–65 pg/mL, functional sensitivity of 6.0 pg/mL, within-run precision of 0.6–2.8%, and total precision of 1.6–3.4%. Concentrations of 25(OH)D were determined using novel assay, liquid chromatography-tandem mass spectrometry (LC-MS/MS) with lower limits of quantification (LLOQ) of 25(OH)D (0.5 ng/mL). Briefly, the assay involves analyte purification using liquid–liquid extraction followed by chromatographical separation using a chiral column in tandem with a rapid resolution microbore column. Full method validation parameters have been previously reported [21,22].

2.3. Demographic and Anthropometric Data

At enrollment, demographic and social characteristics were recorded. Maternal pre-pregnancy BMI was either normal (18–25 kg/m^2) or overweight (25–30 kg/m^2). We collected maternal, infant, and labor data from the medical records, umbilical cord blood samples at the time of delivery, and stored aliquots of plasma and serum at $-70\ ^\circ$C, until assays were performed. We also evaluated neonatal anthropometry at birth. All neonatal anthropometric measurements were performed by the same trained nurse between 12 and 72 h of age according to standard techniques [23]. The following measurements were recorded: birth weight, height, neck–rump length, upper arm, femur, and knee-heel lengths; head, chest, abdominal, upper arm and middle thigh circumferences, and anterior chest and abdominal skinfold thickness. Birth weight of the neonates was obtained naked on regularly calibrated scales on a calibrated infant scale that was verified as accurate with a certified weight (Troemner, Thorofare, NJ, USA). Knee–heel length was measured with a hand-held BK5 infant knemometer (Force Technology, Brondby, Denmark). Instrument software calculated the mean of 10 sequential readings and generated a printed report of all readings and the calculated mean. We also measured the neonatal height to the nearest millimeter using an Ellard newborn lengthboard (Ellard Instrumentation Ltd., Seattle, WA, USA). Abdominal, upper arm and middle thigh head, mid-upper arm, and maximal head circumferences were measured using a plastic encircling tape (Child Growth Foundation, London, UK). Abdominal skin fold was measured using Holtain calipers (Holtain, Crymych, UK).

2.4. Neonatal and Maternal Vitamin D Status Cut-Offs and Combined VDR Polymorphisms Evaluation

Different 25(OH)D cut-offs for neonates and mothers were included, according to their vitamin D status at birth: for neonates, cut-offs of [25(OH)D $\leq$25 and >25 nmol/L] and [25(OH)D $\leq$50] [7] were adopted, whereas for mothers, a 25(OH)D cut-off of [25(OH)D $\leq$50 and >50 nmol/L)] was investigated [24]. Following this classification, maternal and neonatal VDR polymorphisms were assessed at birth to investigate the potential different effects of different neonatal and maternal 25(OH)D cut-offs on neonatal birth anthropometry.

2.5. VDR Analysis

DNA was isolated from peripheral blood samples by a QIAamp DNA Blood Mini Kit (Cat. no. 51304, QIAGEN), according to the manufacturer's protocol. In order to determine the genotypes of rs7975232 (ApaI), rs7731236 (TaqI), rs757343 (Tru9I), and rs1544410 (BsmI) SNPs within the *VDR* gene, polymerase chain reaction (PCR) and restriction fragment length polymorphism (RFLP) methods were performed as previously described [25]. The real-time PCR (RT-PCR) method was used for determining genotypes of rs2228570 (FokI) SNP by using the Simple Probe (LightSNiP, TibMolBiol, Berlin, Germany) and LightCycler Fast Start DNA Master HybProbe Kit (Cat. no. 12239272001, Roche) with a LightCycler 480 Instrument II (Roche). Melting curve analysis was performed for genotyping as previously described [26].

Each SNP allele was named after as follows: for rs7731236 (TaqI), "t" represents C, "T" represents T nucleotide; for rs7975232 (ApaI), "a" represents C, "A" represents A nucleotide; for rs757343 (Tru9I), "u" represents A, "U" represents G nucleotide; for rs1544410 (BsmI), "b" represents G, "B" represents A nucleotide, and for rs2228570 (FokI), "f" represents T, and "F" represents C nucleotide.

2.6. UVB Measurements

UVB radiation includes wavelengths from 280 to 320 nm. UVB data for the broad geographical region of Thessaloniki, Greece were collected at the Laboratory of Atmospheric Physics, School of Physics, Aristotle University of Thessaloniki.

The daily integral of vitamin D effective UVB radiation (09:00 to 16:00 local time) was used as the most representative parameter for UVB exposure. These hours were selected as indicative, since they are related to the beginning and the end of the working period for the majority of the Greek population. Individual sunlight exposure was recorded for each participant during that period. Finally, mean UVB exposure during the previous 45 days (daily integral) before blood sample collection (estimated mean half-life of vitamin D) was calculated for each participant.

2.7. Statistical Analysis

Given that there was no comparison group, the distributions of genotypes of SNPs within the maternal and neonatal VDR genes were given as frequency data. Mean birth neonatal anthropometry data including height (cm), weight (g), head circ/ce (cm), neck rump length (cm), chest circ/ce (cm), abdominal circ/ce (cm), abdominal circ/ce iliac (cm), skin fold abdominal (cm), skin fold high anterior (cm), high thigh circ/ce (cm), middle thigh circ/ce (cm), upper arm length (cm), lower arm radial circ/ce (cm), lower leg calf circ/ce (cm), femur length (cm), and knee-heel length (cm) values of minor allele carriers and homozygote major allele carriers in groups were compared with *T*-Tests. If the Levene's test for equality of variances is $p > 0.05$, then equal variances assumed Sig (2-tailed) p values of *T*-Tests were given. If the Levene's test for equality of variances is $p < 0.05$, then the equal variances not assumed Sig (2-tailed) p values of *T*-Tests were given. The data and p values adjusted for maternal and paternal height (cm), UVB radiation, BMI pre-pregnancy (kg/m^2), BMI terminal (kg/m^2), and weeks of gestation by one-way analysis of covariance (ANCOVA). Corrected effect size was calculated with Hedge's g (Hg) where 0.2 is suggested as a small effect size, 0.5 is the medium effect, and 0.8 is a larger effect [27]. Post-hoc power analysis was performed for significant outcomes. All data were presented as the means $\pm$ SD in the text and figure legends. The tests were performed in groups by stratifying data for maternal or neonatal 25OHD level cutoff values. Statistical analyses were performed by SPSS 24.0 software(IBM, Armonk, New York, NY, USA).

3. Results

Seventy mother–neonate pairs were included in the study. Given four neonates had missing birth neonatal anthropometry data, they were excluded from related analysis. The demographic and laboratory data of mothers and neonates are presented in Table 1. VDR single nucleotide polymorphisms (SNPs) and the genotype distributions of mothers and neonates are presented in Table 2.

Table 1. Maternal and neonatal demographic and anthropometric characteristics.

Maternal	
Number (n)	66
Age (years)	31.92 ± 6.08
Height (cm)	164.85 ± 5.47
Weight; pre-pregnancy (kg)	67.56 ± 14.54
Weight; term (kg)	85.43 ± 14.30
BMI; pre-pregnancy (kg/m^2)	24.91 ± 4.81
BMI; term (kg/m^2)	29.62 ± 5.80
Weeks of gestation (n)	38.80 ± 1.56
Smoking [n (%)]	10 (0.14)
Alcohol consumption [n (%)]	8 (0.11)
Previous live births [n (%)]	26 (0.37)
Daily Calcium Supplementation [n (%)]	37 (0.56)
Daily Calcium Supplementation (mg)	423.07 ± 319.07
Paternal height	177.85 ± 6.14
Neonatal	
Number (n)	66
Gender; Males [n (%)]	38 (0.58)
Height (cm)	50.48 ± 1.96
Weight (g)	3292.12 ± 414.25
Head Circumference (cm)	34.40 ± 2.83
Neck rump length (cm)	17.66 ± 2.16
Chest Circumference (cm)	30.97 ± 1.97
Abdominal Circumference (cm)	28.11 ± 2.03
Abdominal Circumference iliac (cm)	25.94 ± 1.71
Skin fold; abdominal (cm)	2.95 ± 0.50
Upper Arm Circumference (cm)	9.74 ± 0.74
High thigh Circumference (cm)	15.41 ± 1.48
Middle thigh Circumference (cm)	13.36 ± 1.16
Upper Arm Length (cm)	13.65 ± 0.94
Lower Leg Calf Circumference (cm)	10.23 ± 0.83
Femur Length (cm)	9.94 ± 0.57
Knee-Heel Length (cm)	9.15 ± 0.62

Data are presented as mean $\pm$ standard deviation (SD) for continuous variables and frequencies [numbers (%)] for categorical variables. Abbreviations: BMI, body mass index.

Table 2. Vitamin D receptor single nucleotide polymorphisms genotype distributions of mothers and neonates.

SNP	APAI			TAQI			BSMI			FOKI			TRU9I		
Genotype	AA	Aa	aa	TT	Tt	tt	BB	Bb	bb	FF	Ff	ff	UU	Uu	uu
Maternal (n:%)	29 (0.41)	33 (0.47)	8 (0.12)	25 (0.36)	33 (0.47)	12 (0.17)	26 (0.37)	21 (0.30)	23 (0.33)	32 (0.46)	32 (0.46)	6 (0.08)	41 (0.59)	26 (0.37)	3 (0.04)
Neonatal (n:%)	23 (0.33)	39 (0.56)	8 (0.11)	27 (0.38)	32 (0.46)	11 (0.16)	19 (0.27)	27 (0.39)	24 (0.34)	34 (0.49)	31 (0.44)	5 (0.07)	46 (0.66)	22 (0.31)	2 (0.03)

3.1. Birth Neonatal Anthropometry (Neonatal Cut-Offs at Birth >50 nmol/L and <25 and >25 nmol/L) according to Neonatal VDR Polymorphisms

Birth neonatal anthropometry was investigated in neonates whose 25(OH)D at birth was <25 and >25 nmol/L, respectively, and compared according to neonatal VDR polymorphisms. After adjustments, the abdominal circumference (cm) of neonates who had the TAQI SNP TT genotype and 25(OH)D < 25 nmol/L were significantly higher than that of neonates who had Tt or tt genotypes ($p = 0.038$, Hg = 1.138) (Table 3), whereas for neonates with 25(OH)D >25 nmol/L, no significant difference was observed in any birth neonatal anthropometry (Table 4). There was no significant difference in any additional birth neonatal anthropometry parameter, which was investigated in neonates with 25(OH)D > 50 nmol/L, according to neonatal VDR polymorphisms (Table 5).

3.2. Birth Neonatal Anthropometry (Maternal Cut-Offs at Birth <50 and >50 nmol/L) According to Maternal VDR Polymorphisms

Birth neonatal anthropometry was investigated in neonates whose maternal 25(OH)D at birth was <50 and >50 nmol/L, respectively, and compared according to maternal VDR polymorphisms. After adjustments, weight, neck rump length, chest circumference, abdominal circumference, abdominal circumference (iliac), high thigh circumference, middle thigh circumference, lower arm radial circumference, and lower leg calf circumference of neonates who had the TAQl SNP TT genotype and maternal 25(OH)D < 50 nmol/L were significantly higher than that of neonates who had the Tt or tt genotypes ($p = 0.001$, Hg = 1.341, $p = 0.036$, Hg = 0.976, $p = 0.004$, Hg = 1.381, $p = 0.001$, Hg = 1.554, $p = 0.001$, Hg = 1.351, $p = 0.028$, Hg = 0.918, $p = 0.008$, Hg = 1.090, $p = 0.002$, Hg = 1.217, and $p = 0.020$, Hg = 1.263; respectively), (Table 6). Skin fold high anterior was significantly lower in neonates who had the BSMI SNP BB genotype, compared to that of neonates with Bb or bb genotypes ($p = 0.041$, Hg = 0.950) (Table 6), whereas neck rump length was significantly higher in neonates who had the FOKI SNP FF genotype compared to that of neonates who had Ff or ff genotypes ($p = 0.042$, Hg = 1.228) (Table 6).

There was no significant difference in any additional birth neonatal anthropometry parameter, which was investigated in neonates whose maternal 25(OH)D concentration was >50 nmol/L when compared according to maternal VDR polymorphisms, except neonatal height. The height of the neonates with UU was significantly higher than the ones with Uu or uu ($p = 0.032$, Hg = 0.444) (Table 7).

3.3. Birth Neonatal Anthropometry (Maternal Cut-Off at Birth <75 nmol/L) According to Neonatal VDR Polymorphisms

Birth neonatal anthropometry was also investigated in neonates whose maternal 25(OH)D was <75 nmol/L and compared according to neonatal VDR polymorphisms. After adjustments, the lower arm radial circumference of neonates who had the APAl SNP AA genotype was significantly lower than that of neonates who had Aa or aa genotypes ($p = 0.043$, Hg = 0.966) (Table 8), whereas no other significant differences were evident (Table 8).

Table 3. Birth neonatal anthropometry (neonatal vitamin D status at birth <25 nmol/L), according to neonatal VDR polymorphisms.

SNP	Genotype	N	Height (cm)	Weight (g)	Head Circ/ce (cm)	Neck Rump Length (cm)	Chest Circ/ce (cm)	Abdominal Circ/ce (cm)	Abdominal Circ/ce Iliac (cm)	Skin Fold Abdominal (cm)	Skin Fold High Anterior (cm)	High Thigh Circ/ce (cm)	Middle Thigh Circ/ce (cm)	Upper Arm Length (cm)	Lower Arm Radial Circ/ce (cm)	Lower Leg Calf Circ/ce (cm)	Femur Length (cm)	Knee-Heel Length (cm)
APA1	AA	9	49.50 ± 1.6	3261.11 ± 341	34.66 ± 0.7	17.82 ± 2.0	30.57 ± 2.5	27.95 ± 2.3	25.80 ± 1.6	2.71 ± 0.3	3.64 ± 0.5	15.24 ± 1.3	13.43 ± 1.5	13.44 ± 0.5	8.90 ± 0.7	10.33 ± 0.8	10.02 ± 0.4	9.40 ± 0.4
	Aa + aa	16	50.81 ± 1.9	3352.50 ± 372	35.06 ± 5.4	17.45 ± 2.0	31.20 ± 1.8	27.88 ± 1.6	26.01 ± 1.3	2.95 ± 0.2	4.00 ± 0.4	15.61 ± 1.1	13.55 ± 0.8	13.32 ± 0.7	8.96 ± 0.5	10.36 ± 0.6	9.93 ± 0.6	9.24 ± 0.6
	p-value		0.1	0.55	0.83	0.67	0.48	0.92	0.72	0.06	0.09	0.47	0.8	0.69	0.81	0.9	0.73	0.52
TAQ1	TT	8	50.62 ± 2.3	3523.75 ± 532	36.56 ± 7.5	18.95 ± 1.4 (17.00 ± 1.2) *	32.13 ± 2.3 (29.86 ± 1.3) *	29.10 ± 2.4 (26.61 ± 1.3) *	27.06 ± 1.9 (25.13 ± 0.7) *	3.02 ± 0.2	4.07 ± 0.3	16.32 ± 0.9 (14.83 ± 1.0) *	14.30 ± 1.3 (13.11 ± 0.9) *	13.51 ± 0.8	9.41 ± 0.8 (8.58 ± 0.3) *	10.76 ± 0.9	10.06 ± 0.7	9.48 ± 0.9
	Tt + tt	17	50.20 ± 1.7	3223.52 ± 190	34.14 ± 0.8	16.94 ± 1.9 (17.48 ± 2.2) *	30.43 ± 1.7 (31.37 ± 1.4) *	27.34 ± 1.7 (28.09 ± 1.3) *	25.41 ± 0.8 (26.00 ± 1.2) *	2.78 ± 0.3	3.77 ± 0.5	15.08 ± 1.1 (15.76 ± 1.0) *	13.13 ± 0.7 (13.52 ± 0.7) *	13.30 ± 0.6	8.71 ± 0.3 (9.00 ± 0.6) *	10.16 ± 0.4	9.92 ± 0.4	9.21 ± 0.3
	p-value adjusted *p* Effect size Power		0.62	0.16	0.39	**0.017** *0.95* *	**0.054** *0.31* *	**0.092** *0.038* * *1.138* [†] *0.76* ϕ	**0.048** *0.060* *	0.073	0.17	**0.013** *0.25* *	**0.01** *0.49* *	0.49	**0.052** *0.073* *	0.11	0.58	0.43
BSM1	BB	8	49.31 ± 1.6	3165.00 ± 195	34.56 ± 0.7	17.32 ± 1.5	29.95 ± 1.8	27.22 ± 0.9	25.28 ± 0.6	2.65 ± 0.3 (2.93 ± 0.3) *	3.57 ± 0.5 (3.93 ± 0.5) *	14.87 ± 0.8	12.97 ± 0.6	13.40 ± 0.6	8.86 ± 0.2	10.01 ± 0.4	9.96 ± 0.4	9.28 ± 0.2
	Bb + bb	17	50.82 ± 1.8	3392.35 ± 395	35.08 ± 5.2	17.71 ± 2.24	31.46 ± 2.0	28.22 ± 2.1	26.24 ± 1.6	2.96 ± 0.2 (2.91 ± 0.3) *	4.01 ± 0.4 (3.90 ± 0.4) *	15.76 ± 1.2	13.75 ± 1.1	13.35 ± 0.7	9.07 ± 0.7	10.47 ± 0.7	9.97 ± 0.6	9.30 ± 0.6
	p-value		0.066	0.067	0.78	0.66	0.09	0.22	0.12	**0.014** *0.80* *	**0.04** *0.80* *	0.084	0.097	0.88	0.13	0.2	0.97	0.92
FOK1	FF	10	49.90 ± 1.5	3256.00 ± 307	34.25 ± 0.8	18.21 ± 1.5	30.94 ± 2.1	27.75 ± 2.2	25.89 ± 1.4	2.80 ± 0.2	3.70 ± 0.5	15.36 ± 1.2	13.63 ± 1.4	13.34 ± 0.5	8.78 ± 0.7	10.27 ± 0.8	10.05 ± 0.4	9.11 ± 0.6
	Ff + ff	15	50.63 ± 2.1	3362.00 ± 390	35.36 ± 5.5	17.17 ± 2.2	31.00 ± 2.1	28.01 ± 1.6	25.97 ± 1.5	2.90 ± 0.3	3.98 ± 0.4	15.56 ± 1.1	13.42 ± 0.8	13.38 ± 0.7	9.04 ± 0.5	10.41 ± 0.5	9.91 ± 0.6	9.42 ± 0.5
	p-value		0.36	0.47	0.53	0.21	0.93	0.74	0.89	0.41	0.16	0.69	0.66	0.87	0.3	0.61	0.57	0.18
TRU91	UU	18	50.50 ± 2.0	3325.00 ± 359	35.11 ± 5.0	17.32 ± 1.9	30.68 ± 1.9	27.70 ± 1.5	25.81 ± 1.3	2.90 ± 0.2	3.83 ± 0.5	15.30 ± 1.1	13.32 ± 0.8	13.37 ± 0.7	8.88 ± 0.5	10.35 ± 0.5	9.96 ± 0.6	9.34 ± 0.5
	Uu + uu	7	49.92 ± 1.7	3305.71 ± 379	34.42 ± 1.0	18.27 ± 2.0	31.74 ± 2.2	28.44 ± 2.5	26.25 ± 1.7	2.77 ± 0.3	3.97 ± 0.3	15.92 ± 1.3	13.98 ± 1.5	13.35 ± 0.7	9.07 ± 0.8	10.35 ± 0.9	9.98 ± 0.5	9.05 ± 0.7
	p-value		0.51	0.9	0.73	0.29	0.26	0.38	0.51	0.36	0.54	0.25	0.18	0.96	0.52	0.99	0.92	0.19

If the Levene's test for equality of variances $p > 0.05$ then equal variances assumed Sig (2-tailed) p values of T-Tests were given. If the Levene's test for equality of variances $p < 0.05$ then equal variances not assumed Sig (2-tailed) p values of T-Tests were given. * The data and p values adjusted for maternal height (cm), BMI pre-pregnancy (kg/m^2), BMI terminal (kg/m^2), and weeks of gestation by one-way analysis of covariance (ANCOVA). [†] Corrected effect size was calculated with Hedge's g (Hg). ϕ Post-hoc power analysis. Abbreviations: VDR: Vitamin D receptor; 25(OH)D: 25-hydroxy-vitamin D. Digits in bold refer to significant effects.

Table 4. Birth neonatal anthropometry (neonatal vitamin D status at birth >25 nmol/L) according to neonatal VDR polymorphisms.

SNP	Genotype	N	Height (cm)	Weight (g)	Head Circ/ce (cm)	Neck Rump Length (cm)	Chest Circ/ce (cm)	Abdominal Circ/ce (cm)	Abdominal Circ/ce Iliac (cm)	Skin Fold Abdominal (cm)	Skin Fold High Anterior (cm)	High Thigh Circ/ce (cm)	Middle Thigh Circ/ce (cm)	Upper Arm Length (cm)	Lower Arm Radial Circ/ce (cm)	Lower Leg Calf Circ/ce (cm)	Femur Length (cm)	Knee-Heel Length (cm)
APA1	AA	18	50.50 ± 1.8	3241.66 ± 423	34.30 ± 0.8	17.80 ± 2.3	30.98 ± 2.0	28.15 ± 2.1	25.94 ± 2.0	2.93 ± 0.3	3.76 ± 0.4	15.27 ± 1.5	13.15 ± 1.2	13.77 ± 0.5	8.96 ± 0.8	10.16 ± 0.9	9.93 ± 0.6	9.18 ± 0.4
	Aa + aa	23	50.63 ± 2.1	3301.73 ± 475	33.91 ± 1.5	17.65 ± 2.3	30.88 ± 1.9	28.29 ± 2.1	25.93 ± 1.7	3.07 ± 0.7	3.82 ± 0.4	15.43 ± 1.7	13.38 ± 1.2	13.88 ± 1.3	9.04 ± 0.5	10.17 ± 0.9	9.93 ± 0.4	8.98 ± 0.7
	p-value		0.84	0.67	0.33	0.84	0.87	0.84	0.99	0.43	0.65	0.75	0.55	0.75	0.69	0.96	0.99	0.33

Table 4. *Cont.*

SNP	Genotype	N	Height (cm)	Weight (g)	Head Circ/ce (cm)	Neck Rump Length (cm)	Chest Circ/ce (cm)	Abdominal Circ/ce (cm)	Abdominal Circ/ce Iliac (cm)	Skin Fold Abdominal (cm)	Skin Fold High Anterior (cm)	High Thigh Circ/ce (cm)	Middle Thigh Circ/ce (cm)	Upper Arm Length (cm)	Lower Arm Radial Circ/ce (cm)	Lower Leg Calf Circ/ce (cm)	Femur Length (cm)	Knee-Heel Length (cm)
TAQI	TT	17	51.35 ± 2.1 (50.79 ± 2.1) *	3423.52 ± 488	34.44 ± 0.9	17.69 ± 2.9	31.34 ± 1.9	28.83 ± 2.1	26.20 ± 1.8	3.09 ± 0.8	3.90 ± 0.4	15.68 ± 1.7	13.58 ± 1.2	14.10 ± 1.4	9.15 ± 0.6	10.41 ± 0.8	10.01 ± 0.5	9.05 ± 0.8
	Tt + tt	24	50.02 ± 1.7 (50.75 ± 2.1) *	3170.41 ± 395	33.83 ± 1.4	17.73 ± 1.7	30.64 ± 1.9	27.80 ± 2.0	25.74 ± 1.9	2.95 ± 0.3	3.72 ± 0.4	15.14 ± 1.5	13.06 ± 1.1	13.64 ± 0.5	8.90 ± 0.7	9.99 ± 0.9	9.87 ± 0.6	9.08 ± 0.5
	p-value		**0.035** *0.88 **	**0.075**	0.13	0.95	0.26	0.12	0.45	0.47	0.16	0.3	0.18	0.16	0.26	0.14	0.44	0.88
BSMI	BB	16	50.40 ± 1.9	3220.00 ± 420	34.28 ± 0.8	18.06 ± 1.8	30.91 ± 2.1	28.18 ± 2.2	26.08 ± 2.1	2.95 ± 0.3	3.75 ± 0.4	15.43 ± 1.5	13.26 ± 1.2	13.78 ± 0.5	9.01 ± 0.8	10.20 ± 0.9	9.91 ± 0.7	9.16 ± 0.5
	Bb + bb	25	50.68 ± 2.1	3310.80 ± 471	33.96 ± 1.5	17.49 ± 2.5	30.94 ± 1.8	28.26 ± 2.1	25.84 ± 1.7	3.05 ± 0.6	3.83 ± 0.3	15.32 ± 1.7	13.29 ± 1.2	13.86 ± 1.2	9.00 ± 0.5	10.14 ± 0.8	9.94 ± 0.4	9.00 ± 0.7
	p-value		0.67	0.53	0.44	0.44	0.96	0.91	0.69	0.57	0.54	0.84	0.94	0.81	0.98	0.86	0.89	0.44
FOKI	FF	20	50.65 ± 2.2	3370.50 ± 453	33.92 ± 1.7	18.31 ± 2.3	31.31 ± 1.8	28.44 ± 2.1	26.32 ± 2.0	3.08 ± 0.8	3.90 ± 0.3	15.64 ± 1.8	13.44 ± 1.3	14.03 ± 1.2	9.17 ± 0.7	10.34 ± 1.0	9.97 ± 0.5	9.16 ± 0.5
	Ff + ff	21	50.50 ± 1.8	3184.76 ± 435	34.23 ± 0.6	17.15 ± 2.0	30.56 ± 1.9	28.03 ± 2.1	25.58 ± 1.6	2.95 ± 0.2	3.70 ± 0.4	15.10 ± 1.4	13.12 ± 1.0	13.64 ± 0.6	8.85 ± 0.5	10.00 ± 1.0	9.89 ± 0.6	8.98 ± 0.7
	p-value		0.81	0.18	0.45	0.1	0.22	0.54	0.21	0.49	0.13	0.3	0.41	0.23	0.13	0.22	0.66	0.39
TRU9I	UU	20	50.22 ± 1.7	3187.50 ± 459	33.85 ± 1.6	17.31 ± 2.3	30.29 ± 1.7 (30.93 ± 2.3) *	27.44 ± 2.2 (28.02 ± 2.9) *	25.42 ± 1.7	3.11 ± 0.7	3.83 ± 0.4	14.77 ± 1.4 (15.53 ± 1.8) *	12.89 ± 1.1 (13.42 ± 1.5) *	13.80 ± 1.3	8.89 ± 0.6	9.91 ± 0.8	9.97 ± 0.4	9.04 ± 0.7
	Uu + uu	21	50.90 ± 2.1	3359.04 ± 433	34.30 ± 0.7	18.10 ± 2.1	31.53 ± 1.9 (31.05 ± 1.8) *	28.98 ± 1.7 (28.39 ± 1.8) *	26.43 ± 1.8	2.92 ± 0.3	3.77 ± 0.4	15.93 ± 1.6 (15.24 ± 1.5) *	13.65 ± 1.2 (13.30 ± 1.0) *	13.86 ± 0.7	9.11 ± 0.6	10.40 ± 0.8	9.89 ± 0.6	9.09 ± 0.5
	p-value		0.28	0.22	0.25	0.26	**0.038** *0.96 **	**0.019** *0.74 **	**0.084**	0.31	0.65	**0.022** *0.78 **	**0.043** *0.94 **	0.85	0.29	**0.08**	0.66	0.8

If the Levene's test for equality of variances $p > 0.05$ then equal variances assumed Sig (2-tailed) p values of *T*-Tests were given. If the Levene's test for equality of variances $p < 0.05$ then equal variances not assumed Sig (2-tailed) p values of *T*-Tests were given. * The data and p values adjusted for maternal height (cm), BMI pre-pregnancy (kg/m^2), BMI terminal (kg/m^2), and weeks of gestation by one-way analysis of covariance (ANCOVA). Abbreviations: VDR: Vitamin D receptor; 25(OH)D: 25-hydroxy-vitamin D. Digits in bold refer to significant effects.

Table 5. Birth neonatal anthropometry (neonatal vitamin D status at birth >50 nmol/L) according to neonatal VDR polymorphisms.

SNP	Genotype	N	Height (cm)	Weight (g)	Head Circ/ce (cm)	Neck Rump Length (cm)	Chest Circ/ce (cm)	Abdominal Circ/ce (cm)	Abdominal Circ/ce Iliac (cm)	Skin Fold Abdominal (cm)	Skin Fold Thigh Anterior (cm)	High Thigh Circ/ce (cm)	Middle Thigh Circ/ce (cm)	Upper Arm Length (cm)	Lower Arm Radial Circ/ce (cm)	Lower Leg Calf Circ/ce (cm)	Femur Length (cm)	Knee-Heel Length (cm)
APAI	AA	3	51.66 ± 1.4	3336.66 ± 457	35.16 ± 0.2	17.96 ± 1.9	17.96 ± 0.6	32.80 ± 1.0	26.40 ± 0.4	2.73 ± 0.4	3.86 ± 0.2	16.13 ± 0.5	13.90 ± 0.1	14.16 ± 0.4	8.76 ± 0.5	10.06 ± 0.8	8.86 ± 0.4	9.93 ± 1.0
	Aa + aa	11	50.36 ± 2.0	3223.63 ± 542	33.72 ± 1.8	18.06 ± 3.0	18.06 ± 2.0	31.47 ± 2.5	26.66 ± 2.2	2.96 ± 0.4	3.94 ± 0.4	15.80 ± 1.7	13.56 ± 1.3	14.22 ± 1.6	9.21 ± 0.8	10.38 ± 1.1	8.98 ± 0.4	9.83 ± 0.8
	p-value		NA	NA	NA	NA	NA	NA	NA	NA	NA	NA	NA	NA	NA	NA	NA	NA
TAQI	TT	7	50.78 ± 1.7	3218.57 ± 608	34.21 ± 1.1	17.15 ± 2.9	31.47 ± 1.9	29.21 ± 2.4	26.52 ± 2.3	2.82 ± 0.3	3.97 ± 0.5	15.22 ± 1.7	13.30 ± 1.3	14.54 ± 1.9	9.11 ± 0.6	10.22 ± 1.0	9.85 ± 0.4	8.88 ± 1.0
	Tt + tt	7	50.50 ± 2.2	3277.14 ± 438	33.85 ± 2.2	18.92 ± 2.4	32.04 ± 1.9	28.54 ± 2.2	26.68 ± 1.8	3.00 ± 0.4	3.88 ± 0.3	16.52 ± 0.9	13.97 ± 1.1	13.88 ± 0.6	9.11 ± 0.9	10.40 ± 1.0	9.85 ± 0.4	9.02 ± 0.7
	p-value		0.79	0.84	0.71	0.24	0.6	0.6	0.89	0.45	0.71	0.11	0.33	0.41	1	0.76	1	0.77
BSMI	BB	3	51.66 ± 1.5	3336.66 ± 457	35.16 ± 0.2	17.96 ± 1.9	32.80 ± 0.6	29.06 ± 1.06	26.40 ± 0.4	2.73 ± 0.4	3.86 ± 0.2	16.13 ± 0.5	13.90 ± 0.3	14.16 ± 0.45	8.76 ± 0.5	10.06 ± 0.8	9.93 ± 0.4	8.86 ± 1.0
	Bb + bb	11	50.36 ± 2.0	3223.63 ± 542	33.72 ± 1.8	18.06 ± 3.1	31.47 ± 2.0	28.82 ± 2.56	26.66 ± 2.3	2.96 ± 0.3	3.94 ± 0.4	15.81 ± 1.6	13.56 ± 1.3	14.22 ± 1.6	9.21 ± 0.8	10.38 ± 1.0	9.83 ± 0.4	8.98 ± 0.8
	p-value		NA	NA	NA	NA	NA	NA	NA	NA	NA	NA	NA	NA	NA	NA	NA	NA

Table 5. *Cont.*

SNP	Genotype	N	Height (cm)	Weight (g)	Head Circ/ce (cm)	Neck Rump Length (cm)	Chest Circ/ce (cm)	Abdominal Circ/ce (cm)	Abdominal Circ/ce Iliac (cm)	Skin Fold Abdominal (cm)	Skin Fold Anterior (cm)	High Thigh Circ/ce (cm)	Middle Thigh Circ/ce (cm)	Upper Arm Length (cm)	Lower Arm Radial Circ/ce (cm)	Lower Leg Calf Circ/ce (cm)	Femur Length (cm)	Knee-Heel Length (cm)
FOKI	FF	9	50.61 ± 2.3	3294.44 ± 442	33.72 ± 2.0	18.42 ± 1.9	31.54 ± 1.5	28.42 ± 2.4	26.63 ± 2.1	2.93 ± 0.4	4.00 ± 0.5	16.11 ± 1.1	13.66 ± 0.9	14.58 ± 1.6	9.12 ± 0.7	10.32 ± 0.9	9.90 ± 0.4	9.14 ± 0.3
	Ff + ff	5	50.70 ± 1.3	3164.00 ± 664	34.60 ± 0.8	17.36 ± 4.0	32.14 ± 2.8	29.70 ± 1.8	26.56 ± 2.0	2.88 ± 0.2	3.80 ± 0.3	15.46 ± 2.1	13.58 ± 1.8	13.54 ± 0.5	9.10 ± 0.9	10.30 ± 1.2	9.78 ± 0.4	8.62 ± 1.4
	p-value		0.93	0.66	0.38	0.51	0.6	0.33	0.95	0.82	0.4	0.46	0.91	0.19	0.96	0.97	0.62	0.46
TRU9I	UU	10	50.60 ± 2.1	3228.00 ± 554	34.00 ± 1.9	18.63 ± 2.3	31.49 ± 2.1	28.42 ± 2.5	26.25 ± 2.1	3.00 ± 0.3	3.86 ± 0.4	15.78 ± 1.6	13.55 ± 1.3	14.31 ± 1.6	9.06 ± 0.8	10.18 ± 1.1	9.77 ± 0.4	8.88 ± 1.0
	Uu + uu	4	50.75 ± 1.8	3297.50 ± 449	34.12 ± 1.1	16.57 ± 3.5	32.42 ± 0.9	30.02 ± 0.7	27.50 ± 1.6	2.70 ± 0.4	4.10 ± 0.3	16.12 ± 1.3	13.85 ± 1.0	13.97 ± 0.3	9.25 ± 0.6	10.65 ± 0.7	10.07 ± 0.3	9.15 ± 0.2
	p-value		NA	NA	NA	NA	NA	NA	NA	NA	NA	NA	NA	NA	NA	NA	NA	NA

If the Levene's test for equality of variances $p > 0.05$ then equal variances assumed Sig (2-tailed) p values of *T*-Tests were given. If the Levene's test for equality of variances $p < 0.05$ then equal variances not assumed Sig (2-tailed) p values of *T*-Tests were given. Abbreviations: VDR: Vitamin D receptor; 25(OH)D: 25-hydroxy-vitamin D, NA: Not applicable.

Table 6. Birth neonatal anthropometry (maternal vitamin D status at birth <50 nmol/L) according to maternal VDR polymorphisms.

SNP	Genotype	N	Height (cm)	Weight (g)	Head Circ/ce (cm)	Neck Rump Length (cm)	Chest Circ/ce (cm)	Abdominal Circ/ce (cm)	Abdominal Circ/ce Iliac (cm)	Skin Fold Abdominal (cm)	Skin Fold High Anterior (cm)	High Thigh Circ/ce (cm)	Middle Thigh Circ/ce (cm)	Upper Arm Length (cm)	Lower Arm Radial Circ/ce (cm)	Lower Leg Calf Circ/ce (cm)	Femur Length (cm)	Knee-Heel Length (cm)
APA1	AA	18	49.66 ± 1.6	3282.77 ± 375	34.41 ± 0.7	17.93 ± 1.7	30.26 ± 1.9	27.76 ± 2.0	25.50 ± 1.7	2.81 ± 0.3	3.60 ± 0.5 (3.64 ± 0.4) *	15.05 ± 1.2	13.16 ± 1.2	13.56 ± 0.5	8.87 ± 0.8	10.21 ± 0.7	9.96 ± 0.7	9.24 ± 0.4
	Aa + aa	25	50.54 ± 1.9	3320.40 ± 361	34.46 ± 4.4	17.43 ± 1.8	31.17 ± 1.8	28.08 ± 1.6	25.99 ± 1.5	3.08 ± 0.6	3.96 ± 0.4 (3.98 ± 0.4) *	15.55 ± 1.4	13.48 ± 0.9	13.41 ± 0.7	9.04 ± 0.6	10.30 ± 0.7	9.94 ± 0.5	9.24 ± 0.5
	p-value		0.13	0.74	0.96	0.35	0.13	0.57	0.32	0.12	**0.017** *0.097* *	0.22	0.34	0.47	0.18	0.44	0.93	0.99
TAQI	TT	14	50.50 ± 2.2	3520.71 ± 426 *(3503.33 ± 348)* *	35.42 ± 5.7	18.58 ± 1.5 *(18.32 ± 1.1)* *	31.99 ± 2.0 *(31.91 ± 1.4)* *	29.27 ± 1.8 *(29.01 ± 1.3)* *	26.84 ± 1.8 *(26.51 ± 1.6)* *	3.18 ± 0.8	4.04 ± 0.4 *(4.00 ± 0.4)* *	16.16 ± 1.4 *(15.86 ± 1.2)* *	14.07 ± 1.2 *(13.85 ± 1.0)* *	13.60 ± 0.6	9.35 ± 0.7 *(9.17 ± 0.6)* *	10.73 ± 0.8 *(10.56 ± 0.7)* *	10.07 ± 0.6	9.43 ± 0.6
	Tt + tt	29	50.01 ± 1.6	3200.34 ± 281 *(3146.31 ± 218)* *	33.96 ± 0.9	17.18 ± 1.7 *(16.82 ± 1.7)* *	30.21 ± 1.5 *(30.07 ± 1.3)* *	27.31 ± 1.4 *(26.99 ± 1.3)* *	25.27 ± 1.1 *(25.00 ± 0.8)* *	2.86 ± 0.2	3.69 ± 0.5 *(3.80 ± 0.4)* *	14.94 ± 1.1 *(14.88 ± 1.0)* *	12.99 ± 0.7 *(12.97 ± 0.7)* *	13.41 ± 0.6	8.78 ± 0.5 *(8.66 ± 0.3)* *	10.03 ± 0.6 *(9.91 ± 0.4)* *	9.89 ± 0.5	9.15 ± 0.3
	p-value *adjusted p* *Effect size* *Power*		0.43	**0.019** *0.001* * *1.341* † *0.94* φ	0.36	**0.012** *0.036* * *0.976* † *0.94* φ	**0.003** *0.004* * *1.381* † *0.99* φ	**0.001** *0.001* * *1.554* † *1.00* φ	**0.001** *0.001* * *1.351* † *0.92* φ	0.07	**0.031** *0.15* *	**0.003** *0.028* * *0.918* † *0.75* φ	**0.001** *0.008* * *1.090* † *0.84* φ	0.35	**0.018** *0.002* * *1.217* † *0.85* φ	**0.003** *0.020* * *1.263* † *0.90* φ	0.38	0.17
BSMI	BB	16	49.53 ± 1.6	3202.50 ± 310	34.31 ± 0.7	17.58 ± 1.4	29.91 ± 1.5 *(29.84 ± 1.0)* *	27.45 ± 1.5	27.45 ± 1.5	2.78 ± 0.3	3.53 ± 0.5 *(3.60 ± 0.4)* *	14.93 ± 0.9	12.98 ± 0.7	13.54 ± 0.5	8.80 ± 0.6	10.13 ± 0.6	9.90 ± 0.7	9.17 ± 0.4
	Bb + bb	27	50.55 ± 1.8	3365.18 ± 384	34.51 ± 4.2	17.67 ± 1.9	31.30 ± 1.9 *(31.05 ± 1.7)* *	28.24 ± 1.9	28.24 ± 1.9	3.07 ± 0.6	3.97 ± 0.4 *(3.98 ± 0.4)* *	15.58 ± 1.4	13.56 ± 1.1	13.43 ± 0.6	9.06 ± 0.6	10.34 ± 0.8	9.98 ± 0.5	9.28 ± 0.5
	p-value *adjusted p* *Effect size* *Power*		0.081	0.15	0.84	0.87	**0.021** *0.079* *	0.17	0.17	0.1	**0.004** *0.041* * *0.950* † *0.85* φ	0.11	0.082	0.6	0.21	0.37	0.68	0.5

Table 6. *Cont.*

SNP	Genotype	N	Height (cm)	Weight (g)	Head Circ/ce (cm)	Neck Rump Length (cm)	Chest Circ/ce (cm)	Abdominal Circ/ce (cm)	Abdominal Circ/ce Iliac (cm)	Skin Fold Abdominal (cm)	Skin Fold High Anterior (cm)	High Thigh Circ/ce (cm)	Middle Thigh Circ/ce (cm)	Upper Arm Length (cm)	Lower Arm Radial Circ/ce (cm)	Lower Leg Calf Circ/ce (cm)	Femur Length (cm)	Knee-Heel Length (cm)
FOKI	FF	15	49.66 ± 1.6	3373.33 ± 348	33.86 ± 1.2	18.52 ± 1.4 (18.57 ± 1.1) *	30.78 ± 1.7	28.05 ± 2.0	25.89 ± 1.7	3.01 ± 0.8	3.82 ± 0.5	15.24 ± 1.5	15.24 ± 1.5	13.46 ± 0.5	8.98 ± 0.8	10.30 ± 0.8	10.08 ± 0.5	9.16 ± 0.5
	Ff + ff	28	50.44 ± 1.9	3267.85 ± 371	34.75 ± 4.0	17.17 ± 1.7 (16.79 ± 1.6) *	30.79 ± 2.0	27.89 ± 1.7	25.73 ± 1.5	2.94 ± 0.2	3.80 ± 0.5	15.39 ± 1.2	15.39 ± 0.2	13.48 ± 0.6	8.96 ± 0.5	10.24 ± 0.6	9.87 ± 0.6	9.28 ± 0.5
	p-value adjusted *p* *Effect size* *Power*		0.19	0.37	0.41	**0.014** *0.042* * *1.228* † *0.99* φ	0.99	0.79	0.75	0.69	0.86	0.72	0.78	0.94	0.91	0.79	0.3	0.43
TRU9I	UU	26	50.44 ± 1.8	3286.92 ± 368	34.67 ± 4.2	17.28 ± 1.7	30.48 ± 1.7	27.46 ± 1.5 (27.21 ± 1.5) *	25.58 ± 1.4	2.95 ± 0.3	3.86 ± 0.5	15.05 ± 1.1	12.98 ± 1.3	13.40 ± 0.7	8.88 ± 0.5	10.17 ± 0.7	10.04 ± 0.5	9.33 ± 0.4
	Uu + uu	17	49.76 ± 1.8	3331.76 ± 365	34.08 ± 0.8	18.19 ± 1.5	31.26 ± 2.1	28.70 ± 2.1 (28.30 ± 1.5) *	26.10 ± 1.7	2.92 ± 0.4	3.72 ± 0.4	15.78 ± 1.5	13.86 ± 1.3	13.58 ± 0.5	9.10 ± 0.7	10.40 ± 0.8	9.80 ± 0.6	9.10 ± 0.5
	p-value		0.24	0.69	0.58	0.093	0.19	**0.031** *0.083* *	0.28	0.31	0.4	0.078	0.12	0.38	0.27	0.34	0.22	0.15

If the Levene's test for equality of variances $p > 0.05$ then equal variances assumed Sig (2-tailed) p values of T-Tests were given. If the Levene's test for equality of variances $p < 0.05$ then equal variances not assumed Sig (2-tailed) p values of T-Tests were given. * The data and p values adjusted for maternal height (cm), BMI pre-pregnancy (kg/m^2), BMI terminal (kg/m^2) and weeks of gestation by one-way analysis of covariance (ANCOVA). † Corrected effect size was calculated with Hedge's g (Hg). φ Post-hoc power analysis. Abbreviations: VDR: Vitamin D receptor; 25(OH)D: 25-hydroxy-vitamin D. Digits in bold refer to significant effects.

Table 7. Birth neonatal anthropometry (maternal vitamin D status at birth >50 nmol/L) according to maternal VDR polymorphisms.

SNP	Genotype	N	Height (cm)	Weight (g)	Head Circ/ce (cm)	Neck Rump Length (cm)	Chest Circ/ce (cm)	Abdominal Circ/ce (cm)	Abdominal Circ/ce Iliac (cm)	Skin Fold Abdominal (cm)	Skin Fold High Anterior (cm)	High Thigh Circ/ce (cm)	Middle Thigh Circ/ce (cm)	Upper Arm Length (cm)	Lower Arm Radial Circ/ce (cm)	Lower Leg Calf Circ/ce (cm)	Femur Length (cm)	Knee-Heel Length (cm)
APAI	AA	9	51.16 ± 1.8	3178.88 ± 437	34.44 ± 0.8	17.55 ± 3.1	32.03 ± 2.1	28.74 ± 2.3	26.67 ± 2.2	2.95 ± 0.4	3.97 ± 0.2	15.70 ± 1.7	13.41 ± 1.5	25.93 ± 0.8	9.06 ± 0.8	10.23 ± 1.1	9.96 ± 0.4	9.27 ± 0.4
	Aa + aa	14	51.00 ± 2.2	3326.42 ± 550	34.25 ± 1.7	17.82 ± 2.8	30.74 ± 1.8	28.18 ± 2.4	25.93 ± 1.8	2.92 ± 0.3	3.78 ± 0.3	15.42 ± 1.8	13.40 ± 1.3	13.87 ± 0.6	8.97 ± 0.6	10.16 ± 0.9	9.92 ± 0.4	8.81 ± 0.9
	p-value		0.85	0.5	0.75	0.82	0.14	0.59	0.39	0.87	0.19	0.72	0.98	0.72	0.75	0.87	0.82	0.17
TAQI	TT	11	51.90 ± 1.9	3372.72 ± 580	34.72 ± 0.9	17.47 ± 3.5	31.09 ± 1.9	28.46 ± 2.5	26.01 ± 1.9	2.92 ± 0.4	3.85 ± 0.4	15.53 ± 1.7	13.48 ± 1.3	14.30 ± 1.7	9.08 ± 0.6	10.26 ± 0.9	9.98 ± 0.5	8.88 ± 0.9
	Tt + tt	12	50.29 ± 1.9	3173.33 ± 425	33.95 ± 1.7	17.95 ± 2.1	31.39 ± 2.2	28.35 ± 2.2	26.41 ± 2.18	2.95 ± 0.3	3.86 ± 0.3	15.53 ± 1.8	13.33 ± 1.4	13.71 ± 0.6	8.94 ± 0.8	10.12 ± 1.1	9.90 ± 0.4	9.10 ± 0.6
	p-value		**0.06**	0.35	0.21	0.69	0.73	0.91	0.64	0.89	0.93	0.99	0.8	0.28	0.64	0.74	0.69	0.52
BSMI	BB	15	51.06 ± 1.8	3200.00 ± 462	34.50 ± 0.8	18.28 ± 2.3	31.93 ± 2.3	28.70 ± 2.4	26.72 ± 2.4	2.97 ± 0.4	4.00 ± 0.2	15.87 ± 1.8	13.53 ± 1.6	13.88 ± 0.7	9.07 ± 0.9	10.23 ± 1.2	10.00 ± 0.4	9.27 ± 0.4
	Bb + bb	8	51.06 ± 2.2	3305.33 ± 536	34.23 ± 1.6	17.42 ± 3.1	30.88 ± 1.9	28.24 ± 2.3	25.96 ± 1.7	2.92 ± 0.3	3.78 ± 0.4	15.35 ± 1.7	13.33 ± 1.3	14.06 ± 1.5	8.97 ± 0.6	10.16 ± 0.9	9.90 ± 0.4	8.84 ± 0.8
	p-value		0.99	0.64	0.68	0.5	0.25	0.67	0.39	0.75	0.15	0.51	0.74	0.77	0.75	0.87	0.66	0.22
FOKI	FF	15	51.13 ± 2.1	3291.33 ± 470	34.20 ± 1.6	18.03 ± 2.7	31.59 ± 2.0	28.36 ± 2.4	26.46 ± 2.1	2.96 ± 0.4	3.84 ± 0.3	15.84 ± 1.7	13.60 ± 1.4	14.14 ± 1.4	9.10 ± 0.7	10.33 ± 1.1	9.91 ± 0.4	9.12 ± 0.5
	Ff + ff	8	50.93 ± 2.0	3226.25 ± 593	34.56 ± 0.8	17.13 ± 3.2	30.60 ± 2.0	28.47 ± 2.4	25.78 ± 1.8	2.90 ± 0.3	3.90 ± 0.4	14.95 ± 1.7	13.03 ± 1.3	13.73 ± 0.9	8.83 ± 0.6	9.92 ± 0.8	9.98 ± 0.6	8.75 ± 1.0
	p-value		0.83	0.77	0.57	0.48	0.28	0.91	0.45	0.72	0.69	0.25	0.36	0.49	0.39	0.35	0.73	0.28

Table 7. *Cont.*

SNP	Genotype	N	Height (cm)	Weight (g)	Head Circ/ce (cm)	Neck Rump Length (cm)	Chest Circ/ce (cm)	Abdominal Circ/ce (cm)	Abdominal Circ/ce Iliac (cm)	Skin Fold Abdominal (cm)	Skin Fold High Anterior (cm)	High Thigh Circ/ce (cm)	Middle Thigh Circ/ce (cm)	Upper Arm Length (cm)	Lower Arm Radial Circ/ce (cm)	Lower Leg Calf Circ/ce (cm)	Femur Length (cm)	Knee-Heel Length (cm)
	UU	12	50.16 ± 1.9 (50.50 ± 1.5) *	3178.33 ± 512	33.95 ± 1.8	17.39 ± 2.9	30.47 ± 2.0	27.78 ± 2.6	25.67 ± 1.9	2.95 ± 0.3	3.76 ± 0.4	14.95 ± 1.7	12.98 ± 1.3	14.01 ± 1.5	8.91 ± 0.7	10.01 ± 0.9	9.80 ± 0.3	8.94 ± 0.9
TRU9I	Uu + uu	11	52.04 ± 1.7 (49.72 ± 2.0) *	3367.27 ± 499	34.72 ± 0.6	18.08 ± 2.8	32.09 ± 1.7	29.08 ± 1.8	26.82 ± 1.9	2.92 ± 0.4	3.96 ± 0.2	16.16 ± 1.8	13.86 ± 1.3	13.98 ± 0.9	9.11 ± 0.7	10.39 ± 1.0	10.09 ± 0.5	9.05 ± 0.5
	p-value *adjusted p* *Effect size* *Power*		**0.026** *0.032* * *0.444* † *0.18* ϕ	0.38	0.21	0.57	**0.059**	0.19	0.17	0.89	0.17	0.1	0.13	0.95	0.52	0.36	0.14	0.74

If the Levene's test for equality of variances $p > 0.05$ then equal variances assumed Sig (2-tailed) p values of *T*-Tests were given. If the Levene's test for equality of variances $p < 0.05$ then equal variances not assumed Sig (2-tailed) p values of *T*-Tests were given. * The data and p values adjusted for maternal height (cm), BMI pre-pregnancy (kg/m^2), BMI terminal (kg/m^2), and weeks of gestation by one-way analysis of covariance (ANCOVA). † Corrected effect size was calculated with Hedge's g (H*g*). ϕ Post-hoc power analysis. Abbreviations: VDR: Vitamin D receptor; 25(OH)D: 25-hydroxy-vitamin D. Digits in bold refer to significant effects.

Table 8. Birth neonatal anthropometry (maternal vitamin D status at birth <75 nmol/L) according to neonatal VDR polymorphisms.

SNP	Genotype	N	Height (cm)	Weight (g)	Head Circ/ce (cm)	Neck Rump Length (cm)	Chest Circ/ce (cm)	Abdominal Circ/ce (cm)	Abdominal Circ/ce Iliac (cm)	Skin Fold Abdominal (cm)	Skin Fold High Anterior (cm)	High Thigh Circ/ce (cm)	Middle Thigh Circ/ce (cm)	Upper Arm Length (cm)	Lower Arm Radial Circ/ce (cm)	Lower Leg Calf Circ/ce (cm)	Femur Length (cm)	Knee-Heel Length (cm)
	AA	18	50.58 ± 2.1	3251.66 ± 407	35.72 ± 4.8 (34.50 ± 0.5) *	17.30 ± 2.0	30.54 ± 1.7	27.71 ± 1.4	25.43 ± 1.5	2.80 ± 0.2	3.81 ± 0.4	14.80 ± 1.0 (14.97 ± 0.8) *	13.03 ± 0.9	13.58 ± 0.7	8.69 ± 0.5 (8.56 ± 0.2) *	9.95 ± 0.7	9.91 ± 0.7	9.21 ± 0.5
APAI	Aa + aa	37	50.39 ± 1.9	3334.05 ± 397	33.89 ± 1.1 (34.00 ± 0.9) *	17.97 ± 2.0	31.18 ± 2.0	28.32 ± 2.1	26.15 ± 1.6	3.03 ± 0.6	3.87 ± 0.4	15.70 ± 1.8 (15.64 ± 1.5) *	13.57 ± 1.2	13.71 ± 1.1	9.09 ± 0.6 (9.05 ± 0.6) *	10.37 ± 0.8	9.94 ± 0.5	9.18 ± 0.5
	p-value *adjusted p* *Effect size* *Power*		0.74	0.47	**0.032** *0.24* *	0.24	0.25	0.27	0.12	0.12	0.66	**0.029** *0.20* *	0.1	0.67	**0.031** *0.043* * *0.966* † *0.99* ϕ	0.067	0.88	0.81
	TT	21	50.54 ± 1.8	3296.66 ± 405	34.11 ± 0.8	17.55 ± 2.1	31.27 ± 2.0	28.46 ± 2.3	28.46 ± 2.3	3.00 ± 0.7	3.91 ± 0.3	15.51 ± 1.5	13.62 ± 1.3	13.60 ± 1.3	9.02 ± 0.6	10.44 ± 0.7	9.85 ± 0.4	9.10 ± 0.5
TAQI	Tt + tt	34	50.39 ± 1.1	3313.52 ± 400	34.72 ± 3.7	17.88 ± 1.9	30.78 ± 1.8	27.91 ± 1.5	27.91 ± 1.5	2.92 ± 0.3	3.81 ± 0.5	15.34 ± 1.3	1.26 ± 1.0	13.70 ± 0.7	8.92 ± 0.6	10.11 ± 0.8	9.98 ± 0.6	9.24 ± 0.3
	p-value		0.78	0.88	0.47	0.56	0.37	0.3	0.39	0.56	0.43	0.66	0.26	0.72	0.59	0.14	0.43	0.33
	BB	16	50.71 ± 2.2	3275.62 ± 426	35.93 ± 5.1	17.26 ± 2.1	30.53 ± 1.8	27.62 ± 1.5	25.38 ± 1.5	2.82 ± 0.2	3.83 ± 0.4	14.80 ± 1.1 (15.02 ± 0.9) *	13.01 ± 0.9	13.68 ± 0.7	8.71 ± 0.5	9.94 ± 0.8	9.94 ± 0.7	9.25 ± 0.5
BSMI	Bb + bb	39	50.34 ± 1.8	3320.00 ± 391	33.89 ± 1.0	17.95 ± 1.9	31.15 ± 1.7	28.33 ± 2.0	26.14 ± 1.5	3.01 ± 0.5	3.35 ± 0.4	15.65 ± 1.5 (15.58 ± 1.5) *	13.55 ± 1.2	13.66 ± 1.1	9.06 ± 0.6	10.35 ± 0.7	9.92 ± 0.5	9.16 ± 0.5
	p-value		0.53	0.71	0.13	0.25	0.29	0.21	0.11	0.24	0.89	**0.048** *0.25* *	0.11	0.96	0.074	0.087	0.93	0.6

Table 8. *Cont.*

SNP	Genotype	N	Height (cm)	Weight (g)	Head Circ/ce (cm)	Neck Rump Length (cm)	Chest Circ/ce (cm)	Abdominal Circ/ce (cm)	Abdominal Circ/ce Iliac (cm)	Skin Fold Abdominal (cm)	Skin Fold High Anterior (cm)	High Thigh Circ/ce (cm)	Middle Thigh Circ/ce (cm)	Upper Arm Length (cm)	Lower Arm Radial Circ/ce (cm)	Lower Leg Calf Circ/ce (cm)	Femur Length (cm)	Knee-Heel Length (cm)
FOKI	FF	27	50.38 ± 2.2	3426.66 ± 459 (3311.42 ± 479) *	34.92 ± 4.1	18.48 ± 1.6 *(18.52 $\pm$ 1.8)* *	31.15 ± 2.0	28.29 ± 2.2	26.15 ± 1.8	3.02 ± 0.6	3.85 ± 0.4	15.52 ± 1.5	13.58 ± 1.2	13.85 ± 1.2	9.07 ± 0.7	10.32 ± 0.9	9.96 ± 0.5	9.12 ± 0.5
	Ff + ff	28	50.51 ± 1.7	3191.77 ± 293 *(3250.00 $\pm$ 297)* *	34.07 ± 0.9	17.05 ± 2.1 *(17.04 $\pm$ 2.3)* *	30.79 ± 1.8	27.96 ± 1.6	25.69 ± 1.3	2.88 ± 0.2	3.85 ± 0.4	15.29 ± 1.3	13.21 ± 1.0	13.48 ± 0.7	8.86 ± 0.5	10.15 ± 0.6	9.90 ± 0.5	9.25 ± 0.4
	p-value		0.81	**0.03** *0.79* *	0.29	**0.008** *0.14* *	0.49	0.52	0.3	0.31	0.98	0.56	0.23	0.17	0.24	0.46	0.67	0.35
TRU9I	UU	35	50.67 ± 2.1	3325.14 ± 400	34.65 ± 3.7	17.95 ± 1.9	30.95 ± 1.8	27.96 ± 1.8	25.98 ± 1.6	3.04 ± 0.5	3.88 ± 0.5	15.35 ± 1.3	13.33 ± 1.0	13.76 ± 1.2	9.00 ± 0.6	10.25 ± 0.8	10.02 ± 0.5	9.28 ± 0.4
	Uu + uu	20	50.07 ± 1.8	3275.50 ± 404	34.20 ± 0.7	17.49 ± 2.1	31.00 ± 2.0	28.41 ± 2.0	25.81 ± 1.6	2.80 ± 0.3	3.80 ± 0.3	15.51 ± 1.6	13.52 ± 1.3	13.50 ± 0.4	8.90 ± 0.6	10.21 ± 0.8	9.77 ± 0.6	9.03 ± 0.5
	p-value		0.29	0.66	0.59	0.47	0.92	0.4	0.7	0.097	0.54	0.7	0.56	0.26	0.58	0.87	0.12	0.08

If the Levene's test for equality of variances $p > 0.05$ then Equal variances assumed Sig (2-tailed) p values of T-Tests were given. If the Levene's test for equality of variances $p < 0.05$ then equal variances not assumed Sig (2-tailed) p values of T-Tests were given. * The data and p values adjusted for maternal height (cm), BMI pre-pregnancy (kg/m^2), BMI terminal (kg/m^2), and weeks of gestation by one-way analysis of covariance (ANCOVA). † Corrected effect size was calculated with Hedge's g (Hg). ϕ Post-hoc power analysis. Abbreviations: VDR: Vitamin D receptor; 25(OH)D: 25-hydroxy-vitamin D. Digits in bold refer to significant effects.

4. Discussion

This study aimed to evaluate the combined effects of maternal and neonatal VDR polymorphisms (ApaI, TaqI, BsmI, FokI, Tru9I) and different maternal and neonatal 25(OH)D cut-offs on neonatal birth anthropometry at birth including a population from a sunny Mediterranean area in Northern Greece. Results from this maternal–neonatal pair cohort indicate that: (i) the maternal TAQI VDR polymorphism significantly affects neonatal birth anthropometry when maternal 25(OH)D concentrations are <50 nmol/L, but not for a higher cut-off of >50 nmol/L, (ii) neonatal VDR polymorphisms combined with neonatal 25(OH)D > 25 nmol/L have negligible effects on birth anthropometry, whereas this combination exerts a minimal effect, in the presence of neonatal TAQI polymorphism with neonatal 25(OH)D values < 25 nmol/L, and (iii) FOKI and BSMI maternal VDR polymorphisms demonstrate minimal—out of a consistent pattern—effects on skin fold and neck-rump length, which warrant further investigation in future studies with larger samples from mothers and neonates.

These findings are the first to be reported on the combined effects of maternal and neonatal VDR polymorphisms and respective 25(OH)D cut-offs on neonatal birth anthropometry from this region. Moreover, these findings indicate that there are variable degrees of VDR polymorphism functionality depending on maternal and neonatal 25(OH)D concentrations, which result in different anthropometric patterns at birth. In the daily clinical setting, these findings also identify different "safe" maternal 25(OH)D cut-offs (>50 nmol/L for maternal and >25 nmol/L for neonatal vitamin D status), whose attainment practically prevents genetic functional VDR influences on a given neonatal outcome. This is the first mechanistic study of this kind, which combines both aspects of vitamin D physiology, fluctuating concentrations of 25(OH) D, and common VDR polymorphisms with a discourse on the specific neonatal outcome.

Although associations between VDR polymorphisms with a plethora of adverse pregnancy outcomes such as preterm birth and SGA neonates [10–18] have been suggested, evidence is still inconclusive, primarily due to the lack of a pathophysiological connection of individual vitamin D status and functionality of VDR polymorphisms. The most commonly investigated polymorphisms included were the BsmI (rs1544410), ApaI (rs7975232), FokI (rs2228570), and TaqI (rs731236) polymorphisms, while TaqI and FokI consisted of a single base change (A to G and G to A in exons 9 and 2, respectively), and BsmI and ApaI were located in the last intron of the sequence and resulted from a single base change (G to A and A to C, respectively). However, results in the field are highly inconsistent, mainly due to the absence of incorporated standardized thresholds of vitamin D status in the initial study design, which could enable a universal stratification of mothers and neonates. The racial diversity of included populations might also contribute to the inconsistency of the results, underlying the importance of regionally-derived data in the implementation of national health policies [6,28,29].

We have previously highlighted the importance of population-specific genetic profiling in understanding vitamin D deficiency among neonates and their mothers and the protective effect of the maternal FokI FF genotype against the development of neonatal vitamin D deficiency [25(OH)D < 30 nmol/L] [13]. However, there was a lack of simultaneous assessment of both maternal and neonatal VDR polymorphisms in one snapshot, with different 25(OH)D thresholds, oriented toward a detailed evaluation of birth anthropometry as a method of crude estimation of neonatal adiposity [23], which could identify an adverse metabolic offspring profile in later adult life [30].

We included a maternal cut-off of 50 mol/l, but not one of 75 nmol/L, since our cohort did not include women with higher 25(OH)D concentrations, since none of them was supplemented. On the other hand, we considered that including a maternal cut-off of 25 nmol/L would be far from the widely adopted international recommendations for maternal values during pregnancy [2,5,18]. Our goal was to explore vitamin D status and VDR polymorphism interactions in the most common equilibrium pattern observed in non-supplemented women from our region. We followed the same rationale for neonates by including lower cut-offs of 25(OH)D, based on previous observations on maternal–neonatal vitamin D equilibrium at birth [3,4]. Future studies with higher maternal and neonatal cut-offs could be useful in exploring the broader spectrum of maternal–neonatal interactions in this setting.

Mechanistic pathways between VDR expression and offspring outcomes remain largely unclear. Apart from its classical intracellular pathways, which allows the ligand-bound VDR to form heterodimers with nuclear retinoid X receptor (RXR) and recruit co-factors to modulate gene transcription [31], vitamin D can also exert rapid non-genomic effects, probably via VDR located within the plasma membrane [32,33]. However, the functional effects of VDR and its allelic variants on birth anthropometry have not been elucidated, until recently.

Interestingly, we observed that maternal and neonatal TaqI polymorphism is a significant modulator of neonatal birth anthropometry when maternal and neonatal values are in a range of <50 nmol/L and <25 nmol/L, respectively. Results about the effect of VDR polymorphisms including the TaqI polymorphism on neonatal birth anthropometry are currently inconclusive: Swamy and colleagues [16] prospectively evaluated the effect of 38 VDR polymorphisms on several birth outcomes on 615 pregnant women including birth weight. A total of eight out of 38 SNPs examined significantly affected birth weight in black but not in white women, indicating a biologically plausible association that could depend on ethnicity, providing a partial explanation for the observed racial disparity in several pregnancy outcomes.

In a previous study including participants of Caucasian origin, boys with the BB genotype were shorter at birth and grew less from birth to the age of 16.9 than their Bb and bb counterparts. A prediction model including parental height, birth height, birth weight, and VDR alleles could predict up to 39% of the total variation in adult height [34]. Similarly, in a maternal–neonatal cohort from Australia [14], neonates of vitamin D deficient mothers had lower birth weight with FF or Ff, but not ff genotype, whereas thicker subscapular and suprailiac skinfolds with ff, but not the FF or Ff genotype. Placental genetic variations in vitamin D metabolism through investigation of five vitamin D metabolism genes (CUBN, LRP2, VDR, GC, and CYP2R1) was also associated through a sex-specific pattern with birthweight, but not with other neonatal outcomes [17]. In our study, although BSMI SNP BB and FOKI SNP FF genotypes were associated with anterior skin fold and neck rump length and a maternal 25(OH)D cut-off <50 nmol/L, we considered that these findings did not establish a solid biological effect that could identify a genetic variation pattern, as evident for TaqI, where a plethora of birth anthropometry parameters were uniformly affected.

Regarding the effect of TaqI polymorphism on neonatal anthropometry, Barchitta et al. [18] reported that birth weight increased with an increasing number of mutated alleles, concluding that a beneficial effect of TaqI polymorphism in this regard could not be ruled out. However, these results should be interpreted with caution due to the high heterogeneity in design and outcomes, sample size, ethnicity, geographical diversity, sun exposure, dietary calcium and vitamin D intake, and maternal habits [35–37]. This study has certain limitations. First, the sample size was small and not powered to detect additional differences in other maternal-neonatal cut-offs, but it was sufficiently powered to show significant differences regarding the main aim of the study. Second, the cross-sectional design of the study could not prove a causal relationship. Third, all women were Caucasian, so our results cannot be safely generalized to other ethnicities, known to differ at least in the frequency of VDR polymorphisms, indicating that further similar studies from other regions could be useful, in order to elucidate the full extent of the ethnic VDR variation effect in neonatal outcomes. Finally, gender influences on anthropometric parameters in association with specific VDR polymorphisms have not been assessed due to the limited number of neonates included in the study. Therefore, the hypothesis of a gender-specific effect requires future investigation in studies with larger samples.

In conclusion, these results indicate that maternal TAQI VDR polymorphism significantly affects neonatal birth anthropometry, when maternal 25(OH) concentrations are <50 nmol/L, but not for a higher cut-off of >50 nmol/L, whereas this effect is minimally evident in the presence of neonatal TAQI polymorphism with neonatal 25(OH)D values <25 nmol/L.

No other effects of VDR common polymorphisms were evident using specific maternal and neonatal cut-offs. The implications of these findings could be incorporated in daily clinical practice by targeting a maternal 25(OH)D cut-off >50 nmol/L, which could be protective against any potential effect of genetic VDR polymorphism variances on neonatal birth anthropometry.

Author Contributions: S.N.K. designed and conducted the study, interpreted the results, and drafted the original and revised versions. E.D. and D.G.-A. conducted the VDR polymorphism analysis, statistical analysis, and drafted the original and revised versions. M.A. conducted the VDR polymorphism analysis. C.A., D.S., D.E. and D.K. contributed to the data interpretation, statistical analysis, and drafting of the original and revised versions. All authors have read and agreed to the published version of the manuscript.

Funding: This research received no external funding.

Institutional Review Board Statement: The study was conducted according to the guidelines of the Declaration of Helsinki, and approved by the Institutional Review Board (or Ethics Committee) of Aristotle University of Thessaloniki, Greece (Approval number 1/19-12-2011).

Informed Consent Statement: Informed consent was obtained from all subjects involved in the study.

Conflicts of Interest: The authors declare no conflict of interest.

References

1. Tsakiridis, I.; Kasapidou, E.; Dagklis, T.; Leonida, I.; Leonida, C.; Bakaloudi, D.R.; Chourdakis, M. Nutrition in Pregnancy: A Comparative Review of Major Guidelines. *Obstet. Gynecol. Surv.* **2020**, *75*, 692–702. [CrossRef] [PubMed]
2. Palacios, C.; Kostiuk, L.K.; Peña-Rosas, J.P. Vitamin D supplementation for women during pregnancy. *Cochrane Database Syst. Rev.* **2019**, *7*, CD008873. [CrossRef] [PubMed]
3. Karras, S.N.; Wagner, C.L.; Castracane, V.D. Understanding vitamin D metabolism in pregnancy: From physiology to pathophysiology and clinical outcomes. *Metabolism* **2018**, *86*, 112–123. [CrossRef]
4. Karras, S.N.; Shah, I.; Petroczi, A.; Goulis, D.G.; Bili, H.; Papadopoulou, F.; Harizopoulou, V.; Tarlatzis, B.C.; Naughton, D.P. An observational study reveals that neonatal vitamin D is primarily determined by maternal contributions: Implications of a new assay on the roles of vitamin D forms. *Nutr. J.* **2013**, *12*, 77. [CrossRef] [PubMed]
5. Markestad, T.; Aksnes, L.; Ulstein, M.; Aarskog, D. 25-hydroxyvitamin D and 1, 25-dihydroxyvitamin D of D2 and D3 origin in maternal and umbilical cord serum after vitamin D2 supplementation in human pregnancy. *Am. J. Clin. Nutr.* **1984**, *40*, 1057–1063. [CrossRef]
6. Pilz, S.; Trummer, C.; Pandis, M.; Schwetz, V.; Aberer, F.; Grübler, M.; Verheyen, N.; Tomaschitz, A.; März, W. Vitamin D: Current guidelines and future outlook. *Anticancer Res.* **2018**, *38*, 1145–1151.
7. Munns, C.F.; Shaw, N.; Kiely, M.; Specker, B.L.; Thacher, T.D.; Ozono, K.; Michigami, T.; Tiosano, D.; Mughal, M.Z.; Mäkitie, O.; et al. Global consensus recommendations on prevention and management of nutritional rickets. *J. Clin. Endocrinol. Metab.* **2016**, *101*, 394–415. [CrossRef]
8. Karras, S.N.; Anagnostis, P.; Naughton, D.; Annweiler, C.; Petroczi, A.; Goulis, D.G. Vitamin D during pregnancy: Why observational studies suggest deficiency and interventional studies show no improvement in clinical outcomes? A narrative review. *J. Endocrinol. Investig.* **2015**, *38*, 1265–1275. [CrossRef]
9. Wagner, C.L.; Hollis, B.W.; Kotsa, K.; Fakhoury, H.; Karras, S.N. Vitamin D administration during pregnancy as prevention for pregnancy, neonatal and postnatal complications. *Rev. Endocr. Metab. Disord.* **2017**, *18*, 307–322. [CrossRef]
10. Suarez, F.; Zeghoud, F.; Rossignol, C.; Walrant, O.; Garabédian, M. Association between vitamin D receptor gene polymorphism and sex-dependent growth during the first two years of life. *J. Clin. Endocrinol. Metab.* **1997**, *82*, 2966–2970. [CrossRef]
11. Barišić, A.; Pereza, N.; Hodžić, A.; Krpina, M.G.; Ostojić, S.; Peterlin, B. Genetic variation in the maternal vitamin D receptor FokI gene as a risk factor for recurrent pregnancy loss. *J. Matern. Fetal. Neonatal Med.* **2019**, *16*, 1–6. [CrossRef] [PubMed]
12. Zhu, B.; Huang, K.; Yan, S.; Hao, J.; Zhu, P.; Chen, Y.; Ye, A.; Tao, F. VDR Variants rather than Early Pregnancy Vitamin D Concentrations Are Associated with the Risk of Gestational Diabetes: The Ma'anshan Birth Cohort (MABC) Study. *J. Diabetes Res.* **2019**, *2019*, 8313901. [CrossRef] [PubMed]

13. Karras, S.N.; Koufakis, T.; Antonopoulou, V.; Goulis, D.G.; Alaylıoğlu, M.; Dursun, E.; Gezen-Ak, D.; Annweiler, C.; Pilz, S.; Fakhoury, H.; et al. Vitamin D receptor Fokl polymorphism is a determinant of both maternal and neonatal vitamin D concentrations at birth. *J. Steroid Biochem. Mol. Biol.* **2020**, *199*, 105568. [CrossRef]

14. Morley, R.; Carlin, J.B.; Pasco, J.A.; Wark, J.D.; Ponsonby, A.L. Maternal 25-hydroxyvitamin D concentration and offspring birth size: Effect modification by infant VDR genotype. *Eur. J. Clin. Nutr.* **2009**, *63*, 802–804. [CrossRef] [PubMed]

15. Bodnar, L.M.; Catov, J.M.; Zmuda, J.M.; Cooper, M.E.; Parrott, M.S.; Roberts, J.M.; Marazita, M.L.; Simhan, H.N. Maternal serum 25-hydroxyvitamin D concentrations are associated with small-for-gestational age births in white women. *J. Nutr.* **2010**, *140*, 999–1006. [CrossRef]

16. Swamy, G.K.; Garrett, M.E.; Miranda, M.L.; Ashley-Koch, A.E. Maternal vitamin D receptor genetic variation contributes to infant birthweight among black mothers. *Am. J. Med. Genet. A* **2011**, *155A*, 1264–1271. [CrossRef]

17. Workalemahu, T.; Badon, S.E.; Dishi-Galitzky, M.; Qiu, C.; Williams, M.A.; Sorensen, T.; Enquobahrie, A.D. Placental genetic variations in vitamin D metabolism and birthweight. *Placenta* **2017**, *50*, 78–83. [CrossRef]

18. Barchitta, M.; Maugeri, A.; la Rosa, M.C.; Lio, R.M.S.; Favara, G.; Panella, M.; Cianci, A.; Agodi, A. Single Nucleotide Polymorphisms in Vitamin D Receptor Gene Affect Birth Weight and the Risk of Preterm Birth: Results From the "Mamma & Bambino" Cohort and A Meta-Analysis. *Nutrients* **2018**, *10*, 1172.

19. Chun, S.K.; Shin, S.; Kim, M.Y.; Joung, H.; Chung, J. Effects of maternal genetic polymorphisms in vitamin D-binding protein and serum 25-hydroxyvitamin D concentration on infant birth weight. *Nutrition* **2017**, *35*, 36–42. [CrossRef]

20. McGrath, J.J.; Keeping, D.; Saha, S.; Chant, D.C.; Lieberman, D.E.; O'Callaghan, M.J. Seasonal fluctuations in birth weight and neonatal limb length; does prenatal vitamin D influence neonatal size and shape? *Early Hum. Dev.* **2005**, *81*, 609–618. [CrossRef]

21. Shah, I.; James, R.; Barker, J.; Petroczi, A.; Naughton, D.P. Misleading measures in Vitamin D analysis: A novel LC-MS/MS assay to account for epimers and isobars. *Nutr. J.* **2011**, *10*, 46. [CrossRef] [PubMed]

22. Shah, I.; Petroczi, A.; Naughton, D.P. Method for simultaneous analysis of eight analogues of vitamin D using liquid chromatography tandem mass spectrometry. *Chem. Central. J.* **2012**, *6*, 112. [CrossRef] [PubMed]

23. Keeping, J.D. Determinants and Components of Size at Birth. Ph.D. Thesis, University of Aberdeen, Scotland, UK, 1981.

24. Ross, A.C.; Taylor, C.L.; Yaktine, A.L.; del Valle, H.B. (Eds.) *Dietary Reference Intakes for Calcium and Vitamin D.*; The National Academies Press: Washington, DC, USA, 2011; p. 98.

25. Gezen-Ak, D.; Alaylıoğlu, M.; Genç, G.; Gündüz, A.; Candaş, E.; Bilgiç, B.; Atasoy, İ.; Apaydın, H.; Kızıltan, G.; Gürvit, H.; et al. GC and VDR SNPs and Vitamin D Levels in Parkinson's Disease: The relevance to clinical features. *Neuromol. Med.* **2017**, *19*, 24–40. [CrossRef] [PubMed]

26. Gezen-Ak, D.; Dursun, E.; Bilgiç, B.; Hanağasi, H.; Ertan, T.; Gürvit, H.; Emre, M.; Eker, E.; Ulutin, T.; Uysal, O.; et al. Vitamin D receptor gene haplotype is associated with late-onset Alzheimer's disease. *Tohoku J. Exp. Med.* **2012**, *228*, 189–196. [CrossRef] [PubMed]

27. Durlak, J.A. How to select, calculate, and interpret effect sizes. *J. Pediatr. Psychol.* **2009**, *34*, 917–928. [CrossRef]

28. Karras, S.N.; Fakhoury, H.; Muscogiuri, G.; Grant, W.B.; van den Ouweland, J.M.; Colao, A.M.; Kotsa, K. Maternal vitamin D levels during pregnancy and neonatal health: Evidence to date and clinical implications. *Ther. Adv. Musculoskelet. Dis.* **2016**, *8*, 124–135. [CrossRef]

29. Karras, S.N.; Paschou, S.A.; Kandaraki, E.; Anagnostis, P.; Annweiler, C.; Tarlatzis, B.C.; Hollis, B.W.; Grant, W.B.; Goulis, D.G. Hypovitaminosis D in pregnancy in the Mediterranean region: A systematic review. *Eur. J. Clin. Nutr.* **2016**, *70*, 979–986. [CrossRef]

30. Muthayya, S.; Dwarkanath, P.; Thomas, T.; Vaz, M.; Mhaskar, A.; Mhaskar, R.; Thomas, A.; Bhat, S.; Kurpad, A.V. Anthropometry and body composition of south Indian babies at birth. *Public Health Nutr.* **2006**, *9*, 896–903.

31. Freedman, L.P.; Arce, V.; Fernandez, R.P. DNA sequences that act as high affinity targets for the vitamin D3 receptor in the absence of the retinoid x receptor. *Mol. Endocrinol.* **1994**, *8*, 265–273.

32. Gezen-Ak, D.; Dursun, E. Molecular basis of vitamin D action in neurodegeneration: The story of a team perspective. *Hormones* **2019**, *18*, 17–21. [CrossRef]

33. Dursun, E.; Gezen-Ak, D. Vitamin D receptor is present on the neuronal plasma membrane and is co-localized with amyloid precursor protein, ADAM10 or Nicastrin. *PLoS ONE* **2017**, *12*, e0188605. [CrossRef] [PubMed]

34. Lorentzon, M.; Lorentzon, R.; Nordström, P. Vitamin D receptor gene polymorphism is associated with birth height, growth to adolescence, and adult stature in healthy Caucasian men: A cross-sectional and longitudinal study. *J. Clin. Endocrinol. Metab.* **2000**, *85*, 1666–1670. [PubMed]

35. Pereira-Santos, M.; Carvalho, G.Q.; Louro, I.D.; Santos, D.B.D.; Oliveira, A.M. Polymorphism in the vitamin D receptor gene is associated with maternal vitamin D concentration and neonatal outcomes: A Brazilian cohort study. *Am. J. Hum. Biol.* **2019**, *31*, e23250. [CrossRef] [PubMed]

36. Pilz, S.; März, W.; Cashman, K.D.; Kiely, M.E.; Whiting, S.J.; Holick, M.F.; Grant, W.B.; Pludowski, P.; Hiligsmann, M.; Trummer, C.; et al. Rationale and plan for vitamin D food fortification: A review and guidance paper. *Front. Endocrinol.* **2018**, *9*, 373. [CrossRef]

37. Anastasiou, A.; Karras, S.N.; Bais, A.; Grant, W.B.; Kotsa, K.; Goulis, D.G. Ultraviolet radiation and effects on humans: The paradigm of maternal vitamin D production during pregnancy. *Eur. J. Clin. Nutr.* **2017**, *71*, 1268–1272.

 *nutrients*

Review

Does the High Prevalence of Vitamin D Deficiency in African Americans Contribute to Health Disparities?

Bruce N. Ames [1], William B. Grant [2,*] and Walter C. Willett [3,4]

[1] Molecular and Cell Biology, Emeritus, University of California, Berkeley, CA 94720, USA; bnames@berkeley.edu

[2] Sunlight, Nutrition and Health Research Center, San Francisco, CA 94164-1603, USA

[3] Departments of Nutrition and Epidemiology, Harvard T.H. Chan School of Public Health, Boston, MA 02115, USA; wwillett@hsph.harvard.edu

[4] Channing Division of Network Medicine, Department of Medicine, Brigham and Women's Hospital, Harvard Medical School, Boston, MA 02115, USA

* Correspondence: wbgrant@infionline.net

Abstract: African Americans have higher incidence of, and mortality from, many health-related problems than European Americans. They also have a 15 to 20-fold higher prevalence of severe vitamin D deficiency. Here we summarize evidence that: (i) this health disparity is partly due to insufficient vitamin D production, caused by melanin in the skin blocking the UVB solar radiation necessary for its synthesis; (ii) the vitamin D insufficiency is exacerbated at high latitudes because of the combination of dark skin color with lower UVB radiation levels; and (iii) the health of individuals with dark skin can be markedly improved by correcting deficiency and achieving an optimal vitamin D status, as could be obtained by supplementation and/or fortification. Moderate-to-strong evidence exists that high 25-hydroxyvitamin D levels and/or vitamin D supplementation reduces risk for many adverse health outcomes including all-cause mortality rate, adverse pregnancy and birth outcomes, cancer, diabetes mellitus, Alzheimer's disease and dementia, multiple sclerosis, acute respiratory tract infections, COVID-19, asthma exacerbations, rickets, and osteomalacia. We suggest that people with low vitamin D status, which would include most people with dark skin living at high latitudes, along with their health care provider, consider taking vitamin D_3 supplements to raise serum 25-hydroxyvitamin D levels to 30 ng/mL (75 nmol/L) or possibly higher.

Keywords: African American; Hispanic; European American; blacks; whites; health disparities; vitamin D; 25-hydroxyvitamin D; UVB

Citation: Ames, B.N.; Grant, W.B.; Willett, W.C. Does the High Prevalence of Vitamin D Deficiency in African Americans Contribute to Health Disparities?. *Nutrients* **2021**, *13*, 499. https://doi.org/10.3390/nu13020499

Academic Editors: Tyler Barker and Roberto Iacone

Received: 31 December 2020

Accepted: 28 January 2021

Published: 3 February 2021

Publisher's Note: MDPI stays neutral with regard to jurisdictional claims in published maps and institutional affiliations.

1. Introduction

The vitamin D hormone controls the activity of thousands of protein-encoding human genes [1]. Therefore, optimum levels are likely to be important for health. Synthesis of vitamin D in human skin depends on solar UVB radiation, whose levels are low at high latitudes, as in the United States and Europe, and highest at equatorial latitudes [2]. High concentration of melanin, the brown-black pigment in skin, is appropriate for the high UV radiation dose regions such as in the tropical plains as it absorbs UVB radiation absorption, thereby reducing production of free radicals and destruction of folate, but permitting adequate vitamin D production [3]. At higher latitudes, where the UVB radiation dose is lower [2], the rate of synthesis of vitamin D correspondingly decreases, potentially disrupting many metabolic functions that depend on that vitamin and leading to poorer health.

Here we discuss various aspects of such latitude–skin color mismatch and health disparities. By latitude–skin color mismatch, we mean that skin pigmentation is not appropriate for the solar UV doses at various latitudes, either too dark as for African Americans (defined as people living in the United States with some African ancestry) to efficiently produce vitamin D, or too light to protect against the harmful effects of UV radiation as for people living close to the equator, such as those with Anglo-Celtic ancestry in Australia. (Note that people of African descent have dark melanin, called eumelanin, while Anglo-Celtics have yellow-to-reddish melanin called pheomelanin.) This mismatch is particularly impactful in African Americans, whose dark skin is well adapted to the high UVB levels at low equatorial latitudes [3]. However, as a legacy of slavery and more recent migration, African Americans now reside at higher latitudes than in their ancestral environments. That geographic shift is largely responsible for a high prevalence of vitamin D deficiency (serum 25-hydroxyvitamin D (25(OH)D) levels < 20 ng/mL) in African Americans [4,5] independent of diet and other factors [6]. This high prevalence of deficiency potentially contributes to many health disparities. Because vitamin D deficiency can be easily remedied by supplementation or, to a lesser extent, fortification of food, the health implications of this deficiency are important to understand.

Many factors adversely affect the health of African Americans, including high rates of poverty [7], poor housing and residential environments [8], and lack of access to affordable health care. Living in racially segregated, poor neighborhoods also exposes residents to risk of crime [9], thereby limiting time spent outdoors, as well as reducing access to well-stocked grocery stores and pharmacies. Limited educational opportunities frequently result in having jobs with high social interaction and thus greater risk of COVID-19. The high incarceration rate of African American males has resulted in many children being raised by single mothers. While these factors play important roles in racial health disparities and require sustained efforts to correct at individual and societal levels, vitamin D deficiency can be corrected rapidly and inexpensively. In this review we examine the potential health benefits of addressing this deficiency.

This narrative review considers the potential health effects of inadequate vitamin D in humans. Although the motivation for this review is the high prevalence of vitamin D deficiency in African Americans, we draw on the literature from all populations because our underlying biology is similar across all racial groups, even though the prevalence of exposures, here serum levels of 25(OH)D, can differ greatly. Thus, the findings have implications for other groups with darker skin and low 25(OH)D levels, such as US immigrants from Mexico and South Asia, and for European Americans with limited sun exposure. (The term European Americans is used to represent white, non-Hispanic Americans.)

When they are available, we cite meta-analyses or pooled primary data from multiple studies. Because ideal randomized trials are often difficult or impossible to conduct, conclusions regarding causality will usually need to be based on the weight of evidence from multiple types of study [10]. The strengths and limitations of the various approaches to determine relationships between vitamin D and health outcomes are presented in the Appendix A. Ideally, vitamin D's health effects in populations with dark skin would be evaluated directly in such groups, but in most studies the number of such participants has been too small to evaluate separately. Nevertheless, we highlight studies of subgroups of African Americans and Hispanics (17% of the U.S. population) when available. We also pay special attention to subgroups with low baseline serum levels of vitamin D in randomized trials as this where an effect of supplementation may be expected to be seen; failure to do this may lead to misleading negative conclusions. In some randomized trials comparing vitamin D supplements with a placebo, those with low serum levels of 25(OH)D are excluded for ethical reasons and/or are treated, again potentially leading to misleading conclusions.

2. Current Status of Knowledge

2.1. Vitamin D: Synthesis and Metabolism

Vitamin D_3 is synthesized in human skin by the UVB-dependent conversion of 7-dehydrocholesterol to vitamin D_3 (herein referred to as vitamin D when used as a supplement). Vitamin D_3 is then converted to $25(OH)D_3$, a precursor of the crucial vitamin D steroid hormone, 1,25-dihydroxyvitamin D_3, or calcitriol, in a reaction requiring magnesium [11,12], which is widely deficient in the American diet [13]. Calcitriol binds to a specific binding protein, the vitamin D receptor (VDR). The resulting complex interacts with human DNA regulatory sequences known as vitamin D response elements (VDREs; 15 bases long), which reportedly vary in number between a few thousand to ten thousand [14,15]. VDREs respond specifically to calcitriol by activating or inactivating their adjacent genes [16–18]; this response may vary depending on the location of the VDRE and level of 25(OH)D [15]. The unusually large number of calcitriol-responsive DNA sites strongly suggests that sufficiently high 25(OH)D levels, which may vary by outcome are necessary for optimal health and longevity [16,19].

2.2. Evolution of Skin Pigmentation

Skin pigmentation is an evolutionary response to the intense solar UVB at low latitudes, where early humans evolved. Dark skin, through the presence of abundant melanin, protected humans living in Africa, southern India, and other parts of Asia against strong UVB, which causes severe sunburn, damages DNA, and destroys skin folate [3,20–22]. According to a widely accepted hypothesis, people in ancient times moved from low to higher latitudes, and skin pigment evolved (by several mutations) to be lighter, depending on distance from the equator, permitting more efficient production of vitamin D [3,23–26]. Others have suggested that lighter skin resulted from the acquisition of genetic variants from populations that immigrated into northern Europe, but this is still compatible with production of vitamin D being the initial selective factor for these variants [27]. (These authors also hypothesize that variations in genes encoding for proteins responsible for the transport, metabolism and signaling of vitamin D provide alternative mechanisms of adaptation to a life in northern latitudes without suffering the consequences of vitamin D deficiency. However, such mechanisms and loss of melanin are not mutually exclusive, and in either case they would leave people of African descent now living in northern latitudes at risk of vitamin D deficiency.) The importance of solar exposure is illustrated by findings that Africans with dark skin living at low latitudes have levels of 25(OH)D of 29 to 46 ng/mL [28–30], that are substantially higher than those of African Americans (mean 25(OH)D ~16 ng/mL) [31]. These differences, and the similarity in levels of 25(OH)D in European Americans and Africans living in Africa, are shown in Figure 1 [32]. Thus, darker skin pigmentation in Africans living in Africa appears to allow adequate vitamin D synthesis while protecting against sunburn and other damage. Direct genetic evidence that melanin reduces synthesis of vitamin D is provided by findings that Nigerians with albinism have significantly higher 25(OH)D levels than those with normal pigmentation [33]. The interaction between skin melanin and sunlight was further illustrated in a study of pregnant women in the southeastern US; the ratios of winter-to-summer prevalence of vitamin D insufficiency were 3.58 (95% CI 1.64 to 7.81) for European-American, 1.52 (95% CI 1.18 to 1.95) for Hispanic, and 1.14 (95% CI 0.99 to 1.30) for African-American women [34].

In contrast to the slow migration in ancient times, in more recent times there was rapid movement of equatorial Africans to various regions, such as North America, due to slave transport. When the destination is at higher northern latitudes than that of the ancestral country of origin, a mismatch between skin color and UV radiation occurs and lower UVB penetration of the skin to the layer with 7-dehydrocholesterol results in deficient endogenous vitamin D production. The consequent health problems can take years to manifest and thus are both subtle and insidious. A reverse mismatch occurs when light-skinned individuals move to low latitudes (e.g., an Irish person moving to Australia), resulting in increased risk for severe sunburn (and later, high rates of skin cancer). The reverse mismatch is recognized quickly and can be mitigated by using hats and sunscreen.

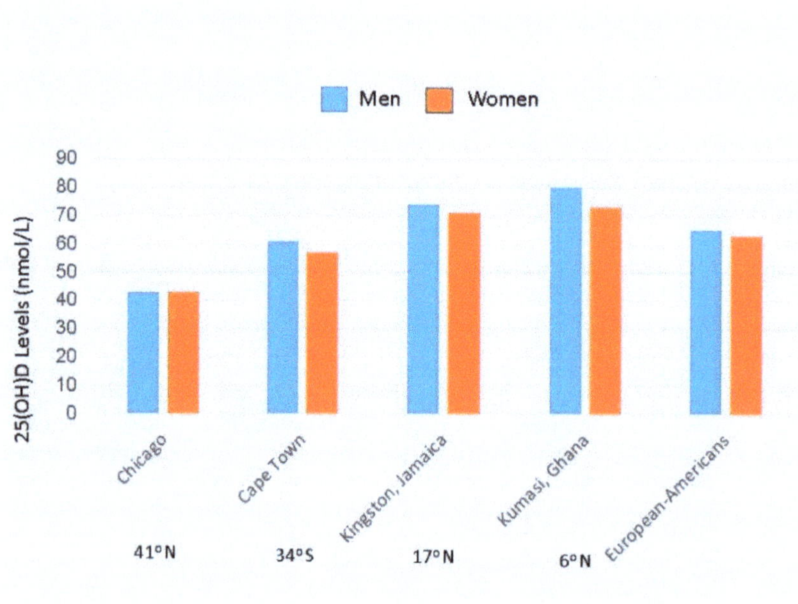

Figure 1. Average serum 25(OH)D levels (nmol/L) in men and women of African Ancestry ages 25 to 45 years living in four sites [32], and European Americans. The latitudes of the cities are given below the names of the cities. Note: Divide by 2.5 to convert nmol/L to ng/mL.

2.3. Prevalence of 25(OH)D Deficiency by Race/ethnicity Group

According to data from National Health and Nutrition Examination Survey (NHANES) 2001–2010, the prevalence of vitamin D deficiency (25(OH)D < 20 ng/mL) among those not taking vitamin D supplements was 75% for non-Hispanic blacks, 44% for Hispanics, and 20% for non-Hispanic whites (Figure 2), whereas severe deficiency (<10 ng/mL) was 17% in non-Hispanic blacks and only 1% in non-Hispanic whites [35]. Although definitions vary [36], there is consensus that levels below 10 ng/mL are a serious concern. A high prevalence of low 25(OH)D levels has also been documented in many other parts of the world [37–41].

2.4. Vitamin D and Health Outcomes

Most epidemiologic studies of vitamin D and health outcomes have used plasma or serum levels of 25(OH)D to measure vitamin D status. That approach has the advantage of integrating intake, solar exposure, skin color, and genetic factors. A single measure of 25(OH)D serves as a good measure of long-term status for an individual; however, the within-person correlation between 25(OH)D levels decreases as follow-up time increases [42]. Downstream metabolites of 25(OH)D are too variable over time to serve as a stable indicator of vitamin D status [43]. Other indicators of vitamin D status, such as parathyroid hormone, may improve our assessment [44], but have not yet been widely used in epidemiologic studies. Some studies have used vitamin D intake calculated from food intake, with or without supplements.

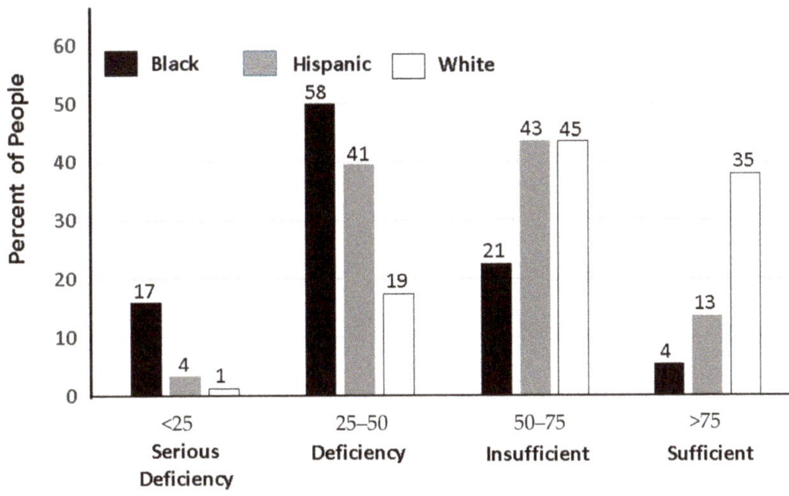

Figure 2. Prevalence of serum 25(OH)D levels in National Health and Nutrition Examination Survey (NHANES) survey, 2001–2010, by race/ethnicity category among nonusers of vitamin D supplements [35]. Additional calculations courtesy of X. Liu. Note: Divide by 2.5 to convert nmol/L to ng/mL.

2.5. Skeletal Health

Adequate vitamin D has long been recognized as essential for bone health, and the 2011 Institute of Medicine (IOM) review of vitamin D requirements concluded that rickets and osteomalacia were the only established consequences of low vitamin D status [36]. Thus, the relation to osteomalacia served to set recommendations for vitamin D intake: the estimated average requirement (EAR—at which half the population is deficient and half is not) for serum 25(OH)D was set at 16 ng/mL. On this basis also, levels below 12 ng/mL were considered deficient, 12 to 20 ng/mL were considered "at risk of inadequacy", and levels above 20 ng/mL were considered sufficient for 97% of the population. Other groups have defined deficiency as levels below 30 ng/mL [45]. Since 2011, much additional evidence has supported the important effects of vitamin D beyond bone health, and the relation between serum levels of 25(OH)D and these health outcomes cannot be assumed to be the same as that with osteomalacia.

Serum levels of 25(OH)D are positively associated with bone mineral density in both European Americans and African Americans [46], but Africans and African Americans have long been known to have higher bone mineral density (BMD) [47] and lower risk of fragility fractures than Europeans [48]. Possible mechanisms may be that African Americans have higher calcium retention, lower calcium excretion, and greater bone resistance to parathyroid hormone than European Americans [47,49,50]. The reason why populations migrating from Africa to higher latitudes evolved to have weaker bones is unclear, but in the context of low UV radiation a trade-off for reductions in pelvic deformity and obstructed labor has been suggested [51]. Whatever the mechanisms, the greater bone strength of African Americans, and the assumption that the only consequence of low 25(OH)D levels is poor bone health, seems to have led many to believe that the low serum levels 25(OH)D in African Americans are not a concern. Notably, the 2011 IOM review of vitamin D did not emphasize the high prevalence of vitamin D deficiency in African Americans, even by their strictest definition of less than 12 ng/mL, and concluded that "requirements are being met by most if not all persons in both countries [US and Canada]".

A finding of low levels of vitamin D-binding protein (VDBP) in African Americans, and thus presumably higher biologically active vitamin D, has been suggested as an explanation for healthy bone mass in African Americans despite low 25(OH(D level [52]. However, the report of low VDBP levels was subsequently shown to be an artifact of the monoclonal antibody assay used in that study; when measured by a polyclonal method or proteomic assay, levels of VDBP were similar in African- and European Americans [53]. This, and findings of much higher 25(OH)D levels in Africans living traditional lifestyles in equatorial regions, support the conclusion that the low levels of African Americans are not "natural" but due to environmental factors, primarily inadequate sun exposure.

2.6. Pregnancy and Early Development

Pregnant African-American women have higher risk of many pregnancy-related complications than European American or Hispanic women (Table 1).

Table 1. Pregnancy and birth outcomes as a function of ethnicity.

Outcome	Ethnicity (%)			Ratio		Ref.
	Black	Hispanic	White	Black/White	Hispanic/White	
Cesarean delivery	35.9	31.7	30.9	1.17	1.03	[54]
In-hospital mother death rate	0.21	0.05	0.05	4.23	0.98	[55]
Preeclampsia,	9.8	7.7	6.7	1.50	1.10	[56]
Low birth weight	13.7	7.3	7.0	2.00	1.04	[54]
Preterm birth	13.9	9.6	9.1	1.50	1.06	[54]
Small for gestational age	10.8	6.5	5.7	1.91	1.14	[57]

Pregnant women with darker skin color have lower 25(OH)D levels than women with lighter skin [34,58–60]. Indirect support for a role of vitamin D in development is provided by the finding that during pregnancy, maternal serum levels of $1,25(OH)_2D$ increase by 75% and those of 25(OH)D by about 30% [61]. The placenta plays a major role regarding these increases [62]. Further, when $1,25(OH)_2D$ stimulates the vitamin D receptors, it can affect the expression of hundreds to thousands of genes [1,15], and fetal development is guided by gene expression.

1. Preeclampsia. In a meta-analysis of data from 27 randomized controlled trials (RCTs) including 4777 participants, vitamin D treatment reduced risk of preeclampsia by 63% (OR = 0.37 (95% CI, 0.26 to 0.52)) [63]. Results were similar with respect to beginning of supplementation, supplementation until delivery, whether or not calcium was also supplemented, and whether the trial was blinded. Increased vitamin D dosage up to 7000 IU/d was associated with reduced risk of preeclampsia.
2. Low birth weight and small for gestational age. In a meta-analysis of 24 RCTs involving 5405 participants, vitamin D supplementation (800 to 7000 IU/day) during pregnancy reduced risk of offspring being small for gestational age by 28% (Risk ratio = 0.72 (95% CI, 0.52 to 0.99, p = 0.04)) [64]. In an observational study conducted in Cincinnati involving 276 black infants and 162 white infants, cord blood vitamin D deficiency was associated with being small for gestational age for black infants (OR = 2.4 (95% CI, 1.0 to 5.8, p = 0.04)) but not white infants (OR = 1.1 (95% CI, 0.3 to 3.9, p = 0.86)) [65]. Vitamin D deficiency was associated with increased risk of preeclampsia among both black and white women: for blacks, OR = 2.3 (95% CI, 1.0 to 5.4, p = 0.04) and for whites, OR = 4.1 (95% CI = 1.0, 16.1, p = 0.05),
3. Preterm birth. In a meta-analysis of 18 observational studies, maternal 25(OH)D level <20 ng/mL versus >20 ng/mL was associated with a pooled OR = 1.25 (95%CI: 1.13 to 1.38) of preterm delivery [66]. In five studies, this was also the case for spontaneous preterm delivery; for 25(OH)D < 20 ng/mL vs. >20 ng/mL pooled OR = 1.45 (95% CI, 1.20 to 1.75). In a meta-analysis of six RCTs, involving 1880 participants with a total of 77 preterm births, vitamin D supplementation reduced preterm delivery

by 43%; the pooled relative risk was 0.57 (95%CI: 0.36 to 0.91)) [66]. The vitamin D dose varied from 400 IU/d to 4000 IU/d. An open-label vitamin D supplementation study involving 1064 pregnant women including African-American and Hispanic and European-American women was conducted in South Carolina [67]. Women were counseled during their first prenatal visit on how to achieve >40 ng/mL 25(OH)D and given free bottles of 5000 IU vitamin D$_3$. In the fully-adjusted model, achieving >40 ng/mL vs. <20 ng/mL resulted in an OR for preterm delivery of 0.41 (95% CI, 0.24 to 0.72). Significantly lower risks of preterm birth were seen for both white and non-white women.

4. Neurologic development. In 2008, vitamin D deficiency during pregnancy was hypothesized to be a risk factor for autism [68]. In a meta-analysis of 25 observational studies, higher vs. lower serum levels of 25(OH)D during pregnancy or in newborn blood at birth were associated with a 28% lower risk of attention deficit–hyperactivity disorder in the offspring [pooled relative risk = 0.72 (95% CI = 0.59 go 0.89, $p = 0.002$)] and a 58% lower risk of autism-related traits (pooled relative risk = 0.42 (95% CI = 0.25 to 0.71, $p = 0.001$)) [69]. Vitamin D supplementation during pregnancy has reduced risk of abnormal neurologic development, and administration of 4000 IU/d during pregnancy caused no adverse effects in a trial conducted in South Carolina [61,70].

5. Cesarean delivery. In an observational study in Boston involving 253 women of whom 43 had a cesarean delivery, women with serum 25(OH)D levels lower than 15 ng/mL at time of delivery had 3.8 times the rate of primary cesarean delivery as compared to women with higher levels [71].

6. The overall evidence strongly supports the harmful nature of vitamin D deficiency among pregnant women for both pregnancy-related outcomes and for fetal development. Further, vitamin D supplementation has reduced the risk of pregnancy-related complications, particularly for women with severe deficiency [72].

2.7. Cancer

For many cancers, African Americans have higher incidence and mortality rates than European Americans; disparities exist for cancers of the bladder, breast, colon, endometrium, lung, ovary, pancreas, prostate, rectum, testes, and vagina, and for Hodgkin's lymphoma [73,74] (Table 2). Higher incidence and lower survival both contribute to some of those differences; for 2008–2012, African American males had a 12% higher overall cancer incidence and a 27% higher mortality rate than white men, whereas African American females had a 4% lower incidence rate but a 14% higher mortality rate than white women [74]. In many of the analyses, these differences in cancer rates were adjusted for a variety of potential confounding variables. Smoking, a major cause of cancer, does not account for the disparities because smoking rates for African Americans and European Americans are similar [74].

In single-country geographical ecological studies, solar UVB doses are inversely associated with mortality rates for many cancers among white people [75], and within the U.S. similar inverse associations are seen among both European Americans and African Americans [76–78]. Variables related to socioeconomic status can be hard to account for completely, especially in ecological studies. However, among male health professionals with similar education and occupation, African Americans with few risk factors for hypovitaminosis D had risks of cancer similar to those of white men; in contrast, African-American men with several risk factors for hypovitaminosis D had a 57% higher total cancer incidence and 127% higher cancer mortality rate [79]. Risk factors for hypovitaminosis D in this population included living in a region with low solar UVB doses, not spending much recreational time out of doors, and not taking vitamin D supplements. The excess risks were greater for digestive-tract cancers. The mechanisms by which vitamin D may reduce risk of cancer incidence and death include effects on cellular differentiation, proliferation, and apoptosis; anti-angiogenesis; and anti-metastasis [80], as well as anti-inflammatory [80,81] and immune-enhancing [82] mechanisms.

Table 2. Incidence and mortality rates for select cancers in the U.S. for males and females, 2008–2012 [74].

Sex and Cancer Type	Incidence *			Mortality **		
	Black	White	Black/White Ratio	Black	White	Black/White Ratio
Male						
Prostate	208.7	123.0	1.70	47.2	19.9	2.38
Lung	93.4	79.3	1.18	74.9	62.2	1.20
Colorectal	60.3	47.4	1.27	27.6	18.2	1.52
Kidney	24.2	21.8	1.11	5.7	5.9	0.97
Liver	16.5	9.3	1.77	12.8	7.6	1.69
Stomach	15.1	7.8	1.93	9.4	3.6	2.58
Female						
Breast	124.3	128.1	0.97	31.0	21.9	1.42
Lung	51.4	58.7	0.87	36.7	41.4	0.89
Colorectal	44.1	36.2	1.22	18.2	12.9	1.41
Kidney	13.0	11.3	1.15	2.6	2.3	1.13
Stomach	8.0	4.3	2.30	4.5	1.8	2.48
Liver	4.8	3.2	1.52	4.4	3.1	1.43

* Age-adjusted cases/100,000/yr; ** Age-adjusted deaths/100,000/yr.

Colorectal cancer. Among various malignancies, low vitamin D status has been most consistently associated with colorectal cancer. In ecological analyses within the United States, colorectal cancer (CRC) mortality among European Americans has been lowest in southwestern states and highest in northeastern states, and lowest in the southern states and highest in the northern states for African Americans (data missing for many states) consistent with the pattern of solar UVB doses in summer [77,78,83]. In an analysis of race and 25(OHD) levels in relation to risk of death due to colorectal cancer [84], a significant two-fold increase in risk was seen among both non-Hispanic white and non-Hispanic black participants when comparing those with 25(OHD) levels less than 20 ng/mL to those with higher levels. Further, adjustment for vitamin levels accounted for almost half of the excess risk of colorectal cancer seen for black compared with white participants. In a recent systematic review and meta-analysis of 11 observational studies involving 7718 patients with CRC, overall survival was 32% greater when comparing high with low levels of 25(OH)D [85]. Thus, substantial evidence suggests a benefit for vitamin D in reducing CRC incidence and mortality.

Bladder and kidney cancers. In a meta-analysis of four prospective studies and one case–control study [86], the risk of urinary bladder cancer was 32% higher when comparing low versus high 25(OH)D level (risk ratio = 1.32 (95% confidence interval (CI), 1.15 to 1.89)). In a meta-analysis of two prospective cohort studies and seven nested case-control studies involving 130,609 participants who developed 1815 cases of kidney cancer, the highest 25(OH)D levels were associated with a significant 21% lower incidence (OR = 0.79, (95% CI, 0.69 to 0.91)) of kidney cancer [87].

Prostate cancer. In contrast to other cancers, higher 25(OH)D levels have been associated with a modestly higher risk of prostate cancer in prospective studies. A meta-analysis of 19 prospective cohort or nested case-control studies with a total of 35,583 participants and 12,786 prostate cancer cases found that higher 25(OH)D level was associated with increased prostate cancer relative risk = 1.15 (95% CI 1.02 to 1.06) [88]. On the other hand, a meta-analysis of six cohorts of 7648 patients with prostate cancer, for prostate cancer-specific mortality the hazard ratio for high vs. low 25(OH)D was 0.91 (95% CI: 0.88–0.95) for prediagnosis studies and 0.84 (95% CI: 0.58–1.21) for postdiagnosis serum levels [89]. In a case–control study, African-American men with a higher intake of vitamin D had a lower risk of total and aggressive prostate cancer; these associations were not seen in European men [90].

Breast cancer. In a meta-analysis of cohort studies, women with higher versus lower baseline serum levels of 25(OH)D had a barely significant 8% lower incidence of breast cancer [91]; the inverse association was limited to premenopausal women. However, in a pooled analysis of cohort studies with 10,353 cases of breast cancer, standardized serum levels of 25(OH)D were not associated with risk of breast cancer overall or by menopausal status. There was also no statistically significant difference by race (P for heterogeneity = 0.90). For the same increment in 25(OH)D levels, the RR was 0.98 (CI, 0.95 to 1.02) in whites (9,579 cases); 1.28 (CI, 0.99 to 1.65) in blacks (290 cases); and 1.13 (CI, 0.76 to 1.68) in Asians (275 cases) [92]. In a cohort of 59,000 African-American women, predicted serum 25(OH)D levels (based on sun exposure, dietary intake, adiposity, and other variables) were inversely associated with risk of breast cancer (1454 cases): risk was 23% higher for the lowest versus the highest quintile [93]. In a recent case–control study among black women, daylight hours spent outdoors per year was inversely associated with lower risk of breast cancer [94].

Total cancer: The VITAL Randomized Trial. In the large VITAL trial [95] participants were randomized to 2000 IU of vitamin D per day and followed for five years. Although vitamin D was interpreted to have no significant overall effect on total cancer incidence, the incidence among African Americans was reduced by 23% (HR = 0.77 (95% CI, 0.59 to 1.01, p = 0.06)). Further, after excluding the first two years of follow-up as part of the planned analysis, total cancer mortality was significantly (p <0.05) reduced by 25% (HR = 0.75 (95% CI, 0.59 to 0.96)) among all participants. Notably, the inclusion of participants with a relatively high baseline serum 25(OH)D level (mean = 31 ng/mL), many of whom also took supplementary vitamin D, plus the limited duration of follow-up, may have obscured benefits of vitamin D for cancer incidence. In a recent meta-analysis of ten RCTs including VITAL, no benefit of vitamin D supplementation was seen for cancer incidence (6537 cases) [96]. However, cancer mortality was reduced by 13% (95% CI, 4% to 21%) in the five available trials (1591 deaths). In another secondary analysis, there was a significant reduction in advanced cancers (metastatic or fatal) for those randomized to vitamin D compared with placebo [97].

Thus, the findings from randomized trials support vitamin D supplementation for reducing cancer mortality among all participants and cancer incidence among African Americans.

2.8. Diabetes Mellitus

In the United States, the age-standardized prevalence of total diabetes is approximately twice as high among non-Hispanic blacks and Hispanics compared to non-Hispanic whites [98]. In a meta-analysis of 28 trials with 3848 participants, vitamin D supplementation reduced HbA1c level by 0.48% (95% CI, 0.18 to 0.79), fasting plasma glucose level by 0.46 mmol/L (95% CI, 0.19 to 0.74), and homeostatic model assessment for insulin resistance (HOMA-IR) level by 0.39 (95% CI, 0.11 to 0.68), in comparison with the control group [99]. Supplemental vitamin D also improved insulin sensitivity in patients with initial low 25(OH)D levels [100].

Several RCTs, each reported as negative, have examined how vitamin D affects risk of diabetes among individuals with prediabetes [101,102]. In the Vitamin D and Type 2 Diabetes (D2d) Study among patients with prediabetes (25% of the 2423 participants were African American) [102], those randomized to vitamin D (4000 IU/d) had a nonsignificant 12% (−25 to +4%) lower progression to type 2 diabetes (T2DM) than those receiving placebo [102]. However, in a post hoc analysis among participants with a baseline 25(OH)D level of less than 12 ng/mL (103 participants), progression to diabetes was 62% lower with vitamin D verses placebo (95% CI, 20 to 0.82%). In two other randomized trials, modest and not statistically significant reductions in risk of T2DM were found with vitamin D supplementation [102,103]. If the results for those three randomized trials are combined, the overall reduction in risk is statistically significant (hazard ratio = 0.88; 95% CI, 0.78 to 0.99; p = 0.04, unpublished analysis).

Subsequently, an additional secondary analysis of the D2d trial was published [104]. The relationship between intra-trial 25(OH)D levels and incidence of T2DM was determined. The HR for T2DM for an increase of 10 ng/mL in intra-trial 25(OH)D level (n = 1074) was 0.75 (95% CI 0.68–0.82) among those assigned to vitamin D and 0.90 (0.80–1.02) among those assigned to placebo. The HRs for T2DM among participants treated with vitamin D who maintained intra-trial 25(OH)D levels of 40–50 (n = 319) and $\geq$50 ng/mL (n = 430) were 0.48 (0.29–0.80) and 0.29 (0.17–0.50), respectively, compared with those who maintained a level of 20–30 ng/mL (n = 78). In a recent Mendelian randomization (MR) analysis, genetically predicted 25(OHD)D levels, and particularly alleles in genes involved in vitamin D synthesis) were inversely associated with incidence of type 2 diabetes [105].

Thus, the available evidence supports a modest overall benefit of vitamin D in reducing risk of T2DM [102] and possibly a substantial benefit among people with low serum 25(OH)D levels, such as African Americans.

2.9. Cardiovascular Disease

Inverse associations have been reported in studies of serum levels of 25(OH)D with risk of cardiovascular disease [106], including analyses specifically among African Americans [107–109]. However, no association was seen in MR studies [110] and in the large VITAL trial [95], including in participants with serum levels of 25(OH)D below 20 ng/mL and in African Americans. No association was also seen in a meta-analysis of vitamin D RCTs [111].

2.10. Alzheimer's Disease and Dementia and Cognitive Function

Multiple lines of evidence support a role of vitamin D in lowering risk of Alzheimer's disease (AD) [112]. In a meta-analysis of seven prospective studies and one retrospective cohort study (1953 cases of dementia and 1607 cases of AD), a serum level of 25(OH)D <10 ng/mL was associated with a 31% higher risk of dementia and a 33% higher risk of AD when compared with levels >20 ng/mL [113]. In one prospective study, 30% of participants were African American; higher baseline levels of 25(OH)D were associated with lower rates of cognitive decline but the numbers of were not large enough for race-specific analyses [114].

In MR analyses using the International Genomic of Alzheimer's Project (IGAP) dataset, risk of AD was found to be lower for individuals with genetic variants predicting higher levels of serum 25(OH)D compared to those without these variants [115]. In the most recent analysis using six such alleles (21,982 cases of Alzheimer's disease and 41,944 controls), the relative risk per allele was 0.62 (95% confidence interval 0.46 to 0.84) [115]. Together with the data based on serum levels of 25(OH)D, these findings provide substantial evidence that adequate vitamin D will reduce risk of dementia.

2.11. Multiple Sclerosis

In ecological studies, consistent with animal models [116], low solar UVB exposure is strongly associated with greater risk of multiple sclerosis (MS) [117] in whites, African Americans, and Hispanic Americans [118]. The inverse association between UVB exposure and MS also was seen in studies of individuals [119,120]. Solar UVB exposure in winter appears especially important.

In cohort studies, serum levels of 25(OH)D have been inversely associated with risk of MS [121]. In U.S. military recruits, levels greater than 40 ng/mL were associated with the lowest risk of MS, a level few African Americans attained [121]. In a Swedish study, 25(OH)D levels in the highest quintile were associated with a 32% lower incidence of MS [122]. Low levels of 25(OH)D in neonatal blood spots were strongly associated with MS later in life [123], supporting the importance of maternal vitamin D status during pregnancy. In a large cohort of women, use of vitamin D supplements greater than or equal to 400 IU/day was associated with lower risk of MS, but intake from diet, which rarely exceeds 400 IU/day, was not [124]. A cohort study conducted in southern California included a modest number of Black, Hispanic, and White participants with MS and matched controls. An inverse association between serum 25(OH)D level and incidence of MS was seen in Whites, but not among Blacks or Hispanics [118], but in all three groups a careful assessment of lifetime solar exposure was inversely associated with risk of MS. The authors suggested that something about solar exposure independent of vitamin D may be protective for MS, but an alternative explanation could be that their lifetime solar exposure assessment provided a better indication of long term vitamin D status (which would have been correlated with solar exposure over this period) than a single blood measurement collected in midlife.

A review of vitamin D supplementation in MS found little benefit even at high vitamin D doses [125]. The reasons for lack of benefit of vitamin D supplement suggested by the authors included the number of participants being too low, the length of the trial too short, baseline serum 25(OH)D levels too high, and other treatments being administered reducing the potential of vitamin D to help.

In a MR analysis of two large cohorts including 7391 cases of MS and 14,777 controls, a genetic risk score comprised of three alleles known to be associated with higher plasma 25(OH)D predicted levels was associated with lower risk of MS [126]. In the meta-analyses of these cohorts, the relative risk per allele was 0.85 (95% CI, 0.76 to 0.94, $p = 0.003$). This result, in combination with the other extensive observational study evidence, supports a protective role of adequate vitamin D intake for incidence of MS.

2.12. Acute Respiratory Tract Infections and COVID-19

Substantial evidence indicates that higher serum 25(OH)D levels can reduce the risk or severity of acute respiratory tract infections, possibly including COVID-19. Potential mechanisms include role of vitamin D in innate and acquired immunity [127,128]. This relation was suggested by observations that the seasonal increase in influenza infections corresponds with lower solar UVB doses and 25(OH)D levels [129] and that in the 1918–1919 influenza pandemic the case-fatality rates were much lower in the southwestern U.S. states than in the northeastern states [130]. In a meta-analysis of 25 RCTs involving 10933 participants, vitamin D supplementation (daily or weekly) reduced risk of acute respiratory tract infections by 19% [131]. For participants with baseline 25(OH)D level <10 ng/mL, the reduction was 70%. In a post hoc analysis of an RCT conducted among 208 post-menopausal African-American women living in New York, supplementation with 1000 or 2000 IU/day of vitamin D_3 compared to a placebo significantly reduced rates of influenza and colds [132].

Adequate vitamin D supplementation has also been hypothesized to decrease incidence and death from COVID-19; in addition to reducing viral replication, vitamin D may limit excess production of pro-inflammatory cytokines underlying the "cytokine storm" that damages the lungs and other organs [130,133]. Incidence and mortality of COVID have been far higher in African Americans than in European Americans [134]; after adjustment for age, African Americans are 4.5 times more likely to die from COVID-19 than European Americans [135]. Much of this could be due to more crowded housing, riskier jobs, dependence on public transportation, and higher prevalence of existing cardiometabolic conditions, but the high prevalence of vitamin D deficiency may also contribute.

Among patients with COVID-19, the disease resulting from SARS-CoV-2 infection followed by a dysregulated immune response, low levels of 25(OH)D at time of diagnosis have been associated with more severe illness [136–138]. Reverse causation cannot be excluded because serum 25(OH)D level decreases in response to acute inflammatory disease [139,140], but serum 25(OH)D levels have been associated with SARS-CoV-2 virus positivity using seasonally-adjusted 25(OH)D levels from the preceding 12 months [141,142].

The strong suggestion of benefits of vitamin D supplementation for preventing or treating COVID-19 has encouraged the initiation of supplementation trials [143]. In a non-randomized intervention study conducted in Spain among hospitalized patients hospitalized for COVID-19 [144], high doses of vitamin D (as 25(OH) D_3) were administered in combination with standard care; only 1/50 required admission to the intensive care unit compared to 13/26 comparable control patients. An RCT conducted in India involving 40 SARS-CoV-2 positive patients with serum 25(OH)D with mean values near

9 ng/mL were randomized into high-dose vitamin D treatment ($n = 16$) and control ($n = 24$) groups [145]. Ten (63%) participants in the intervention group and five (21%) participants in the control arm ($p < 0.02$) became SARS-CoV-2 RNA negative. According to the registry of clinical trials [146] as of 30 December 2020, there were at least 35 RCTs registered examining the role of vitamin D supplementation in prevention or treatment of COVID-19. In these trials, it will be important to distinguish among those with low versus adequate vitamin D status at baseline.

2.13. Asthma Exacerbations

A combined analysis of two RCTs conducted with pregnant women found that vitamin D supplementation (2400 IU/d and 4000 IU/d) reduced risk of asthma/recurrent wheeze from 0–3 years by 24% (aOR = 0.74 (95% CI, 0.57 to 0.96)) [147]. The effect was strongest for those with baseline 25(OH)D level $\geq$30 ng/mL (aOR = 0.54 (95% CI, 0.33 to 0.88)). A secondary analysis of the 4000 IU/d vitamin D RCT found that there was no difference with respect to race for African American vs. non-African American [148]. In a meta-analysis of individual participant data from seven RCTs with high-quality evidence, vitamin D supplementation reduced by 26% the risk of asthma exacerbation that required treatment with systemic corticosteroids [149]. The reduction was 67% for individuals with initial 25(OH)D level <10 ng/mL (92 participants). Thus, there is good evidence that raising serum 25(OH)D levels reduces risk of asthma or its exacerbation.

2.14. All-Cause Mortality

A comparison of death rates for African Americans with those of European Americans shows a large disparity for many diseases [150]. For example, the disparity in death rates from all causes for people aged 50 to 64 years is 45% higher in African American than in European Americans (Figure 3). Differences in multiple factors, such as hypertension, obesity, diet, income, education, and lower access to medical care, may contribute to some of those disparities, but they do not fully explain the differences [151–155].

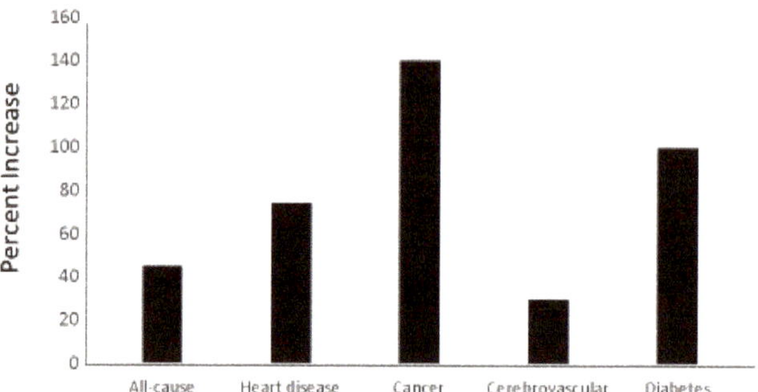

Figure 3. Percentage Increase in Cause-Specific Mortality for Black compared to White Americans, Ages 50–64 years—United States, 2015 [150].

In a meta-analysis (~29,000 subjects from five Northern European countries), serum levels of 25(OH)D below 32 ng/mL were associated with the highest mortality [156]. In another meta-analysis of 32 studies, mortality for those in the lowest (<9 ng/mL) was about double compared with those with the highest serum 25(OH)D level. The lowest mortality was reached near 40 ng/mL and plateaued above this level [157]. Most of those analyses accounted for potentially confounding variables such as adiposity, physical activity, and smoking, but some residual confounding could not be excluded.

More direct evidence for causality comes from a MR analysis documenting an association between genetically determined serum 25(OH)D levels and both total and cancer-specific mortality [110]. Although in a MR analysis from the UK Biobank genetically determined 25(OH)D level was not associated with all-cause mortality rate [158], that study was underpowered according to the authors, especially given that the association appears nonlinear.

A US cohort of 3075 adults aged 70–79 years of age was followed for 8.5 years [159]. Although the prevalence of vitamin D deficiency was much higher in Blacks than in Whites, lower baseline serum levels of 25(OH)D were similarly associated with higher mortality in both Black and White participants (Figure 4). Because of the large Black/White difference in vitamin D status, 25(OH)D levels below 30 ng/mL statistically accounted for 38% of mortality in Blacks and 11% in Whites. In a multivariate model without 25(OH)D levels, Blacks had 22% higher mortality than Whites, but after inclusion of 25(OH)D in the model the excess mortality in Blacks was only 9% and not statistically significant.

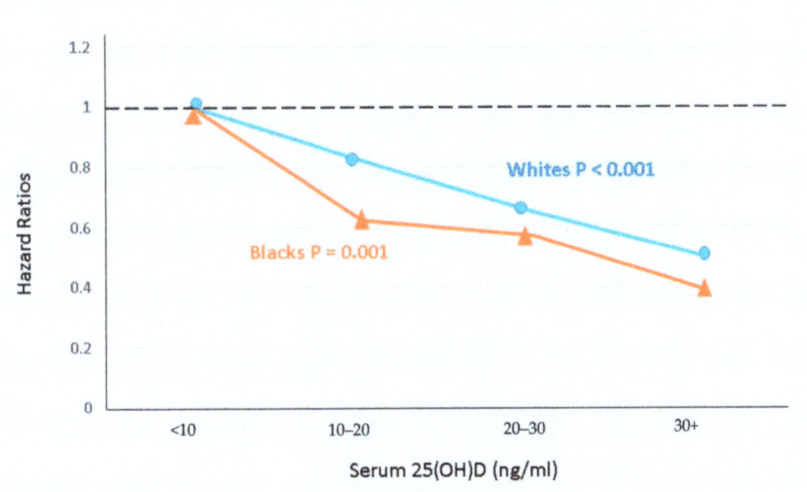

Figure 4. Serum 25(OH)D levels and all-cause mortality for elderly Black (*n* = 1023) and White (*n* = 1615) men and women followed for up to 8.5 years [159]. Hazard ratios with < 10 ng/mL serum 25(OH)D as the reference were adjusted for other predictors of mortality.

Similar findings were seen in a nested case-control analysis from a cohort of largely African-American participants, in which 1852 cohort members who died were matched to a similar number of participants who remained alive. Using baseline serum samples, the multivariate OR for death for those in lowest quartile compared with the highest quartile was 1.60 (95% confidence interval (CI): 1.20, 2.14, p_{trend} = 0.003) for African Americans and 2.11 (95% CI: 1.39, 3.21, p_{trend} < 0.001) for non-African Americans; the adjusted mortality rate became flat above approximately 30 ng/mL for both groups [160]. In the VITAL trial, no significant effect of vitamin D supplements on total mortality was seen in black or white participants, but the study was limited by duration and relatively high baseline levels [95].

Thus, there is good evidence from observational studies that all-cause mortality rate is inversely correlated with serum 25(OH)D concentrations up to about 30 to 40 ng/mL and then the association becomes flat.

3. Discussion

This review calls attention to the health-related consequences of low 25(OH)D levels for people with dark skin living at high northern latitudes, in both annual average and winter or summer [31,161]. Serum 25(OH)D levels are much lower among people with dark skin than among those with light skin living at similar latitudes. The prevalence of a serious deficiency value of 10 ng/mL (or less) is particularly high among African Americans. Regardless of skin color, low 25(OH)D levels are associated with higher incidence or poorer outcomes for many diseases. Evidence is particularly strong for several complications of pregnancy, multiple sclerosis, dementia, type 2 diabetes, colorectal cancer, total cancer mortality, and acute respiratory tract infections. For several of these diseases, causality is supported by either RCTs (such as for cancer mortality [95], diabetes mellitus [102], and acute respiratory tract infections [131]) or by the combination of prospective cohort and MR studies (such as for MS and dementia). Even if not all those relationships are ultimately determined to be causal, the consequences of vitamin D deficiency on the remaining diseases are important. On the basis of that evidence, vitamin D deficiency is highly likely to contribute to disparities in health status between people with dark and light skin at high latitudes. Hopefully, this review, by assembling the latest information on vitamin D for many health outcomes, will motivate physicians and patients to consider improving vitamin D status as an efficient way to improve health regardless of skin type.

Our study contrasts with the interpretations of several recent major RCTs on vitamin D supplementation [95,102] which reported no benefit of vitamin D supplementation. Although RCTs of vitamin D supplementation can be useful or definitive if a clear effect is seen, they also can be misinterpreted or misleading if no statistically significant effect is seen for various reasons, such as that the baseline and achieved 25(OH)D levels were either not measured or not considered in designing the trial [162,163]. Therefore, important benefits would be missed if many participants had levels high enough before randomization. For ethical or practical reasons, many RCTs do not focus on participants with vitamin D deficiency [95,102,164] who would be the people who would benefit most from supplementation. Also, some trials permit or encourage all participants to take additional vitamin D (400 to 800 IU/d), thereby reducing the risk of disease in the control group [95]. For some trial outcomes, especially cancer incidence, supplementation for long periods may be needed, but the effort is complicated by declining adherence. Our reasons for reaching a different conclusion include consideration of a broader literature, including important recent studies, and of secondary analyses for subpopulations, such as African Americans, most likely to benefit [165].

Observational cohort studies can circumvent some of those problems, but some residual confounding may be hard to exclude. Community-based observational studies in which participants take a vitamin D dose of their choice, have 25(OH)D levels measured semiannually, and report any changes in health status such as those conducted by GrassrootsHealth.net, e.g., [166], can play a role. No single type of study will provide the best evidence for all hypotheses, and the greatest insights will come from a thoughtful combination of research strategies. Those studies should especially include people with dark skin, in particular African Americans.

Preventing Vitamin D Deficiency

Few foods, mainly fish and fish liver, have substantial amounts of vitamin D, which is primarily synthesized in the skin. Therefore, low 25(OH)D levels are largely determined by melanin levels and contemporary lifestyles (including getting little sun exposure by staying indoors, covering the body, and using sunscreen extensively) and excess body fat [167]. For people with dark skin, especially if living at northern latitudes and in winter, typical sun exposure will usually not be adequate to prevent deficiency. Notably, while leisure-time sun exposure contributes to serum levels of 25(OH)D in European Americans, it does so minimally in African Americans [168]. Vitamin D supplementation can effectively prevent deficiency. Fortifying milk with vitamin D has prevented rickets in children, but the amount of vitamin D (100 IU per 8 oz [240 mL]) would have only small effects on serum levels in adults. In addition, milk consumption has decreased over time and lactose intolerance is common, especially among African Americans [169]. Thus, intakes may need to be increased primarily by fortifying additional foods, including non-dairy, or use of supplements; issues of dose and frequency of administration suggest that levels of at least 30 ng/mL would be a reasonable target. Notably, while solar UVB exposure is an important source of vitamin D, African Americans have high prevalence of 25(OH)D levels below 30 ng/mL in both summer (88%) and winter (93%), which contrasts for European Americans (61% in winter and 49% in summer) and Hispanics (86% in winter and 57% in summer) [161].

One strategy would be to screen routinely and to supplement people with low serum levels, but ideally a safe dose could be identified that yields near-optimal serum levels for almost everyone. A level of 20 ng/mL or higher was considered sufficient by an IOM committee in 2011 [36]; however, levels between 20 and 30 ng/mL have also been associated with lower risks of colorectal cancer [170], total mortality [157], dementia [113], multiple sclerosis [122] and bone mineral density [49,171].

Other researchers have suggested that 40 to 60 ng/mL is optimal based on results of small observational studies with participants taking high vitamin D$_3$ doses [172–174]. Three-quarters of African Americans not already taking supplements have levels that do not ensure adequacy even by the IOM definition (20 ng/mL), and 96% have levels below 30 ng/mL. Notably, for African Americans living in Boston, 4000 IU/day was required to achieve serum levels of 30 ng/mL [175]. Similarly, among men with early prostate cancer, serum levels of 25(OH)D were much lower in African Americans compared with European Americans, but after supplementation of 4000 IU/day for one year, levels increased in both groups and were nearly identical [176]. In 2012, vitamin D supplements were used by only 12% of African Americans and 22% of European Americans [177]. Noting the between-person variation in response to the same dose of vitamin D, some have suggested the desirability of monitoring indicators of biological function [178]. While this deserves consideration, it would add greatly to costs, and the specific variables to monitor are not clear at this time [179].

Vitamin D supplementation up to 4000 IU of vitamin D daily was considered to be safe by the 2011 IOM review. Further assurance comes from the trial using 4000 IU/day for 2.5 years [102]. An intake of 4000 IU/day has been used without adverse effects during pregnancy [61,67]. Higher intakes may also be safe [170,180] and warrant further study. At very high doses, such as in accidental exposures, vitamin D can produce death, neurological symptoms and serious damage (e.g., 1 million IU/day for several weeks, although the damage can sometimes be reversed [181]); at less extreme doses, a primary concern has been hypercalciuria and kidney stones [182], although any excess risk of kidney stones appears to be minimal when taking vitamin D supplements up to 4000 IU per day [183].

4. Conclusions

Together, ecological studies, prospective cohort studies based on blood levels, MR studies, and randomized trials provide moderate-to-strong evidence that low levels of 25(OH)D have many adverse health consequences. In addition, the fact that the vitamin D hormone, calcitriol, controls a considerable percentage of the human genome [1] indicates that it must be of huge general importance for health.

Much evidence is at present derived from studies of people of European descent. However, the benefits of supplementation for most health outcomes (other than skeletal effects) probably apply to all groups but are likely to be greatest for people with dark skin living at higher latitudes, such as African Americans, as well as most people in winter and those spending little time in the sun during summers. Many of the health disparities we discuss also have a basis in income inequality, poorer education and employment opportunities, poor housing, food insecurity, and other social inequalities; efforts to improve 25(OH)D levels will not lessen the need to address these factors but should improve health outcomes. While further research is needed to identify the optimal strategy for vitamin D supplementation and fortification, no reason exists to delay addressing vitamin D deficiency among populations with high prevalence of deficiency such as African Americans. The potential benefits promise to be large, and much evidence indicates that the risks of supplementation up to 4000 IU per day vitamin D are minimal.

Author Contributions: Conceptualization, B.N.A.; Methodology, B.N.A., W.B.G. and W.C.W. Writing—Original Draft Preparation, B.N.A. and W.B.G.; Writing—Review & Editing, B.N.A., W.B.G. and W.C.W. All authors have read and agreed to the published version of the manuscript.

Funding: This research received no external funding.

Acknowledgments: B.N.A. thanks Giovanna Ferro-Luzzi Ames for editing and helpful comment; Dylan Moss for early exhaustive literature research; Joyce McCann, Ken Beckman, Rhonda Patrick, Mark Shigenaga for helpful comments.

Conflicts of Interest: W.B.G. receives funding from Bio-Tech Pharmacal, Inc. (Fayetteville, AR). The other authors have no conflicts of interest to declare.

Appendix A

Many types of epidemiological study are used to determine the extent to which vitamin D affects health outcomes. The strengths and weaknesses of the types of study used to determine the relationship between vitamin D and health outcomes are outlined in Table 1.

Table 1. Comparison of epidemiological approaches to determine relationships between vitamin D and health outcomes.

Approach	Method	Strengths	Weaknesses
Ecological, geographical	Compare health outcomes with indices of solar UVB doses and other risk-modifying factors	Can include large numbers of participants	Subject to confounding factors; indices used may not apply to those with health outcomes
Observational, prospective	Enroll participants, draw blood, obtain information, follow for a long period.	25(OH)D has inputs from solar UVB, diet, and supplements.	25(OH)D changes with time including season so effect decreases as follow-up time increases; control of confounding may not be complete.
Observational, case-control	Measure 25(OH)D near time of diagnosis, match with controls.	Appropriate when health outcome is affected by recent 25(OH)D or 25(OH)D changes little over time.	Disease status may affect 25(OH)D; control choice may be biased.
Observational, cross-sectional	Measure 25(OH)D and health status of a representative number of a population	Many health outcomes can be studied.	25(OH)D may not be similar to that prior to health outcome; health status may affect 25(OH)D.
Non-randomized vitamin D supplementation	Enroll participants, measure parameters, instruct.	High vitamin D doses can be used;	Confounding by self-selected use of supplements
RCT	Enroll participants, randomize to vitamin D or placebo supplementation, follow.	Effects found are likely due to vitamin D.	False negatives are possible because enrollees often have high 25(OH)D or may be given low vitamin D doses; compliance issues; other sources of vitamin D occur. Initiation may be after onset of disease and duration may be too short to have an effect.
MR	Measure alleles of genes that affect 25(OH)D levels.	Independent of many factors including actual 25(OH)D.	The alleles may not reflection biological activity, and confounding is still possible. Can be misleading when relationship of vitamin D is nonlinear, and statistical power can be low.

MR—Mendelian randomization; RCT—Randomized controlled trial.

References

1. Ramagopalan, S.V.; Heger, A.; Berlanga, A.J.; Maugeri, N.J.; Lincoln, M.R.; Burrell, A.; Handunnetthi, L.; Handel, A.E.; Disanto, G.; Orton, S.M.; et al. A ChIP-seq defined genome-wide map of vitamin D receptor binding: Associations with disease and evolution. *Genome Res.* **2010**, *20*, 1352–1360. [CrossRef] [PubMed]
2. Engelsen, O. The relationship between ultraviolet radiation exposure and vitamin D status. *Nutrients* **2010**, *2*, 482–495. [CrossRef] [PubMed]

3. Jablonski, N.G.; Chaplin, G. Colloquium paper: Human skin pigmentation as an adaptation to UV radiation. *Proc. Natl. Acad. Sci. USA* **2010**, *107* (Suppl. 2), 8962–8968. [CrossRef]
4. Weishaar, T.; Rajan, S.; Keller, B. Probability of Vitamin D Deficiency by Body Weight and Race/Ethnicity. *J. Am. Board Fam. Med.* **2016**, *29*, 226–232. [CrossRef] [PubMed]
5. Andersen, T.B.; Dalgaard, C.J.; Skovsgaaaard, C.V.; Selaya, P. Historical migration and contemporary health. *Oxf. Econ. Pap.* **2021**, 1–27. [CrossRef]
6. Forrest, K.Y.; Stuhldreher, W.L. Prevalence and correlates of vitamin D deficiency in US adults. *Nutr. Res.* **2011**, *31*, 48–54. [CrossRef]
7. Williams, D.R.; Yan, Y.; Jackson, J.S.; Anderson, N.B. Racial Differences in Physical and Mental Health: Socio-economic Status, Stress and Discrimination. *J. Health Psychol.* **1997**, *2*, 335–351. [CrossRef]
8. Williams, D.R.; Collins, C. Racial residential segregation: A fundamental cause of racial disparities in health. *Public Health Rep.* **2001**, *116*, 404–416. [CrossRef]
9. Sampson, R.J.; Raudenbush, S.W.; Earls, F. Neighborhoods and violent crime: A multilevel study of collective efficacy. *Science* **1997**, *277*, 918–924. [CrossRef]
10. Moukayed, M.; Grant, W.B. The roles of UVB and vitamin D in reducing risk of cancer incidence and mortality: A review of the epidemiology, clinical trials, and mechanisms. *Rev. Endocr. Metab. Disord.* **2017**, *18*, 167–182. [CrossRef]
11. Holick, M.F. Vitamin D deficiency. *N. Engl. J. Med.* **2007**, *357*, 266–281. [CrossRef] [PubMed]
12. Bikle, D.D. Vitamin D and cancer: The promise not yet fulfilled. *Endocrine* **2014**, *46*, 29–38. [CrossRef] [PubMed]
13. Dai, Q.; Zhu, X.; Manson, J.E.; Song, Y.; Li, X.; Franke, A.A.; Costello, R.B.; Rosanoff, A.; Nian, H.; Fan, L.; et al. Magnesium status and supplementation influence vitamin D status and metabolism: Results from a randomized trial. *Am. J. Clin. Nutr.* **2018**, *108*, 1249–1258. [CrossRef] [PubMed]
14. Carlberg, C. Vitamin D Genomics: From In Vitro to In Vivo. *Front. Endocrinol.* **2018**, *9*, 250. [CrossRef]
15. Shirvani, A.; Kalajian, T.A.; Song, A.; Holick, M.F. Disassociation of Vitamin D's Calcemic Activity and Non-calcemic Genomic Activity and Individual Responsiveness: A Randomized Controlled Double-Blind Clinical Trial. *Sci. Rep.* **2019**, *9*, 17685. [CrossRef]
16. Pike, J.W.; Christakos, S. Biology and Mechanisms of Action of the Vitamin D Hormone. *Endocrinol. Metab. Clin. North Am.* **2017**, *46*, 815–843. [CrossRef]
17. Patrick, R.P.; Ames, B.N. Vitamin D hormone regulates serotonin synthesis. Part 1: Relevance for autism. *FASEB J. Off. Publ. Fed. Am. Soc. Exp. Biol.* **2014**, *28*, 2398–2413. [CrossRef]
18. Patrick, R.P.; Ames, B.N. Vitamin D and the omega-3 fatty acids control serotonin synthesis and action, part 2: Relevance for ADHD, bipolar disorder, schizophrenia, and impulsive behavior. *FASEB J. Off. Publ. Fed. Am. Soc. Exp. Biol.* **2015**, *29*, 2207–2222. [CrossRef]
19. Ames, B.N. Prolonging healthy aging: Longevity vitamins and proteins. *Proc. Natl. Acad. Sci. USA* **2018**, *115*, 10836–10844. [CrossRef]
20. Jablonski, N.G.; Chaplin, G. The evolution of skin pigmentation and hair texture in people of African ancestry. *Dermatol. Clin.* **2014**, *32*, 113–121. [CrossRef]
21. Jablonski, N.G.; Chaplin, G. The roles of vitamin D and cutaneous vitamin D production in human evolution and health. *Int. J. Paleopathol.* **2018**, *23*, 54–59. [CrossRef] [PubMed]
22. Chaplin, G. Geographic distribution of environmental factors influencing human skin coloration. *Am. J. Phys. Anthropol.* **2004**, *125*, 292–302. [CrossRef] [PubMed]
23. Clemens, T.L.; Adams, J.S.; Henderson, S.L.; Holick, M.F. Increased skin pigment reduces the capacity of skin to synthesise vitamin D3. *Lancet* **1982**, *1*, 74–76. [CrossRef]
24. Matsuoka, L.Y.; Wortsman, J.; Haddad, J.G.; Kolm, P.; Hollis, B.W. Racial pigmentation and the cutaneous synthesis of vitamin D. *Arch. Dermatol.* **1991**, *127*, 536–538. [CrossRef]
25. Adhikari, K.; Mendoza-Revilla, J.; Sohail, A.; Fuentes-Guajardo, M.; Lampert, J.; Chacon-Duque, J.C.; Hurtado, M.; Villegas, V.; Granja, V.; Acuna-Alonzo, V.; et al. A GWAS in Latin Americans highlights the convergent evolution of lighter skin pigmentation in Eurasia. *Nat. Commun.* **2019**, *10*, 358. [CrossRef]
26. Campbell, P.A.; Young, M.W.; Lee, R.C. Vitamin D Clinical Pharmacology: Relevance to COVID-19 Pathogenesis. *J. Natl. Med. Assoc.* **2020**. [CrossRef]
27. Hanel, A.; Carlberg, C. Skin colour and vitamin D: An update. *Exp. Dermatol.* **2020**, *29*, 864–875. [CrossRef]
28. Luxwolda, M.F.; Kuipers, R.S.; Kema, I.P.; Dijck-Brouwer, D.A.; Muskiet, F.A. Traditionally living populations in East Africa have a mean serum 25-hydroxyvitamin D concentration of 115 nmol/l. *Br. J. Nutr.* **2012**, *108*, 1557–1561. [CrossRef]
29. Luxwolda, M.F.; Kuipers, R.S.; Kema, I.P.; van der Veer, E.; Dijck-Brouwer, D.A.; Muskiet, F.A. Vitamin D status indicators in indigenous populations in East Africa. *Eur. J. Nutr.* **2013**, *52*, 1115–1125. [CrossRef]
30. Chiang, D.; Kramer, H.; Luke, A.; Cooper, R.; Aloia, J.; Bovet, P.; Plange-Rhule, J.; Forrester, T.; Lambert, V.; Camacho, P.; et al. 25-Hydroxyvitamin D and blood pressure: A plateau effect in adults with African ancestry living at different latitudes. *J. Hypertens.* **2017**, *35*, 968–974. [CrossRef]
31. Ginde, A.A.; Liu, M.C.; Camargo, C.A., Jr. Demographic differences and trends of vitamin D insufficiency in the US population, 1988-2004. *Arch. Intern. Med.* **2009**, *169*, 626–632. [CrossRef] [PubMed]

32. Kramer, H.; Camacho, P.; Aloia, J.; Luke, A.; Bovet, P.; Rhule, J.P.; Forrester, T.; Lambert, V.; Harders, R.; Dugas, L.; et al. Association between 25-Hydroxyvitamin D and Intact Parathyroid Hormone Levels across Latitude among Adults with African Ancestry. *Endocr. Pr.* **2016**, *22*, 911–919. [CrossRef] [PubMed]

33. Enechukwu, N.; Cockburn, M.; Ogun, G.; Ezejiofor, O.I.; George, A.; Ogunbiyi, A. Higher vitamin D levels in Nigerian albinos compared with pigmented controls. *Int. J. Dermatol.* **2019**, *58*, 1148–1152. [CrossRef] [PubMed]

34. Chawla, D.; Daniels, J.L.; Benjamin-Neelon, S.E.; Fuemmeler, B.F.; Hoyo, C.; Buckley, J.P. Racial and ethnic differences in predictors of vitamin D among pregnant women in south-eastern USA. *J. Nutr. Sci.* **2019**, *8*, e8. [CrossRef] [PubMed]

35. Liu, X.; Baylin, A.; Levy, P.D. Vitamin D deficiency and insufficiency among US adults: Prevalence, predictors and clinical implications. *Br. J. Nutr.* **2018**, *119*, 928–936. [CrossRef]

36. Ross, A.C.; Manson, J.E.; Abrams, S.A.; Aloia, J.F.; Brannon, P.M.; Clinton, S.K.; Durazo-Arvizu, R.A.; Gallagher, J.C.; Gallo, R.L.; Jones, G.; et al. The 2011 report on dietary reference intakes for calcium and vitamin D from the Institute of Medicine: What clinicians need to know. *J. Clin. Endocrinol. Metab.* **2011**, *96*, 53–58. [CrossRef]

37. Arabi, A.; El Rassi, R.; El-Hajj Fuleihan, G. Hypovitaminosis D in developing countries-prevalence, risk factors and outcomes. *Nat. Rev. Endocrinol.* **2010**, *6*, 550–561. [CrossRef]

38. Zhang, W.; Stoecklin, E.; Eggersdorfer, M. A glimpse of vitamin D status in Mainland China. *Nutrition* **2013**, *29*, 953–957. [CrossRef]

39. Gupta, A. Vitamin D deficiency in India: Prevalence, causalities and interventions. *Nutrients* **2014**, *6*, 729–775. [CrossRef]

40. Vatandost, S.; Jahani, M.; Afshari, A.; Amiri, M.R.; Heidarimoghadam, R.; Mohammadi, Y. Prevalence of vitamin D deficiency in Iran: A systematic review and meta-analysis. *Nutr. Health* **2018**, 260106018802968. [CrossRef]

41. Pilz, S.; Marz, W.; Cashman, K.D.; Kiely, M.E.; Whiting, S.J.; Holick, M.F.; Grant, W.B.; Pludowski, P.; Hiligsmann, M.; Trummer, C.; et al. Rationale and Plan for Vitamin D Food Fortification: A Review and Guidance Paper. *Front. Endocrinol.* **2018**, *9*, 373. [CrossRef] [PubMed]

42. Grant, W.B. Effect of follow-up time on the relation between prediagnostic serum 25-hydroxyvitamin D and all-cause mortality rate. *Dermatoendocrinol* **2012**, *4*, 198–202. [CrossRef] [PubMed]

43. Bailie, G.R.; Johnson, C.A. Comparative review of the pharmacokinetics of vitamin D analogues. *Semin. Dial.* **2002**, *15*, 352–357. [CrossRef] [PubMed]

44. Peiris, A.N.; Youssef, D.; Grant, W.B. Secondary hyperparathyroidism: Benign bystander or culpable contributor to adverse health outcomes? *South Med. J.* **2012**, *105*, 36–42. [CrossRef] [PubMed]

45. Holick, M.F.; Binkley, N.C.; Bischoff-Ferrari, H.A.; Gordon, C.M.; Hanley, D.A.; Heaney, R.P.; Murad, M.H.; Weaver, C.M.; Endocrine, S. Evaluation, treatment, and prevention of vitamin D deficiency: An Endocrine Society clinical practice guideline. *J. Clin. Endocrinol. Metab.* **2011**, *96*, 1911–1930. [CrossRef] [PubMed]

46. Bischoff-Ferrari, H.A.; Dietrich, T.; Orav, E.J.; Dawson-Hughes, B. Positive association between 25-hydroxy vitamin D levels and bone mineral density: A population-based study of younger and older adults. *Am. J. Med.* **2004**, *116*, 634–639. [CrossRef] [PubMed]

47. Bell, N.H. Bone and mineral metabolism in African Americans. *Trends Endocrinol. Metab.* **1997**, *8*, 240–245. [CrossRef]

48. Aloia, J.F. African Americans, 25-hydroxyvitamin D, and osteoporosis: A paradox. *Am. J. Clin. Nutr.* **2008**, *88*, 545S–550S. [CrossRef]

49. Gutierrez, O.M.; Farwell, W.R.; Kermah, D.; Taylor, E.N. Racial differences in the relationship between vitamin D, bone mineral density, and parathyroid hormone in the National Health and Nutrition Examination Survey. *Osteoporos. Int.* **2011**, *22*, 1745–1753. [CrossRef]

50. Redmond, J.; Palla, L.; Yan, L.; Jarjou, L.M.; Prentice, A.; Schoenmakers, I. Ethnic differences in urinary calcium and phosphate excretion between Gambian and British older adults. *Osteoporos. Int.* **2015**, *26*, 1125–1135. [CrossRef]

51. Vieth, R. Weaker bones and white skin as adaptions to improve anthropological "fitness" for northern environments. *Osteoporos. Int.* **2020**, *31*, 617–624. [CrossRef] [PubMed]

52. Powe, C.E.; Evans, M.K.; Wenger, J.; Zonderman, A.B.; Berg, A.H.; Nalls, M.; Tamez, H.; Zhang, D.; Bhan, I.; Karumanchi, S.A.; et al. Vitamin D-binding protein and vitamin D status of black Americans and white Americans. *N. Engl. J. Med.* **2013**, *369*, 1991–2000. [CrossRef] [PubMed]

53. Hollis, B.W.; Bikle, D.D. Vitamin D-binding protein and vitamin D in blacks and whites. *N. Engl. J. Med.* **2014**, *370*, 879–880. [CrossRef] [PubMed]

54. Martin, J.A.; Hamilton, B.E.; Osterman, M.J.K.; Driscoll, A.K.; Drake, P. Births: Final Data for 2016. *Natl. Vital Stat. Rep. Cent. Dis. Control Prev. Natl. Cent. Health Stat. Natl. Vital Stat. Syst.* **2018**, *67*, 1–55.

55. Tangel, V.; White, R.S.; Nachamie, A.S.; Pick, J.S. Racial and Ethnic Disparities in Maternal Outcomes and the Disadvantage of Peripartum Black Women: A Multistate Analysis, 2007–2014. *Am. J. Perinatol.* **2018**. [CrossRef]

56. Gyamfi-Bannerman, C.; Pandita, A.; Miller, E.C.; Boehme, A.K.; Wright, J.D.; Siddiq, Z.; D'Alton, M.E.; Friedman, A.M. Preeclampsia outcomes at delivery and race. *J. Matern. Fetal Neonatal Med.* **2019**, 1–8. [CrossRef]

57. Bediako, P.T.; BeLue, R.; Hillemeier, M.M. A Comparison of Birth Outcomes Among Black, Hispanic, and Black Hispanic Women. *J. Racial Ethn. Health Disparities* **2015**, *2*, 573–582. [CrossRef]

58. Johnson, D.D.; Wagner, C.L.; Hulsey, T.C.; McNeil, R.B.; Ebeling, M.; Hollis, B.W. Vitamin D deficiency and insufficiency is common during pregnancy. *Am. J. Perinatol.* **2011**, *28*, 7–12. [CrossRef]

59. Luque-Fernandez, M.A.; Gelaye, B.; VanderWeele, T.; Ferre, C.; Siega-Riz, A.M.; Holzman, C.; Enquobahrie, D.A.; Dole, N.; Williams, M.A. Seasonal variation of 25-hydroxyvitamin D among non-Hispanic black and white pregnant women from three US pregnancy cohorts. *Paediatr. Perinat. Epidemiol.* **2014**, *28*, 166–176. [CrossRef]

60. Tian, Y.; Holzman, C.; Siega-Riz, A.M.; Williams, M.A.; Dole, N.; Enquobahrie, D.A.; Ferre, C.D. Maternal Serum 25-Hydroxyvitamin D Concentrations during Pregnancy and Infant Birthweight for Gestational Age: A Three-Cohort Study. *Paediatr. Perinat. Epidemiol.* **2016**, *30*, 124–133. [CrossRef]

61. Hollis, B.W.; Johnson, D.; Hulsey, T.C.; Ebeling, M.; Wagner, C.L. Vitamin D supplementation during pregnancy: Double-blind, randomized clinical trial of safety and effectiveness. *J. Bone Miner. Res. Off. J. Am. Soc. Bone Miner. Res.* **2011**, *26*, 2341–2357. [CrossRef] [PubMed]

62. Liu, N.Q.; Hewison, M. Vitamin D, the placenta and pregnancy. *Arch. Biochem. Biophys.* **2012**, *523*, 37–47. [CrossRef] [PubMed]

63. Fogacci, S.; Fogacci, F.; Banach, M.; Michos, E.D.; Hernandez, A.V.; Lip, G.Y.H.; Blaha, M.J.; Toth, P.P.; Borghi, C.; Cicero, A.F.G.; et al. Vitamin D supplementation and incident preeclampsia: A systematic review and meta-analysis of randomized clinical trials. *Clin. Nutr.* **2020**, *39*, 1742–1752. [CrossRef] [PubMed]

64. Bi, W.G.; Nuyt, A.M.; Weiler, H.; Leduc, L.; Santamaria, C.; Wei, S.Q. Association Between Vitamin D Supplementation During Pregnancy and Offspring Growth, Morbidity, and Mortality: A Systematic Review and Meta-analysis. *JAMA Pediatr.* **2018**, *172*, 635–645. [CrossRef]

65. Seto, T.L.; Tabangin, M.E.; Langdon, G.; Mangeot, C.; Dawodu, A.; Steinhoff, M.; Narendran, V. Racial disparities in cord blood vitamin D levels and its association with small-for-gestational-age infants. *J. Perinatol. Off. J. Calif. Perinat. Assoc.* **2016**, *36*, 623–628. [CrossRef]

66. Zhou, S.S.; Tao, Y.H.; Huang, K.; Zhu, B.B.; Tao, F.B. Vitamin D and risk of preterm birth: Up-to-date meta-analysis of randomized controlled trials and observational studies. *J. Obs. Gynaecol. Res.* **2017**, *43*, 247–256. [CrossRef]

67. McDonnell, S.L.; Baggerly, K.A.; Baggerly, C.A.; Aliano, J.L.; French, C.B.; Baggerly, L.L.; Ebeling, M.D.; Rittenberg, C.S.; Goodier, C.G.; Mateus Nino, J.F.; et al. Maternal 25(OH)D concentrations ≥40 ng/mL associated with 60% lower preterm birth risk among general obstetrical patients at an urban medical center. *PLoS ONE* **2017**, *12*, e0180483. [CrossRef]

68. Cannell, J.J. Autism and vitamin D. *Med Hypotheses* **2008**, *70*, 750–759. [CrossRef]

69. Garcia-Serna, A.M.; Morales, E. Neurodevelopmental effects of prenatal vitamin D in humans: Systematic review and meta-analysis. *Mol. Psychiatry* **2020**, *25*, 2468–2481. [CrossRef]

70. Wagner, C.L.; Hollis, B.W. The Implications of Vitamin D Status During Pregnancy on Mother and her Developing Child. *Front. Endocrinol.* **2018**, *9*, 500. [CrossRef]

71. Merewood, A.; Mehta, S.D.; Chen, T.C.; Bauchner, H.; Holick, M.F. Association between vitamin D deficiency and primary cesarean section. *J. Clin. Endocrinol. Metab.* **2009**, *94*, 940–945. [CrossRef] [PubMed]

72. Rostami, M.; Tehrani, F.R.; Simbar, M.; Bidhendi Yarandi, R.; Minooee, S.; Hollis, B.W.; Hosseinpanah, F. Effectiveness of Prenatal Vitamin D Deficiency Screening and Treatment Program: A Stratified Randomized Field Trial. *J. Clin. Endocrinol. Metab.* **2018**, *103*, 2936–2948. [CrossRef] [PubMed]

73. Grant, W.B.; Peiris, A.N. Differences in vitamin D status may account for unexplained disparities in cancer survival rates between African and white Americans. *Dermatoendocrinology* **2012**, *4*, 85–94. [CrossRef] [PubMed]

74. DeSantis, C.E.; Siegel, R.L.; Sauer, A.G.; Miller, K.D.; Fedewa, S.A.; Alcaraz, K.I.; Jemal, A. Cancer statistics for African Americans, 2016: Progress and opportunities in reducing racial disparities. *CA A Cancer J. Clin.* **2016**, *66*, 290–308. [CrossRef] [PubMed]

75. Moukayed, M.; Grant, W.B. Molecular link between vitamin D and cancer prevention. *Nutrients* **2013**, *5*, 3993–4021. [CrossRef]

76. Grant, W.B. An estimate of premature cancer mortality in the U.S. due to inadequate doses of solar ultraviolet-B radiation. *Cancer* **2002**, *94*, 1867–1875. [CrossRef]

77. Grant, W.B.; Garland, C.F. The association of solar ultraviolet B (UVB) with reducing risk of cancer: Multifactorial ecologic analysis of geographic variation in age-adjusted cancer mortality rates. *Anticancer. Res.* **2006**, *26*, 2687–2699.

78. Grant, W.B. Lower vitamin-D production from solar ultraviolet-B irradiance may explain some differences in cancer survival rates. *J. Natl. Med. Assoc.* **2006**, *98*, 357–364.

79. Giovannucci, E.; Liu, Y.; Willett, W.C. Cancer incidence and mortality and vitamin D in black and white male health professionals. *Cancer Epidemiol. Biomark. Prev.* **2006**, *15*, 2467–2472. [CrossRef]

80. Krishnan, A.V.; Feldman, D. Mechanisms of the anti-cancer and anti-inflammatory actions of vitamin D. *Annu. Rev. Pharm. Toxicol.* **2011**, *51*, 311–336. [CrossRef]

81. Christakos, S.; Dhawan, P.; Verstuyf, A.; Verlinden, L.; Carmeliet, G. Vitamin D: Metabolism, Molecular Mechanism of Action, and Pleiotropic Effects. *Physiol. Rev.* **2016**, *96*, 365–408. [CrossRef] [PubMed]

82. Song, M.; Nishihara, R.; Wang, M.; Chan, A.T.; Qian, Z.R.; Inamura, K.; Zhang, X.; Ng, K.; Kim, S.A.; Mima, K.; et al. Plasma 25-hydroxyvitamin D and colorectal cancer risk according to tumour immunity status. *Gut* **2016**, *65*, 296–304. [CrossRef] [PubMed]

83. Garland, C.F.; Garland, F.C. Do sunlight and vitamin D reduce the likelihood of colon cancer? *Int. J. Epidemiol.* **1980**, *9*, 227–231. [CrossRef] [PubMed]

84. Fiscella, K.; Winters, P.; Tancredi, D.; Hendren, S.; Franks, P. Racial disparity in death from colorectal cancer: Does vitamin D deficiency contribute? *Cancer* **2011**, *117*, 1061–1069. [CrossRef] [PubMed]

85. Maalmi, H.; Walter, V.; Jansen, L.; Boakye, D.; Schottker, B.; Hoffmeister, M.; Brenner, H. Association between Blood 25-Hydroxyvitamin D Levels and Survival in Colorectal Cancer Patients: An Updated Systematic Review and Meta-Analysis. *Nutrients* **2018**, *10*, 896. [CrossRef]

86. Zhang, H.; Zhang, H.; Wen, X.; Zhang, Y.; Wei, X.; Liu, T. Vitamin D Deficiency and Increased Risk of Bladder Carcinoma: A Meta-Analysis. *Cell Physiol. Biochem.* **2015**, *37*, 1686–1692. [CrossRef]

87. Lin, G.; Ning, L.; Gu, D.; Li, S.; Yu, Z.; Long, Q.; Hou, L.N.; Tan, W.L. Examining the association of circulating 25-hydroxyvitamin D with kidney cancer risk: A meta-analysis. *Int. J. Clin. Exp. Med.* **2015**, *8*, 20499–20507.

88. Gao, J.; Wei, W.; Wang, G.; Zhou, H.; Fu, Y.; Liu, N. Circulating vitamin D concentration and risk of prostate cancer: A dose-response meta-analysis of prospective studies. *Clin. Risk Manag.* **2018**, *14*, 95–104. [CrossRef]

89. Song, Z.Y.; Yao, Q.; Zhuo, Z.; Ma, Z.; Chen, G. Circulating vitamin D level and mortality in prostate cancer patients: A dose-response meta-analysis. *Endocr. Connect* **2018**, *7*, R294–R303. [CrossRef]

90. Batai, K.; Murphy, A.B.; Ruden, M.; Newsome, J.; Shah, E.; Dixon, M.A.; Jacobs, E.T.; Hollowell, C.M.; Ahaghotu, C.; Kittles, R.A. Race and BMI modify associations of calcium and vitamin D intake with prostate cancer. *BMC Cancer* **2017**, *17*, 64. [CrossRef]

91. Estebanez, N.; Gomez-Acebo, I.; Palazuelos, C.; Llorca, J.; Dierssen-Sotos, T. Vitamin D exposure and Risk of Breast Cancer: A meta-analysis. *Sci. Rep.* **2018**, *8*, 9039. [CrossRef] [PubMed]

92. Visvanathan, K.; Mondul, A.; Zeleniuch-Jacquotte, A.; Mukhtar, T.K.; Smith-Warner, S.A.; Ziegler, R.G. A pooled analysis of 17 cohorts [abstract]. In Proceedings of the Thirty-Seventh Annual CTRC-AACR San Antonio Breast Cancer Symposium, San Antonio, TX, USA, 9–13 December 2014.

93. Palmer, J.R.; Gerlovin, H.; Bethea, T.N.; Bertrand, K.A.; Holick, M.F.; Ruiz-Narvaez, E.N.; Wise, L.A.; Haddad, S.A.; Adams-Campbell, L.L.; Kaufman, H.W.; et al. Predicted 25-hydroxyvitamin D in relation to incidence of breast cancer in a large cohort of African American women. *Breast Cancer Res. BCR* **2016**, *18*, 86. [CrossRef] [PubMed]

94. Qin, B.; Xu, B.; Ji, N.; Yao, S.; Pawlish, K.; Llanos, A.A.M.; Lin, Y.; Demissie, K.; Ambrosone, C.B.; Hong, C.C.; et al. Intake of vitamin D and calcium, sun exposure, and risk of breast cancer subtypes among black women. *Am. J. Clin. Nutr.* **2020**, *111*, 396–405. [CrossRef] [PubMed]

95. Manson, J.E.; Cook, N.R.; Lee, I.M.; Christen, W.; Bassuk, S.S.; Mora, S.; Gibson, H.; Gordon, D.; Copeland, T.; D'Agostino, D.; et al. Vitamin D Supplements and Prevention of Cancer and Cardiovascular Disease. *N. Engl. J. Med.* **2019**, *380*, 33–44. [CrossRef]

96. Keum, N.; Lee, D.H.; Greenwood, D.C.; Manson, J.E.; Giovannucci, E. Vitamin D supplementation and total cancer incidence and mortality: A meta-analysis of randomized controlled trials. *Ann. Oncol.* **2019**, *30*, 733–743. [CrossRef]

97. Chandler, P.D.; Chen, W.Y.; Ajala, O.N.; Hazra, A.; Cook, N.; Bubes, V.; Lee, I.M.; Giovannucci, E.L.; Willett, W.; Buring, J.E.; et al. Effect of Vitamin D3 Supplements on Development of Advanced Cancer: A Secondary Analysis of the VITAL Randomized Clinical Trial. *JAMA Netw. Open* **2020**, *3*, e2025850. [CrossRef]

98. Menke, A.; Casagrande, S.; Geiss, L.; Cowie, C.C. Prevalence of and Trends in Diabetes Among Adults in the United States, 1988–2012. *JAMA* **2015**, *314*, 1021–1029. [CrossRef]

99. Mirhosseini, N.; Vatanparast, H.; Mazidi, M.; Kimball, S.M. Vitamin D Supplementation, Glycemic Control, and Insulin Resistance in Prediabetics: A Meta-Analysis. *J. Endocr. Soc.* **2018**, *2*, 687–709. [CrossRef]

100. Niroomand, M.; Fotouhi, A.; Irannejad, N.; Hosseinpanah, F. Does high-dose vitamin D supplementation impact insulin resistance and risk of development of diabetes in patients with pre-diabetes? A double-blind randomized clinical trial. *Diabetes Res. Clin. Pract.* **2019**, *148*, 1–9. [CrossRef]

101. Kawahara, T. Eldecalcitol, a Vitamin D Analog, for Diabetes Prevention in Impaired Glucose Tolerance (DPVD Study). *Diabetes* **2018**, *67* (Suppl. 1). [CrossRef]

102. Pittas, A.G.; Dawson-Hughes, B.; Sheehan, P.; Ware, J.H.; Knowler, W.C.; Aroda, V.R.; Brodsky, I.; Ceglia, L.; Chadha, C.; Chatterjee, R.; et al. Vitamin D Supplementation and Prevention of Type 2 Diabetes. *N. Engl. J. Med.* **2019**. [CrossRef] [PubMed]

103. Jorde, R.; Sollid, S.T.; Svartberg, J.; Schirmer, H.; Joakimsen, R.M.; Njolstad, I.; Fuskevag, O.M.; Figenschau, Y.; Hutchinson, M.Y. Vitamin D 20,000 IU per Week for Five Years Does Not Prevent Progression From Prediabetes to Diabetes. *J. Clin. Endocrinol. Metab.* **2016**, *101*, 1647–1655. [CrossRef] [PubMed]

104. Dawson-Hughes, B.; Staten, M.A.; Knowler, W.C.; Nelson, J.; Vickery, E.M.; LeBlanc, E.S.; Neff, L.M.; Park, J.; Pittas, A.G.; Investigators, D.d.R.G. Intratrial Exposure to Vitamin D and New-Onset Diabetes Among Adults With Prediabetes: A Secondary Analysis From the Vitamin D and Type 2 Diabetes (D2d) Study. *Diabetes Care* **2020**. [CrossRef] [PubMed]

105. Xu, Y.; Zhou, Y.; Liu, J.; Wang, C.; Qu, Z.; Wei, Z.; Zhou, D. Genetically increased circulating 25(OH)D level reduces the risk of type 2 diabetes in subjects with deficiency of vitamin D: A large-scale Mendelian randomization study. *Medicine* **2020**, *19*, e23672. [CrossRef] [PubMed]

106. Gholami, F.; Moradi, G.; Zareei, B.; Rasouli, M.A.; Nikkhoo, B.; Roshani, D.; Ghaderi, E. The association between circulating 25-hydroxyvitamin D and cardiovascular diseases: A meta-analysis of prospective cohort studies. *BMC Cardiovasc. Disord.* **2019**, *19*, 248. [CrossRef]

107. Schneider, A.L.; Lutsey, P.L.; Selvin, E.; Mosley, T.H.; Sharrett, A.R.; Carson, K.A.; Post, W.S.; Pankow, J.S.; Folsom, A.R.; Gottesman, R.F.; et al. Vitamin D, vitamin D binding protein gene polymorphisms, race and risk of incident stroke: The Atherosclerosis Risk in Communities (ARIC) study. *Eur. J. Neurol.* **2015**, *22*, 1220–1227. [CrossRef] [PubMed]

108. Judd, S.E.; Morgan, C.J.; Panwar, B.; Howard, V.J.; Wadley, V.G.; Jenny, N.S.; Kissela, B.M.; Gutierrez, O.M. Vitamin D deficiency and incident stroke risk in community-living black and white adults. *Int. J. Stroke* **2016**, *11*, 93–102. [CrossRef]

109. Paul, S.; Judd, S.E.; Howard, V.J.; Safford, M.S.; Gutierrez, O.M. Association of 25-hydroxyvitamin D with incident coronary heart disease in the Reasons for Geographic and Racial Differences in Stroke (REGARDS) study. *Am. Heart J.* **2019**, *217*, 140–147. [CrossRef]

110. Afzal, S.; Brondum-Jacobsen, P.; Bojesen, S.E.; Nordestgaard, B.G. Genetically low vitamin D concentrations and increased mortality: Mendelian randomisation analysis in three large cohorts. *BMJ* **2014**, *349*, g6330. [CrossRef]

111. Orkaby, A.R.; Djousse, L.; Manson, J.E. Vitamin D supplements and prevention of cardiovascular disease. *Curr. Opin. Cardiol.* **2019**, *34*, 700–705. [CrossRef]

112. Larsson, S.C.; Traylor, M.; Markus, H.S.; Michaelsson, K. Serum Parathyroid Hormone, 25-Hydroxyvitamin D, and Risk of Alzheimer's Disease: A Mendelian Randomization Study. *Nutrients* **2018**, *10*, 1501. [CrossRef] [PubMed]

113. Jayedi, A.; Rashidy-Pour, A.; Shab-Bidar, S. Vitamin D status and risk of dementia and Alzheimer's disease: A meta-analysis of dose-response. *Nutr. Neurosci.* **2019**, *22*, 750–759. [CrossRef] [PubMed]

114. Miller, J.W.; Harvey, D.J.; Beckett, L.A.; Green, R.; Farias, S.T.; Reed, B.R.; Olichney, J.M.; Mungas, D.M.; DeCarli, C. Vitamin D Status and Rates of Cognitive Decline in a Multiethnic Cohort of Older Adults. *JAMA Neurol.* **2015**, *72*, 1295–1303. [CrossRef]

115. Wang, L.; Qiao, Y.; Zhang, H.; Zhang, Y.; Hua, J.; Jin, S.; Liu, G. Circulating Vitamin D Levels and Alzheimer's Disease: A Mendelian Randomization Study in the IGAP and UK Biobank. *J. Alzheimers Dis.* **2020**, *73*, 609–618. [CrossRef] [PubMed]

116. Irving, A.A.; Marling, S.J.; Seeman, J.; Plum, L.A.; DeLuca, H.F. UV light suppression of EAE (a mouse model of multiple sclerosis) is independent of vitamin D and its receptor. *Proc. Natl. Acad. Sci. USA* **2019**, *116*, 22552–22555. [CrossRef]

117. Orton, S.M.; Wald, L.; Confavreux, C.; Vukusic, S.; Krohn, J.P.; Ramagopalan, S.V.; Herrera, B.M.; Sadovnick, A.D.; Ebers, G.C. Association of UV radiation with multiple sclerosis prevalence and sex ratio in France. *Neurology* **2011**, *76*, 425–431. [CrossRef]

118. Langer-Gould, A.; Lucas, R.; Xiang, A.H.; Chen, L.H.; Wu, J.; Gonzalez, E.; Haraszti, S.; Smith, J.B.; Quach, H.; Barcellos, L.F. MS Sunshine Study: Sun Exposure But Not Vitamin D Is Associated with Multiple Sclerosis Risk in Blacks and Hispanics. *Nutrients* **2018**, *10*, 268. [CrossRef]

119. Gallagher, L.G.; Ilango, S.; Wundes, A.; Stobbe, G.A.; Turk, K.W.; Franklin, G.M.; Linet, M.S.; Freedman, D.M.; Alexander, B.H.; Checkoway, H. Lifetime exposure to ultraviolet radiation and the risk of multiple sclerosis in the US radiologic technologists cohort study. *Mult. Scler.* **2019**, *25*, 1162–1169. [CrossRef]

120. Hedstrom, A.K.; Olsson, T.; Kockum, I.; Hillert, J.; Alfredsson, L. Low sun exposure increases multiple sclerosis risk both directly and indirectly. *J. Neurol.* **2019**, *267*, 1045–1052. [CrossRef]

121. Munger, K.L.; Levin, L.I.; Hollis, B.W.; Howard, N.S.; Ascherio, A. Serum 25-hydroxyvitamin D levels and risk of multiple sclerosis. *JAMA* **2006**, *296*, 2832–2838. [CrossRef]

122. Bistrom, M.; Alonso-Magdalena, L.; Andersen, O.; Jons, D.; Gunnarsson, M.; Vrethem, M.; Hultdin, J.; Sundstrom, P. High serum concentration of vitamin D may protect against multiple sclerosis. *Mult. Scler. J. Exp. Transl. Clin.* **2019**, *5*, 2055217319892291. [CrossRef] [PubMed]

123. Nielsen, N.M.; Munger, K.L.; Koch-Henriksen, N.; Hougaard, D.M.; Magyari, M.; Jorgensen, K.T.; Lundqvist, M.; Simonsen, J.; Jess, T.; Cohen, A.; et al. Neonatal vitamin D status and risk of multiple sclerosis: A population-based case-control study. *Neurology* **2017**, *88*, 44–51. [CrossRef] [PubMed]

124. Munger, K.L.; Zhang, S.M.; O'Reilly, E.; Hernan, M.A.; Olek, M.J.; Willett, W.C.; Ascherio, A. Vitamin D intake and incidence of multiple sclerosis. *Neurology* **2004**, *62*, 60–65. [CrossRef] [PubMed]

125. Yeh, W.Z.; Gresle, M.; Jokubaitis, V.; Stankovich, J.; van der Walt, A.; Butzkueven, H. Immunoregulatory effects and therapeutic potential of vitamin D in multiple sclerosis. *Br. J. Pharmacol.* **2020**, *177*, 4113–4133. [CrossRef] [PubMed]

126. Rhead, B.; Baarnhielm, M.; Gianfrancesco, M.; Mok, A.; Shao, X.; Quach, H.; Shen, L.; Schaefer, C.; Link, J.; Gyllenberg, A.; et al. Mendelian randomization shows a causal effect of low vitamin D on multiple sclerosis risk. *Neurol. Genet.* **2016**, *2*, e97. [CrossRef]

127. Baeke, F.; Takiishi, T.; Korf, H.; Gysemans, C.; Mathieu, C. Vitamin D: Modulator of the immune system. *Curr. Opin. Pharm.* **2010**, *10*, 482–496. [CrossRef]

128. Meftahi, G.H.; Jangravi, Z.; Sahraei, H.; Bahari, Z. The possible pathophysiology mechanism of cytokine storm in elderly adults with COVID-19 infection: The contribution of "inflame-aging". *Inflamm. Res.* **2020**, *69*, 825–839. [CrossRef]

129. Cannell, J.J.; Vieth, R.; Umhau, J.C.; Holick, M.F.; Grant, W.B.; Madronich, S.; Garland, C.F.; Giovannucci, E. Epidemic influenza and vitamin D. *Epidemiol. Infect.* **2006**, *134*, 1129–1140. [CrossRef]

130. Grant, W.B.; Giovannucci, E. The possible roles of solar ultraviolet-B radiation and vitamin D in reducing case-fatality rates from the 1918-1919 influenza pandemic in the United States. *Dermatoendocrinology* **2009**, *1*, 215–219. [CrossRef]

131. Martineau, A.R.; Jolliffe, D.A.; Hooper, R.L.; Greenberg, L.; Aloia, J.F.; Bergman, P.; Dubnov-Raz, G.; Esposito, S.; Ganmaa, D.; Ginde, A.A.; et al. Vitamin D supplementation to prevent acute respiratory tract infections: Systematic review and meta-analysis of individual participant data. *BMJ* **2017**, *356*, i6583. [CrossRef]

132. Aloia, J.F.; Li-Ng, M. Re: Epidemic influenza and vitamin D. *Epidemiol. Infect.* **2007**, *135*, 1095–1096. [CrossRef] [PubMed]

133. Mercola, J.; Grant, W.B.; Wagner, C.L. Evidence Regarding Vitamin D and Risk of COVID-19 and Its Severity. *Nutrients* **2020**, *12*, 3361. [CrossRef] [PubMed]

134. Millett, G.A.; Jones, A.T.; Benkeser, D.; Baral, S.; Mercer, L.; Beyrer, C.; Honermann, B.; Lankiewicz, E.; Mena, L.; Crowley, J.S.; et al. Assessing Differential Impacts of COVID-19 on Black Communities. *Ann. Epidemiol.* **2020**. [CrossRef] [PubMed]

135. CDC. Health Equity Considerations and Racial and Ethnic Minority Groups. Available online: https://www.cdc.gov/coronavirus/2019-ncov/need-extra-precautions/racial-ethnic-minorities.html (accessed on 6 September 2020).

136. Baktash, V.; Hosack, T.; Patel, N.; Shah, S.; Kandiah, P.; Van Den Abbeele, K.; Mandal, A.K.J.; Missouris, C.G. Vitamin D status and outcomes for hospitalised older patients with COVID-19. *Postgrad. Med. J.* **2020**. [CrossRef] [PubMed]

137. Panagiotou, G.; Tee, S.A.; Ihsan, Y.I.; Athar, W.; Marchitelli, G.; Kelly, D.; Boot, C.S.; Stock, N.; Macfarlane, J.; Martineau, A.R.; et al. Low serum 25-hydroxyvitamin D (25[OH]D) levels in patients hospitalised with COVID-19 are associated with greater disease severity: Results of a local audit of practice. *Clin. Endocrinol.* **2020**. [CrossRef]

138. Radujkovic, A.; Hippchen, T.; Tiwari-Heckler, S.; Dreher, S.; Boxberger, M.; Merle, U. Vitamin D Deficiency and Outcome of COVID-19 Patients. *Nutrients* **2020**, *12*, 2757. [CrossRef] [PubMed]

139. French, C.B.; McDonnell, S.L.; Vieth, R. 25-Hydroxyvitamin D variability within-person due to diurnal rhythm and illness: A case report. *J. Med. Case Rep.* **2019**, *13*, 29. [CrossRef]

140. Smolders, J.; van den Ouweland, J.; Geven, C.; Pickkers, P.; Kox, M. Letter to the Editor: Vitamin D deficiency in COVID-19: Mixing up cause and consequence. *Metabolism* **2020**, 154434. [CrossRef]

141. Meltzer, D.O.; Best, T.J.; Zhang, H.; Vokes, T.; Arora, V.; Solway, J. Association of Vitamin D Status and Other Clinical Characteristics With COVID-19 Test Results. *JAMA Netw. Open* **2020**, *3*, e2019722. [CrossRef]

142. Kaufman, H.W.; Niles, J.K.; Kroll, M.H.; Bi, C.; Holick, M.F. SARS-CoV-2 positivity rates associated with circulating 25-hydroxyvitamin D levels. *PLoS ONE* **2020**, *15*, e0239252. [CrossRef]

143. Wang, R.; DeGruttola, V.; Ler, K.H.; Mayer, K.H.; Redline, S.; Hazra, A.; Mora, S.; Willett, W.C.; Gammaa, D.; Manson, J.E. The Vitamin D for COVID-19 (VIVID) Trial: A Pragmatic Cluster-Randomized Design. *Contemp. Clin. Trials* **2021**, *100*, 106176. [CrossRef] [PubMed]

144. Entrenas Castillo, M.; Entrenas Costa, L.M.; Vaquero Barrios, J.M.; Alcala Diaz, J.F.; Miranda, J.L.; Bouillon, R.; Quesada Gomez, J.M. "Effect of Calcifediol Treatment and best Available Therapy versus best Available Therapy on Intensive Care Unit Admission and Mortality Among Patients Hospitalized for COVID-19: A Pilot Randomized Clinical study". *J. Steroid. Biochem. Mol. Biol.* **2020**, 105751. [CrossRef] [PubMed]

145. Rastogi, A.; Bhansali, A.; Khare, N.; Suri, V.; Yaddanapudi, N.; Sachdeva, N.; Puri, G.D.; Malhotra, P. Short term, high-dose vitamin D supplementation for COVID-19 disease: A randomised, placebo-controlled, study (SHADE study). *Postgrad. Med. J.* **2020**. [CrossRef] [PubMed]

146. ClinicalTrials.Gov. Studies for Vitamin D, COVID19. Available online: https://clinicaltrials.gov/ct2/results?cond=COVID19&term=vitamin+D&cntry=&state=&city=&dist= (accessed on 30 December 2020).

147. Wolsk, H.M.; Chawes, B.L.; Litonjua, A.A.; Hollis, B.W.; Waage, J.; Stokholm, J.; Bonnelykke, K.; Bisgaard, H.; Weiss, S.T. Prenatal vitamin D supplementation reduces risk of asthma/recurrent wheeze in early childhood: A combined analysis of two randomized controlled trials. *PLoS ONE* **2017**, *12*, e0186657. [CrossRef] [PubMed]

148. Wolsk, H.M.; Harshfield, B.J.; Laranjo, N.; Carey, V.J.; O'Connor, G.; Sandel, M.; Strunk, R.C.; Bacharier, L.B.; Zeiger, R.S.; Schatz, M.; et al. Vitamin D supplementation in pregnancy, prenatal 25(OH)D levels, race, and subsequent asthma or recurrent wheeze in offspring: Secondary analyses from the Vitamin D Antenatal Asthma Reduction Trial. *J. Allergy Clin. Immunol.* **2017**, *140*, 1423–1429 e1425. [CrossRef] [PubMed]

149. Jolliffe, D.A.; Greenberg, L.; Hooper, R.L.; Griffiths, C.J.; Camargo, C.A., Jr.; Kerley, C.P.; Jensen, M.E.; Mauger, D.; Stelmach, I.; Urashima, M.; et al. Vitamin D supplementation to prevent asthma exacerbations: A systematic review and meta-analysis of individual participant data. *Lancet. Respir. Med.* **2017**, *5*, 881–890. [CrossRef]

150. Cunningham, T.J.; Croft, J.B.; Liu, Y.; Lu, H.; Eke, P.I.; Giles, W.H. Vital Signs: Racial Disparities in Age-Specific Mortality Among Blacks or African Americans—United States, 1999–2015. *Mmwr. Morb. Mortal. Wkly. Rep.* **2017**, *66*, 444–456. [CrossRef]

151. Ford, M.E.; Magwood, G.; Brown, E.T.; Cannady, K.; Gregoski, M.; Knight, K.D.; Peterson, L.L.; Kramer, R.; Evans-Knowell, A.; Turner, D.P. Disparities in Obesity, Physical Activity Rates, and Breast Cancer Survival. *Adv. Cancer Res.* **2017**, *133*, 23–50. [CrossRef]

152. Musemwa, N.; Gadegbeku, C.A. Hypertension in African Americans. *Curr. Cardiol. Rep.* **2017**, *19*, 129. [CrossRef]

153. Reeder-Hayes, K.E.; Troester, M.A.; Meyer, A.M. Reducing Racial Disparities in Breast Cancer Care: The Role of 'Big Data'. *Oncology (Williston Park)* **2017**, *31*, 756–762.

154. Singh, G.K.; Daus, G.P.; Allender, M.; Ramey, C.T.; Martin, E.K.; Perry, C.; Reyes, A.A.L.; Vedamuthu, I.P. Social Determinants of Health in the United States: Addressing Major Health Inequality Trends for the Nation, 1935-2016. *Int. J. MCH AIDS* **2017**, *6*, 139–164. [CrossRef] [PubMed]

155. Wilcox, S.; Sharpe, P.A.; Liese, A.D.; Dunn, C.G.; Hutto, B. Socioeconomic factors associated with diet quality and meeting dietary guidelines in disadvantaged neighborhoods in the Southeast United States. *Ethn. Health* **2018**, 1–17. [CrossRef]

156. Gaksch, M.; Jorde, R.; Grimnes, G.; Joakimsen, R.; Schirmer, H.; Wilsgaard, T.; Mathiesen, E.B.; Njolstad, I.; Lochen, M.L.; Marz, W.; et al. Vitamin D and mortality: Individual participant data meta-analysis of standardized 25-hydroxyvitamin D in 26916 individuals from a European consortium. *PLoS ONE* **2017**, *12*, e0170791. [CrossRef] [PubMed]

157. Garland, C.F.; Kim, J.J.; Mohr, S.B.; Gorham, E.D.; Grant, W.B.; Giovannucci, E.L.; Baggerly, L.; Hofflich, H.; Ramsdell, J.W.; Zeng, K.; et al. Meta-analysis of all-cause mortality according to serum 25-hydroxyvitamin D. *Am. J. Public Health* **2014**, *104*, e43–e50. [CrossRef] [PubMed]

158. Meng, X.; Li, X.; Timofeeva, M.N.; He, Y.; Spiliopoulou, A.; Wei, W.Q.; Gifford, A.; Wu, H.; Varley, T.; Joshi, P.; et al. Phenome-wide Mendelian-randomization study of genetically determined vitamin D on multiple health outcomes using the UK Biobank study. *Int. J. Epidemiol.* **2019**, *48*, 1425–1434. [CrossRef]

159. Kritchevsky, S.B.; Tooze, J.A.; Neiberg, R.H.; Schwartz, G.G.; Hausman, D.B.; Johnson, M.A.; Bauer, D.C.; Cauley, J.A.; Shea, M.K.; Cawthon, P.M.; et al. 25-Hydroxyvitamin D, parathyroid hormone, and mortality in black and white older adults: The health ABC study. *J. Clin. Endocrinol. Metab.* **2012**, *97*, 4156–4165. [CrossRef]

160. Signorello, L.B.; Han, X.; Cai, Q.; Cohen, S.S.; Cope, E.L.; Zheng, W.; Blot, W.J. A prospective study of serum 25-hydroxyvitamin d levels and mortality among African Americans and non-African Americans. *Am. J. Epidemiol.* **2013**, *177*, 171–179. [CrossRef]

161. Schleicher, R.L.; Sternberg, M.R.; Looker, A.C.; Yetley, E.A.; Lacher, D.A.; Sempos, C.T.; Taylor, C.L.; Durazo-Arvizu, R.A.; Maw, K.L.; Chaudhary-Webb, M.; et al. National Estimates of Serum Total 25-Hydroxyvitamin D and Metabolite Concentrations Measured by Liquid Chromatography-Tandem Mass Spectrometry in the US Population during 2007–2010. *J. Nutr.* **2016**, *146*, 1051–1061. [CrossRef]

162. Heaney, R.P. Guidelines for optimizing design and analysis of clinical studies of nutrient effects. *Nutr. Rev.* **2014**, *72*, 48–54. [CrossRef]

163. Grant, W.B.; Boucher, B.J.; Bhattoa, H.P.; Lahore, H. Why vitamin D clinical trials should be based on 25-hydroxyvitamin D concentrations. *J. Steroid Biochem. Mol. Biol.* **2018**, *177*, 266–269. [CrossRef]

164. Rollins, L.; Sy, A.; Crowell, N.; Rivers, D.; Miller, A.; Cooper, P.; Teague, D.; Jackson, C.; Henry Akintobi, T.; Ofili, E. Learning and Action in Community Health: Using the Health Belief Model to Assess and Educate African American Community Residents about Participation in Clinical Research. *Int. J. Environ. Res. Public Health* **2018**, *15*, 1862. [CrossRef] [PubMed]

165. Grant, W.B.; Boucher, B.J. Why Secondary Analyses in Vitamin D Clinical Trials Are Important and How to Improve Vitamin D Clinical Trial Outcome Analyses-A Comment on "Extra-Skeletal Effects of Vitamin D". *Nutrients* **2019**, *11*, 2182. [CrossRef]

166. McDonnell, S.L.; Baggerly, C.A.; French, C.B.; Baggerly, L.L.; Garland, C.F.; Gorham, E.D.; Hollis, B.W.; Trump, D.L.; Lappe, J.M. Breast cancer risk markedly lower with serum 25-hydroxyvitamin D concentrations ≥ 60 vs <20 ng/mL (150 vs 50 nmol/L): Pooled analysis of two randomized trials and a prospective cohort. *PLoS ONE* **2018**, *13*, e0199265. [CrossRef]

167. Migliaccio, S.; Di Nisio, A.; Mele, C.; Scappaticcio, L.; Savastano, S.; Colao, A.; Obesity Programs of nutrition, E.R.; Assessment, G. Obesity and hypovitaminosis D: Causality or casualty? *Int. J. Obes. Suppl.* **2019**, *9*, 20–31. [CrossRef] [PubMed]

168. Freedman, D.M.; Cahoon, E.K.; Rajaraman, P.; Major, J.M.; Doody, M.M.; Alexander, B.H.; Hoffbeck, R.W.; Kimlin, M.G.; Graubard, B.I.; Linet, M.S. Sunlight and other determinants of circulating 25-hydroxyvitamin D levels in black and white participants in a nationwide U.S. study. *Am. J. Epidemiol.* **2013**, *177*, 180–192. [CrossRef]

169. Bailey, R.K.; Fileti, C.P.; Keith, J.; Tropez-Sims, S.; Price, W.; Allison-Ottey, S.D. Lactose intolerance and health disparities among African Americans and Hispanic Americans: An updated consensus statement. *J. Natl. Med. Assoc.* **2013**, *105*, 112–127. [CrossRef]

170. McCullough, M.L.; Zoltick, E.S.; Weinstein, S.J.; Fedirko, V.; Wang, M.; Cook, N.R.; Eliassen, A.H.; Zeleniuch-Jacquotte, A.; Agnoli, C.; Albanes, D.; et al. Circulating Vitamin D and Colorectal Cancer Risk: An International Pooling Project of 17 Cohorts. *J. Natl. Cancer Inst.* **2019**, *111*, 158–169. [CrossRef]

171. Bischoff-Ferrari, H.A.; Kiel, D.P.; Dawson-Hughes, B.; Orav, J.E.; Li, R.; Spiegelman, D.; Dietrich, T.; Willett, W.C. Dietary calcium and serum 25-hydroxyvitamin D status in relation to BMD among U.S. adults. *J. Bone Miner. Res. Off. J. Am. Soc. Bone Miner. Res.* **2009**, *24*, 935–942. [CrossRef]

172. McDonnell, S.L.; Baggerly, C.; French, C.B.; Baggerly, L.L.; Garland, C.F.; Gorham, E.D.; Lappe, J.M.; Heaney, R.P. Serum 25-Hydroxyvitamin D Concentrations ≥40 ng/ml Are Associated with >65% Lower Cancer Risk: Pooled Analysis of Randomized Trial and Prospective Cohort Study. *PLoS ONE* **2016**, *11*, e0152441. [CrossRef]

173. Mirhosseini, N.; Vatanparast, H.; Kimball, S.M. The Association between Serum 25(OH)D Status and Blood Pressure in Participants of a Community-Based Program Taking Vitamin D Supplements. *Nutrients* **2017**, *9*, 1244. [CrossRef]

174. Raed, A.; Bhagatwala, J.; Zhu, H.; Pollock, N.K.; Parikh, S.J.; Huang, Y.; Havens, R.; Kotak, I.; Guo, D.H.; Dong, Y. Dose responses of vitamin D3 supplementation on arterial stiffness in overweight African Americans with vitamin D deficiency: A placebo controlled randomized trial. *PLoS ONE* **2017**, *12*, e0188424. [CrossRef] [PubMed]

175. Ng, K.; Scott, J.B.; Drake, B.F.; Chan, A.T.; Hollis, B.W.; Chandler, P.D.; Bennett, G.G.; Giovannucci, E.L.; Gonzalez-Suarez, E.; Meyerhardt, J.A.; et al. Dose response to vitamin D supplementation in African Americans: Results of a 4-arm, randomized, placebo-controlled trial. *Am. J. Clin. Nutr.* **2014**, *99*, 587–598. [CrossRef] [PubMed]

176. Garrett-Mayer, E.; Wagner, C.L.; Hollis, B.W.; Kindy, M.S.; Gattoni-Celli, S. Vitamin D3 supplementation (4000 IU/d for 1 y) eliminates differences in circulating 25-hydroxyvitamin D between African American and white men. *Am. J. Clin. Nutr.* **2012**, *96*, 332–336. [CrossRef] [PubMed]

177. Kantor, E.D.; Rehm, C.D.; Du, M.; White, E.; Giovannucci, E.L. Trends in Dietary Supplement Use Among US Adults From 1999–2012. *JAMA* **2016**, *316*, 1464–1474. [CrossRef]

178. Carlberg, C.; Haq, A. The concept of the personal vitamin D response index. *J. Steroid Biochem. Mol. Biol.* **2018**, *175*, 12–17. [CrossRef]

179. Carlberg, C. Nutrigenomics of Vitamin D. *Nutrients* **2019**, *11*, 676. [CrossRef]

180. Malihi, Z.; Wu, Z.; Lawes, C.M.M.; Scragg, R. Adverse events from large dose vitamin D supplementation taken for one year or longer. *J. Steroid Biochem. Mol. Biol.* **2019**, *188*, 29–37. [CrossRef]

181. Araki, T.; Holick, M.F.; Alfonso, B.D.; Charlap, E.; Romero, C.M.; Rizk, D.; Newman, L.G. Vitamin D intoxication with severe hypercalcemia due to manufacturing and labeling errors of two dietary supplements made in the United States. *J. Clin. Endocrinol. Metab.* **2011**, *96*, 3603–3608. [CrossRef]

182. Taylor, P.N.; Davies, J.S. A review of the growing risk of vitamin D toxicity from inappropriate practice. *Br. J. Clin. Pharmacol.* **2018**, *84*, 1121–1127. [CrossRef]

183. Schulster, M.L.; Goldfarb, D.S. Vitamin D and Kidney Stones. *Urology* **2020**, *139*, 1–7. [CrossRef]

Article

Effect of Monthly Vitamin D Supplementation on Preventing Exacerbations of Asthma or Chronic Obstructive Pulmonary Disease in Older Adults: Post Hoc Analysis of a Randomized Controlled Trial

Carlos A. Camargo, Jr. [1,*], Les Toop [2], John Sluyter [3], Carlene M. M. Lawes [3], Debbie Waayer [3], Kay-Tee Khaw [4], Adrian R. Martineau [5] and Robert Scragg [3]

1   Department of Emergency Medicine, Massachusetts General Hospital, Harvard Medical School, Boston, MA 02114, USA
2   Department of General Practice, University of Otago, Christchurch 8140, New Zealand; les.toop@otago.ac.nz
3   School of Population Health, University of Auckland, Auckland 1142, New Zealand; j.sluyter@auckland.ac.nz (J.S.); Carlene.Lawes@waitematadhb.govt.nz (C.M.M.L.); d.waayer@auckland.ac.nz (D.W.); r.scragg@auckland.ac.nz (R.S.)
4   Department of Public Health and Primary Care, University of Cambridge, Cambridge CB2 2QQ, UK; kk101@cam.ac.uk
5   Institute for Population Health Sciences, Barts and The London School of Medicine and Dentistry, Queen Mary University of London, London E1 2AB, UK; a.martineau@qmul.ac.uk
*   Correspondence: ccamargo@partners.org

Citation: Camargo, C.A., Jr.; Toop, L.; Sluyter, J.; Lawes, C.M.M.; Waayer, D.; Khaw, K.-T.; Martineau, A.R.; Scragg, R. Effect of Monthly Vitamin D Supplementation on Preventing Exacerbations of Asthma or Chronic Obstructive Pulmonary Disease in Older Adults: Post Hoc Analysis of a Randomized Controlled Trial. *Nutrients* **2021**, *13*, 521. https://doi.org/10.3390/nu13020521

Academic Editors: Spyridon N. Karras and Pawel Pludowski
Received: 24 December 2020
Accepted: 1 February 2021
Published: 6 February 2021

Publisher's Note: MDPI stays neutral with regard to jurisdictional claims in published maps and institutional affiliations.

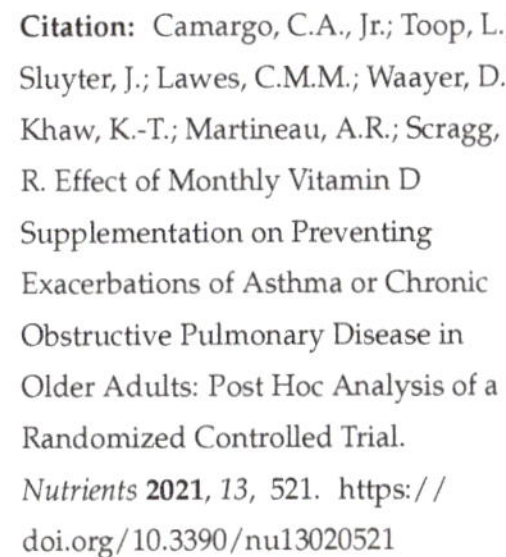

Abstract: Randomized controlled trials have suggested that vitamin D supplementation can prevent asthma and chronic obstructive pulmonary disease (COPD) exacerbations. For COPD, the benefit appears to be limited to individuals with baseline 25-hydroxyvitamin D (25OHD) levels <25 nmol/L. We performed a post hoc analysis of data from a randomized, double-blinded, placebo-controlled trial to investigate the effect that monthly, high-dose vitamin D supplementation (versus placebo) had on older adults with asthma and/or COPD. Specifically, we investigated whether vitamin D supplementation prevented exacerbations of these conditions. Participants were randomly assigned either to an initial oral dose of 200,000 IU vitamin D3 followed by 100,000 IU monthly or to placebo, with an average follow-up period of 3.3 years. Among the 5110 participants, 775 had asthma or COPD at the beginning of the study, and were eligible for inclusion in this analysis. Exacerbations were defined by the prescription of a short-burst of oral corticosteroids. The mean age of the participants was 67 years old, and 56% were male. The mean baseline blood 25OHD level was 63 nmol/L; 2.3% were <25 nmol/L. Overall, we found that vitamin D supplementation did not affect the exacerbation risk (hazard ratio 1.08; 95%CI 0.84–1.39). Among those with baseline 25OHD <25 nmol/L, however, the hazard ratio was 0.11 (95%CI 0.02–0.51); *p* for interaction = 0.001. Although monthly vitamin D supplementation had no overall impact on risk of exacerbations of asthma or COPD, we found evidence of a probable benefit among those with severe vitamin D deficiency.

Keywords: asthma; chronic obstructive pulmonary disease; exacerbations; oral corticosteroids; randomized controlled trial; supplement; vitamin D

1. Introduction

Among older adults, asthma and chronic obstructive pulmonary disease (COPD) are two major obstructive airway diseases that can be difficult to distinguish, and that may coexist as asthma-COPD overlap syndrome [1]. Both are chronic diseases characterized by periodic exacerbations, which are usually triggered by respiratory virus infections [2]. The treatment of asthma and COPD exacerbations is addressed in international management guidelines [3,4]. The guidelines also address interventions (such as inhaled corticosteroids) that aim to prevent exacerbations, which remain a major cause of morbidity, mortality, and increased healthcare costs.

Given the beneficial effect that vitamin D supplementation has on acute respiratory infections (ARI) [5], a growing number of researchers have investigated whether vitamin D supplements could potentially prevent exacerbations of asthma or COPD [6]. In a 2017 individual participant data (IPD) meta-analysis of all known randomized controlled trials (RCTs) worldwide, Jolliffe and colleagues demonstrated that vitamin D supplements can reduce the risk of asthma exacerbations requiring systemic corticosteroids, especially among those with baseline serum hydroxyvitamin D (25OHD) levels <25 nmol/L [7]. In contrast, vitamin D supplementation had no overall effect on the risk of COPD exacerbations among older individuals [8]. However, a prespecified subgroup analysis did show benefits from vitamin D supplements among COPD patients with baseline 25OHD levels of <25 nmol/L. Indeed, all three trials in this IPD meta-analysis reported the same subgroup finding [9–11]. A fourth trial with only 88 participants reported an overall protective effect of vitamin D supplementation [12].

Nevertheless, questions remain. We performed a large RCT of monthly high-dose vitamin D supplements (versus placebo) in >5000 older adults in New Zealand. In the current, prespecified analysis, we investigated the effect of vitamin D supplementation on the subgroup of participants with asthma or COPD. Specifically, we investigated whether taking the monthly vitamin D supplement reduced exacerbations overall, or at least in the subgroup with baseline 25OHD <25 nmol/L.

2. Materials and Methods

Older adults with asthma or COPD came from a large, population-based RCT of vitamin D supplementation called the Vitamin D Assessment (ViDA) study; this trial was carried out in Auckland, New Zealand. Full details of the study methods [13], and the findings on the main outcomes of cardiovascular disease [14], falls/fractures [15], and ARI [16] have been published. The New Zealand Multi-Region Ethics Committee, Wellington, approved the trial in October 2010 (MEC/09/08/082), and the main outcomes were registered with the Australian New Zealand Clinical Trials Registry in April 2011 (ACTRN12611000402943). All participants gave their written informed consent.

The trial investigators invited 5250 people, mainly from family practices in Auckland ($n = 5107$) and a small number from community groups ($n = 143$), who underwent baseline assessments at the School of Population Health, Tamaki Campus, University of Auckland, between 5 April, 2011 and 6 November, 2012. The participants were asked questions about their demographic status, lifestyle (including smoking and physical activity), intake of vitamin D supplements within the study inclusion criteria ($\leq$600 IU per day if aged 50–70 years; $\leq$800 IU per day if aged 71–84 years), and past medical history, as informed by a doctor (including asthma, COPD, and other medical disorders) [13]. The weight (to the nearest 0.1 kg) and height (to the nearest 0.1 cm) of the participants were measured in the research clinic.

Spirometry was performed using a KoKo Trek spirometer (nSpire Health; Longmont, CO, USA) with participants in a seated position, maximally inhaling and then forcibly exhaling while watching a clock on a computer screen for at least 6 s. Only three efforts were performed due to time constraints as a result of the large sample size, and to avoid exhaustion in the elderly participants. All other spirometry recommendations were fulfilled [17]. The maximum values of forced expiratory volume in 1 s (FEV1; in L), forced vital capacity (FVC; in L), and their ratio (FEV1/FVC) were derived from the three efforts.

A blood sample was collected and immediately centrifuged for an initial measurement of corrected serum calcium (those >2.50 mmol/L were excluded). The remaining serum was stored at $-80\ °C$ for later measurement of serum 25OHD, using liquid chromatography–tandem mass spectrometry (API 4000 by SCIEX; Framingham, MA, USA) with 12.7% interassay coefficient of variation, by a local laboratory participating in the Vitamin D

External Quality Assessment Scheme program (www.deqas.org). The baseline 25OHD was deseasonalized using standard methods [18].

After the baseline assessment, participants were mailed a run-in questionnaire with a masked placebo capsule. On return of the questionnaire within four weeks, 5110 participants were randomized by the study statistician—within random block sizes of 8, 10, or 12, and stratified by ethnic origin (Māori, Pacific Island, South Asian, European, or other) and 5-year age groups—to receive identical-looking softgel capsules containing either vitamin D_3 (100,000 IU) or placebo. The capsules were provided by Tishcon Corporation (Westbury, NY, USA).

The capsules were mailed monthly to the homes of the participants, with two capsules in the first letter (an initial bolus of 200,000 IU vitamin D_3 or placebo), and thereafter, one capsule monthly (100,000 IU vitamin D_3 or placebo). Two participants (both from the placebo arm) withdrew during the follow-up period, which ended on 31 July 2015, so that 2558 participants received vitamin D_3 and 2550 received placebo capsules.

New Zealand residents have a unique National Health Index (NHI) number. This number was used to link each participant to information held by the Ministry of Health on all dispensed prescriptions from two years before the assessment until the end of the follow-up on 31 July 2015. This included information on the medication, dose, number of tablets or volume dispensed, and date of dispensing. Data on all medications used for the management of asthma or COPD were extracted; these included short-acting β-agonists (SABA)—inhaled or not inhaled, inhaled long-acting β-agonists (LABA), inhaled short or long-acting muscarinic antagonists (SAMA or LAMA), inhaled combination SABA and SAMA agents, inhaled mast cell stabilizers, inhaled corticosteroids (ICS) with or without LABA, and oral corticosteroids. No participants with asthma or COPD received a prescription for inhaled orciprenaline, injected bronchodilators (such as aminophylline), or injected corticosteroids. The Ministry of Health also provided information about any deaths in the cohort during the follow-up period.

Asthma identification: Participants were asked at their baseline assessment, "Have you ever been told by a doctor that you have asthma?" Those who answered "Yes" and had been dispensed a prescription for ICS, SABA, or LABA at any time from 12 months before randomization to 36 months after, were defined as having asthma and were included in the current analysis.

COPD identification: Given the well-known underdiagnosis of COPD in the general population [19,20], spirometry was used to identify participants who had a ratio of forced expiratory volume in one second (FEV_1), divided by forced vital capacity (FVC) of <0.70, and who had smoked >100 cigarettes in their whole life, regardless of whether they were current or former smokers. These individuals were defined as having COPD, and were included in the current analysis.

Primary outcome: Exacerbations of asthma or COPD were identified by any prescription of oral corticosteroids more than 20 days apart for a short period (e.g., several days), consistent with dosing regimens recommended in international management guidelines [3,4], after joint adjudication by two senior clinicians (CAC, LT) who were blinded to the treatment allocation. Any participants selected according to the above asthma or COPD identification criteria were excluded if they had medical conditions commonly treated by oral corticosteroids—e.g., doctor-diagnosed arthritis reported at baseline, except for osteoarthritis; and doctor-diagnosed inflammatory bowel disease or multiple sclerosis reported at baseline or in the final end-of-study questionnaire in July 2015.

Figure 1 shows that 1420 participants were selected by the initial screen of either having doctor-diagnosed asthma or a FEV_1/FVC ratio <0.70. After excluding participants with missing FEV_1/FVC ratios, asthma participants not dispensed an inhaled asthma medication, and COPD participants that had missing information on their smoking status or those who had never been smokers, 983 participants remained eligible. Of these, a further 208 were excluded for having conditions commonly treated by oral corticosteroids. This left an analytic sample of 775 participants for the post hoc analysis, with 214 having asthma only, 356 having COPD only, and 205 having both conditions; 30 of these participants died during follow-up. This selection was made without knowledge of the treatment allocation.

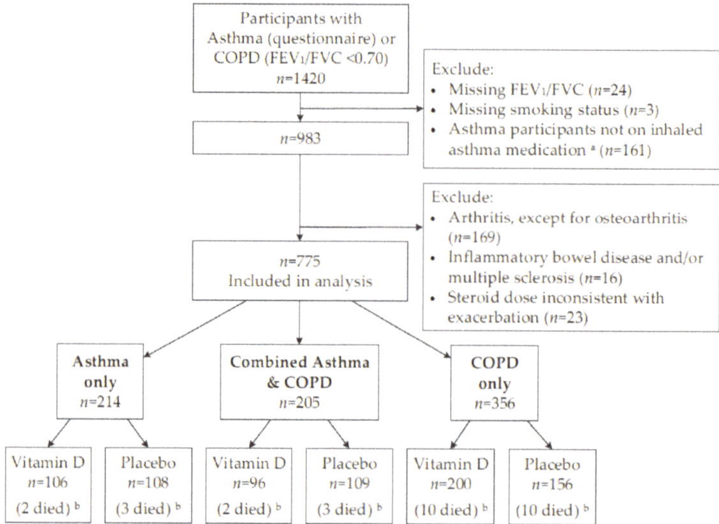

Figure 1. Flow diagram for selecting participants with asthma or chronic obstructive pulmonary disease in the ViDA study. COPD = chronic obstructive pulmonary disease; FEV_1 = forced expiratory volume in 1 s; FVC = forced vital capacity; SABA = short-acting β-agonist; LABA = long-acting β-agonist. ᵃ Dispensed inhaled corticosteroids, SABA or LABA at any time from 12 months before randomisation, to 36 months after. ᵇ Died by 31 July 2015.

Statistical Analysis

Data were analyzed using SAS (Cary, NC, USA, version 9.4). Chi-square tests and t-tests were used to compare proportions and means, respectively. The PHREG procedure was used to calculate hazard ratios (HR) for repeated prescriptions of oral corticosteroids, using the mean cumulative function overlay for comparison groups, adjusted for covariates. The mean cumulative function is the average cumulative number of prescriptions at a time point during follow-up. Effect modification was assessed by the creation of interaction terms in models that also included the main effects for both variables in the regression model. Incidence rates were calculated by dividing the number of oral steroid prescriptions by person-time. *p*-values were not corrected to account for the multiple hypothesis tests because, given the known heterogeneity in the effectiveness of vitamin D supplementation in preventing ARI and asthma or COPD exacerbations [5,7,8], we did not want to miss any potentially important findings [21]. A two-sided $p < 0.05$ was considered statistically significant.

3. Results

The selected sample of participants had a mean (SD) age of 66.6 (8.3) years and 56.4% were men. Most participants were of European/Other ethnicity (82.7%) and well-educated, with 49.7% having attended tertiary education and 48.7% still in paid employment. Only 12.4% were current tobacco smokers, and a high proportion were former smokers (62.7%). Overall, the mean (SD) FEV_1 was 2.16 (0.70) L, FVC was 3.24 (0.96) L, and ratio of FEV_1/FVC was 0.67 (0.10); the observed baseline 25OHD was 62.5 (23.7) nmol/L.

With regard to disease severity, 74% of participants with asthma only were taking a long-term controller medication (e.g., inhaled corticosteroids) at baseline. Among all participants with COPD, the baseline mean (SD) FEV_1 was 2.09 (0.69) L, FVC was 3.31

(0.96) L, and ratio of FEV_1/FVC was 0.63 (0.07). Most COPD patients were in GOLD stages 1 and 2: 226 (40%) in stage 1, 271 (48%) in stage 2, 58 (10%) in stage 3, and 6 (1%) in stage 4.

Table 1 compares baseline characteristics between the vitamin D and placebo groups. There were no between-group differences in distributions of demographic or lifestyle variables, asthma or COPD status, or in mean levels of spirometry and anthropometry ($p > 0.05$). However, participants in the vitamin D group had slightly higher observed and deseasonalized mean 25OHD concentrations than those in the placebo group ($p < 0.05$).

Table 1. Baseline comparison of the vitamin D supplemented and placebo groups.

Variable	Vitamin D ($n = 402$)	Placebo ($n = 373$)	*p*-Value
Age (years)			
50–59	84 (20.9)	75 (20.1)	
60–69	169 (42.0)	152 (40.8)	0.77
70–79	128 (31.8)	120 (32.2)	
80–84	21 (5.2)	26 (7.0)	
Sex–male	227 (56.5)	210 (56.3)	0.96
Ethnicity			
Māori	30 (7.5)	38 (10.2)	
Pacific Island	15 (3.7)	22 (5.9)	0.25
South Asian	16 (4.0)	13 (3.5)	
European/Other	341 (84.8)	300 (80.4)	
Education (highest level) [a]			
Primary school	8 (2.0)	5 (1.3)	
Secondary school	198 (49.3)	179 (48.0)	0.71
Tertiary	196 (48.8)	189 (50.7)	
In paid employment [a]			
Yes	203 (50.5)	174 (46.8)	
No			0.45
Retired	172 (42.8)	166 (44.6)	
Other	27 (6.7)	32 (8.6)	
Tobacco smoking			
Current	50 (12.4)	46 (12.3)	
Ex	250 (62.2)	236 (63.3)	0.95
Never	102 (25.4)	91 (24.4)	
Vigorous physical activity (hours per week) [a]			
None	149 (38.6)	163 (46.4)	
1–2	102 (26.4)	76 (21.7)	0.09
>2	135 (35.0)	112 (31.9)	
Take vitamin D supplements [b]	25 (6.2)	31 (8.3)	0.26
Type of asthma/COPD			
Asthma only	106 (26.4)	108 (29.0)	
COPD only	200 (49.8)	156 (41.8)	0.07
Combined asthma & COPD	96 (23.9)	109 (29.2)	
Spirometry, mean (SD)			
FEV_1 (L)	2.18 (0.71)	2.15 (0.70)	0.53
FEV_1, % predicted	78 (19)	79 (19)	0.98
FVC (L)	3.27 (0.98)	3.21 (0.93)	0.37
FVC, % predicted	90 (17)	91 (17)	0.79
Ratio FEV_1/FVC	0.67 (0.09)	0.67 (0.10)	0.69
FEV_1/FVC, % predicted	86 (12)	86 (11)	0.70
PEF (L/min)	353 (116)	360 (118)	0.43

Table 1. *Cont.*

Variable	Vitamin D ($n = 402$)	Placebo ($n = 373$)	*p*-Value
Anthropometry, mean (SD)			
Height (cm)	168.9 (9.3)	168.6 (9.2)	0.70
Weight (kg)	80.3 (16.2)	80.6 (17.3)	0.81
Body mass index (kg/m^2)	28.1 (5.1)	28.3 (5.6)	0.57
Corrected serum calcium, mean (SD) (mmol/L)	2.28 (0.07)	2.27 (0.07)	0.10
25-hydroxyvitamin D (nmol/L)			
Observed, mean (SD)	64.5 (23.1)	60.4 (24.2)	0.02
Deseasonalized, mean (SD)	66.7 (22.3)	63.1 (22.9)	0.03
Deseasonalized category			
<25.0	8 (2.0)	10 (2.7)	
25.0–49.9	82 (20.4)	109 (29.2)	0.03
50.0–74.9	172 (42.8)	134 (35.9)	
≥75.0	140 (34.8)	120 (32.2)	

Results are number (%) unless otherwise indicated. Abbreviations: FEV$_1$ = forced expiratory volume in 1 s; FVC = forced vital capacity; PEF = peak expiratory flow rate. [a] Numbers do not add to total for column because of missing/don't know responses. [b] ≤600 IU per day if aged 50–70 years; ≤800 IU per day if aged 71–84 years.

Table 2 shows the number and incidence rates of oral corticosteroid prescriptions for all participants, and for demographic and 25OHD categories within the vitamin D and placebo groups; it also shows HRs for repeated prescriptions in the vitamin D group compared to the placebo group, adjusted for age, sex, and ethnicity. For all participants in the analysis, the incidence rate for oral corticosteroid prescriptions was 0.40 per person-year in the vitamin D group and 0.39 per person-year in the placebo group. The HR was 1.08 (95%CI 0.84–1.39) adjusting for age, sex, and ethnicity, with the mean cumulative function curves being similar in the two treatment groups (Figure 2).

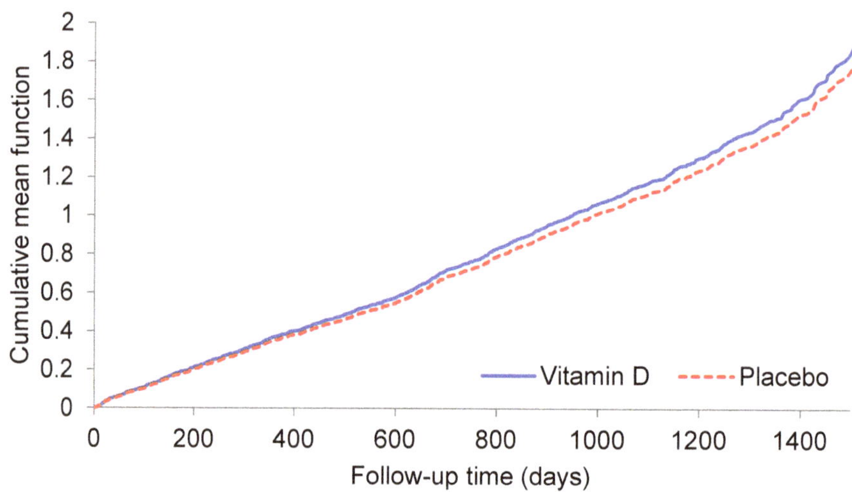

Figure 2. Mean cumulative number of exacerbations of asthma or chronic obstructive pulmonary disease during follow-up to 31 July 2015, by study treatment.

Table 2. Numbers, incidence rates, and adjusted hazard ratios of oral corticosteroid prescriptions (placebo as reference), by treatment group.

Participants	Vitamin D			Placebo			Hazard Ratio (95%CI) [a]	p-Value (Wald X^2)
	Number of Participants	Number of Prescriptions /p-Y	Incidence Rate per p-Y	Number of Participants	Number of Prescriptions /p-Y	Incidence Rate per p-Y		
All	402	536/1324	0.40	373	469/1218	0.39	1.08 (0.84, 1.39)	0.55
Asthma-COPD category								
Asthma only	106	172/352	0.49	108	173/367	0.47	1.18 (0.80, 1.74)	0.41
COPD only	200	176/650	0.27	156	124/494	0.25	1.07 (0.69, 1.67)	0.75
Combined asthma & COPD	96	188/320	0.59	109	172/357	0.48	1.21 (0.78, 1.88)	0.39
Age (years)								
50–59	84	155/299	0.52	75	117/270	0.43	1.23 (0.73, 2.07)	0.43
60–69	169	220/550	0.40	152	206/494	0.42	0.99 (0.68, 1.45)	0.97
70–79	128	139/410	0.34	120	108/374	0.29	1.21 (0.73, 2.03)	0.46
80–84	21	22/65	0.34	26	38/80	0.48	0.54 (0.19, 1.52)	0.24
Sex								
Male	227	242/744	0.33	210	270/676	0.40	0.83 (0.58, 1.18)	0.30
Female	175	294/580	0.51	163	199/542	0.37	1.46 (1.03, 2.06)	0.03
Ethnicity								
Māori	30	41/100	0.41	38	72/127	0.57	0.68 (0.33, 1.42)	0.31
Pacific Island	15	48/53	0.91	22	49/73	0.67	1.37 (0.56, 3.31)	0.49
South Asian	16	33/51	0.65	13	20/43	0.47	1.73 (0.51, 5.92)	0.38
European/Other	341	414/1120	0.37	300	328/976	0.34	1.10 (0.82, 1.46)	0.54
25OHD category (nmol/L) [b]								
<25.0	8	2/27	0.07	10	23/33	0.70	0.11 (0.02, 0.51)	0.005
25.0–49.9	82	152/265	0.57	109	150/360	0.42	1.45 (0.92, 2.30)	0.11
50.0–74.9	172	232/569	0.41	134	168/437	0.38	1.09 (0.76, 1.58)	0.64
≥75.0	140	150/463	0.32	120	128/389	0.33	1.03 (0.62, 1.70)	0.92

Abbreviations: 25OHD = 25-hydroxyvitamin D; 95%CI, 95% confidence interval; COPD = chronic obstructive pulmonary disease; p-Y = person-years. [a] Adjusted for age, sex, and ethnicity (as appropriate). [b] Based on deseasonalized concentrations.

The incidence rate was lower in both treatment arms for participants with COPD only, compared to those with asthma only or with combined asthma/COPD. However, vitamin D had no effect in preventing prescriptions within each of these asthma-COPD categories, as HRs were not different from 1.00 ($p > 0.05$). Unexpectedly, the incidence rate was increased in women given vitamin D compared to placebo (HR 1.46; 95%CI 1.03–2.06; p for interaction = 0.04). This finding prompted a post hoc analysis to examine for possible baseline imbalance in lung function, a major predictor of exacerbations. Indeed, among women, the vitamin D group had worse lung function (e.g., FEV_1 % predicted 71% in vitamin D group vs. 75% in placebo group; p = 0.09). Adjusting for this modest difference attenuated the original HR from 1.46 to 1.28 (95%CI 0.92–1.79).

When participants were analyzed by baseline 25OHD category, there was a strong protective effect of vitamin D supplementation in those with deseasonalized 25OHD <25 nmol/L (HR 0.11, 95%CI 0.02–0.51; p = 0.005; adjusting for age, sex, and ethnicity), along with a highly significant interaction when comparing the effect of vitamin D in participants with a baseline 25OHD below and above 25 nmol/L (p for interaction = 0.001). Further adjustment for baseline FEV_1 % predicted had no material impact on the effect of vitamin D in participants with baseline 25OH <25 nmol/L (HR 0.10; p = 0.004). The HR reduction was slightly attenuated after adjusting for asthma-COPD status (HR 0.24; p = 0.06), but the interaction comparing participants with baseline 25OHD below and above 25 nmol/L remained highly significant (p for interaction = 0.002). To explore if there were potentially important differences in the distribution of baseline variables from participants with baseline 25OHD <25 nmol/L, we repeated Table 1 in this subgroup; randomization yielded two similar groups (Supplementary Materials, Table S1). Figure 3 shows the mean cumulative function for each treatment group by baseline 25OHD category.

Adherence-related data support fidelity to the protocol. For example, 98% of the 775 participants reported taking the study capsule over the study period. In prior publications, we have documented, in the random subset of participants who underwent multiple blood testing as part of a safety evaluation, that the observed blood 25OHD levels in the intervention group increased from approximately 63 nmol/L to 135 nmol/L, consistent with their vitamin D supplementation, while the 25OHD in the placebo group did not change [14–16]. Participant retention was also high; for example, 77% of the 775 participants returned the final monthly questionnaire (July 2015). Lastly, we have previously reported that the vitamin D intervention did not affect participant-reported adverse events [22,23].

To provide better context for the mostly null RCT findings, we also examined the *observational* (noninterventional) association between baseline 25OHD levels and future risk of asthma or COPD exacerbation in the placebo group only (Supplementary Materials, Table S2). Participants with asthma (with or without COPD) had significantly increased HR compared to those with COPD only. The HR also was increased in older participants, men, and Māori or Pacific Island participants. Despite the limited statistical power, there was a borderline significant (p = 0.08) increase in HR among participants with baseline 25OHD <25 nmol/L compared to those ≥75 nmol/L.

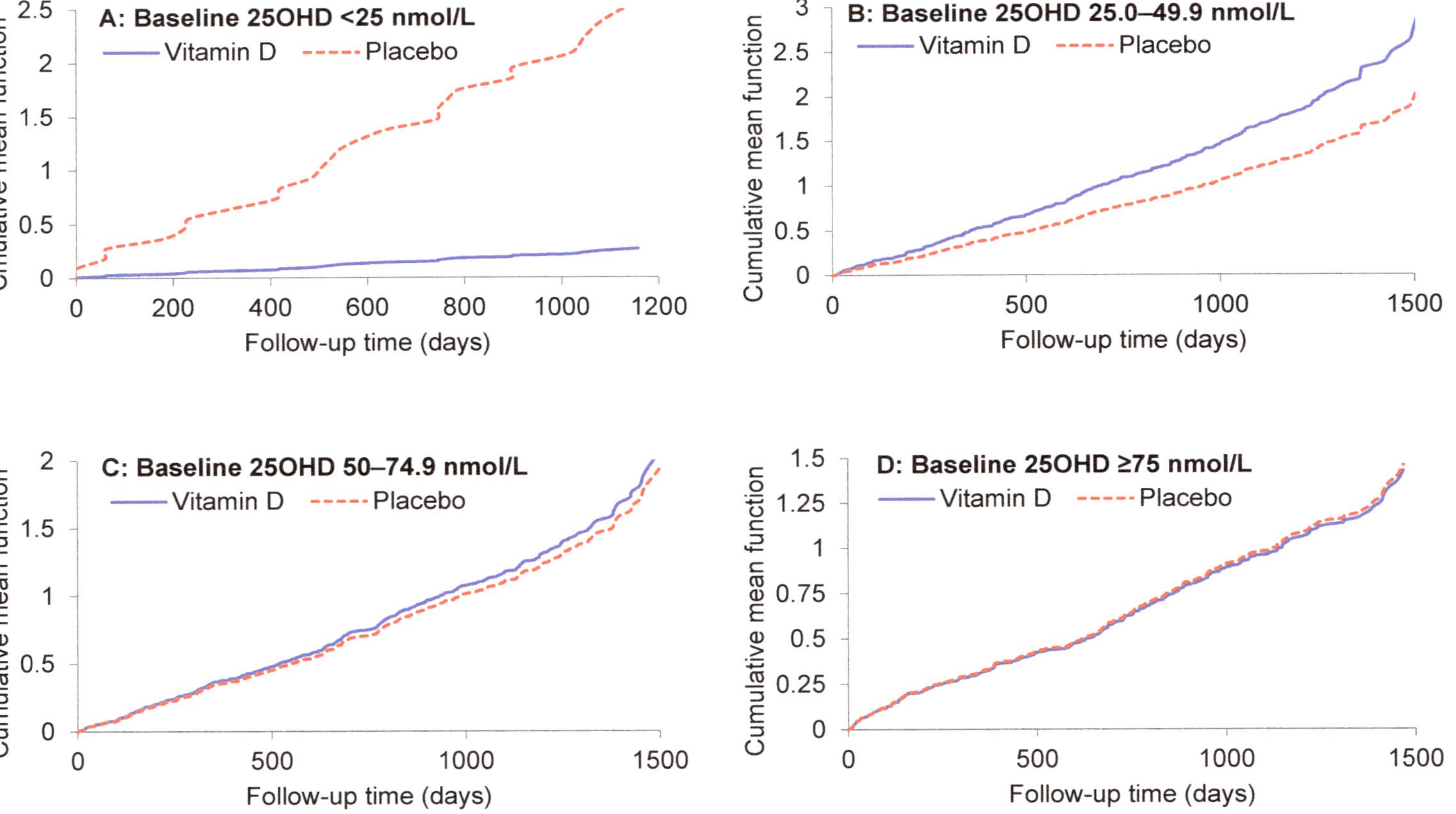

Figure 3. Mean cumulative number of exacerbations of asthma or chronic obstructive pulmonary disease during follow-up to 31 July 2015 for deseasonalized baseline serum 25-hydroxyvitamin D subgroups (panels **A–D**), by study treatment.

4. Discussion

In this post hoc analysis of RCT data from 775 older adults with asthma or COPD, monthly high-dose vitamin D supplementation (compared to placebo) did not prevent exacerbations of asthma or COPD. However, in the prespecified subgroup with baseline 25OHD <25 nmol/L, the observed benefit was striking (HR 0.11; 95%CI 0.02–0.51). The consistency of this subgroup finding with prior RCTs focused on either asthma or COPD [7,8] suggests that the finding is not due to chance alone.

Recent IPD meta-analyses of RCTs on the effects of vitamin D supplementation on ARI [5], asthma exacerbations [7], and COPD exacerbations [8] provide an excellent overview of the most relevant RCT literature. Briefly, the ARI meta-analysis showed an overall benefit of vitamin D supplements, but with substantial heterogeneity according to study population and dosing regimen; those who received the most benefit had baseline 25OHD <25 nmol/L and did not receive bolus dosing [5]. The asthma exacerbation meta-analysis showed an overall benefit, and suggested potentially greater benefit among those with baseline 25OHD <25 nmol/L [7]. Lastly, the COPD exacerbation meta-analysis showed no benefit overall, but consistent benefit for those with baseline 25OHD <25 nmol/L [8]. The current RCT findings are consistent with the results of the asthma and COPD meta-analyses. The new findings extend the results of earlier studies by having a much longer duration (average of 3.3 years) and by showing, in the same RCT, that despite no apparent effect on ARI [16], monthly vitamin D supplementation prevents asthma/COPD exacerbations among those with vitamin D deficiency.

The apparent discordance between the effects of vitamin D supplementation on the prevention of ARI *per se* (avoid bolus dosing for benefit) versus on asthma/COPD exacerbations (bolus dosing works in specific patient populations) merits further study. The results raise the intriguing possibility that, while viral respiratory infections trigger most exacerbations, the beneficial effects of vitamin D on asthma and COPD exacerbations may involve other effects, such as its anti-inflammatory actions [24]—which may occur regardless of the exact dosing regimen. Although ViDA is underpowered to look at the effect of vitamin D supplementation among vitamin D-deficient adults with asthma only, COPD only, or asthma-COPD overlap, we look forward to contributing our data to future IPD meta-analyses. Regardless, the emerging differences in how patient population and dosing regimen can modify the effects of vitamin D supplements suggest that ARI, asthma exacerbation, and COPD exacerbation outcomes are not truly interchangeable.

Although the ViDA trial suggests that monthly high-dose vitamin D supplementation is safe, we did note a 46% elevated risk of exacerbation among women assigned to the vitamin D supplement group (Table 2). We are not aware of any prior observational or interventional study that suggests that women experience more asthma or COPD exacerbations at higher levels of circulating 25OHD, or in response to taking vitamin D supplements [6]. Accordingly, we believe these results are most likely due to chance. We encourage further research to further investigate this subgroup finding.

The current report has several major strengths, including its study design (randomized, double-blinded, placebo-controlled trial) with high protocol adherence and the clearly demonstrated effect of the intervention on blood 25OHD levels [14–16]. Nevertheless, this RCT, like all RCTs, has potential limitations. First, we remind readers that we tested one vitamin D regimen (initial bolus, then monthly high-dose boluses) in one population (older adults); therefore, the relevance of the current study to, for example, daily vitamin D dosing in children with asthma is unclear. Second, the primary outcome was based on the prescribing of systemic corticosteroids, which requires clinician recognition of the exacerbation and prescribing of the appropriate treatment. While it is likely that we missed exacerbation events, we assume that these events were equally distributed across the two randomly assigned groups; this would tend to bias results toward the null. Third, despite starting with 775 participants, the few trial participants with baseline 25OHD <25 nmol/L (n = 18, 2%) precluded analyses within the asthma only, COPD only, or asthma-COPD overlap groups. By contributing our trial data to future IPD meta-analyses, we hope to compile sufficiently large numbers of participants in these important patient groups to pursue the issues further.

In summary, although monthly high-dose vitamin D supplementation had no overall impact on exacerbations of asthma or COPD in these older adults, we found evidence of probable benefit among those with severe vitamin D deficiency (baseline 25OHD < 25 nmol/L). While it remains possible that the subgroup finding was due to chance, it was very similar to results from other vitamin D trials in the literature, particularly those focused on the prevention of COPD exacerbations [8]. The exact threshold (and mechanism) for the observed benefit requires further study. To maximize the scientific yield of future RCTs, we encourage that vitamin D researchers carefully consider the importance of the trial population, vitamin D dosing regimen, and other factors when both designing and interpreting the current results, and those from future trials [25].

Supplementary Materials: The following tables are available online at https://www.mdpi.com/2072-6643/13/2/521/s1: Table S1: Baseline comparison of the vitamin D supplemented and placebo groups among participants with 25-hydroxyvitamin D < 25 nmol/L. Table S2: Hazard ratios of oral corticosteroid prescriptions, adjusted for other variables in the table, in the placebo group only ($n = 373$).

Author Contributions: Conceptualization, C.A.C., R.S.; methodology, C.A.C., L.T., J.S., A.R.M., R.S.; statistical analysis, J.S., R.S.; resources, K.-T.K.; data curation, J.S., D.W.; writing—original draft preparation, C.A.C., R.S.; writing—review and editing, C.A.C., L.T., J.S., C.M.M.L., D.W., K.-T.K., A.R.M., R.S.; project administration, D.W., R.S.; funding acquisition, C.A.C., R.S. All authors have read and agreed to the published version of the manuscript.

Funding: This research was funded by the Health Research Council of New Zealand (grant 10/400), and by the Accident Compensation Corporation of New Zealand.

Institutional Review Board Statement: The study was conducted according to the guidelines of the Declaration of Helsinki, and approved by the New Zealand Multi-Region Ethics Committee, Wellington (MEC/09/08/082, October 2010).

Informed Consent Statement: Informed consent was obtained from all subjects involved in the study.

Data Availability Statement: The data presented in this study are available on reasonable request from the corresponding author. The data are not publicly available due to privacy restriction.

Acknowledgments: We thank the participants and the ViDA study staff for their outstanding contributions.

Conflicts of Interest: The authors declare no conflict of interest. The funders had no role in the design of the study; in the collection, analyses, or interpretation of data; in the writing of the manuscript; or in the decision to publish the results.

References

1. Alshabanat, A.; Zafari, Z.; Albanyan, O.; Dairi, M.; Fitzgerald, J.M. Asthma and COPD Overlap Syndrome (ACOS): A Systematic Review and Meta Analysis. *PLoS ONE* **2015**, *10*, e0136065. [CrossRef]
2. Ritchie, A.I.; Farne, H.; Singanayagam, A.; Jackson, D.J.; Mallia, P.; Johnston, S.L. Pathogenesis of Viral Infection in Exacerbations of Airway Disease. *Ann. Am. Thorac. Soc.* **2015**, *12* (Suppl. 2), S115-32. [CrossRef]
3. Global Initiative for Asthma (GINA). Global strategy for asthma management and prevention. Available online: http://www.ginasthma.com/ (accessed on 30 October 2020).
4. Global Initiative for Chronic Obstructive Lung Disease (GOLD). Global strategy for the diagnosis, management, and prevention of COPD. Available online: http://www.goldcopd.com/ (accessed on 30 October 2020).
5. Martineau, A.R.; Jolliffe, D.A.; Hooper, R.L.; Greenberg, L.; Aloia, J.F.; Bergman, P.; Dubnov-Raz, G.; Esposito, S.; Ganmaa, D.; Ginde, A.A.; et al. Vitamin D supplementation to prevent acute respiratory tract infections: Systematic review and meta-analysis of individual participant data. *BMJ* **2017**, *356*, i6583. [CrossRef] [PubMed]
6. Camargo, C.A., Jr. Vitamin D, acute respiratory infection, and asthma/chronic obstructive pulmonary disease. In *Vitamin D*, 4th ed.; Feldman, D., Pike, J.W., Bouillon, R., Giovannucci, E., Goltzman, D., Hewison, M., Eds.; Elsevier Academic Press: Cambridge, MA, USA, 2018; pp. 1096–1120.

7. Jolliffe, D.A.; Greenberg, L.; Hooper, R.L.; Griffiths, C.J.; Camargo, C.A.; Kerley, C.P.; Jensen, M.E.; Mauger, D.; Stelmach, I.; Urashima, M.; et al. Vitamin D supplementation to prevent asthma exacerbations: A systematic review and meta-analysis of individual participant data. *Lancet Respir. Med.* **2017**, *5*, 881–890. [CrossRef]

8. Jolliffe, D.A.; Greenberg, L.; Hooper, R.L.; Mathyssen, C.; Rafiq, R.; De Jongh, R.T.; Camargo, C.A.; Griffiths, C.J.; Janssens, W.; Martineau, A.R. Vitamin D to prevent exacerbations of COPD: Systematic review and meta-analysis of individual participant data from randomised controlled trials. *Thorax* **2019**, *74*, 337–345. [CrossRef] [PubMed]

9. Lehouck, A.; Mathieu, C.; Carremans, C.; Baeke, F.; Verhaegen, J.; Van Eldere, J.; Decallonne, B.; Bouillon, R.; Decramer, M.; Janssens, W. High doses of vitamin D to reduce exacerbations in chronic obstructive pulmonary disease: A randomized trial. *Ann. Intern. Med.* **2012**, *156*, 105–114. [CrossRef]

10. Martineau, A.R.; James, W.Y.; Hooper, R.L.; Barnes, N.C.; Jolliffe, D.A.; Greiller, C.L.; Islam, K.; McLaughlin, D.; Bhowmik, A.; Timms, P.M.; et al. Vitamin D3 supplementation in patients with chronic obstructive pulmonary disease (ViDiCO): A multicentre, double-blind, randomised controlled trial. *Lancet Respir. Med.* **2015**, *3*, 120–130. [CrossRef]

11. Rafiq, R.; Prins, H.J.; Boersma, W.G.; Daniels, J.M.; Heijer, M.D.; Lips, P.; De Jongh, R.T. Effects of daily vitamin D supplementation on respiratory muscle strength and physical performance in vitamin D-deficient COPD patients: A pilot trial. *Int. J. Chronic Obstr. Pulm. Dis.* **2017**, *12*, 2583–2592. [CrossRef]

12. Zendehdel, A.; Gholami, M.R.; Anbari, K.; Ghanadi, K.; Bachari, E.C.; Azargoon, A. Effects of Vitamin D Intake on FEV1 and COPD Exacerbation: A Randomized Clinical Trial Study. *Glob. J. Heal. Sci.* **2014**, *7*, 243–248. [CrossRef]

13. SScragg, R.; Waayer, D.; Stewart, A.W.; Lawes, C.M.; Toop, L.; Murphy, J.; Khaw, K.-T.; Camarago, C.A., Jr. The Vitamin D Assessment (ViDA) Study: Design of a randomized controlled trial of vitamin D supplementation for the prevention of cardiovascular disease, acute respiratory infection, falls and non-vertebral fractures. *J. Steroid Biochem. Mol. Biol.* **2016**, *164*, 318–325. [CrossRef] [PubMed]

14. Scragg, R.; Stewart, A.W.; Waayer, D.; Lawes, C.M.M.; Toop, L.; Sluyter, J.; Murphy, J.; Khaw, K.-T.; Camargo, C.A. Effect of Monthly High-Dose Vitamin D Supplementation on Cardiovascular Disease in the Vitamin D Assessment Study: A Randomized Clinical Trial. *JAMA Cardiol.* **2017**, *2*, 608–616. [CrossRef] [PubMed]

15. Khaw, K.-T.; Stewart, A.W.; Waayer, D.; Lawes, C.M.M.; Toop, L.; Camargo, C.A.; Scragg, R. Effect of monthly high-dose vitamin D supplementation on falls and non-vertebral fractures: Secondary and post-hoc outcomes from the randomised, double-blind, placebo-controlled ViDA trial. *Lancet Diabetes Endocrinol.* **2017**, *5*, 438–447. [CrossRef]

16. Camargo, C.A.; Sluyter, J.; Stewart, A.W.; Khaw, K.-T.; Lawes, C.M.M.; Toop, L.; Waayer, D.; Scragg, R. Effect of Monthly High-Dose Vitamin D Supplementation on Acute Respiratory Infections in Older Adults: A Randomized Controlled Trial. *Clin. Infect. Dis.* **2019**, *71*, 311–317. [CrossRef] [PubMed]

17. Miller, M.R.; Hankinson, J.; Brusasco, V.; Burgos, F.; Casaburi, R.; Coates, A.; Crapo, R.; Enright, P.; Van Der Grinten, C.P.M.; Gustafsson, P.; et al. Standardisation of spirometry. *Eur. Respir. J.* **2005**, *26*, 319–338. [CrossRef]

18. Sachs, M.C.; Shoben, A.; Levin, G.P.; Robinson-Cohen, C.; Hoofnagle, A.N.; Swords-Jenny, N.; Ix, J.H.; Budoff, M.; Lutsey, P.L.; Siscovick, D.S.; et al. Estimating mean annual 25-hydroxyvitamin D concentrations from single measurements: The Multi-Ethnic Study of Atherosclerosis. *Am. J. Clin. Nutr.* **2013**, *97*, 1243–1251. [CrossRef]

19. Hill, K.; Goldstein, R.S.; Guyatt, G.H.; Blouin, M.; Tan, W.C.; Davis, L.L.; Heels-Ansdell, D.M.; Erak, M.; Bragaglia, P.J.; Tamari, I.E.; et al. Prevalence and underdiagnosis of chronic obstructive pulmonary disease among patients at risk in primary care. *Can. Med Assoc. J.* **2010**, *182*, 673–678. [CrossRef]

20. Martinez, C.H.; Mannino, D.M.; Jaimes, F.A.; Curtis, J.L.; Han, M.K.; Hansel, N.N.; Diaz, A.A. Undiagnosed Obstructive Lung Disease in the United States. Associated Factors and Long-term Mortality. *Ann. Am. Thorac. Soc.* **2015**, *12*, 1788–1795. [CrossRef]

21. Rothman, K.J. No adjustments are needed for multiple comparisons. *Epidemiology* **1990**, *1*, 43–46. [CrossRef] [PubMed]

22. Malihi, Z.; Lawes, C.M.; Wu, Z.; Huang, Y.; Waayer, D.; Toop, L.; Khaw, K.-T.; Camargo, C.A.; Scragg, R. Monthly high-dose vitamin D3 supplementation and self-reported adverse events in a 4-year randomized controlled trial. *Clin. Nutr.* **2019**, *38*, 1581–1587. [CrossRef] [PubMed]

23. Malihi, Z.; Lawes, C.M.M.; Wu, Z.; Huang, Y.; Waayer, D.; Toop, L.; Khaw, K.-T.; Camargo, C.A.; Scragg, R. Monthly high-dose vitamin D supplementation does not increase kidney stone risk or serum calcium: Results from a randomized controlled trial. *Am. J. Clin. Nutr.* **2019**, *109*, 1578–1587. [CrossRef] [PubMed]

24. Sassi, F.; Tamone, C.; D'Amelio, P. Vitamin D: Nutrient, Hormone, and Immunomodulator. *Nutrients* **2018**, *10*, 1656. [CrossRef] [PubMed]

25. Camargo, C.A., Jr.; Martineau, A.R. Vitamin D to prevent COVID-19: Recommendations for the design of clinical trials. *FEBS J.* **2020**, *287*, 3689–3692. [CrossRef] [PubMed]

Article

Effects of Vitamin D Supplementation on Surrogate Markers of Fertility in PCOS Women: A Randomized Controlled Trial

Elisabeth Lerchbaum [1,*], Verena Theiler-Schwetz [1], Martina Kollmann [2], Monika Wölfler [2], Stefan Pilz [1], Barbara Obermayer-Pietsch [1] and Christian Trummer [1]

[1] Department of Internal Medicine, Division of Endocrinology and Diabetology, Medical University of Graz, Auenbruggerplatz 15, 8036 Graz, Austria; verena.schwetz@medunigraz.at (V.T.-S.); stefan.pilz@medunigraz.at (S.P.); barbara.obermayer@medunigraz.at (B.O.-P.); christian.trummer@medunigraz.at (C.T.)

[2] Department of Obstetrics and Gynecology, Division of Gynecological Endocrinology and Reproductive Medicine, Medical University of Graz, Auenbruggerplatz 14, 8036 Graz, Austria; martina.kollmann@medunigraz.at (M.K.); monika.woelfler@medunigraz.at (M.W.)

* Correspondence: elisabeth.lerchbaum@medunigraz.at; Tel.: +43-316-385-82383

Abstract: Vitamin D (VD) might play an important role in polycystic ovary syndrome (PCOS) and female fertility. However, evidence from randomized controlled trials (RCT) is sparse. We examined VD effects on anti-Müllerian hormone (AMH) and other endocrine markers in PCOS and non-PCOS women. This is a post hoc analysis of a single-center, double-blind RCT conducted between December 2011 and October 2017 at the endocrine outpatient clinic at the Medical University of Graz, Austria. We included 180 PCOS women and 150 non-PCOS women with serum 25-hydroxyvitamin D (25(OH)D) concentrations <75 nmol/L in the trial. We randomized subjects to receive 20,000 IU of VD3/week (119 PCOS, 99 non-PCOS women) or placebo (61 PCOS, 51 non-PCOS women) for 24 weeks. Outcome measures were AMH, follicle-stimulating hormone (FSH), luteinizing hormone (LH), estradiol, dehydroepiandrosterone sulfate, and androstenedione. In PCOS women, we observed a significant treatment effect on FSH (mean treatment effect 0.94, 95% confidence interval [CI] 0.087 to 1.799, $p = 0.031$) and LH/FSH ratio (mean treatment effect -0.335, 95% CI -0.621 to 0.050, $p = 0.022$), whereas no significant effect was observed in non-PCOS women. In PCOS women, VD treatment for 24 weeks had a significant effect on FSH and LH/FSH ratio but no effect on AMH levels.

Keywords: vitamin D; polycystic ovary syndrome; anti-Müllerian hormone; follicle-stimulating hormone; randomized controlled trial

Citation: Lerchbaum, E.; Theiler-Schwetz, V.; Kollmann, M.; Wölfler, M.; Pilz, S.; Obermayer-Pietsch, B.; Trummer, C. Effects of Vitamin D Supplementation on Surrogate Markers of Fertility in PCOS Women: A Randomized Controlled Trial. *Nutrients* **2021**, *13*, 547. https://doi.org/10.3390/nu13020547

Academic Editors: Luigi Barrea and Giovanna Muscogiuri

Received: 21 December 2020
Accepted: 2 February 2021
Published: 7 February 2021

Publisher's Note: MDPI stays neutral with regard to jurisdictional claims in published maps and institutional affiliations.

1. Introduction

Vitamin D (VD) is a steroid hormone with well-known effects on calcium and bone metabolism [1]. Accumulating evidence from cross-sectional studies indicates an association of low 25-hydroxyvitamin D (25(OH)D) concentrations with various conditions including obesity, metabolic disorders [2,3], cardiovascular disease [4], hypogonadism [5], polycystic ovary syndrome (PCOS) [6], and decreased female fertility [7]. It has been hypothesized that a possible VD effect on ovarian anti-Müllerian hormone (AMH) might be a putative component explaining the complex relationship of VD and human reproduction [8]. AMH is an ovarian biomarker playing a central role in folliculogenesis and ovarian dysfunction. Several in vitro as well as in vivo studies examined the potential effects of vitamin D on ovarian function [9,10]. Kinuta et al. [9] found that VD receptor null female mice suffer from ovarian insufficiency that is characterized by impaired follicular development. A recent meta-analysis assessed the reproductive outcomes of 2700 subfertile women and found a significant association of favorable outcomes with replete vitamin D status [10]. It has been hypothesized that VD acts upon the ovarian follicle and may improve oocyte quality [10]. As impaired ovarian function is also related to obesity, it should be mentioned that obesity is associated with low VD status due to decreased physical activity, low sun exposure, and sequestration in the adipose tissue [11,12]. Furthermore, it has been hypothesized that low 25(OH)D concentrations are involved in the development of obesity by influencing adipogenesis [12].

PCOS is the most common endocrine disorder among women of childbearing age [13]. Of note, PCOS has a very high prevalence and up to 10% of women of reproductive age are affected by PCOS [13]. In addition to hyperandrogenemia and metabolic disturbances such as obesity and insulin resistance, affected women frequently suffer from decreased fertility due to anovulation [13,14]. Moreover, alterations in lipid pattern are associated with obstetric complications in PCOS women [15]. Diet plays an important role in the pathogenesis of PCOS and obesity is related to the severity of the syndrome [16,17]. An increasing number of studies have examined the association of VD status with various features of PCOS. Whereas the majority of observational studies point towards a link of deficient VD status with obesity, metabolic disturbances, and anovulation, data derived from randomized controlled trials (RCTs) are limited [6]. Compared to healthy women, PCOS women have higher AMH levels and AMH is considered as an important diagnostic and prognostic marker in PCOS [18]. Existing cross-sectional studies on the association of 25(OH)D concentrations and AMH levels have reported inconsistent results [19]. Although a small RCT among VD-deficient infertile PCOS women reported a positive VD effect on AMH levels [20], data from large RCTs are lacking. Therefore, a recent systematic review and meta-analysis on VD and AMH concluded that large RCTs of VD supplementation are necessary to elucidate the complex relationship of VD and AMH [19].

Consequently, we performed a post hoc analysis of our RCT that was designed to examine VD effects on endocrine and metabolic parameters in PCOS and non-PCOS women. We aim to investigate VD effects on AMH levels as well as on other endocrine parameters involved in reproduction, including follicle-stimulating hormone (FSH), luteinizing hormone (LH), and estradiol in PCOS as well as in healthy premenopausal women without PCOS. Furthermore, we analyze VD effects on dehydroepiandrosterone sulfate (DHEAS) and androstenedione levels.

2. Materials and Methods

This study is a post hoc analysis of a single-center, placebo-controlled, double-blind, parallel-group study performed at the Medical University of Graz (MUG), Austria. We designed our study to examine VD effects on endocrine and metabolic parameters in PCOS as well as in healthy women without PCOS.

We have published details on the study design and methods previously [21,22]. The design, conducting, and publication of this study adhere to the recommendations of the CONSORT Statement (http://www.consort-statement.org/). We registered the trial at http://www.clinicaltrialsregister.eu (EudraCT number, 2011-000994-30) and at clinicaltrials.gov (ClinicalTrials.gov Identifier NCT01721915). The local ethics committee approved the study protocol (EK 23-300 ex 10/11).

2.1. Subjects

Premenopausal women aged ≥18 years with 25(OH)D concentrations <75 nmol/L were eligible for our study. In the PCOS group, we established a diagnosis of PCOS using the Rotterdam criteria [23] if two out of the following three features were met: clinical and/or biochemical hyperandrogenism, polycystic ovaries, and/or oligo-/anovulation. We excluded disorders with similar clinical features before we made the diagnosis of PCOS. Non-PCOS women were required to show none of the Rotterdam PCOS criteria.

Exclusion criteria in both groups were hypercalcemia (defined as plasma calcium concentrations >2.65 mmol/L), regular vitamin D supplementation within 3 months prior to study inclusion, prevalent type 2 diabetes mellitus, use of insulin-sensitizing drugs (i.e., metformin, incretin mimetic drugs, thiazolidinedione, sulfonylurea) within 6 months prior to study inclusion, hormonal contraception within 3 months prior to study inclusion, use of lipid-lowering drugs or other drugs affecting insulin sensitivity or serum androgens (e.g., niacin, corticosteroids, beta-blockers, calcium channel blockers, thiazide diuretics) as well as disorders apart from PCOS associated with irregular menses and/or androgen excess.

We recruited PCOS and non-PCOS women from patients of the Division of Endocrinology and Diabetology and the Division of Gynecological Endocrinology and Reproductive Medicine at the MUG by conversation during routine visits in the outpatient clinics. Furthermore, we recruited participants from female hospital staff and female family members of hospital staff, and written information about the study was posted in the outpatient clinic. We informed all study participants during recruitment about the possibility of receiving a placebo.

Healthy Women

We included not only PCOS but also healthy women in our RCT as vitamin D might have varying effects among women with and without PCOS. As outlined above, the relationship of vitamin D and AMH is complex. Vitamin D might increase AMH levels in healthy women [24] but decrease AMH levels in PCOS women [20]. Therefore, to examine whether vitamin D effects vary depending on the respective group, we included PCOS as well as healthy women without PCOS in our analyses.

2.2. Intervention

We allocated subjects to the placebo (PBO) or VD group according to a computer-generated randomization list using a ratio of 2:1. We placed study medication into numbered bottles according to this list.

The VD group received an oral dose of 20,000 IU VD3 per week (equivalent to 2857 IU/day) as 50 oily drops per week (Oleovit D3 drops; Fresenius Kabi Austria GmbH, Linz) for 24 weeks. Our PBO group received 50 oily drops without VD for 24 weeks. PBO oil contained the same oil as Oleovit D3 drops (without VD content). Fresenius Kabi Austria GmbH, Linz delivered the PBO oil. All investigators involved in the enrollment of study subjects, data collection as well as assignment to intervention were masked to participant allocation. In order to improve and verify compliance, we asked study participants to return full as well as empty study medication bottles at the end of the study.

2.3. Outcome Measures

This is a post hoc analysis of our RCT including PCOS and non-PCOS women. We investigate VD effects on endocrine parameters including AMH, FSH, LH, estradiol, DHEAS, and androstenedione levels.

2.4. Procedures

We collected basal blood samples for measurement of 25(OH)D, AMH, FSH, LH, estradiol, DHEAS, and androstenedione between 8.00 and 9.00 a.m. after an overnight fast. We used 25(OH)D concentrations determined by immunoassay for evaluation of inclusion criteria. We performed biobanking of all remaining blood samples by freezing and storing at $-80\,^{\circ}$C until analysis. In addition, we measured serum concentrations of 25(OH)D by well-adjusted isotope dilution–liquid chromatography tandem mass spectrometry (ID-LC-MS/MS) methods in 2018 [21,22].

We measured FSH, LH, and estradiol levels on a daily basis. LH and FSH were measured using Access® hLH and hFSH CLIA (Beckman Coulter Inc., Brea, CA, USA), respectively. 17β-estradiol was determined using IMMULATE® CLIA assays (Siemens Healthcare Diagnostics Products Ltd., Glyn Rhonwy, UK). We measured AMH, androstenedione, and DHEAS levels once weekly, and blood samples were frozen and stored at $-40\,^{\circ}$C until analysis. We measured DHEAS (Labor Diagnostika Nord, Nordhorn, Germany) and androstenedione (Siemens Healthcare Diagnostics Products Ltd.) via enzyme-linked immunosorbent assay (ELISA), with intra-assay and interassay coefficients of variation (CV) of <10%. In our laboratory, the assay for AMH was changed in November 2014 from the ultra-sensitive anti-Müllerian hormone/Müllerian-inhibiting substance enzyme-linked immunosorbent assay (ELISA) kit (Ansh Labs, Webster, TX, USA) to the Access 2 immunosorbent assay system (Beckmann Coulter). We compared both assays and found a good correlation (r = 0.95). Both AMH assays show intra-assay and interassay CV of <10%. Laboratory kits and assays did not change between 2011 and 2017 for the remaining outcome parameters.

Vitamin D intake was assessed by questionnaires.

2.5. Statistical Analyses

We performed sample size calculation based on the data derived from a pilot study conducted among PCOS women [25]. In detail, we found a reduction area under the curve (AUC) glucose from 115 ± 17 at baseline to 103 ± 18 at the end of the study after 24 weeks VD supplementation. We therefore calculated a sample size of 92 participants to detect a treatment difference at two-sided 0.05 significance levels with a probability of 90%, if the true difference between treatments is 12 with a standard deviation of 17. As the analyses of VD effects according to genotype profile were a secondary outcome measure (results have been published previously [26]), we randomized study participants 2:1 (VD:PBO) in order to increase the sample size in the VD group. The number of enrolled PCOS subjects was increased from 150 to 180 to ensure an adequate power to detect differences regarding AUCgluc.

We used descriptive statistics as well as the Kolmogorov–Smirnov test to analyze the distribution of data. We present continuous data with normal distribution as means with SD and continuous data following a skewed distribution as median with interquartile range. We performed log transformation of skewed variables and rechecked log transformed data for normal distribution before parametric tests were performed. We used Student's T-test and χ^2-test for comparisons of baseline characteristics between groups. Delta (Δ) values (value at the end of the study minus baseline value) were calculated for 25(OH)D and outcome measures. We used Pearson correlation analysis to determine relationships between variables. We performed multivariable stepwise linear regression analysis with LH/FSH ratio and androstenedione as the dependent variables, and with BMI, age, and 25(OH)D as independent variables.

We executed analyses of outcome variables according to the intention-to-treat principle. In these analyses, we included all subjects with baseline and follow-up values. We applied analysis of covariance and adjusted our analyses for baseline values to test for differences in the respective outcome variables between the VD and the PBO group at the end of the study. We performed subgroup analyses of PCOS women with irregular menses. All statistical procedures were performed with SPSS version 26 (SPSS Inc., Chicago, IL, USA). We considered a *p*-value of <0.05 as statistically significant.

3. Results

We screened ~500 PCOS women and ~300 healthy women without PCOS who routinely visited the endocrine outpatient clinic or responded to written information material for study eligibility. We randomized and enrolled 180 PCOS women and 150 healthy women in the study (participant flow charts have been published previously [21,22]).We randomized the first subject in December 2011 and we performed the last follow-up in October 2017.

In Table 1, we display the baseline characteristics of all study subjects. In PCOS women, we observed significantly higher BMI (*p* = 0.001), AMH levels (*p* < 0.001), LH levels (*p* = 0.02), LH/FSH ratio (*p* < 0.001), DHEAS (*p* < 0.001), and androstenedione levels (*p* < 0.001), whereas age (*p* < 0.001), 25(OH)D (*p* = 0.019), FSH (*p* < 0.001), and estradiol levels (*p* < 0.001) were lower compared to healthy women without PCOS. PCOS women in the VD group were significantly younger compared to PCOS women in the PBO group. In healthy women without PCOS, baseline estradiol levels were significantly lower in the VD group compared to the PBO group. We found no significant differences among the remaining baseline characteristics between VD and PBO groups in PCOS as well as in healthy women.

Table 1. Baseline characteristics of study subjects. Data are shown as means with standard deviation, median, and interquartile range or proportion as appropriate. PCOS—polycystic ovary syndrome; VD—vitamin D; PBO—placebo; BMI—body mass index; 25(OH)D—25-hydroxyvitamin D; AMH—anti-Müllerian hormone; FSH—follicle-stimulating hormone; LH—luteinizing hormone; DHEAS—dehydroepiandrosterone sulfate. We performed comparisons of baseline characteristics between women in the VD and the PBO groups using Student's *t*-test and χ^2-test. Season 1: January–March; season 2: April–June; season 3: July–September; season 4: October–December.

	All PCOS Women **(*n* = 180)**	**VD (*n* = 119)**	**PBO (*n* = 61)**	***p*-Value**
Age (years)	26.0 ± 5.0	25.4 ± 4.6	27.2 ± 5.5	0.022
BMI (kg/m^2)	27.6 ± 7.5	27.3 ± 7.4	28.3 ± 7.8	0.453
25(OH)D * (nmol/L)	50.4 ± 19.0	50.7 ± 19.5	49.9 ± 18.3	0.798
AMH (ng/mL)	7.67 (4.09–15.0)	7.62 (4.23–15.0)	7.71 (3.15–15.0)	0.547
FSH (µU/mL)	5.97 ± 2.41	6.04 ± 2.59	5.94 ± 2.33	0.783
LH (µU/mL)	8.88 (4.26–14.5)	8.89 (4.20–15.34)	8.86 (3.82–14.18)	0.830
LH/FSH ratio	1.48 (0.88–2.30)	1.52 (0.88–2.54)	1.38 (0.68–2.55)	0.530
Estradiol (pg/mL)	60.6 (41.0–122.0)	59.1 (39.3–123.0)	64.0 (43.5–158.0)	0.311
DHEAS (µg/mL)	1.90 (1.24–2.97)	1.94 (1.16–3.22)	1.9 (1.28–3.07)	0.789
Androstenedione (ng/mL)	3.36 (2.26–4.87)	2.4 (1.48–4.24)	2.61 (1.79–3.96)	0.937
Vitamin D intake (IU/day)	31 (14–76)	31 (16–67)	31 (13–78)	0.582
Season of recruitment				
Season 1	38.3%	36.1%	42.6%	0.442
Season 2	26.1%	26.1%	26.2%	
Season 3	17.8%	21.0%	11.5%	
Season 4	17.8%	16.8%	19.7%	
	All Healthy Women **(*n* = 150)**	**VD (*n* = 99)**	**PBO (*n* = 51)**	
Age (years)	35.8 ± 8.7	35.7 ± 8.9	36.1 ± 8.4	0.826
BMI (kg/m^2)	25.2 ± 5.5	25.5 ± 5.3	24.7 ± 5.8	0.398
25(OH)D * (nmol/L)	55.4 ± 18.9	55.4 ± 18.9	55.3 ± 18.9	0.996
AMH (ng/mL)	1.97 (0.32–4.38)	1.89 (0.29–5.2)	2.41 (0.32–5.30)	0.546
FSH (µU/mL)	9.86 ± 13.11	9.67 ± 12.05	9.96 ± 13.69	0.898
LH (µU/mL)	6.28 (3.72–11.0)	6.28 (3.24–11.50)	6.48 (4.04–14.20)	0.119
LH/FSH ratio	0.93 (0.51–1.59)	0.87 (0.48–1.57)	1.12 (10.51–2.03)	0.242
Estradiol (pg/mL)	92.6 (50.5–156.0)	83.4 (41.5–145)	114 (61.1–212.0)	0.006
DHEAS (µg/mL)	1.21 (0.78–2.0)	1.20 (0.75–2.03)	1.23 (0.76–2.19)	0.508
Androstenedione (ng/mL)	2.50 (1.56–3.96)	2.4 (1.48–4.24)	2.61 (1.79–3.96)	0.642
Vitamin D intake (IU/day)	50 (26–77)	50 (22–80)	50 (27–72)	0.471
Season of recruitment				
Season 1	30.7%	29.3%	33.3%	0.942
Season 2	32.7%	32.3%	33.3%	
Season 3	10.0%	10.1%	9.8%	
Season 4	26.7%	28.3%	23.5%	

* We measured 25(OH)D by liquid chromatography tandem mass spectrometry.

3.1. Cross-Sectional Analyses

In PCOS women, we found a significant correlation of 25(OH)D levels with LH/FSH ratio ($r = -0.195$, $p = 0.009$) as well as with androstenedione levels ($r = 0.15$, $p = 0.043$). We observed no significant correlation of 25(OH)D levels with the remaining endocrine parameters (AMH, LH, FSH, estradiol, and DHEAS). In analyses adjusted for age and BMI, the correlation of 25(OH)D with LH/FSH ratio ($p = 0.011$) remained stable but was attenuated for androstenedione ($p = 0.070$).

In healthy women, we observed no significant correlation of 25(OH)D with endocrine parameters.

3.2. Outcome Analyses

3.2.1. PCOS Women

In PCOS women, the mean ($\pm$SD) overall treatment period was 176 ± 23 days in the VD group and 176 ± 21 days in the PBO group ($p = 0.906$). A total of 123 study participants completed both study visits.

In Table 2, we display results of outcome analyses. In PCOS women, we found a significant VD effect on FSH levels as well as on LH/FSH ratio. We found no significant effect on the remaining outcome parameters. After exclusion of PCOS women with regular menses ($n = 19$), VD effects on FSH levels (mean treatment effect 0.271, 95% CI 0.27 to 2.06, $p = 0.011$) and LH/FSH ratio (mean treatment effect -0.401, 95% CI -0.705 to -0.097, $p = 0.010$) remained stable.

Table 2. Continuous outcome variables at baseline and end of the study in PCOS women with available values at both study visits. We display data as means with standard deviation or medians and interquartile range as appropriate. We calculated treatment effects with 95% confidence interval and *p*-values by analysis of covariance for group differences at the end of the study. Analyses were adjusted for baseline values. IQR—interquartile range; AMH—anti-Müllerian hormone; VD—vitamin D; PBO—placebo; FSH—follicle-stimulating hormone; LH—luteinizing hormone; DHEAS—dehydroepiandrosterone sulfate.

	Baseline Visit	Study End	Treatment Effect (95% Confidence Interval)	*p*-Value
		AMH * (ng/mL)		
VD ($n = 80$)	7.6 (4.2–15.0)	7.0 (4.2–15.5)	0.097 (-0.081 to 0.276)	0.282
PBO ($n = 40$)	7.7 (3.2–15.0)	7.6 (2.8–14.4)		
		FSH (µU/mL)		
VD ($n = 81$)	6.04 ± 2.59	6.16 ± 2.46	0.94 (0.087 to 1.799)	0.031
PBO ($n = 41$)	5.94 ± 2.33	5.23 ± 1.78		
		LH * (µU/mL)		
VD ($n = 79$)	8.9 (4.2–15.3)	9.4 (3.4–15.2)	-0.184 (-0.497 to 0.129)	0.248
PBO ($n = 41$)	8.9 (3.8–14.2)	8.8 (4.1–14.7)		
		Estradiol * (pg/mL)		
VD ($n = 81$)	59.1 (39.3–123.0)	59.4 (33.9–169.0)	-0.096 (-0.351 to 0.159)	0.460
PBO ($n = 41$)	64.0 (43.5–158.0)	73.8 (44.2–193.0)		
		LH/FSH ratio *		
VD ($n = 79$)	1.52 (0.88–2.54)	1.45 (0.79–2.73)	-0.335 (-0.621 to -0.050)	0.022
PBO ($n = 41$)	1.38 (0.69–2.55)	1.73 (0.76–3.32)		
		DHEAS * (µg/mL)		

Table 2. Cont.

	Baseline Visit	**Study End**	**Treatment Effect (95% Confidence Interval)**	***p*-Value**
VD ($n = 81$)	1.94 (1.16–3.22)	1.96 (1.06–3.12)	-0.016 (-0.142 to 0.11)	0.805
PBO ($n = 41$)	1.9 (1.28–3.07)	2.12 (1.31–3.23)		
		Androstenedione * (ng/mL)		
VD ($n = 80$)	3.41 (2.24–4.95)	3.68 (2.55–6.0)	0 (-0.131 to 0.130)	0.996
PBO ($n = 40$)	3.32 (2.05–5.58)	3.86 (2.33–7.11)		

* Skewed variables for which logarithmic transformed values were used in ANCOVA, but untransformed values are shown in the table.

We observed a significant negative correlation of Δ25(OH)D levels with ΔLH/FSH ratio ($r = -0.208$, $p = 0.024$) and a trend with ΔFSH ($r = 0.169$, $p = 0.066$). We observed no significant correlation of Δ25(OH)D with ΔAMH, ΔLH, Δestradiol, Δandrostenedione, and ΔDHEAS ($p > 0.05$ for all).

Table 3 shows 25(OH)D concentrations at baseline and the end of the study in PCOS women. VD supplementation significantly increased 25(OH)D concentrations.

Table 3. 25(OH)D concentrations at baseline and at the end of the study in subjects with available values at both study visits. Data are shown as means with standard deviation. Treatment effects with 95% confidence interval and *p*-values were calculated by ANCOVA for group differences at follow-up with adjustment for baseline value.

	Baseline	**Follow-Up (24 Weeks)**	**Treatment Effect (95% Confidence Interval)**	***p*-Value**
		PCOS women		
		25(OH)D (nmol/L)		
VD ($n = 79$)	48.8 ± 16.8	90.2 ± 20.1	33.4 (24.5 to 42.2)	<0.001
PBO ($n = 44$)	48.8 ± 17.5	56.8 ± 29.5		
		Healthy women		
		25(OH)D (nmol/L)		
VD ($n = 82$)	55.8 ± 19.9	95.3 ± 26.2	28.5 (19.3 to 37.7)	<0.001
PBO ($n = 44$)	56.2 ± 19.3	67.0 ± 24.8		

PCOS—polycystic ovary syndrome; 25(OH)D—25-hydroxyvitamin D; VD—vitamin D; PBO—placebo.

VD effects on metabolic parameters are shown in supplemental Tables S1 and S2. In PCOS women, we found a significant beneficial VD effect on glucose levels at 60 min during the oral glucose tolerance test (Supplementary Table S1). In non-PCOS women, VD treatment had a significant unfavorable effect on insulin resistance and insulin sensitivity (Supplementary Table S2).

3.2.2. Non-PCOS Women

In healthy women, the mean (± SD) treatment duration was 174 ± 44 days in the VD group and 173 ± 23 days in the PBO group ($p = 0.884$). In total, 127 participants completed the entire study including the last follow-up visit after 24 weeks.

In healthy women without PCOS, we found no significant VD effect on outcome measures ($p > 0.05$ for all, data not shown). Furthermore, we observed no significant correlation of Δ25(OH)D with changes in outcome measures ($p > 0.05$ for all).

Table 3 shows 25(OH)D concentrations at baseline and the end of the study in healthy women. We found a significant VD effect on 25(OH)D concentrations in women without PCOS.

4. Discussion

In our RCT in PCOS women with baseline 25(OH)D concentrations <75 nmol/L, VD treatment had a significant effect on FSH levels and LH/FSH ratio. We found, however, no significant VD effect on AMH levels and the remaining endocrine parameters. In healthy women with serum 25(OH)D <75 nmol/L at baseline, we observed no significant VD effect on outcome measures.

Interestingly, we observed a significant VD effect on FSH levels and LH/FSH ratio as well as a significant correlation between Δ25(OH)D and ΔLH/FSH ratio in PCOS women. In the pathophysiology of PCOS, abnormalities of the hypothalamic–pituitary–ovarian axis play an important role [13]. A relative increase in LH to FSH release is caused by a disturbance in the secretion pattern of the gonadotrophin-releasing hormone [27]. Furthermore, ovarian estrogen is responsible for causing an abnormal feedback mechanism that results in increased LH release [28]. An elevated LH/FSH ratio is a common finding in PCOS and as a result, ovulation does not occur in many PCOS patients [29]. It has been reported that VD alters FSH sensitivity, indicating a possible physiological role for VD in the development and luteinization of the ovarian follicle [8]. Among induced PCOS rats, VD treatment increases the normal follicle number through increasing FSH and estradiol and decreasing LH [30]. Furthermore, Kinuta et al. [9] demonstrated that VD promoted folliculogenesis and follicular development in PCOS rats by increasing progesterone and estrogen levels and regulating the LH/FSH ratio.

Our results contribute to the mounting evidence from cross-sectional and interventional studies on favorable VD effects on reproduction [7,31,32]. It has been hypothesized that physiological levels of VD might have a beneficial role in ovulation and endometrial receptivity [33]. Consistently, findings from systematic reviews and meta-analyses analyzing the association of VD and assisted reproduction outcomes suggest that women with replete VD status have more live births, more positive pregnancy tests, and more clinical pregnancies compared to women with deficient or insufficient 25(OH)D [31,32]. Recently, Butts et al. [31] reported that VD deficiency in PCOS women who underwent ovarian stimulation for infertility treatment was linked with significantly diminished rates of ovulation, of pregnancy, and ultimately, a reduced chance of live birth. Of note, there was no significant association of VD deficiency with ovulation, pregnancy, or live birth in non-PCOS women with unexplained infertility [31]. In light of the high prevalence of insufficient VD levels in PCOS women [6] and the significant burden of decreased fertility in affected women, our findings deserve investigation in future large RCTs including PCOS women as well as women without PCOS. Considering the fact that VD supplementation is a safe and cheap treatment, our findings might be of high clinical interest. It should, however, be emphasized that the clinical relevance of our findings regarding reproduction remains to be determined as we investigated only surrogate parameters involved in fertility.

We failed to find a significant VD effect on AMH levels. Existing evidence on the relationship of VD and AMH levels is conflicting [19]. It has been shown that VD regulates AMH levels in vitro, both directly through the AMH promoter [34] and indirectly by regulating the number of granulosa cells and AMH signaling in cultures of ovarian follicles [35]. In contrast to the consistency of the in vitro data, the evidence of a link between VD and AMH in women is contentious. The majority of cross-sectional studies failed to find a significant correlation of 25(OH)D levels and AMH [19]. In contrast, a prospective study including PCOS women observed an association of VD supplementation with a decrease in serum AMH levels [36]. Furthermore, positive VD effects on AMH levels were found in a prospective study including infertile women with diminished ovarian reserve [24]. To date, there are only two small RCTs investigating VD effects on AMH levels in women [20,37]. In a study among VD-deficient infertile PCOS women, participants received either 50,000 IU VD/week ($n = 17$) or PBO ($n = 17$) for 8 weeks. The authors found a significant decrease in AMH levels in the VD group compared with PBO [20]. Dennis et al. [37] conducted an RCT in 49 young women with regular menses to evaluate the effects of a single high dose of VD (50,000 IU, taken on the first day of the menstrual cycle) versus PBO on AMH levels during the following week. Interestingly, the authors observed a significant progressive increase in AMH levels in the following week after VD supplementation. In our RCT, we found no significant VD effect on AMH levels in PCOS or non-PCOS women. These different results might be related to varying VD doses, study duration, baseline 25(OH)D levels, age, and sample size. However, as 25(OH)D concentrations at the end of the study were high in both groups, it is unlikely that the lack of a significant VD effect on AMH levels in our study is related to insufficient vitamin D doses.

Our study has several limitations. First, as we investigated women with relatively high baseline 25(OH)D levels, we cannot exclude significant VD effects on AMH levels in women with lower baseline 25(OH)D levels. Another possible limitation is the relatively high drop-out rate in PCOS women. Furthermore, since blood samples were collected regardless of the participants' menstrual cycle, the results regarding some of the measured parameters (e.g., FSH, LH, estradiol) should be interpreted with caution. As gonadotropins vary consistently during the phases of the menstrual cycle, our results regarding vitamin D effects on FSH and LH/FSH ratio should be interpreted in light of this limitation. We cannot rule out that blood sampling in the first week of the menstrual cycle in PCOS women with a regular menstrual cycle would provide different results. Nevertheless, only a small number of PCOS women had a regular menstrual cycle ($n = 19$) and the exclusion of these PCOS women from our analyses did not materially change our results. Moreover, AMH levels are stable across the menstrual cycle and typically demonstrate minimal intercycle and intracycle variability [38,39]. As we did not assess data on sun exposure, we were not able to adjust our analyses for this potential confounder. Furthermore, we did not assess the dietary pattern of study participants. This limitation might influence our results, as it has been demonstrated that specific diets such as the Mediterranean diet are associated with circulating 25(OH)D concentrations [40]. Finally, our findings should be interpreted with caution because our results derive from a post hoc analysis and we did not adjust for multiple testing as our analyses were all based on a priori pre-specified hypotheses. Nevertheless, we cannot exclude that our statistical analyses revealed a false-positive finding. The strengths of our study include its design as an RCT, the large sample size as well as the inclusion of PCOS and non-PCOS women.

5. Conclusions

In summary, we found no significant VD effect on AMH levels but a significant effect on FSH levels and LH/FSH ratio in PCOS women. Our results, therefore, support the idea that VD may be involved in reproductive function in PCOS women. In light of previous data suggesting a possible favorable VD effect on female fertility, further adequately powered RCTs are of clinical importance to clarify the potential positive effects of VD on reproductive function in PCOS women.

Supplementary Materials: The following are available online at https://www.mdpi.com/2072-6643/13/2/547/s1, Table S1: Outcome variables at baseline and follow-up at study end in PCOS women with available values at both study visits, Table S2: Outcome variables at baseline and follow-up at study end in non-PCOS women with available values at both study visits.

Author Contributions: Conceptualization, E.L., B.O.-P.; methodology, E.L., S.P.; formal analysis, E.L.; investigation, M.K., M.W., B.O.-P.; resources, M.K., M.W., B.O.-P.; data curation, E.L., C.T., V.T.-S.; writing—original draft preparation, E.L.; writing—review and editing, S.P., C.T., V.T.-S.; supervision, B.O.-P.; project administration, E.L., V.T.-S., B.O.-P.; funding acquisition, E.L. All authors have read and agreed to the published version of the manuscript.

Funding: This study was supported by funding from the Austrian Science Fund (FWF), project no.: KLI 274.

Informed Consent Statement: Informed consent was obtained from all subjects involved in the study.

Data Availability Statement: The data presented in this study are available on request from the corresponding author.

Acknowledgments: First, we thank all study subjects for participating in our study. Roswitha Gumpold recruited study participants. Furthermore, the Endocrinology Lab platform and Cornelia Missbrenner continuously supported our study and Fresenius Kabi provided the study medication. Open Access Funding by the Austrian Science Fund (FWF).

Conflicts of Interest: The authors declare no conflict of interest.

References

1. Holick, M.F. Vitamin D deficiency. *N. Engl. J. Med.* **2007**, *357*, 266–281. [CrossRef]
2. Wehr, E.; Pilz, S.; Schweighofer, N.; Giuliani, A.; Kopera, D.; Pieber, T.R.; Obermayer-Pietsch, B. Association of hypovitaminosis D with metabolic disturbances in polycystic ovary syndrome. *Eur. J. Endocrinol.* **2009**, *161*, 575–582. [CrossRef]
3. Trummer, C.; Pilz, S.; Schwetz, V.; Obermayer-Pietsch, B.; Lerchbaum, E. Vitamin D, PCOS and androgens in men: A systematic review. *Endocr. Connect.* **2018**, *7*, R95–R113. [CrossRef]
4. Pilz, S.; März, W.; Wellnitz, B.; Seelhorst, U.; Fahrleitner-Pammer, A.; Dimai, H.P.; Boehm, B.O.; Dobnig, H. Association of vitamin D deficiency with heart failure and sudden cardiac death in a large cross-sectional study of patients referred for coronary angiography. *J. Clin. Endocrinol. Metab.* **2008**, *93*, 3927–3935. [CrossRef]
5. Lerchbaum, E.; Pilz, S.; Trummer, C.; Rabe, T.; Schenk, M.; Heijboer, A.C.; Obermayer-Pietsch, B. Serum vitamin D levels and hypogonadism in men. *Andrology* **2014**, *2*, 748–754. [CrossRef]
6. Mu, Y.; Cheng, D.; Yin, T.L.; Yang, J. Vitamin D and polycystic ovary syndrome: A narrative review. *Reprod. Sci.* **2020**. [CrossRef]
7. Lerchbaum, E.; Rabe, T. Vitamin D and female fertility. *Curr. Opin. Obstet. Gynecol.* **2014**, *26*, 145–150. [CrossRef]
8. Irani, M.; Merhi, Z. Role of vitamin D in ovarian physiology and its implication in reproduction: A systematic review. *Fertil. Steril.* **2014**, *102*, 460–468. [CrossRef]
9. Kinuta, K.; Tanaka, H.; Moriwake, T.; Aya, K.; Kato, S.; Seino, Y. Vitamin D is an important factor in estrogen biosynthesis of both female and male gonads. *Endocrinology* **2000**, *141*, 1317–1324. [CrossRef]
10. Chu, J.; Gallos, I.; Tobias, A.; Tan, B.; Eapen, A.; Coomarasamy, A. Vitamin D and assisted reproductive treatment outcome: A systematic review and meta-analysis. *Hum. Reprod.* **2018**, *33*, 65–80. [CrossRef]
11. Savastano, S.; Barrea, L.; Savanelli, M.C.; Nappi, F.; Di Somma, C.; Orio, F.; Colao, A. Low vitamin D status and obesity: Role of nutritionist. *Rev. Endocr. Metab. Disord.* **2017**, *18*, 215–225. [CrossRef] [PubMed]
12. Barrea, L.; Frias-Toral, E.; Pugliese, G.; Garcia-Velasquez, E.; De Los Angeles Carignano, M.; Savastano, S.; Colao, A.; Muscogiuri, G. Vitamin D in obesity and obesity-related diseases: An overview. *Minerva Endocrinol.* **2020**. [CrossRef]
13. Goodarzi, M.O.; Dumesic, D.A.; Chazenbalk, G.; Azziz, R. Polycystic ovary syndrome: Etiology, pathogenesis and diagnosis. *Nat. Rev. Endocrinol.* **2011**, *7*, 219–231. [CrossRef]
14. Norman, R.J.; Dewailly, D.; Legro, R.S.; Hickey, T.E. Polycystic ovary syndrome. *Lancet* **2007**, *370*, 685–697. [CrossRef]
15. Palomba, S.; Falbo, A.; Chiossi, G.; Muscogiuri, G.; Fornaciari, E.; Orio, F.; Tolino, A.; Colao, A.; La Sala, G.B.; Zullo, F. Lipid profile in nonobese pregnant women with polycystic ovary syndrome: A prospective controlled clinical study. *Steroids* **2014**, *88*, 36–43. [CrossRef]
16. Barrea, L.; Arnone, A.; Annunziata, G.; Muscogiuri, G.; Laudisio, D.; Salzano, C.; Pugliese, G.; Colao, A.; Savastano, S. Adherence to the Mediterranean diet, dietary patterns and body composition in women with Polycystic Ovary Syndrome (PCOS). *Nutrients* **2019**, *11*, 2278. [CrossRef]
17. Barrea, L.; Marzullo, P.; Muscogiuri, G.; Di Somma, C.; Scacchi, M.; Orio, F.; Aimaretti, G.; Colao, A.; Savastano, S. Source and amount of carbohydrate in the diet and inflammation in women with polycystic ovary syndrome. *Nutr. Res. Rev.* **2018**, *31*, 291–301. [CrossRef]
18. Qi, X.; Pang, Y.; Qiao, J. The role of anti-Müllerian hormone in the pathogenesis and pathophysiological characteristics of polycystic ovary syndrome. *Eur. J. Obstet. Gynecol. Reprod. Biol.* **2016**, *199*, 82–87. [CrossRef]
19. Moridi, I.; Chen, A.; Tal, O.; Tal, R. The Association between Vitamin D and Anti-Müllerian Hormone: A systematic review and meta-analysis. *Nutrients* **2020**, *12*, 1567. [CrossRef] [PubMed]
20. Dastorani, M.; Aghadavod, E.; Mirhosseini, N.; Foroozanfard, F.; Zadeh Modarres, S.; Amiri Siavashani, M.; Asemi, Z. The effects of vitamin D supplementation on metabolic profiles and gene expression of insulin and lipid metabolism in infertile polycystic ovary syndrome candidates for in vitro fertilization. *Reprod. Biol. Endocrinol.* **2018**, *16*, 94. [CrossRef]
21. Trummer, C.; Schwetz, V.; Kollmann, M.; Wölfler, M.; Münzker, J.; Pieber, T.R.; Pilz, S.; Heijboer, A.C.; Obermayer-Pietsch, B.; Lerchbaum, E. Effects of vitamin D supplementation on metabolic and endocrine parameters in PCOS: A randomized-controlled trial. *Eur. J. Nutr.* **2019**, *58*, 2019–2028. [CrossRef]
22. Trummer, C.; Theiler-Schwetz, V.; Kollmann, M.; Wölfler, M.; Münzker, J.; Pilz, S.; Pieber, T.R.; Heijboer, A.C.; Obermayer-Pietsch, B.; Lerchbaum, E. Effects of vitamin D supplementation on metabolic and endocrine parameters in healthy premenopausal women: A randomized controlled trial. *Clin. Nutr.* **2020**, *39*, 718–726. [CrossRef]
23. Rotterdam ESHRE/ASRM-Sponsored PCOS Consensus Workshop Group. Revised 2003 consensus on diagnostic criteria and long-term health risks related to polycystic ovary syndrome (PCOS). *Hum. Reprod.* **2004**, *19*, 41–47. [CrossRef]
24. Naderi, Z.; Kashanian, M.; Chenari, L.; Sheikhansari, N. Evaluating the effects of administration of 25-hydroxyvitamin D supplement on serum anti-mullerian hormone (AMH) levels in infertile women. *Gynecol. Endocrinol.* **2018**, *34*, 409–412. [CrossRef] [PubMed]

25. Wehr, E.; Pieber, T.R.; Obermayer-Pietsch, B. Effect of vitamin D3 treatment on glucose metabolism and menstrual frequency in polycystic ovary syndrome women: A pilot study. *J. Endocrinol. Investig.* **2011**, *34*, 757–763. [PubMed]
26. Trummer, O.; Schweighofer, N.; Haudum, C.W.; Trummer, C.; Pilz, S.; Theiler-Schwetz, V.; Keppel, M.H.; Grübler, M.; Pieber, T.R.; Renner, W.; et al. Genetic components of 25-Hydroxyvitamin D increase in three randomized controlled trials. *J. Clin. Med.* **2020**, *9*, 570. [CrossRef]
27. Yen, S.S. The polycystic ovary syndrome. *Clin. Endocrinol.* **1980**, *12*, 177–207. [CrossRef]
28. McKenna, T.J. Pathogenesis and treatment of polycystic ovary syndrome. *N. Engl. J. Med.* **1988**, *318*, 558–562. [CrossRef]
29. Johansson, J.; Stener-Victorin, E. Polycystic ovary syndrome: Effect and mechanisms of acupuncture for ovulation induction. *Evid. Based Complement. Alternat. Med.* **2013**, *2013*, 762615. [CrossRef]
30. Behmanesh, N.; Abedelahi, A.; Charoudeh, H.N.; Alihemmati, A. Effects of vitamin D supplementation on follicular development, gonadotropins and sex hormone concentrations, and insulin resistance in induced polycystic ovary syndrome. *Turk. J. Obstet. Gynecol.* **2019**, *16*, 143–150. [CrossRef] [PubMed]
31. Butts, S.F.; Seifer, D.B.; Koelper, N.; Senapati, S.; Sammel, M.D.; Hoofnagle, A.N.; Kelly, A.; Krawetz, S.A.; Santoro, N.; Zhang, H.; et al. Eunice kennedy shriver national institute of child health and human development reproductive medicine network. Vitamin D deficiency is associated with poor ovarian stimulation outcome in PCOS but not unexplained infertility. *J. Clin. Endocrinol. Metab.* **2019**, *104*, 369–378. [CrossRef]
32. Pilz, S.; Zittermann, A.; Obeid, R.; Hahn, A.; Pludowski, P.; Trummer, C.; Lerchbaum, E.; Pérez-López, F.R.; Karras, S.N.; März, W. The role of Vitamin D in fertility and during pregnancy and lactation: A review of clinical data. *Int. J. Environ. Res. Public Health* **2018**, *15*, 2241. [CrossRef]
33. Firouzabadi, R.D.; Rahmani, E.; Rahsepar, M.; Firouzabadi, M.M. Value of follicular fluid vitamin D in predicting the pregnancy rate in an IVF program. *Arch. Gynecol. Obstet.* **2014**, *289*, 201–206. [CrossRef]
34. Malloy, P.J.; Peng, L.; Wang, J.; Feldman, D. Interaction of the vitamin D receptor with a vitamin D response element in the Mullerian-inhibiting substance (MIS) promoter: Regulation of MIS expression by calcitriol in prostate cancer cells. *Endocrinology* **2009**, *150*, 1580–1587. [CrossRef]
35. Merhi, Z.; Doswell, A.; Krebs, K.; Cipolla, M. Vitamin D alters genes involved in follicular development and steroidogenesis in human cumulus granulosa cells. *J. Clin. Endocrinol. Metab.* **2014**, *99*, E1137–E1145. [CrossRef] [PubMed]
36. Irani, M.; Minkoff, H.; Seifer, D.B.; Merhi, Z. Vitamin D increases serum levels of the soluble receptor for advanced glycation end products in women with PCOS. *J. Clin. Endocrinol. Metab.* **2014**, *99*, E886–E890. [CrossRef]
37. Dennis, N.A.; Houghton, L.A.; Pankhurst, M.W.; Harper, M.J.; McLennan, I.S. Acute supplementation with high dose Vitamin D3 Increases serum anti-müllerian hormone in young women. *Nutrients* **2017**, *9*, 719. [CrossRef] [PubMed]
38. Dorgan, J.F.; Spittle, C.S.; Egleston, B.L.; Shaw, C.M.; Kahle, L.L.; Brinton, L.A. Assay reproducibility and within-person variation of Müllerian inhibiting substance. *Fertil. Steril.* **2010**, *94*, 301–304. [CrossRef]
39. Tal, R.; Seifer, D.B. Ovarian reserve testing: A user's guide. *Am. J. Obstet. Gynecol.* **2017**, *217*, 129–140. [CrossRef]
40. Barrea, L.; Muscogiuri, G.; Laudisio, D.; Pugliese, G.; de Alteriis, G.; Colao, A.; Savastano, S. Influence of the Mediterranean diet on 25- Hydroxyvitamin D levels in adults. *Nutrients* **2020**, *12*, 1439. [CrossRef]

 *nutrients*

Review

Medication-Related Osteonecrosis of the Jaw (MRONJ): Are Antiresorptive Drugs the Main Culprits or Only Accomplices? The Triggering Role of Vitamin D Deficiency

Luca Dalle Carbonare [1,*], Monica Mottes [2] and Maria Teresa Valenti [1]

[1] Department of Medicine, Section of Internal Medicine, University of Verona, 37134 Verona, Italy; mariateresa.valenti@univr.it
[2] Department of Neurosciences, Biomedicine and Movement Sciences, Section of Biology and Genetics, University of Verona, 37134 Verona, Italy; monica.mottes@univr.it
* Correspondence: luca.dallecarbonare@univr.it; Tel.: +39-045-8126062

Abstract: Osteonecrosis of the jaw (ONJ) is a severe clinical condition characterized mostly but not exclusively by an area of exposed bone in the mandible and/or maxilla that typically does not heal over a period of 6–8 weeks. The diagnosis is first of all clinical, but an imaging feedback such as Magnetic Resonance is essential to confirm clinical suspicions. In the last few decades, medication-related osteonecrosis of the jaw (MRONJ) has been widely discussed. From the first case reported in 2003, many case series and reviews have appeared in the scientific literature. Almost all papers concerning this topic conclude that bisphosphonates (BPs) can induce this severe clinical condition, particularly in cancer patients. Nevertheless, the exact mechanism by which amino-BPs would be responsible for ONJ is still debatable. Recent findings suggest a possible alternative explanation for BPs role in this pattern. In the present work we discuss how a condition of osteomalacia and low vitamin D levels might be determinant factors.

Keywords: aminobisphosphonates; BRONJ; denosumab; MRONJ; osteomalacia; osteonecrosis; jaw; pathophysiology

Citation: Dalle Carbonare, L.; Mottes, M.; Valenti, M.T. Medication-Related Osteonecrosis of the Jaw (MRONJ): Are Antiresorptive Drugs the Main Culprits or Only Accomplices? The Triggering Role of Vitamin D Deficiency. *Nutrients* **2021**, *13*, 561. https://doi.org/10.3390/nu13020561

Academic Editors: Spyridon N. Karras and Pawel Pludowski
Received: 5 December 2020
Accepted: 2 February 2021
Published: 8 February 2021

Publisher's Note: MDPI stays neutral with regard to jurisdictional claims in published maps and institutional affiliations.

1. Introduction

Osteonecrosis of the jaw (ONJ) is a severe clinical condition characterized by an area of exposed bone in the mandible and/or maxilla that typically does not heal over a period of 6–8 weeks. The diagnostic criteria were updated in 2014 by the American Association of Oral and Maxillofacial Surgeons and based on clinical features and radiologic imaging in presence of pharmacological history or ongoing use of antiresorptive agents, in particular bisphosphonates (BPs) or antiangiogenic agents such as monoclonal antibodies targeting vascular endothelial growth factor (VEGF) receptors [1]. A special committee of the American Association of Oral and Maxillofacial Surgeons (AAOMS) suggested changing the nomenclature of bisphosphonate-related osteonecrosis of the jaw (BRONJ) to medication-related osteonecrosis of the jaw (MRONJ) as a consequence of increasing cases of osteonecrosis due to the association with other antiresorptive and antiangiogenic therapies [1]. However, the nomenclature concerning this pathology has been and still is the subject of debate [2–5]. MRONJ classification considers four disease stages. Stage 0—the prodromal period. No clinical evidence of necrotic bone, and nonspecific clinical findings, radiographic changes, and symptoms. Radiographically, it can reveal an unexplained bone loss not attributed to periodontal inflammation with a change in trabecular bone pattern; Stage 1—Exposed and necrotic bone, or fistulae that probe to the bone in asymptomatic patients who have no evidence of infection. These patients may also present with the radiographic findings mentioned for Stage 0 which are localized to the alveolar bone region; Stage 2—Exposed and necrotic bone, or fistulae that probe to the bone, associated with infection as evidenced by pain and erythema in the region of the exposed bone, with or without purulent drainage. These patients are typically symptomatic; Stage 3—Exposed and necrotic bone, or fistulae that probe to bone, with evidence of infection, and one or more of the following: (1) exposed necrotic bone extending beyond the region of alveolar bone, i.e., inferior border and ramus in the mandible, maxillary sinus and zygoma in the maxilla; (2) pathologic fracture; (3) extraoral fistula; (4) oral antral/oral nasal communication; (5) osteolysis extending to the inferior border of the mandible or sinus floor [1].

The diagnosis is first of all clinical, but 3D imaging techniques (CT, cone beam), Single-Photon Emission Tomography (SPECT) and Magnetic Resonance (MR), are important to confirm the clinical suspicions. In the last decade, the problem of ONJ has been widely discussed. From the first case reported in 2003, many additional case series and reviews appeared in the scientific literature. Almost all publications concerning this topic conclude that BPs, antiresorptive and antiangiogenic drugs can induce this severe clinical condition, particularly in cancer patients. Generally, antiresorptive drugs are bone targeting agents used to prevent skeletal resorption following different pathologies such as metabolic and degenerative diseases. In addition to bone targeting drugs, medications without antiresorptive properties such as angiogenic inhibitors, tyrosine kinase inhibitors or inhibitors of the mammalian target of rapamycin (mTOR) and cytotoxic molecules used for chemotherapy may also increase the risk of osteonecrosis of the jaw [6].

Nevertheless, the exact mechanism by which these drugs, in particular amino-BPs, would be responsible for MRONJ is still subject to discussion. However, many hypotheses have been proposed and different pathophysiological mechanisms have been investigated, supporting the main role of drugs in the pathogenesis of this severe condition.

Among other hypotheses, osteomalacia following vitamin D deficiency has been considered an important factor in the pathogenesis of ONJ.

Therefore, with the aim to describe the ONJ problem and the medical context of this pathology, in this review we discuss recent studies related to ONJ and antiresorptive drugs, as well as the involvement of osteomalacia due to low vitamin D levels as a triggering factor for ONJ.

2. Epidemiology

ONJ lesions occur more commonly in the mandible than in the maxilla (65% mandible, 28.4% maxilla, 6.5% both mandible and maxilla, 0.1% other locations). ONJ incidence in patients who are prescribed oral BPs for the treatment of osteoporosis is very low and ranges from 1.04 to 69 per 100,000 patient-years. The incidence of ONJ in patients prescribed intravenous (i.v.) BPs for the treatment of ONJ ranges from 0 to 90 per 100,000 patient-years. With regard to denosumab, ONJ incidence ranges from 0 to 30.2 per 100,000 patient-years [7]. Based on different national surveys the incidence of ONJ in osteoporotic patients receiving BPs ranged from 0.01% to 0.07% [8,9]. On the basis of these epidemiologic data, ONJ impact on the osteoporotic population appears to be very rare and therefore negligible.

In cancer patients treated with i.v. BPs the incidence of ONJ is higher, ranging from 0 to 12,222 per 100,000 patient-years [7]; recently, an incidence of about 0.8% (48 out of 6018) in breast cancer patients has been observed [10]. However, ONJ incidence in this particular setting may be influenced by the malignancy type/severity as well as by the assumption of other drugs that may impact bone health, such as glucocorticoids. In addition, in the presence of bone metastases, the doses of drugs used for the management of bone disease are significantly higher compared to those used in osteoporosis, therefore the oncologic setting appears to be very peculiar compared to other clinical conditions involving the skeleton.

Considering the epidemiological data discussed so far and the prognostic clinical impact of fragility fractures increasing morbidity and disability, as well as mortality, the precautionary interruption of an antifracture treatment should be carefully evaluated. If we assume that an antiresorptive therapy may grant a long-term fracture risk reduction of around 30%, the benefit/risk ratio (prevented fracture/adverse skeletal event), particularly in high-risk subjects, would be at least 100:1 [11]. In addition, as has been recently observed, the interruption of the antiresorptive therapy, in particular denosumab treatment, is associated with a significant fracture rate increase [12].

However, it is important to highlight that, even in long-term treatments, serious adverse event rates are generally stable over time, varying between 11.5 and 14.4 per 100 participant-years, against a 10.9 to 11.7 withdrawal per 100 participant-years in placebo [11,13]. These data further support the pursuance of antifracture therapy in high-risk patients. From another point of view, emerging evidence confirms that antiresorptive drug treatment discontinuation aimed at ONJ risk reduction is unneeded [14].

3. Clinical and Genetic Risk Factors for ONJ

Many clinical factors have been considered in the pathogenesis of ONJ, particularly dental surgery. A recent study reported that in 48 patients ONJ triggers were: dental extraction in 20 of them (35.1%), periodontal disease in 14 (24.6%), denture trauma in 6 (10.5%), other dental surgery in 2 (3.5%). Spontaneous ONJ was observed in 20 patients (35.1%). Infection was present in 13/27 (48.1%) induced ONJ and in 7/18 (38.9%) spontaneous ONJ cases [10].

The patients' features are also important: immunodeficiency, comorbidities such as diabetes as well as the presence of autoimmune diseases have been suggested as risk factors for ONJ [15]. The local triggering factors were examined in a recent review: tooth extraction was reckoned in 46% of individuals, implant placement in 14%, prosthetic trauma in 14% [16].

Genetic and epigenetic studies have been performed to evaluate individual risks of developing ONJ in patients treated with antiresorptive drugs. It has been reported that the A allele frequency of the A/C rs2297480 polymorphism of farnesyl pyrophosphate synthase (FDPS), an enzymatic target of BPs, correlates positively with ONJ after 18–24 months of zoledronate treatment [17]. A genome-wide association study (GWAS), has reported that a single nucleotide polymorphism (SNP) occurring in Cytochrome p450 CYP2C8 is associated with a higher risk to develop ONJ in patients affected by multiple myeloma treated with BP therapy [18]. An exome-wide association analysis (ExWAS), highlighted two SNPs on chromosome 10 occurring in two promoter sequences of the SIRT1/HERC4 locus which seemed to be associated with MRONJ [19]. On the other hand, the promoter SNP rs932658 regulates the expression of SIRT1 and presumably lowers the risk of MRONJ by increasing SIRT1 expression [20]. According to this hypothesis, in the presence of high concentrations of BP in bone, or with frequent intravenous dosing, toxicity to other bone cells including soft tissue might occur. Concerning this aspect, the potential role of cumulative doses of BPs in fostering the onset of bone alterations seems unlikely, particularly for zoledronic acid [8]. In fact, ONJ has been observed with a wide range of BP doses, varying from a single dose of zoledronic acid (4 mg) to 60 administrations [8]; risk increases dramatically with higher cumulative doses, higher administrations, and longer observation time.

In such a complex scenario the precise role and action course of BPs, denosumab or antiangiogenic drugs in MRONJ is still under discussion. It is important to ascertain whether they are the main culprits or rather detrimental factors, among others, in the pathogenesis of ONJ. Such elucidation would improve the management of patients at high fracture risk requiring long-term antifracture treatments.

4. Bone Remodeling Impairment

Bone remodeling is a crucial lifelong process that allows old bone tissue removal from the skeleton and its replacement with new bone. It also ensures bone reshaping/replacement following fractures and microdamage. Osteoclasts (OC) literally "bone-breaking cells", perform bone resorption, while osteoblasts (OB) produce the collagen rich extracellular matrix and participate in its mineralization. Osteoclast and osteoblast activity must be balanced through coupling in order to maintain skeletal mass throughout the lifespan; however, certain diseases and aging itself lead to unbalanced pathological conditions. Regarding the topic under discussion in the present review, impairment of osteoclast-mediated bone remodeling and angiogenesis have been reckoned to play a major pathogenetic role in MRONJ [21,22]. The site-specific effect, restricted to the jaw bone, is ascribed to a differentiated proliferation and osseous response to BPs by craniofacial bones, due to their different embryonic origin (i.e., from the cranial neural crest). Antiresorptive drugs (BPs and Denosumab) target OCs but affect OBs as well. Bone homeostasis depends on OC/OB crosstalk which is regulated by the RANK-RANKL-OPG network; osteoclasts targeting drugs might favor ONJ by disrupting the coupling process [23]. Osteocytes, mature osteoblasts embedded into the mineralized matrix, which are the most abundant and long-lived cells in bone, play an important role in bone remodeling control, by secreting Sclerostin and DKK1, two inhibitors of WNT signaling pathway, and RANKL, which reduce bone formation (Figure 1). In vitro experiments on MLO-Y4, an osteocyte-like cell line, have shown that Zoledronate administration significantly enhanced both RANKL and Sclerostin expression [24]. These data demonstrate that BPs also exert their influence on osteocytes, suggesting osteocytes' potential role in MRONJ development. Jaw predisposition to MRONJ is justified by the very rapid turnover rate in alveolar bone [25].

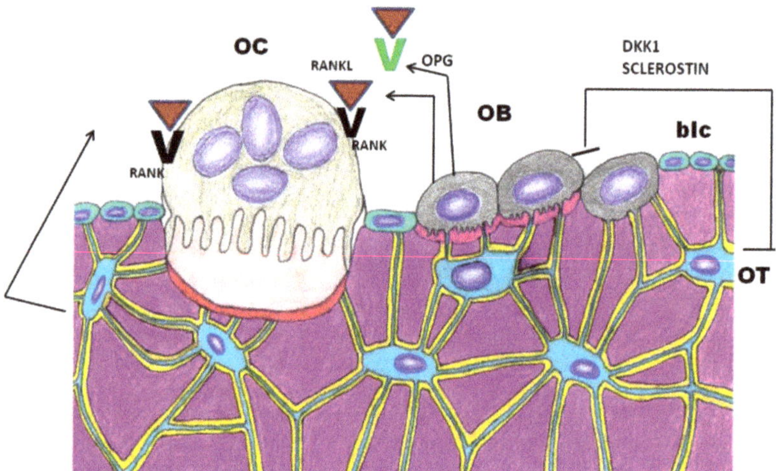

Figure 1. Schematic representation of Osteoclast/Osteoblast/Osteocyte cross-talk mediators involved in bone remodeling. OC: osteoclast; OB: osteoblast; OT: osteocyte; blc: bone lining cell. DKK1: Dickkopf WNT signaling pathway inhibitor 1; RANKL: Receptor activator of nuclear factor kappa-B ligand; RANK: Receptor activator of nuclear factor kappa-B; OPG: osteoprotegerin, a RANKL decoy receptor.

5. How Do Different Antiresorptive Drugs Interfere with Bone Turnover

In respect to turnover suppression, we have previously observed that zoledronic acid increases the anabolic window preserving bone formation activity compared to other less powerful BPs such as risedronate, avoiding the so-called frozen bone [26,27]. From another point of view, Reid in 2009 suggested that MRONJ is caused by powerful BPs direct toxicity to bone and soft tissue cells, probably deriving from their effects on the mevalonate pathway [28]. BPs concentration in the jaw can be higher compared to other skeleton areas [29]. In fact, BPs preferably affect this area in consequence of its higher remodeling and turnover rate. By suppressing bone metabolism, BPs may induce physiological microdamage in the jaw affecting its biomechanical abilities. In addition, a lower pH consequent to oral invasive procedures allows BPs accumulation, i.e., toxic concentrations. It has been suggested that the fostering factors for MRONJ are: BPs potency, treatment duration, concomitant oral surgery [30,31]. In addition to BPs, other therapeutic molecules can inhibit osteoclasts like denosumab, an anti-RANKL monoclonal antibody is currently used for the treatment of osteoporosis, primary and metastatic bone cancer as well as rheumatoid arthritis [32–34]. However, cases of ONJ have been reported in patients receiving denosumab [35,36].

6. Animal Models Contribution to ONJ Studies

A suitable animal model is necessary to better understand the pathophysiology of ONJ. The challenging task is to generate animal models showing signs similar to ONJ clinical picture [37]. The in vivo model should expose the oral cavity bone following bisphosphonate treatment in association with other factors occurring in humans such as dental trauma or immunosuppression [38,39]. It is important to consider that ONJ occurs in humans after at least 8 weeks exposure. Timing may vary for animal models. Therefore, establishing the correct timeline for the observation of ONJ effects represents a starting point for studies related to the physiology and pathophysiology of the jaw.

Studies performed in dogs demonstrated that the bone turnover rate in the jaw is 6/10-fold higher than in long bones. Such bone turnover might increase 10-fold further upon dental extraction [40,41].

Since BPs affect bone by suppressing its turnover, the suppression/reduction of bone turnover might be considered the main cause of ONJ pathophysiology [42,43]. As intra-cortical remodeling suppression is a favoring factor for ONJ, it has been hypothesized that animal species with intracortical remodeling may render ONJ effects more appropri-ately [37]. Allen et al. have chosen dogs, characterized by intracortical remodeling in the skeleton and, in particular, in the jaw. For this purpose, the authors used intact female beagles treated daily with vehicle or alendronate (0.20 or 1.0 mg/kg/day) and the duration of this treatment was one or three years. During this study the authors reported exposed oral bone absence in all animals; jaw matrix necrosis areas were observed in 25% of dogs treated with the lower doses, in 17% or 33% of dogs treated with the higher dose after 1 year or 3 years, respectively [37].

Another suitable animal model for studying ONJ is the rodent. Rodents are widely used for studies related to skeletal diseases. However, the absence in rodents of intracortical remodeling, a favoring factor for ONJ, generally limits their use. However, it has been demonstrated that intracortical turnover occurs in C3H mice long bones [44] suggesting that selected mice strains may be useful for studying ONJ.

Recently, Holtmann et al. consulted Embase, Medline, and the Cochrane Library in order to identify the appropriate model for MRONJ [45]. In this retrospective study, the authors found that rats, mice, dogs, minipigs, sheep and rabbits were the most used animal models. In particular, studies performed on the rat model focused on BPs' effects on the jaw after tooth extraction. According to Vasconcelos et al. clodronate (a nonamino-BPs), was less likely to induce ONJ than zoledronate [46]. However, most of the other studies em-ployed amino-BPs such as zoledronate, alendronate or pamidronate. Studies performed by using zoledronate in rats clearly showed the effects of ONJ [47], while the administration of alendronate showed controversial results. The combined use of an amino-bisphosphonate plus a corticosteroid led to the appearance of ONJ-like lesions [48]; Sonis et al. observed in rats treated with bisphosphonate and corticosteroid more relevant ONJ lesions than in zoledronate-only administration [49]. Aghaloo et al. observed that periodontitis is a triggering factor for the development of ONJ with high-dose administration of zoledronate; other studies confirm these results [50].

In studies employing murine models the effects of bisphosphonate in association with corticosteroids compared to the effects of zoledronate alone have also been investigated. The combination of bisphosphonate together with corticosteroids seems to enhance the development of ONJ lesions following tooth extraction in mice. However, other authors did not observe these effects [51]. In addition, it has been demonstrated that the presence of a periapical disease in mice promotes ONJ following zoledronate administration or treatment with the anti-receptor activator of nuclear factor kappa beta ligand antibody (anti-RANKL Ab) [52]. Such a finding has not been observed in rats. In studies using a pig model or a sheep model treated with zoledronate alone or zoledronate in association with corticosteroids, respectively, the presence of ONJ lesions was observed [53–55]. Pig is a very useful model for studying skeletal diseases as its bone regeneration pattern is similar to what is expected in humans [53]. In particular, the minipig is considered the most reliable model for ONJ pathophysiology investigations [53,54]. Yet, due to actual bone physiology differences, the direct translation of animal model findings to human ONJ pathophysiology is questionable.

7. Osteomalacia and Vitamin D

Osteomalacia is characterized by low phosphate levels causing impaired bone mineralization, bone pains, myopathies and enthesopathies. In addition to hypophosphatemia, biochemical aspects include normal or low levels of serum calcium, normal or high levels or alkaline phosphatase, low or insufficient levels of serum 1, 25 dihydroxy vitamin D as well as normal serum intact parathormone levels and alterations related to the maximum tubular resorptive capacity for phosphorus/glomerular filtration rate [56,57]. The causes of osteomalacia may be identified in underlying mechanisms such as vitamin D deficiency/resistance, vitamin D-independent low calcium serum levels, hypophosphatemic diseases, mineralization impairment due to aluminum toxicity (antacids, dialysis fluid), fluorosis (i.e., endemic fluorosis from borehole water) iron (in dialysis patients, or patients with FGF23 mediated hypophosphatemia), etidronate overdose (in Paget's disease), or environmental intoxication with cadmium [58]. In addition, metabolic acidosis occurring in gastrointestinal or renal disorders may contribute to bone mineralization disruption [58]. Recently, the involvement of FGF23, an osteocyte-borne hormone, in osteomalacia has been suggested [59]. Osteomalacia can also be related to congenital connective tissue disorders such as osteogenesis imperfecta type VI [60] or the rare fibrogenesis imperfecta ossium [61]. However, the actual prevalence of osteomalacia is elusive. In the Middle East and Asia low calcium intake and severe vitamin D deficiency are common; in Pakistan a prevalence of 2% to 3.6% of diagnosed osteomalacia has been reported in young women [62]. In Western countries elderly people are at high risk of osteomalacia: a 2% to 5% prevalence for this disorder has been reported in different studies [63,64]. Interestingly, a larger number of biopsies revealed a 4.9% prevalence of individuals with osteomalacic features in Germany [65]. In addition, osteomalacia is present in patients with gastrointestinal disorders (i.e., celiac disease) [66]. After gastric bypass surgery, patients may develop vitamin D deficiency even if only 25% of these bariatric patients with suspected osteomalacia will be actually confirmed as osteomalacic by histomorphometric analyses [67,68]. Hypovitaminosis D has been found in prostate, multiple myeloma, colorectal and breast cancer patients [69]. In particular, Nogues et al. found vitamin D deficiency in 85–92% of breast cancer patients [70] whereas Neuhouser et al. reported a prevalence of 76.8% of vitamin D insufficiency in a study conducted on 426 breast cancer survivors [71]. Other studies conducted on breast cancer patients confirmed a prevalence >70% of vitamin D deficiency [72–74]. Fakih et al. reported in a study performed on colorectal cancer patients that 21% stage I-III patients and 32% stage IV patients had very low vitamin D serum levels (<15 ng/mL) [75]. Trump et al., performing a case control study in prostate cancer patients reported vitamin D deficiency (<20 ng/mL) and insufficiency (20–31 ng/mL) in 40% and 32% of cases respectively; notably, the authors found 31% vitamin D deficiency and 40% insufficiency among controls [76]. Finally, an alarming study reported vitamin D deficiency in metastatic bone disease and multiple myeloma patients [77]. In this study the authors reported that serum 25-OH-D levels are rarely sufficient in breast, prostate or MM bone metastatic patients.

8. Vitamin D and Oral Pathology

Various studies highlight the role of vitamin D in oral pathology. Vitamin D plays an important role in periodontology as it contributes to maintaining healthy periodontal tissues, reducing the risks of gingivitis and chronic periodontitis by activating the immune response [78]. A study performed in 562 senior citizens demonstrated that subjects receiving a high vitamin D dose (>800 IU daily) showed a lower risk of developing a severe form of chronic periodontitis compared to those receiving a lower vitamin D daily dose (<400 IU) [79]. Furthermore, the association between low levels of vitamin D intake and increased caries risk has been reported in children in different studies [80–83]. The alleged association between vitamin D deficiency and ONJ is an intriguing topic. In fact, some researchers did not find such an association [84], while others have observed that low vitamin D levels are risk factors for the development of ONJ [85,86]. In particular, in a randomized study performed in osteoporotic patients no correlation between vitamin D intake and ONJ was found [84]. On the contrary, MRONJ prevalence was reported in patients with low vitamin D levels in a two-year retrospective study performed in 63 patients treated with antiresorptive medication [85]. Recently, Demircan et al. performed a case control study (20 patients with ONJ and 20 healthy controls) in order to evaluate bone marker levels in bisphosphonate-induced-ONJ [86]. Interestingly, the researchers found higher PTH levels and lower TSH, Vit-D, osteocalcin and NTX levels in ONJ patients compared to controls.

9. Histomorphometric Study

Some years ago, the results of a histomorphometric study performed in our laboratory suggested a possible novel explanation for BPs' role in this pattern [87]. In the cited study, we performed jaw bone biopsies in patients treated with BPs with or without ONJ and we found a mineralization defect in the jaws of all ONJ patients, highlighting the presence of osteomalacia at the histological level (Figure 2). On the contrary, control subjects did not show any osteomalacic pattern in jaw biopsies. Furthermore, control subjects, who had been followed up as part of a cohort study, did not develop any sign of MRONJ up to one year after bone sampling. Interestingly, four patients who had been excluded from the study because of osteomyelitis, turned out to be osteomalacic upon histomorphometric evaluation of jaw biopsies and developed clinical and/or radiological signs of MRONJ within six months, suggesting the mineralization defect to be a pivotal factor in the pathogenesis of ONJ.

From the histological point of view, osteomalacia is characterized by inadequate or delayed mineralization of osteoid in mature cortical and spongy bone, leading to bone softening and sclerosis. When these aspects are referred to the jaw, they appear consistent with the ONJ stage 0 characteristics. Furthermore, the osteomalacic condition might be a necessary but not sufficient prodromal condition for ONJ development. This histologic pattern may facilitate inflammatory/infective processes preventing complete bone restoration, which may be further biased by BP administration. Whether osteomalacia in ONJ patients is a local phenomenon or a systemic condition is still questionable. However, this is not a relevant issue, since even focal osteomalacia can lead to the previously described bone alterations characterizing ONJ.

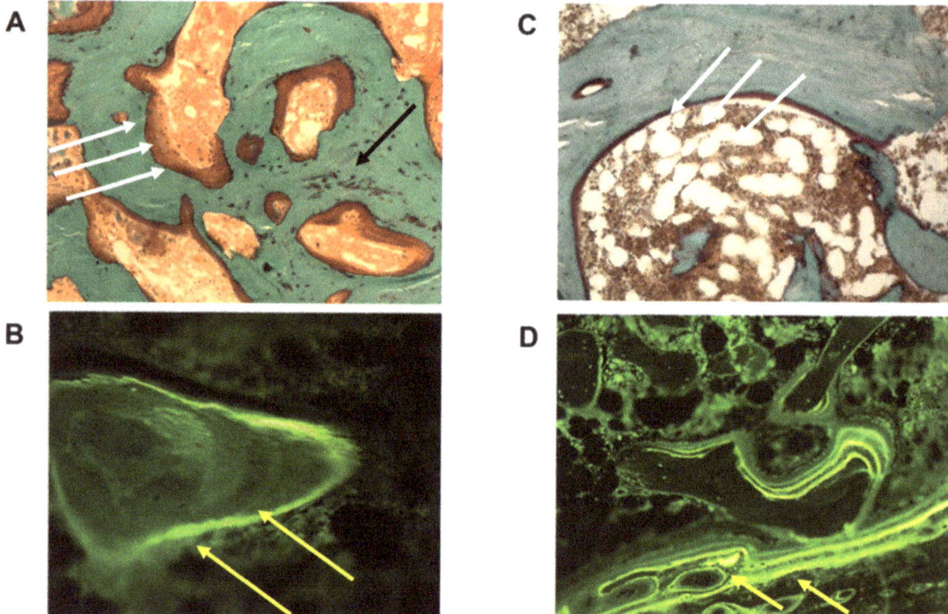

Figure 2. Histological section of the jaw. Left panels show biopsy from patient with Osteonecrosis of the jaw (ONJ), Goldner-stained (**A**) and tetracycline double labeling sections (**B**). Right panels represent an age-matched control subject. Note the large quantity of unmineralized osteoid ((**A**), red color, white arrows), area of woven bone ((**A**), black arrow), lacking the double labeling ((**B**), yellow arrows) in ONJ compared to control (**C,D**). These findings represent the histological pattern of a mineralization defect (magnification 200×).

10. Discussion and Conclusions

On the basis of the histomorphometric results discussed above, BPs' role in the pathogenesis of ONJ should be reviewed. In particular, as has been described previously for patients with bone diseases, the treatment with BPs in the presence of osteomalacia can emphasize a mineralization defect [88]. Consequently, BPs' contribution to the pathogenesis of ONJ might be secondary to the osteomalacic condition. The finding of an impaired turnover, consequent to the mineralization defect, rather than an excessive osteoclasts inhibition induced by BPs or by other antiresorptive agents, represents a new insight in the pathogenesis of ONJ. Moreover, it is worth stressing that more powerful antiresorptive agents (e.g., zoledronic acid and denosumab) contribute more to MRONJ pathogenesis than less powerful and structurally different molecules such as clodronate or oral formulations [89,90]. Recently, it has been observed that nonamino-BPs could also prevent ONJ due to their most potent antioxidant and anti-inflammatory activities [91]. On the other hand, an osteomalacic pattern, frequently secondary to vitamin D deficiency, rather than BPs potency or cumulative doses, may explain the high incidence of ONJ in cancer patients (Figure 3).

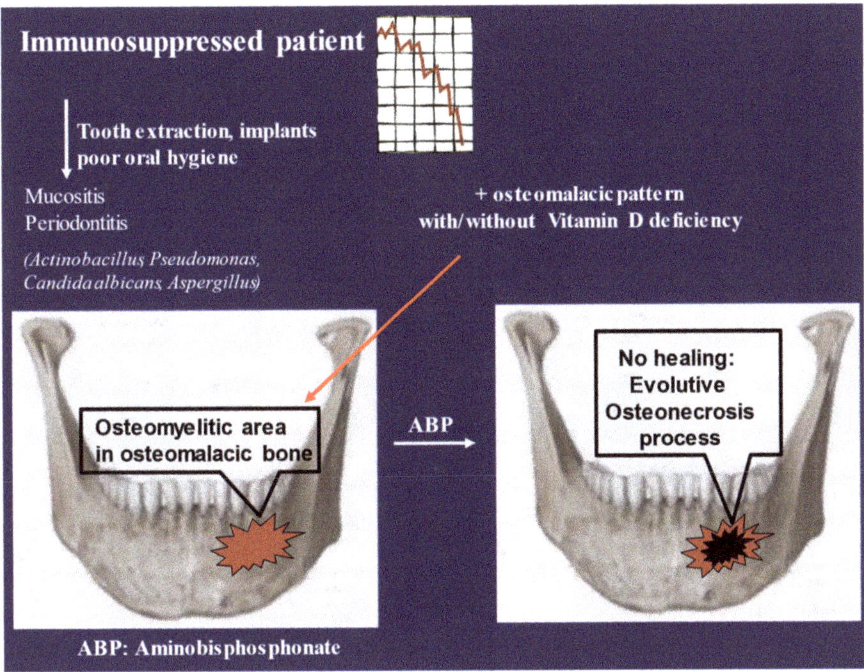

Figure 3. Graphical updated pathophysiologic mechanism of medication-related osteonecrosis of the jaw (MRONJ): In immunosuppressed patients (causes: cancer, chemotherapy, corticoids), surgical interventions and/or poor oral hygiene promote an osteomyelitis complication. In the presence of osteomalacic bone, the use of aminobisphosphonates (ABPs) contributes, through several pathways, to hamper bone healing and to promote the osteonecrosis process.

In fact, cancer patients and immunocompromised patients in general, show a high prevalence of hypovitaminosis D, which is very difficult to correct; they actually need higher doses of cholecalciferol than healthy subjects [72]. In these patients the hormone deprivation or, in general hypogonadism, may promote hypovitaminosis D [92]. Notably, not all patients with an osteomalacic histological pattern show hypovitaminosis D: this suggests the presence of focal osteomalacia in some of them. This situation has been described already in kidney transplant patients characterized by immunocompromised status, low bone turnover and osteomalacic pattern, suggesting vitamin D resistance [93]. In such cases, higher doses or alternatively vitamin D active metabolites should be administered in order to overcome this condition, so as to improve the safety target value of serum vitamin D. Osteomalacia in the jaws might be a pivotal factor in MRONJ pathogenesis and should be considered before starting BPs treatment.

In summary: ONJ is a severe and multifactorial clinical condition; its incidence is low in cancer, almost irrelevant in osteoporosis. ONJ results from a combination of different concomitant factors: none of these is singly sufficient to induce ONJ. The main incident factors beside the presence of an immunosuppressive status, osteomalacia and the use of antiresorptive agents, are: concomitant assumption of drugs such as steroids, dental interventions, oral and gingival infections. Moreover, the role played by antiresorptive drugs has not been completely understood yet, but they do not appear to be the main culprits in ONJ pathogenesis; the pre-existing condition of general/local osteomalacia might be a pivotal factor for drugs involvement in the pathogenetic mechanism. Vitamin D plays an important role in the prevention of ONJ; the safety level of 25-OH vitamin D has to be investigated (Table 1).

Table 1. The key points of this review are summarized.

1.	ONJ is a severe and multifactorial clinical condition
2.	ONJ incidence is low in cancer, almost irrelevant in osteoporosis
3.	The incident factors are: immunodeficiency, assumption of drugs such as glucocorticoids, dental interventions, oral and gingival infections
4.	A general or local osteomalacia condition may be the main factor in the pathogenetic involvement of antiresorptive drugs
5.	Vitamin D is important in the prevention of ONJ
6.	The safety levels of 25-OH vitamin D in this pattern need to be specifically investigated

ONJ, osteonecrosis of the jaw.

These considerations point towards some important clinical implications: firstly, BPs should not be considered direct pathogenetic factors for ONJ; secondly, hypovitaminosis D correction in immunocompromised patients in view of a dental intervention should be considered as the antibiotic prophylaxis, before starting a BP treatment. The effective safety level of serum 25-OH vitamin D in this particular setting should be determined by ad hoc studies.

Such precautions seem to be more effective for these patients rather than BPs treatment discontinuation, considering the dramatic impact of fragility fractures.

Further studies are needed to confirm the actual interplay occurring between osteomalacia and BPs in ONJ pathogenesis, although the results of the abovementioned histomorphometric study head towards the acquittal of BPs as the main culprits.

Author Contributions: Conceptualization, L.D.C.; methodology, L.D.C. and M.T.V.; validation, L.D.C., M.M., M.T.V.; data curation, M.M., M.T.V.; writing—original draft preparation, L.D.C., M.T.V.; writing—review and editing, L.D.C., M.M., M.T.V. All authors have read and agreed to the published version of the manuscript.

Funding: This research received no external funding.

Conflicts of Interest: The authors declare no conflict of interest.

References

1. Ruggiero, S.L.; Dodson, T.B.; Fantasia, J.; Goodday, R.; Aghaloo, T.; Mehrotra, B.; O'Ryan, F. American Association of Oral and Maxillofacial Surgeons position paper on medication-related osteonecrosis of the jaw—2014 update. *J. Oral Maxillofac. Surg.* **2014**, *72*, 1938–1956.
2. Colella, G.; Fusco, V.; Campisi, G. American Association of Oral and Maxillofacial Surgeons position paper: Bisphosphonate-Related Osteonecrosis of the Jaws-2009 update: The need to refine the BRONJ definition. *J. Oral Maxillofac. Surg.* **2009**, *67*, 2698–2699.
3. Otto, S.; Marx, R.E.; Tröltzsch, M.; Ristow, O.; Ziebart, T.; Al-Nawas, B.; Groetz, K.A.; Ehrenfeld, M.; Mercadante, V.; Porter, S.; et al. Comments on "diagnosis and management of osteonecrosis of the jaw: A systematic review and international consensus". *J. Bone Miner. Res.* **2015**, *30*, 1113–1115.
4. Schiodt, M.; Otto, S.; Fedele, S.; Bedogni, A.; Nicolatou-Galitis, O.; Guggenberger, R.; Herlofson, B.B.; Ristow, O.; Kofod, T. Workshop of European task force on medication-related osteonecrosis of the jaw—Current challenges. *Oral Dis.* **2019**, *10*, 1815–1821.
5. Yarom, N.; Peterson, D.E.; Bohlke, K.; Saunders, D.P. Reply to Fusco et al. *JCO Oncol. Pract.* **2020**, *16*, 145–146.
6. Nicolatou-Galitis, O.; Kouri, M.; Papadopoulou, E.; Vardas, E.; Galiti, D.; Epstein, J.B.; Elad, S.; Campisi, G.; Tsoukalas, N.; Bektas-Kayhan, K.; et al. Osteonecrosis of the jaw related to non-antiresorptive medications: A systematic review. *Support. Care Cancer* **2019**, *27*, 383–394.
7. Khan, A.A.; Morrison, A.; Hanley, A.D.; Felsenberg, D.; McCauley, L.K.; O'Ryan, F.; Reid, I.R.; Ruggiero, S.L.; Taguchi, A.; Tetradis, S.; et al. Diagnosis and management of osteonecrosis of the jaw: A systematic review and international consensus. *J. Bone Miner. Res.* **2015**, *30*, 3–23.
8. Mavrokokki, T.; Cheng, A.; Stein, B.; Goss, A. Nature and frequency of bisphosphonate-associated osteonecrosis of the jaws in Australia. *J. Oral Maxillofac. Surg.* **2007**, *65*, 415–423.
9. Ulmner, M.; Jarnbring, F.; Törring, O. Osteonecrosis of the jaw in Sweden associated with the oral use of bisphosphonate. *J. Oral Maxillofac. Surg.* **2014**, *72*, 76–82.

10. Kizub, D.A.; Miao, J.; Schubert, M.M.; Paterson, A.H.G.; Clemons, M.; Dees, E.C.; Ingle, J.N.; Falkson, C.I.; Barlow, W.E.; Hortobagyi, G.N.; et al. Risk factors for bisphosphonate-associated osteonecrosis of the jaw in the prospective randomized trial of adjuvant bisphosphonates for early-stage breast cancer (SWOG 0307). *Support Care Cancer* **2020**. [CrossRef]
11. Dennison, E.; Cooper, C.; Kanis, J.; Bruyère, O.; Silverman, S.; McCloskey, E.; Abrahamsen, B.; Prieto-Alhambra, D.; Ferrari, S.; On behalf of the IOF Epidemiology/Quality of Life Working Group. Fracture risk following intermission of osteoporosis therapy. *Osteoporos. Int.* **2019**, *30*, 1733–1743.
12. Cummings, S.R.; Ferrari, S.; Eastell, R.; Gilchrist, N.; Jensen, J.-E.B.; McClung, M.; Roux, C.; Törring, O.; Valter, I.; Wang, A.T.; et al. Vertebral fractures after discontinuation of denosumab: A post hoc analysis of the randomized placebo-controlled FREEDOM trial and its extension. *J. Bone Miner. Res.* **2018**, *33*, 190–198.
13. Ferrari, S.L.; Lewiecki, E.M.; Butler, P.W.; Kendler, D.L.; Napoli, N.; Huang, S.; Crittenden, D.B.; Pannacciulli, N.; Siris, E.; Binkley, N. Favorable skeletal benefit/risk of long-term denosumab therapy: A virtual-twin analysis of fractures prevented relative to skeletal safety events observed. *Bone* **2020**, *134*, 115287.
14. Ottesen, C.; Schiodt, M.; Gotfredsen, K. Efficacy of a high-dose antiresorptive drug *holiday* to reduce the risk of medication-related osteonecrosis of the jaw (MRONJ): A systematic review. *Heliyon* **2020**, *6*, e03795.
15. Fujieda, Y.; Doi, M.; Asaka, T.; Ota, M.; Hisada, R.; Ohnishi, N.; Kono, M.; Kameda, H.; Nakazawa, D.; Kato, M.; et al. Incidence and risk of antiresorptive agent-related osteonecrosis of the jaw (ARONJ) after tooth extraction in patients with autoimmune disease. *J. Bone Miner. Metab.* **2020**, *38*, 1–8.
16. Maciel, A.P.; Quispe, R.A.; Martins, L.J.O.; Caldas, R.J.; Santos, P.S.D.S. Clinical profile of individuals with bisphosphonate-related osteonecrosis of the jaw: An integrative review. *Sao Paulo Med. J.* **2020**, *138*, 326–335.
17. Marini, F.; Tonelli, P.; Cavalli, L.; Cavalli, T.; Masi, L.; Falchetti, A.; Brandi, M.L. Pharmacogenetics of bisphosphonate-associated osteonecrosis of the jaw. *Front. Biosci. (Elite Ed.)* **2011**, *3*, 364–370.
18. Sarasquete, M.E.; García-Sanz, R.; Marín, L.; Alcoceba, M.; Chillón, M.D.C.; Balanzategui, A.; Santamaria, C.; Rosiñol, L.; De La Rubia, J.; Hernandez, M.T.; et al. Bisphosphonate-related osteonecrosis of the jaw is associated with polymorphisms of the cytochrome P450 CYP2C8 in multiple myeloma: A genome-wide single nucleotide polymorphism analysis. *Blood* **2008**, *112*, 2709–2712.
19. Yang, G.; Hamadeh, I.S.; Katz, J.; Riva, A.; Lakatos, P.; Balla, B.; Kosa, J.; Vaszilko, M.; Pelliccioni, G.A.; Davis, N.; et al. SIRT1/HERC4 locus associated with bisphosphonate-induced osteonecrosis of the jaw: An exome-wide association analysis. *J. Bone Miner. Res.* **2018**, *33*, 91–98.
20. Yang, G.; Collins, J.M.; Rafiee, R.; Singh, S.; Langaee, T.; McDonough, C.W.; Holliday, L.S.; Wang, D.; Lamba, J.K.; Kim, Y.S.; et al. *SIRT1* Gene SNP rs932658 Is Associated With Medication-Related Osteonecrosis of the Jaw. *J. Bone Miner. Res.* **2020**. [CrossRef]
21. Allen, R.M.; Burr, D.B. The pathogenesis of bisphosphonate-related osteonecrosis of the jaw: So many hypotheses, so few data. *J. Oral Maxillofac. Surg.* **2009**, *67*, 61–70.
22. Heim, N.; Götz, W.; Kramer, F.-J.; Faron, A. Antiresorptive drug-related changes of the mandibular bone densitiy in medication-related osteonecrosis of the jaw patients. *Dentomaxillofac. Radiol.* **2019**, *48*, 20190132.
23. Anesi, A.; Generali, L.; Sandoni, L.; Pozzi, S.; Grande, A. From osteoclast differentiation to osteonecrosis of the jaw: Molecular and clinical insights. *Int. J. Mol. Sci.* **2019**, *20*, 4925.
24. Kim, H.J.; Kim, H.J.; Choi, Y.; Bae, M.-K.; Hwang, D.S.; Shin, S.-H.; Lee, J.-Y. Zoledronate enhances osteocyte-mediated osteoclast differentiation by IL-6/RANKL axis. *Int. J. Mol. Sci.* **2019**, *20*, 1467.
25. Kim, J.; Lee, D.-H.; Dziak, R.; Ciancio, S. Bisphosphonate-related osteonecrosis of the jaw: Current clinical significance and treatment strategy review. *Am. J. Dent.* **2020**, *33*, 115–128.
26. Carbonare, L.D.; Zanatta, M.; Gasparetto, A.; Valenti, M.T. Safety and tolerability of zoledronic acid and other bisphosphonates in osteoporosis management. *Drug Healthc. Patient Saf.* **2010**, *2*, 121.
27. Carbonare, L.D.; Mottes, M.; Malerba, G.; Mori, A.; Zaninotto, M.; Plebani, M.; Dellantonio, A.; Valenti, M.T. Enhanced osteogenic differentiation in zoledronate-treated osteoporotic patients. *Int. J. Mol. Sci.* **2017**, *18*, 1261.
28. Reid, I.R. Osteonecrosis of the jaw—Who gets it, and why? *Bone* **2009**, *44*, 4–10.
29. Drake, M.T.; Clarke, B.L.; Khosla, S. Bisphosphonates: Mechanism of action and role in clinical practice. *Mayo Clin. Proc.* **2008**, *83*, 1032–1045.
30. Benlidayi, C.I.; Guzel, R. Oral bisphosphonate related osteonecrosis of the jaw: A challenging adverse effect. *ISRN Rheumatol.* **2013**, *2013*, 1–6.
31. Paulo, S.; Abrantes, A.M.; Laranjo, M.; Carvalho, F.L.; Serra, A.C.; Botelho, M.F.; Ferreira, M.M. Bisphosphonate-related osteonecrosis of the jaw: Specificities. *Oncol. Rev.* **2014**, *8*, 254.
32. Lewiecki, E.M. Denosumab update. *Curr. Opin. Rheumatol.* **2009**, *21*, 369–373.
33. Lewiecki, E.M. Denosumab in postmenopausal osteoporosis: What the clinician needs to know. *Ther. Adv. Musculoskelet. Dis.* **2009**, *1*, 13–26.
34. Fizazi, K.; Lipton, A.; Mariette, X.; Body, J.-J.; Rahim, Y.; Gralow, J.R.; Gao, G.M.; Wu, L.; Sohn, W.; Jun, S. Randomized phase II trial of denosumab in patients with bone metastases from prostate cancer, breast cancer, or other neoplasms after intravenous bisphosphonates. *J. Clin. Oncol.* **2009**, *27*, 1564–1571.
35. Aghaloo, L.T.; Felsenfeld, A.L.; Tetradis, S. Osteonecrosis of the jaw in a patient on Denosumab. *J. Oral Maxillofac. Surg.* **2010**, *68*, 959–963.

36. Lima, M.V.D.S.; Rizzato, J.; Marques, D.V.G.; Kitakawa, D.; Carvalho, L.F.D.C.E.S.D.; Scherma, A.P.; Carvalho, L.F.C.S. Denosumab related osteonecrosis of jaw: A case report. *J. Oral Maxillofac. Res.* **2018**, *9*, e1.

37. Allen, M. Animal models of osteonecrosis of the jaw. *J. Musculoskelet. Neuronal Interact.* **2007**, *7*, 358.

38. Ruggiero, S.L.; Dodson, T.B.; Assael, L.A.; Landesberg, R.; Marx, R.E.; Mehrotra, B. American Association of Oral and Maxillofacial Surgeons position paper on bisphosphonate-related osteonecrosis of the jaw—2009 update. *Aust. Endod. J.* **2009**, *35*, 119–130.

39. Khosla, S.; Burr, D.; Cauley, J.; Dempster, D.W.; Ebeling, P.R.; Felsenberg, D.; Gagel, R.F.; Gilsanz, V.; Guise, T.; Koka, S.; et al. Bisphosphonate-associated osteonecrosis of the jaw: Report of a task force of the American Society for Bone and Mineral Research. *J. Bone Miner. Res.* **2007**, *22*, 1479–1491.

40. Garetto, L.P.; Tricker, N.D. Remodeling of bone surrounding the implant interface. In Bridging the Gap between Dental and Orthopaedic Implants, Proceedings of the 3rd Annual Indiana Conference, Indianapolis, IN, USA, 13–16 May 1998; Garetto, L.P., Turner, C.H., Duncan, R.L., Burr, D.B., Eds.; Indiana University School of Dentistry: Indianapolis, IN, USA, 2002.

41. Huja, S.S.; Fernandez, S.A.; Hill, K.J.; Li, Y. Remodeling dynamics in the alveolar process in skeletally mature dogs. *Anat. Rec. A Discov. Mol. Cell. Evol. Biol.* **2006**, *288*, 1243–1249.

42. Van Poznak, C.; Estilo, C. Osteonecrosis of the jaw in cancer patients receiving IV bisphosphonates. *Oncology* **2006**, *20*, 1053–1062.

43. Ruggiero, S.L.; Fantasia, J.; Carlson, E. Bisphosphonate-related osteonecrosis of the jaw: Background and guidelines for diagnosis, staging and management. *Oral Surg. Oral Med. Oral Pathol. Oral Radiol. Endod.* **2006**, *102*, 433–441.

44. Li, C.Y.; Schaffler, M.B.; Wolde-Semait, H.T.; Hernandez, C.J.; Jepsen, K.J. Genetic background influences cortical bone response to ovariectomy. *J. Bone Miner. Res.* **2005**, *20*, 2150–2158.

45. Holtmann, H.; Lommen, J.; Kübler, N.R.; Sproll, C.; Rana, M.; Karschuck, P.; Depprich, R. Pathogenesis of medication-related osteonecrosis of the jaw: A comparative study of in vivo and in vitro trials. *J. Int. Med. Res.* **2018**, *46*, 4277–4296.

46. Vasconcelos, A.C.U.; Berti-Couto, S.A.; Azambuja, A.A.; Salum, F.G.; Figueiredo, M.A.; Da Silva, V.D.; Cherubini, K. Comparison of effects of clodronate and zoledronic acid on the repair of maxilla surgical wounds–histomorphometric, receptor activator of nuclear factor-kB ligand, osteoprotegerin, von Willebrand factor, and caspase-3 evaluation. *J. Oral Pathol. Med.* **2012**, *41*, 702–712.

47. Janovszky, Á.; Szabó, A.; Varga, R.; Garab, D.; Borós, M.; Mester, C.; Beretka, N.; Zombori, T.; Wiesmann, H.-P.; Bernhardt, R.; et al. Periosteal microcirculatory reactions in a zoledronate-induced osteonecrosis model of the jaw in rats. *Clin. Oral Investig.* **2015**, *19*, 1279–1288.

48. Abtahi, J.; Agholme, F.; Sandberg, O.; Aspenberg, P. Effect of local vs. systemic bisphosphonate delivery on dental implant fixation in a model of osteonecrosis of the jaw. *J. Dent. Res.* **2013**, *92*, 279–283.

49. Sonis, S.T.; Watkins, B.A.; Lyng, G.D.; Lerman, M.A.; Anderson, K.C. Bony changes in the jaws of rats treated with zoledronic acid and dexamethasone before dental extractions mimic bisphosphonate-related osteonecrosis in cancer patients. *Oral Oncol.* **2009**, *45*, 164–172.

50. Song, M.; AlShaikh, A.; Kim, T.; Kim, S.; Dang, M.; Mehrazarin, S.; Shin, K.-H.; Kang, M.; Park, N.-H.; Kim, R.H. Preexisting periapical inflammatory condition exacerbates tooth extraction–induced bisphosphonate-related osteonecrosis of the jaw lesions in mice. *J. Endod.* **2016**, *42*, 1641–1646.

51. Kuroshima, S.; Yamashita, J. Chemotherapeutic and antiresorptive combination therapy suppressed lymphangiogenesis and induced osteonecrosis of the jaw-like lesions in mice. *Bone* **2013**, *56*, 101–109.

52. Aghaloo, T.L.; Cheong, S.; Bezouglaia, O.; Kostenuik, P.; Atti, E.; Dry, S.M.; Pirih, F.Q.; Tetradis, S. RANKL inhibitors induce osteonecrosis of the jaw in mice with periapical disease. *J. Bone Miner. Res.* **2014**, *29*, 843–854.

53. Pautke, C.; Kreutzer, K.; Weitz, J.; Knödler, M.; Münzel, D.; Wexel, G.; Otto, S.; Hapfelmeier, A.; Sturzenbaum, S.; Tischer, T. Bisphosphonate related osteonecrosis of the jaw: A minipig large animal model. *Bone* **2012**, *51*, 592–599.

54. Li, Y.; Xu, J.; Mao, L.; Liu, Y.; Gao, R.; Zheng, Z.; Chen, W.; Le, A.; Shi, S.; Wang, S. Allogeneic mesenchymal stem cell therapy for bisphosphonate-related jaw osteonecrosis in swine. *Stem Cells Dev.* **2013**, *22*, 2047–2056.

55. Voss, P.J.; Stoddart, M.J.; Bernstein, A.; Schmelzeisen, R.; Nelson, K.; Stadelmann, V.A.; Ziebart, T.; Poxleitner, P. Zoledronate induces bisphosphonate-related osteonecrosis of the jaw in osteopenic sheep. *Clin. Oral Investig.* **2016**, *20*, 31–38.

56. Moreira, C.A.; Costa, T.M.R.L.; Marques, J.V.O.; Sylvestre, L.; Almeida, A.C.R.; Maluf, E.M.C.P.; Borba, V.Z.C. Prevalence and clinical characteristics of X-linked hypophosphatemia in Paraná, southern Brazil. *Arch. Endocrinol. Metab.* **2020**, *64*, 796–802.

57. Bhadada, S.K.; Bhansali, A.; Upreti, V.; Dutta, P.; Santosh, R.; Das, S.; Nahar, U. Hypophosphataemic rickets/osteomalacia: A descriptive analysis. *Indian J. Med Res.* **2010**, *131*, 399.

58. Laurent, M.R.; Bravenboer, N.; van Schoor, N.M.; Bouillon, R.; Pettifor, J.M.; Lips, P. Rickets and osteomalacia. In *Primer on the Metabolic Bone Diseases and Disorders of Mineral Metabolism*; John Wiley & Sons: Newark, NJ, USA, 2018; pp. 684–694.

59. Feng, J.Q.; Clinkenbeard, E.L.; Yuan, B.; White, K.E.; Drezner, M.K. Osteocyte regulation of phosphate homeostasis and bone mineralization underlies the pathophysiology of the heritable disorders of rickets and osteomalacia. *Bone* **2013**, *54*, 213–221.

60. Homan, E.P.; Rauch, F.; Grafe, I.; Lietman, C.; Doll, J.A.; Dawson, B.; Bertin, T.; Napierala, D.; Morello, R.; Gibbs, R.; et al. Mutations in SERPINF1 cause osteogenesis imperfecta type VI. *J. Bone Miner. Res.* **2011**, *26*, 2798–2803.

61. Frame, B.; Frost, H.M.; Pak, C.Y.; Reynolds, W.; Argen, R.J. Fibrogenesis imperfecta ossium: A collagen defect causing osteomalacia. *N. Engl. J. Med.* **1971**, *285*, 769–772.

62. Herm, F.; Killguss, H.; Stewart, A. Osteomalacia in Hazara District, Pakistan. *Trop. Dr.* **2005**, *35*, 8–10.

63. Campbell, A.G.; Hosking, D.J.; Kemm, J.R.; Boyd, R.V. How common is osteomalacia in the elderly? *Lancet* **1984**, *324*, 386–388.

64. Lips, P. Vitamin D deficiency and secondary hyperparathyroidism in the elderly: Consequences for bone loss and fractures and therapeutic implications. *Endocr. Rev.* **2001**, *22*, 477–501.
65. Priemel, M.; Von Domarus, C.; Klatte, T.O.; Kessler, S.; Schlie, J.; Meier, S.; Proksch, N.; Pastor, F.; Netter, C.; Streichert, T.; et al. Bone mineralization defects and vitamin D deficiency: Histomorphometric analysis of iliac crest bone biopsies and circulating 25-hydroxyvitamin D in 675 patients. *J. Bone Miner. Res.* **2010**, *25*, 305–312.
66. Rabelink, N.M.; Westgeest, H.M.; Bravenboer, N.; Jacobs, M.A.J.M.; Lips, P. Bone pain and extremely low bone mineral density due to severe vitamin D deficiency in celiac disease. *Arch. Osteoporos.* **2011**, *6*, 209–213.
67. Parfitt, A.; Pødenphant, J.; Villanueva, A.; Frame, B. Metabolic bone disease with and *without osteomalacia* after intestinal bypass surgery: A bone histomorphometric study. *Bone* **1985**, *6*, 211–220.
68. Bisballe, S.; Eriksen, E.F.; Melsen, F.; Mosekilde, L.; Sorensen, O.H.; Hessov, I. Osteopenia and osteomalacia after gastrectomy: Interrelations between biochemical markers of bone remodelling, vitamin D metabolites, and bone histomorphometry. *Gut* **1991**, *32*, 1303–1307.
69. Gupta, D.; Vashi, P.G.; Trukova, K.; Lis, C.G.; Lammersfeld, C.A. Prevalence of serum vitamin D deficiency and insufficiency in cancer: Review of the epidemiological literature. *Exp. Ther. Med.* **2011**, *2*, 181–193.
70. Nogués, X.; Servitja, S.; Peña, M.J.; Prieto-Alhambra, D.; Nadal, R.; Mellibovsky, L.; Albanell, J.; Diez-Perez, A.; Tusquets, I. Vitamin D deficiency and bone mineral density in postmenopausal women receiving aromatase inhibitors for early breast cancer. *Maturitas* **2010**, *66*, 291–297.
71. Neuhouser, M.L.; Bernstein, L.; Hollis, B.W.; Xiao, L.; Ambs, A.; Baumgartner, K.; Baumgartner, R.; McTiernan, A.; Ballard-Barbash, R. Serum vitamin D and breast density in breast cancer survivors. *Cancer Epidemiol. Prev. Biomark.* **2010**, *19*, 412–417.
72. Crew, K.D.; Shane, E.; Cremers, S.; McMahon, D.J.; Irani, D.; Hershman, D.L. High prevalence of vitamin D deficiency despite supplementation in premenopausal women with breast cancer undergoing adjuvant chemotherapy. *J. Clin. Oncol.* **2009**, *27*, 2151.
73. Rainville, C.; Khan, Y.; Tisman, G. Triple negative breast cancer patients presenting with low serum vitamin D levels: A case series. *Cases J.* **2009**, *2*, 1–5.
74. Neuhouser, M.L.; Sorensen, B.; Hollis, B.W.; Ambs, A.; Ulrich, C.M.; McTiernan, A.; Bernstein, L.; Wayne, S.; Gilliland, F.; Baumgartner, K.; et al. Vitamin D insufficiency in a multiethnic cohort of breast cancer survivors. *Am. J. Clin. Nutr.* **2008**, *88*, 133–139.
75. Fakih, M.G.; Trump, D.L.; Johnson, C.S.; Tian, L.; Muindi, J.; Sunga, A.Y. Chemotherapy is linked to severe vitamin D deficiency in patients with colorectal cancer. *Int. J. Colorectal Dis.* **2009**, *24*, 219–224.
76. Trump, D.L.; Chadha, M.K.; Sunga, A.Y.; Fakih, M.G.; Ashraf, U.; Silliman, C.G.; Hollis, B.W.; Nesline, M.K.; Tian, L.; Tan, W.; et al. Vitamin D deficiency and insufficiency among patients with prostate cancer. *BJU Int.* **2009**, *104*, 909.
77. Maier, G.S.; Horas, K.; Kurth, A.A.; Lazovic, D.; Seeger, J.B.; Maus, U. Prevalence of vitamin D deficiency in patients with bone metastases and multiple myeloma. *Anticancer Res.* **2015**, *35*, 6281–6285.
78. Jagelavičienė, E.; Vaitkevičienė, I.; Šilingaitė, D.; Šinkūnaitė, E.; Daugėlaitė, G. The relationship between vitamin D and periodontal pathology. *Medicina* **2018**, *54*, 45.
79. Alshouibi, E.; Kaye, E.; Cabral, H.; Leone, C.; Garcia, R. Vitamin D and periodontal health in older men. *J. Dent. Res.* **2013**, *92*, 689–693.
80. Zhou, F.; Zhou, Y.; Shi, J. The association between serum 25-hydroxyvitamin D levels and dental caries in US adults. *Oral Dis.* **2020**, *26*, 1537–1547.
81. Kim, I.-J.; Lee, H.-S.; Ju, H.-J.; Na, J.-Y.; Oh, H.-W. A cross-sectional study on the association between vitamin D levels and caries in the permanent dentition of Korean children. *BMC Oral Health* **2018**, *18*, 1–6.
82. Chhonkar, A.; Gupta, A.; Arya, V. Comparison of vitamin D level of children with severe early childhood caries and children with no caries. *Int. J. Clin. Pediatric Dent.* **2018**, *11*, 199.
83. Deane, S.; Schroth, R.J.; Sharma, A.; Rodd, C. Combined deficiencies of 25-hydroxyvitamin D and anemia in preschool children with severe early childhood caries: A case–control study. *Paediatr. Child Health* **2018**, *23*, e40–e45.
84. Danila, M.I.; Outman, R.C.; Rahn, E.J.; Mudano, A.S.; Redden, D.T.; Li, P.; Allison, J.J.; Anderson, F.A.; Wyman, A.; Greenspan, S.L.; et al. Evaluation of a multimodal, direct-to-patient educational intervention targeting barriers to osteoporosis care: A randomized clinical trial. *J. Bone Miner. Res.* **2018**, *33*, 763–772.
85. Heim, N.; Warwas, F.B.; Wilms, C.T.; Reich, R.H.; Martini, M. Vitamin D (25-OHD) deficiency may increase the prevalence of medication-related osteonecrosis of the jaw. *J. Cranio-Maxillofac. Surg.* **2017**, *45*, 2068–2074.
86. Demircan, S.; Isler, S. Changes in serological bone turnover markers in bisphosphonate induced osteonecrosis of the jaws: A case control study. *Niger. J. Clin. Pract.* **2020**, *23*, 154.
87. Bedogni, A.; Saia, G.; Bettini, G.; Tronchet, A.; Totola, A.; Bedogni, G.; Tregnago, P.; Valenti, M.T.; Bertoldo, F.; Ferronato, G.; et al. Osteomalacia: The missing link in the pathogenesis of bisphosphonate-related osteonecrosis of the jaws? *Oncologist* **2012**, *17*, 1114.
88. Boyce, B.F.; Adamson, B.B.; Gallacher, S.J.; Byars, J.; Ralston, S.H.; Boyle, I.T. Mineralisation defects after pamidronate for Paget's disease. *Lancet* **1994**, *343*, 1231–1232.
89. Jakob, T.; Tesfamariam, Y.M.; Macherey, S.; Kuhr, K.; Adams, A.; Monsef, I.; Heidenreich, A.; Skoetz, N. Bisphosphonates or RANK-ligand-inhibitors for men with prostate cancer and bone metastases: A network meta-analysis. *Cochrane Database Syst. Rev.* **2020**, *12*, CD013020.

90. Endo, Y.; Funayama, H.; Yamaguchi, K.; Monma, Y.; Yu, Z.; Deng, X.; Oizumi, T.; Shikama, Y.; Tanaka, Y.; Okada, S.; et al. Basic Studies on the Mechanism, Prevention, and Treatment of Osteonecrosis of the Jaw Induced by Bisphosphonates. *Yakugaku zasshi: J. Pharm. Soc. Japan* **2020**, *140*, 63–79.
91. Suzuki, K.; Takeyama, S.; Murakami, S.; Nagaoka, M.; Chiba, M.; Igarashi, K.; Shinoda, H. Structure-Dependent Effects of Bisphosphonates on Inflammatory Responses in Cultured Neonatal Mouse Calvaria. *Antioxidants* **2020**, *9*, 503.
92. D'Andrea, S.; Martorella, A.; Coccia, F.; Castellini, C.; Minaldi, E.; Totaro, M.; Parisi, A.; Francavilla, F.; Barbonetti, A. Relationship of Vitamin D status with testosterone levels: A systematic review and meta-analysis. *Endocrine* **2020**. [CrossRef]
93. Monier-Faugere, M.C.; Mawad, H.; Qi, Q.; Friedler, R.M.; Malluche, H.H. High prevalence of low bone turnover and occurrence of osteomalacia after kidney transplantation. *J. Am. Soc. Nephrol.* **2000**, *11*, 1093–1099.

nutrients

Article

Effect of Selected Factors on the Serum 25(OH)D Concentration in Women Treated for Breast Cancer

Agnieszka Radom [1], Andrzej Wędrychowicz [2], Stanisław Pieczarkowski [2], Szymon Skoczeń [3] and Przemysław Tomasik [4,*]

[1] Medical Laboratory Diagmed, Lwowska 20, 33-300 Nowy Sącz, Poland; a_radom@interia.pl
[2] Department of Pediatrics, Gastroenterology and Nutrition, Faculty of Medicine, Jagiellonian University Medical College, Wielicka 265, 30-663 Kraków, Poland; andrzej.wedrychowicz@uj.edu.pl (A.W.); stpiecz@wp.pl (S.P.)
[3] Department of Oncology and Hematology, Pediatric Institute, Faculty of Medicine, Jagiellonian University Medical College, Wielicka 265, 30-663 Kraków, Poland; szymon.skoczen@uj.edu.pl
[4] Department of Clinical Biochemistry, Pediatric Institute, Faculty of Medicine, Jagiellonian University Medical College, Wielicka 265, 30-663 Kraków, Poland
* Correspondence: p.tomasik@uj.edu.pl

Abstract: Maintaining an optimal vitamin D concentration reduces the risk of recurrence and extends survival time in patients after breast cancer treatment. Data on vitamin D deficiency among Polish women after breast cancer therapy are limited. Thus, the aim of the study was the analysis of vitamin D status in post-mastectomy patients, considering such factors as seasons, social habits, vitamin D supplementation and its measurements. The study involved 94 women after breast cancer treatment. Serum vitamin D concentration was measured, and a questionnaire, gathering demographic and clinical data regarding cancer, diet, exposure to sun radiation, and knowledge of recommendations on vitamin D supplementation, was delivered twice, in both winter and in summer. The control group consisted of 94 age-matched women with no oncological history. In women after breast cancer treatment, 25-hydroxyvitamin D (25(OH)D) deficiency was much more frequent than in the general population. Only about half of the patients supplemented vitamin D at the beginning of the study. After the first test and the issuing of recommendations on vitamin D supplementation, the percentage of vitamin D supplemented patients increased by about 30% in study groups. The average dose of supplement also increased. None of the women that were not supplementing vitamin D and were tested again in winter had optimal 25(OH)D concentration. It was concluded that vitamin deficiency is common in women treated for breast cancer. Medical advising about vitamin D supplementation and monitoring of 25(OH)D concentration should be improved.

Keywords: vitamin D deficiency; vitamin D measurement; vitamin D supplementation; breast cancer

Citation: Radom, A.; Wędrychowicz, A.; Pieczarkowski, S.; Skoczeń, S.; Tomasik, P. Effect of Selected Factors on the Serum 25(OH)D Concentration in Women Treated for Breast Cancer. *Nutrients* **2021**, *13*, 564. https://doi.org/10.3390/nu13020564

Academic Editor: Spyridon N. Karras
Received: 17 December 2020
Accepted: 3 February 2021
Published: 9 February 2021

Publisher's Note: MDPI stays neutral with regard to jurisdictional claims in published maps and institutional affiliations.

1. Introduction

Breast cancer is one of the biggest problems in public health. Globally, more than one million cases are diagnosed annually [1]. In terms of cancer mortality in women, breast cancer ranks in second position. In 2015, in the Polish population, breast cancer was the cause of 14.1% of deaths due to cancer; in 2016 it was already 14.5%, while in 2017 this percentage was 14.8% [2–4]. In the world, these proportions differ, varying significantly depending on race, latitude, and socioeconomic status [5].

The main biological function of vitamin D is the maintenance of homeostasis of calcium-phosphate management and the regulation of bone metabolism. It has been proven to play a role in the proper functioning of the immune, cardiovascular, and reproductive systems. In addition, vitamin D deficiency is associated with an increased incidence of type 1 and 2 diabetes, obesity, asthma, inflammatory bowel disease and cancer [6,7]. In 1990, Garland et al. were the first to demonstrate a negative relationship between total, average annual exposure to solar radiation and age-dependent mortality of breast cancer patients [8]. Many studies and meta-analyses have shown a relationship between vitamin D status and breast cancer risk [9–12]. However, in some of them, the negative correlation of breast cancer risk with 25-hydroxyvitamin D (25(OH)D) concentration was only found in retrospective studies and or in specific subpopulations of women with regard to menopausal status and ethnicity [13–15]. The discovery of the nuclear vitamin D receptor (VDR) and the demonstration of its presence in cancer cells gave rise to research into the role of vitamin D in the development and course of cancer [16,17]. Vitamin D has anticancer properties by affecting inflammation, cell's growth, maturation, and proliferation. It inhibits angiogenesis and metastasis ability, reduces the number of estrogen receptors, inhibits the expression of adhesion molecules, regulates miRNA expression, modulates the hedgehog signaling pathway, and induces breast cancer cell apoptosis in vitro and in vivo [18–20]. In vitamin D deficiency, there is a dysregulation of cells' growth and proliferation, and facilitation of neoangiogenesis and carcinogenesis. At least 35 genes are regulated with vitamin D in breast tissue, and their activity is associated with invasiveness and cancer metastasis [21].

In recent years, great attention has been paid to maintaining proper 25(OH)D concentration in a healthy population as well as especially in people after cancer treatment. Despite numerous reports on the relationship between serum 25(OH)D concentration and the risk of breast cancer, its progression and distant prognosis, there are no uniform guidelines on what doses of vitamin D and what serum concentrations should be considered appropriate in both healthy and oncological patients. Guidelines of American Cancer Society/American Society of Clinical Oncology recommended calcium (1200 mg/d) and vitamin D (600–1000 IU/d) supplementation in breast cancer patients from 50 years of age to reduce bone loss-related mortality [22]. In turn, the guidelines of the European Society of Medical Oncology (ESMO) state that the daily supply of vitamin D in breast cancer patients should be in the amount of 1000–2000 IU [23]. Polish guidelines for supplementation and treatment of vitamin D for healthy people and risk groups of deficiencies recommend that adults should take 800–2000 IU vitamin D per day depending on body weight. In the risk groups of vitamin D deficiency to which cancer patients belong, supplementation should be carried out under the control of laboratory determinations 25(OH)D to maintain an optimal concentration between 30 and 50 ng/mL [24]. According to Polish Standards of Nutritional Treatment in Oncology (2015), indications for supplementation of vitamin D include a documented deficiency in the blood or typical clinical characteristics of vitamin D deficiency [25]. There are also no global uniform laboratory criteria for determining vitamin D deficiency based on its blood concentrations. According to the recommendation of the US National Academy of Medicine, adults should supplement vitamin D to maintain serum concentrations of 25(OH)D above 20 ng/mL (50 nmol/L) [26,27]. The American Association of Clinical Endocrinology (AACE) and the Endocrine Society recommend serum concentrations of 25(OH)D $\geq$30 ng/mL (75 nmol/L) as sufficient [28]. The 2016 Guidelines of the German Food Society state that the desired serum vitamin D concentration is >20 ng/mL (>50 nmol/L) [29]. Although prevention and treatment of vitamin D deficiencies is recommended in the daily practice of doctors and clinical nutritionists, this problem is often overlooked in oncological patients due to the lack of uniform guidelines for such patients, and determining which specialist is responsible for implementing and monitoring vitamin D supplementation. Therefore, the aim of the study was to evaluate the serum 25(OH)D concentrations of women after being treated with breast cancer according to the seasons, eating and social habits, vitamin D supplementation, and the recommendations of the attending physicians and the effect of vitamin D determinations on improving vitamin D status in subsequent testing and changing behavior to obtain and maintain the recommended serum vitamin D concentration.

2. Materials and Methods

2.1. Studied Groups

The patients who underwent radical treatment of breast cancer (mastectomy) were included in the study. The groups were separated depending on the season in which the patients were included in the study. Group A—62 women, in whom the first questionnaire and vitamin D concentration determination was carried out in winter (December 2016–January 2017). In these patients, the research procedure was repeated in the summer (July–August 2017). Group B—32 women with the first survey and determination of vitamin D concentration made in the summer (July–August 2017) and repeated in the winter (December 2017–January 2018). Patients from groups A and B, after first tests, received laboratory interpretation of the 25(OH)D concentration result, and additionally, in the case of results beyond the reference values, patients were informed about the need to obtain doctor's advice. Control group (93 women) was recruited among the participants of the regional screening vitamin D testing program for people at risk—at the age of 40 and above. The program was carried out in November–December 2017. Detailed data are available in Supplementary material File S1: Recruitment of patients.

2.2. Survey

The study used an indirect research method in the form of a questionnaire, which each of the respondents completed twice, in winter and summer, before or after (not longer than a month) the determination of vitamin 25(OH)D. The survey asked, inter alia, about basic demographic and anthropometric data (age, height, and body mass); clinical, cancer-focused (age at which breast cancer was diagnosed, presence of neoplasms including breast cancer among relatives, methods of breast cancer treatment, radiation therapy in the past), data on the reproductive system (age at the first menstruation, age at menopause, birth of the first child, use of hormone replacement therapy (HRT)); eating habits (frequency of consumption of vitamin D-rich products); exposure to natural solar/UV radiation (average daily exposure time, protection against UV radiation); knowledge of recommendations on vitamin D supplementation in oncological patients and recommendations provided by the attending physician on vitamin D supplementation, use of vitamin D supplementation—the dose taken all year round. Wallace's rule of nines was used to estimate the percentage of the body area that patients expose to sunlight. Food products rich in vitamin D were selected based on the Nutrition Standards for the Polish population [30] and Polish recommendations for the prevention of vitamin D deficiency [31]. Additional data are available in Supplementary Materials File S2: Survey validation; File S3: Personal data anonymization.

2.3. Laboratory Determinations

Venous blood was collected from all subjects fasting, in the morning. Determinations of vitamin D concentration were performed using the Elecsys Vitamin D total II tests (Roche Diagnostics, Mannheim, Germany) on the Integra cobas e411 analyzer (Roche Diagnostics, Mannheim, Germany) according to the manufacturer manual.

2.4. Statistics

The distribution was verified by the D'Agostino–Pearson normality test. The results are presented using the descriptive statistics. To evaluate relationships between continuous variables with normal distributions, the parametric Pearson correlation test was used. In other cases, the non-parametric test was used—Spearman's rank correlation. The analysis of variance (ANOVA) or Friedman's rank test was used to compare the groups in terms of a measurable feature. To assess the differences between the selected factors in the group A and B, the student's t-test was used for normally distributed variables, and the Mann–Whitney U-test (both for independent variables) for variables with a different distribution. In the case of comparing the dependent variables, the paired t-test was used and Wilcoxon signed-rank test for a non-parametric data. The chi square test was used to compare the distribution of data categorized in studied groups. The results in which $p < 0.05$ were considered statistically significant. Statistica 13 (TIBCO Software Inc., Palo Alto, CA, USA) and MedCalc 15.8 (MedCalc Software Ltd., Ostend Belgium) were used for analysis.

Sample Size Calculation

The sample size for laboratory determinations of vitamin D was calculated from data obtained from the winter and summer tests of first studied group A (women treated from breast cancer), using "Sample size for before-after study (Paired *t*-test)" calculator from the University of California San Francisco, Clinical & Translational Science Institute website. The calculated effect size was 10.7, the standard deviation of the change in the outcome was 15.1 with alpha (type I error rate) set at 5% and a beta (type II error rate) of 10%. Calculation using the T statistic and non-centrality parameter showed a sample size of 23, and approximation using the Z statistic instead of the T statistic showed required a sample size of 21 paired measurements.

The recommended sample size for the survey is 214 and above. With our studied group (size *n* = 92; combined group A and B), the margin of error was 8.17% instead of the recommended 5%, with a confidence level of 90%.

3. Results

3.1. Demographical and Anthropometrical Characteristic of Studied Groups

Study groups A and B as well as the control group were similar in terms of age, body mass and BMI. Detailed data are presented in Table S1, in Supplementary Materials. Also important clinical data are available in Supplementary Materials Table S2: Characteristics of patients related to the reproductive system and Table S3: Characteristics of patients related to the cancer disease.

3.2. Vitamin D Concentration in Studied Patients

The mean concentration of vitamin D in group A from the first sampling was 27.6 ± 14.1 ng/mL, and six months later, the mean concentration was 38.3 ± 12.2 ng/mL ($p = 0.000$ paired *t*-test). In group B, in the first sampling, the concentration of vitamin D was 29.6 ± 13.6 ng/mL, and after six months, it increased to 32.4 ± 13.3 ng/mL ($p = 0.340$ paired *t*-test). In the control group, the mean vitamin D concentration (determined in winter) was 32.2 ± 14.4 ng/mL (Figure 1).

In group A, in the results obtained in winter, the serum concentration of 25(OH)D in 33.9% of patients was below 20 ng/mL (below the reference range), in 25.8% it was within 20–30 ng/mL (suboptimal level), while in 40.3% of patients it was above 30 ng/mL (normal value). In the results obtained in the summer (second sampling), serum 25(OH)D concentration below 20 ng/mL was observed in 7.1% of patients, in 21.4% it was in the range of 20–30 ng/mL, and in 71.4% of patients, it was above 30 ng/mL (Figure 2A).

In group B, in the results obtained from the first sampling (summer), serum concentration of 25(OH)D in 33.3% of patients was below 20 ng/mL, in 24.2% it was within 20–30 ng/mL range, while in 42.5% of patients it was above 30 ng/mL. In the results obtained from the second sampling (winter), serum 25(OH)D concentration below 20 ng/mL was found in 14.3% of patients, in 39.3% it was within the range of 20–30 ng/mL, and in 46,4% of patients were above 30 ng/mL (Figure 2B).

In the control group (sampling in winter), the serum concentration of 25(OH)D in 19.4% of patients was below 20 ng/mL, in 28.0% it was within the range of 20–30 ng/mL, while 52.7% were above 30 ng/mL (Figure 2C).

3.3. The Concentration of Vitamin D and Diet

In the group of patients examined in winter, significantly lower concentrations of vitamin D were noted in women who consumed milk 1–2 times a week compared to those who did not drink milk at all or occasionally (respectively, 23.1 ng/mL and 30.7 ng/mL, $p = 0.0145$). On the other hand, in the study group tested for the first time in the summer, patients who consumed eggs 1–2 times a week had higher vitamin D values compared to those who did not eat eggs or ate them sporadically (29.4 ng/mL and 17.4 ng/mL, respectively p = 0.0522). None of the other analyzed diet components had a statistically significant effect on the 25(OH)D concentration in the studied patients. The detailed information about consumption of food rich in vitamin D in studied groups is delivered in Supplementary material Table S4: The frequency of consumption of foods rich in vitamin D per week in the combined group A + B and control group, before entering the study and Table S5: Number of patients consumed food rich in vitamin D before first and second testing in combined A + B groups.

3.4. Supplementation of Vitamin D in Studied Patients

At the time of enrollment in the study, more than half of patients after breast cancer treatment did not supplement with vitamin D. In group A, it was 51.6% (n = 32), and, in group B, 56.2% (n = 18). According to the data from the questionnaires performed during the second sampling, after obtaining the vitamin D concentration result from the first testing, the percentage of patients supplemented vitamin D in group A increased to 75.8% (n = 47), also, in group B, the percentage increased to 75.0%. The average dose of supplemented vitamin D also increased. In group A, the average dose of vitamin D taken before the first test (in winter) was, on average, 1500 units per day (range from 200 to 4000 units). Before the next test, the average intake of vitamin D was slightly above 1700 units per day (range from 500 to 4000 units). In group B, the average intake of vitamin D before the first examination was nearly 2000 units per day (range from 1000 to 8000 units). Before the next sampling, it increased, on average, almost to 2500 units per day. Detailed data about the concentration of vitamin D depending on supplementation are presented in Table 1.

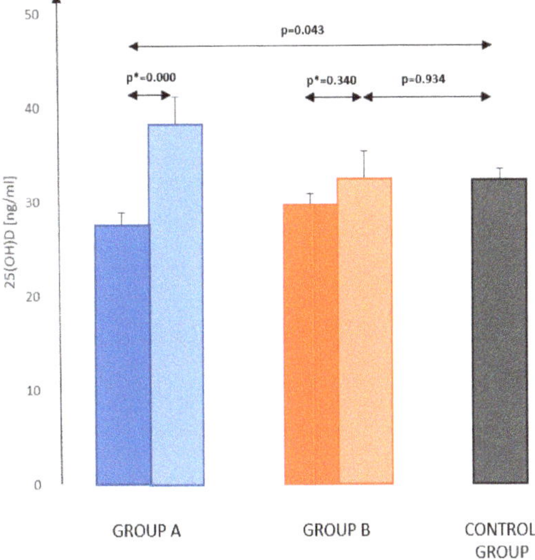

Figure 1. Mean concentrations of 25(OH)D (±SD—error bars) in the healthy control group and the groups of women treated for breast cancer. Values before inclusion into study—dark (left) bars in the pair and during second testing half year later—pale bars (right) in the pair. Group A—women treated for breast cancer tested the first time in winter; Group B—women treated for breast cancer tested the first time in summer. Comparison of the results in the control group (obtained in winter) with the results of studied groups in winter. p*—paired t-test; p—t-test.

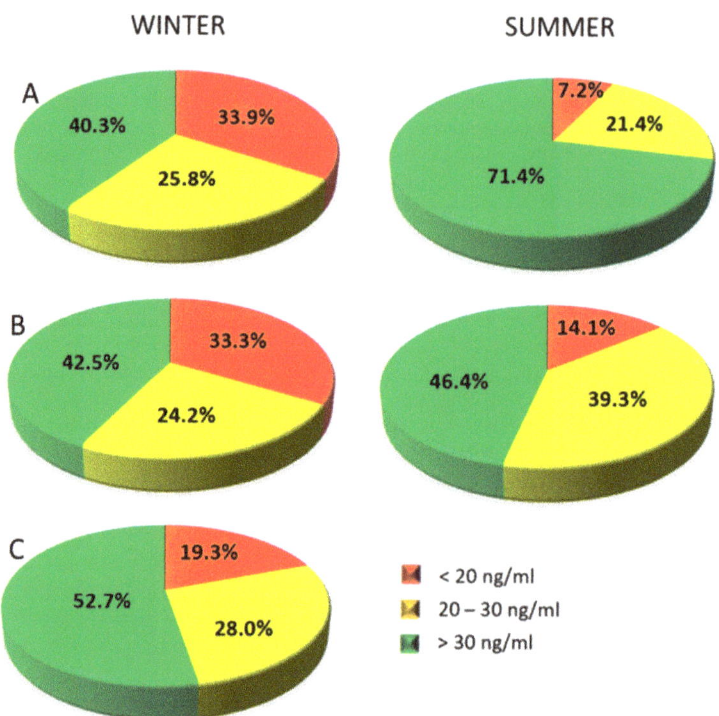

Figure 2. Percentage of women with vitamin 25(OH)D concentrations below the reference value range, with suboptimal and optimal values in Group A—women treated for breast cancer tested the first time in winter; testing during winter—left disc, testing during summer—right disc. (**A**), Group B—women treated for breast cancer tested the first time in summer; testing during winter—left disc, testing during summer—right disc (**B**), and the control group (**C**) (testing during winter only).

Table 1. Average 25(OH)D concentration in women treated for breast cancer and in healthy persons in relation to vitamin D supplementation and season.

	n Supplementing/ *n* Non-Supplementing	Patients Supplementing Vitamin D (ng/mL)	Patients Non-Supplementing Vitamin D (ng/mL)	*p*
Group A winter	30/32	34.6 ± 14.2	22.4 ± 11.9	$p = 0.0006$
Group A summer	47/15	38.9 ± 12.2	34.8 ± 12.1	$p = 0.24$
Group B summer	16/18	36.8 ± 15.4	24.0 ± 9.0	$p = 0.012$
Group B winter	24/8	35.3 ± 13.7	22.8 ± 5.3	$p = 0.001$
Control group winter	41/36	38.3 ± 16.3	27.6 ± 10.9	$p = 0.0006$

Notes: Group A—women treated for breast cancer tested first time in winter; Group B—women treated for breast cancer tested the first time in summer; *n*—sample size; the variables are presented as mean ± SD.

In group A, during the first part of the study (winter), the deficit of 25(OH)D was found in 10.0% (*n* = 3) of patients who took the pharmacopeial form of vitamin D and in 59.4% (*n* = 19) of patients who did not use supplementation. The suboptimal level was found in 26.7% (*n* = 8) of patients taking vitamin D preparations and in 25.0% (*n* = 8) of patients not taking pharmacopeial vitamin D. Recommended vitamin D concentration values (>30 ng/mL) were found in 19 patients (63.3%) taking pharmacopeial vitamin D and 5 people (15.6%) not taking any vitamin D supplementation. In the results obtained in the second part of the study (summer; group A), 25(OH)D concentrations below 20 ng/mL were found in 6.4% (*n* = 3) of patients who took the pharmacopeial form of vitamin D, and in 26.7% (*n* = 4) of non-supplementing patients; the suboptimal level was found in 23.4% (*n* = 11) of people taking vitamin D supplements and in 26.7% (*n* = 4) of patients not taking the pharmacopeial vitamin D. Optimal vitamin D concentration was found in 70.2% (*n* = 33) of people using pharmacopeial vitamin D and in 46.6% (*n* = 7) of people not using vitamin D supplementation (Figure 3A).

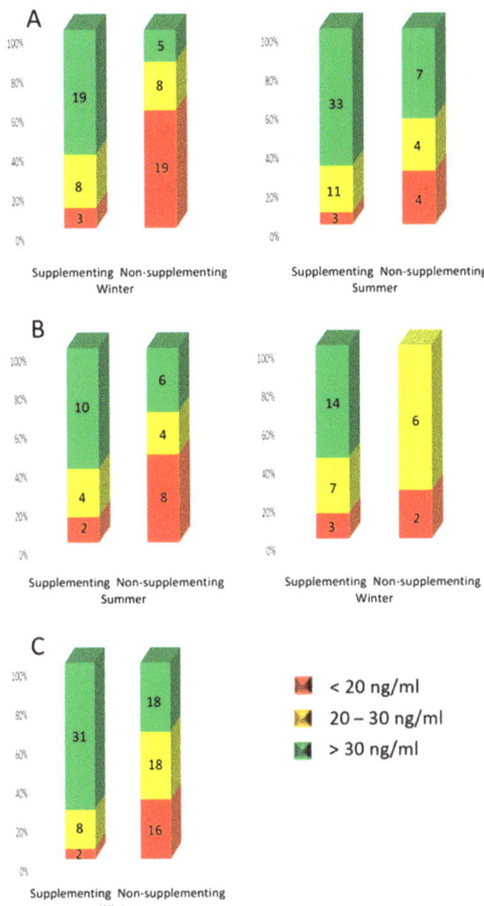

Figure 3. Percentage of patients in the healthy control group and the groups of women treated for breast cancer, with deficiency, suboptimal and optimal vitamin D levels depending on supplementation of vitamin D and season—winter/summer. Group A—women treated for breast cancer tested the first time in winter (**A**); Group B—women treated for breast cancer tested the first time in summer (**B**), and the control group tested during winter only (**C**). The number of persons in every subgroup is listed on the bars.

In group B, in the results of first sampling (summer), the vitamin D deficit was found in 12.5% (*n* = 2) patients who took the pharmacopeial form of vitamin D, and 44.5% (*n* = 8) in the group non-supplementing pharmacopeial vitamin D; the suboptimal level was found in 25.0% (*n* = 4) of people taking and 22.2% (*n* = 4) of those not taking vitamin D preparations. The optimal value of vitamin D concentration was found in 62.5% (*n* = 10) of people using pharmacopeial vitamin D and in 33.3% (*n* = 6) of people not supplementing vitamin D. In the results obtained in the second sampling (winter), the 25(OH)D deficiency was found in three patients (12.5%) who took the pharmacopeial form of vitamin D, and two non-supplementing (25.0%). The suboptimal level was found in seven (29.2%) patients taking vitamin D preparations and in six (75.0%) patients not taking the pharmacopeial vitamin D. The optimal vitamin D concentration was found in 14 patients (58.3%) taking the pharmacopeial vitamin D and in none of the patients who did not supplement vitamin D (Figure 3B).

In the control group, the vitamin D deficient were two (4.9%) persons who supplemented with vitamin D and 16 (30.8%) who did not supplement. The suboptimal level was found in eight (19.5%) people using vitamin D supplementation and in eighteen (34.6%) patients without such a support. The optimal values of vitamin D concentration were found in thirty-one (75.6%) people taking the pharmacopeial vitamin D and in eighteen (34.6%) people who did not supplement vitamin D (Figure 3C).

The percentages of women supplementing vitamin D in groups A and B at the first sampling with deficiency, suboptimal concentration, and optimal values, were similar (respectively, group A—10.0%, 26.7%, 63.3%; group B—12.5%, 25%, 62.6%). In the control group, among women supplementing vitamin D, the percentage of persons with the optimal vitamin D results was significantly higher (75.6%). On the other hand, among the non-supplementing patients in group B, there was a significantly lower percentage of results indicating vitamin D deficiency (44.5%) and a higher percentage indicating suboptimal and optimal (22.2 and 33.3%, respectively) than in group A (59.4%, 25.0%, and 15.6%, respectively).

3.5. Knowledge of Recommendations on Vitamin D Supplementation

In group A, 14 (22.6%) patients declared that they knew the recommendations regarding vitamin D supplementation in women with breast cancer before entering the study. Among them, 10 (71.42%) declared the use of vitamin D supplementation. However, only five patients (35.7%) had a determined 25(OH)D concentration in the past. The average value of vitamin D concentration in group A, among those who knew the recommendations and used vitamin D supplementation, was 39.3 ng/mL, and, among women who knew the recommendation, but did not use vitamin D supplementation, the average concentration was 24.0 ng/mL, while the average concentration of vitamin D, in women who did not know the recommendations and did not use vitamin D supplements, was lower—20.1 ng/mL (Figure 4A).

In the second series of studies carried out in the summer, 39 (62.9%) patients declared knowledge of the recommendations regarding vitamin D supplementation in women with breast cancer, of which 92.3% ($n = 36$) declared the vitamin D supplementation. The average value of vitamin D concentration in the group of respondents knowing the recommendations and using vitamin D supplementation was 38.1 ng/mL, and among women who knew the recommendation, but did not use vitamin D supplements, this value was, on average, 30.3 ng/mL. Similarly, the average concentration of vitamin D in women who still did not know the recommendation and did not use vitamin D supplementation was 32.4 ng/mL (Figure 4A).

In group B, four patients (12.5%) declared knowledge of the recommendations regarding vitamin D supplementation in women with breast cancer before entering the study (summer). All the patients who knew the recommendations stated that they were using vitamin D supplements, and three of them (75%) had measured 25(OH)D concentration in the past. The mean vitamin D concentration among patients familiar with the recommendations and taking vitamin D supplements was 46.2 ng/mL. The mean concentration of vitamin D in women who did not know the recommendations and did not use vitamin D supplements was 24.6 ng/mL (Figure 4B).

In the next series of studies carried out in winter, 18 (56.3%) patients declared that they knew the recommendations regarding vitamin D supplementation in women with breast cancer, of which 15 (83.3%) used vitamin D supplements. The average value of vitamin D concentration in those who knew the recommendations and used vitamin D supplements was 36.6 ng/mL, and among women who knew the recommendation, but did not use vitamin D supplements, this value was, on average, 26.3 ng/mL. The mean concentration of vitamin D in women who did not know and did not use vitamin D supplements was 19.8 ng/mL (Figure 4B).

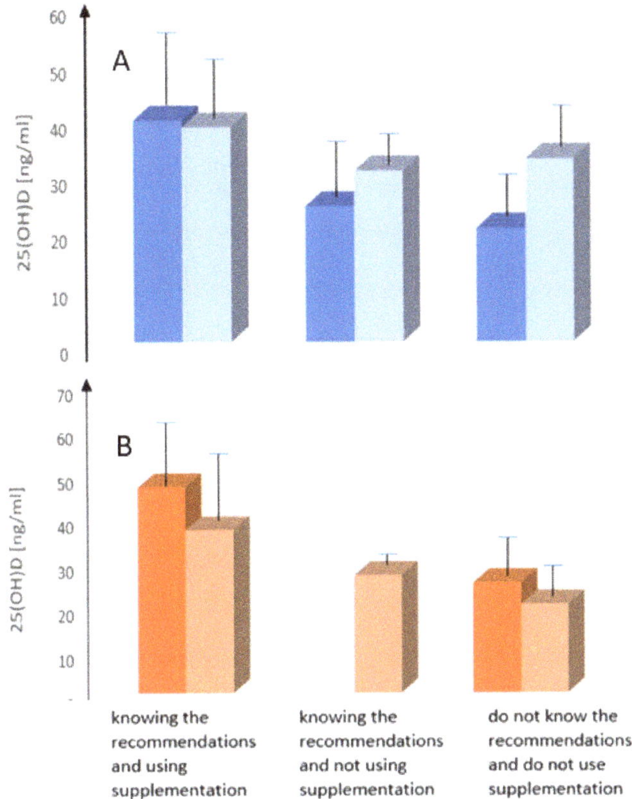

Figure 4. Mean serum 25(OH)D concentration ($\pm$SD—error bars) in the groups of women treated for breast cancer depending on the knowledge of vitamin D supplementation guidelines and supplementation of vitamin D. (**A**)—Group A—women treated for breast cancer tested the first time in winter. (**B**)—Group B—women treated for breast cancer tested the first time in summer. Dark (**left**) bars in the pair—values and data before inclusion into the study; pale bars (**right**) in the pair values and data during second testing half year later.

Only 15 women from group A and B had determined vitamin D which constitutes 15.9% of all examined patients. In addition, a similar percentage of women from the control group had previously undergone such a laboratory test (15.1%, $n = 14$).

Vitamin D supplementation was recommended to patients in group A and B by GPs in 35.0% of cases ($n = 14$), specialists in 30.0% of cases ($n = 12$) (including oncologists, neurologists, endocrinologists, gynecologists, diabetologists, internists), and pharmacists in 10.0% of cases ($n = 4$). Some patients made their own decisions to start supplementation (25.0%; $n = 10$) without any medical professional advice. In the control group, vitamin D intake was most often recommended by a physician (63.8%; $n = 30$) or a pharmacist (8.5%; $n = 4$), and was also, in some cases, an independent decision of the patients (27.7%; $n = 13$) based on information obtained from media broadcasts and friends (Figure 5).

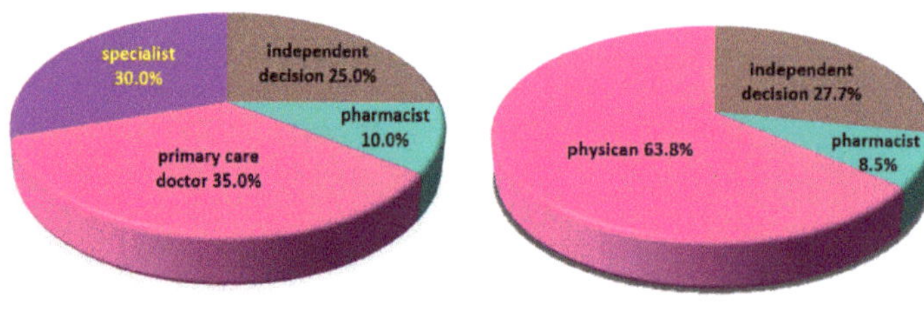

Figure 5. Sources of information on vitamin D supplementation in women treated for breast cancer before joining the program (combined group A in winter and group B in summer)—**left** disc; and in healthy control group—**right** disc. The detailed percentages are listed in the figure.

3.6. Exposure of Patients to Solar Radiation

In groups A and B, most of the patients declared that they protected themselves from the sun in summer (72.6% ($n = 45$) and 71.8% ($n = 23$), respectively); in the control group, the percentage of patients protecting themselves from the sun was lower—59.1% ($n = 55$); ($p = 0.27$, chi^2 test). Most patients, after breast cancer treatment, protected themselves from the sun by wearing appropriate clothing (73.4%); 26.6% used only—or additionally—cosmetics with UV filters. In the control group, more than half of the patients used cosmetics with UV filters (50.7%) as basic or additional protection.

Patients from both studied groups (A + B), who did not protect themselves against solar radiation, according to their assessment, spent, on average, 151 ± 101 min in the sun daily during the summer, and patients from the control group spent 108 ± 101 min in the sun daily ($p = 0.13$).

There were no differences in vitamin D concentrations measured during the summer between the group of patients avoiding the sun (wearing appropriate clothing and/or using cosmetics with UV filters; 34.9 ± 13.9 ng/mL) and those who declared that they did not use any protection and even willingly sunbathed (34.1 ± 11.1 ng/mL).

3.7. Correlation of Vitamin D Concentration with Age

There was no statistically significant correlation between the age of the patients and the vitamin D concentration determined at the first measurement in the combined both studied groups (A + B) ($R = 0.010$, $p = 0.91$) and in the control group ($R = 0.19$, $p = 0.068$).

3.8. Correlation of Serum Vitamin D Concentration with Diet

To assess the effect of diet on 25(OH)D concentration, the data obtained from all patients during the winter period were analyzed to avoid bias due to endogenous production after exposure to sunlight. The correlation of vitamin D concentration in winter in all subjects with the total consumption of foods rich in vitamin D (fish, eggs, milk) was not statistically significant ($R = 0.041$; $p = 0.585$). Additionally, the correlation was assessed in the group of people who did not supplement vitamin D in order to ignore the effect of supplementation on the concentration of vitamin D. Again, in this case, no significant correlation was found ($R = 0.004$, $p = 0.969$).

3.9. Correlation of the Declared Tanning Time with the Concentration of Vitamin D

To analyze the influence of sunlight on vitamin D concentration, such a relationship was analyzed in patients who did not supplement vitamin D by correlating the time and area of skin exposed to the sun with the concentration of vitamin D determined in the summer months. The result was statistically significant—$R = 0.584$, $p = 0.0014$ (Figure 6).

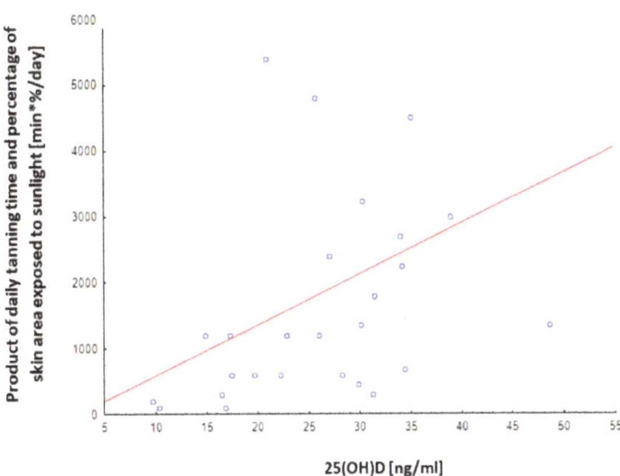

Figure 6. Correlation of the product of the declared tanning time and the percentage of the skin area exposed to sunlight with the concentration of vitamin D determined in the summer period in all surveyed people without vitamin D supplementation. $n = 27$, Pearson's R = 0.584, $p = 0.0014$.

4. Discussion

Recently, hypovitaminosis D has been recognized as one of the risk factors for breast cancer, and the concentration of vitamin D in the blood becoming a prognostic factor [21]. The results of meta-analyses show that higher serum 25(OH)D$_3$ concentration determined after breast cancer diagnosis correlates with lower mortality due to breast cancer. In particular, patients with 25(OH)D$_3$ in the highest quartile had approximately half the mortality rate compared to those with vitamin D in the lowest quartile [32]. There are also many reports on the relationship between the serum concentration of vitamin D and the achieved clinical results of breast cancer treatment and risk of recurrence [33,34]. The above-described relationship between the concentration of vitamin D and the risk of breast cancer or recurrence, as well as therapeutic success in the event of this disease, indicates the need to pay attention also to this factor. It can be consciously easily modulated by the time spent in the sun and diet, and it is also possible to administer this vitamin as pharmacopeial medication [32].

Vitamin D deficiency (serum 25(OH)D concentration < 20 ng/mL or <50 nmol/L) in the European population is a common phenomenon. The reports of the European Calcified Tissue Society estimate that 30–40% of the population of Central, Eastern, and Southeastern Europe is vitamin D deficient [35]. National studies showed even worse data [36,37]. Vitamin D deficiency is also common in the population of breast cancer patients, occurring in 23.0% to 95.6% of patients [38]. In the present study, a deficiency defined as 25(OH)D concentration below 20 ng/mL was demonstrated in the first sampling in 33.9% of patients in group A and 33.3% of patients in group B. Suboptimal concentration at the same time was found in 25.8% of patients in group A and 24.2% of patients in group B. In the control group, vitamin D deficiency was found in 19.3% patients, and suboptimal levels were observed in 28.0%. Andersen et al. showed that 30% of females in a group of American patients, after breast cancer treatment, were vitamin D deficient [39]. In addition, Apoe in a similar group showed that 62% of patients had 25(OH)D concentration below 30 ng/mL [40]. Mechado et al. tested 209 Brazilian women after breast cancer treatment and 26.2% had vitamin D deficiency and 55.6% had suboptimal levels [41]. These results confirm several published sets of data about the higher prevalence of vitamin D deficiency in the population of people with cancer. Furthermore, it should be remembered that serum 25(OH)D concentration above 20 ng/mL ensures the proper functioning of the calcium-phosphate metabolism and bone density, but only concentrations above 30 ng/mL ensure the optimal amount of vitamin D for other bodily functions, including proper functioning of immune mechanisms, as well as obtaining a protective effect in the context of neoplastic diseases [42].

Endogenous production stimulated by solar radiation is the primary source of vitamin D and is determined by geographical factors (latitude, the ozone layer, cloud cover, albedo), season, and individual factors (genetic, including skin phototype, time spent outdoors, body area exposed to the sunlight, applying cosmetics with UV filters, etc.) [43–45]. It is assumed that over 90% of the vitamin D requirement in summer is covered by the endogenous conversion of previtamin into provitamin D in the skin due to UVB radiation [46,47].

The present studies confirmed the seasonality of changes in 25(OH)D concentration in patients after breast cancer treatment. After excluding from the analysis the patients who declared supplementation with vitamin D, the tests performed for the first time in winter showed a higher percentage of patients with vitamin D deficiency compared to the tests performed in summer (58% and 44%, respectively). Additionally, in the group tested firstly in the summer, a significantly higher percentage of patients (33%) with optimal vitamin D concentrations was found than in patients tested first time in winter (16%). A similar seasonality in 25(OH)D concentration in patients after treatment for breast cancer has been demonstrated in other studies. Acevedo et al. found that the concentrations of 25(OH)D in women with breast cancer in summer were significantly higher compared to the tests performed in winter (p = 0.0322) [48]. Eliassen et al. also demonstrated the dependence of 25(OH)D concentration, in patients with breast cancer, on the time of year in which the measurement was performed [49]. A study of serum 25(OH)D concentration in 1940 women by Shi et al. during the first 6 months after breast cancer diagnosis showed higher 25(OH)D concentration in summer and fall [34].

Exposition to sunlight is an easily modifiable factor, but is season dependent. In the latitude of central Europe (between 30 and 55 degrees latitude), approximately half an hour of exposure to solar radiation in the mid-afternoon period during the summer, three times a week with bare limbs, is sufficient to obtain a serum concentration of 25(OH)D equal to or higher than 20 ng/mL (50 nmol/L) in 90% of the white population. The solar UVB radiation from October to March is weakened by an extended path through the atmosphere. As a result, in this period of the year, the skin synthesis of vitamin D is weaker and insufficient to cover human needs [45]. In addition, the change in lifestyle in recent decades, which has diminished exposure to sunlight, is of great importance in this respect [50]. According to the generally prevailing opinion, "excessive exposure" (for which there is no definition) to the sun is a factor causing skin cancer. That has led to a situation where most people believe that they should avoid sun radiation and over-shield the body or use cosmetics with UV filters [51]. The results of my research indicate that the majority of patients from groups A and B avoid sun radiation (72% (n = 45) and 73% (n = 23), respectively): two-thirds in the form of covered clothing, and the rest only, or additionally, used creams with protective filters. Despite such protection, there were no differences in vitamin D concentrations determined in summer between the group of patients declaring no use of protection from the sun and those who avoided the sun. That may result from a longer overall time spent outdoors by the latter women. The percentage of the body area protected with a UV filter, the SPF factor and the thickness of the cream layer applied to the skin should be also taken into account. At the same time, a statistically significant correlation was demonstrated (R = 0.584, p = 0.0014) between the time and area of the skin exposed to sunlight in patients who did not supplement vitamin D and the vitamin D concentration determined in the summer.

It is estimated that an average of 100–200 IU (2.5–5.0 µg) of vitamin D is consumed daily with the diet, which is 15–30% of the requirement [52,53]. Therefore analyzing data from the survey, it was not surprising that none of the women (n = 6) who decided not to take pharmacological supplementation, only to enrich their diet with natural products rich in vitamin D during a second sampling in a winter, did obtain an optimal vitamin D level. The results significantly confirm that it is impossible to supply an adequate amount of vitamin D in the latitude of Central Europe only through the diet in the winter months when there is no skin synthesis of vitamin D.

The lack of a statistically significant correlation between the diet and the concentration of vitamin D in the serum (R = 0.041; p = 0.585) may result from the widespread enrichment of various foods, which the study participants may not have pointed out or may not know.

Supplementation with pharmacopeial preparations is a relatively easy and inexpensive method of delivering vitamin D. The importance of vitamin D supplementation in long-term follow-up of breast cancer patients was shown in several papers. For instance, it was shown that the mortality due to breast cancer in patients who started supplementation after cancer diagnosis was 20% lower compared to the group not supplementing vitamin D, and the reduction in mortality was 49% in patients who started supplementing with such supplementation within no more than 6 months after diagnosis [54].

In the present study, during enrollment in the study, vitamin D supplementation was used by less than half of the patients (group A 48.4%; n = 30, group B 43.8%; n = 14). Patients who did not supplement vitamin D had significantly lower 25(OH)D concentration, regardless of the season. On the other hand, supplementation increased the percentage of women whose 25(OH)D concentration exceeded 20 ng/mL. Similar results were found in the group of 332 Swiss women after treatment for breast cancer. Only 133 patients took calcium supplementation with vitamin D or vitamin D alone. However, in many patients, despite the supplementation (800 IU), the vitamin D level remained suboptimal [55].

There are no uniform global guidelines on what vitamin D doses and serum concentrations should be considered as a target in the group of patients after breast cancer treatment. The National Comprehensive Cancer Network recommends that cancer patients should maintain 25(OH)D concentration above 30 ng/mL [56]. The American Cancer Society/American Society of Clinical Oncology recommends vitamin D supplementation in patients with breast cancer starting from the age of 50 at a dose of 600–1000 IU/day [22]; in turn, ESMO recommends higher daily doses of vitamin D in this group of patients, amounting to 1000–2000 IU/day [23]. It should be noted that both societies recommend the above doses to prevent bone loss related to the nature of the disease and the side effects of treatment, and not to prevent a recurrence. Since 2018, the Polish group of experts recommends vitamin D supplementation under the control of serum concentrations so that it should be between 30 and 50 ng/mL in groups with the risk of vitamin D deficiency, including, among others, the population of breast cancer patients. Supplementation should be carried out using daily doses of vitamin D between 800 and 2000 IU for adults, differentiated depending on its supply in the diet and current weight [24]. In addition, several scientists, based on the analysis of epidemiological studies of cancer risk, its progression, recurrence, and mortality, recommend higher doses, to maintain higher 25(OH)D concentration between 30-40 ng/mL, up to 60 ng/mL [56–58].

The percentage of patients declaring knowledge of the recommendations in the field of vitamin D supplementation in women with breast cancer before inclusion in this study was low—22.6% in group A and 12.5% in group B. During the second test, 62.9% and 56.3% of women in groups A and B declared knowledge of the recommendations regarding vitamin D supplementation. In addition, the percentage of people using vitamin D supplements increased in the group where knowledge of the recommendations was declared (during the first sampling in group A—71.42%; in the second—92.3%). Patients who did not know the recommendations and did not use supplementation had lower mean serum concentrations of 25(OH)D compared to those who knew the recommendations and supplemented, and even those who knew the recommendations but did not supplement vitamin D. Most of the surveyed women obtained information on the recommended principles of vitamin D supplementation from doctors; 35% from specialists and 30% from primary care physicians. Pharmacists advised supplementation in every tenth patient. Overall, 25% of patients made decisions about supplementation independently. More than half of the respondents (56.6%) had the dosage determined by a doctor, every fourth patient was advised by a pharmacist, and the remaining 23.3% of patients had chosen a vitamin D dose themselves.

Similar data were obtained in other countries. In Ireland, in 2005 only 15.5% of patients with breast cancer were instructed by a doctor about the need for supplementation and received a prescription for vitamin D. Six years later, this percentage increased to 36.9% of patients and the average daily dose of prescribed vitamin D was 857 IU/day [59]. In the USA, at the beginning of the 21st century, only 56% of breast cancer patients after chemotherapy-induced menopause received information on the recommended supplementation with vitamin D and calcium [60]. The best results in communicating the need for vitamin D supplementation in women after breast cancer treatment were reported in Croatia. According to a prospective 3.5-year study by Bosković et al., 75.7% of patients with breast cancer received a prescription for vitamin D and calcium from a medical oncologist. Unfortunately, a large group of patients, after breast cancer treatment (40–80% depending on the center where the women were treated), do not follow medical recommendations on vitamin D supplementation [61].

The presented data show that, in Poland and many other countries, a lot of patients diagnosed with breast cancer do not obtain information on the supplementation of vitamin D from attending doctors (oncologists, endocrinologists, gynecologists and GPs). Kimiafar et al. unequivocally showed that patients with breast cancer expect information about the disease, its course, prognosis, rehabilitation process, and variants of diagnostic and therapeutic procedures. However, they would most like to understand the possible side effects therapy, and allowed diets. Despite the growing awareness of physicians about the needs of patients with cancer, many patients still feel that they receive insufficient information or information that is unclear or incomprehensible. That is why patients often use alternative sources of information, such as the internet, books, and other patients, to obtain information on how to use supplements or compose the right diet [62].

All these data indicate the need for education of both patients and doctors about the benefits of maintaining optimal 25(OH)D concentration, especially in the group of patients treated for breast cancer. Whether vitamin D supplementation in oncological patients is the domain of oncology specialists or primary care physicians should be clearly defined.

Some studies show that patients, after breast cancer treatment, more often use vitamin D supplements when they are tested for the 25(OH)D concentration. In the large prospective The Sister Study, at the first testing, the regular use of vitamin D supplement was declared by 56% of women. Along with testing, participating women were informed about the planned vitamin D determination after a few years. During the second determination of vitamin D, regular supplementation was declared by 84% of women, and mean 25(OH)D concentrations were higher, due to the greater number of women who reported using vitamin D supplements [63].

The present study also showed an increased number of women treated for breast cancer supplemented vitamin D, after testing the 25(OH)D. The patients received laboratory interpretation of the result and were advised to obtain detailed recommendations, including recommended doses of supplementation and laboratory monitoring of vitamin D concentrations, from a doctor. When enrolled in the study, more than half of the patients from groups A and B did not supplement vitamin D, while, during the second testing, the percentage of people supplementing vitamin D increased to over 75% in both study groups.

According to this study and reports of other authors, vitamin D supplementation must be adapted to the season, lifestyle, and individual characteristics of a person, including the individual ability to absorb vitamin D [35]. Even advanced algorithms that allow individualized dosages of vitamin D, e.g., depending on age, body weight, or menopausal status, do not always bring the expected effects [64]. It is difficult to determine the optimal dosage due to the supply of vitamin D in the diet and endogenous production after exposure to solar radiation, the use of cosmetics with filters, the amount of fat tissue, skin pigmentation, and air pollution [65]. Therefore, in the Polish Standards of Nutritional Treatment in Oncology (2015), vitamin D supplementation should be ordered based on the result of laboratory determination of this component in the blood [25]. Similar recommendations were issued in 2018 by the Spanish Society of Oncological Medicine (SEOM) [66].

The data of the present study show that only about 15% of women from the study and control groups had their 25(OH)D concentration determined before being included in the research program. Similar results were obtained by Andersen et al. in a population of 553 American women diagnosed with breast cancer no later than 2 years before study enrollment. Interestingly, it was assessed whether conventional medicine physicians and practitioners of alternative medicine monitor the concentration of 25(OH)D in the group of patients treated conventionally, but supplemented with alternative medicine therapies. It was found that women who used complementary therapies in addition to conventional medicine had more frequently measured blood vitamin D compared to patients using only conventional medicine (30% vs. 16%) [39]. It can be assumed that doctors of natural medicine attach more importance to supplementation and holistic compensation than doctors ordering targeted pharmacological treatment.

The first measurement of 25(OH)D concentration during this study, and discussing the result with the patients and indicating the necessity to contact the attending physician in the case of deficiencies, prompted the patients to compensate for the deficits. There was a 26.73% reduction in the number of vitamin D deficient women in the next testing in group A and by 19.04% in group B, regardless of the season of the first and second tests. It was associated with an increase in the percentage of patients who used vitamin D supplementation by half and with higher mean concentration of 25(OH)D in the second measurement compared to the first.

Although supplementation with vitamin D under the control of laboratory determinations is the best solution, it should be remembered that laboratory tests are inconvenient for the patient and increase the cost of treatment. However, the price of vitamin D determinations is almost traceable to the costs of oncological treatment in the case of recurrence, and social costs of the patient's death.

The limitation of the study may result from the use of the questionnaire in research proceedings. Patients participating in the study may have given the wrong answer due to difficulties in understanding the questions, or may have knowingly concealed the truth. Bearing in mind the above issues, a pilot was conducted along with the validation of the survey, the aim of which was to eliminate the factors affecting the credibility of the survey by using control questions and adjusting the questions contained in the survey so that they were as understandable as possible for the respondents (data available in supplements). A lack of control group analysis in summer slightly increases the uncertainty of some comparisons; however, the most important are relationships inside the same study group (tested twice). The influence of tumor location and type as well as the applied additional treatment during and after mastectomy was not analyzed and may be a source of potential bias. The number of patients in the study is sufficient for comparisons of laboratory data inside groups. The relatively low number of survey participants could be a source of uncertainty of data from the questionnaire.

5. Conclusions

- Women with treated breast cancer are more likely to develop vitamin D deficiency than the healthy population.
- The seasons, and the time and area of the skin exposed to solar radiation, influence the concentration of 25(OH)D in women after treatment for breast cancer.
- Diet is not important in maintaining the proper concentration of vitamin D in women after treatment for breast cancer.
- Supplementation with vitamin D significantly improves its status in women after treatment for breast cancer; therefore, oncologists and GPs should recommend it to patients.
- Knowledge of recommendations on vitamin D supplementation among patients after treatment for breast cancer is directly related to the more frequent use of supplementation and less frequent 25(OH)D deficiency.
- Laboratory monitoring of vitamin D concentration has a positive effect on maintaining optimal vitamin D concentration in women after breast cancer treatment.

- Routine vitamin D testing should be introduced to the screening panel in follow-up patients after breast cancer treatment.

Supplementary Materials: The following are available online at https://www.mdpi.com/2072-6643/13/2/564/s1, File S1: Recruitment of patients; File S2: Survey validation; File S3: Personal data anonymization; Table S1: Anthropometric characteristics of the patients; Table S2: Characteristics of patients related to the reproductive system; Table S3: Characteristics of patients related to the cancer disease; Table S4: The frequency of consumption of foods rich in vitamin D per week in the combined group A + B and control group, before entering the study; Table S5: Number of patients consumed food rich in vitamin D before first and second testing in combined A + B groups.

Author Contributions: Conceptualization P.T. and S.S.; methodology, A.R., P.T. and A.W.; software, A.R. and S.P.; validation, A.R. and P.T.; formal analysis, A.R. and S.P.; investigation, A.R.; data analysis, A.R., S.S. and P.T.; writing—original draft preparation A.R., P.T. and S.P.; writing—review and editing, P.T.; visualization, A.R. and A.W.; supervision, P.T. All authors have read and agreed to the published version of the manuscript.

Funding: This research received no external funding.

Institutional Review Board Statement: The study was conducted according to the guidelines of the Declaration of Helsinki, and approved by the Jagiellonian University Ethics Committee.

Informed Consent Statement: Informed consent was obtained from all subjects involved in the study.

Data Availability Statement: The data presented in this study are available on request from the corresponding author. The data are not publicly available due to General Data Protection Regulation (GDPR) and lack of such a permission from participants.

Conflicts of Interest: The authors declare no conflict of interest.

References

1. Janssens, J.; Vandeloo, M. Rak piersi: Bezpośrednie i pośrednie czynniki ryzyka związane z wiekiem i stylem życia. *Nowotw. J. Oncol.* **2009**, *59*, 159–167.
2. Wojciechowska, U.; Czaderny, K.; Ciuba, A.; Olasek, P.; Didkowska, J. Nowotwory Złośliwe w Polsce w 2016 Roku. Krajowy Rejestr Nowotworów. Available online: http://onkologia.org.pl/wp-content/uploads/Nowotwory_2016.pdf (accessed on 11 October 2019).
3. Didkowska, J.; Wojciechowska, U.; Olasek, P. Nowotwory Złośliwe w Polsce w 2015 Roku. Krajowy Rejestr Nowotworów. Available online: http://onkologia.org.pl/wp-content/uploads/Nowotwory_2015.pdf (accessed on 11 October 2019).
4. Didkowska, J.; Wojciechowska, U.; Czaderny, K.; Olasek, P.; Ciuba, A. Nowotwory Złośliwe w Polsce w 2017 Roku. Krajowy Rejestr Nowotworów. Available online: http://onkologia.org.pl/wp-content/uploads/Nowotwory_2017.pdf (accessed on 11 October 2019).
5. Romieu, I.I.; Amadou, A.; Chajes, V. The Role of Diet, Physical Activity, Body Fatness, and Breastfeeding in Breast Cancer in Young Women: Epidemiological Evidence. *Rev. Investig. Clin.* **2017**, *69*, 193–203. [CrossRef]
6. Chun, R.F.; Shieh, A.; Gottlieb, C.; Yacoubian, V.; Wang, J.; Hewison, M.; Adams, J.S. Vitamin D Binding Protein and the Biological Activity of Vitamin D. *Front. Endocrinol.* **2019**, *10*, 718. [CrossRef]
7. De La Puente-Yagüe, M.; Cuadrado-Cenzual, M.A.; Ciudad-Cabañas, M.J.; Hernández-Cabria, M.; Collado, L. Vitamin D: And its role in breast cancer. *Kaohsiung J. Med. Sci.* **2018**, *34*, 423–427. [CrossRef]
8. Garland, F.C.; Garland, C.F.; Gorham, E.D.; Young, J.F. Geographic variation in breast cancer mortality in the United States: A hypothesis involving exposure to solar radiation. *Prev. Med.* **1990**, *19*, 614–622. [CrossRef]
9. Wu, X.; Zhou, T.; Cao, N.; Ni, J.; Wang, X. Role of Vitamin D Metabolism and Activity on Carcinogenesis. *Oncol. Res. Featur. Preclin. Clin. Cancer Ther.* **2015**, *22*, 129–137. [CrossRef]
10. Chen, G.-C.; Zhang, Z.-L.; Wan, Z.; Wang, L.; Weber, P.; Eggersdorfer, M.; Qin, L.-Q.; Zhang, W. Circulating 25-hydroxyvitamin D and risk of lung cancer: A dose–response meta-analysis. *Cancer Causes Control* **2015**, *26*, 1719–1728. [CrossRef]
11. Bauer, S.R.; Hankinson, S.E.; Bertone-Johnson, E.R.; Ding, E.L. Plasma Vitamin D Levels, Menopause, and Risk of Breast Cancer. *Medicine* **2013**, *92*, 123–131. [CrossRef]
12. Shao, T.; Klein, P.; Grossbard, M.L. Vitamin D and Breast Cancer. *Oncologist* **2011**, *17*, 36–45. [CrossRef]
13. Gandini, S.; Boniol, M.; Haukka, J.; Byrnes, G.; Cox, B.; Sneyd, M.J.; Mullie, P.; Autier, P. Meta-analysis of observational studies of serum 25-hydroxyvitamin D levels and colorectal, breast and prostate cancer and colorectal adenoma. *Int. J. Cancer* **2011**, *128*, 1414–1424. [CrossRef]
14. Yin, L.; Grandi, N.; Raum, E.; Haug, U.; Arndt, V.; Brenner, H. Meta-analysis: Serum vitamin D and breast cancer risk. *Eur. J. Cancer* **2010**, *46*, 2196–2205. [CrossRef]

15. Mondul, A.M.; Weinstein, S.J.; Layne, T.M.; Albanes, D. Vitamin D and Cancer Risk and Mortality: State of the Science, Gaps, and Challenges. *Epidemiol. Rev.* **2017**, *39*, 28–48. [CrossRef] [PubMed]
16. Atoum, M.; Alzoughool, F. Vitamin D and Breast Cancer: Latest Evidence and Future Steps. *Breast Cancer Basic Clin. Res.* **2017**, *11*, 1–8. [CrossRef]
17. Pandolfi, F.; Franza, L.; Mandolini, C.; Conti, P. Immune Modulation by Vitamin D: Special Emphasis on Its Role in Prevention and Treatment of Cancer. *Clin. Ther.* **2017**, *39*, 884–893. [CrossRef] [PubMed]
18. Cadeau, C.; Fournier, A.; Mesrine, S.; Clavel-Chapelon, F.; Fagherazzi, G.; Boutron-Ruault, M.-C. Interaction between current vitamin D supplementation and menopausal hormone therapy use on breast cancer risk: Evidence from the E3N cohort. *Am. J. Clin. Nutr.* **2015**, *102*, 966–973. [CrossRef]
19. Fathi, N.; Ahmadian, E.; Shahi, S.; Roshangar, L.; Khan, H.; Kouhsoltani, M.; Dizaj, S.M.; Sharifi, S. Role of vitamin D and vitamin D receptor (VDR) in oral cancer. *Biomed. Pharmacother.* **2019**, *109*, 391–401. [CrossRef]
20. O'Brien, K.M.; Sandler, D.P.; Taylor, J.A.; Weinberg, C.R. Serum Vitamin D and Risk of Breast Cancer within Five Years. *Environ. Health Perspect.* **2017**, *125*, 077004. [CrossRef]
21. Takalkar, U.; Asegaonkar, S.; Advani, S. Vitamin D and prevention of breast cancer. *J. Cancer Res. Forecast* **2018**, *1*, 1–4.
22. Runowicz, C.D.; Leach, C.R.; Henry, N.L.; Henry, K.S.; Mackey, H.T.; Cowens-Alvarado, R.L.; Cannady, R.S.; Pratt-Chapman, M.L.; Edge, S.B.; Jacobs, L.A.; et al. American Cancer Society/American Society of Clinical Oncology Breast Cancer Survivorship Care Guideline. *CA Cancer J. Clin.* **2015**, *66*, 43–73. [CrossRef] [PubMed]
23. Coleman, R.; Body, J.J.; Aapro, M.; Hadji, P.; Herrstedt, J. Bone health in cancer patients: ESMO Clinical Practice Guidelines. *Ann. Oncol.* **2014**, *25*, iii124–iii137. [CrossRef]
24. Rusińska, A.; Pludowski, P.; Walczak, M.; Borszewska-Kornacka, M.K.; Bossowski, A.; Chlebna-Sokół, D.; Czech-Kowalska, J.; Dobrzanska, A.; Franek, E.; Helwich, E.; et al. Vitamin D Supplementation Guidelines for General Population and Groups at Risk of Vitamin D Deficiency in Poland—Recommendations of the Polish Society of Pediatric Endocrinology and Diabetes and the Expert Panel with Participation of National Specialist Consultants and Representatives of Scientific Societies—2018 Update. *Front. Endocrinol.* **2018**, *9*, 246. [CrossRef]
25. Kłek, S.; Jankowski, M.; Kruszewski, W.J.; Fijuth, J.; Kapała, A.; Kabata, P.; Wysocki, P.J.; Krzakowski, M.; Rutkowski, P. Standardy leczenia żywieniowego w onkologii. *Nowotw. J. Oncol.* **2015**, *65*, 320–337. [CrossRef]
26. Pilz, S.; Hahn, A.; Schön, C.; Wilhelm, M.; Obeid, R. Effect of Two Different Multimicronutrient Supplements on Vitamin D Status in Women of Childbearing Age: A Randomized Trial. *Nutrients* **2017**, *9*, 30. [CrossRef] [PubMed]
27. Kotlarczyk, M.P.; Perera, S.; Ferchak, M.A.; Nace, D.A.; Resnick, N.M.; Greenspan, S.L. Vitamin D deficiency is associated with functional decline and falls in frail elderly women despite supplementation. *Osteoporos. Int.* **2017**, *28*, 1347–1353. [CrossRef]
28. Camacho, P.M.; Petak, S.M.; Binkley, N.; Clarke, B.L.; Harris, S.T.; Hurley, D.L.; Kleerekoper, M.; Lewiecki, E.M.; Miller, P.D.; Narula, H.S.; et al. American Association of Clinical Endocrinologists and American College of Endocrinology Clinical Practice Guidelines for the Diagnosis and Treatment of Postmenopausal Osteoporosis—2016. *Endocr. Pract.* **2016**, *22*, 1–42. [CrossRef]
29. German Nutrition Society. New reference values for vitamin D. *Ann. Nutr. Metab.* **2012**, *60*, 241–246. [CrossRef]
30. Jarosz, M. *Normy Żywienia dla Populacji Polskiej—Nowelizacja*; Wydawnictwa—Instytut Żywności i Żywienia: Warsaw, Poland, 2012.
31. Stanowisko Zespołu Ekspertów. Polskie zalecenia dotyczące profilaktyki niedoborów witaminy D—2009. *Ginekol. Pol.* **2010**, *81*, 149–153.
32. Mohr, S.B.; Gorham, E.D.; Kim, J.; Hofflich, H.; Garland, C.F. Meta-analysis of vitamin D sufficiency for improving survival of patients with breast cancer. *Anticancer Res.* **2014**, *34*, 1163–1166.
33. Kuzmickiene, I.; Atkocius, V.; Aleknavicius, E.; Ostapenko, V. Impact of season of diagnosis on mortality among breast cancer survivors. *J. Cancer Res. Ther.* **2018**, *14*, 1091. [CrossRef]
34. Shi, L.; Nechuta, S.; Gao, Y.-T.; Zheng, Y.; Dorjgochoo, T.; Wu, J.; Cai, Q.; Zheng, W.; Lu, W.; Shu, X.O. Correlates of 25-Hydroxyvitamin D among Chinese Breast Cancer Patients. *PLoS ONE* **2014**, *9*, e86467. [CrossRef]
35. Lips, P.; Cashman, K.D.; Lamberg-Allardt, C.; Bischoff-Ferrari, H.A.; Obermayer-Pietsch, B.; Bianchi, M.L.; Stepan, J.; Fuleihan, G.E.-H.; Bouillon, R. Current vitamin D status in European and Middle East countries and strategies to prevent vitamin D deficiency: A position statement of the European Calcified Tissue Society. *Eur. J. Endocrinol.* **2019**, *180*, P23–P54. [CrossRef]
36. Płudowski, P.; Ducki, C.; Konstantynowicz, J.; Jaworski, M. Vitamin D status in Poland. *Pol. Arch. Intern. Med.* **2016**, *126*, 530–539. [CrossRef] [PubMed]
37. Rabenberg, M.; Scheidt-Nave, C.; A Busch, M.; Rieckmann, N.; Hintzpeter, B.; Mensink, G.B. Vitamin D status among adults in Germany—Results from the German Health Interview and Examination Survey for Adults (DEGS1). *BMC Public Health* **2015**, *15*, 1–15. [CrossRef]
38. Foumani, R.S.; Khodaie, F. The Correlation of Plasma 25-Hydroxyvitamin D Deficiency with Risk of Breast Neoplasms: A Systematic Review. *Iran. J. Cancer Prev.* **2016**, *9*, 1–7. [CrossRef]
39. Andersen, M.R.; Sweet, E.; Hager, S.; Gaul, M.; Dowd, F.; Standish, L.J. Effects of Vitamin D Use on Health-Related Quality of Life of Breast Cancer Patients in Early Survivorship. *Integr. Cancer Ther.* **2019**, *18*, 1–12. [CrossRef]
40. Apoe, O.; Jung, S.-H.; Liu, H.; Seisler, D.K.; Charlamb, J.; Zekan, P.; Wang, L.X.; Unzeitig, G.W.; Garber, J.; Marshall, J.; et al. Effect of Vitamin D Supplementation on Breast Cancer Biomarkers: CALGB 70806 (Alliance) Study Design and Baseline Data. *Am. J. Hematol.* **2016**, *12*, 4–9.

41. Machado, M.R.M.; Almeida-Filho, B.D.S.; Vespoli, H.D.L.; Schmitt, E.B.; Nahas-Neto, J.; Nahas, E.A. Low pretreatment serum concentration of vitamin D at breast cancer diagnosis in postmenopausal women. *Menopause* **2019**, *26*, 293–299. [CrossRef] [PubMed]

42. Valverde, C.N.; Gómez, J.Q. Vitamin D, determinant of bone and extrabone health. Importance of vitamin D supplementation in milk and dairy products. *Nutr. Hosp.* **2015**, *31*, 18–25.

43. Woon, F.C.; Chin, Y.S.; Ismail, I.H.; Batterham, M.; Latiff, A.H.A.; Gan, W.Y.; Appannah, G.; Hussien, S.H.M.; Edi, M.; Tan, M.L.; et al. Vitamin D deficiency during pregnancy and its associated factors among third trimester Malaysian pregnant women. *PLoS ONE* **2019**, *14*, e0216439. [CrossRef] [PubMed]

44. Sediyama, C.M.O. Lifestyle and vitamin D dosage in women with breast cancer. *Nutr. Hosp.* **2016**, *33*, 1179–1186. [CrossRef]

45. Krzyścin, J.W.; Jarosławski, J.; Sobolewski, P. A mathematical model for seasonal variability of vitamin D due to solar radiation. *J. Photochem. Photobiol. B Biol.* **2011**, *105*, 106–112. [CrossRef]

46. Guo, J.; Lovegrove, J.A.; Givens, D.I. A Narrative Review of The Role of Foods as Dietary Sources of Vitamin D of Ethnic Minority Populations with Darker Skin: The Underestimated Challenge. *Nutrients* **2019**, *11*, 81. [CrossRef] [PubMed]

47. Zhang, X.; Harbeck, N.; Jeschke, U.; Sixou, S. Influence of vitamin D signaling on hormone receptor status and HER2 expression in breast cancer. *J. Cancer Res. Clin. Oncol.* **2016**, *143*, 1107–1122. [CrossRef]

48. Acevedo, F.; Pérez, V.; Pérez-Sepúlveda, A.; Florenzano, P.; Artigas, R.; Medina, L.; Sánchez, C. High prevalence of vitamin D deficiency in women with breast cancer: The first Chilean study. *Breast* **2016**, *29*, 39–43. [CrossRef] [PubMed]

49. Eliassen, A.H.; Warner, E.T.; Rosner, B.; Collins, L.C.; Beck, A.H.; Quintana, L.M.; Tamimi, R.M.; Hankinson, S.E. Plasma 25-Hydroxyvitamin D and Risk of Breast Cancer in Women Followed over 20 Years. *Cancer Res.* **2016**, *76*, 5423–5430. [CrossRef] [PubMed]

50. O'Sullivan, F.; Raftery, T.; Van Weele, M.; Van Geffen, J.; McNamara, D.; O'Morain, C.; Mahmud, N.; Kelly, D.; Healy, M.; O'Sullivan, M.; et al. Sunshine is an Important Determinant of Vitamin D Status Even among High-dose Supplement Users: Secondary Analysis of a Randomized Controlled Trial in Crohn's Disease Patients. *Photochem. Photobiol.* **2019**, *95*, 1060–1067. [CrossRef]

51. Hoel, D.G.; De Gruijl, F. Sun Exposure Public Health Directives. *Int. J. Environ. Res. Public Health* **2018**, *15*, 2794. [CrossRef]

52. Directive 2002/46/EC of the European Parliament and of the Council of 10 June 2002. Available online: https://eur-lex.europa.eu/legal-content/EN/ALL/?uri=celex%3A32002L0046 (accessed on 7 May 2020).

53. Dymkowska-Malesa, M.; Walczak, Z. Suplementacja w sporcie. *Now. Lek.* **2011**, *80*, 3, 199–204.

54. Madden, J.M.; Murphy, L.; Zgaga, L.; Bennett, K. De novo vitamin D supplement use post-diagnosis is associated with breast cancer survival. *Breast Cancer Res. Treat.* **2018**, *172*, 179–190. [CrossRef]

55. Baumann, M.; Dani, S.; Dietrich, D.; Hochstrasser, A.; Klingbiel, D.; Mark, M.T.; Riesen, W.F.; Ruhstaller, T.; Templeton, A.J.; Thürlimann, B. Vitamin D levels in Swiss breast cancer survivors. *Swiss Med. Wkly.* **2018**, *148*, w14576. [CrossRef]

56. Griffin, N.; Dowling, M. Vitamin D supplementation and clinical outcomes in cancer survivorship. *Br. J. Nurs.* **2018**, *27*, 1121–1128. [CrossRef] [PubMed]

57. Grant, W.B. A review of the evidence supporting the Vitamin D-cancer related hypothesis in 2017. *Anticancer Res.* **2018**, *38*, 1121–1236. [CrossRef] [PubMed]

58. Karthikayan, A.; Sureshkumar, S.; Kadambari, D.; Vijayakumar, C. Low serum 25-hydroxy vitamin D levels are associated with aggressive breast cancer variants and poor prognostic factors in patients with breast carcinoma. *Arch. Endocrinol. Metab.* **2018**, *62*, 452–459. [CrossRef] [PubMed]

59. Madden, J.M.; Duffy, M.J.; Zgaga, L.; Bennett, K. Trends in vitamin D supplement use in a general female and breast cancer population in Ireland: A repeated cross-sectional study. *PLoS ONE* **2018**, *13*, e0209033. [CrossRef] [PubMed]

60. Tham, Y.-L.; Sexton, K.; Weiss, H.L.; Elledge, R.M.; Friedman, L.C.; Kramer, R.M. The adherence to practice guidelines in the assessment of bone health in women with chemotherapy-induced menopause. *J. Support. Oncol.* **2006**, *4*, 295–298.

61. Bošković, L.; Gašparić, M.; Petković, M.; Gugić, D.; Lovasić, I.B.; Soldić, Ž.; Miše, B.P.; Dabelić, N.; Vazdar, L.; Vrdoljak, E. Bone health and adherence to vitamin D and calcium therapy in early breast cancer patients on endocrine therapy with aromatase inhibitors. *Breast* **2017**, *31*, 16–19. [CrossRef]

62. Kimiafar, K.; Sarbaz, M.; Sales, S.S.; Esmaeili, M.; Ghazvini, Z.J. Breast cancer patients' information needs and information-seeking behavior in a developing country. *Breast* **2016**, *28*, 156–160. [CrossRef]

63. O'Brien, K.M.; Sandler, D.P.; House, M.; A Taylor, J.; Weinberg, C.R. The Association of a Breast Cancer Diagnosis with Serum 25-Hydroxyvitamin D Concentration over Time. *Am. J. Epidemiol.* **2019**, *188*, 637–645. [CrossRef]

64. Heaney, R.P. Vitamin D in Health and Disease. *Clin. J. Am. Soc. Nephrol.* **2008**, *3*, 1535–1541. [CrossRef]

65. Gallagher, J.C.; Sai, A.; Templin, T.; Smith, L. Dose Response to Vitamin D Supplementation in Postmenopausal Women. *Ann. Intern. Med.* **2012**, *156*, 425–437. [CrossRef]

66. Peñas, R.D.L.; Majem, M.; Perez, J.; Virizuela, J.A.; Cancer, E.; Taín, P.D.; Donnay, O.; Hurtado, A.; Jiménez-Fonseca, P.; Ocon, M.J. SEOM clinical guidelines on nutrition in cancer patients (2018). *Clin. Transl. Oncol.* **2019**, *21*, 87–93. [CrossRef] [PubMed]

Article

Adequate Vitamin D Intake Cannot Be Achieved within Carbon Emission Limits Unless Food Is Fortified: A Simulation Study

Maaike J. Bruins * and Ulla Létinois

DSM Nutrition Products, Wurmisweg 576, CH-4303 Kaiseraugst, Switzerland; ulla.letinois@DSM.com
* Correspondence: maaike.bruins@DSM.com

Abstract: This study applied linear programming using a Dutch "model diet" to simulate the dietary shifts needed in order to optimize the intake of vitamin D and to minimize the carbon footprint, considering the popularity of the diet. Scenarios were modelled without and with additional fortified bread, milk, and oil as options in the diets. The baseline diet provided about one fifth of the adequate intake of vitamin D from natural food sources and voluntary vitamin D-fortified foods. Nevertheless, when optimizing this diet for vitamin D, these food sources together were insufficient to meet the adequate intake required, unless the carbon emission and calorie intake were increased almost 3-fold and 2-fold, respectively. When vitamin D-fortified bread, milk, and oil were added as options to the diet, along with increases in fish consumption, and decreases in sugar, snack, and cake consumption, adequate intakes for vitamin D and other nutrients could be met within the 2000 kcal limits, along with a relatively unchanged carbon footprint. Achieving vitamin D goals while reducing the carbon footprint by 10% was only possible when compromising on the popularity of the diet. Adding vitamin D to foods did not contribute to the total carbon emissions. The modelling study shows that it is impossible to obtain adequate vitamin D through realistic dietary shifts alone, unless more vitamin D-fortified foods are a necessary part of the diet.

Keywords: dietary modelling; sustainable diet; vitamin D intake; fortification; carbon emission

1. Introduction

Food production has a considerable impact on greenhouse gas emissions [1]. The planet cannot sustain a continuation of the current dietary habits, especially when it comes to feeding the 10 billion people living on the planet by 2050. There is a growing understanding of the types of diets and food patterns that can be part of the solution in order to reduce environmental impact, while optimizing health in terms of nutrient adequacies when shifting dietary patterns [2]. Some governments have already incorporated sustainability into their national dietary guidelines [3]. Even though sustainability and health considerations are increasingly driving consumer purchasing decisions, consumers still face challenges when changing dietary habits in order to improve their nutrition and sustainability [4,5].

Vitamin D deficiency is among the most neglected major public health problems worldwide [6]. Surveys show that vitamin D deficiency is highly prevalent among all population groups, with severe deficiency (<25 nmol/L) and deficiency (<50 nmol/L) rates estimated to be 7% and 37% globally, respectively [6], and the vitamin D requirements are largely unmet in most populations [7,8]. In the Netherlands, one study found that no adults met the estimated average requirement for vitamin D [8]. Food provides a relatively small proportion of the vitamin D supply, while vitamin D produced in the skin from UVB light makes the greatest contribution [9]. An adequate intake of vitamin D-rich food is not the only difficulty, as adequate sunlight exposure can be a challenge with sun avoidance and less time spent outdoors. The high prevalence (83%) of low serum 25(OH)D levels <50 nmol/L in Dutch adults suggests that vitamin D from diet and UVB exposure combined are not adequate [10]. To ensure that individuals consume adequate vitamin D, irrespective

Citation: Bruins, M.J.; Létinois, U. Adequate Vitamin D Intake Cannot Be Achieved within Carbon Emission Limits Unless Food Is Fortified: A Simulation Study. *Nutrients* **2021**, *13*, 592. https://doi.org/10.3390/nu13020592

Academic Editor: Spyridon N. Karras

Received: 30 December 2020
Accepted: 8 February 2021
Published: 11 February 2021

of their exposure to sunlight, the Institute of Medicine (IOM) and the European Food Safety Authority (EFSA) set the adequate intake for vitamin D based on assumed low sun exposure and the intake needed in order to achieve a serum 25(OH)D of $\geq$50 nmol/L, a level unlikely to pose adverse musculoskeletal health outcomes [11,12]. Moreover, experts have highlighted the potential immunomodulant, anti-inflammatory, and anti-infective roles of vitamin D beyond bone and muscle health [9,13].

However, obtaining an adequate intake of vitamin D from the diet alone is difficult, as only few foods naturally contain significant amounts of vitamin D [14]. As vitamin D food sources include mainly oily fish, meat, dairy, and eggs, shifting to more plant-based diets is likely to further aggravate the risk of vitamin D deficiency. It remains controversial among professionals whether sufficient vitamin D can be obtained from a healthy diet. This study simulated the shifts needed within a Dutch "model diet" to overcome vitamin D shortfalls, as well as the consequences for calorie intake and carbon emissions. In addition, dietary shifts were modelled by extending the diet with fortified milk, bread, and vegetable oils optimizing for vitamin D, as well as the vitamin D and carbon footprint combined.

2. Materials and Methods

2.1. Linear Modelling Methods

The linear modelling program Optimeal® 2.0 (Blonk Consultants, Gouda, the Netherlands) was used to model scenarios of dietary shifts in the Netherlands. The program can propose dietary shifts from the current diet based on nutritional or environmental dietary goals, optimizing for popularity, by searching for scenarios of foods that resemble current diets as closely as possible. The program can set boundaries (constraints) for 36 nutrients, and for energy to be fulfilled or limited through the upper boundaries. In all scenarios, the recommended nutrient intake (RNI) and adequate intake (AI) were set as the lower boundaries, and tolerable upper intake level (UL) and maximum reference value (MRV) were set as the upper boundaries. We combined the data from the Dutch National Food Survey and Food Composition Database and the Life Cycle Assessment databases, resulting in a consolidated dataset with daily amounts consumed, nutritional value, popularity estimates, and carbon footprint estimates for 251 food items. The intake frequency of food items were used as a proxy for food popularity. This Dutch model diet was optimized for vitamin D (and carbon footprint) through linear modelling using the following scenarios.

2.2. Scenario 1: Optimizing the Current Diet for Vitamin D without Energy Constraints

In the first scenario, the baseline diet was optimized for an adequate intake of vitamin D. The baseline diet included some voluntary vitamin D-fortified foods, such as juices, fat spreads, soy-drinks, and breakfast cereals. The recommendations for nutrients had to be fulfilled, while the upper limits for calories were removed to allow the optimized diet to reach the adequate intake of vitamin D (13.4 µg/d).

2.3. Scenario 2: Optimizing the Current Diet for Vitamin D within Energy Constraints

In the second scenario, the diet was optimized for vitamin D, limiting energy intake to 2000 kcal and fulfilling nutrient recommendations. The diet was optimized to reach 9.6 µg/d, the maximum achievable amount of vitamin D, within a 2000 kcal constraint.

2.4. Scenario 3: Optimizing the Current Diet with Additional Fortified Foods for Vitamin D

In this scenario, vitamin D-fortified whole grain breads, semi-skimmed milk, and oil (soy, arachidic, and sunflower) were added to the food repertoire. These diets were optimized for a vitamin D intake of 13.4 µg/d, meeting nutrient recommendations, within a 2000 kcal limit. Combinations of two or three vitamin D-fortified foods were modelled.

2.5. Scenario 4: Optimizing the Current Diet with Additional Fortified Foods for Vitamin D and CO_2

Like the previous scenario, vitamin D-fortified bread, milk, and oil were added to the diet. Both an adequate vitamin D intake and carbon footprint were set as the goals, while fulfilling nutrient references within a 2000 kcal limit. Either a capped (limit at 3.9 kg CO_2 eq) or a 10% reduced footprint (limit at 3.5 kg CO_2 eq) was simulated.

2.6. Food and Nutrition Data Used in the Model

Chronic food consumption (food records taking into account different survey periods) from the 24-h dietary recall Dutch National Food Consumption Survey (DNFCS) of 2003 was retrieved from the EFSA Comprehensive European Food Consumption Database [15] at food classification system "FoodEx" level 3, i.e., food category sub-items, such as type of cheese. The nutritional composition of foods was defined using the Dutch Food Composition Database (NEVO) food composition tables of 2016 [16]. If the food item was not available in the NEVO database, the nutritional profile was selected from the United States Department of Agriculture (USDA) food composition database. After adding hypothetical fortified foods (see Section 2.8), this resulted in a set of 251 food items.

The nutrition goals were based on the RNI's and AI's for vitamins and minerals, as recommended by the EFSA for adults [17]. The adequate intake for vitamin D is based on minimal exposure to sunlight [11]. For modelling purposes, reference values were averaged for men and women. Assuming an average 2243 kcal consumption for adult women and men at a moderate physical activity level [18], the dietary reference values were also adjusted to 2000 kcal. For instance, the vitamin D reference value of 15 μg/d was adjusted to 13.4 μg/d per 2000 kcal (Table A1). The upper bounds or maximum reference values (MRVs) for carbohydrates, free sugars, total fat, saturated fatty acids, trans-fatty acids, cholesterol, and sodium were based on reference values from the World Health Organization and the Food and Agriculture Organization [19] (Table A1).

2.7. Nutrient Density of the Diet

The baseline and modelled diets were standardized to 2000 kcal in order to calculate the nutrient density per 2000 kcal diet, which allowed for comparisons between countries or genders, irrespective of calorie intake or reporting, providing a good reflection of the diet quality. We calculated the mean adequacy ratio (MAR) as an overall measure of the nutrient adequacy of the diet. The MAR was calculated as desired nutrients in a 2000 kcal diet as a percent of the RNI or AI, truncated at 100%, and averaged for 26 qualifying nutrients [20]. The mean excess ratio (MER) was calculated as percent of the MRV, and was averaged for 6 undesired nutrients (total fats, saturated fatty acids, trans fatty acids, cholesterol, added sugars, and sodium) [20]. The added sugars were estimated from the sum of the food categories, in which mono- and di-saccharides almost exclusively represented the added sugars, and by subtracting the estimated lactose content in dairy foods from the total mono- and di-saccharides.

2.8. Fortified Foods Used in the Model

The current diet in the Netherlands already includes some vitamin D-fortified products, such as breakfast and porridge cereals, on average fortified at 4.2 μg/100 g and 16.5 μg/100 g, respectively, as well as fat spreads, fortified on average at 7.5 μg/100 g of vitamin D. Bread and vegetable oil offer a suitable opportunity for improving vitamin D intake through fortification, as they are consumed by a large proportion of the population in fairly constant amounts [21], are among the categories considered acceptable by consumers in Nordic countries [22], and their fortification with vitamin D is technically feasible. Fortified milk offers another option to voluntary fortify food, but this may reach less people, as some population groups do not consume milk. Vitamin D levels of 2, 6, and 15 μg/100 g in semi-skimmed milk, whole grain bread, and vegetable oil, respectively, were selected. Fortified bread and milk were constrained to two servings daily, so as to avoid proposing an unrealistically high consumption of these food items in the simulated diet.

2.9. Carbon Footprint Data Used in the Model

Optimeal® 2.0 contains the environmental impact data of more than 200 food products, including carbon footprints. If the carbon footprint data were not available, they were obtained from the Agri-footprint® 3.0 Life Cycle Inventory food database (SimaPro Life Cycle Analysis software 8, Amersfoort, the Netherlands). The implementation of the impact assessment methods in SimaPro were used without modification. The carbon footprint was calculated using the IPCC 2013 GWP 100a assessment method and the results were expressed as kg CO_2 equivalents, using the associated characterization factors for the relevant greenhouse gases. This modelling study uses an attributional life cycle assessment estimating what share of the global environmental burdens belongs to a product. It was estimated that one kilogram of vitamin D3 can have a carbon footprint of less than 200 kg CO_2 equivalent based on primary data for the production of vitamin D3 (internal data). The carbon footprints of the whole grain bread, semi-skimmed milk, and oil were about 0.09, 0.12 kg, and 0.2–0.4 CO_2 equivalent per 100 g, respectively. Adding 2, 6, and 15 µg of vitamin D per 100 g of bread, milk, and oil, respectively, added ~0.001% CO_2 to the total CO_2 footprint of the food product.

3. Results

3.1. Scenario 1: Optimizing the Current Diet for Vitamin D without Energy Constraints

Fish; meat products; dairy; eggs; and some voluntary fortified foods such as juice, fat spreads, and breakfast cereals, are the main sources of vitamin D in the Dutch diet (Figure 1A). The baseline diet provided about 3 µg/d of vitamin D per 2000 kcal, contributing 21% of the adequate intake of vitamin D per 2000 kcal. Animal-source products provided 2 µg/d of vitamin D, fortified foods provided 1 µg/d, and mushrooms provided 0.01 µg/d. Achieving 13.4 µg/d of vitamin D was not possible with the current diet within the 2000 kcal intake limit. Therefore, the upper constraints for energy intake were removed. An adequate vitamin D intake could be reached when increasing the carbon footprint 2.8-fold (Figure 1B) and increasing the calorie consumption two-fold (Figure 1C). The increase in the carbon footprint of the optimized diet compared with the baseline diet was mainly attributable to an increase in the carbon footprints of meat products (3-fold); dairy (4-fold); oils, fat, and fat spreads (7-fold); egg products (11-fold); fish (15-fold); and legumes (17-fold), respectively (Figure 1B).

3.2. Scenario 2: Optimizing the Current Diet for Vitamin D within Energy Constraints

Despite the inclusion of voluntary vitamin D-fortified foods in the Dutch baseline diet, only 9.6 µg/d instead of the adequate 13.4 µg/d vitamin D could be achieved within the energy constraints of 2000 kcal (Figure 2A). To achieve 9.6 µg/d of vitamin D, fish (smoked herring and eel, and fish fingers), egg products (fried and boiled eggs), meat products (minced meat balls, meat soup, pate, and lean sausages), fortified breakfast cereals (cornflakes), fortified margarine, butter cakes, and vegetables (fried mushrooms) provided most of the vitamin D in the vitamin D-optimized diet (Figure 2B). Optimizing the diet for vitamin D within the 2000 kcal boundary increased the carbon footprint 1.7-fold compared with the baseline diet (Figure 2B). The carbon footprint increased 11-fold for egg products, 7-fold for fish, 6-fold for vegetables, and 2-fold for meat products, relative to the baseline. To achieve the vitamin D goals while meeting the nutrient recommendations, calorie consumption from egg products, fish, vegetables, and meat products would need to increase 11-, 10-, 6-, and 2-fold, respectively, while reducing calories from most other food categories (Figure 2C).

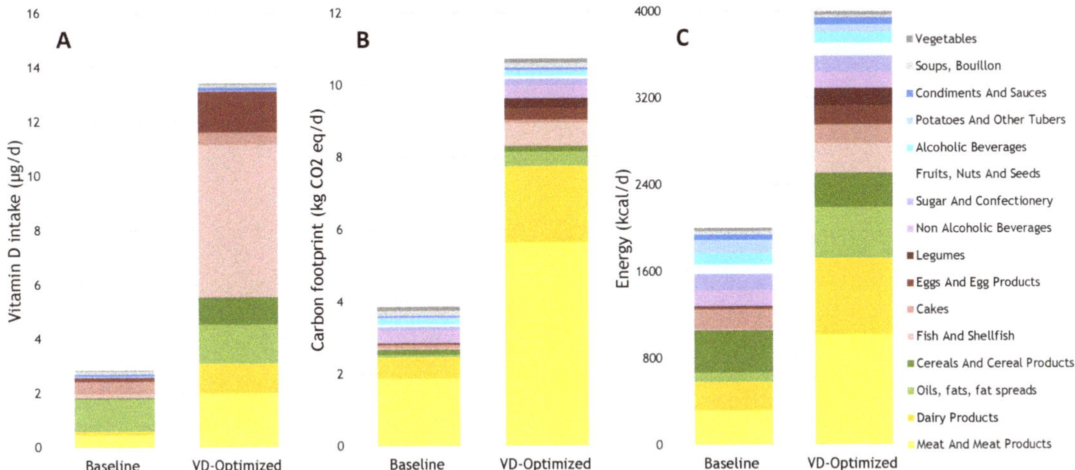

Figure 1. Daily contributions to (**A**) vitamin D intake, (**B**) carbon footprint, and (**C**) energy intake from the baseline diet and the diet optimized for vitamin D assuming no energy intake restrictions.

3.3. Scenario 3: Optimizing the Current Diet with Additional Fortified Foods for Vitamin D

When adding vitamin D-fortified bread, milk, and oil to the Dutch baseline diet, it was possible to optimize the diet with an adequate vitamin D intake of 13.4 µg/d, meeting the other nutrient requirements while remaining within the 2000 kcal consumption constraint (Figure 3A). Vitamin D from fish increased 22-fold from baseline, and from fortified bread and breakfast cereals it increased 170-fold from baseline (Figure 3A). Fortified bread was proposed over fortified milk or oil as source of vitamin D. When fortified bread was excluded as a dietary option, fortified oil was proposed over fortified milk (data not shown). When fortified oil was also excluded, fortified milk could fulfill the vitamin D requirements adequately (data not shown). Optimizing the baseline diet to meet the adequate intake of vitamin D involved an 8% increase in the total diet carbon footprint, coming mostly from fish and vegetables, of which the carbon footprints increased 6- and 2-fold compared with baseline, respectively (Figure 3B). To achieve vitamin D goals while also meeting the other nutrient recommendations, calorie consumption from fish and vegetables would need to increase 8- and 2-fold, respectively (Figure 3C). In exchange, calories from cakes, sugar, snacks, potatoes, and tubers would need to decrease.

3.4. Scenario 4: Optimizing the Current Diet with Additional Fortified Foods for Vitamin D and CO_2

Optimizing the diet for an adequate vitamin D intake and capped CO_2 emission, while satisfying nutrient recommendations, was feasible through a small shift from animal-source foods to fortified cereals (data not shown). A 10% CO_2 footprint reduction could only be achieved when removing the minimum nutrient recommendations or significantly shifting to less popular food items. In the latter scenario, vitamin D was obtained from an increased consumption of fish and fortified bread and breakfast cereals (Figure 4A), and from a shift from unfortified to fortified foods. The net 10% reduction in CO_2 was a result of less CO_2 (−33%) from meat, dairy, non-alcoholic drinks, cakes, sugar, and snacks, and a smaller increase in CO_2 from the total of legumes, fruits, nuts, seeds, fish, and vegetables (Figure 4B). Meeting the vitamin D and CO_2 goals while the satisfying nutrient recommendations required a significant shift in calorie intake (−33%), moving from meat, dairy, non-alcoholic drinks, cakes, sugar, and snacks, towards eggs (2-fold), vegetables (3-fold), and fish (8-fold; Figure 4C).

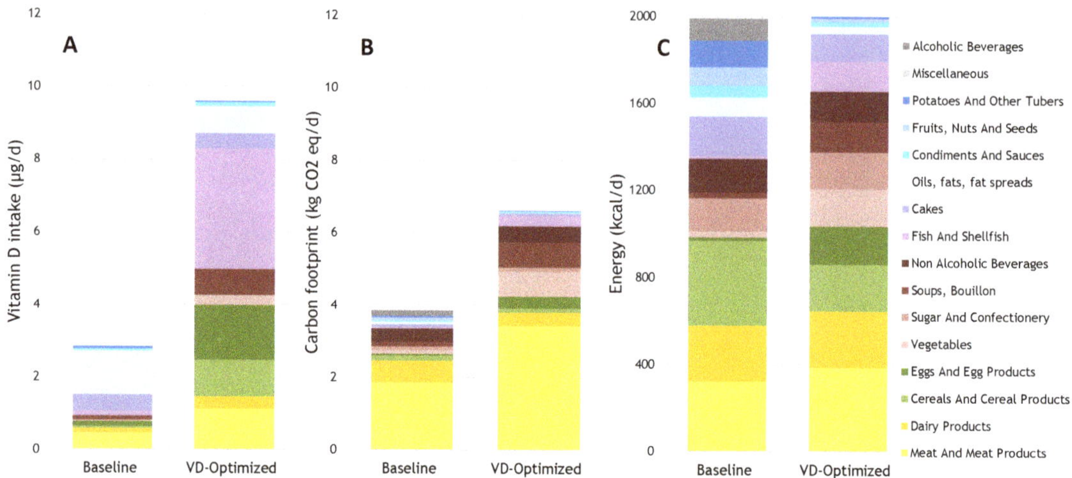

Figure 2. Daily contributions to (**A**) vitamin D intake, (**B**) carbon footprint, and (**C**) energy intake from the baseline diet and the diet optimized for vitamin D within a 2000 kcal boundary.

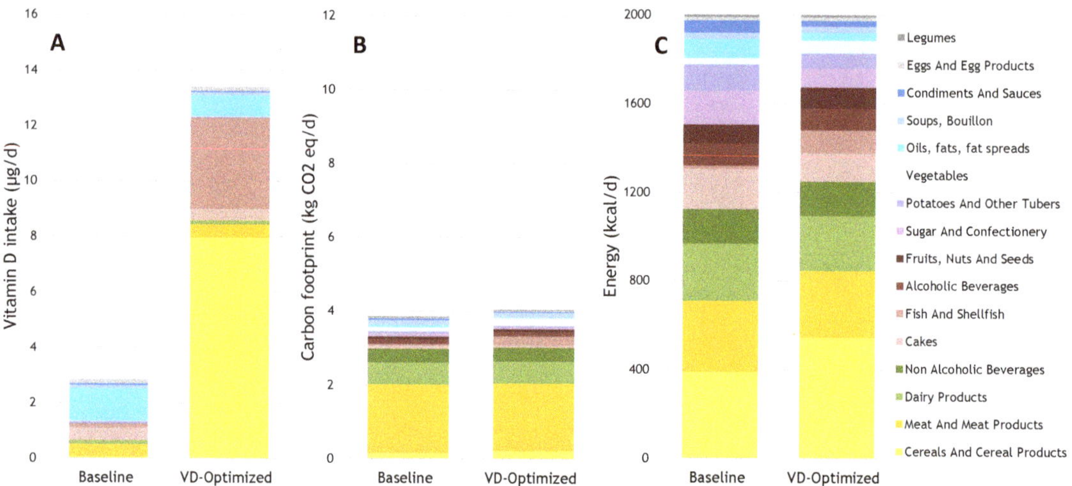

Figure 3. Daily contributions to (**A**) vitamin D intake; (**B**) carbon footprint; and (**C**) energy intake from the baseline diet and diet with additional fortified bread, milk, and oil optimized for vitamin D within a 2000 kcal boundary.

3.5. Nutrient Density of the Current Diet and the Optimized Diets

The calculated MAR and MER of the usual and optimized diet are shown in Table 1. The MAR of the desired nutrients in the Dutch baseline diet was 86%, with vitamin D being the first limiting nutrient (followed by seafood omega-3 fatty acids and fiber). The MER of the nutrients overconsumed relative to the maximum reference values was 120% (20% excess). Optimizing the diet for vitamin D and satisfying the nutrient recommendations without energy constraints increased the MAR to 100%, but increased the MER 2.4-fold. After adding vitamin D-fortified whole grain bread, milk, and oil, the MAR increased to 100% and the MER decreased to 112%. Setting additional goals to reduce CO_2 by 10% by compromising on popularity reduced the MER to 100%

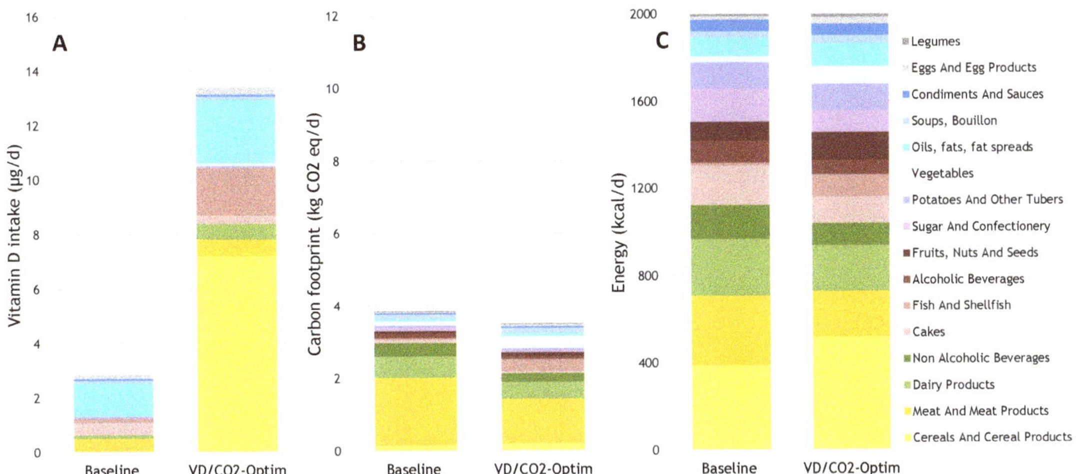

Figure 4. Daily contributions to (**A**) vitamin D intake; (**B**) carbon footprint; and (**C**) energy intake from the baseline diet and the diet with additional fortified bread, milk, and oil optimized for vitamin D and CO_2 within a 2000 kcal boundary.

Table 1. The mean adequacy ratio (MAR) and mean excess ratio (MER) of the Dutch diet: (1) baseline, (2) after the inclusion of fortified milk and bread, and (3) optimizing for vitamin D and (4) vitamin D and CO_2.

	Current	Scenario 1 Usual Diet Vitamin D Goals No Energy Limits	Scenario 2 Usual Diet Maximum Vitamin D 2000 kcal Limits	Scenario 3 Extra Fortified Foods Vitamin D Goals 2000 kcal Limits	Scenario 4 Extra Fortified Foods Vitamin D Goals CO_2 Goals 2000 kcal Limits
Mean adequacy ratio (MAR)	86%	100%	100%	100%	100%
Mean excess ratio (MER)	120%	242%	154%	112%	100%

4. Discussion

This simulation study demonstrates that even with a diet that is relatively abundant in vitamin D-rich foods, it is not possible to achieve an adequate intake of vitamin D without greatly increasing the carbon emission and calorie intake. Adding vitamin D-fortified options to the diet allowed for achieving the adequate intake of vitamin D and nutrient recommendations without sacrificing the carbon footprint and popularity of the diet.

The adequate vitamin D intake of 15 µg/d set by the EFSA and IOM represents the average adequate intake to achieve a serum 25(OH)D of $\geq$50 nmol/L [11,12]. The assumed low average year-round sun exposure in these dietary guidelines is realistic for the northern latitude of the Netherlands, with a high prevalence of vitamin D deficiency [10]. The Dutch model diet contributed approximately 3 µg/d per 2000 kcal (i.e., 20% of the adequate intake for vitamin D). This is comparable to the average vitamin D intake of 4.1 µg/d reported for European countries [23]. Two-thirds of the vitamin D in the Dutch model diet came from animal-source foods, one-third from voluntary vitamin D-fortified foods, and mushrooms contributed marginally.

In this study, the Dutch model diet was optimized to meet the adequate intake for vitamin D. This was only achievable when the calorie intake increased 2-fold and the carbon footprint increased almost 3-fold. However, the inclusion of additional vitamin D-fortified bread, milk, and oil in the diet, along with shifts in energy consumption towards fish and more plant-based nutrient-dense food sources, allowed for achieving an adequate vitamin D intake with minor compromises on the carbon emission and popularity of the diet within 2000 kcal limits. Clearly, the improvement in vitamin D adequacy (from 21% to 100%) and average nutrient adequacy (from 86% to 100%) was larger than the 8% increase in the carbon footprint. As only μg amounts of vitamin D are added to foods, vitamin D contributes only 1 permille to the carbon footprint of a food product and not to the total diet. A 10% reduction in carbon emissions while meeting the nutrient recommendations was feasible when shifting the intake of popular products such as meat, dairy, sugar, snacks, cakes, and non-alcoholic drinks more towards fish, fruits, nuts, vegetables, and eggs. However, these dietary changes may be less acceptable. Large reductions in meat, fish, eggs, and dairy products are not an option, as they provide essential or important sources of calcium; iodine; zinc; iron; and vitamins B2, B3, B5, B6, B12, and D [24–26].

Our study has various limitations; first, the food survey used in the model was from 2003, whereas food patterns likely changed over recent years. Second, the study focused solely on vitamin D intake relative to dietary references. Future work could consider integrating sun exposure as a source of vitamin D status in the model. Third, only the carbon footprint was selected as indicator of environmental impact, but other aspects such as land occupation and water use were not considered. The main strength of the study was the integral consideration of the popularity, nutrition, and climate aspects of the diets. Additional drivers of dietary choices, such as price, could be addressed in future research.

Previous studies concluded that without the universal fortification of staple foods or a dramatic increase in fish consumption, the current vitamin D intakes are too low to meet the recommendations or to sustain a healthy vitamin D status in the population [27–30]. This is substantiated by our findings, showing that an unrealistic increase in animal-source foods and the consequent carbon footprint is needed in order to meet the adequate vitamin D intake. Fortified whole grain bread was proposed over other fortified foods as a source of vitamin D, probably because it contributes to filling the fiber intake gap in the Netherlands, is popular, and has a relatively favorable carbon footprint. When fortifying foods, acceptable foods with a low carbon footprint addressing a nutrient gap should be considered.

Achieving sufficient vitamin D from the sun has become an increasing challenge, with more sun avoidance, time spent indoors, and a narrowing gap between beneficial and harmful UV exposure time to obtain desirable vitamin D. Various simulation studies show that the inclusion of vitamin D-fortified foods in the diet can be a viable and safe approach to improve intakes or reduce the prevalence of inadequate intakes [30,31]. Vitamin D-fortified bread and milk were able to reduce low vitamin D status in the winter season [32]. Food fortification with vitamin D in order to improve public health has been shown to be a cost-effective approach [33]. In voluntary fortification approaches, it is important that it is well-accepted by the population itself [33]. Consumers' perceived health benefits and the appropriateness of the product are important drivers of purchasing and consumption [22]. In Finland, voluntary vitamin D fortification of milk products and fat spreads has been well-accepted since 2003, and helped the Finnish population reach vitamin D levels ≥50 nmol/L in 2011 [34]. Enriching the vitamin D content of eggs, milk, and meat by adding vitamin D to feed represents another potential complementary approach to address inadequate vitamin D intake at a population level [35]. Animal-source foods continue to be an important part of diets, as they provide micronutrients that are difficult to obtain in adequate quantities from plant-source foods alone. Vitamin D supplement intakes and recommendations have also shown to contribute significantly to achieving sufficient vitamin D status [34,36].

The present study shows that adequate intakes for vitamin D cannot be achieved with the current diet alone within realistic calorie and carbon emission limits, and additional vitamin D sources are needed to overcome the shortfalls. Universal fortification along with small dietary shifts represents an approach to improve the vitamin D status of the general population, at a high acceptability without affecting the carbon footprint.

Author Contributions: M.J.B. contributed to food and nutrition data collection, and U.L. contributed to the LCA data collection. Both authors contributed to the linear modelling, writing, interpretation, and discussion of the manuscript. Both of the authors have read and agreed to the published version of the manuscript.

Funding: This research received no external funding.

Institutional Review Board Statement: Not applicable.

Informed Consent Statement: Not applicable.

Acknowledgments: The authors thank Marcelo Tyszler from Blonk Consultants for advice on modelling using Optimeal®.

Conflicts of Interest: M.J.B. and U.L. are employed by DSM Nutritional Products, a manufacturer of nutritional ingredients.

Appendix A

Table A1. Dietary reference values averaged for men and women adjusted for 2000 kcal energy consumption applied in the dietary modelling.

		RNI [1] or AI [2]	UL [3] or MRV [4]
Energy	kcal	2000	2000
Protein [5]	g	58.1	125
Polyunsaturated fatty acids	g	13.3	26.6
Linoleic acid	g	8.9	19.4
α-Linolenic acid	g	0.9	4.8
Fiber	g	25	
Water	g	2300	3800
Alcohol	g	0	10
DHA+EPA [6]	mg	250	1000
Vitamin A	μg RAE	624	3000
Thiamin (B1)	mg	0.84	
Riboflavin (B2)	mg	1.43	
Niacin (B3)	mg NE	13.4	
Vitamin B6	mg	1.52	25
Folate (B9)	μg FE	294	1000
Vitamin B12	μg	3.57	
Vitamin C	mg	91	
Vitamin D	μg	13.4	100
Vitamin E	mg	10.7	300
Vitamin K	μg	62	
Calcium	mg	847	2500
Copper	mg	1.3	5
Iodine	μg	156	600
Iron	mg	10.9	70

Table A1. *Cont.*

		RNI [1] or AI [2]	UL [3] or MRV [4]
Magnesium	mg	290	530
Phosphorus	mg	490	3000
Potassium	mg	3121	
Selenium	μg	62	300
Zinc	mg	10.2	25
Tryptophan	g	0.3	
Threonine	g	1.1	
Isoleucine	g	1.5	
Leucine	g	3.0	
Lysine	g	2.3	
Methionine	g	0.8	
Valine	g	2.0	
Histidine	g	0.8	
Carbohydrates	g		300
Added sugar	g		50
Total fat	g		78
Saturated fatty acids	g		22
Trans-fatty acids	g		2.2
Cholesterol	mg		300
Sodium	mg		2000

[1] Recommended nutrient intake (RNI); [2] adequate intake (AI); [3] tolerable upper intake level (UL); [4] maximum reference value (MRV); [5] at body weight of 70 kg; [6] DHA—docosahexaenoic acid; EPA—eicosapentaenoic acid.

References

1. Poore, J.; Nemecek, T. Reducing food's environmental impacts through producers and consumers. *Science* **2018**, *360*, 987–992. [CrossRef]
2. van Dooren, C. A Review of the Use of Linear Programming to Optimize Diets, Nutritiously, Economically and Environmentally. *Front. Nutr.* **2018**, *5*, 48. [CrossRef]
3. Lang, T. Re-Fashioning Food Systems with Sustainable Diet Guidelines: Towards a SDG Strategy. Available online: https://foodresearch.org.uk/publications/re-fashioning-food-systems-with-sustainable-diet-guidelines/ (accessed on 10 February 2021).
4. Macdiarmid, J.I.; Douglas, F.; Campbell, J. Eating like there's no tomorrow: Public awareness of the environmental impact of food and reluctance to eat less meat as part of a sustainable diet. *Appetite* **2016**, *96*, 487–493. [CrossRef] [PubMed]
5. Tyszler, M.; Kramer, G.; Blonk, H. Just eating healthier is not enough: Studying the environmental impact of different diet scenarios for Dutch women (31–50 years old) by linear programming. *Int. J. Life Cycle Assess* **2016**. [CrossRef]
6. Hilger, J.; Friedel, A.; Herr, R.; Rausch, T.; Roos, F.; Wahl, D.A.; Pierroz, D.D.; Weber, P.; Hoffmann, K. A systematic review of vitamin D status in populations worldwide. *Br. J. Nutr.* **2014**, *111*, 23–45. [CrossRef] [PubMed]
7. Mertens, E.; Kuijsten, A.; Dofkova, M.; Mistura, L.; D'Addezio, L.; Turrini, A.; Dubuisson, C.; Favret, S.; Havard, S.; Trolle, E.; et al. Geographic and socioeconomic diversity of food and nutrient intakes: A comparison of four European countries. *Eur. J. Nutr.* **2018**, *58*, 1475–1493. [CrossRef] [PubMed]
8. Spiro, A.; Buttriss, J.L. Vitamin D: An overview of vitamin D status and intake in Europe. *Nutr. Bull.* **2014**, *39*, 322–350. [CrossRef]
9. Amrein, K.; Scherkl, M.; Hoffmann, M.; Neuwersch-Sommeregger, S.; Kostenberger, M.; Tmava Berisha, A.; Martucci, G.; Pilz, S.; Malle, O. Vitamin D deficiency 2.0: An update on the current status worldwide. *Eur. J. Clin. Nutr.* **2020**, *74*, 1498–1513. [CrossRef]
10. Meems, L.M.; de Borst, M.H.; Postma, D.S.; Vonk, J.M.; Kremer, H.P.; Schuttelaar, M.L.; Rosmalen, J.G.; Weersma, R.K.; Wolffenbuttel, B.H.; Scholtens, S.; et al. Low levels of vitamin D are associated with multimorbidity: Results from the LifeLines Cohort Study. *Ann. Med.* **2015**, *47*, 474–481. [CrossRef]
11. European Food Safety Authority (EFSA). Dietary reference values for vitamin D. *EFSA J.* **2016**, *14*, 4547.
12. Ross, A.C.; Manson, J.E.; Abrams, S.A.; Aloia, J.F.; Brannon, P.M.; Clinton, S.K.; Durazo-Arvizu, R.A.; Gallagher, J.C.; Gallo, R.L.; Jones, G.; et al. The 2011 report on dietary reference intakes for calcium and vitamin D from the Institute of Medicine: What clinicians need to know. *J. Clin. Endocrinol. Metab.* **2011**, *96*, 53–58. [CrossRef] [PubMed]

13. Charoenngam, N.; Holick, M.F. Immunologic Effects of Vitamin D on Human Health and Disease. *Nutrients* **2020**, *12*, 2097. [CrossRef] [PubMed]
14. United States Department of Agriculture (USDA). Scientific Report of the 2015 Dietary Guidelines Advisory Committee. In *Appendix E-3.3: Meeting Vitamin D Recommended Intakes in USDA Food PatternsI*; United States Department of Agriculture (USDA), Agricultural Research Service: Washington, DC, USA, 2015.
15. European Food Safety Authority (EFSA). The EFSA Comprehensive European Food Consumption Database. Available online: https://www.efsa.europa.eu/en/food-consumption/comprehensive-database (accessed on 28 March 2015).
16. Dutch Food Composition Database. NEVO online Version 2016/5.0. Available online: https://nevo-online.rivm.nl/ (accessed on 28 March 2016).
17. European Food Safety Authority (EFSA). Dietary Reference Values for nutrients Summary report. *EFSA Supporting Publ.* **2017**, e15121. [CrossRef]
18. European Food Safety Authority (EFSA). Scientific Opinion on Dietary Reference Values for energy. *EFSA J.* **2013**, *11*, 3005. [CrossRef]
19. World Health Organization (WHO); Food and Agriculture Organization of the United Nations (FAO). *Joint WHO/FAO Expert Consultation on Diet, Nutrition and the Prevention of Chronic Diseases*; WHO: Geneva, Switzerland, 2003.
20. Fern, E.B.; Watzke, H.; Barclay, D.V.; Roulin, A.; Drewnowski, A. The Nutrient Balance Concept: A New Quality Metric for Composite Meals and Diets. *PLoS ONE* **2015**, *10*, e0130491. [CrossRef] [PubMed]
21. Cashman, K.D.; O'Dea, R. Exploration of strategic food vehicles for vitamin D fortification in low/lower-middle income countries. *J. Steroid Biochem. Mol. Biol.* **2019**, *195*, 105479. [CrossRef]
22. Jahn, S.; Tsalis, G.; Lahteenmaki, L. How attitude towards food fortification can lead to purchase intention. *Appetite* **2019**, *133*, 370–377. [CrossRef]
23. Jenab, M.; Salvini, S.; van Gils, C.H.; Brustad, M.; Shakya-Shrestha, S.; Buijsse, B.; Verhagen, H.; Touvier, M.; Biessy, C.; Wallstrom, P.; et al. Dietary intakes of retinol, beta-carotene, vitamin D and vitamin E in the European Prospective Investigation into Cancer and Nutrition cohort. *Eur. J. Clin. Nutr.* **2009**, *63* (Suppl. 4), S150–S178. [CrossRef]
24. Sobiecki, J.G.; Appleby, P.N.; Bradbury, K.E.; Key, T.J. High compliance with dietary recommendations in a cohort of meat eaters, fish eaters, vegetarians, and vegans: Results from the European Prospective Investigation into Cancer and Nutrition-Oxford study. *Nutr. Res.* **2016**, *36*, 464–477. [CrossRef] [PubMed]
25. White, R.R.; Hall, M.B. Nutritional and greenhouse gas impacts of removing animals from US agriculture. *Proc. Natl. Acad. Sci. USA* **2017**, *114*, E10301–E10308. [CrossRef]
26. Springmann, M.; Wiebe, K.; Mason-D'Croz, D.; Sulser, T.B.; Rayner, M.; Scarborough, P. Health and nutritional aspects of sustainable diet strategies and their association with environmental impacts: A global modelling analysis with country—level detail. *Lancet Planet Health* **2018**, *2*, e451–e461. [CrossRef]
27. Black, L.J.; Walton, J.; Flynn, A.; Kiely, M. Adequacy of vitamin D intakes in children and teenagers from the base diet, fortified foods and supplements. *Public Health Nutr.* **2014**, *17*, 721–731. [CrossRef] [PubMed]
28. Lehmann, U.; Gjessing, H.R.; Hirche, F.; Mueller-Belecke, A.; Gudbrandsen, O.A.; Ueland, P.M.; Mellgren, G.; Lauritzen, L.; Lindqvist, H.; Hansen, A.L.; et al. Efficacy of fish intake on vitamin D status: A meta-analysis of randomized controlled trials. *Am. J. Clin. Nutr.* **2015**, *102*, 837–847. [CrossRef] [PubMed]
29. Laaksi, I.T.; Ruohola, J.P.; Ylikomi, T.J.; Auvinen, A.; Haataja, R.I.; Pihlajamaki, H.K.; Tuohimaa, P.J. Vitamin D fortification as public health policy: Significant improvement in vitamin D status in young Finnish men. *Eur. J. Clin. Nutr.* **2006**, *60*, 1035–1038. [CrossRef] [PubMed]
30. Allen, R.E.; Dangour, A.D.; Tedstone, A.E.; Chalabi, Z. Does fortification of staple foods improve vitamin D intakes and status of groups at risk of deficiency? A United Kingdom modeling study. *Am. J. Clin. Nutr.* **2015**, *102*, 338–344. [CrossRef]
31. Harika, R.K.; Dotsch-Klerk, M.; Zock, P.L.; Eilander, A. Compliance with Dietary Guidelines and Increased Fortification Can Double Vitamin D Intake: A Simulation Study. *Ann. Nutr. Metab.* **2016**, *69*, 246–255. [CrossRef] [PubMed]
32. Madsen, K.H.; Rasmussen, L.B.; Andersen, R.; Molgaard, C.; Jakobsen, J.; Bjerrum, P.J.; Andersen, E.W.; Mejborn, H.; Tetens, I. Randomized controlled trial of the effects of vitamin D-fortified milk and bread on serum 25-hydroxyvitamin D concentrations in families in Denmark during winter: The VitmaD study. *Am. J. Clin. Nutr.* **2013**, *98*, 374–382. [CrossRef]
33. Pilz, S.; Marz, W.; Cashman, K.D.; Kiely, M.E.; Whiting, S.J.; Holick, M.F.; Grant, W.B.; Pludowski, P.; Hiligsmann, M.; Trummer, C.; et al. Rationale and Plan for Vitamin D Food Fortification: A Review and Guidance Paper. *Front. Endocrinol. (Lausanne)* **2018**, *9*, 373. [CrossRef]
34. Jaaskelainen, T.; Itkonen, S.T.; Lundqvist, A.; Erkkola, M.; Koskela, T.; Lakkala, K.; Dowling, K.G.; Hull, G.L.; Kroger, H.; Karppinen, J.; et al. The positive impact of general vitamin D food fortification policy on vitamin D status in a representative adult Finnish population: Evidence from an 11-y follow-up based on standardized 25-hydroxyvitamin D data. *Am. J. Clin. Nutr.* **2017**, *105*, 1512–1520. [CrossRef]
35. Guo, J.; Lovegrove, J.A.; Givens, D.I. 25(OH)D3-enriched or fortified foods are more efficient at tackling inadequate vitamin D status than vitamin D3. *Proc. Nutr. Soc.* **2018**, *77*, 282–291. [CrossRef]
36. Pittaway, J.K.; Ahuja, K.D.; Beckett, J.M.; Bird, M.L.; Robertson, I.K.; Ball, M.J. Make vitamin D while the sun shines, take supplements when it doesn't: A longitudinal, observational study of older adults in Tasmania, Australia. *PLoS ONE* **2013**, *8*, e59063. [CrossRef] [PubMed]

Article

Vitamin D Status, Bone Mineral Density, and *VDR* Gene Polymorphism in a Cohort of Belarusian Postmenopausal Women

Pavel Marozik [1,2,*], Alena Rudenka [3], Katsiaryna Kobets [1] and Ema Rudenka [4]

[1] Laboratory of Human Genetics, Institute of Genetics and Cytology of the National Academy of Sciences of Belarus, 220072 Minsk, Belarus; E.Kobets@igc.by

[2] Department of General Biology and Genetics, International Sakharov Environmental Institute of the Belarusian State University, 220070 Minsk, Belarus

[3] Department of Cardiology and Rheumatology, Belarusian Medical Academy of Post-Graduate Education, 220013 Minsk, Belarus; alenka.v.ru@gmail.com

[4] Department of Cardiology and Internal Diseases, Belarusian State Medical University, 220116 Minsk, Belarus; rudenka.ema@gmail.com

[*] Correspondence: P.Marozik@igc.by; Tel.: +375-17-364-1614

Abstract: Vitamin D plays an important role in bone metabolism and is important for the prevention of multifactorial pathologies, including osteoporosis (OP). The biological action of vitamin is realized through its receptor, which is coded by the *VDR* gene. *VDR* gene polymorphism can influence individual predisposition to OP and response to vitamin D supplementation. The aim of this work was to reveal the effects of *VDR* gene ApaI rs7975232, BsmI rs1544410, TaqI rs731236, FokI rs2228570, and Cdx2 rs11568820 variants on bone mineral density (BMD), 25-hydroxyvitamin D level, and OP risk in Belarusian women. Methods. The case group included 355 women with postmenopausal OP, and the control group comprised 247 women who met the inclusion criteria. TaqMan genotyping assay was used to determine *VDR* gene variants. Results. Rs7975232 A/A, rs1544410 T/T, and rs731236 G/G single variants and their A-T-G haplotype showed a significant association with increased OP risk (for A-T-G, OR = 1.8, p = 0.0001) and decreased BMD (A-T-G, -0.09 g/cm^2, p = 0.0001). The rs11568820 A-allele showed a protective effect on BMD (+0.22 g/cm^2, p = 0.027). A significant dose effect with 25(OH)D was found for rs1544410, rs731236, and rs11568820 genotypes. Rs731236 A/A was associated with the 25(OH)D deficiency state. Conclusion. Our novel data on the relationship between *VDR* gene variants and BMD, 25(OH)D level, and OP risk highlights the importance of genetic markers for personalized medicine strategy.

Keywords: vitamin D; *VDR* gene; polymorphism; predisposition; osteoporosis; bone mineral density

Citation: Marozik, P.; Rudenka, A.; Kobets, K.; Rudenka, E. Vitamin D Status, Bone Mineral Density, and *VDR* Gene Polymorphism in a Cohort of Belarusian Postmenopausal Women. *Nutrients* **2021**, *13*, 837. https://doi.org/10.3390/nu13030837

Academic Editor: Karras Spyridon

Received: 31 December 2020
Accepted: 17 February 2021
Published: 4 March 2021

Publisher's Note: MDPI stays neutral with regard to jurisdictional claims in published maps and institutional affiliations.

1. Introduction

Osteoporosis (OP) is defined as a systemic skeletal disease characterized by low bone mineral density (BMD) and a microarchitectural deterioration of bone tissue, leading to increased bone fragility and susceptibility to fracture [1]. Postmenopausal osteoporosis (PMO) is the most common form of primary OP, affecting menopausal women. The clinical significance of OP lies in its serious complications such as low-energy fractures, causing an increased risk of morbidity and mortality, especially in the elderly [2]. The impact of OP and fragility fractures on human health is huge: more than 9 million osteoporotic fractures are registered annually in the world [2]. Such fractures are associated with 26,300 life years lost and 1.16 million quality-adjusted life years (QALYs) lost yearly in EU countries, and the costs of treatment of osteoporotic fractures in 2010 has been estimated at €37 billion [3]. Projected demographic changes will cause an increase in fracture burden in coming decades [4].

The pathogenesis of OP is complex and includes many factors, among which genetic factors are of particular importance, as up to 90% of susceptibility to OP may be genetically determined [5]. However, the influence of environmental factors on the risk of OP should not be underestimated. In accordance with modern concepts, vitamin D is a steroid prohormone, which, along with parathyroid hormone (PTH) and calcitonin, plays a major role in the regulation of genes involved in calcium–phosphorus and bone metabolism [6]. Vitamin D effects are mediated by its binding to a specific steroid receptor (vitamin D receptor, VDR) that has a transcription factor activity [7]. The formation of the vitamin D steroid receptor complex results in the activation or silencing of numerous target genes, regulating bone remodeling, calcium homeostasis, and immune response.

The human *VDR* gene is located on the 12th chromosome (12q12-14) and consists of 14 exons spanning about 75 kb: eight protein-coding exons (2–9), six untranslated exons (1A–1F), located on the non-coding 5′region, and several promoter regions that are DNA sequences recognized by RNA polymerase as a launching pad for the initiation of specific transcription [8]. Even a small modification in a gene may affect the structure and functional activity of the receptor. Common single nucleotide variations (SNVs) in the *VDR* gene may be associated with different biological responses to vitamin D. The *VDR* ApaI (rs7975232, c.1025-49C>A), BsmI (rs1544410, 1024+443C>T), and TaqI (rs731236, c.1056A>G) SNVs are located at the 3′-untranslated end. They do not alter the amino acid sequence of the encoded protein but influence gene expression, regulating mRNA stability [9]. The *VDR* FokI variant (rs2228570, c.2T(A, f)>C(G, F), p.Met1Arg) is located in the coding region of the *VDR* gene (exon 2) and leads to the loss of the ATG translation initiation region, resulting in a shorter and more active receptor protein [10]. *VDR* Cdx2 G-to-A (rs11568820) substitution is located in the promoter region and causes 30% increased transcriptional activity [11].

Although several studies on different populations revealed an association of *VDR* gene variants with BMD [12,13] and serum 25(OH)D [14,15], many issues in this area are not fully understood. An investigation of *VDR* gene polymorphisms may help clarify criteria for the identification of individuals with high risk of PMO and thus conduct a timely set of preventive measures in target risk groups as well as evaluate effectiveness of therapy [16].

The aim of our study was to investigate the relationship between *VDR* gene single variants and haplotypes and PMO risk, BMD, and serum 25(OH)D level in Belarusian postmenopausal women.

2. Materials and Methods

2.1. Study Subjects

This study was a cross-sectional cohort study conducted at the outpatient department and inpatient clinic. Patients were recruited at the Minsk City Center for Osteoporosis and Bone-Muscular Diseases Prevention and Rheumatologic Department of 1st Minsk city clinic (Minsk, Belarus). The study protocol was approved by the Local Research Ethics Committee of Belarusian Medical Academy of Postgraduate Education. White Caucasian women were screened for participation. Inclusion criteria were wiliness to participate in the study, female sex, duration of menopause at least 3 years, and established diagnosis of OP according to World Health Organization Diagnostic Criteria [17]. Exclusion criteria: presence of other metabolic bone diseases (such as Paget's disease and osteomalacia), diseases, affecting bone metabolism (such as endocrine osteopathy, renal failure, Cron's disease, rheumatic diseases etc.), malignant tumors, use of medications likely influencing BMD. After assessing compliance with inclusion and exclusion criteria, all the enrolled women signed written informed consent for participation in the study in accordance with the declaration of Helsinki (as revised in 2013). Participants of the study have filled out questionnaires to identify clinical risk factors for OP (age of menopause, history of fractures etc.).

2.2. Clinical Evaluation

BMD was evaluated by DXA (GE Lunar, Madison, WI, USA). Calibration of the device was performed daily using a standard spine phantom provided by the manufacturer. Lumbar spine (LS, L1–L4) and femoral neck (FN) BMD (g/cm^2) was measured on the same machine. Diagnosis of OP was established on the basis of T-score criteria for Caucasian women [17].

Fasting blood samples for biochemical and electrochemiluminescence blood tests were obtained from the cubital vein in the morning, not earlier than 10–12 h after the last meal, into a sterile vacuum Vacutainer tube without additives. Determination of serum vitamin D was performed by electrochemiluminescence immunoassay on the Cobas e411 analyzer (Roche Diagnostic, Rotkreuz, Switzerland). All patients were concealed about adequate calcium dietary intake and were taking calcium supplementation in the form of calcium carbonate equivalent to 500 mg elemental calcium. All subjects were supplemented with a daily dose of 400–800 international units (IU) of cholecalciferol according to European guidance [18]. In accordance with international recommendations, the level of vitamin D was considered appropriate at 25(OH)D value > 30 ng/mL, insufficiency was diagnosed at rates of 20-30 ng/mL, and 25(OH)D concentration less than 20 ng/mL was considered as vitamin D deficiency [19].

2.3. Genotyping

Genomic DNA was isolated from whole blood using the standard phenol–chloroform extraction, the concentration and purity were measured using a NanoDrop 8000 spectrophotometer (Thermo Fisher Scientific, Wilmington, NC, USA). Information on *VDR* gene variants was obtained from the Entrez Gene database (www.ncbi.nlm.nih.gov/gene, accessed on 5 June 2020). Selected SNVs (ApaI rs7975232, BsmI rs1544410, TaqI rs731236, FokI rs2228570 and Cdx2 rs11568820) were determined using the quantitative polymerase chain reaction (PCR) with TaqMan Probes (Thermo Fisher Scientific, Waltham, MA, USA) in the CFX96™ Touch Real-Time PCR Detection Systems (Bio-Rad Laboratories, Hercules, NY, USA) as previously described [20,21]. The whole reacting volume in PCR tubes was 10 µL, including 5 µL iTaq™ Universal Probes Supermix (Bio-Rad Laboratories, Hercules, NY, USA), 3.75 µL of mQ water, 0.25 µL × 40 TaqMan™ SNP Genotyping Assay, and 1 µL of genomic DNA (15 ng). The reactions were performed with an initial denaturation at 95 °C for 10 min, followed by 40 cycles of denaturation at 95 °C for 15 s, annealing, and synthesis at 60 °C for 30 s. The final extension was performed at 72 °C for 1 min. Negative and positive controls were randomly included across each PCR run, and several samples were randomly re-genotyped for quality control purposes.

2.4. Statistical Analysis

Data analysis was performed using the programming language R. Continuous variables presented as median (25%, 75% interquartile range) and compared using Mann–Whitney U-test. The deviation from Hardy–Weinberg equilibrium was assessed by the chi-square (χ^2) test. The genetic risk of pathology was estimated using odds ratios, with 95% confidence intervals (CI) and calculated in comparison to reference (major homozygous) genotype. The codominant model was defined and tested for all SNVs. Logistic regression models were used to assess the difference between the characteristics of analyzed groups for categorical data and for comparison of genotype frequencies between these groups. A Multivariate Linear Regression model was used to adjust for confounding factors, and an ANOVA test was used for analysis of continuous variables distribution between genotypes. Beta (β) measures the difference in quantitative traits between genotypes. Pairwise linkage disequilibrium (LD) and haplotype analysis were performed using the R-packages "haplo.stats" (v.1.7.9) and "SNPassoc" (v.1.9-2); the programs used likelihood ratio tests in a generalized linear model and the expectation-maximization algorithm. The differences between the groups were considered statistically significant at $p < 0.05$. p-values corrected for multiple testing using the False Discovery Rate (*p* FDR) with Benjamini and Hochberg procedure (n = 5, multiple comparisons).

3. Results

In total, 927 subjects were screened for participation in the study, 325 of them were excluded due to eligibility, and 602 of them met inclusion criteria and were allocated to patients with PMO (355 women) and control (247 women) groups, followed by clinical examinations and genetic testing.

3.1. Study Subjects Charactetistic

The clinical characteristics of the analyzed cohort are summarized in Table 1. The mean age of all individuals was 62.4 years, the mean weight, height, and BMI were 72.4 kg, 159.8 cm, and 28.4 kg/m^2, respectively. All participants were ethnic Belarusians.

Table 1. Clinical characteristics of study subjects.

Clinical Characteristic	PMO Patients	Control	*p*-Value
n (%)	355 (59.0)	247 (41.0)	
Age, years	63.0 (57.0; 70.0)	62.0 (58.0; 67.0)	0.12
Age at menopause, years	50.0 (47.0; 52.0)	50.0 (48.0; 52.0)	0.32
Weight, kg	66.0 (58.0; 74.0)	81.0 (73.0; 93.0)	**0.0001**
Height, cm	160.0 (156.0; 165.0)	159.0 (155.0; 164.0)	0.20
BMI, kg/m^2	25.5 (22.5; 28.7)	31.6 (28.3; 36.2)	**0.0001**
LS BMD, g/cm^2	0.9 (0.8; 0.9)	1.3 (1.2; 1.4)	**0.0001**
LS T-score	−2.6 (-3.2; −2.0)	0.7 (−0.1; 1.3)	**0.0001**
LS Z-score	−1.1 (-1.7; −0.5)	1.3 (0.6; 2.1)	**0.0001**
FN BMD, g/cm^2	0.8 (0.7; 0.9)	1.1 (1.0; 1.2)	**0.0001**
FN T-score	−1.7 (−2.4; −1.1)	0.4 (−0.1; 1.0)	**0.0001**
FN Z-score	−0.7 (−1.2; −0.1)	1.0 (0.4; 1.6)	**0.0001**
25-hydroxyvitamin D (25(OH)D), ng/mL	27.0 (21.1; 35.8)	19.9 (15.2; 26.4)	**0.0001**
Fractures in history	49 (13.8%)	7 (2.8%)	

The data are presented as median (25%; 75% interquartile range). BMI, body mass index; LS, lumbar spine; FN, femur neck; BMD, bone mineral density; PMO, postmenopausal osteoporosis. The values highlighted in bold indicate a significant association.

Cases and controls did not differ in age, age at menopause, and height (Mann-Whitney *U*-test did not reveal difference, *p* > 0.1). A strong difference between groups was revealed for weight, BMI, lumbar spine, and femoral neck BMD, T-, and Z-scores. The weight and BMI variables were considered potential confounding factors and were adjusted in analysis of association between groups. In the study cohort, 49 patients with PMO had a fracture history (at least one) compared to seven individuals from the control group. The baseline serum 25(OH)D level ranged from 7.4 to 70 ng/mL, and the mean level in all individuals was 25.5 ng/mL. There was a statistically significant difference between analyzed groups in the plasma 25(OH)D level.

3.2. The Relationship between VDR Gene Variants and PMO Risk

All subjects were genotyped in the study; the genotype frequencies distribution is presented in Table 2. The five most common polymorphic loci of the *VDR* gene were selected from key publications and studied as candidate markers of PMO. These SNVs with previously established involvement in vitamin D and bone tissue metabolisms were included to the study to validate their effect by analysis of combinations of genetic variants on independent cohorts.

Table 2. The Hardy–Weinberg equilibrium *p*-values and distribution of genotype frequencies of *VDR* gene variants in patients with postmenopausal osteoporosis and control.

Gene Variant	Genotype	PMO *n* = 355		CON *n* = 247		OR (95% CI)	Adjusted OR (95% CI) [1]
		%	HWE	%	HWE		
rs7975232	C/C	23.7		31.2		Ref.	Ref.
	C/A	45.9	0.17	48.2	0.7	1.3 (0.9–1.9)	1.3 (0.8–2.1)
	A/A	30.4		20.6		**1.9 (1.2–3.1) ***	**2.1 (1.3–3.6) ***
rs1544410	C/C	25.4		36.4		Ref.	Ref.
	C/T	46.7	0.24	46.6	0.6	1.5 (0.9–2.1)	**1.6 (1.0–2.4) ***
	T/T	27.9		17.0		**2.4 (1.5–3.8) ****	**2.1 (1.2–3.6) ***
rs731236	A/A	24.2		37.2		Ref.	Ref.
	A/G	47.6	0.4	45.8	0.51	**1.6 (1.1–2.3) ****	**2.1 (1.3–3.3) ****
	G/G	28.2		17.0		**2.6 (1.6–4.1) ****	**2.3 (1.4–4.0) ****
rs2228570	G/G	27.9		26.7		Ref.	Ref.
	A/G	46.2	0.17	50.2	1.0	0.9 (0.6–1.3)	0.9 (0.6–1.5)
	A/A	25.9		23.1		1.1 (0.7–1.7)	1.0 (0.6–1.8)
rs11568820	C/C	69.0		66.4		Ref.	Ref.
	C/T	30.1	0.02	29.6	0.68	0.9 (0.6–1.3)	1.2 (0.8–1.8)
	T/T	0.9		4.0		0.2 (0.1–0.7)	0.2 (0.1–1.0)
	C/T+T/T [2]	31.0		33.6		0.8 (0.6–1.3)	1.1 (0.7–1.6)

PMO, postmenopausal osteoporosis; CON, control; HWE, Hardy–Weinberg equilibrium; OR, odds ratio; CI, confidence interval; Ref., referent value; * *p* FDR < 0.05, ** *p* FDR < 0.01; [1] Adjusted by confounding factors (BMI); [2] Dominant model of inheritance used due to low minor allele frequency. The values highlighted in bold indicate a significant association.

The minor allele frequencies of all analyzed SNVs were not significantly different from those taken from GnomAD data for Europeans [22]. By the analysis, the genotyping data were found to be in correspondence to the expected Hardy–Weinberg equilibrium at the 5% level in the control group ($p > 0.1$). In case group, the deviation from the Hardy–Weinberg equilibrium was revealed only for the rs7975232 variant ($p = 0.02$), which was possibly due to the very low frequency of the T/T homozygous genotype.

The analysis of genotype frequencies of *VDR* gene variants, presented in Table 2, showed significant differences in their distribution between both groups. The most frequent homozygous genotype was taken for reference. Comparing the genotype frequencies between PMO and CON groups, statistically significant differences after FDR correction for multiple testing were found for rs7975232, rs1544410, and rs731236 variants of the *VDR* gene. The PMO group individuals were more likely to carry the rs7975232 A/A genotype (30.4%) compared to the CON group (20.6%, OR = 1.9, 95% CI 1.2-3.1, *p* FDR = 0.0175). The rs1544410 T/T genotype was significantly over-represented in PMO patients (27.9%) compared to control group (17.0%, OR = 2.4, 95% CI 1.5-3.8, *p* FDR = 0.0028). For the bearers of the rs731236 G/G homozygous genotype, the risk of osteoporosis was increased (OR = 2.6, 95% CI 1.6-4.1, *p* FDR = 0.0015). An increased risk of PMO was also revealed for the bearers of the heterozygous genotype A/G, OR = 1.6 (95% CI 1.1-2.3).

To reduce the potential impact of confounding factors, the association analysis was further adjusted for BMI. When adjusted, the associated gene variants remained significant; additionally, a statistically significant association was revealed for the rs1544410 heterozygous C/T genotype (Table 2).

There was no association with osteoporosis risk found for rs2228570 and rs11568820 variants. Since the rs11568820 T-allele frequency was very low, we used a dominant model of inheritance and merged C/T+T/T genotypes. Despite the absence of significant association after the Yates correction, it can be noted that the frequency of the rs11568820 T/T-genotype is significantly higher in the CON group (4.0%) compared to PMO patients (0.9%).

Next, we analyzed the pairwise linkage disequilibrium between *VDR* gene variants. An LD plot was constructed using combined genotype data from both groups of individuals (Figure 1).

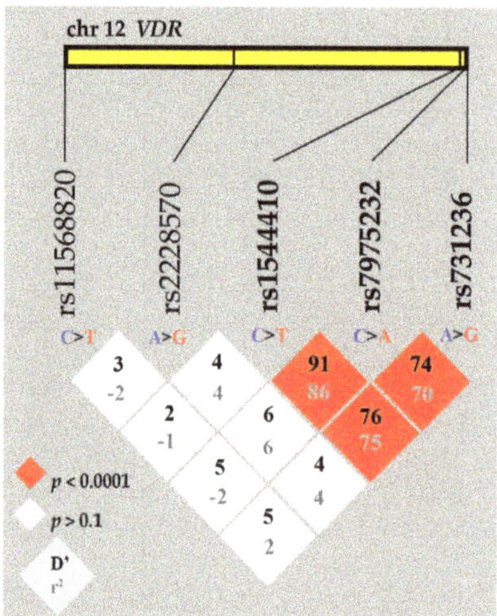

Figure 1. Linkage disequilibrium (LD) plot for rs7975232, rs1544410, rs731236, rs2228570, and rs11568820 variants of the *VDR* gene. LD is displayed as pairwise D′ values multiplied by 100 and given for each SNV combination within each cell. Red cells correspond to a very strong LD; rs7975232, rs1544410, and rs731236 variants are in the same LD block.

Performed LD analysis identified one haplotype block, which was composed of *VDR* rs7975232, rs1544410, and rs731236 variants. The measure of linkage strength D′ between rs7975232 and rs1544410 was 91, $p < 0.0001$. The positive coefficient of correlation r^2 suggests that major alleles of rs7975232, rs1544410, and rs731236 variants are likely to be inherited together, as well as minor alleles. The rs2228570 and rs11568820 variants do not exhibit significant LD, the D′ coefficient ranged for them from 2 to 6; therefore, they were removed from further analysis.

Based on LD data, in further analysis, we combined three *VDR* gene variants from the same block and performed the haplotype analysis. Haplotypes were constructed from all possible allelic combinations and compared between the PMO and control (CON) groups (Table 3).

Table 3. Haplotype analysis of *VDR* rs7975232, rs1544410, and rs731236 gene variants in patients with postmenopausal osteoporosis (PMO) and control (CON) group.

Haplotype	Frequency, %		Haplotype Score	*p* FDR	Logistic Regression	
	PMO	CON			OR (95% CI)	*p*
C-C-A	38.6	49.6	−3.53	0.001	Ref.	-
A-T-G	44.9	31.7	**4.41**	0.00005	**1.8 (1.4–2.3)**	**0.0001**
A-T-A	4.0	7.1	−1.78	0.12	0.9 (0.6–1.4)	0.63
C-C-G	5.7	4.1	0.84	0.50	1.6 (0.9–2.7)	0.11
A-C-A	4.2	3.5	0.15	0.88	1.3 (0.8–2.2)	0.34
Rare *	2.6	4.0	-	-	-	-

OR, odds ratio; Ref., referent value; CI, confidence interval; FDR, false discovery rate. * Rare haplotypes—other possible haplotypes with total frequency less than 3%. The values highlighted in bold indicate a significant association.

The haplotype analysis revealed five combinations (C-C-A, A-T-G, A-T-A, C-C-G, and A-C-A) of the possible eight with a frequency greater than 3%. These five haplotypes jointly presented in 96.5% of study subjects. Statistically significant differences between analyzed groups were revealed in the global distribution of allelic combinations (global $p = 0.00023$), suggesting an association of analyzed haplotypes with the risk of PMO. A statistically significant difference was revealed in the distribution of the most frequent C-C-A and A-T-G haplotypes between the PMO and CON groups even after FDR correction for multiple testing. The C-C-A haplotype, constructed from three wild-type alleles, was the most frequent (total frequency 43.4%). This haplotype frequency was significantly higher among controls (49.6%) than among cases (38.6%, p FDR = 0.001). The negative haplotype score value of -3.53 suggests that this combination is associated with a decreased risk of PMO. The total frequency of the A-T-G haplotype was 39.4%; it was significantly under-represented in the CON group (31.7%) compared to the PMO group (44.9%, p FDR = 0.00005), suggesting that this allelic combination might confer a greater susceptibility to PMO (the highest haplotype score of 4.41 points). In comparison with the most frequent reference (wild-type) haplotype C-C-A, for the bearers of the A-T-G haplotype, the risk of PMO was significantly higher (OR = 1.8, 95% CI 1.4-2.3, $p = 0.0001$). No significant association was found for other constructed haplotypes.

3.3. The Relationship between VDR Gene Variants and Lumbar Spine BMD Level

The association analysis between *VDR* gene single variants and haplotypes and LS BMD level was performed using linear regression on the combined cohort of study subjects (Figure 2).

The analysis revealed four *VDR* gene variants associated with L1-L4 BMD level. The observed difference in BMD level for rs7975232, rs1544410, and rs731236 minor homozygous genotypes compared to reference genotypes was almost the same ($\beta = -0.13$ g/cm^2, p FDR = 0.0003; $\beta = -0.15$ g/cm^2, p FDR = 0.0005, and $\beta = -0.13$ g/cm^2, p FDR = 0.00025, respectively). Such a similar effect may be explained by the previously observed high LD between rs7975232, rs1544410, and rs731236. Interestingly, for all these variants, there was a gene/dose response revealed: the highest level of BMD was found for the bearers of major homozygous genotype, the intermediate level was found in heterozygotes, and the lowest level was found in minor homozygotes (Figure 2A–C). Quantitative analysis of the rs11568820 variant revealed a more significant association with LS BMD level (Figure 2E). The substitution of G to A was associated with much higher LS BMD values ($\beta = 0.22$, 95% CI 0.07-0.38, p FDR = 0.027), suggesting that the rs11568820 A-allele has a protective effect. No significant association for rs2228570 was found with LS BMD (Figure 2D). As there was a strong positive LD between rs7975232, rs1544410, and rs731236 variants, we performed an analysis of association of their haplotypes with LS BMD level (Figure 2F). The bearers of the A-T-G haplotype showed a higher decrease in LS BMD compared to reference C-C-A ($\beta = -0.09$, 95% CI $-0.13 \ldots -0.06$, p FDR = 0.0001). No significant association with BMD was found for other haplotypes.

3.4. The Relationship between VDR Gene Variants and Serum 25(OH)D Level

The effect of vitamin D is mediated through binding to a specific steroid receptor with the activity of a transcription factor, thus regulating the synthesis of protein actively participating in bone metabolism and maintaining calcium homeostasis. Variation in *VDR* gene may alter receptor functions, suggesting possible changes in serum 25(OH)D concentration. The relationship between *VDR* gene variants and serum 25(OH)D level is presented in Figure 3.

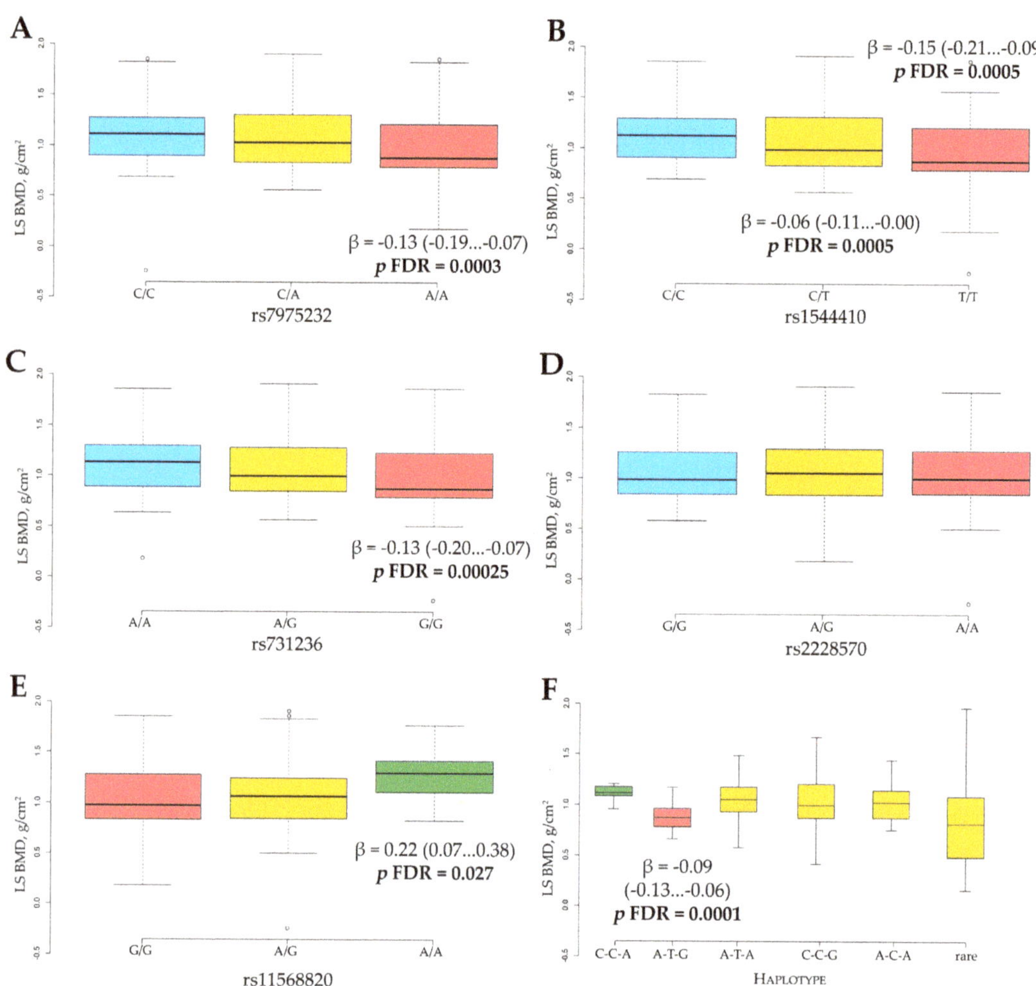

Figure 2. Lumbar spine (LS) BMD level in relation to *VDR* gene variants rs7975232 (**A**), rs1544410 (**B**), rs731236 (**C**), rs2228570 (**D**), rs11568820 (**E**), and rs7975232, rs1544410, and rs731236 haplotypes (**F**). For rs7975232, rs1544410, rs731236, and rs11568820 variants, gene/dose dependence was revealed. The rs11568820 was the only *VDR* gene variant with protective effect. *p*-values corrected for multiple testing using the FDR, β is the difference compared to the reference value. The data are presented as β (95% CI).

We revealed a statistically significant association of rs1544410, rs731236, and rs11568820 gene variants with the serum 25(OH)D level. The genetic effects of these three markers on baseline serum 25(OH)D level were gene/dose dependent. Interestingly, for rs1544410 (Figure 3B) and rs731236 (Figure 3C), the lowest 25(OH)D level was typical for the reference genotype, while it was intermediate for heterozygotes and the highest for the bearers of minor homozygous genotypes (ANOVA test $p = 0.006$ and $p = 0.0005$, respectively). For the bearers of the rs1544410 T/T genotype, the 25(OH)D concentration was 3.6 ng/mL higher compared to C/C genotypes (p FDR = 0.015), whereas for the bearers of the rs731236 G/G genotype, it was 4.6 ng/mL higher (p FDR = 0.0002). The opposite gene/dose relationship was revealed for the rs11568820 variant, when wild-type G/G homozygotes showed a higher 25(OH)D increase as compared to the A/G and A/A genotypes (Figure 3E).

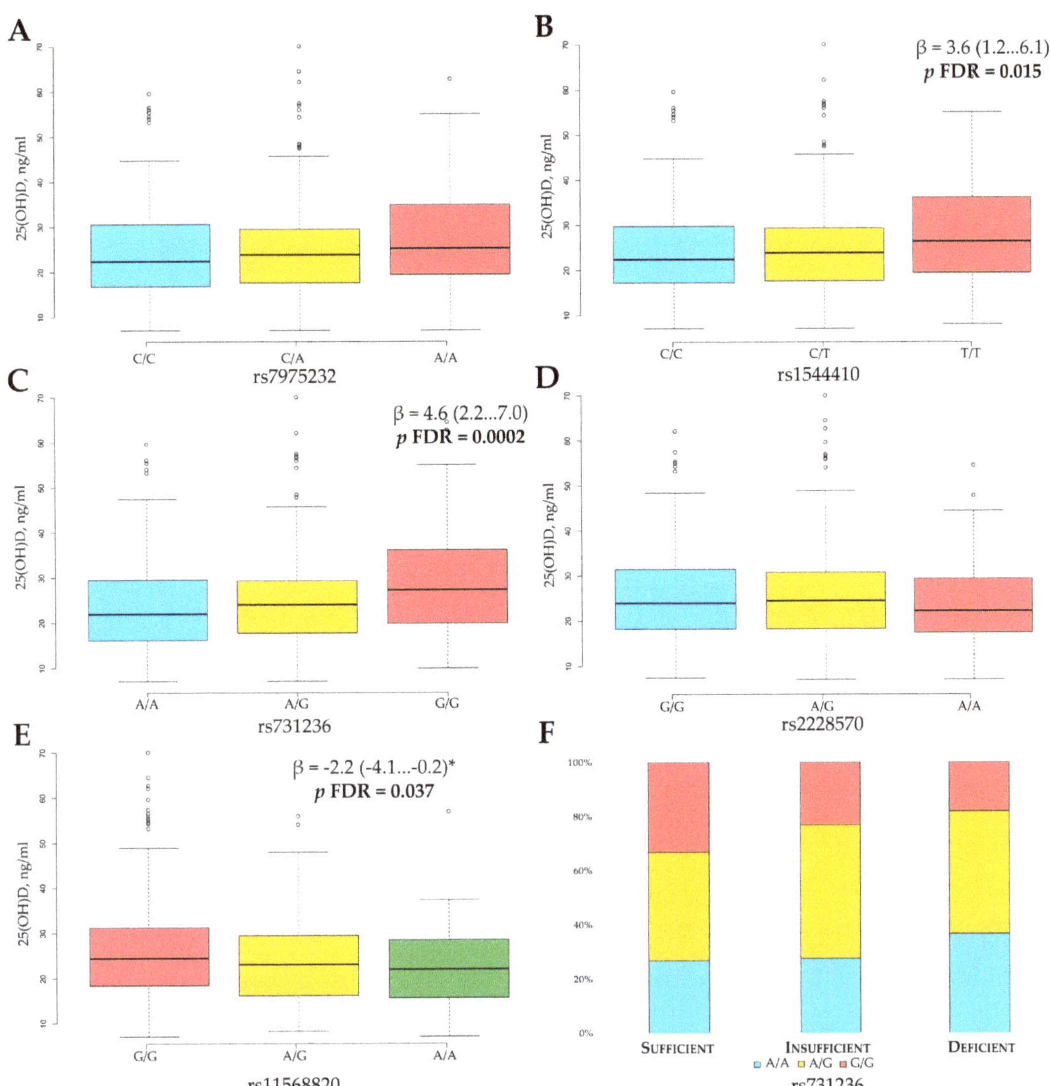

Figure 3. The association of serum 25-hydroxyvitamin D levels with *VDR* gene variants rs7975232 (**A**), rs1544410 (**B**), rs731236 (**C**), rs2228570 (**D**), rs11568820 (**E**), and rs731236 genotype distribution in groups based on Vitamin D status (**F**). For rs1544410, rs731236, and rs11568820 variants, gene/dose dependence was revealed. The rs11568820 was the only *VDR* gene variant with protective effect. *p*-values corrected for multiple testing using the FDR, *β* is the difference compared to the reference value. The data are presented as *β* (95% CI). * Dominant model of inheritance.

We also assessed the distribution of each *VDR* variant of genotype in different groups of study participants according to vitamin D level (sufficient, insufficient, deficient). Using a two-tailed χ^2 test, a statistically significant difference in genotype distribution between groups was revealed only for the rs731236 variant ($\chi^2 = 12.8$, $p = 0.012$, Figure 3F). The G/G genotype was over-represented in a group of participants with a "sufficient" state, while the A/A genotype was associated with vitamin D deficiency.

In order to increase the statistical power, we analyzed the association of serum 25(OH)D level with rs1544410, rs731236, and rs11568820 haplotypes (Table 4).

Table 4. The association of *VDR* rs7975232, rs1544410, and rs731236 haplotypes with serum 25(OH)D levels.

Haplotype	Frequency, %	25(OH)D, ng/mL (mean ± SE)	Linear Regression	
			β (95% CI)	*p* FDR
C-C-A	44.8	21.7 ± 0.7	Ref.	-
A-T-G	39.8	23.7 ± 0.8	**2.0 (0.7 … 3.4)**	**0.017**
A-T-A	4.4	25.4 ± 1.4	−1.7 (−4.4 … 1.0)	0.35
C-C-G	4.2	23.5 ± 1.5	0.2 (−2.8 … 3.1)	0.91
A-C-A	3.6	24.0 ± 1.5	−0.3 (−3.2 … 2.6)	1.0
Rare *	3.2	27.9 ± 4.3	4.2 (1.1 … 7.4)	0.02

Ref., referent value; SE, standard error; β, difference compared to reference value; CI, confidence interval. * Rare haplotypes—other possible haplotypes with total frequency less than 3%. The values highlighted in bold indicate a significant association.

The haplotype analysis revealed five combinations (C-C-A, A-T-G, A-T-A, C-C-G, and A-C-A) with a frequency greater than 3%. The C-C-A haplotype was the most frequent (total frequency 44.8%) and was taken for reference. The total frequency of the A-T-G haplotype was 39.8%. For the bearers of the A-T-G haplotype, the 25(OH)D level was significantly higher compared to the reference haplotype (β = 2.0, 95% CI 0.7–3.4, *p* FDR = 0.017). No significant association with BMD level was found for other haplotypes.

4. Discussion

PMO is the most widespread type of disease, which causes serious medical, social, and economic difficulties for society. Research in the area of such complex (multifactorial) pathology aims to reveal both the environmental and genetic factors affecting disease development. According to numerous epidemiological studies, family and twin observations, up to 90% of OP cases are genetically determined [23]. The early detection of genetic factors, associated with predisposition to PMO, may help increase the prophylaxis and treatment effectiveness, although the evaluation of genetic background is complicated due to the involvement of multiple gene networks and their interaction with various environmental factors.

Vitamin D effects have been widely investigated in various populations with regard to its possible effect on PMO risk. The huge interest in vitamin D is explained primarily by its activity in calcium homeostasis, bone formation, and the regulation of bone mineral density. Vitamin D binds to a specific steroid receptor, which is coded by the *VDR* gene. VDR has a transcription factor activity and influences the expression activity of numerous target genes. The intensive study of the *VDR* gene made it possible to identify polymorphic variants that may lead to structural or functional changes in protein expression. Therefore, these variants may serve as potential clinical and diagnostic markers of bone muscular pathology. However, *VDR* gene studies continue and in a number of research studies, conflicting data on the distribution of genotype frequencies of the different loci of this gene were shown [12,13,23], thus providing a basis for further work in the area.

In the present study, we analyzed the association of the five most commonly studied *VDR* gene variants with PMO risk, BMD level, and serum 25(OH)D concentration in the Belarusian population. These SNVs were selected based on their previously established role in bone metabolism, modulation of *VDR* expression, and activity. Several previous research studies have suggested that *VDR* gene variants may influence the BMD level; this effect is variable and population-dependent [12,13], but complex studies on SNVs association with vitamin D level are still lacking.

The observed differences in body weight and BMI between groups are quite expected, as decreased BMI and body weight are well-known risk factors of OP. The revealed average level of BMI in the control group can be explained by the population features of the Belarusian cohort.

We compared the allele distribution of analyzed *VDR* gene variants with that of the populations included in the gnomAD database [22]. For rs7975232, rs1544410, rs731236, and rs11568820, our study revealed no significant difference in minor allele frequency (MAF) as compared to European cohort. However, MAF (allele A) for rs2228570 was significantly higher (48%) as compared to the gnomAD European cohort (37%). This difference may be due to the cohort size, ethnicity, or gender specificity.

The analysis of our data revealed a strong association of rs7975232, rs1544410, and rs731236 variants with PMO risk, where A/A, T/T, and G/G genotypes, correspondingly, were over-represented in the patients group. This association remained even after correction for cofounding factors, suggesting the predominant contribution of hereditary factors and their role as potential markers of PMO risk. For all these three *VDR* gene variants located in the ligand-binding domain at the 3′-end of the gene, a strong linkage disequilibrium was found. The located upstream rs2228570 in the DNA-binding domain and rs11568820 in the promoter region did not show LD. These data are in agreement with previous studies on European populations [24,25]. Despite the fact that a strong LD was revealed between rs7975232, rs1544410, and rs731236 markers, this linkage is not full, and it is not possible to predict the alleles of one SNV, knowing the alleles of other SNVs. Perhaps, this explains that the risk of PMO in haplotype analysis (for the bearers of A-T-G haplotype) was lower compared to single gene analysis (for the bearers of rs7975232 A/A, rs1544410 T/T, and rs731236 G/G genotypes).

When examined with respect to the interaction with LS BMD, in addition to rs7975232, rs1544410, and rs731236, a statistically significant association was also revealed for rs11568820. However, there was no effect revealed for rs2228570. Interestingly, a dose effect was observed for all these variants, which is very important. Women with the rs7975232, rs1544410, and rs731236 minor homozygous genotypes had the lowest BMD level compared to bearers of other genotypes. In contrast, the rs11568820 A/A genotype was associated with increased BMD level, suggesting its protective effect. Unlike other gene variants, rs11568820 is located in the promoter region of the *VDR* gene, and previously, it was reported that the A-allele relates to higher promoter activity in vitro [26]. The same effect on BMD was observed in Slovenian women with PMO [27]. Previously, various studies also reported a significant association with LS BMD of rs7975232 [28] and rs1544410 [27] in Caucasian women. The most unexpected result of our study was the absence of a statistically significant association of rs2228570 with BMD level, although it was previously reported in Caucasian women with PMO [28]. On the other hand, our results are consistent with a huge meta-analysis, which was performed recently [12]. Such differences within the same populations may be explained by the involved environmental factors, gene–gene and gene–environment interactions or sample sizes. The rs2228570 variant plays an important role in message stability and post-transcriptional processes in the *VDR* gene, and due to its functions, more extensive research is required. Interestingly, our results showed consistency: rs7975232, rs1544410, and rs731236 were associated both with a decreased risk of PMO and high levels of LS BMD. This result may be explained by strong LD between variants. Moreover, the simultaneous increase of PMO risk and decrease in the lumbar BMD level for the bearers of the A-T-G haplotype was observed. The same association was previously revealed in Polish [28] and Dutch [29] populations. High LD creates additional difficulties, since the exact gene variant with a causative effect is unknown.

Possibly, the most interesting part of the study is the analysis of the association of *VDR* gene variants with vitamin D status. Such an interest in vitamin D is explained by importance for bone health and by its pleiotropic biological action, which mediates predisposition not only to bone muscular diseases but also to many other complex diseases. Recent studies have suggested that SNVs within the *VDR* gene may influence the level or activity of vitamin D. In the present study, we revealed a statistically significant gene/dose association of rs1544410, rs731236, and rs11568820 variants with plasma 25(OH)D level. The rs1544410 T/T and rs731236 G/G genotypes are associated with an increased level of plasma vitamin D compared to the reference genotype, while rs11568820 combined

A/G+A/A genotype bearers were characterized by a decrease. The same association was previously revealed in Caucasian [30,31], Chinese [32], and Egyptian [33] populations. In the analysis of *VDR* gene variants distribution within groups according to vitamin D status (sufficient, insufficient, deficient), only rs731236 revealed a significant association, whereas the bearers of A/A genotypes were over-represented in the group of patients with deficiency. These data are confirmed by a recent study on response to vitamin D supplementation [15]. In haplotype analysis, which was performed to test whether combinations of different *VDR* gene variants may predict serum 25(OH)D concentration, the highest level of serum vitamin was found in bearers of A-T-G alleles, which are associated with PMO risk and low LS BMD. Located at the 3′-end of the *VDR* gene, rs7975232, rs1544410, and rs731236 variants are associated with the different-length polyadenylate sequences and affect the stability of mRNA, while the rs11568820 variant could change the transcription activity of the promoter region of the gene [34]. Thus, we may hypothesize that the increased level of circulating 25(OH)D in patients compared to controls (Table 1) may be explained by the fact that unfavorable *VDR* genotypes were over-represented in the PMO group, possibly altering metabolic feedback loops or the effectiveness of vitamin metabolism. This hypothesis is confirmed by the fact that the level of VDR mRNA was remarkably reduced in bearers of the rs1544410 T/T genotype compared to individuals with the C/C genotype [35]. The bearers of favorable rs7975232 C-, rs1544410 C-, and rs731236 A-alleles have higher VDR receptor expression, thus leading to increased vitamin D metabolism. The possible mechanism may include an alteration of vitamin D-mediated gene expression by differential activity of the VDR receptor. Specific SNVs may decrease the activity of a wide range of enzymes involved in the production and elimination of 25(OH)D, promoting an increase in circulating serum 25(OH)D level and a simultaneous decrease in intercellular 1,25-dihydroxyvitamin D concentration, resulting in adverse health effects caused by vitamin D deficiency despite higher circulating 25(OH)D concentration [14]. This hypothesis requires further confirmation in other studies.

The remaining rs2228570 with a previously established role in vitamin D-related pathways revealed no association with plasma 25(OH)D level in our cohort. This variant, which is located in the second exon, forms a second methionine start site, producing a shorter protein receptor, which displayed higher transcriptional activity. Our results are contrary to studies where in bearers of the rs2228570 G/G genotype, an increase in 25(OH)D concentration [36] and affected 25(OH)D hydroxylation [37] was reported. Although we found no significant association for this variant with PMO risk, LS BMD, and 25(OH)D level in the present study, broader research with a bigger cohort may confirm this variant as a marker for personalized medicine purposes.

Nevertheless, a very interesting result is that the variants associated with low LS BMD levels were also associated with high 25(OH)D levels and vice versa. Such a tendency is found for all analyzed SNVs, suggesting the differential action of vitamin D on the local cell level in bones and circulating in plasma. At least rs7975232, rs1544410, rs731236, and rs11568820 might help to identify individuals with increased PMO risk and vitamin D status. The revealed considerable variation in serum 25(OH)D in individuals with different *VDR* genotypes further suggest that a one-size-fits-all approach to vitamin D supplementation may not be appropriate. For more accurate evaluation of the association of *VDR* gene polymorphism with predisposition to PMO, the analysis of various environmental factors, such as diet, sun exposure, exercise, and other is required. The relationship between gene variation and vitamin D status also requires further investigation. In addition, other gene variants within vitamin D pathway, as well as epigenetic factors can also play a significant role in the pathogenesis of disease.

5. Conclusions

We investigated the association of five major *VDR* gene variants with PMO risk, BMD level, and serum 25(OH)D concentration. Our study shows novel data on vitamin D genetics and homeostasis, particularly on the significant association of four markers with

BMD and 25(OH)D levels. *VDR* rs7975232, rs1544410, rs731236, and rs11568820 might be taken into consideration for individual PMO risk assessment and the development of personalized recommendations for the optimization of vitamin D supplementation.

Author Contributions: Conceptualization, E.R.; formal analysis, P.M.; investigation, A.R. and K.K.; writing—original draft preparation, P.M. and A.R.; writing—review and editing, E.R.; supervision, E.R.; project administration, E.R.; funding acquisition, P.M. and A.R. All authors have read and agreed to the published version of the manuscript.

Funding: This research was funded by NATIONAL ACADEMY OF SCIENCES OF BELARUS, scientific and technical program of the Union State "DNA-identification", grant number 6.3.

Institutional Review Board Statement: The study was conducted according to the guidelines of the Declaration of Helsinki, and approved by the Ethical Committee of Belarusian Medical Academy of Post-Graduate Education, Minsk, Belarus (protocol 4 issued on 4 December 2017).

Informed Consent Statement: Informed consent was obtained from all subjects involved in the study.

Data Availability Statement: The data supporting the conclusions of this manuscript will be made available by the authors to any qualified researcher.

Conflicts of Interest: The authors declare no conflict of interest. The funders had no role in the design of the study; in the collection, analyses, or interpretation of data; in the writing of the manuscript, or in the decision to publish the results.

References

1. NIH Consensus Development Panel on Osteoporosis Prevention, Diagnosis, and Therapy. Osteoporosis prevention, diagnosis, and therapy. *JAMA* **2001**, *285*, 785–795. [CrossRef]
2. Curtis, E.M.; Moon, R.J.; Harvey, N.C.; Cooper, C. The impact of fragility fracture and approaches to osteoporosis risk assessment worldwide. *Bone* **2017**, *104*, 29–38. [CrossRef] [PubMed]
3. Hernlund, E.; Svedbom, A.; Ivergård, M.; Compston, J.; Cooper, C.; Stenmark, J.; McCloskey, E.V.; Jönsson, B.; Kanis, J.A. Osteoporosis in the European Union: Medical management, epidemiology and economic burden. A report prepared in collaboration with the International Osteoporosis Foundation (IOF) and the European Federation of Pharmaceutical Industry Associations (EFPIA). *Arch. Osteoporos.* **2013**, *8*, 136. [CrossRef]
4. Borgström, F.; Karlsson, L.; Ortsäter, G.; Norton, N.; Halbout, P.; Cooper, C.; Lorentzon, M.; McCloskey, E.V.; Harvey, N.C.; Javaid, M.K.; et al. Fragility fractures in Europe: Burden, management and opportunities. *Arch. Osteoporos.* **2020**, *15*, 59. [CrossRef] [PubMed]
5. Stewart, T.L.; Ralston, S.H. Role of genetic factors in the pathogenesis of osteoporosis. *J. Endocrinol.* **2000**, *166*, 235–245. [CrossRef] [PubMed]
6. Bhattoa, H.P.; Konstantynowicz, J.; Laszcz, N.; Wojcik, M.; Pludowski, P. Vitamin D: Musculoskeletal health. *Rev. Endocr. Metab. Disord.* **2017**, *18*, 363–371. [CrossRef] [PubMed]
7. Pike, J.W.; Meyer, M.B. The vitamin D receptor: New paradigms for the regulation of gene expression by 1,25-dihydroxyvitamin D(3). *Endocrinol. Metab. Clin. N. Am.* **2010**, *39*, 255–269. [CrossRef]
8. Miyamoto, K.; Kesterson, R.A.; Yamamoto, H.; Taketani, Y.; Nishiwaki, E.; Tatsumi, S.; Inoue, Y.; Morita, K.; Takeda, E.; Pike, J.W. Structural organization of the human vitamin D receptor chromosomal gene and its promoter. *Mol. Endocrinol.* **1997**, *11*, 1165–1179. [CrossRef]
9. Jurutka, P.W.; Whitfield, G.K.; Hsieh, J.C.; Thompson, P.D.; Haussler, C.A.; Haussler, M.R. Molecular nature of the vitamin D receptor and its role in regulation of gene expression. *Rev. Endocr. Metab. Disord.* **2001**, *2*, 203–216. [CrossRef] [PubMed]
10. Gross, C.; Krishnan, A.V.; Malloy, P.J.; Eccleshall, T.R.; Zhao, X.Y.; Feldman, D. The vitamin D receptor gene start codon polymorphism: A functional analysis of FokI variants. *J. Bone Miner. Res. Off. J. Am. Soc. Bone Miner. Res.* **1998**, *13*, 1691–1699. [CrossRef]
11. Yamamoto, H.; Miyamoto, K.; Li, B.; Taketani, Y.; Kitano, M.; Inoue, Y.; Morita, K.; Pike, J.W.; Takeda, E. The caudal-related homeodomain protein Cdx-2 regulates vitamin D receptor gene expression in the small intestine. *J. Bone Miner. Res. Off. J. Am. Soc. Bone Miner. Res.* **1999**, *14*, 240–247. [CrossRef] [PubMed]
12. Zhang, L.; Yin, X.; Wang, J.; Xu, D.; Wang, Y.; Yang, J.; Tao, Y.; Zhang, S.; Feng, X.; Yan, C. Associations between VDR Gene Polymorphisms and Osteoporosis Risk and Bone Mineral Density in Postmenopausal Women: A systematic review and Meta-Analysis. *Sci. Rep.* **2018**, *8*, 981. [CrossRef]
13. Wang, S.; Ai, Z.; Song, M.; Yan, P.; Li, J.; Wang, S. The association between vitamin D receptor FokI gene polymorphism and osteoporosis in postmenopausal women: A meta-analysis. *Climacteric J. Int. Menopause Soc.* **2020**, 1–6. [CrossRef]
14. McGrath, J.J.; Saha, S.; Burne, T.H.; Eyles, D.W. A systematic review of the association between common single nucleotide polymorphisms and 25-hydroxyvitamin D concentrations. *J. Steroid Biochem. Mol. Biol.* **2010**, *121*, 471–477. [CrossRef]

15. Tomei, S.; Singh, P.; Mathew, R.; Mattei, V.; Garand, M.; Alwakeel, M.; Sharif, E.; Al Khodor, S. The Role of Polymorphisms in Vitamin D-Related Genes in Response to Vitamin D Supplementation. *Nutrients* **2020**, *12*, 2608. [CrossRef]
16. Marozik, P.; Alekna, V.; Rudenko, E.; Tamulaitiene, M.; Rudenka, A.; Mastaviciute, A.; Samokhovec, V.; Cernovas, A.; Kobets, K.; Mosse, I. Bone metabolism genes variation and response to bisphosphonate treatment in women with postmenopausal osteoporosis. *PLoS ONE* **2019**, *14*, e0221511. [CrossRef]
17. Cosman, F.; de Beur, S.J.; LeBoff, M.S.; Lewiecki, E.M.; Tanner, B.; Randall, S.; Lindsay, R. Clinician's Guide to Prevention and Treatment of Osteoporosis. *Osteoporos. Int.* **2014**, *25*, 2359–2381. [CrossRef]
18. Kanis, J.A.; Burlet, N.; Cooper, C.; Delmas, P.D.; Reginster, J.Y.; Borgstrom, F.; Rizzoli, R. European guidance for the diagnosis and management of osteoporosis in postmenopausal women. *Osteoporos. Int.* **2008**, *19*, 399–428. [CrossRef]
19. Holick, M.F.; Binkley, N.C.; Bischoff-Ferrari, H.A.; Gordon, C.M.; Hanley, D.A.; Heaney, R.P.; Murad, M.H.; Weaver, C.M. Evaluation, treatment, and prevention of vitamin D deficiency: An Endocrine Society clinical practice guideline. *J. Clin. Endocrinol. Metab.* **2011**, *96*, 1911–1930. [CrossRef] [PubMed]
20. Marozik, P.M.; Tamulaitiene, M.; Rudenka, E.; Alekna, V.; Mosse, I.; Rudenka, A.; Samokhovec, V.; Kobets, K. Association of Vitamin D Receptor Gene Variation with Osteoporosis Risk in Belarusian and Lithuanian Postmenopausal Women. *Front. Endocrinol.* **2018**, *9*, 305. [CrossRef] [PubMed]
21. Rudenka, A.V.; Rudenka, E.V.; Samokhovec, V.Y.; Kobets, K.V.; Marozik, P.M. Association of vitamin D receptor gene polymorphism with a bone mineral density level in postmenopausal women. *Proc. Natl. Acad. Sci. Belarus Med. Ser.* **2019**, *16*, 192–201. (in Russian). [CrossRef]
22. Karczewski, K.J.; Francioli, L.C.; Tiao, G.; Cummings, B.B.; Alföldi, J.; Wang, Q.; Collins, R.L.; Laricchia, K.M.; Ganna, A.; Birnbaum, D.P.; et al. The mutational constraint spectrum quantified from variation in 141,456 humans. *Nature* **2020**, *581*, 434–443. [CrossRef]
23. Urano, T.; Inoue, S. Genetics of osteoporosis. *Biochem. Biophys. Res. Commun.* **2014**, *452*, 287–293. [CrossRef]
24. Rukin, N.J.; Strange, R.C. What are the frequency, distribution, and functional effects of vitamin D receptor polymorphisms as related to cancer risk? *Nutr. Rev.* **2007**, *65*, S96–S101. [CrossRef]
25. Thakkinstian, A.; D'Este, C.; Attia, J. Haplotype analysis of VDR gene polymorphisms: A meta-analysis. *Osteoporos. Int.* **2004**, *15*, 729–734. [CrossRef] [PubMed]
26. Arai, H.; Miyamoto, K.I.; Yoshida, M.; Yamamoto, H.; Taketani, Y.; Morita, K.; Kubota, M.; Yoshida, S.; Ikeda, M.; Watabe, F.; et al. The polymorphism in the caudal-related homeodomain protein Cdx-2 binding element in the human vitamin D receptor gene. *J. Bone Miner. Res. Off. J. Am. Soc. Bone Miner. Res.* **2001**, *16*, 1256–1264. [CrossRef] [PubMed]
27. Mencej-Bedrac, S.; Prezelj, J.; Kocjan, T.; Teskac, K.; Ostanek, B.; Smelcer, M.; Marc, J. The combinations of polymorphisms in vitamin D receptor, osteoprotegerin and tumour necrosis factor superfamily member 11 genes are associated with bone mineral density. *J. Mol. Endocrinol.* **2009**, *42*, 239–247. [CrossRef] [PubMed]
28. Horst-Sikorska, W.; Kalak, R.; Wawrzyniak, A.; Marcinkowska, M.; Celczynska-Bajew, L.; Slomski, R. Association analysis of the polymorphisms of the VDR gene with bone mineral density and the occurrence of fractures. *J. Bone Miner. Metab.* **2007**, *25*, 310–319. [CrossRef]
29. Fang, Y.; van Meurs, J.B.; Arp, P.; van Leeuwen, J.P.; Hofman, A.; Pols, H.A.; Uitterlinden, A.G. Vitamin D binding protein genotype and osteoporosis. *Calcif. Tissue Int.* **2009**, *85*, 85–93. [CrossRef] [PubMed]
30. Marini, F.; Falcini, F.; Stagi, S.; Fabbri, S.; Ciuffi, S.; Rigante, D.; Cerinic, M.M.; Brandi, M.L. Study of vitamin D status and vitamin D receptor polymorphisms in a cohort of Italian patients with juvenile idiopathic arthritis. *Sci. Rep.* **2020**, *10*, 17550. [CrossRef]
31. Barry, E.L.; Rees, J.R.; Peacock, J.L.; Mott, L.A.; Amos, C.I.; Bostick, R.M.; Figueiredo, J.C.; Ahnen, D.J.; Bresalier, R.S.; Burke, C.A.; et al. Genetic variants in CYP2R1, CYP24A1, and VDR modify the efficacy of vitamin D3 supplementation for increasing serum 25-hydroxyvitamin D levels in a randomized controlled trial. *J. Clin. Endocrinol. Metab.* **2014**, *99*, E2133–E2137. [CrossRef]
32. Ling, Y.; Lin, H.; Aleteng, Q.; Ma, H.; Pan, B.; Gao, J.; Gao, X. Cdx-2 polymorphism in Vitamin D Receptor gene was associated with serum 25-hydroxyvitamin D levels, bone mineral density and fracture in middle-aged and elderly Chinese women. *Mol. Cell. Endocrinol.* **2016**, *427*, 155–161. [CrossRef] [PubMed]
33. Elhoseiny, S.M.; Morgan, D.S.; Rabie, A.M.; Bishay, S.T. Vitamin D Receptor (VDR) Gene Polymorphisms (FokI, BsmI) and their Relation to Vitamin D Status in Pediatrics βeta Thalassemia Major. *Indian J. Hematol. Blood Transfus. Off. J. Indian Soc. Hematol. Blood Transfus.* **2016**, *32*, 228–238. [CrossRef] [PubMed]
34. Fang, Y.; van Meurs, J.B.; d'Alesio, A.; Jhamai, M.; Zhao, H.; Rivadeneira, F.; Hofman, A.; van Leeuwen, J.P.; Jehan, F.; Pols, H.A.; et al. Promoter and 3'-untranslated-region haplotypes in the vitamin d receptor gene predispose to osteoporotic fracture: The rotterdam study. *Am. J. Hum. Genet.* **2005**, *77*, 807–823. [CrossRef] [PubMed]
35. Luo, X.Y.; Yang, M.H.; Wu, F.X.; Wu, L.J.; Chen, L.; Tang, Z.; Liu, N.T.; Zeng, X.F.; Guan, J.L.; Yuan, G.H. Vitamin D receptor gene BsmI polymorphism B allele, but not BB genotype, is associated with systemic lupus erythematosus in a Han Chinese population. *Lupus* **2012**, *21*, 53–59. [CrossRef]
36. Monticielo, O.A.; Brenol, J.C.; Chies, J.A.; Longo, M.G.; Rucatti, G.G.; Scalco, R.; Xavier, R.M. The role of BsmI and FokI vitamin D receptor gene polymorphisms and serum 25-hydroxyvitamin D in Brazilian patients with systemic lupus erythematosus. *Lupus* **2012**, *21*, 43–52. [CrossRef] [PubMed]
37. Smolders, J.; Damoiseaux, J.; Menheere, P.; Tervaert, J.W.; Hupperts, R. Fok-I vitamin D receptor gene polymorphism (rs10735810) and vitamin D metabolism in multiple sclerosis. *J. Neuroimmunol.* **2009**, *207*, 117–121. [CrossRef] [PubMed]

 *nutrients*

Review

Lung-Centric Inflammation of COVID-19: Potential Modulation by Vitamin D

Hana. M. A. Fakhoury [1,*,†], Peter R. Kvietys [2,†], Ismail Shakir [2], Hashim Shams [2], William B. Grant [3] and Khaled Alkattan [4]

[1] Department of Biochemistry and Molecular Medicine, College of Medicine, Alfaisal University, P.O. Box 50927, Riyadh 11533, Saudi Arabia
[2] Department of Physiology, College of Medicine, Alfaisal University, P.O. Box 50927, Riyadh 11533, Saudi Arabia; pkvietys@alfaisal.edu (P.R.K.); ishakir@alfaisal.edu (I.S.); hshams@alfaisal.edu (H.S.)
[3] Sunlight, Nutrition, and Health Research Center, P.O. Box 641603, San Francisco, CA 94164-1603, USA; wbgrant@infionline.net
[4] Department of Surgery, College of Medicine, Alfaisal University, P.O. Box 50927, Riyadh 11533, Saudi Arabia; KKattan@alfaisal.edu
* Correspondence: hana.fakhoury@gmail.com
† These authors contributed equally to this work.

Abstract: SARS-CoV-2 infects the respiratory tract and leads to the disease entity, COVID-19. Accordingly, the lungs bear the greatest pathologic burden with the major cause of death being respiratory failure. However, organs remote from the initial site of infection (e.g., kidney, heart) are not spared, particularly in severe and fatal cases. Emerging evidence indicates that an excessive inflammatory response coupled with a diminished antiviral defense is pivotal in the initiation and development of COVID-19. A common finding in autopsy specimens is the presence of thrombi in the lungs as well as remote organs, indicative of immunothrombosis. Herein, the role of SARS-CoV-2 in lung inflammation and associated sequelae are reviewed with an emphasis on immunothrombosis. In as much as vitamin D is touted as a supplement to conventional therapies of COVID-19, the impact of this vitamin at various junctures of COVID-19 pathogenesis is also addressed.

Keywords: acute respiratory distress syndrome (ARDS); coronavirus; COVID-19; cytokine storm; inflammasome; neutrophil extracellular traps (NETs); SARS-CoV-2; vitamin D

Citation: Fakhoury, H..M.A.; Kvietys, P.R.; Shakir, I.; Shams, H.; Grant, W.B.; Alkattan, K. Lung-Centric Inflammation of COVID-19: Potential Modulation by Vitamin D. *Nutrients* **2021**, *13*, 2216. https://doi.org/10.3390/nu13072216

Academic Editor: Carsten Carlberg

Received: 15 June 2021
Accepted: 25 June 2021
Published: 28 June 2021

Publisher's Note: MDPI stays neutral with regard to jurisdictional claims in published maps and institutional affiliations.

1. Introduction

The mechanism of infection, transmission, and clinical presentations of SARS-CoV-2 are qualitatively similar to those of its predecessor, SARS-CoV [1–3]. Notably, both SARS-CoV-2 and SARS-CoV highjack angiotensin-converting enzyme 2 (ACE2) on the membranes of host cells to gain entry [4]. ACE2 is expressed on apical membranes of human respiratory and gastrointestinal epithelial cells [5,6], accounting for proposed means of transmission and clinical manifestations of the current disease, COVID-19. Specifically, human-to-human transmission of SARS-CoV-2 may occur via expired air droplets or a fecal–oral route; the former well documented [3], while the latter has been posited [7]. The clinical presentations include respiratory problems (e.g., cough, dyspnea) and to a lesser extent intestinal complaints (e.g., diarrhea) [8,9]. Most infected individuals develop mild symptoms that resolve without the need for hospitalization. However, a small number of patients develop pneumonia and require hospitalization, and eventually recover. Others worsen, progressing to acute respiratory distress syndrome (ARDS) and require aggressive treatment in intensive care units. The major cause of death is respiratory failure [10–13], although multiple organ damage and sepsis can occur in severe COVID-19 cases [2,11]. In accordance with the clinical presentation, microscopic analyses of autopsy specimens indicate that lungs bear the greatest pathologic burden [10,14–17]. The predominant histologic features include diffuse alveolar damage, thrombosis, and inflammatory infiltrates consisting of macrophages, lymphocytes, and neutrophils [10,15–23]. Organs remote from the initial site of infection (e.g., heart, kidney, brain) may also exhibit pathology (e.g., thrombi) [10,14,17,24,25]. Of particular significance, thrombi, macrophage recruitment, and diminished T and B lymphocytes are noted in hilar lymph nodes and the spleen [15].

The mechanisms involved in the pathogenesis of COVID-19 are the subject of an intense research effort. The information emerging indicates that, in addition to viral virulence, the host's immune response appears to play a major role. Specifically, an excessive inflammatory response, coupled with an impaired antiviral (e.g., interferon) response, are currently touted as causative [26–34]. A common characteristic of severe COVID-19 patients is lymphopenia; markers of T cell exhaustion are also reported in some [15,27,28,30,33,35–41]. An inadequate interferon response would impede the eradication of the virus thereby exacerbating and prolonging the inflammatory response and associated sequelae [27–29,42]. As a case in point, a significant contribution to the lethality of COVID-19 is the inflammation-induced formation of microvascular thrombi, referred to as immunothrombosis [43–49].

Vitamin D can impact numerous pathways involved in host immune responses to viral infections [50,51]. The dietary or skin-derived vitamin D precursors are sequentially hydroxylated to form the active vitamin D [1,25(OH)2D]. A variety of immune and non-immune cells possess the enzymatic machinery to generate (e.g., CYP27B1) or inactivate (e.g., CYP24A1) vitamin D [50,52–55]. Vitamin D can exert both genomic and non-genomic effects by binding to the vitamin D receptor (VDR). A well-documented genomic function is the generation of the antiviral peptide, LL-37 [50,51,56]. A unique regulatory feature of vitamin D is the ability to induce an appropriate inflammatory response and suppress an excessive one [51].

A significant majority of COVID-19 patients have vitamin D insufficiency, based on cut-off values for 25-OHD $\leq$ 10–20 ng/mL [57–61]. A recent meta-analysis indicates that low levels of vitamin D (20–30 ng/mL) are associated with a greater susceptibility to SARS-CoV-2 infection, as well as severity and mortality of COVID-19 [62]. Conversely, in a multicenter retrospective study, supplemental 25-OHD during the first month of hospitalization reduced in-hospital mortality [63]. Thus, it is not surprising that vitamin D has been touted as a potential therapeutic adjunct to conventional approaches in the treatment of COVID-19 [56,57,60,64]. However, based on some controversial issues [52], caution is recommended pending the outcomes of clinical trials targeting the therapeutic efficacy of vitamin D in COVID-19 patients [60,64].

Herein, a narrative approach will be used to address the role of SARS-CoV-2 in lung inflammation that can lead to the manifestations of severe COVID-19 such as ARDS, coagulopathy, and multiorgan dysfunction. The potential impact of vitamin D at various stages of COVID-19 pathogenesis will be addressed in an evidence-based manner (Figure 1). To this end, the PubMed database was mined for vitamin D/VDR data relevant to the aberrant immune response of COVID-19 and yielded the following. Studies addressing the prophylactic or therapeutic efficacy of vitamin D for the ARDS of COVID-19 are limited [57,63,65–67]. A transgenic murine model of COVID-19 that mimics the disease in humans is available [68,69]; however, it has not been used for interventional studies of vitamin D/VDR signaling. An additional issue complicating animal studies is the potential for species-specific inflammatory signaling pathways [50]. Finally, the bulk of the information is derived from cell-based studies using tractable immune cells (e.g., circulating monocytes), which may not reflect the responses in relevant lung cells (e.g., alveolar macrophages). With these limitations in mind, we address the most salient features of vitamin D/VDR signaling relevant to the innate immune response of ARDS. Wherever possible, reviews are cited to direct the reader to relevant original studies.

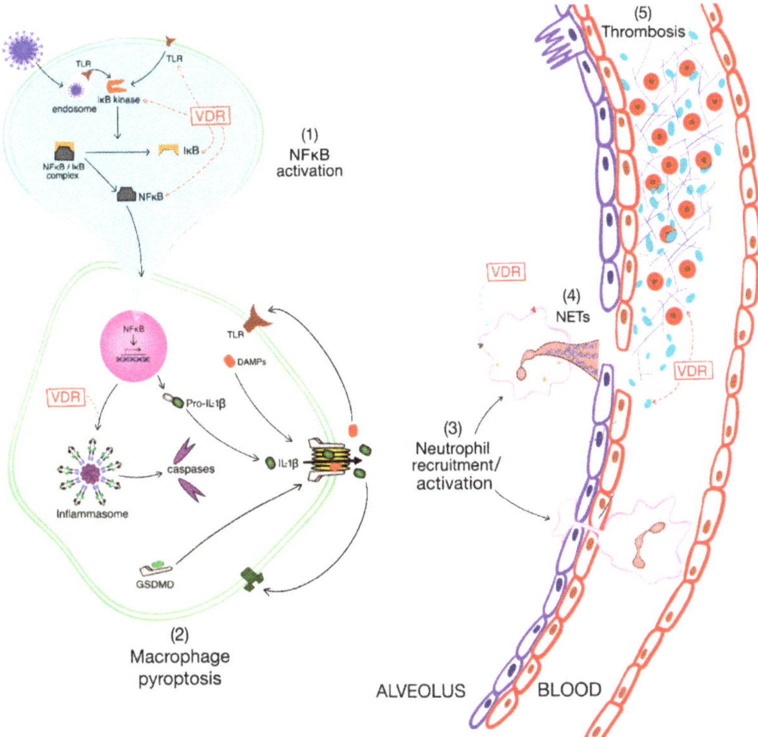

Figure 1. Impact of vitamin D/VDR signaling at various junctures of lung-centric inflammation and immunothrombosis. Experimentally verified inhibitory pathways are indicated with red dashed arrows while an ambiguous pathway is indicated with a blue dashed arrow. As addressed in the text, VDR signaling inhibits the response of alveolar macrophages to viral infection at the level of (1) NFκB signaling and (2) inflammasome activation, thereby preventing neutrophil recruitment and activation (3). The impact of VDR signaling on NET generation (4) may be context-dependent; VDR can promote or inhibit NETs. VDR signaling inhibits thrombosis (5) by interfering with platelet function and fibrin generation. VDR: Vitamin D Receptor; NETs: Neutrophil Extracellular Traps.

2. Current Status of Knowledge

2.1. Intrapulmonary Tropism

SARS-CoV-2 productively infects the human nasal and bronchiolar epithelial cells; primarily ciliated epithelia and, to a lesser extent, goblet cells [4,5,12,20,70–73]. The initial event is an adhesive interaction between the spike (S) protein of the virus with ACE2 on apical membranes of lung epithelia. Subsequently, the S protein is proteolytically activated (e.g., TMPRSS2, furin) allowing for fusion-induced entry [4,74]. After replicating its genome, SARS-CoV-2 preferentially exits via the apical membrane [20,71]. This mode of entry and exit would ensure infection of downstream lung epithelial cells while limiting remote organ involvement. The epithelium remains intact for up to 2–4 days post infection (dpi); progressive infection eventually results in epithelial permeability [71,72]. Of note is that little injury is incurred by SARS-CoV-2 replication within epithelial cells until 3–4 dpi, after which epithelial cell injury and death occur [70–72]. Cell death is a result of apoptosis [71,72], presumably as an antiviral defense mechanism [75,76].

Within the alveolar compartment, type II epithelial cells, endothelial cells, as well as macrophages and dendritic cells, express the requisite machinery for SARS-CoV-2 infection (e.g., ACE2, TMPRSS2, and/or furin) [10,15,26,77]. Infection of type II epithelial cells is productive and utilizes an apical entry and exit pathway [78]. Infection results in the upregulation of pro-inflammatory and antiviral transcriptional pathways [73,78]. The pro-inflammatory pathways (e.g., NFκB) are dominant in the early stages of infection (1–2 dpi); whereas antiviral interferon signaling (e.g., STAT) is delayed (3–4 dpi). As the infection progresses, apoptotic pathways become activated [73,78]. Loss of type II pneumocytes is particularly detrimental, since they generate surfactant, reabsorb fluid from the airspace, and serve as progenitors for the repair of epithelial damage [79].

SARS-CoV-2 infection of endothelium is a matter of debate. Capillary organoids are permissive for SARS-CoV-2 infection and replication [80]. However, lung autopsies of COVID-19 patients are equivocal; some report infection of the endothelium [10], while others highlight the lack of endothelial infection [18,19]. Irrespective of this, endothelial dysfunction and/or injury is common, as evidenced by microvascular thrombi, inflammatory cell infiltration, and capillary sprouting [10,18,19]. A probable scenario holds that endothelial dysfunction contributes to the formation of occlusive emboli resulting in hypoxia, a powerful stimulus for angiogenesis [10].

The major sentinel immune cells of the lung are the alveolar macrophages and dendritic cells. There seems to be little doubt that macrophages can be infected by SARS-CoV-2 [12,24,26,77,81–83]. However, as compared to pneumocytes, fewer resident or infiltrated macrophages are infected [18,19]. While phenotypically quite diverse [84], alveolar macrophages are generally classified as either pro-inflammatory (M1) or pro-resolving (M2) [77]. SARS-CoV-2 can infect both M1 and M2 macrophages; M1 being more permissive [82]. Bronchoalveolar lavage fluid (BALF) of severe COVID-19 patients is characterized by a diminished resident M2 population in favor of infiltrated M1 macrophages [85]. Of note is that SARS-CoV-2 infection of macrophages does not yield viable progeny [81,82]. Despite an abortive infection, the macrophages can generate pro-inflammatory cytokines/chemokines. Dendritic cells, major antigen-presenting cells, can also be infected by SARS-CoV-2. As was the case in macrophages, the infection is abortive [81]. Further, their interferon response is diminished; an effect attributed to viral antagonism of signaling pathways (e.g., STAT). Dendritic cells isolated from COVID-19 patients exhibit impaired maturation and functionality, as evidenced by an inability to stimulate CD4 and CD8 T cell proliferation [37].

In summary, SARS-CoV-2 can productively infect nasal, bronchial, and alveolar epithelial cells, while infection of macrophages and dendritic cells is abortive. This cell-specific differential infection (productive vs. abortive) is the same as noted with SARS-CoV [77]. Since nearly peak viral titers are incurred prior to discernible cytopathic effects, epithelial cell death (primarily, apoptosis) is not considered to be a direct effect of the virus; rather, it is attributed to the host immune response [10,70,73,78]. The net effect of SARS-CoV-2 induced alveolar damage (epithelial and endothelial) is a breakdown of the air-blood barrier, thereby limiting oxygen exchange and eventually culminating in respiratory failure.

2.2. Intrapulmonary Tropism: Impact of Vitamin D

Apart from being a receptor for SARS-CoV-2, ACE2 has a homeostatic function in the lungs by regulating the local renin–angiotensin system (RAS) [86,87]. In brief, renin-derived angiotensin I (AngI) is converted to angiotensin II (Ang II) by ACE. Ang II interacts with its receptor (AT1R) which triggers downstream pathways that are detrimental to lung function (e.g., pro-oxidant, pro-inflammatory). As a countermeasure, ACE2 nullifies the effects of Ang II by cleaving it to the heptapeptide Ang 1–7 which interacts with its cognate receptor (MasR) to exert beneficial effects (e.g., antioxidant, anti-inflammatory). Thus, an imbalance in the relative activity of the two converting enzymes that favor ACE over ACE2 promotes lung injury and vice-versa. Vitamin D is a negative regulator of the local RAS (increasing ACE2/ACE ratio) and thereby protects against acute lung injury (ALI) in rodents. For example, ALI induced by local (LPS, acid) or remote (peritonitis) challenges increases lung inflammation and injury as well as systemic hypoxia; effects attributed to an increased local RAS and ACE/ACE2 ratio [88–91]. Vitamin D/VDR signaling suppresses lung inflammation and injury by inhibiting Ang II/AT1R signaling and promoting Ang

1–7/MasR signaling [88,90]. Based on these and other documented effects of vitamin/VDR signaling on the local RAS, vitamin D has been touted as a potential therapeutic approach to treat ARDS of COVID-19 [92]. As a caveat, SARS-CoV-2 uses ACE2 to infect human lung cells [4,74,78]. Thus, it is unclear how a vitamin D-induced increase in the ACE2/ACE ratio will impact lung injury or disease progression induced by SARS-CoV-2.

At mucosal sites exposed to the external environment (e.g., gut, bronchi), antimicrobial peptides (AMPs) represent the first line of defense against pathogens [53,93]. LL37 is an AMP that can be detected in isolated lung epithelial cells and alveolar macrophages [94]. The cathelicidin gene encoding LL37 contains a vitamin D response element and can be regulated by vitamin D/VDR signaling [50]. Human bronchial epithelial cells constitutively express the requisite machinery (e.g., CYP27B1, VDR) to ensure intracrine activation of vitamin D/VDR signaling in response to the exogenous inactive vitamin D precursor, 25(OH)D. VDR-induced transcription generates LL37 in isolated human airway epithelial cells even in the absence of infection; however, viral infection has a potentiating effect [54,95–97]. The antiviral effects of LL37 include both extracellular (e.g., destruction of the viral envelope) and intracellular (e.g., inhibition of viral replication) modalities [54]. Based on its broad antiviral activity, it has been proposed that the vitamin D–LL37 axis may be effective against SARS-CoV-2 [56]. LL37 may also inhibit binding of SARS-CoV-2 to ACE2. In silico structural studies predict binding sites for LL37 on the viral S protein [98] and in a cell-free system this interaction prevents the binding of the S protein to ACE2 [99]. Of note, in a small safety and efficacy trial in COVID-19 patients, oral administration of *L. lactis*, genetically modified to produce LL37, was deemed safe and alleviated respiratory symptoms such as cough and shortness of breath [100]. However, the enrolled cohort were only mildly symptomatic and firm conclusions of therapeutic efficacy await controlled larger scale clinical trials.

COVID-19 lung histopathology is characterized by inflammatory cell infiltration and diffuse alveolar damage, with the blood–air barrier defect ultimately causing systemic hypoxia. In general, neither dietary depletion, genetic blockade, nor supplementation of vitamin D appreciably affects the inflammatory status or epithelial integrity of the unstressed lungs of rodents [88,90,101–104]. However, in models of ALI (e.g., LPS), either vitamin D or VDR deficiency exacerbates lung inflammation, barrier dysfunction, and systemic oxygenation [102,104]; meanwhile, supplementation with vitamin D is protective [88,104–106]. Although there are detractors from this paradigm [101,107], these detractors may not be anomalies. Seemingly paradoxical roles of vitamin D are most likely context-dependent (e.g., species, models, cell types) [54,55,96].

A context-dependent vitamin D/VDR signaling is also operative in the immune sentinel cells of the lung, such as alveolar macrophages and dendritic cells. A common cell-based model employs either bone marrow- or monocyte-derived macrophages and dendritic cells. Ex vivo induction of human macrophage differentiation in the presence of vitamin D does not appear to affect their polarization to either M1 or M2 phenotypes [108,109]. Further, vitamin D/VDR signaling in differentiated macrophages is either pro- or anti-inflammatory depending on the existing infectious/inflammatory milieu [108–114]. Consensus holds that, in response to viral infection, macrophage vitamin D/VDR signaling initially activates pro-inflammatory pathways (e.g., increased LL37, IL-8), while a more delayed anti-inflammatory response (e.g., decreased IL-8, increased IL-10) serves to limit immune-mediated injury [54,55,112,114]. With respect to human dendritic cells, supplementation with vitamin D during or after differentiation renders them tolerogenic [53,55,115,116]. Tolerogenic dendritic cells generate an anti-inflammatory milieu by secreting less pro-inflammatory cytokines, inhibiting effector T cell function (both CD4+ and CD8+ T cells), and promoting regulatory T cell conversion [55,115,117]. The vitamin D-induced tolerogenic response is delayed, presumably due to a delay in upregulation of CYP27B1 and VDR expression in both dendritic and effector T cells. It has been proposed that this delay allows for the clearance of invading microbes and subsequently quiets the immune response to avoid collateral tissue damage [55].

2.3. The Inflammatory Response

Transcriptomic [118] and proteomic [33,40,85,119,120] analyses of bronchoalveolar lavage fluid (BALF) of COVID-19 patients indicate that a pro-inflammatory environment is present in their lungs. Their BALF contains high levels of chemokines and cytokines, with the former detected earlier in longitudinal sampling [85]. Correspondingly, the BALF was enriched with innate immune cells such as neutrophils, monocytes, and to a lesser extent, dendritic cells [33,40,85,119]. The neutrophils and macrophages exhibited an activated phenotype in comparison to their circulating counterparts. The generation of a pro-inflammatory milieu is most likely initiated by either infected epithelial cells or resident macrophages [18,19,119]. These cells detect specific molecular features of inhaled virions (e.g., RNAs, proteins) referred to as pathogen-associated molecular patterns (PAMPs). Different PAMPs are recognized by an array of pattern recognition receptors (PRRs) that activate various signaling pathways, most of which converge to activate the transcription factor, NFκB [26,27,78]. NFκB transactivates various pro-inflammatory genes, generating chemokines (e.g., IL-8) and cytokines (e.g., IL-6, TNFα) [26,27,78,121]. As the infection progresses, leading to tissue injury [15], PRRs on macrophages recognize material released by damaged cells (damage-associated molecular patterns; DAMPs) and mount signaling cascades that also activate NFκB and thereby amplify the inflammation. The PRR/NFκB pathway has been proposed as a potential therapeutic target for COVID-19 [122,123].

A major function of NFκB is to initiate the assembly and activation of the NLRP3 inflammasome; a multiprotein complex that generates IL-1β. This cytokine lacks a signal sequence, so secretion to extracellular space occurs through pores in the plasma membrane formed by gasdermin D (GSDMD) [124,125]. Multiple GSDMDs are inserted into the plasma membrane and oligomerize to form pores [126], thereby allowing IL-1β to leave the cell. Excessive GSDMD pores can rupture the plasma membrane and induce a lytic form of cell death, termed "pyroptosis" [124,125].

Emerging evidence indicates that an NLRP3 inflammasome is formed in COVID-19 patients and may predict the disease trajectory. Human monocytes infected by SARS-CoV-2 secrete IL-1β and undergo pyroptosis [127,128], indicating that the virus can induce a functional inflammasome. Sera of COVID-19 patients contain active caspase; higher levels are prevalent in more severe cases [127]. Furthermore, lung tissues of fatal cases contain the fully assembled NLRP3 inflammasome [127,129]. The inflammasome components are localized to the resident or recruited monocytes/macrophages and, to a lesser extent, alveolar epithelial cells. It has been proposed that the enhanced lethality of COVID-19 in older patients is a result of age-related hyperactivation of the NLRP3 inflammasome [130].

2.4. The Inflammatory Response: Impact of Vitamin D

A major family of PRRs are the toll-like receptors (TLRs), membrane spanning glycoproteins that can detect viral PAMPs [131] and host DAMPs [122]. SARS-CoV-2, like other coronaviruses, are most likely sensed via their nucleic acids (e.g., ssRNA, dsRNA) by endosomal PRRs (e.g., TLR3 and TLR7). In addition, viral membrane proteins as well as various DAMPs from injured cells can be detected by plasma membrane PRRs (e.g., TLR4). Agonists of TLR3 and TLR4 increase cytokine (e.g., TNFα, IL-1β) production by human lung macrophages, with TLR4 agonists being the most potent [132]. In human monocytes, vitamin D/VDR signaling reduces surface levels of TLR4, while not affecting intracellular TLR3 [133,134]. Thus, while the vitamin D/VDR axis may not impact viral-mediated TLR signaling, it may downregulate DAMP-mediated TLR pathways.

In quiescent cells, NFκB is inhibited by IκB which binds to the NFκB dimer and prevents its translocation to the nucleus [121]. Pro-inflammatory stimuli activate IκB kinase which phosphorylates IκB. Subsequent ubiquitination targets IκB for proteasomal degradation. The loss of IκB frees the NFκB dimer to enter the nucleus and transcribe relevant pro-inflammatory genes. Several lines of evidence indicate that the nuclear translocation of NFκB is impeded by vitamin D/VDR signaling. In VDR deficient fibroblasts, there is a reduction in basal levels of IκB and an increase in nuclear levels of NFκB [135]. Exogenous vitamin D increases IκB and decreases NFκB translocation to the nucleus of human lung epithelial cells or murine macrophages [136,137]. In a similar vein, vitamin D or VDR overexpression inhibits IκB kinase activity in fibroblasts; an effect mediated by the physical interaction of the VDR with the kinase [138]. Finally, VDR can also physically interact with NFκB, as demonstrated in murine tissues [139,140] and macrophages [141]. However, the precise docking sites involved in VDR interactions with IκB kinase and NFκB have not been identified.

Vitamin D/VDR signaling inhibits tissue inflammation and injury mediated by the NLRP3 inflammasome in various in vivo murine models [142–144]. Loss and gain of function approaches support a role for the vitamin D/VDR pathway to dampen activation and the function of the NLRP3 inflammasome. For example, VDR inhibits caspase activation, generation of mature IL-1β, and GSDMD-mediated pyroptosis in a murine model of kidney injury, as well as human tubular epithelial cells [143,144]. VDR can physically interact with NLRP3 [144,145]; the ligand-binding domain of VDR and the amino-terminal pyrin domain of NLRP3 are required for complex formation [144]. The VDR-NLRP3 interaction prevents the inflammasome function in both murine and human macrophages [144].

An increase in intracellular oxidant stress has been implicated in the activation of the NLRP3 inflammasome [146–148]. Oxidant stress occurs when the generation of ROS exceeds the antioxidant capacity of the cell. An important transcription factor that enhances the antioxidant status of cells is Nrf2 [149]. The promoter region of the Nrf2 gene contains a response element that binds VDR [150]. In human epithelial cells, vitamin D/VDR signaling blunts ROS-mediated activation of the NLRP3 inflammasome by promoting Nrf2 translocation to the nucleus where its transcriptional activity increases cellular antioxidant enzymes [151].

2.5. Immunothrombosis and Remote Organ Dysfunction

Organs remote from the initial site of SARS-CoV-2 infection can exhibit pathology, particularly in severe cases [2,10,14,15,17,24,83,152]. It has been proposed that the excessive inflammatory response within the lungs results in the spill-over of cytokines into the systemic circulation causing a "cytokine storm" syndrome [12,26,29,153,154]. However, a more likely scenario is the generation of a localized cytokine storm within the lungs of COVID-19 patients [120,155].

The subsequent recruitment and hyper-activation of neutrophils results in the generation of neutrophil extracellular traps (NETs) [156,157]. NET components have been detected in tracheal aspirates [43,158] of COVID patients. Lung tissue from fatal cases contains NETs in close association with diffuse alveolar damage [43,44,158]. Importantly, NETs decorated with platelets as well as occlusive thrombi have been noted within the lung microvasculature [43–45]. These observations are in accordance with immunothrombosis, a process linking innate immunity to thrombosis for defense against pathogens [23,49,159]. However, excessive immunothrombosis leads to occlusion of numerous pulmonary blood vessels and precipitates ARDS.

Whether the immunothrombosis of COVID-19 is confined to the lungs or can impact remote organs is still controversial. Activated neutrophils and platelets, as well as platelet–neutrophil aggregates, are present in the systemic circulation of patients [45,160]. Further, sera from COVID-19 patients can induce NET formation in neutrophils isolated from healthy donors [161]. Circulating markers of fibrin degradation (e.g., D-dimers), and NET remnants are elevated in COVID-19, with higher levels in more severe or fatal cases [45,161]. However, while NET formation has been consistently noted in the lungs of COVID-19, their presence in remote organs has either been noted [45] or not detected [44]. Nonetheless, in fatal cases of COVID-19, microvascular thrombi as well as ischemic infarcts are present in multiple organs [10,14,17,24,152].

2.6. Immunothrombosis and Remote Organ Dysfunction: Impact of Vitamin D

While clinical studies indicate an inverse relationship between vitamin D status and thrombotic events [162], the impact of VDR signaling on specific steps involved in the development of immunothrombosis is less clear. For example, the effects of vitamin D/VDR signaling on the generation of NETs is ambiguous [163,164]. Platelet activation is increased in blood samples from vitamin D deficient individuals [165], while vitamin D inhibits platelet aggregation in vitro [166]. The antithrombin gene has multiple vitamin D response elements, and paricalcitol increases antithrombin expression in, and secretion from, cultured cells [167]. Further, a transcriptomic analysis of data derived from human monocytes identified the thrombomodulin gene as a target of vitamin D/VDR signaling [168]. Collectively, these observations predict an antithrombotic function of vitamin D/VDR signaling. Unexpectedly, however, correcting vitamin D deficiency in otherwise healthy individuals does not consistently affect their blood thrombogenic profile. Vitamin D supplementation of deficient subjects either increases [169] or reduces [170] thrombogenicity.

Further work is warranted to systematically assess the potential benefit of vitamin D in immunothrombosis of COVID-19. This is particularly important since anticoagulants (e.g., heparinoids) are currently advocated to alleviate hypercoagulation in these patients [171] and therapeutic vitamin D may increase the probability of bleeding events [172].

3. Conclusions

The anti-inflammatory and anti-thrombotic effects of vitamin D are promising features that suggest efficacy against immunothrombosis of COVID-19. Results of ongoing clinical trials should either validate or refute a beneficial role for vitamin D in alleviating the ARDS of COVID-19 and associated sequelae.

Author Contributions: Conceptualization, H.M.A.F., P.R.K. and K.A.; methodology, H.M.A.F., P.R.K. and W.B.G.; writing—original draft preparation, H.M.A.F. and P.R.K.; writing—review and editing, P.R.K., H.M.A.F., I.S., H.S., W.B.G. and K.A.; visualization, H.M.A.F., P.R.K. and I.S. All authors have read and agreed to the published version of the manuscript.

Funding: This research received no external funding.

Institutional Review Board Statement: Not applicable.

Informed Consent Statement: Not applicable.

Data Availability Statement: Not applicable.

Conflicts of Interest: W.B.G. receives funding from Bio-Tech Pharmacal, Inc. (Fayetteville, AR, USA). The other authors have no conflict of interest to declare.

References

1. Fung, S.-Y.; Yuen, K.-S.; Ye, Z.-W.; Chan, C.-P.; Jin, D.-Y. A tug-of-war between severe acute respiratory syndrome Vitamin D and host antiviral defence: Lessons from other pathogenic viruses. *Emerg. Microbes Infect.* **2020**, *9*, 558–570. [CrossRef]
2. Tang, D.; Comish, P.; Kang, R. The hallmarks of COVID-19 disease. *PLoS Pathog.* **2020**, *16*, e1008536. [CrossRef]
3. Petrosillo, N.; Viceconte, G.; Ergonul, O.; Ippolito, G.; Petersen, E. COVID-19, SARS and MERS: Are they closely related? *Clin. Microbiol. Infect.* **2020**, *26*, 729–734. [CrossRef]
4. Hoffmann, M.; Kleine-Weber, H.; Schroeder, S.; Krüger, N.; Herrler, T.; Erichsen, S.; Schiergens, T.S.; Herrler, G.; Wu, N.-H.; Nitsche, A. SARS-CoV-2 cell entry depends on ACE2 and TMPRSS2 and is blocked by a clinically proven protease inhibitor. *Cell* **2020**, *181*, 271–280.e8. [CrossRef] [PubMed]
5. Lamers, M.M.; Beumer, J.; van der Vaart, J.; Knoops, K.; Puschhof, J.; Breugem, T.I.; Ravelli, R.B.G.; Paul van Schayck, J.; Mykytyn, A.Z.; Duimel, H.Q.; et al. SARS-CoV-2 productively infects human gut enterocytes. *Science* **2020**, *369*, 50–54. [CrossRef]
6. Ziegler, C.G.; Allon, S.J.; Nyquist, S.K.; Mbano, I.M.; Miao, V.N.; Tzouanas, C.N.; Cao, Y.; Yousif, A.S.; Bals, J.; Hauser, B.M. SARS-CoV-2 receptor ACE2 is an interferon-stimulated gene in human airway epithelial cells and is detected in specific cell subsets across tissues. *Cell* **2020**, *181*, 1016–1035.e19. [CrossRef] [PubMed]
7. Ding, S.; Liang, T.J. Is SARS-CoV-2 also an enteric pathogen with potential fecal-oral transmission? A COVID-19 virological and clinical review. *Gastroenterology* **2020**, *159*, 53–61. [CrossRef]
8. Xu, Z.; Shi, L.; Wang, Y.; Zhang, J.; Huang, L.; Zhang, C.; Liu, S.; Zhao, P.; Liu, H.; Zhu, L.; et al. Pathological findings of COVID-19 associated with acute respiratory distress syndrome. *Lancet Respir. Med.* **2020**, *8*, 420–422. [CrossRef]

9. Klopfenstein, T.; Kadiane-Oussou, N.J.; Royer, P.Y.; Toko, L.; Gendrin, V.; Zayet, S. Diarrhea: An underestimated symptom in Coronavirus disease 2019. *Clin. Res. Hepatol. Gastroenterol.* **2020**, *44*, 282–283. [CrossRef] [PubMed]

10. Ackermann, M.; Verleden, S.E.; Kuehnel, M.; Haverich, A.; Welte, T.; Laenger, F.; Vanstapel, A.; Werlein, C.; Stark, H.; Tzankov, A.; et al. Pulmonary vascular endothelialitis, thrombosis, and angiogenesis in Covid-19. *N. Engl. J. Med.* **2020**, *383*, 120–128. [CrossRef]

11. Li, X.; Ma, X. Acute respiratory failure in COVID-19: Is it "typical" ARDS? *Crit. Care* **2020**, *24*, 1–5. [CrossRef]

12. Tay, M.Z.; Poh, C.M.; Renia, L.; MacAry, P.A.; Ng, L.F.P. The trinity of COVID-19: Immunity, inflammation and intervention. *Nat. Rev. Immunol.* **2020**, *20*, 363–374. [CrossRef]

13. Phua, J.; Weng, L.; Ling, L.; Egi, M.; Lim, C.M.; Divatia, J.V.; Shrestha, B.R.; Arabi, Y.M.; Ng, J.; Gomersall, C.D.; et al. Intensive care management of coronavirus disease 2019 (COVID-19): Challenges and recommendations. *Lancet Respir. Med.* **2020**, *8*, 506–517. [CrossRef]

14. Deshmukh, V.; Motwani, R.; Kumar, A.; Kumari, C.; Raza, K. Histopathological observations in COVID-19: A systematic review. *J. Clin. Pathol.* **2021**, *74*, 76–83. [CrossRef] [PubMed]

15. Liu, Q.; Shi, Y.; Cai, J.; Duan, Y.; Wang, R.; Zhang, H.; Ruan, Q.; Li, J.; Zhao, L.; Ping, Y. Pathological changes in the lungs and lymphatic organs of 12 COVID-19 autopsy cases. *Natl. Sci. Rev.* **2020**, *7*, 1868–1878. [CrossRef]

16. Lax, S.F.; Skok, K.; Zechner, P.; Kessler, H.H.; Kaufmann, N.; Koelblinger, C.; Vander, K.; Bargfrieder, U.; Trauner, M. Pulmonary arterial thrombosis in COVID-19 with fatal outcome: Results from a prospective, single-center, clinicopathologic case series. *Ann. Intern. Med.* **2020**, *173*, 350–361. [CrossRef] [PubMed]

17. Fox, S.E.; Akmatbekov, A.; Harbert, J.L.; Li, G.; Quincy Brown, J.; Vander Heide, R.S. Pulmonary and cardiac pathology in African American patients with COVID-19: An autopsy series from New Orleans. *Lancet Respir. Med.* **2020**, *8*, 681–686. [CrossRef]

18. Carsana, L.; Sonzogni, A.; Nasr, A.; Rossi, R.S.; Pellegrinelli, A.; Zerbi, P.; Rech, R.; Colombo, R.; Antinori, S.; Corbellino, M.; et al. Pulmonary post-mortem findings in a series of COVID-19 cases from northern Italy: A two-centre descriptive study. *Lancet Infect. Dis.* **2020**, *20*, 1135–1140. [CrossRef]

19. Schaefer, I.M.; Padera, R.F.; Solomon, I.H.; Kanjilal, S.; Hammer, M.M.; Hornick, J.L.; Sholl, L.M. In Situ detection of SARS-CoV-2 in lungs and airways of patients with COVID-19. *Mod. Pathol.* **2020**, *33*, 2104–2114. [CrossRef] [PubMed]

20. Milewska, A.; Kula-Pacurar, A.; Wadas, J.; Suder, A.; Szczepanski, A.; Dabrowska, A.; Owczarek, K.; Marcello, A.; Ochman, M.; Stacel, T.; et al. Replication of severe acute respiratory syndrome coronavirus 2 in human respiratory epithelium. *J. Virol.* **2020**, *94*, e00957-20. [CrossRef]

21. Klok, F.; Kruip, M.; Van der Meer, N.; Arbous, M.; Gommers, D.; Kant, K.; Kaptein, F.; van Paassen, J.; Stals, M.; Huisman, M. Incidence of thrombotic complications in critically ill ICU patients with COVID-19. *Thromb. Res.* **2020**, *191*, 145–147. [CrossRef] [PubMed]

22. Poissy, J.; Goutay, J.; Caplan, M.; Parmentier, E.; Duburcq, T.; Lassalle, F.; Jeanpierre, E.; Rauch, A.; Labreuche, J.; Susen, S.; et al. Pulmonary embolism in patients with COVID-19: Awareness of an increased prevalence. *Circulation* **2020**, *142*, 184–186. [CrossRef] [PubMed]

23. Iba, T.; Levy, J.H.; Levi, M.; Connors, J.M.; Thachil, J. Coagulopathy of coronavirus disease 2019. *Crit. Care Med.* **2020**, *48*, 1358–1364. [CrossRef]

24. Rapkiewicz, A.V.; Mai, X.; Carsons, S.E.; Pittaluga, S.; Kleiner, D.E.; Berger, J.S.; Thomas, S.; Adler, N.M.; Charytan, D.M.; Gasmi, B. Megakaryocytes and platelet-fibrin thrombi characterize multi-organ thrombosis at autopsy in COVID-19: A case series. *EClinicalMedicine* **2020**, *24*, 100434. [CrossRef]

25. Song, E.; Zhang, C.; Israelow, B.; Lu-Culligan, A.; Prado, A.V.; Skriabine, S.; Lu, P.; Weizman, O.E.; Liu, F.; Dai, Y.; et al. Neuroinvasion of SARS-CoV-2 in human and mouse brain. *J. Exp. Med.* **2021**, *218*, e20202135. [CrossRef] [PubMed]

26. Merad, M.; Martin, J.C. Pathological inflammation in patients with COVID-19: A key role for monocytes and macrophages. *Nat. Rev. Immunol.* **2020**, *20*, 355–362. [CrossRef] [PubMed]

27. Lee, S.; Channappanavar, R.; Kanneganti, T.D. Coronaviruses: Innate immunity, inflammasome activation, inflammatory cell death, and cytokines. *Trends Immunol.* **2020**, *41*, 1083–1099. [CrossRef] [PubMed]

28. Lucas, C.; Wong, P.; Klein, J.; Castro, T.B.R.; Silva, J.; Sundaram, M.; Ellingson, M.K.; Mao, T.; Oh, J.E.; Israelow, B.; et al. Longitudinal analyses reveal immunological misfiring in severe COVID-19. *Nature* **2020**, *584*, 463–469. [CrossRef]

29. Fajgenbaum, D.C.; June, C.H. Cytokine Storm. *N. Engl. J. Med.* **2020**, *383*, 2255–2273. [CrossRef]

30. Chen, Z.; John Wherry, E. T cell responses in patients with COVID-19. *Nat. Rev. Immunol.* **2020**, *20*, 529–536. [CrossRef]

31. Mudd, P.A.; Crawford, J.C.; Turner, J.S.; Souquette, A.; Reynolds, D.; Bender, D.; Bosanquet, J.P.; Anand, N.J.; Striker, D.A.; Martin, R.S.; et al. Distinct inflammatory profiles distinguish COVID-19 from influenza with limited contributions from cytokine storm. *Sci. Adv.* **2020**, *6*, eabe3024. [CrossRef]

32. Blanco-Melo, D.; Nilsson-Payant, B.E.; Liu, W.C.; Uhl, S.; Hoagland, D.; Moller, R.; Jordan, T.X.; Oishi, K.; Panis, M.; Sachs, D.; et al. Imbalanced host response to SARS-CoV-2 drives development of COVID-19. *Cell* **2020**, *181*, 1036–1045.e9. [CrossRef] [PubMed]

33. Ronit, A.; Berg, R.M.G.; Bay, J.T.; Haugaard, A.K.; Ahlstrom, M.G.; Burgdorf, K.S.; Ullum, H.; Rorvig, S.B.; Tjelle, K.; Foss, N.B.; et al. Compartmental immunophenotyping in COVID-19 ARDS: A case series. *J. Allergy Clin. Immunol.* **2021**, *147*, 81–91. [CrossRef] [PubMed]

34. Hadjadj, J.; Yatim, N.; Barnabei, L.; Corneau, A.; Boussier, J.; Smith, N.; Péré, H.; Charbit, B.; Bondet, V.; Chenevier-Gobeaux, C.; et al. Impaired type I interferon activity and inflammatory responses in severe COVID-19 patients. *Science* **2020**, *369*, 718–724. [CrossRef]

35. Diao, B.; Wang, C.; Tan, Y.; Chen, X.; Liu, Y.; Ning, L.; Chen, L.; Li, M.; Liu, Y.; Wang, G.; et al. Reduction and functional exhaustion of T cells in patients with coronavirus disease 2019 (COVID-19). *Front. Immunol.* **2020**, *11*, 827. [CrossRef]

36. Giamarellos-Bourboulis, E.J.; Netea, M.G.; Rovina, N.; Akinosoglou, K.; Antoniadou, A.; Antonakos, N.; Damoraki, G.; Gkavogianni, T.; Adami, M.E.; Katsaounou, P.; et al. Complex immune dysregulation in Covid-19 patients with severe respiratory failure. *Cell Host Microbe* **2020**, *27*, 992–1000.e3. [CrossRef] [PubMed]

37. Zhou, R.; To, K.K.; Wong, Y.C.; Liu, L.; Zhou, B.; Li, X.; Huang, H.; Mo, Y.; Luk, T.Y.; Lau, T.T.; et al. Acute SARS-CoV-2 infection impairs dendritic cell and T cell responses. *Immunity* **2020**, *53*, 864–877.e5. [CrossRef] [PubMed]

38. Zheng, H.-Y.; Zhang, M.; Yang, C.-X.; Zhang, N.; Wang, X.-C.; Yang, X.-P.; Dong, X.-Q.; Zheng, Y.-T. Elevated exhaustion levels and reduced functional diversity of T cells in peripheral blood may predict severe progression in COVID-19 patients. *Cell. Mol. Immunol.* **2020**, *17*, 541–543. [CrossRef] [PubMed]

39. Zhao, Q.; Meng, M.; Kumar, R.; Wu, Y.; Huang, J.; Deng, Y.; Weng, Z.; Yang, L. Lymphopenia is associated with severe coronavirus disease 2019 (COVID-19) infections: A systemic review and meta-analysis. *Int. J. Infect. Dis.* **2020**, *96*, 131–135. [CrossRef] [PubMed]

40. De Biasi, S.; Meschiari, M.; Gibellini, L.; Bellinazzi, C.; Borella, R.; Fidanza, L.; Gozzi, L.; Iannone, A.; Lo Tartaro, D.; Mattioli, M.; et al. Marked T cell activation, senescence, exhaustion and skewing towards TH17 in patients with COVID-19 pneumonia. *Nat. Commun.* **2020**, *11*, 3434. [CrossRef] [PubMed]

41. Zheng, M.; Gao, Y.; Wang, G.; Song, G.; Liu, S.; Sun, D.; Xu, Y.; Tian, Z. Functional exhaustion of antiviral lymphocytes in COVID-19 patients. *Cell Mol. Immunol.* **2020**, *17*, 533–535. [CrossRef]

42. Manjili, R.H.; Zarei, M.; Habibi, M.; Manjili, M.H. COVID-19 as an acute inflammatory disease. *J. Immunol.* **2020**, *205*, 12–19. [CrossRef]

43. Middleton, E.A.; He, X.Y.; Denorme, F.; Campbell, R.A.; Ng, D.; Salvatore, S.P.; Mostyka, M.; Baxter-Stoltzfus, A.; Borczuk, A.C.; Loda, M.; et al. Neutrophil extracellular traps contribute to immunothrombosis in COVID-19 acute respiratory distress syndrome. *Blood* **2020**, *136*, 1169–1179. [CrossRef]

44. Radermecker, C.; Detrembleur, N.; Guiot, J.; Cavalier, E.; Henket, M.; d'Emal, C.; Vanwinge, C.; Cataldo, D.; Oury, C.; Delvenne, P.; et al. Neutrophil extracellular traps infiltrate the lung airway, interstitial, and vascular compartments in severe COVID-19. *J. Exp. Med.* **2020**, *217*, e20201012. [CrossRef]

45. Leppkes, M.; Knopf, J.; Naschberger, E.; Lindemann, A.; Singh, J.; Herrmann, I.; Sturzl, M.; Staats, L.; Mahajan, A.; Schauer, C.; et al. Vascular occlusion by neutrophil extracellular traps in COVID-19. *EBioMedicine* **2020**, *58*, 102925. [CrossRef]

46. McDonald, B.; Davis, R.P.; Kim, S.J.; Tse, M.; Esmon, C.T.; Kolaczkowska, E.; Jenne, C.N. Platelets and neutrophil extracellular traps collaborate to promote intravascular coagulation during sepsis in mice. *Blood* **2017**, *129*, 1357–1367. [CrossRef]

47. Yadav, V.; Chi, L.; Zhao, R.; Tourdot, B.E.; Yalavarthi, S.; Jacobs, B.N.; Banka, A.; Liao, H.; Koonse, S.; Anyanwu, A.C. ENTPD-1 disrupts inflammasome IL-1β–driven venous thrombosis. *J. Clin. Investig.* **2019**, *129*, 2872–2877. [CrossRef] [PubMed]

48. Gupta, N.; Sahu, A.; Prabhakar, A.; Chatterjee, T.; Tyagi, T.; Kumari, B.; Khan, N.; Nair, V.; Bajaj, N.; Sharma, M.; et al. Activation of NLRP3 inflammasome complex potentiates venous thrombosis in response to hypoxia. *Proc. Natl. Acad. Sci. USA* **2017**, *114*, 4763–4768. [CrossRef] [PubMed]

49. Gaertner, F.; Massberg, S. Blood coagulation in immunothrombosis—At the frontline of intravascular immunity. *Semin. Immunol.* **2016**, *28*, 561–569. [CrossRef]

50. Bishop, E.L.; Ismailova, A.; Dimeloe, S.; Hewison, M.; White, J.H. Vitamin D and immune regulation: Antibacterial, antiviral, anti-inflammatory. *JBMR Plus* **2021**, *5*, e10405. [CrossRef] [PubMed]

51. Weiss, S.T.; Litonjua, A.A. Vitamin D in host defense: Implications for future research. *Am. J. Respir. Cell Mol. Biol.* **2017**, *56*, 692–693. [CrossRef]

52. Lee, C. Controversial effects of Vitamin D and related genes on viral infections, pathogenesis, and treatment outcomes. *Nutrients* **2020**, *12*, 962. [CrossRef] [PubMed]

53. Fakhoury, H.M.A.; Kvietys, P.R.; AlKattan, W.; Anouti, F.A.; Elahi, M.A.; Karras, S.N.; Grant, W.B. Vitamin D and intestinal homeostasis: Barrier, microbiota, and immune modulation. *J. Steroid Biochem. Mol. Biol.* **2020**, *200*, 105663. [CrossRef] [PubMed]

54. Teymoori-Rad, M.; Shokri, F.; Salimi, V.; Marashi, S.M. The interplay between Vitamin D and viral infections. *Rev. Med. Virol.* **2019**, *29*, e2032. [CrossRef]

55. Cantorna, M.T.; Rogers, C.J.; Arora, J. Aligning the paradoxical role of Vitamin D in gastrointestinal immunity. *Trends Endocrinol. Metab. TEM* **2019**, *30*, 459–466. [CrossRef] [PubMed]

56. Crane-Godreau, M.A.; Clem, K.J.; Payne, P.; Fiering, S. Vitamin D deficiency and air pollution exacerbate COVID-19 through suppression of antiviral peptide LL37. *Front. Public Health* **2020**, *8*, 232. [CrossRef] [PubMed]

57. Grant, W.B.; Lahore, H.; McDonnell, S.L.; Baggerly, C.A.; French, C.B.; Aliano, J.L.; Bhattoa, H.P. Evidence that Vitamin D supplementation could reduce risk of influenza and COVID-19 infections and deaths. *Nutrients* **2020**, *12*, 988. [CrossRef]

58. Panagiotou, G.; Tee, S.A.; Ihsan, Y.; Athar, W.; Marchitelli, G.; Kelly, D.; Boot, C.S.; Stock, N.; Macfarlane, J.; Martineau, A.R. Low serum 25-Hydroxyvitamin D (25 [OH] D) levels in patients hospitalized with COVID-19 are associated with greater disease severity. *Clin. Endocrinol.* **2020**, *93*, 508–511. [CrossRef]

59. Carpagnano, G.E.; Di Lecce, V.; Quaranta, V.N.; Zito, A.; Buonamico, E.; Capozza, E.; Palumbo, A.; Di Gioia, G.; Valerio, V.N.; Resta, O. Vitamin D deficiency as a predictor of poor prognosis in patients with acute respiratory failure due to COVID-19. *J. Endocrinol. Investig.* **2020**, *44*, 765–771. [CrossRef] [PubMed]
60. Ferder, L.; Martín Giménez, V.M.; Inserra, F.; Tajer, C.; Antonietti, L.; Mariani, J.; Manucha, W. Vitamin D supplementation as a rational pharmacological approach in the Covid-19 pandemic. *Am. J. Physiol. Lung Cell. Mol. Physiol.* **2020**, *319*, L941–L948. [CrossRef] [PubMed]
61. Griffin, G.; Hewison, M.; Hopkin, J.; Kenny, R.A.; Quinton, R.; Rhodes, J.; Subramanian, S.; Thickett, D. Preventing Vitamin D deficiency during the COVID-19 pandemic: UK definitions of Vitamin D sufficiency and recommended supplement dose are set too low. *Clin. Med.* **2021**, *21*, e48–e51. [CrossRef]
62. Akbar, M.R.; Wibowo, A.; Pranata, R.; Setiabudiawan, B. Low serum 25-hydroxyvitamin D (Vitamin D) level is associated with susceptibility to COVID-19, severity, and mortality: A systematic review and meta-analysis. *Front. Nutr.* **2021**, *8*, 660420. [CrossRef]
63. Alcala-Diaz, J.F.; Limia-Perez, L.; Gomez-Huelgas, R. Calcifediol treatment and hospital mortality due to COVID-19: A cohort study. *Nutrients* **2021**, *13*, 1760. [CrossRef]
64. Rhodes, J.M.; Subramanian, S.; Laird, E.; Griffin, G.; Kenny, R.A. Perspective: Vitamin D deficiency and COVID-19 severity— plausibly linked by latitude, ethnicity, impacts on cytokines, ACE2 and thrombosis. *J. Intern. Med.* **2021**, *289*, 97–115. [CrossRef]
65. Bilezikian, J.P.; Bikle, D.; Hewison, M.; Lazaretti-Castro, M.; Formenti, A.M.; Gupta, A.; Madhavan, M.V.; Nair, N.; Babalyan, V.; Hutchings, N. Mechanisms in endocrinology: Vitamin D and COVID-19. *Eur. J. Endocrinol.* **2020**, *183*, R133–R147. [CrossRef] [PubMed]
66. Vimaleswaran, K.S.; Forouhi, N.G.; Khunti, K. Vitamin D and covid-19. *BMJ* **2021**, *372*, n544. [CrossRef] [PubMed]
67. Mercola, J.; Grant, W.B. Evidence regarding Vitamin D and risk of COVID-19 and its severity. *Nutrients* **2020**, *12*, 3361. [CrossRef]
68. Winkler, E.S.; Bailey, A.L.; Kafai, N.M.; Nair, S.; McCune, B.T.; Yu, J. SARS-CoV-2 infection of human ACE2-transgenic mice causes severe lung inflammation and impaired function. *Nat. Immunol.* **2020**, *21*, 1327–1335. [CrossRef] [PubMed]
69. Yinda, C.K.; Port, J.R. K18-hACE2 mice develop respiratory disease resembling severe COVID-19. *PLoS Pathog.* **2021**, *17*, e1009195. [CrossRef] [PubMed]
70. Chu, H.; Chan, J.F.; Yuen, T.T.; Shuai, H.; Yuan, S.; Wang, Y.; Hu, B.; Yip, C.C.; Tsang, J.O.; Huang, X.; et al. Comparative tropism, replication kinetics, and cell damage profiling of SARS-CoV-2 and SARS-CoV with implications for clinical manifestations, transmissibility, and laboratory studies of COVID-19: An observational study. *Lancet Microbe* **2020**, *1*, e14–e23. [CrossRef]
71. Zhu, N.; Wang, W.; Liu, Z.; Liang, C.; Wang, W.; Ye, F.; Huang, B.; Zhao, L.; Wang, H.; Zhou, W.; et al. Morphogenesis and cytopathic effect of SARS-CoV-2 infection in human airway epithelial cells. *Nat. Commun.* **2020**, *11*, 3910. [CrossRef] [PubMed]
72. Pizzorno, A.; Padey, B.; Julien, T.; Trouillet-Assant, S.; Traversier, A.; Errazuriz-Cerda, E.; Fouret, J.; Dubois, J.; Gaymard, A.; Lescure, F.X.; et al. Characterization and treatment of SARS-CoV-2 in nasal and bronchial human airway epithelia. *Cell Rep. Med.* **2020**, *1*, 100059. [CrossRef] [PubMed]
73. Mulay, A.; Konda, B.; Garcia, G.; Yao, C.; Beil, S.; Sen, C.; Purkayastha, A.; Kolls, J.K.; Pociask, D.A.; Pessina, P.; et al. SARS-CoV-2 infection of primary human lung epithelium for COVID-19 modeling and drug discovery. *bioRxiv Prepr. Serv. Biol.* **2020**. [CrossRef]
74. Hoffmann, M.; Kleine-Weber, H.; Pohlmann, S. A multibasic cleavage site in the spike protein of SARS-CoV-2 is essential for infection of human lung cells. *Mol. Cell* **2020**, *78*, 779–784.e5. [CrossRef] [PubMed]
75. Ren, Y.; Shu, T.; Wu, D.; Mu, J.; Wang, C.; Huang, M.; Han, Y.; Zhang, X.Y.; Zhou, W.; Qiu, Y.; et al. The ORF3a protein of SARS-CoV-2 induces apoptosis in cells. *Cell Mol. Immunol.* **2020**, *17*, 881–883. [CrossRef]
76. Orzalli, M.H.; Kagan, J.C. Apoptosis and necroptosis as host defense strategies to prevent viral infection. *Trends Cell Biol.* **2017**, *27*, 800–809. [CrossRef] [PubMed]
77. Abassi, Z.; Knaney, Y.; Karram, T.; Heyman, S.N. The lung macrophage in SARS-CoV-2 infection: A friend or a foe? *Front. Immunol.* **2020**, *11*, 1312. [CrossRef]
78. Huang, J.; Hume, A.J.; Abo, K.M.; Werder, R.B.; Villacorta-Martin, C.; Alysandratos, K.-D.; Beermann, M.L.; Simone-Roach, C.; Lindstrom-Vautrin, J.; Olejnik, J. SARS-CoV-2 infection of pluripotent stem cell-derived human lung alveolar type 2 cells elicits a rapid epithelial-intrinsic inflammatory response. *Cell Stem Cell* **2020**, *27*, 962–973.e7. [CrossRef]
79. Olajuyin, A.M.; Zhang, X.; Ji, H.L. Alveolar type 2 progenitor cells for lung injury repair. *Cell Death Discov.* **2019**, *5*, 63. [CrossRef]
80. Monteil, V.; Kwon, H.; Prado, P.; Hagelkruys, A.; Wimmer, R.A.; Stahl, M.; Leopoldi, A.; Garreta, E.; Hurtado Del Pozo, C.; Prosper, F.; et al. Inhibition of SARS-CoV-2 infections in engineered human tissues using clinical-grade soluble human ACE2. *Cell* **2020**, *181*, 905–913.e7. [CrossRef]
81. Yang, D.; Chu, H.; Hou, Y.; Chai, Y.; Shuai, H.; Lee, A.C.-Y.; Zhang, X.; Wang, Y.; Hu, B.; Huang, X. Attenuated interferon and proinflammatory response in SARS-CoV-2–infected human dendritic cells is associated with viral antagonism of STAT1 phosphorylation. *J. Infect. Dis.* **2020**, *222*, 734–745. [CrossRef] [PubMed]
82. Boumaza, A.; Gay, L.; Mezouar, S.; Diallo, A.B.; Michel, M.; Desnues, B.; Raoult, D.; La Scola, B.; Halfon, P.; Vitte, J. Monocytes and macrophages, targets of SARS-CoV-2: The clue for Covid-19 immunoparalysis. *bioRxiv Prepr. Serv. Biol.* **2020**. [CrossRef]
83. Wang, C.; Xie, J.; Zhao, L.; Fei, X.; Zhang, H.; Tan, Y.; Nie, X.; Zhou, L.; Liu, Z.; Ren, Y.; et al. Alveolar macrophage dysfunction and cytokine storm in the pathogenesis of two severe COVID-19 patients. *EBioMedicine* **2020**, *57*, 102833. [CrossRef]

84. Evren, E.; Ringqvist, E.; Willinger, T. Origin and ontogeny of lung macrophages: From mice to humans. *Immunology* **2020**, *160*, 126–138. [CrossRef] [PubMed]

85. Liao, M.; Liu, Y.; Yuan, J.; Wen, Y.; Xu, G.; Zhao, J.; Cheng, L.; Li, J.; Wang, X.; Wang, F.; et al. Single-cell landscape of bronchoalveolar immune cells in patients with COVID-19. *Nat. Med.* **2020**, *26*, 842–844. [CrossRef]

86. Gheblawi, M.; Wang, K.; Viveiros, A.; Nguyen, Q.; Zhong, J.-C.; Turner, A.J.; Raizada, M.K.; Grant, M.B.; Oudit, G.Y. Angiotensin-converting enzyme 2: SARS-CoV-2 receptor and regulator of the renin-angiotensin system: Celebrating the 20th anniversary of the discovery of ACE2. *Circ. Res.* **2020**, *126*, 1456–1474. [CrossRef]

87. Groß, S.; Jahn, C.; Cushman, S.; Bär, C.; Thum, T. SARS-CoV-2 receptor ACE2-dependent implications on the cardiovascular system: From basic science to clinical implications. *J. Mol. Cell. Cardiol.* **2020**, *144*, 47–53. [CrossRef]

88. Xu, J.; Yang, J.; Chen, J.; Luo, Q.; Zhang, Q.; Zhang, H. Vitamin D alleviates lipopolysaccharide-induced acute lung injury via regulation of the renin-angiotensin system. *Mol. Med. Rep.* **2017**, *16*, 7432–7438. [CrossRef]

89. Li, Y.; Zeng, Z.; Cao, Y.; Liu, Y.; Ping, F.; Liang, M.; Xue, Y.; Xi, C.; Zhou, M.; Jiang, W. Angiotensin-converting enzyme 2 prevents lipopolysaccharide-induced rat acute lung injury via suppressing the ERK1/2 and NF-κB signaling pathways. *Sci. Rep.* **2016**, *6*, 27911. [CrossRef]

90. Kong, J.; Zhu, X.; Shi, Y.; Liu, T.; Chen, Y.; Bhan, I.; Zhao, Q.; Thadhani, R.; Li, Y.C. VDR attenuates acute lung injury by blocking Ang-2-Tie-2 pathway and renin-angiotensin system. *Mol. Endocrinol.* **2013**, *27*, 2116–2125. [CrossRef]

91. Imai, Y.; Kuba, K.; Rao, S.; Huan, Y.; Guo, F.; Guan, B.; Yang, P.; Sarao, R.; Wada, T.; Leong-Poi, H.; et al. Angiotensin-converting enzyme 2 protects from severe acute lung failure. *Nature* **2005**, *436*, 112–116. [CrossRef]

92. Malek Mahdavi, A. A brief review of interplay between vitamin D and angiotensin-converting enzyme 2: Implications for a potential treatment for COVID-19. *Rev. Med Virol.* **2020**, *30*, e2119. [CrossRef]

93. Geitani, R.; Moubareck, C.A.; Xu, Z.; Karam Sarkis, D.; Touqui, L. Expression and roles of antimicrobial peptides in innate defense of airway mucosa: Potential implication in cystic fibrosis. *Front. Immunol.* **2020**, *11*, 1198. [CrossRef]

94. Rivas-Santiago, B.; Hernandez-Pando, R.; Carranza, C.; Juarez, E.; Contreras, J.L.; Aguilar-Leon, D.; Torres, M.; Sada, E. Expression of cathelicidin LL-37 during Mycobacterium tuberculosis infection in human alveolar macrophages, monocytes, neutrophils, and epithelial cells. *Infect. Immun.* **2008**, *76*, 935–941. [CrossRef]

95. Brockman-Schneider, R.A.; Pickles, R.J.; Gern, J.E. Effects of Vitamin D on airway epithelial cell morphology and rhinovirus replication. *PLoS ONE* **2014**, *9*, e86755. [CrossRef]

96. Hansdottir, S.; Monick, M.M.; Hinde, S.L.; Lovan, N.; Look, D.C.; Hunninghake, G.W. Respiratory epithelial cells convert inactive vitamin D to its active form: Potential effects on host defense. *J. Immunol.* **2008**, *181*, 7090–7099. [CrossRef] [PubMed]

97. Telcian, A.G.; Zdrenghea, M.T.; Edwards, M.R.; Laza-Stanca, V.; Mallia, P.; Johnston, S.L.; Stanciu, L.A. Vitamin D increases the antiviral activity of bronchial epithelial cells in vitro. *Antivir. Res.* **2017**, *137*, 93–101. [CrossRef] [PubMed]

98. Bharat, K.; Lokhande, T.; Banerjee, K.V.; Swamy, M. An in Silico Scientific Basis for LL-37 as a Therapeutic and Vitamin D as Preventive for Covid-19. *ChemRxiv* **2020**, preprint.

99. Roth, A.; Luetke, S.; Meinberger, D.; Hermes, G.; Sengle, G.; Koch, M.; Streichert, T.; Klatt, A.R. LL-37 fights SARS-CoV-2: The Vitamin D-inducible peptide LL-37 inhibits binding of SARS-CoV-2 spike protein to its cellular receptor angiotensin converting enzyme 2 In Vitro. *bioRxiv Prepr. Serv. Biol.* **2020**. [CrossRef]

100. Zhang, H.; Zhao, Y.; Jiang, X.; Zhao, Y.; Li, Y.; Li, C.; Dong, M.; Luan, Z.; Yan, C.; Jiao, J. Preliminary evaluation of the safety and efficacy of oral human antimicrobial peptide LL-37 in the treatment of patients of COVID-19, a small-scale, single-arm, exploratory safety study. *medRxiv* **2020**. [CrossRef]

101. Niederstrasser, J.; Herr, C.; Wolf, L.; Lehr, C.M.; Beisswenger, C.; Bals, R. Vitamin D deficiency does not result in a breach of host defense in murine models of pneumonia. *Infect. Immun.* **2016**, *84*, 3097–3104. [CrossRef]

102. Dancer, R.C.; Parekh, D.; Lax, S.; D'Souza, V.; Zheng, S.; Bassford, C.R.; Park, D.; Bartis, D.G.; Mahida, R.; Turner, A.M.; et al. Vitamin D deficiency contributes directly to the acute respiratory distress syndrome (ARDS). *Thorax* **2015**, *70*, 617–624. [CrossRef] [PubMed]

103. Gorman, S.; Buckley, A.G.; Ling, K.M.; Berry, L.J.; Fear, V.S.; Stick, S.M.; Larcombe, A.N.; Kicic, A.; Hart, P.H. Vitamin D supplementation of initially Vitamin D-deficient mice diminishes lung inflammation with limited effects on pulmonary epithelial integrity. *Physiol. Rep.* **2017**, *5*, e13371. [CrossRef]

104. Shi, Y.Y.; Liu, T.J.; Fu, J.H.; Xu, W.; Wu, L.L.; Hou, A.N.; Xue, X.D. Vitamin D/VDR signaling attenuates lipopolysaccharide-induced acute lung injury by maintaining the integrity of the pulmonary epithelial barrier. *Mol. Med. Rep.* **2016**, *13*, 1186–1194. [CrossRef]

105. Zhang, M.; Dong, M.; Liu, W.; Wang, L.; Luo, Y.; Li, Z.; Jin, F. 1α, 25-dihydroxyvitamin D3 ameliorates seawater aspiration-induced acute lung injury via NF-κB and RhoA/Rho kinase pathways. *PLoS ONE* **2014**, *9*, e104507. [CrossRef] [PubMed]

106. Zheng, S.; Yang, J.; Hu, X.; Li, M.; Wang, Q.; Dancer, R.C.A.; Parekh, D.; Gao-Smith, F.; Thickett, D.R.; Jin, S. Vitamin D attenuates lung injury via stimulating epithelial repair, reducing epithelial cell apoptosis and inhibits TGF-beta induced epithelial to mesenchymal transition. *Biochem. Pharmacol.* **2020**, *177*, 113955. [CrossRef] [PubMed]

107. Chen, H.; Lu, R.; Zhang, Y.G.; Sun, J. Vitamin D receptor deletion leads to the destruction of tight and adherens junctions in lungs. *Tissue Barriers* **2018**, *6*, 1–13. [CrossRef]

108. Dionne, S.; Duchatelier, C.F.; Seidman, E.G. The influence of Vitamin D on M1 and M2 macrophages in patients with Crohn's disease. *Innate Immun.* **2017**, *23*, 557–565. [CrossRef] [PubMed]

109. Rao Muvva, J.; Parasa, V.R.; Lerm, M.; Svensson, M.; Brighenti, S. Polarization of human monocyte-derived cells with Vitamin D promotes control of mycobacterium tuberculosis infection. *Front. Immunol.* **2019**, *10*, 3157. [CrossRef] [PubMed]

110. Liang, S.; Cai, J.; Li, Y.; Yang, R. 1, 25-Dihydroxy-Vitamin D3 induces macrophage polarization to M2 by upregulating T-cell Ig-mucin-3 expression. *Mol. Med. Rep.* **2019**, *19*, 3707–3713.

111. Chen, L.; Eapen, M.S.; Zosky, G.R. Vitamin D both facilitates and attenuates the cellular response to lipopolysaccharide. *Sci. Rep.* **2017**, *7*, 45172. [CrossRef]

112. Dauletbaev, N.; Herscovitch, K.; Das, M.; Chen, H.; Bernier, J.; Matouk, E.; Bérubé, J.; Rousseau, S.; Lands, L. Down-regulation of IL-8 by high-dose vitamin D is specific to hyperinflammatory macrophages and involves mechanisms beyond up-regulation of DUSP1. *Br. J. Pharmacol.* **2015**, *172*, 4757–4771. [CrossRef]

113. Tulk, S.E.; Liao, K.C.; Muruve, D.A.; Li, Y.; Beck, P.L.; MacDonald, J.A. Vitamin D3 metabolites enhance the NLRP3-dependent secretion of IL-1β from human THP-1 monocytic cells. *J. Cell. Biochem.* **2015**, *116*, 711–720. [CrossRef]

114. Ryynänen, J.; Carlberg, C. Primary 1,25-dihydroxyvitamin D3 response of the interleukin 8 gene cluster in human monocyte- and macrophage-like cells. *PLoS ONE* **2013**, *8*, e78170. [CrossRef] [PubMed]

115. Bscheider, M.; Butcher, E.C. Vitamin D immunoregulation through dendritic cells. *Immunology* **2016**, *148*, 227–236. [CrossRef]

116. Vanherwegen, A.S.; Eelen, G.; Ferreira, G.B.; Ghesquiere, B.; Cook, D.P.; Nikolic, T.; Roep, B.; Carmeliet, P.; Telang, S.; Mathieu, C.; et al. Vitamin D controls the capacity of human dendritic cells to induce functional regulatory T cells by regulation of glucose metabolism. *J. Steroid Biochem. Mol. Biol.* **2019**, *187*, 134–145. [CrossRef] [PubMed]

117. Švajger, U.; Rozman, P.J. Synergistic effects of interferon-γ and vitamin D3 signaling in induction of ILT-3highPDL-1high tolerogenic dendritic cells. *Front. Immunol.* **2019**, *10*, 2627. [CrossRef]

118. Zhou, Z.; Ren, L.; Zhang, L.; Zhong, J.; Xiao, Y.; Jia, Z.; Guo, L.; Yang, J.; Wang, C.; Jiang, S.; et al. Heightened innate immune responses in the respiratory tract of COVID-19 patients. *Cell Host Microbe* **2020**, *27*, 883–890.e2. [CrossRef] [PubMed]

119. Pandolfi, L.; Fossali, T.; Frangipane, V.; Bozzini, S.; Morosini, M.; D'Amato, M.; Lettieri, S.; Urtis, M.; Di Toro, A.; Saracino, L.; et al. Broncho-alveolar inflammation in COVID-19 patients: A correlation with clinical outcome. *BMC Pulm. Med.* **2020**, *20*, 301. [CrossRef] [PubMed]

120. Wang, C.; Kang, K.; Gao, Y.; Ye, M.; Lan, X.; Li, X.; Zhao, M.; Yu, K. Cytokine levels in the body fluids of a patient with COVID-19 and acute respiratory distress syndrome: A case report. *Ann. Intern. Med.* **2020**, *173*, 499–501. [CrossRef]

121. Liu, T.; Zhang, L.; Joo, D.; Sun, S.-C. NF-κB signaling in inflammation. *Signal Transduct. Target. Ther.* **2017**, *2*, 17023. [CrossRef]

122. Carty, M.; Guy, C.; Bowie, A.G. Detection of viral infections by innate immunity. *Biochem. Pharmacol.* **2021**, *183*, 114316. [CrossRef]

123. Hariharan, A.; Hakeem, A.R. The role and therapeutic potential of NF-kappa-B pathway in severe COVID-19 patients. *Inflammopharmacology* **2021**, *29*, 91–100. [CrossRef]

124. Tsuchiya, K. Inflammasome-associated cell death: Pyroptosis, apoptosis, and physiological implications. *Microbiol. Immunol.* **2020**, *64*, 252–269. [CrossRef] [PubMed]

125. Broz, P.; Pelegrín, P. The gasdermins, a protein family executing cell death and inflammation. *Nat. Rev. Immunol.* **2020**, *20*, 143–157. [CrossRef] [PubMed]

126. Mulvihill, E.; Sborgi, L.; Mari, S.A.; Pfreundschuh, M.; Hiller, S. Mechanism of membrane pore formation by human gasdermin-D. *EMBO J.* **2018**, *37*, e98321. [CrossRef] [PubMed]

127. Rodrigues, T.S.; de Sa, K.S.G.; Ishimoto, A.Y.; Becerra, A.; Oliveira, S.; Almeida, L.; Goncalves, A.V.; Perucello, D.B.; Andrade, W.A.; Castro, R.; et al. Inflammasomes are activated in response to SARS-CoV-2 infection and are associated with COVID-19 severity in patients. *J. Exp. Med.* **2021**, *218*, e20201707. [CrossRef] [PubMed]

128. Ferreira, A.C.; Soares, V.C.; de Azevedo-Quintanilha, I.G.; Dias, S.; Fintelman-Rodrigues, N.; Sacramento, C.Q.; Mattos, M.; de Freitas, C.S.; Temerozo, J.R.; Teixeira, L.; et al. SARS-CoV-2 engages inflammasome and pyroptosis in human primary monocytes. *Cell Death Discov.* **2021**, *7*, 43. [CrossRef] [PubMed]

129. Toldo, S.; Bussani, R.; Nuzzi, V.; Bonaventura, A.; Mauro, A.G.; Cannata, A.; Pillappa, R.; Sinagra, G.; Nana-Sinkam, P.; Sime, P.; et al. Inflammasome formation in the lungs of patients with fatal COVID-19. *Inflamm. Res. Off. J. Eur. Histamine Res. Soc.* **2021**, *70*, 7–10. [CrossRef]

130. Lara, P.C.; Macias-Verde, D.; Burgos-Burgos, J. Age-induced NLRP3 inflammasome over-Activation increases lethality of SARS-CoV-2 pneumonia in elderly patients. *Aging Dis.* **2020**, *11*, 756–762. [CrossRef]

131. Kawasaki, T.; Kawai, T. Discrimination between self and non-self-nucleic acids by the innate immune system. *Int. Rev. Cell Mol. Biol.* **2019**, *344*, 1–30. [CrossRef]

132. Grassin-Delyle, S.; Abrial, C.; Salvator, H.; Brollo, M.; Naline, E.; Devillier, P. The role of toll-like receptors in the production of cytokines by human lung macrophages. *J. Innate Immun.* **2020**, *12*, 63–73. [CrossRef] [PubMed]

133. Sadeghi, K.; Wessner, B.; Laggner, U.; Ploder, M.; Tamandl, D.; Friedl, J.; Zügel, U.; Steinmeyer, A.; Pollak, A.; Roth, E.; et al. Vitamin D3 down-regulates monocyte TLR expression and triggers hyporesponsiveness to pathogen-associated molecular patterns. *Eur. J. Immunol.* **2006**, *36*, 361–370. [CrossRef] [PubMed]

134. Dickie, L.J.; Church, L.D.; Coulthard, L.R.; Mathews, R.J.; Emery, P.; McDermott, M.F. Vitamin D3 down-regulates intracellular Toll-like receptor 9 expression and Toll-like receptor 9-induced IL-6 production in human monocytes. *Rheumatology* **2010**, *49*, 1466–1471. [CrossRef] [PubMed]

135. Sun, J.; Kong, J.; Duan, Y.; Szeto, F.L.; Liao, A.; Madara, J.L.; Li, Y.C. Increased NF-κB activity in fibroblasts lacking the vitamin D receptor. *Am. J. Physiol. Endocrinol. Metab.* **2006**, *291*, E315–E322. [CrossRef]

136. Cohen-Lahav, M.; Shany, S.; Tobvin, D.; Chaimovitz, C.; Douvdevani, A. Vitamin D decreases NFκB activity by increasing IκBα levels. *Nephrol. Dial. Transplant.* **2006**, *21*, 889–897. [CrossRef] [PubMed]

137. Hansdottir, S.; Monick, M.M.; Lovan, N.; Powers, L.; Gerke, A.; Hunninghake, G.W. Vitamin D decreases respiratory syncytial virus induction of NF-kappaB-linked chemokines and cytokines in airway epithelium while maintaining the antiviral state. *J. Immunol.* **2010**, *184*, 965–974. [CrossRef]

138. Chen, Y.; Zhang, J.; Ge, X.; Du, J.; Deb, D.K.; Li, Y.C. Vitamin D receptor inhibits nuclear factor κB activation by interacting with IκB kinase β protein. *J. Biol. Chem.* **2013**, *288*, 19450–19458. [CrossRef] [PubMed]

139. Chen, Y.H.; Yu, Z.; Fu, L.; Wang, H.; Chen, X.; Zhang, C.; Lv, Z.M.; Xu, D.X. Vitamin D3 inhibits lipopolysaccharide-induced placental inflammation through reinforcing interaction between vitamin D receptor and nuclear factor kappa B p65 subunit. *Sci. Rep.* **2015**, *5*, 10871. [CrossRef]

140. Xu, S.; Chen, Y.H.; Tan, Z.X.; Xie, D.D.; Zhang, C.; Zhang, Z.H.; Wang, H.; Zhao, H.; Yu, D.X.; Xu, D.X. Vitamin D3 pretreatment regulates renal inflammatory responses during lipopolysaccharide-induced acute kidney injury. *Sci. Rep.* **2015**, *5*, 18687. [CrossRef]

141. Ma, D.; Zhang, R.N.; Wen, Y.; Yin, W.N.; Bai, D.; Zheng, G.Y.; Li, J.S.; Zheng, B.; Wen, J.K. 1, 25(OH)(2)D(3)-induced interaction of vitamin D receptor with p50 subunit of NF-κB suppresses the interaction between KLF5 and p50, contributing to inhibition of LPS-induced macrophage proliferation. *Biochem. Biophys. Res. Commun.* **2017**, *482*, 366–374. [CrossRef]

142. Li, H.; Zhong, X.; Li, W.; Wang, Q. Effects of 1, 25-dihydroxyvitamin D3 on experimental periodontitis and AhR/NF-κB/NLRP3 inflammasome pathway in a mouse model. *J. Appl. Oral Sci.* **2019**, *27*, e20180713. [CrossRef] [PubMed]

143. Jiang, S.; Zhang, H.; Li, X.; Yi, B.; Huang, L.; Hu, Z.; Li, A.; Du, J.; Li, Y.; Zhang, W. Vitamin D/VDR attenuate cisplatin-induced AKI by down-regulating NLRP3/Caspase-1/GSDMD pyroptosis pathway. *J. Steroid Biochem. Mol. Biol.* **2021**, *206*, 105789. [CrossRef] [PubMed]

144. Rao, Z.; Chen, X.; Wu, J.; Xiao, M.; Zhang, J.; Wang, B.; Fang, L.; Zhang, H.; Wang, X.; Yang, S.; et al. Vitamin D receptor inhibits NLRP3 activation by impeding its BRCC3-mediated deubiquitination. *Front. Immunol.* **2019**, *10*, 2783. [CrossRef] [PubMed]

145. Huang, H.; Hong, J.Y.; Wu, Y.J.; Wang, E.Y.; Liu, Z.Q.; Cheng, B.H.; Mei, L.; Liu, Z.G.; Yang, P.C.; Zheng, P.Y. Vitamin D receptor interacts with NLRP3 to restrict the allergic response. *Clin. Exp. Immunol.* **2018**, *194*, 17–26. [CrossRef]

146. Kelley, N.; Jeltema, D.; Duan, Y.; He, Y. The NLRP3 inflammasome: An overview of mechanisms of activation and regulation. *Int. J. Mol. Sci.* **2019**, *20*, 3328. [CrossRef]

147. Swanson, K.V.; Deng, M.; Ting, J.P. The NLRP3 inflammasome: Molecular activation and regulation to therapeutics. *Nat. Rev. Immunol.* **2019**, *19*, 477–489. [CrossRef]

148. Zhao, C.; Zhao, W. NLRP3 Inflammasome—A key player in antiviral responses. *Front. Immunol.* **2020**, *11*, 211. [CrossRef]

149. Bousquet, J.; Cristol, J.P.; Czarlewski, W.; Anto, J.M.; Martineau, A.; Haahtela, T.; Fonseca, S.C.; Iaccarino, G.; Blain, H.; Fiocchi, A.; et al. Nrf2-interacting nutrients and COVID-19: Time for research to develop adaptation strategies. *Clin. Transl. Allergy* **2020**, *10*, 58. [CrossRef]

150. Chen, L.; Yang, R.; Qiao, W.; Zhang, W.; Chen, J.; Mao, L.; Goltzman, D.; Miao, D. 1,25-dihydroxyvitamin D exerts an antiaging role by activation of Nrf2-antioxidant signaling and inactivation of p16/p53-senescence signaling. *Aging Cell* **2019**, *18*, e12951. [CrossRef]

151. Dai, Y.; Zhang, J.; Xiang, J.; Li, Y.; Wu, D.; Xu, J. Calcitriol inhibits ROS-NLRP3-IL-1β signaling axis via activation of Nrf2-antioxidant signaling in hyperosmotic stress stimulated human corneal epithelial cells. *Redox Biol.* **2019**, *21*, 101093. [CrossRef] [PubMed]

152. Song, E.; Zhang, C.; Israelow, B.; Lu-Culligan, A.; Prado, A.V.; Skriabine, S.; Lu, P.; Weizman, O.E.; Liu, F.; Dai, Y.; et al. Neuroinvasion of SARS-CoV-2 in human and mouse brain. *bioRxiv Prepr. Serv. Biol.* **2020**, *218*. [CrossRef]

153. Barnes, B.J.; Adrover, J.M.; Baxter-Stoltzfus, A.; Borczuk, A.; Cools-Lartigue, J.; Crawford, J.M.; Dassler-Plenker, J.; Guerci, P.; Huynh, C.; Knight, J.S.; et al. Targeting potential drivers of COVID-19: Neutrophil extracellular traps. *J. Exp. Med.* **2020**, *217*, e20200652. [CrossRef]

154. Mehta, P.; McAuley, D.F.; Brown, M.; Sanchez, E.; Tattersall, R.S.; Manson, J.J.; HLH Across Speciality Collaboration, UK. COVID-19: Consider cytokine storm syndromes and immunosuppression. *Lancet* **2020**, *395*, 1033–1034. [CrossRef]

155. Kvietys, P.R.; Fakhoury, H.M.A.; Kadan, S.; Yaqinuddin, A.; Al-Mutairy, E.; Al-Kattan, K. COVID-19: Lung-Centric immunothrombosis. *Front. Cell. Infect. Microbiol.* **2021**, *11*, 534. [CrossRef]

156. Chen, K.W.; Monteleone, M.; Boucher, D.; Sollberger, G.; Ramnath, D.; Condon, N.D.; von Pein, J.B.; Broz, P.; Sweet, M.J.; Schroder, K. Noncanonical inflammasome signaling elicits gasdermin D-dependent neutrophil extracellular traps. *Sci. Immunol.* **2018**, *3*, aar6676. [CrossRef]

157. Sollberger, G.; Choidas, A.; Burn, G.L.; Habenberger, P.; Di Lucrezia, R.; Kordes, S.; Menninger, S.; Eickhoff, J.; Nussbaumer, P.; Klebl, B.; et al. Gasdermin D plays a vital role in the generation of neutrophil extracellular traps. *Sci. Immunol.* **2018**, *3*, aar6689. [CrossRef]

158. Veras, F.P.; Pontelli, M.C.; Silva, C.M.; Toller-Kawahisa, J.E.; de Lima, M.; Nascimento, D.C.; Schneider, A.H.; Caetité, D.; Tavares, L.A.; Paiva, I.M.; et al. SARS-CoV-2-triggered neutrophil extracellular traps mediate COVID-19 pathology. *J. Exp. Med.* **2020**, *217*, e20201129. [CrossRef]

159. Nakazawa, D.; Ishizu, A. immunothrombosis in severe COVID-19. *EBioMedicine* **2020**, *59*, 102942. [CrossRef]

160. Manne, B.K.; Denorme, F.; Middleton, E.A.; Portier, I.; Rowley, J.W.; Stubben, C.; Petrey, A.C.; Tolley, N.D.; Guo, L.; Cody, M.; et al. Platelet gene expression and function in patients with COVID-19. *Blood* **2020**, *136*, 1317–1329. [CrossRef]
161. Zuo, Y.; Yalavarthi, S.; Shi, H.; Gockman, K.; Zuo, M.; Madison, J.A.; Blair, C.; Weber, A.; Barnes, B.J.; Egeblad, M.; et al. Neutrophil extracellular traps in COVID-19. *JCI Insight* **2020**, *5*. [CrossRef]
162. Mohammad, S.; Mishra, A.; Ashraf, M.Z. Emerging role of Vitamin D and its associated molecules in pathways related to pathogenesis of thrombosis. *Biomolecules* **2019**, *9*, 649. [CrossRef] [PubMed]
163. Agraz-Cibrian, J.M.; Giraldo, D.M.; Urcuqui-Inchima, S. 1,25-Dihydroxyvitamin D3 induces formation of neutrophil extracellular trap-like structures and modulates the transcription of genes whose products are neutrophil extracellular trap-associated proteins: A pilot study. *Steroids* **2019**, *141*, 14–22. [CrossRef]
164. Chen, C.; Weng, H.; Zhang, X.; Wang, S.; Lu, C.; Jin, H.; Chen, S.; Liu, Y.; Sheng, A.; Sun, Y. Low-Dose Vitamin D Protects Hyperoxia-Induced Bronchopulmonary Dysplasia by Inhibiting Neutrophil Extracellular Traps. *Front. pediatrics* **2020**, *8*, 335. [CrossRef] [PubMed]
165. Park, Y.C.; Kim, J.; Seo, M.S.; Hong, S.W.; Cho, E.S.; Kim, J.-K. Inverse relationship between vitamin D levels and platelet indices in Korean adults. *Hematology* **2017**, *22*, 623–629. [CrossRef] [PubMed]
166. Sultan, M.; Twito, O.; Tohami, T.; Ramati, E.; Neumark, E.; Rashid, G. Vitamin D diminishes the high platelet aggregation of type 2 diabetes mellitus patients. *Platelets* **2019**, *30*, 120–125. [CrossRef]
167. Toderici, M.; de la Morena-Barrio, M.E.; Padilla, J.; Miñano, A.; Antón, A.I.; Iniesta, J.A.; Herranz, M.T.; Fernández, N.; Vicente, V.; Corral, J. Identification of regulatory mutations in SERPINC1 affecting vitamin D response elements associated with antithrombin deficiency. *PLoS ONE* **2016**, *11*, e0152159.
168. Koivisto, O.; Hanel, A.; Carlberg, C. Key Vitamin D target genes with functions in the immune system. *Nutrients* **2020**, *12*, 1140. [CrossRef]
169. Saliba, W.; Awad, K.; Ron, G.; Elias, M. The effect of Vitamin D supplementation on thrombin generation assessed by the calibrated automated thrombogram. *Clin. Appl. Thromb./Hemost. Off. J. Int. Acad. Clin. Appl. Thromb./Hemost.* **2016**, *22*, 340–345. [CrossRef] [PubMed]
170. Blondon, M.; Biver, E.; Braillard, O.; Righini, M.; Fontana, P.; Casini, A. Thrombin generation and fibrin clot structure after vitamin D supplementation. *Endocr. Connect.* **2019**, *8*, 1447–1454. [CrossRef]
171. Leentjens, J.; van Haaps, T.F.; Wessels, P.F.; Schutgens, R.E.G.; Middeldorp, S. COVID-19-associated coagulopathy and antithrombotic agents-lessons after 1 year. *Lancet Haematol.* **2021**. [CrossRef]
172. Keskin, Ü.; Basat, S. The effect of vitamin D levels on gastrointestinal bleeding in patients with warfarin therapy. *Blood Coagul. Fibrinolysis Int. J. Haemost. Thromb.* **2019**, *30*, 331–336. [CrossRef] [PubMed]

Article

Investigating the Role of Functional Polymorphism of Maternal and Neonatal Vitamin D Binding Protein in the Context of 25-Hydroxyvitamin D Cutoffs as Determinants of Maternal-Neonatal Vitamin D Status Profiles in a Sunny Mediterranean Region

Spyridon N. Karras [1,*], Erdinç Dursun [2,3], Merve Alaylıoğlu [2], Duygu Gezen-Ak [2], Cedric Annweiler [4,5], Fatme Al Anouti [6], Hana M. A. Fakhoury [7], Alkiviadis Bais [8] and Dimitrios Kiortsis [9]

1   National Scholarship Foundation, 55535 Thessaloniki, Greece
2   Brain and Neurodegenerative Disorders Research Laboratories, Department of Medical Biology, Cerrahpasa Faculty of Medicine, Istanbul University-Cerrahpasa, 34381 Istanbul, Turkey; erdincdu@gmail.com (E.D.); merve.alaylioglu@hotmail.com (M.A.); duygugezenak@gmail.com (D.G.-A.)
3   Department of Neuroscience, Institute of Neurological Sciences, Istanbul University-Cerrahpasa, 34381 Istanbul, Turkey
4   Department of Geriatric Medicine and Memory Clinic, Research Center on Autonomy and Longevity, University Hospital, 49035 Angers, France; CeAnnweiler@chu-angers.fr
5   Department of Medical Biophysics, Robarts Research Institute, Schulich School of Medicine and Dentistry, The University of Western Ontario, London, ON N6A 3K7, Canada
6   Department of Health Sciences, College of Natural and Health Sciences, Zayed University, Abu Dhabi 144534, United Arab Emirates; Fatme.AlAnouti@zu.ac.ae
7   Department of Biochemistry and Molecular Biology, College of Medicine, AlFaisal University, Riyadh 11533, Saudi Arabia; hana.fakhoury@gmail.com
8   Laboratory of Atmospheric Physics, Aristotle University of Thessaloniki, 54124 Thessaloniki, Greece; abais@auth.gr
9   Department of Nuclear Medicine, University of Ioannina, 45110 Ioannina, Greece; dkiorts@uoi.gr
*   Correspondence: karraspiros@yahoo.gr

Citation: Karras, S.N.; Dursun, E.; Alaylıoğlu, M.; Gezen-Ak, D.; Annweiler, C.; Al Anouti, F.; Fakhoury, H.M.A.; Bais, A.; Kiortsis, D. Investigating the Role of Functional Polymorphism of Maternal and Neonatal Vitamin D Binding Protein in the Context of 25-Hydroxyvitamin D Cutoffs as Determinants of Maternal-Neonatal Vitamin D Status Profiles in a Sunny Mediterranean Region. *Nutrients* **2021**, *13*, 3082. https://doi.org/10.3390/nu13093082

Academic Editor: Lutz Schomburg

Received: 31 July 2021
Accepted: 27 August 2021
Published: 1 September 2021

Publisher's Note: MDPI stays neutral with regard to jurisdictional claims in published maps and institutional affiliations.

Abstract: Recent results indicate that dysregulation of vitamin D-binding protein (VDBP) could be involved in the development of hypovitaminosis D, and it comprises a risk factor for adverse fetal, maternal and neonatal outcomes. Until recently, there was a paucity of results regarding the effect of maternal and neonatal VDBP polymorphisms on vitamin D status during pregnancy in the Mediterranean region, with a high prevalence of hypovitaminosis D. We aimed to evaluate the combined effect of maternal and neonatal VDBP polymorphisms and different maternal and neonatal 25-hydroxyvitamin D (25(OH)D) cut-offs on maternal and neonatal vitamin D profile. Blood samples were obtained from a cohort of 66 mother–child pairs at birth. Our results revealed that: (i) Maternal VDBP polymorphisms do not affect neonatal vitamin D status at birth, in any given internationally adopted maternal or neonatal cut-off for 25(OH)D concentrations; (ii) neonatal VDBP polymorphisms are not implicated in the regulation of neonatal vitamin D status at birth; (iii) comparing the distributions of maternal VDBP polymorphisms and maternal 25(OH)D concentrations, with cut-offs at birth, revealed that mothers with a CC genotype for rs2298850 and a CC genotype for rs4588 tended to demonstrate higher 25(OH)D ($\geq$75 nmol/L) during delivery (p = 0.05 and p = 0.04, respectively), after adjustments for biofactors that affect vitamin D equilibrium, including UVB, BMI and weeks of gestation. In conclusion, this study from Southern Europe indicates that maternal and neonatal VDBP polymorphisms do not affect neonatal vitamin D status at birth, whereas mothers with CC genotype for rs2298850 and CC genotype for rs4588 demonstrate higher 25(OH)D concentrations. Future larger studies are required to establish a causative effect of these specific polymorphisms in the attainment of an adequate ($\geq$75 nmol/L) maternal vitamin D status during pregnancy.

Keywords: vitamin D; pregnancy; neonatal health; functional polymorphism

1. Introduction

Vitamin D has gained a tremendous width of ongoing scientific research during the past two decades [1–3]. Its undisputed role in bone mineralization has been expanded to a widely adopted hypothesis, associating maternal hypovitaminosis D during pregnancy with an increased risk of the development of adverse pregnancy outcomes and impairment of future offspring's metabolic health [4,5]. Mechanistic evidence reported that maternal 25-hydroxyvitamin D (25(OH)D) correlates strongly with neonatal 25(OH)D concentrations at birth [6–8]. On the other hand, there is a continuing scientific debate and differing criteria of maternal and neonatal vitamin D deficiency worldwide [9,10]. The main reasons for this wide controversy might implicate individual genetic and regional characteristics [11,12], including ethnic variations of vitamin D receptor (VDR) polymorphisms [13], ultraviolet B (UVB) radiation [14] and country-specific dietary patterns [15]. The potential influence of the specific genetic background of each individual for decreasing pregnancy complications and optimizing neonatal health could provide a holistic and personalized clinical approach in daily practice and future vitamin D supplementation practices during pregnancy.

Vitamin D binding protein (VDBP) comprises one of the most important factors for vitamin D metabolism [16–19]. Previous results outline that VDBP metabolism disorders comprise a risk factor for adverse maternal and neonatal outcomes [20–24]. These reports vary according to regional and population parameters, including most European countries, with different public health strategies.

The prevailing view for most eastern European and Mediterranean pregnant populations has been, for decades, that casual exposure to sunlight provides enough vitamin D. Observational data from this region reported a high prevalence of vitamin D deficiency during pregnancy [25,26]. Until recently, there was a paucity of results regarding the effect of maternal and neonatal VDBP polymorphisms on vitamin D metabolism during pregnancy within this region. In addition, so far a combined clinical (in terms of various maternal/neonatal 25(OH) D cut-offs) and genetic (including different maternal and neonatal VDBP polymorphisms) approach has not been investigated.

We aimed to evaluate the combined effect of maternal and neonatal VDBP polymorphisms and different maternal and neonatal 25(OH) D cut-offs on the profiles of maternal and neonatal vitamin D status, within a sunny region of the Mediterranean basin.

2. Methods

2.1. Inclusion and Exclusion Criteria

The study included a cohort of mother–child pairs at birth. Inclusion and exclusion criteria have been previously described [7,13]. Use of vitamin D supplements was also an exclusion criterion. Daily calcium (Ca) supplement use was also recorded. Informed consent was obtained. The study was conducted from January 2018 to September 2018. The study was approved by the Bioethics Committee of the Aristotle University of Thessaloniki, Greece (approval number 1/19-12-2011).

2.2. Demographic and Dietary Data—Biochemical and Hormonal Assays

At enrollment, demographic and social characteristics were recorded. Ca and vitamin D dietary intake during the last month of pregnancy were assessed through a validated semi-quantitative food frequency questionnaire that includes 150 food and beverages [27]. From these data, calculations were made for estimations of consumed quantities (in g per day) based on a food composition database, based on the Greek diet [27] for estimating daily dietary Ca and vitamin D intake. Maternal alcohol use during pregnancy was defined either as none (subdivided into never drinking alcohol or drinking alcohol but not during pregnancy), light (1–2 units per week or at any one time during pregnancy) or moderate (3–6 units per week or at any one time during pregnancy) [28].

Blood samples were obtained from mothers 30–60 min before delivery. Umbilical cord blood was collected, immediately after clamping, from the umbilical vein. Concentrations of 25(OH)D2 and 25(OH)D were determined using liquid chromatography–tandem mass spectrometry (LC–MS/MS), with lower limits of quantification (LLOQ); 25(OH)D2 (0.5 ng/mL), 25(OH)D (0.5 ng/mL) and the sum of both vitamin D forms is provided as total 25 (OH)D [29,30].

2.3. Neonatal and Maternal Vitamin D Status Cut-Offs and Combined VDBP Polymorphisms Evaluation

Differences in the frequency of vitamin D status according to neonatal and maternal VDBP polymorphisms were determined according to their vitamin D status at birth: 25(OH)D $\leq$ 25 nmol/L (deficiency), 25–50 nmol/L (insufficiency) and 25(OH)D $\geq$ 50 nmol/L (sufficiency) [2,4–6]. Following this classification, maternal and neonatal VDBP polymorphisms were assessed at birth to investigate potential differences of maternal and neonatal vitamin D status.

2.4. VDBP Analysis

DNA isolation was performed by QIAamp DNA Blood Mini Kit (Cat. No. 51304, QIAGEN, Hilden, Germany) according to the manufacturer's protocol. Genotypes of VDBP rs2298850, rs4588 and rs7041 SNPs were determined by LightSNiP assay using simple probes (LightSNiP, TibMolBiol, Berlin, Germany) and LightCycler Fast Start DNA Master HybProbe Kit (Cat. No.12239272001, Roche Diagnostics, Mannheim, Germany). Real-time PCR (RT-PCR) was performed with LightCycler 480 Instrument II (Roche Diagnostics, Mannheim, Germany), and genotyping was done by using melting curve analysis as previously described [31].

2.5. UVB Measurements

UVB data for the broad geographical region of Thessaloniki, Greece, were collected at the Laboratory of Atmospheric Physics, School of Physics, Aristotle University of Thessaloniki. The daily integral of vitamin D effective UVB radiation (09:00 to16:00 local time) was expressed as the amount of sunlight hitting a horizontal surface, updated every five minutes, in watts per hour square meter (wh/m^2). Mean UVB exposure during the previous 45 days (daily integral) before blood sample collection (estimated mean half-life of vitamin D) was calculated.

2.6. Statistical Analysis

All analyses that involve the distributions of genotypes of VDBP polymorphisms were analyzed with the chi-square (χ^2) test, df:2 for genotypes. Significance was also confirmed with Cramer's V/Kendall's tau-c. The comparisons between mean values of the groups were performed with one-way ANOVA followed by multiple comparison tests, either Tukey HSD or Dunett C, depending on the normality of the data set. Homogeneity of variances was checked with Levene's Test for homogeneity of variances. When required, the data and p values were adjusted for maternal height (cm), BMI pre-pregnancy (kg/m^2), BMI terminal (kg/m^2), UVB and weeks of gestation by one-way analysis of covariance (ANCOVA). All data are presented as the mean $\pm$ SD in the text and figure legends. p values lower than 0.05 were considered statistically significant. "SPSS 24.0" software was used in these comparisons.

3. Results

Seventy mother–neonate dyads were initially included. Given four neonates had missing birth biochemical data, they were excluded from the related analysis. Demographic, dietary and biochemical data are presented in Table 1.

Table 1. Maternal and neonatal demographic and anthropometric characteristics.

Maternal	
Number (*n*)	66
Age (years)	31.92 ± 6.08
Height (cm)	164.85 ± 5.47
Weight; pre-pregnancy (kg)	67.56 ± 14.54
Weight; term (kg)	85.43 ± 14.30
BMI; pre-pregnancy (kg/m^2)	24.91 ± 4.81
BMI; term (kg/m^2)	29.62 ± 5.80
Weeks of gestation (*n*)	38.80 ± 1.56
Smoking [*n* (%)]	10 (0.14)
Alcohol consumption [*n* (%)]	8 (0.11)
Previous live births [*n* (%)]	26 (0.37)
Daily calcium supplementation [*n* (%)]	37 (0.56)
Daily calcium supplementation (mg)	423.07 ± 319.07
Daily dietary calcium intake during 3rd trimester (mg)	792.5 ± 334.0
Daily dietary vitamin D intake during 3rd trimester (mcg)	2.9 ± 1.2
UVB	0.2 ± 0.1
Paternal height (cm)	177.85 ± 6.14
Neonatal	
Number (*n*)	66
Gender; Males [*n* (%)]	38 (0.58)
Height (cm)	50.48 ± 1.96
Weight (g)	3292.12 ± 414.25

VDBP single nucleotide polymorphisms (SNPs) and the genotype percentage distributions of mothers and neonates are presented in Table 2.

Table 2. Vitamin D binding protein single nucleotide polymorphisms genotype distributions of mothers and neonates (%).

SNP	rs2298850			rs4588			rs7041		
Genotype	CC	CG	GG	CC	CA	AA	GG	GT	TT
Maternal (*n:*%)	33 (0.47)	32 (0.46)	5 (0.07)	32 (0.46)	31 (0.44)	7 (0.10)	19 (0.27)	39 (0.56)	12 (0.17)
Neonatal (*n:*%)	35 (0.50)	28 (0.40)	7 (0.10)	33 (0.47)	30 (0.43)	7 (0.10)	18 (0.26)	38 (0.54)	14 (0.20)

3.1. Distribution of Neonatal and Maternal Vitamin D Status According to VDBP Polymorphisms

Distributions of vitamin D status of maternal–neonatal dyads according to VDBP polymorphisms are presented in Tables 3–6. Data and *p* values were adjusted for maternal height (cm), BMI pre-pregnancy (kg/m^2), BMI terminal (kg/m^2), UVB and weeks of gestation by one-way analysis of covariance (ANCOVA). Covariates appearing in the model are evaluated at the following values: BMI pre-pregnancy (kg/m^2) = 25.09, BMI terminal (kg/m^2) = 29.73 and weeks of gestation = 38.81. No significant difference was observed in any comparisons after adjustment. Mean concentrations of maternal and neonatal vitamin D status (total 25(OH)D), according to maternal and neonatal VDBP polymorphisms, as well as distribution of different states of maternal and neonatal vitamin D equilibrium, were not different among different genotype profiles of VDBP.

Table 3. Mean concentrations of maternal and neonatal 25(OH)D according to VDBP polymorphisms.

Polymorphism	Maternal Genotype	n	Maternal 25OHD Level (nmol/L) Mean ± SD	p Value	Neonatal Geno-type	n	Neonatal 25OHD Level (nmol/L) Mean ± SD	p Value
rs2298850	CC	33	54.14 ± 30.6	0.96	CC	35	48.84 ± 31.4	0.70
	CG	32	51.35 ± 75.1		CG	28	58.90 ± 78.8	
	GG	5	47.10 ± 5.9		GG	7	43.83 ± 18.4	
rs4588	CC	32	55.26 ± 30.4	0.92	CC	33	50.86 ± 31.5	0.77
	CA	31	50.06 ± 76.3		CA	30	56.75 ± 76.5	
	AA	7	49.14 ± 15.7		AA	7	40.66 ± 19.2	
rs7041	GG	19	55.24 ± 34.2	0.67	GG	18	49.12 ± 30.8	0.63
	GT	39	54.97 ± 68.5		GT	38	57.75 ± 69.9	
	TT	12	39.32 ± 18.6		TT	14	41.91 ± 20.2	

Table 4. Distribution of neonatal vitamin D status according to neonatal VDBP genotype polymorphisms.

Polymorphism	Genotype	Deficient $n = 26$ (37%)	Insufficient $n = 29$ (41.5%)	Sufficient $n = 15$ (21.5%)	p Value
rs2298850	CC	14 (54%)	11 (38%)	10 (67%)	
	CG	10 (38%)	13 (45%)	5 (33%)	0.26
	GG	2 (8%)	5 (17%)	0 (0%)	
rs4588	CC	13 (50%)	10 (34%)	10 (67%)	
	CA	10 (39%)	15 (52%)	5 (33%)	0.27
	AA	3 (11%)	4 (14%)	0 (0%)	
rs7041	GG	7 (27%)	8 (28%)	3 (20%)	
	GT	14 (54%)	13 (44%)	11 (73%)	0.42
	TT	5 (19%)	8 (28%)	1 (7%)	

Table 5. Distribution of neonatal vitamin D status according to maternal VDBP genotype polymorphisms.

Polymorphism	Genotype	Deficient $n = 26$ (37%)	Insufficient $n = 29$ (41.5%)	Sufficient $n = 15$ (21.5%)	p Value
rs2298850	CC	13 (50%)	11 (38%)	9 (60%)	
	CG	13 (50%)	14 (48%)	5 (33%)	0.25
	GG	0 (0%)	4 (14%)	1 (7%)	
rs4588	CC	12 (46%)	11 (38%)	9 (60%)	
	CA	14 (54%)	12 (41%)	5 (33%)	0.09
	AA	0 (0%)	6 (21%)	1 (7%)	
rs7041	GG	8 (31%)	6 (21%)	5 (33%)	
	GT	14 (54%)	17 (58%)	8 (54%)	0.87
	TT	4 (15%)	6 (21%)	2 (13%)	

Table 6. Distribution of maternal vitamin D status according to maternal VDBP polymorphisms.

Polymorphism	Genotype	Deficient $n = 18$ (26%)	Insufficient $n = 27$ (39%)	Sufficient $n = 25$ (35%)	p Value
rs2298850	CC	8 (44%)	10 (37%)	15 (60%)	
	CG	10 (56%)	14 (52%)	8 (32%)	0.28
	GG	0 (0%)	3 (11%)	2 (8%)	
rs4588	CC	7 (39%)	10 (37%)	15 (60%)	
	CA	11 (61%)	13 (48%)	7 (28%)	0.14
	AA	0 (0%)	4 (15%)	3 (12%)	
rs7041	GG	5 (28%)	6 (22%)	8 (32%)	
	GT	10 (56%)	15 (56%)	14 (56%)	0.87
	TT	3 (17%)	6 (22%)	3 (12%)	

3.2. Neonatal Cut-Offs at Birth ($\geq$50 nmol/L vs. $\leq$50 nmol/L and $\geq$25 vs. $\leq$25 nmol/L), According to Neonatal VDBP Polymorphisms

Genotype distribution of neonatal VDBP polymorphisms, using different neonatal cut-offs for 25(OH)D at birth, revealed that no significant differences were evident regarding neonatal vitamin D cut-offs of 25 and 50 nmol/L (Table 7).

Table 7. Neonatal vitamin D status at birth (cut-offs at birth $\leq$25 vs. $\geq$25 nmol/L and $\leq$50 vs. $\geq$50 nmol/L) according to neonatal VDBP polymorphisms.

Polymorphism	Genotype	$\leq$50 nmol/L n = 55 (79%)	$\geq$50 nmol/L n = 15 (21%)	p Value	$\leq$25 nmol/L n = 26 (37%)	$\geq$25 nmol/L n = 44 (63%)	p Value
rs2298850	CC	25 (45%)	10 (67%)	0.20	14 (54%)	21 (48%)	0.83
	CG	23 (42%)	5 (33%)		10 (38%)	18 (41%)	
	GG	7 (13%)	0 (0%)		2 (8%)	5 (11%)	
rs4588	CC	23 (42%)	10 (67%)	0.15	13 (50%)	20 (45.5%)	0.83
	CA	25 (45%)	5 (33%)		10 (39%)	20 (45.5%)	
	AA	7 (13%)	0 (0%)		3 (11%)	4 (9%)	
rs7041	GG	15 (27%)	3 (20%)	0.20	7 (27%)	11 (25%)	0.98
	GT	27 (49%)	11 (73%)		14 (54%)	24 (55%)	
	TT	13 (24%)	1 (7%)		5 (19%)	9 (20%)	

3.3. Maternal Vitamin D Status at Birth (Cut-Offs at Birth $\leq$25 vs. $\geq$25 nmol/L, $\leq$50 vs. $\geq$50 nmol/L and $\geq$75 nmol/L vs. $\leq$75 nmol/L) According to Maternal VDBP Polymorphisms

By comparing the distributions of maternal VDBP polymorphisms and maternal 25(OH)D concentrations with cut-offs at birth, we revealed that mothers with CC genotype for rs2298850 and CC genotype for rs4588 tended to demonstrate higher 25(OH)D ($\geq$75 nmol/L) during delivery (p = 0.05 and p = 0.04, respectively), as viewed in Table 8.

Table 8. Maternal vitamin D status at birth (cut-offs at birth $\leq$25 vs. $\geq$25 nmol/L, $\leq$50 vs. $\geq$50 nmol/L and $\geq$75 nmol/L vs. $\leq$75 nmol/L) according to maternal VDBP polymorphisms.

Polymorphism	Genotype	$\leq$25 nmol/L n = 18 (26%)	$\geq$25 nmol/L n = 52 (74%)	p Value	$\leq$50 nmol/L n = 44 (63%)	$\geq$50 nmol/L n = 26 (37%)	p Value	$\leq$75 nmol/L n = 57 (37%)	$\geq$75 nmol/L n = 13 (63%)	p Value
rs2298850	CC	8 (44%)	25 (48%)	0.32	18 (41%)	15 (58%)	0.35	23 (40%)	10 (77%)	0.05
	CG	10 (56%)	22 (42%)		23 (52%)	9 (34%)		29 (51%)	3 (23%)	
	GG	0 (0%)	5 (10%)		3 (7%)	2 (8%)		5 (9%)	0 (0%)	
rs4588	CC	7 (39%)	25 (48%)	0.12	17 (39%)	15 (58%)	0.21	22 (39%)	10 (77%)	0.04
	CA	11 (61%)	20 (39%)		23 (52%)	8 (31%)		29 (51%)	2 (15%)	
	AA	0 (0%)	7 (13%)		4 (9%)	3 (11%)		6 (10%)	1 (8%)	
rs7041	GG	5 (27%)	14 (27%)	0.99	11 (25%)	8 (31%)	0.61	13 (23%)	6 (46%)	0.20
	GT	10 (56%)	29 (56%)		24 (55%)	15 (58%)		33 (58%)	6 (46%)	
	TT	3 (17%)	9 (17%)		9 (20%)	3 (11%)		11 (19%)	1 (8%)	

3.4. Neonatal Vitamin D Status at Birth, According to Maternal VDBP Polymorphisms

There were no significant differences between neonatal 25(OH)D concentrations, with respect to maternal VDBP genotype distribution, using cut-offs of 25 and 50 nmol/L at birth (Tables 9 and 10).

Table 9. Neonatal vitamin D status (25(OH)D $\leq$25 nmol/L vs. 25(OH)D $\geq$25 nmol/L) according to genotype distribution of maternal VDBP polymorphisms.

Polymorphism	Genotype	Neonatal Vitamin D Status		*p* Value
		25(OH)D $\leq$ 25 nmol/L *n* = 44 (63%)	25(OH)D $\geq$ 25 nmol/L *n* = 26 (37%)	
rs2298850	CC	13 (50%)	20 (46%)	
	CG	13 (50%)	19 (43%)	0.20
	GG	0 (0%)	5 (11%)	
rs4588	CC	12 (46%)	20 (46%)	
	CA	14 (54%)	17 (39%)	0.08
	AA	0 (0%)	7 (16%)	
rs7041	GG	8 (31%)	11 (25%)	
	GT	14 (54%)	25 (57%)	0.86
	TT	4 (15%)	8 (138%)	

Table 10. Genotype distribution of maternal VDBP polymorphisms according to neonatal vitamin D status (25(OH)D $\leq$50 nmol/L vs. 25(OH)D $\geq$ 50 nmol/L).

Polymorphism	Genotype	Neonatal Vitamin D Status		*p* Value
		25(OH)D $\leq$50 nmol/L *n* = 55 (79%)	25(OH)D $\geq$ 50 nmol/L *n* = 15 (21%)	
rs2298850	CC	24 (44%)	9 (60%)	
	CG	27 (49%)	5 (33%)	0.52
	GG	4 (7%)	1 (7%)	
rs4588	CC	23 (42%)	9 (60%)	
	CA	26 (47%)	5 (33%)	0.45
	AA	6 (11%)	1 (7%)	
rs7041	GG	14 (26%)	5 (33%)	
	GT	31 (56%)	8 (54%)	0.80
	TT	10 (18%)	2 (13%)	

4. Discussion

Apart from its well-established role as the major plasma carrier protein of vitamin D and its metabolites, VDBP is also considered a critical bioregulator of vitamin D equilibrium during pregnancy [16,17]. Fluctuations of VDBP concentrations during pregnancy, resulting from adaptive changes on the maternal–neonatal interface, have been reported to exert significant effects on the vitamin D profile [16–18]. However, the effects of the specific genetic profile of VDBP polymorphisms on maternal–neonatal vitamin D status, based on widely adopted 25(OH)D cut-offs at term, have not been investigated previously in the region of southern Europe. Our results revealed the following:

(i) maternal VDBP polymorphisms do not affect neonatal vitamin D concentrations at birth, in any given internationally adopted maternal or neonatal cut-off for 25(OH)D concentrations;

(ii) neonatal VDBP polymorphisms are not implicated in the regulation of neonatal vitamin D status at birth;

(iii) in a maternal cohort not affected by vitamin D supplementation during pregnancy, mothers with CC genotype for rs2298850 and CC genotype for rs4588 tended to demonstrate higher 25(OH)D ($\geq$75 nmol/L) concentrations, after adjustments for biofactors that affect vitamin D equilibrium, including UVB, BMI and weeks of gestation. The fact that this finding was evident in a small cohort implies that a biologically plausible basis, which could explain the profound differences of maternal vitamin D status, was observed in our region [25,26], as well as adding to existing genetic influences on maternal hypovitaminosis D during pregnancy [31].

Available studies regarding the interplay of VDBP polymorphisms with 25(OH)D concentrations are conflicting. In specific, rs12512631 and rs7041 were found to affect maternal and cord-blood concentrations of 25(OH)D [32,33]. Insufficient 25(OH)D concentrations were reported in infants of mothers carrying the rs12512631 "C" allele [30].

In addition, GC rs2282679 polymorphism was associated with achieved 25(OH)D concentrations after cholecalciferol supplementation and during pregnancy [32]. The minor allele for rs7041 was also associated with higher 25(OH)D and rs4588 was associated with lower 25(OH)D levels during pregnancy [34], whereas Chinese pregnant women, with VDBP Gc-1f and Gc-1s genotypes, manifested higher plasma 25(OH)D status compared to women with Gc-2 [35].

VDBP concentrations manifest a variable longitudinal increase during pregnancy [18,36], observed only in women with rs7041 GG or GT genotypes [35,37]. Genetic variations of VDBP polymorphisms could partly explain different supplementation responses during pregnancy, regarding clinical outcomes [6,25,26]. Of major interest in a recent cohort with 815 Chinese women, the influence of variants of rs17467825, rs4588, rs2282679 and rs2298850 on maternal 25(OH)D has been reported to be modified by vitamin D supplementation and sunshine exposure [38]. It is interesting to note that significantly higher levels of serum 25(OH)D in homozygous major allele carriers for the rs2298850 of GC gene were also observed in Parkinson's disease cases with slower progression [39]. It becomes evident that a country-specific clinical approach and a tailored approach, according to specific lifestyle and genetic profiles of pregnant women, could result in a more pragmatic approach, in terms of vitamin D supplementation and prevention of maternal and neonatal adverse outcomes [40,41].

This study has several limitations. First, the sample size was small and not powered to detect additional differences in other maternal–neonatal cut-offs, but it was sufficiently powered to show significant differences regarding the main aim of the study. Second, the cross-sectional design of the study could not prove a causal relationship. Third, all women were Caucasian, so our results cannot be safely generalized to other ethnicities. On the other hand, the inclusion of both maternal and neonatal polymorphisms, as well as assessment of different cut-offs, could provide a realistic overview of maternal–neonatal dynamics, which is absent in most previous studies of similar design.

In conclusion, this study, from southern Europe, indicates that maternal and neonatal VDBP polymorphisms do not affect neonatal vitamin D status at birth, whereas mothers with CC genotype for rs2298850 and CC genotype for rs4588 demonstrate higher 25(OH)D concentrations. Future larger studies are required to establish a causative effect of these specific polymorphisms, in the attainment of an adequate ($\geq$75 nmol/L) maternal vitamin D status during pregnancy.

Author Contributions: S.N.K. designed and conducted the study, interpreted the results, and drafted the original and revised versions. E.D. and D.G.-A. conducted the VDBP polymorphism analysis, statistical analysis, and drafted the original and revised versions. M.A. conducted the VDBP polymorphism analysis. C.A. and D.K. contributed to the data interpretation, statistical analysis, and drafting of the original and revised versions. F.A.A., H.M.A.F. and A.B. contributed to the discussion and final edits of the manuscript. All authors have read and agreed to the published version of the manuscript.

Funding: This research received no external funding.

Institutional Review Board Statement: The study was conducted according to the guidelines of the Declaration of Helsinki, and approved by the Institutional Review Board (or Ethics Committee) of Aristotle University of Thessaloniki, Greece (Approval number 1/19-12-2011).

Informed Consent Statement: Informed consent was obtained from all subjects involved in the study.

Data Availability Statement: The data presented in this study are available on request from the corresponding author.

Conflicts of Interest: The authors declare no conflict of interest.

References

1. Dusso, A.S.; Brown, A.J.; Slatopolsky, E. Vitamin D. *Am. J. Physiol. Ren. Physiol.* **2005**, *289*, F8F28. [CrossRef] [PubMed]
2. Holick, M.F. Vitamin D deficiency. *N. Engl. J. Med.* **2007**, *357*, 266–281. [CrossRef]
3. Rajakumar, K.; Greenspan, S.L.; Thomas, S.B.; Holick, F.M. Solar ultraviolet radiation and vitamin D: A historical perspective. *Am. J. Public Health* **2007**, *97*, 17461754. [CrossRef]
4. Palacios, C.; Kostiuk, L.K.; Peña-Rosas, J.P. Vitamin D supplementation for women during pregnancy. *Cochrane Database Syst. Rev.* **2019**, *7*, CD008873. [CrossRef]
5. Pilz, S.; Zittermann, A.; Obeid, R.; Hahn, A.; Pludowski, P.; Trummer, C.; Lerchbaum, E.; Pérez-López, F.R.; Karras, S.N.; März, W. The Role of Vitamin D in Fertility and during Pregnancy and Lactation: A Review of Clinical Data. *Int. J. Environ. Res. Public Health* **2018**, *15*, 2241. [CrossRef]
6. Karras, S.N.; Wagner, C.L.; Castracane, V.D. Understanding vitamin D metabolism in pregnancy: From physiology to pathophysiology and clinical outcomes. *Metabolism* **2018**, *86*, 112–123. [CrossRef]
7. Karras, S.N.; Shah, I.; Petroczi, A.; Goulis, D.G.; Bili, H.; Papadopoulou, F.; Harizopoulou, V.; Tarlatzis, B.C.; Naughton, D.P. An observational study reveals that neonatal vitamin D is primarily determined by maternal contributions: Implications of a new assay on the roles of vitamin D forms. *Nutr. J.* **2013**, *12*, 77. [CrossRef]
8. Markestad, T.; Aksnes, L.; Ulstein, M.; Aarskog, D. 25-Hydroxyvitamin D and 1,25-dihydroxyvitamin D of D2 and D3 origin in maternal and umbilical cord serum after vitamin D2 supplementation in human pregnancy. *Am. J. Clin. Nutr.* **1984**, *40*, 1057–1063. [CrossRef]
9. Holick, M.F.; Binkley, N.C.; Bischoff-Ferrari, H.A.; Gordon, C.M.; Hanley, D.A.; Heaney, R.P.; Murad, M.H.; Weaver, C.M. Evaluation, treatment, and prevention of vitamin D deficiency: An Endocrine Society clinical practice guideline. *J. Clin. Endocrinol. Metab.* **2011**, *96*, 1911–1930. [CrossRef] [PubMed]
10. Munns, C.F.; Shaw, N.; Kiely, M.; Specker, B.L.; Thacher, T.; Ozono, K.; Michigami, T.; Tiosano, D.; Mughal, M.Z.; Mäkitie, O.; et al. Global Consensus Recommendations on Prevention and Management of Nutritional Rickets. *J. Clin. Endocrinol. Metab.* **2016**, *101*, 394–415. [CrossRef] [PubMed]
11. Karras, S.N.; Anagnostis, P.; Naughton, D.; Annweiler, C.; Petroczi, A.; Goulis, D.G. Vitamin D during pregnancy: Why observational studies suggest deficiency and interventional studies show no improvement in clinical outcomes? A narrative review. *J. Endocrinol. Investig.* **2015**, *38*, 1265–1275. [CrossRef] [PubMed]
12. Wagner, C.L.; Hollis, B.W.; Kotsa, K.; Fakhoury, H.; Karras, S.N. Vitamin D administration during pregnancy as prevention for pregnancy, neonatal and postnatal complications. *Rev. Endocr. Metab. Disord.* **2017**, *18*, 307–322. [CrossRef]
13. Karras, S.N.; Dursun, E.; Alaylıoğlu, M.; Gezen-Ak, D.; Annweiler, C.; Skoutas, D.; Evangelidis, D.; Kiortsis, D. Diverse Effects of Combinations of Maternal-Neonatal VDR Polymorphisms and 25-Hydroxyvitamin D Concentrations on Neonatal Birth Anthropometry: Functional Phenocopy Variability Dependence, Highlights the Need for Targeted Maternal 25-Hydroxyvitamin D Cut-Offs during Pregnancy. *Nutrients* **2021**, *13*, 443. [CrossRef] [PubMed]
14. Anastasiou, S.N.; Karras, A.; Bais, W.B.; Grant, K.; Kotsa, K.; Goulis, D.G. Ultraviolet radiation and effects on humans: The paradigm of maternal vitamin D productionduring pregnancy. *Eur. J. Clin. Nutr.* **2017**, *71*, 1268–1272. [CrossRef]
15. Karras, S.N.; Anagnostis, P.; Paschou, S.A.; Kandaraki, E.; Goulis, D.G. Vitamin D statusduring pregnancy: Time for a more unified approach beyond borders? *Eur. J. Clin. Nutr.* **2015**, *69*, 874–877. [CrossRef]
16. Karras, S.N.; Koufakis, T.; Fakhoury, H.; Kotsa, K. Deconvoluting the Biological Roles of Vitamin D-Binding Protein During Pregnancy: A Both Clinical and Theoretical Challenge. *Front. Endocrinol.* **2018**, *9*, 259. [CrossRef] [PubMed]
17. Jassil, N.K.; Sharma, A.; Bikle, D.; Wang, X. Vitamin D binding protein and 25-hydroxyvitamin D levels: Emerging clinical ap-plications. *Endocr. Pract.* **2017**, *23*, 605–613. [CrossRef] [PubMed]
18. Bouillon, R.; Van Assche, F.A.; Van Baelen, H.; Heyns, W.; De Moor, P. Influence of the Vitamin D-binding Protein on the Serum Concentration of 1,25-Dihydroxyvitamin D3. *J. Clin. Investig.* **1981**, *67*, 589–596. [CrossRef] [PubMed]
19. Bikle, D.D.; Gee, E.; Halloran, B.; Haddad, J.G. Free 1,25-dihydroxyvitamin D levels in serum from normal subjects, pregnant subjects, and subjects with liver disease. *J. Clin. Investig.* **1984**, *74*, 1966–1971. [CrossRef]
20. Ma, R.; Gu, Y.; Zhao, S.; Sun, J.; Groome, L.J.; Wang, Y. Expressions of vitamin D metabolic components VDBP, CYP2R1, CYP27B1, CYP24A1, and VDR in placentas from normal and preeclamptic pregnancies. *Am. J. Physiol. Metab.* **2012**, *303*, E928–E935. [CrossRef]
21. Sørensen, I.M.; Joner, G.; Jenum, P.A.; Eskild, A.; Brunborg, C.; Torjesen, P.A.; Stene, L.C. Vitamin D-binding protein and 25-hydroxyvitamin D during pregnancy in mothers whose children later developed type 1 diabetes. *Diabetes Metab. Res. Rev.* **2016**, *32*, 883–890. [CrossRef]
22. Wang, Y.; Wang, O.; Li, W.; Ma, L.; Ping, F.; Chen, L.; Nie, M. Variants in vitamin D bind-ing protein gene are associated with gestational diabetes mellitus. *Medicine* **2015**, *94*, e1693. [CrossRef]
23. Blanton, D.; Han, Z.; Bierschenk, L.; Linga-Reddy, M.P.; Wang, H.; Clare-Salzler, M.; Haller, M.; Schatz, D.; Myhr, C.; She, J.X.; et al. Reduced serum vitamin D-binding protein levels are associated with type 1 diabetes. *Diabetes* **2011**, *60*, 2566–2570. [CrossRef]
24. Kodama, K.; Zhao, Z.; Toda, K.; Yip, L.; Fuhlbrigge, R.; Miao, N.; Fathman, C.G.; Yamada, S.; Butte, A.J.; Yu, L. Expression-Based Genome-Wide Association Study Links Vitamin D–Binding Protein With Autoantigenicity in Type 1 Diabetes. *Diabetes* **2016**, *65*, 1341–1349. [CrossRef] [PubMed]

25. Karras, S.; Paschou, S.A.; Kandaraki, E.; Anagnostis, P.; Annweiler, C.; Tarlatzis, B.C.; Hollis, B.W.; Grant, W.B.; Goulis, D. Hypovitaminosis D in pregnancy in the Mediterranean region: A systematic review. *Eur. J. Clin. Nutr.* **2016**, *70*, 979–986. [CrossRef] [PubMed]
26. Karras, S.; Anagnostis, P.; Annweiler, C.; Naughton, D.P.; Petroczi, A.; Bili, E.; Harizopoulou, V.; Tarlatzis, B.C.; Persinaki, A.; Papadopoulou, F.; et al. Maternal vitamin D status during pregnancy: The Mediterranean reality. *Eur. J. Clin. Nutr.* **2014**, *68*, 864–869. [CrossRef]
27. Trichopoulou, A.; Georga, K. *Composition Tables of Foods and Greek Dishes*; Parisianos: Athens, Greece, 2004.
28. Connelly, R.; Platt, L. Cohort profile: UK millennium cohort study (MCS). *Int. J. Epidemiol.* **2014**, *43*, 1719–1725. [CrossRef] [PubMed]
29. Shah, I.; James, B.; Barker, J.; Petroczi, A.; Naughton, D.P. Misleading measures in Vitamin D analysis: A novel LC-MS/MS assay to account for epimers and isobars. *Nutr. J.* **2011**, *10*, 46. [CrossRef]
30. Shah, I.; Petroczi, A.; Naughton, D.P. Method for simultaneous analysis of eight analogues of vitamin D using liquid chromatography tandem mass spectrometry. *Chem. Central J.* **2012**, *6*, 112. [CrossRef]
31. Gezen-Ak, D.; Dursun, E.; Bilgiç, B.; Hanağasi, H.; Ertan, T.; Gürvit, H.; Emre, M.; Eker, E.; Ulutin, T.; Uysal, O.; et al. Vitamin D receptor gene haplotype is associated with late-onset Alzheimer's disease. *Tohoku J. Exp. Med.* **2012**, *228*, 189–196. [CrossRef] [PubMed]
32. Karras, S.N.; Koufakis, T.; Antonopoulou, V.; Goulis, D.G.; Alaylıoğlu, M.; Dursun, E.; Gezen-Ak, D.; Annweiler, C.; Pilz, S.; Fakhoury, H.; et al. Vitamin D receptor Fokl polymorphism is a determinant of both maternal and neonatal vitamin D concentrations at birth. *J. Steroid Biochem. Mol. Biol.* **2019**, *199*, 105568. [CrossRef] [PubMed]
33. Chun, S.K.; Shin, S.; Kim, M.Y.; Joung, H.; Chung, J. Effects of maternal genetic polymorphisms in vitamin D-binding protein and serum 25-hydroxyvi-tamin D concentration on infant birth weight. *Nutrition* **2017**, *35*, 36–42. [CrossRef]
34. Moon, R.J.; Harvey, N.C.; Cooper, C.; D'Angelo, S.; Curtis, E.M.; Crozier, S.R.; Barton, S.J.; Robinson, S.M.; Godfrey, K.M.; Graham, N.J.; et al. Response to antenatal cholecalciferol supple-mentation is with common vitamin D-related genetic variants. *J. Clin. Endocrinol. Metab.* **2017**, *102*, 2941–2949. [CrossRef] [PubMed]
35. Baca, K.M.; Govil, M.; Zmuda, J.M.; Simhan, H.N.; Marazita, M.L.; Bodnar, L.M. Vitamin D metabolic loci and vitamin D status in Black and White pregnant women. *Eur. J. Obstet. Gynecol. Reprod. Biol.* **2018**, *220*, 61–68. [CrossRef] [PubMed]
36. Shao, B.; Jiang, S.; Muyiduli, X.; Wang, S.; Mo, M.; Li, M.; Wang, Z.; Yu, Y. Vitamin D pathway gene polymorphisms influenced vitamin D level among pregnant women. *Clin. Nutr.* **2018**, *37*, 2230–2237. [CrossRef]
37. Zhang, J.Y.; Lucey, A.; Horgan, R.; Kenny, L.; Kiely, M. Impact of pregnancy on vitamin D status: A longitudinal study. *Br. J. Nutr.* **2014**, *112*, 1081–1087. [CrossRef]
38. Ganz, A.B.; Park, H.; Malysheva, O.V.; Caudill, M.A. Vitamin D binding protein rs7041 genotype alters vitamin D metabolism in pregnant women. *FASEB J.* **2018**, *32*, 2012–2020. [CrossRef]
39. Gezen-Ak, D.; Alaylıoğlu, M.; Genç, G.; Gündüz, A.; Candaş, E.; Bilgiç, B.; Atasoy, I.L.; Apaydın, H.; Kızıltan, G.; Gürvit, H.; et al. GC and VDR SNPs and Vitamin D Levels in Parkinson's Disease: The Relevance to Clinical Features. *Neuro Mol. Med.* **2017**, *19*, 24–40. [CrossRef]
40. Dong, J.; Zhou, Q.; Wang, J.; Lu, Y.; Li, J.; Wang, L.; Wang, L.; Meng, P.; Li, F.; Zhou, H.; et al. As-sociation between variants in vitamin D-binding protein gene and vitamin D deficiency among pregnant women in china. *J. Clin. Lab. Anal.* **2020**, *34*, e23376. [CrossRef] [PubMed]
41. Xie, Z.; Wang, X.; Bikle, D.D. Editorial: Vitamin D Binding Protein, Total and Free Vitamin D Levels in Different Physiological and Pathophysiological Conditions. *Front. Endocrinol.* **2020**, *11*, 317. [CrossRef]

Article

Rapid and Effective Vitamin D Supplementation May Present Better Clinical Outcomes in COVID-19 (SARS-CoV-2) Patients by Altering Serum INOS1, IL1B, IFNg, Cathelicidin-LL37, and ICAM1

Mustafa Sait Gönen [1], Merve Alaylıoğlu [2], Emre Durcan [1], Yusuf Özdemir [3], Serdar Şahin [1], Dildar Konukoğlu [4], Okan Kadir Nohut [5], Seval Ürkmez [6], Berna Küçükece [7], İlker İnanç Balkan [3], H. Volkan Kara [8], Şermin Börekçi [9], Hande Özkaya [1], Zekayi Kutlubay [10], Yalım Dikmen [6], Yılmaz Keskindemirci [11,12], Spyridon N. Karras [13,*], Cedric Annweiler [14,15], Duygu Gezen-Ak [2,*] and Erdinç Dursun [2,16,*]

1   Endocrinology and Metabolism Unit, Department of Internal Medicine, Cerrahpasa Faculty of Medicine, Istanbul University-Cerrahpasa, Istanbul 34098, Turkey; sait.gonen@iuc.edu.tr (M.S.G.); dr.durcan@hotmail.com (E.D.); srdr_shn@hotmail.com (S.Ş.); hmba@iuc.edu.tr (H.Ö.)
2   Brain and Neurodegenerative Disorders Research Laboratories, Department of Medical Biology, Cerrahpasa Faculty of Medicine, Istanbul University-Cerrahpasa, Istanbul 34098, Turkey; merve.alaylioglu@hotmail.com
3   Department of Infectious Diseases and Clinical Microbiology, Cerrahpasa Faculty of Medicine, Istanbul University-Cerrahpasa, Istanbul 34098, Turkey; ozdemiryusufemre1990@gmail.com (Y.Ö.); ilkerinancbalkan@hotmail.com (İ.İ.B.)
4   Department of Medical Biochemistry, Cerrahpasa Faculty of Medicine, Istanbul University-Cerrahpasa, Istanbul 34098, Turkey; dkonuk@istanbul.edu.tr
5   Fikert Biyal Biochemistry Laboratory, Cerrahpasa Faculty of Medicine, Istanbul University-Cerrahpasa, Istanbul 34098, Turkey; okannohut@hotmail.com
6   Department of Anesthesiology and Reanimation, Cerrahpasa Faculty of Medicine, Istanbul University-Cerrahpasa, Istanbul 34098, Turkey; seval.urkmez@istanbul.edu.tr (S.Ü.); ydikmen@iuc.edu.tr (Y.D.)
7   Cerrahpasa Hospital Pharmacy Unit, Cerrahpasa Faculty of Medicine, Istanbul University-Cerrahpasa, Istanbul 34098, Turkey; bernakucukece11@gmail.com
8   Department of Thoracic Surgery, Cerrahpasa Faculty of Medicine, Istanbul University-Cerrahpasa, Istanbul 34098, Turkey; volkan_kara@yahoo.com
9   Department of Pulmonary Diseases, Cerrahpasa Faculty of Medicine, Istanbul University-Cerrahpasa, Istanbul 34098, Turkey; borekcisermin@gmail.com
10 Dermatology and Venerology, Cerrahpasa Faculty of Medicine, Istanbul University-Cerrahpasa, Istanbul 34098, Turkey; zekayikutlubay@hotmail.com
11 General Directorate of Hospitals, Istanbul University-Cerrahpasa, Istanbul 34098, Turkey; yilmaz.keskindemirci@isuzem.com
12 Department of Medical Services and Techniques, Health Services Vocational School, Istanbul University-Cerrahpasa, Istanbul 34098, Turkey
13 National Scholarship Foundation, 55535 Thessaloniki, Greece
14 Division of Geriatric Medicine, Department of Neuroscience, Angers University Hospital, 49035 Angers, France; CeAnnweiler@chu-angers.fr
15 Department of Medical Biophysics, Robarts Research Institute, Schulich School of Medicine and Dentistry, The University of Western Ontario, London, ON N6A 3K7, Canada
16 Department of Neuroscience, Institute of Neurological Sciences, Istanbul University-Cerrahpasa, Istanbul 34098, Turkey
*   Correspondence: karraspiros@yahoo.gr (S.N.K.); duygugezenak@iuc.edu.tr or duygugezenak@gmail.com (D.G.-A.); erdinc.dursun@iuc.edu.tr or erdincdu@gmail.com (E.D.); Tel.: +90-212-414-30-00 (ext. 68016) or +90-533-339-98-82 (E.D.)

Citation: Gönen, M.S.; Alaylıoğlu, M.; Durcan, E.; Özdemir, Y.; Şahin, S.; Konukoğlu, D.; Nohut, O.K.; Ürkmez, S.; Küçükece, B.; Balkan, İ.İ.; et al. Rapid and Effective Vitamin D Supplementation May Present Better Clinical Outcomes in COVID-19 (SARS-CoV-2) Patients by Altering Serum INOS1, IL1B, IFNg, Cathelicidin-LL37, and ICAM1. *Nutrients* 2021, 13, 4047. https://doi.org/10.3390/nu13114047

Academic Editors: Maria Luz Fernandez, Andrea Fabbri and Sara Gandini

Received: 15 September 2021
Accepted: 11 November 2021
Published: 12 November 2021

Publisher's Note: MDPI stays neutral with regard to jurisdictional claims in published maps and institutional affiliations.

Abstract: Background: We aimed to establish an acute treatment protocol to increase serum vitamin D, evaluate the effectiveness of vitamin D3 supplementation, and reveal the potential mechanisms in COVID-19. Methods: We retrospectively analyzed the data of 867 COVID-19 cases. Then, a prospective study was conducted, including 23 healthy individuals and 210 cases. A total of 163 cases had vitamin D supplementation, and 95 were followed for 14 days. Clinical outcomes, routine blood biomarkers, serum levels of vitamin D metabolism, and action mechanism-related parameters were evaluated. Results: Our treatment protocol increased the serum 25OHD levels significantly to above

30 ng/mL within two weeks. COVID-19 cases (no comorbidities, no vitamin D treatment, 25OHD <30 ng/mL) had 1.9-fold increased risk of having hospitalization longer than 8 days compared with the cases with comorbidities and vitamin D treatment. Having vitamin D treatment decreased the mortality rate by 2.14 times. The correlation analysis of specific serum biomarkers with 25OHD indicated that the vitamin D action in COVID-19 might involve regulation of INOS1, IL1B, IFNg, cathelicidin-LL37, and ICAM1. Conclusions: Vitamin D treatment shortened hospital stay and decreased mortality in COVID-19 cases, even in the existence of comorbidities. Vitamin D supplementation is effective on various target parameters; therefore, it is essential for COVID-19 treatment.

Keywords: SARS-CoV-2; COVID-19; vitamin D; cytokine; cathelicidin-LL37; acute respiratory failure

1. Introduction

Since December 2019, the world has been experiencing one of the most striking outbreaks in human history—the COVID-19 pandemic. The main route of COVID-19 transmission was reported as being respiratory droplets and direct contact [1]. It was observed that patients hospitalized in intensive care units (ICU) had high plasma levels of IL-2, IL-7, IL-10, GSCF, IP10, MCP1, MIP1A, and TNFα [2]. Given the natural three-stage clinical course of the disease, inadequate innate immune response in the first stage and immune-mediated damage due to dysregulated immune response in the second stage are considered to be the major determinants of poor outcomes [3]. Several classes of drugs and supplements, including vitamin D, are being evaluated for the treatment of COVID-19, based on the growing evidence regarding the natural history and evolution of the infection obtained from patients [4].

Vitamin D is a secosteroid hormone that has existed on the Earth's surface for 750 million years and regulates many cellular mechanisms [5,6]. After being produced in the skin by sunlight or dietary intake, it is converted to biologically active 1,25-dihydroxyvitamin D in the liver and kidneys, respectively [7,8]. Although the effects of vitamin D on skeletal and bone metabolism have been well recognized for a long time, its extra-skeletal effects have gradually come into prominence within the last 20 years. In addition, its effects on the regulation of the immune response, oxidative stress, cancer biology, and the nervous system are particularly substantial. [6,9–13].

Vitamin D was used to treat tuberculosis even before anti-mycobacterial drugs were introduced [14]. Numerous cross-sectional studies have been reporting the association between low vitamin D levels and increased rates or severity of various infections, or both, such as influenza [15], bacterial vaginosis [16], and human immunodeficiency virus (HIV) infection [17,18]. The ability of vitamin D to regulate immune response and mitigate the course of acute infections has been highlighted in recent years [11,19–22].

Vitamin D_3 replacement is hypothesized to reduce infection-related mortality in intensive care units (ICUs) via increasing hemoglobin concentrations, reducing serum hepcidin concentrations, improving oxygenation on the cellular level, and reversing lung damage [23–29]. Recently, studies have demonstrated an association between vitamin D deficiency and the severity and increased mortality of COVID-19. Vitamin D deficiency has been associated with more severe clinical forms of COVID-19 [30–33]. A study reported that patients supplemented with 10,000 IU/daily vitamin D in COVID-19 presented fewer symptoms than non-supplemented patients [34].

In this study, we aimed to: (1) investigate whether vitamin D deficiency is a risk factor in the clinical course of COVID-19 infection; (2) establish an acute (bolus) treatment protocol to increase serum vitamin D (25 hydroxy-vitamin D-25OHD) to sufficient levels (>30 ng/mL); (3) evaluate the effectiveness of vitamin D_3 supplementation in the COVID-19 treatment, and develop a recommendation for routine treatment of patients in varying clinical severities; (4) reveal the novel potential mechanisms that vitamin D acts on modulating COVID-19 immune response and augment treatment success.

2. Materials and Methods

2.1. Study Design and Patient Groups

The study was conducted in two stages. The flow chart of patient recruitment is shown in Figure 1, in a consort diagram. In the retrospective part, data of 867 patients admitted to Istanbul University-Cerrahpasa (Cerrahpasa Faculty of Medicine) Faculty Hospital between 7 March and 22 May 2020, with a confirmed diagnosis of COVID-19, based on clinical and PCR findings, were analyzed. Considering that other diseases may affect the vitamin D status, severity, or progression of COVID-19 infection, cases with comorbidities such as cancer, thyroid or kidney disease, or cardiovascular or autoimmune diseases were excluded. This left 162 cases in the first part of the study (Figure 1). All patients received anti-virals (hydroxychloroquine, azithromycin, oseltamivir, and favipiravir) and some received anti-cytokine (tocilizumab) treatment, in case of indication, according to current national guidelines. The first stage of the study was conducted to evaluate the effect of serum vitamin D (25OHD) on status in COVID-19.

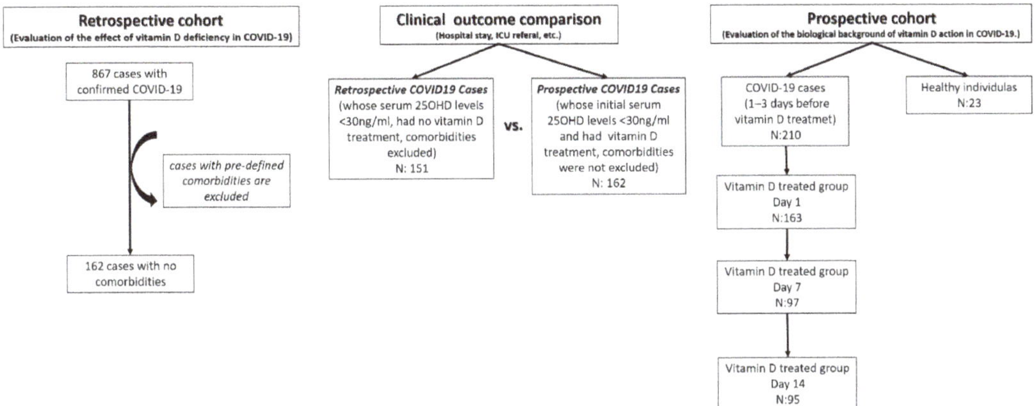

Figure 1. The study design and patient groups.

The second part, which was designed as a prospective randomized controlled study, involved 210 individuals diagnosed with COVID-19 and 23 healthy individuals (mean age 35.5 ± 8.2; range 26–48; 65.2% female). A total of 163 COVID-19 cases whose serum 25OHD levels were less than 30 ng/mL received vitamin D3 (cholecalciferol) treatment, according to the protocol (Table 1), which was created by compiling evidence-based data from the literature [23–26], while 47 cases had no vitamin D treatment at all. A total of 95 out of 163 cases who had vitamin D supplementation were followed for at least 14 days. We should note that the patients that were treated with vitamin D were vitamin D deficient or insufficient (serum 25OHD levels < 30 ng/mL). The safety of the treatment was checked by monitoring serum 25OHD and Ca^{2+} levels (for toxicity and calcification) weekly. In this second part, peripheral blood samples were collected from all patients 1–3 days before treatment and from patients who received vitamin D treatment on day 7 (D7) and day 14 (D14) of the treatment (Figure 1). The second stage of the study was conducted to evaluate the biological background of the effect of vitamin D treatment in COVID-19.

Clinical outcomes, such as hospital stays and ICU referrals, were evaluated in a retrospective cohort to assess the effect of serum vitamin D status, and in both retrospective and prospective cases to evaluate the effect of vitamin D treatment (Figure 1).

Table 1. Vitamin D3 (cholecalciferol) treatment protocol.

	Patient Definition	DAY 1	DAY 2	DAY 3	DAY 4	DAY 5	DAY 6	DAY 7	TOTAL PERIOD	TOTAL DOSE
INPATIENT	Serum 25OHD level < 12 ng/mL	100.000 IU	10.000 IU	10.000 IU	10.000 IU	10.000 IU	10.000 IU	10.000 IU	14 Days	320.000 IU
	Serum 25OHD level 20–12 ng/mL	100.000 IU	5.000 IU	5.000 IU	5.000 IU	5.000 IU	5.000 IU	5.000 IU	14 Days	260.000 IU
	Serum 25OHD level 20–30 ng/mL	100.000 IU	2.000 IU	2.000 IU	2.000 IU	2.000 IU	2.000 IU	2.000 IU	14 Days	224.000 IU
ICU PATIENT	Serum 25OHD level < 12 ng/mL	100.000 IU	100.000 IU	100.000 IU	100.000 IU	100.000 IU			5 Days	500.000 IU
	Serum 25OHD level 20–12 ng/mL	100.000 IU	100.000 IU	100.000 IU	100.000 IU				4 Days	400.000 IU
	Serum 25OHD level 20–30 ng/mL	100.000 IU	100.000 IU	50.000 IU					3 Days	250.000 IU

Participants in the present study were treated according to the current national COVID-19 guidelines, which did not have any recommendation regarding vitamin D supplementation at the time of study or during the manuscript writing process. The study adhered to the ethical principles for medical research involving human participants, described in the World Medical Association's Declaration of Helsinki. The study was approved by the Ethics Committee of Istanbul University, Cerrahpasa, and Republic of Turkey Ministry of Health (Approval Number: Mustafa Sait Gönen-2020-05-06T19_51_05). Signed informed consent was obtained from all study participants.

2.2. Target Parameters

The relation between vitamin D supplementation and disease parameters, such as gender, age, hospitalization time, ICU (intensive care unit) stay, CBC (Complete blood count), Urea, Creatinine, Sodium, Potassium, Chlorine, AST, ALT, Total Bilirubin, LDH, CPK, D-dimer, Ferritin, troponin, and CRP were noted in hospital records and gathered electronically. The analysis was based on comparing these between 2 groups. The data for the aforementioned parameters was gathered from the database of Hospitals General Directorate of Cerrahpasa Faculty of Medicine.

The molecular infrastructure of vitamin D's effectiveness in the COVID-19 treatment protocol was investigated with vitamin D metabolism (25OHD, vitamin D binding protein-DBP, parathormone-PTH, and Ca^{2+}), immune response (cathelicidin-LL-37, Interleukin-IL1b, IL6, IL17, Interferon gamma-INFg, and calcium binding protein B-S100B), and endothelial function (Intercellular Adhesion Molecule 1-ICAM1, Vascular cell adhesion protein 1-VCAM1, nitric oxide-NO, and Nitric Oxide Synthase 1-NOS1)-related parameters. DBP, cathelicidin LL-37, IL1b, IL6, IL17, INFg, S100B, ICAM, VCAM, NO, and NOS parameters were investigated by ELISA, 25OHD, PTH, and Ca^{2+} with CLIA methods. The kits that were used were the following: Elecsys Vitamin D total II (7464215190, Roche, detection range: 3–100 ng/mL, sample dilution factor (SDF): 2); Elecsys PTH (11972103122, Roche, detection range: 1.20–5000 pg/mL, sensitivity: 6.0 pg/mL, SDF: 1); Calcium Gen.2 (05061482190, Roche, detection range: 0.20–5.0 mmol/L); Human LL-37 (Antibacterial Protein LL-37) ELISA Kit (E-EL-H2438, Elabscience, detection range: 1.56–100 ng/mL, sensitivity: 0.94 ng/mL, sample dilution factor (SDF): 1); IL-1 beta Human ELISA Kit (BMS224-2, Thermo, detection range: 3.9–250 pg/mL, sensitivity: 0.3 pg/mL, SDF: 2); Human IL-6 ELISA Kit (BMS213-2, Thermo, detection range: 1.56–100 pg/mL, sensitivity: 0.92 pg/mL, SDF: 2); Human IL-17(Interleukin 17) ELISA Kit (E-EL-H0105, Elabscience, detection range: 31.25–2000 pg/mL, sensitivity: 18.75 pg/mL, SDF: 1); Human IFN-gamma ELISA Kit (BMS228, Thermo, detection range: 1.6–100 pg/mL, sensitivity: 0.99 pg/mL, SDF: 2); Human S100B(S100 Calcium Binding Protein B) ELISA Kit (E-EL-H1297, Elabscience, detection range: 31.25–2000 pg/mL, sensitivity: 18.75 pg/mL, SDF: 1); Human ICAM-1(intercellular adhesion molecule 1) ELISA Kit (E-EL-H6114, Elabscience, detection range: 0.31–20 ng/mL, sensitivity: 0.19 ng/mL, SDF: 1); Human VCAM-1/CD106 (Vascuolar Cell Adhesion Molecule 1) ELISA Kit (E-EL-H5587, Elabscience, detection range: 1.56–100 ng/mL, sensitivity: 0.94 ng/mL, SDF: 1); nitrate–nitrite (index of total NO production) Colorimetric Assay Kit (780001, Cayman, detection limit: 2.5 μM, SDF: 2); Human NOS1/nNOS (Nitric Oxide Synthase 1, Neuronal) ELISA Kit (E-EL-H0742, Elabscience, detection range: 0.16–10 ng/mL, sensitivity: 0.10 ng/mL, SDF: 1); Human DBP (Vitamin D Binding Protein) ELISA Kit (E-EL-H1604, Elabscience, detection range: 3.91–250 ng/mL, sensitivity: 2.35 ng/mL, SDF: 1).

2.3. Statistics

We used the SPSS 24 or GraphPad Prism 7.0a (GraphPad Software, Inc. San Diego, CA, USA) program for the biostatistical analysis of this study. For pairwise comparison, the data were compared using the independent sample t-test when the data were normally distributed and the Mann–Whitney U test when the data were not normally distributed. $p < 0.05$ was accepted as statistically significant. In comparisons of more than two groups, whether the data is normally distributed and whether the difference between the obtained standard deviations is significant were determined firstly by one-way ANOVA, followed by Tukey–Kramer multiple comparison tests, or, for multiple comparisons, Kruskal Wallis then Dunn's multiple comparison tests were used. $p < 0.05$ was accepted as statistically significant. The effect of age or gender difference on categorized data was adjusted with binary logistic regression analysis. When required, the corrected effect size was calculated with Glass' delta (GΔ), where 0.2 is suggested as a small effect size, 0.5 as medium, and 0.8 is a larger effect [35,36]. The overall corrected effect size for multiple comparisons was calculated as the average of individual GΔs determined for each significant outcome [36]. In the prospective study, age and sex adjustment was performed with one way analysis of covariance (ANCOVA) and the observed power was stated. Pearson correlation was used in normally distributed groups, and Spearman correlation was used in non-normally distributed groups, for the correlation analysis between parameters.

3. Results

3.1. The Effect of Serum Vitamin D Status on Clinical Outcomes of Retrospective Cases

The rate of ICU admission was 17.53% (152 out of 867) in the whole cohort and 4.94% (8 out of 162) in the sub-group had no comorbidities. Co-existing diseases increased the risk of ICU admission by 3.6 times ($p = 0.0007$, 95%CI: 1.7100 to 7.3705, OR: 3.55, post-hoc power: 99.9%). The rate of ICU admission was not significantly different in cases with serum 25OHD levels either lower or higher than 12 ng/mL ($p = 0.502$), regardless of comorbidity (Table 2). ICU admission was not significantly different between COVID-19 cases with no comorbidities and COVID-19 cases with no comorbidities but having serum 25OHD levels higher than 12 ng/mL ($p = 0.7459$, 95% CI: 0.3228 to 4.8481, OR: 1.25).

Mean ICU stay in COVID-19 cases, including those with co-existing diseases, was 7.47 ± 7.35, N:152. Mean ICU stay in COVID-19 cases excluding those with co-existing diseases while having serum 25OHD levels lower than 12 ng/mL, was 17.80 ± 6.91, N:5. The ICU stay duration of this group was significantly higher than that of COVID-19 cases including co-existing diseases ($p = 0.0042$, 95% CI: 3.736 to 16.916, post hoc power: 90.7%, Glass' Δ: 1.41). Given the number of COVID-19 cases, excluding those with co-existing diseases whose serum 25OHD levels were higher than 12 ng/mL and who went into ICU, were less than five, we were not able to analyze the ICU stay in this group.

The rate of mortality was 11.19% (97 out of 867) in the whole cohort, including patients with comorbidities. The mortality rate of prospective cases who also had comorbidities but received vitamin D treatment was 5.5% (9 out of 162). Having vitamin D treatment decreased the mortality rate 2.14 times ($p = 0.03$, 95%CI: 1.0585 to 4.3327, OR: 2.14, post-hoc power: 61.0%).

Table 2. Retrospective study. Demographics, routine blood biomarkers, and the serum levels of the targets in key pathways of COVID-19 cases that had no vitamin D treatments, which were separated into four groups according to serum 25OHD levels (<12 ng/mL, 12–20 ng/mL, 20–30 ng/mL, and >30 ng/mL).

		Serum 25OHD Levels				
		<12 ng/mL (L1)	**12–20 ng/mL (L2)**	**20–30 ng/mL (L3)**	**>30 ng/mL (L4)**	*p* **Value**
		n (%)	*n* (%)	*n* (%)	*n* (%)	
Sex	Female	31 (37.8%)	10 (24.4%)	11 (39.3%)	6 (54.5%)	0.23
	Male	51 (62.2%)	31 (75.6%)	17 (60.7%)	5 (45.5%)	
Hospital stay	<8 days	29 (35.4%)	20 (48.8%)	14 (50.0%)	6 (54.5%)	0.30
	>8 days	53 (64.6%)	21 (51.2%)	14 (50.0%)	5 (45.5%)	
ICU referral	Yes	5 (6.1%)	2 (5.0%)	1 (3.6%)	0 (0.0%)	0.82
	No	77 (93.9%)	38 (95.0%)	27 (96.4%)	11 (100%)	
		<12 ng/mL	**12–20 ng/mL**	**20–30 ng/mL**	**>30 ng/mL**	*p* **value for MCT**
	n	82	41	28	11	
Age	Mean ± SD	49.70 ± 13.45	46.75 ± 11.27	54.25 ± 12.35	52.18 ± 12.01	*p* > 0.05 for all groups
Hospital stay (days)	Mean ± SD	9.40 ± 4.78	8.95 ± 4.13	8.39 ± 4.14	6.91 ± 3.36	*p* > 0.05 for all groups
Serum 25OHD levels (ng/mL)	Mean ± SD	8.16 ± 2.22	15.27 ± 2.13	23.80 ± 2.87	44.12 ± 12.87	*p* **< 0.001 for all groups,** overall Post hoc power: 100%, overall Glass' Δ: 6.84
ALT (IU/L)	Mean ± SD	32.53 ± 26.07	43.66 ± 79.13	32.45 ± 17.08	24.01 ± 15.08	*p* > 0.05 for all groups
AST (IU/L)	Mean ± SD	34.72 ± 28.79	36.67 ± 35.91	35.71 ± 18.88	27.54 ± 14.02	*p* > 0.05 for all groups
CRP (mg/L)	Mean ± SD	55.36 ± 70.44	40.85 ± 64.49	49.84 ± 53.85	25.75 ± 26.49	*p* > 0.05 for all groups
Creatinine (mg/dL)	Mean ± SD	0.84 ± 0.19	0.90 ± 0.22	0.91 ± 0.22	0.90 ± 0.25	*p* > 0.05 for all groups
Ca^{2+} (mg/dL)	Mean ± SD	8.75 ± 0.48	8.83 ± 0.53	8.89 ± 0.51	9.22 ± 0.67	**L1 vs. L4** *p* **< 0.05**; *p* > 0.05 for other groups Post hoc power: 61.4%, %, Glass' Δ: 0.98
Sodium (mmol/L)	Mean ± SD	137.76 ± 3.09	138.28 ± 3.20	136.96 ± 3.00	137.73 ± 4.47	*p* > 0.05 for all groups
Urea (mg/dL)	Mean ± SD	27.78 ± 12.46	25.67 ± 6.89	25.75 ± 8.23	26.64 ± 10.14	*p* > 0.05 for all groups
Ferritin (ng/mL)	Mean ± SD	407.55 ± 418.19	322.83 ± 304.59	455.10 ± 442.27	394.76 ± 318.01	*p* > 0.05 for all groups
Hemoglobine (g/dL)	Mean ± SD	13.48 ± 1.54	13.53 ± 1.57	13.51 ± 1.35	13.24 ± 1.23	*p* > 0.05 for all groups
Lymphocyte ($\times 10^3$/μL)	Mean ± SD	1.61 ± 1.00	1.59 ± 0.82	1.45 ± 0.78	1.75 ± 0.93	*p* > 0.05 for all groups
Platelet ($\times 10^3$/μL)	Mean ± SD	217.70 ± 78.02	224.95 ± 76.72	211.07 ± 54.47	210.49 ± 72.20	*p* > 0.05 for all groups
Leukocyte ($\times 10^3$/μL)	Mean ± SD	6.94 ± 2.96	6.72 ± 3.80	5.99 ± 2.09	5.62 ± 1.75	*p* > 0.05 for all groups
D-dimer (mg/L)	Mean ± SD	2.80 ± 12.62	0.62 ± 0.55	2.49 ± 10.06	0.57 ± 0.38	*p* > 0.05 for all groups
Fibrinogen (mg/dL)	Mean ± SD	485.21 ± 178.26	426.99 ± 176.84	464.54 ± 155.50	407.79 ± 167.58	*p* > 0.05 for all groups
	n	18	18	16	*n* < 3	
PTH (pg/mL)	Mean ± SD	37.68 ± 22.87	27.10 ± 10.15	23.48 ± 11.25	-	*p* > 0.05 for all groups
Nitrate–Nitrite (μM)	Mean ± SD	12.35 ± 6.77	10.50 ± 3.89	16.11 ± 5.64	-	**L2 vs. L3** *p* **< 0.05;** *p* **> 0.05** for other groups Post hoc power: 91.6%, Glass' Δ: 1.44
NOS1 (ng/mL)	Mean ± SD	3.00 ± 0.85	3.73 ± 1.22	3.42 ± 1.07	-	*p* > 0.05 for all groups
DBP (ng/mL)	Mean ± SD	450.64 ± 182.61	586.10 ± 221.10	547.78 ± 174.04	-	*p* > 0.05 for all groups
IL1B (pg/mL)	Mean ± SD	6.08 ± 0.94	5.98 ± 1.44	6.34 ± 1.36	-	*p* > 0.05 for all groups
IL6 (pg/mL)	Mean ± SD	17.33 ± 33.40	14.81 ± 27.31	4.60 ± 3.33	-	*p* > 0.05 for all groups
IFNg (pg/mL)	Mean ± SD	6.08 ± 7.72	4.65 ± 4.30	3.87 ± 4.54	-	*p* > 0.05 for all groups
IL17 (pg/mL)	Mean ± SD	2.68 ± 0.57	2.56 ± 0.73	2.84 ± 0.78	-	*p* > 0.05 for all groups
LL37 (ng/mL)	Mean ± SD	19.01 ± 8.22	22.52 ± 9.49	19.33 ± 4.79	-	*p* > 0.05 for all groups
S100B (pg/mL)	Mean ± SD	6.37 ± 8.64	5.84 ± 8.94	7.86 ± 15.17	-	*p* > 0.05 for all groups
ICAM1 (ng/mL)	Mean ± SD	98.03 ± 25.50	103.89 ± 66.33	72.11 ± 23.84	-	*p* > 0.05 for all groups
VCAM1 (ng/mL)	Mean ± SD	578.17 ± 560.15	402.15 ± 302.33	370.82 ± 163.75	-	*p* > 0.05 for all groups

Bold letters indicating the group names or the significant data.

3.2. Retrospective Study

The study samples were investigated in 4 groups: the cases with serum 25OHD levels <12 ng/mL (L1), 12–20 ng/mL (L2), 20–30 ng/mL (L3), or >30 ng/mL (L4), first. The results indicated that, besides serum 25OHD levels, the parameters that were significantly different between groups were serum Ca^{2+} and nitrate–nitrite (Table 2). When study samples were dichotomized according to serum 25OHD levels, we created two groups—the cases with serum 25OHD levels <12 ng/mL and >12 ng/mL—in order to increase the power of the study. We observed that serum DBP and NOS1 levels were significantly high and PTH levels was significantly low in cases whose serum 25OHD levels were >12 ng/mL. The differences between the two groups were the nearly significant Ca^{2+} and creatinine levels (Table 3).

Table 3. Retrospective study. Demographics, routine blood biomarkers, and the serum levels of the targets in key pathways of COVID-19 cases that had no vitamin D treatments, which were separated into two groups according to serum 25OHD levels (<12 ng/mL, >12 ng/mL).

		Serum 25OHD Levels		
		<12 ng/mL	**>12 ng/mL**	***p* Value**
		***n* (%)**	***n* (%)**	
Sex	Female	31 (37.8%)	27 (33.8%)	0.60
	Male	51 (62.2%)	53 (66.2%)	
Hospital stay	<8 days	28 (35%)	38 (49%)	0.08
	>8 days	52 (65%)	40 (51%)	Post hoc power: 42.9%
ICU referral	Yes	5 (6%)	4 (5%)	0.776
	No	75 (94%)	73 (95%)	
Mortality		3 (3.7%)	1 (1.3%)	0.33
		Serum 25OHD levels		
		<12 ng/mL	**>12 ng/mL**	***p* value**
	n	82	79	
Age	Mean ± SD	49.71 ± 13.45	50.16 ± 12.14	0.82
Duration of hospital stay (days)	Mean ± SD	9.40 ± 4.78	8.47 ± 4.05	0.18
Serum 25OHD levels (ng/mL)	Mean ± SD	8.16 ± 2.21	22.22 ± 10.90	**<0.0001** Post hoc power: 100%, Glass' Δ: 6.36
ALT (IU/L)	Mean ± SD	32.53 ± 26.07	36.95 ± 57.57	0.53
AST (IU/L)	Mean ± SD	34.72 ± 28.79	35.06 ± 28.34	0.94
CRP (mg/L)	Mean ± SD	55.36 ± 70.44	41.93 ± 56.86	0.19
Creatinine (mg/dL)	Mean ± SD	0.84 ± 0.19	0.90 ± 0.22	0.056 Post hoc power: 45.6%, Glass' Δ: 0.32
Ca²⁺ (mg/dL)	Mean ± SD	8.75 ± 0.48	8.90 ± 0.55	0.057 Post hoc power: 45.3%, Glass' Δ: 0.31
Sodium (mmol/L)	Mean ± SD	137.76 ± 3.09	137.73 ± 3.34	0.96
Urea (mg/dL)	Mean ± SD	27.78 ± 12.46	25.83 ± 7.79	0.24
Ferritin (ng/mL)	Mean ± SD	407.55 ± 418.19	384.72 ± 367.76	0.74
Hemoglobine (g/dL)	Mean ± SD	13.48 ± 1.53	13.48 ± 1.44	0.99
Lymphocyte ($\times 10^3$/μL)	Mean ± SD	1.61 ± 1.00	1.56 ± 0.82	0.75
Platelet ($\times 10^3$/μL)	Mean ± SD	217.70 ± 78.02	218.02 ± 68.47	0.98
Leukocyte ($\times 10^3$/μL) D-dimer (mg/L)	Mean ± SD Mean ± SD	6.94 ± 2.96 2.80 ± 12.62	6.31 ± 3.05 1.31 ± 6.16	0.19 0.36
Fibrinogen (mg/dL)	Mean ± SD	485.21 ± 178.26	437.49 ± 166.76	0.12
	n	18	34	***p* value**
PTH (pg/mL)	Mean ± SD	37.68 ± 22.87	25.40 ± 10.68	**0.04** Post hoc power: 57.8%, Glass' Δ: 0.54
Nitrate–Nitrite (μM)	Mean ± SD	12.35 ± 6.77	13.14 ± 5.51	0.65
NOS1 (ng/mL)	Mean ± SD	3.00 ± 0.85	3.59 ± 1.14	0.06 Post hoc power: 55.9%, Glass' Δ: 0.69
DBP (ng/mL)	Mean ± SD	450.64 ± 182.61	568.07 ± 198.32	**0.04** Post hoc power: 57.2%, Glass' Δ: 0.64
IL1B (pg/mL)	Mean ± SD	6.08 ± 0.94	6.15 ± 1.39	0.85
IL6 (pg/mL)	Mean ± SD	17.33 ± 33.40	10.00 ± 20.40	0.40
IFNg (pg/mL)	Mean ± SD	6.08 ± 7.72	4.28 ± 4.36	0.37
IL17 (pg/mL)	Mean ± SD	2.68 ± 0.57	2.69 ± 0.76	0.98
LL37 (ng/mL)	Mean ± SD	19.01 ± 8.22	21.02 ± 7.71	0.39
S100B (pg/mL)	Mean ± SD	6.37 ± 8.64	6.79 ± 12.12	0.90
ICAM1 (ng/mL)	Mean ± SD	98.03 ± 25.50	88.93 ± 52.76	0.50
VCAM1 (ng/mL)	Mean ± SD	575.17 ± 560.15	386.96 ± 241.91	0.19

Bold letters indicating the group names or the significant data.

3.3. The Effect of Vitamin D Treatment on Clinical Outcomes: Untreated Retrospective Cases vs. Vitamin D Treated Prospective Cases

Descriptive analyses of age, sex, hospitalization (stay) period, and admission to ICU in COVID-19 cases that had or did not have vitamin D treatment are shown in Table 4. The cases that stayed in hospital longer than 8 days were significantly less in COVID-19 cases that had vitamin D treatment compared with the ones that had no vitamin D treatment ($p = 0.02$) (Table 4); however, the retrospective cohort and prospective cohort differed by means of age gender distribution ($p = 0.004$, $p = 0.008$; respectively), given that the data adjusted for age and sex. The binary logistic regression analysis indicated that the significance of hospital stay (< or >8 days) did not depend on gender. Retrospective COVID-19 cases (without additional disease, without vitamin D treatment, and serum 25OHD <30 ng/mL) had the 1.9-fold increased risk of hospitalization longer than 8 days ($p = 0.007$, OR: 1.91, 95%CI: 1.19–3.06). Increased age was also a risk factor for hospitalization longer than 8 days ($p = 0.023$, OR: 1.03, 95%CI: 1.00–1.06) (Table 4).

3.4. Prospective Study (the Biological Background of Vitamin D Treatment)

3.4.1. Vitamin D Treatment Formula

After following the treatment protocols (Table 1) given in this article, the increase in a patient serum 25OHD levels within 14 days might be predicted with the formula "y = 8.63 ln(x) + 13.66", where x = the initial level of serum 25OHD and y = the predicted serum 25OHD levels 14 days after treatment. The formula was extracted from the graphics of the COVID-19 cases that include the serum 25OHD levels in days 1, 7, and 14 of the treatment protocol. The predicted values of serum 25OHD (n: 142, 34.59 ± 5.27) indicated no significant difference for the comparison with the D14 measured serum 25OHD levels (n: 95, 35.46 ± 10.92), ($p > 0.05$, 95%CI: −1.521 to 3.251). The serum 25OHD levels of COVID-19 cases (day 14 of vitamin D treatment—D14) was significantly higher than that of COVID-19 cases (1–3 days before vitamin D treatment -C), ($p < 0.001$, Table 5).

3.4.2. Mean Comparisons

The serum 25OHD levels of healthy individuals were higher than those in COVID-19 cases that did not receive vitamin D treatment and those who received vitamin D treatment for 14 days. On the other hand, the serum 25 OHD levels of the COVID-19 cases on the 7th and 14th days were higher than the COVID-19 cases 1–3 days before the treatment, which did not receive vitamin D treatment. The Ca^{2+} level of cases was relatively increased on the 14th day after treatment, yet it was statistically significant. Given that the fact that the mean value of serum 25OHD levels begin with 16.62 ± 11.85 and only reached 35.46 ± 10.93, which is far below the possible toxic dose of 100 ng/mL within two weeks, and the serum Ca2+ levels did not increase significantly on the 14th day, the treatment protocol was accepted as safe. Considering the PTH level, it was observed that, although the PTH levels of COVID-19 cases that did not receive vitamin D supplementation were relatively high, this level came close to healthy individuals in COVID-19 cases on the 14th day of vitamin D supplementation. It was determined that serum nitrate–nitrite levels were higher in COVID-19 cases on the 7th and 14th day of the treatment, compared with controls. A similar situation was observed for NOS1 as well. While the DBP level was higher in the cases that did not receive supplementation, compared with the controls, it was observed that the cases that received the supplement gradually decreased and regressed to the control levels on the 7th and 14th days. IL1B level was higher in all case groups compared with controls. Although this was not statistically significant, the IL6 level on the 14th day was found to be lower than the cases that did not take vitamin D supplements. IFNg level remained high in all cases compared with controls. IL17 level was lower in all cases compared with controls. Although the LL37 level remained high in all case groups compared with controls, it was significantly reduced on the 7th and 14th days of supplementation compared with the non-supplemented subjects. S100B level was found to be high in cases that did not take vitamin D supplements compared with controls. It was observed that ICAM1 levels were higher in COVID-19 cases on the 7th and 14th day of the treatment compared with controls. Moreover, cases on the 14th days of the treatment had higher ICAM1 levels than cases that did not receive supplementation.

Table 4. Descriptive analysis of age, sex, hospital stay period and going into ICU in retrospective COVID-19 cases (without additional disease, without vitamin D treatment, and serum 25OHD <30 ng/mL) and prospective COVID-19 cases that were treated with vitamin D.

		Retrospective COVID-19 Cases (without Additional Disease, without Vitamin D Treatment, and Serum 25OHD < 30 ng/mL)	Prospective COVID-19 Cases (with Vitamin D Treatment, and Initial Serum 25OHD < 30 ng/mL)	
		n (%)	*n* (%)	*p* Value
Sex	Female	52 (34.4%)	80 (49.4%)	0.008
	Male	99 (65.6%)	82 (50.6%)	
Hospital stay	<8 days	63 (41.7%)	89 (54.9%)	0.02 *
	>8 days	88 (58.3%)	73 (45.1%)	
ICU referral	Yes	8 (5.3%)	18 (11.0%)	0.07
	No	143 (94.7%)	145 (89.0%)	
Mortality		4 (2.7%)	9 (5.5%)	0.22
	n	151	163	
Age	Mean ± SD	50.23 ± 12.36	55.00 ± 16.45	0.004
Hospital stay (days)	Mean ± SD	8.91 ± 4.35	9.23 ± 6.54	0.30

The data was adjusted for age and sex. * The binary logistic regression analysis indicated that the significance in hospital stay (< or >8 days) did not depend on gender. Retrospective COVID-19 cases (without additional disease, without vitamin D treatment, and serum 25OHD < 30 ng/mL) had the 1.9-fold increased risk of having hospitalization longer than 8 days (p = 0.007, OR: 1.91, 95%CI: 1.19–3.06). Increased age was also a risk factor for hospitalization longer than 8 days (p = 0.023, OR: 1.03, 95% CI: 1.00–1.06). Bold letters indicating the group names or the significant data.

The routine blood parameters were analyzed only in cases of COVID-19 that did not take vitamin D supplements and did take supplements, given they were not followed in healthy subjects. It was observed that the ALT level remained higher on the 7th and 14th days compared with those who did not take supplements. No such change was observed for AST. While the CRP level was high in the cases who did not take the supplement and, in the cases on the 1st day of the supplementation, it was observed that it decreased significantly in the cases on the 7th and 14th days. No change in serum creatinine levels was observed. It was observed that the sodium level remained high on the 7th and 14th days. There was no significant difference between the case groups regarding urea, ferritin, hemoglobin, and D-dimer levels. However, it was observed that the leukocyte and platelet levels were high on the 14th day of the cases that received vitamin D supplements, while the fibrinogen level was significantly lower. Detailed statistical analyses with numbers are mentioned in Table 6.

3.4.3. Correlation Analysis

While a positive correlation was observed between serum 25OHD level and serum Ca^{2+} level in COVID-19 cases that did not receive vitamin D supplementation, no such correlation was observed in healthy controls and cases on the 7th and 14th days of supplementation. While a negative correlation was observed between serum 25OD level and serum PTH level in healthy controls, in cases that did not receive supplementation, and on the 7th day of supplementation, it was observed that this correlation disappeared on the 14th day of supplementation. A negative correlation was observed between serum 25OD level and serum nitrate–nitrite levels, only in cases that did not receive supplementation. When NOS1 was examined, it was observed that serum 25OHD level and NOS1 level were not correlated in healthy controls but negatively correlated in cases that did not receive supplementation and positively correlated in cases that received supplementation. While DBP was not correlated with 25OHD in healthy subjects, it was found to be positively correlated in all case groups. While serum 25OHD level and serum IL1B level were not correlated with the control group in the cases who received supplementation, it was observed that they were positively correlated in the cases who did not receive the supplement. No correlation was detected between IL6 and serum 25OHD levels in any group. While serum 25OHD level and serum IFNg level were not correlated in the control group or the cases receiving supplementation, it was negatively correlated in those who did not receive the supplement. No correlation was detected between IL17, S100B, VCAM1, and serum 25OHD levels in any group. While serum 25OHD level and serum LL37 level were not correlated in the control group or in cases that did not receive supplementation, they were positively correlated in vitamin D supplemented cases. While serum 25OHD level and serum ICAM1 level were not correlated in the control group or in the cases who received supplementation, they were negatively correlated in those who did not receive the supplement. Detailed statistical analyses with numbers are mentioned in Table 7.

Table 5. Prospective study. Serum levels of routine blood biomarkers and key proteins of target pathways in healthy subjects, COVID-19 cases (1–3 days before vitamin D treatment) (C), COVID-19 cases in day 7 (D7), and in day 14 (D14) of vitamin D treatment.

	GROUPS				
	Healthy Subjects (H) ($n = 23$)	**COVID-19 (1–3 Days before Vitamin D Treatment) (C)** ($n = 210$)	**COVID-19 Cases (Day 7 of vit D) (D7)** ($n = 97$)	**COVID-19 Cases (Day 14 of Vit D) (D14)** ($n = 95$)	***p* Value for MCT (Multiple Comparison Test) Age and Sex Adjusted**
Serum 25OHD levels (ng/mL) Mean $\pm$ SD	23.44 ± 9.10	16.62 ± 11.85	31.73 ± 12.29	35.46 ± 10.93	H vs. C $p < 0.05$; H vs. D14 $p < 0.001$; C vs. D7 or D14 $p < 0.001$; $p > 0.05$ for other groups Post hoc power: 100%
Ca^{2+} (mg/dL) Mean $\pm$ SD	8.80 ± 0.41	8.49 ± 0.87	9.06 ± 0.90	9.52 ± 0.72	$p > 0.05$ for all groups Post hoc power: 37%
PTH (pg/mL) Mean $\pm$ SD	28.97 ± 12.14	53.67 ± 114.78	49.92 ± 124.34	33.93 ± 40.15	$p > 0.05$ for all groups Post hoc power: 24%
Nitrate–Nitrite (μM) Mean $\pm$ SD	10.18 ± 6.62	16.58 ± 10.89	17.83 ± 11.67	18.53 ± 10.76	H vs. D7 or D14 $p < 0.05$; $p > 0.05$ for other groups Post hoc power: 62%
NOS1 (ng/mL) Mean $\pm$ SD	0.81 ± 0.35	3.93 ± 2.45	3.56 ± 2.41	2.89 ± 2.00	H vs. C $p < 0.001$; H vs. D7 $p < 0.05$; C vs. D14 $p < 0.01$; $p > 0.05$ for other groups Post hoc power: 98%
DBP (ng/mL) Mean $\pm$ SD	258.16 ± 92.86	416.64 ± 279.55	307.67 ± 258.36	289.74 ± 270.07	C vs. D7 or D14 $p < 0.001$; $p > 0.05$ for other groups Post hoc power: 95%
IL1B (pg/mL) Mean $\pm$ SD	4.44 ± 0.75	7.30 ± 3.00	7.54 ± 4.19	7.07 ± 3.49	H vs. C or D7 $p < 0.05$; H vs. D14 $p < 0.001$; C vs. D14 $p < 0.05$; $p > 0.05$ for other groups Post hoc power: 86%
IL6 (pg/mL) Mean $\pm$ SD	0.86 ± 0.34	19.27 ± 41.66	27.57 ± 64.32	17.82 ± 43.20	$p > 0.05$ for all groups Post hoc power: 22%
IFNg (pg/mL) Mean $\pm$ SD	1.10 ± 0.23	28.01 ± 24.63	35.66 ± 23.34	37.05 ± 21.52	H vs. all groups $p < 0.0001$; C vs. D7 $p < 0.001$; C vs. D14 $p < 0.0001$; $p > 0.05$ for other groups Post hoc power: 100%
IL17 (pg/mL) Mean $\pm$ SD	3.06 ± 1.03	2.09 ± 0.80	1.98 ± 1.21	2.11 ± 1.28	H vs. all groups $p < 0.0001$; $p > 0.05$ for other groups Post hoc power: 99%
LL37 (ng/mL) Mean $\pm$ SD	4.81 ± 2.69	18.51 ± 9.65	15.97 ± 9.23	14.76 ± 6.78	H vs. all groups $p < 0.0001$; C vs. D7 $p < 0.05$; C vs. D14 $p < 0.01$; $p > 0.05$ for other groups Post hoc power: 100%

Table 5. *Cont.*

	Healthy Subjects (H) (*n* = 23)	COVID-19 (1–3 Days before Vitamin D Treatment) (C) (*n* = 210)	COVID-19 Cases (Day 7 of vit D) (D7) (*n* = 97)	COVID-19 Cases (Day 14 of Vit D) (D14) (*n* = 95)	*p* Value for MCT (Multiple Comparison Test) Age and Sex Adjusted
			GROUPS		
S100B (pg/mL) Mean ± SD	1.43 ± 0.25	3.96 ± 6.28	3.03 ± 3.21	3.00 ± 2.56	**H vs. C** $p < 0.05$; $p > 0.05$ for other groups Post hoc power: 57%
ICAM1 (ng/mL) Mean ± SD	71.97 ± 37.92	130.48 ± 84.74	144.15 ± 77.14	145.33 ± 73.56	**H vs. D7** $p < 0.05$; **H vs. D14** $p < 0.01$; **C vs. D14** $p < 0.05$; $p > 0.05$ for other groups Post hoc power: 71%
VCAM1 (ng/mL) Mean ± SD	319.84 ± 138.14	496.33 ± 354.93	571.24 ± 371.16	666.65 ± 463.34	$p > 0.05$ for all groups Post hoc power: 13%

Bold letters indicating the group names or the significant data.

Table 6. Prospective study. Serum levels of routine biomarkers in COVID-19 cases without vitamin D treatment (C), COVID-19 cases in day 7 (D7), and in day 14 (D14) of vitamin D treatment.

	COVID-19 Cases (1–3 Days before Vitamin D Treatment) (C) (*n* = 209)	COVID-19 Cases (Day 7 of Vit D) (D7) (*n* = 99)	COVID-19 Cases (Day 14 of Vit D) (D14) (*n* = 86)	*p* Value for MCT
ALT (IU/L) Mean ± SD	29.08 ± 21.42	49.23 ± 44.76	53.22 ± 62.64	**C vs. D7 or D14** $p < 0.001$; $p > 0.05$ for other groups
AST (IU/L) Mean ± SD	31.44 ± 23.41	35.61 ± 26.62	31.68 ± 29.86	$p > 0.05$ for all groups
CRP (mg/L) Mean ± SD	50.68 ± 66.41	28.13 ± 49.08	10.96 ± 27.27	**C vs. D7 or D14** $p < 0.001$; **D7 vs. D14** $p < 0.001$; $p > 0.05$ for other groups
Creatinine (mg/dL) Mean ± SD	1.03 ± 0.65	1.08 ± 1.02	0.87 ± 0.27	$p > 0.05$ for all groups
Sodium (mmol/L) Mean ± SD	137.08 ± 8.51	139.28 ± 3.69	139.63 ± 3.24	**C vs. D7 or D14** $p < 0.001$; $p > 0.05$ for other groups
Urea (mg/dL) Mean ± SD	35.46 ± 22.64	40.77 ± 28.98	32.22 ± 16.44	$p > 0.05$ for all groups
Ferritin (ng/mL) Mean ± SD	408.15 ± 474.26	421.19 ± 498.75	252.52 ± 299.45	$p > 0.05$ for all groups
Hemoglobi Mean ± SD	12.43 ± 1.89	12.27 ± 1.80	12.69 ± 1.75	$p > 0.05$ for all groups
Lymphocyte ($\times 10^3/\mu$L) Mean ± SD	1.56 ± 0.82	1.60 ± 0.86	1.84 ± 0.65	**C vs. D14** $p < 0.001$; **D7 vs. D14** $p < 0.05$; $p > 0.05$ for other groups
Platelet ($\times 10^3/\mu$L) Mean ± SD	210.80 ± 81.10	296.25 ± 124.71	296.67 ± 91.07	**C vs. D7 or D14** $p < 0.001$; $p > 0.05$ for other groups
Leukocyte ($\times 10^3/\mu$L) Mean ± SD	7.51 ± 7.55	8.31 ± 6.63	7.60 ± 3.02	**C vs. D14** $p < 0.01$; $p > 0.05$ for other groups
D-dimer (mg/L) Mean ± SD	0.99 ± 1.21	1.08 ± 1.22	0.76 ± 0.83	$p > 0.05$ for all groups
Fibrinogen (mg/dL) Mean ± SD	469.60 ± 172.43	449.55 ± 148.01	375.42 ± 116.03	**C vs. D14** $p < 0.001$; **D7 vs. D14** $p < 0.001$; $p > 0.05$ for other groups

Bold letters indicating the group names or the significant data.

Table 7. Prospective study. Serum levels of biomarkers of vit D metabolism and inflammation in healthy subjects, COVID-19 cases (1–3 days before vitamin D treatment), COVID-19 cases in day 7 and in day 14 of vitamin D treatment.

Groups	Ca2+	PTH	Nitrate–Nitrite	NOS1	DBP	IL1B	IL6	IFNg	IL17	LL37	S100B	ICAM1	VCAM1
Healthy subjects ($n = 23$)	NC	$p = 0.08$ 95% CI: -0.68 to 0.05, $r^2 = 0.14$	NC	NC	NC	NC	NC	NC	NC	NC	NC	NC	NC
COVID-19 cases (1–3 days before vitamin D treatment) ($n = 210$)	$p = 0.049$ 95% CI: 0.0006 to 0.27, $r^2 = 0.02$	$p = 0.02$ 95% CI: -0.29 to -0.03, $r^2 = 0.03$	$p = 0.047$ 95% CI: -0.27 to -0.002, $r^2 = 0.02$	$p = 0.06$ 95% CI: -0.26 to 0.005, $r^2 = 0.02$	$p = 0.03$ 95% CI: 0.01 to 0.28, $r^2 = 0.02$	$p < 0.0001$ 95% CI: 0.15 to 0.40, $r^2 = 0.08$	NC	$p = 0.06695$ 95% CI: -0.026 to 0.0009, $r^2 = 0.016$	NC	NC	NC	$p = 0.0003$ 95% CI: -0.37 to -0.12, $r^2 = 0.06$	NC
COVID-19 cases (day 7 of vit D treatment) ($n = 97$)	NC	$p = 0.074$ 95% CI: -0.37 to 0.02, $r^2 = 0.03$	NC	$p = 0.043$ 95% CI: 0.007 to 0.39, $r^2 = 0.04$	$p = 0.043$ 95% CI: 0.007 to 0.39, $r^2 = 0.04$	NC	NC	NC	NC	$p = 0.005$ 95% CI: 0.09 to 0.46, $r^2 = 0.08$	NC	NC	NC
COVID-19 cases (day 14 of vit D treatment) ($n = 95$)	NC	NC	NC	$p = 0.023$ 95% CI: 0.03 to 0.42, $r^2 = 0.05$	$p = 0.033$ 95% CI: 0.02 to 0.41, $r^2 = 0.05$	NC	NC	NC	NC	$p = 0.008$ 95% CI: 0.07 to 0.45, $r^2 = 0.07$	NC	NC	NC

NC: No correlation.

4. Discussion

The present study aimed to evaluate the effectiveness of vitamin D3 supplementation in COVID-19 treatment and reveal the potential mechanisms of vitamin D on COVID-19. Our results indicated that vitamin D treatment shortened the hospitalization period, decreased the mortality rate, and that the effect of vitamin D in COVID-19 might involve regulation of INOS1, IL1B, IFNg, cathelicidin-LL37, and ICAM1.

4.1. The Efficiency of Vitamin D Supplementation

Although vitamin D supplementation is a well-established subject in bone health and bone-related diseases, the knowledge on its effects on extra-skeletal functions is not well established. When vitamin D deficiency was reported to increase the risk of COVID-19 disease [30], we established a vitamin D supplementation protocol from the existing literature, that focused on lung damage, reduced oxygen saturation, and sepsis [23–26]. Our treatment protocol increased the serum 25OHD levels significantly to above 30 ng/mL within two weeks. The Ca^{2+} level of cases was relatively increased on the 14th day after treatment, yet it was statistically significant after age and sex adjustment. PTH levels of COVID-19 cases who did not receive vitamin D supplementation were relatively high; moreover, this level came close to healthy individuals in COVID-19 cases on the 14th day of vitamin D supplementation. DBP level was higher in the cases that did not receive supplementation compared with the controls. However, the cases that received the supplement gradually decreased and regressed to the control levels on the 7th and 14th days. Therefore, we may conclude that the treatment protocol was safe, efficient, and functioning effectively. This protocol might be presented as a way of safe, fast, and significant elevation of serum vitamin D levels in adults in 14 days.

4.2. Vitamin D, Iron, and Hemoglobin

The relationship between iron and vitamin D has been evaluated in three studies [23–26]. Two studies found a significant positive correlation between serum iron and basal vitamin D concentration, hematocrit, and transferrin saturation [24,26]. In another study, low hemoglobin (Hb) and transferrin saturation was observed in babies with low 25(OH)D and low 24.25(OH)2D [25]. On the other hand, anemia is quite common in critical illnesses. Approximately two-thirds of ICU adolescent patients develop anemia in the first week of admission and anemia at admission to ICU [37,38]. Anemia is associated with an increased low oxygen-carrying capacity and cardiovascular morbidity, potentially prolonging mechanical ventilation duration, thus increasing the total risk for mortality [27]. A study of 475 patients hospitalized in intensive care units showed that, in patients with severe vitamin D deficiency (<12 ng/mL), an oral or nasogastric-mediated single dose of 540,000 IU vitamin D3 administration significantly decreased mortality compared with the placebo group. This effect was not observed in those with low vitamin D levels (20–13 ng/mL) [39]. In another study, it was shown that in adults hospitalized in ICU, 100,000 IU daily for five days and a total of 500,000 IU vitamin D3 treatment increased hemoglobin concentrations over time and acutely decreased serum hepcidin concentrations. This effect was not observed in patients receiving 50,000 IU per day, totaling 250,000 IU [40].

Either retrospective or prospective part of our study, there was no significant difference between case groups regarding urea, ferritin, hemoglobin, and D-dimer levels.

4.3. Vitamin D and Sepsis

Sepsis is a life-threatening organ dysfunction caused by the host in response to infection and is still the leading cause of death in critically ill patients [28]. In recent years, studies have shown that vitamin D deficiency or insufficiency is common in critically ill patients, particularly in severe sepsis cases [29]. It is thought that the relationship between vitamin D and sepsis can be explained by mechanisms that work through regulation of the immune system and inflammation, endothelial cell protection, and carbon monoxide regulation [28].

Results from a meta-analysis examining twenty-four studies showed that cases of sepsis had significantly lower vitamin D levels in all populations, especially in Caucasians and Africans, compared with cases without sepsis. Vitamin D levels in sepsis cases were not associated with ALB, PLT, WBC, mortality, PCT, BMI, male to female ratio, IL-6, and CRP levels, nor were they associated with death due to sepsis. However, the meta-analysis suggests that vitamin D deficiency may be a biomarker of sepsis risk in all populations, independent of other variables [29]. Vitamin D administration has been shown to reverse lung injury and reduce the decrease in oxygen saturation in animals with an intratracheal lipopolysaccharide (IT-LPS) sepsis model [29].

In our study, while the CRP level was high in the cases that did not receive vitamin D treatment and in the cases on the 1st day of the treatment, it decreased significantly in the cases on the 7th and 14th days. However, the leukocyte and platelet levels were high on the 14th day of the cases that received vitamin D treatment, whereas the fibrinogen level was significantly lower. It was observed that the ALT level remained higher on the 7th and 14th days compared with those who did not take supplements. No such change was observed for AST.

4.4. Vitamin D and COVID-19

In a study conducted on 212 COVID-19 cases, the probability of having a mild disease is correlated to high levels of vitamin D. On the contrary, as the vitamin D levels decrease, the risk of severe disease increases [30]. Another study demonstrates an association between vitamin D deficiency and severity and increased mortality of COVID-19 [31]. A study reported that supplementation of 10,000 IU/daily vitamin D in COVID-19 patients presented fewer symptoms compared with those non-supplemented on the 7th and 14th day of follow-up, and 10,000 IU/daily vitamin D supplementation for 14 days was sufficient to increase vitamin D serum concentrations in a western Mexican population [34]. A retrospective study done in the United Arab Emirates showed that vitamin D levels lower than 12 ng/mL were significantly associated with a higher risk of COVID-19 severity and of death [32]. A systematic review and meta-analysis study indicated a link between serum vitamin D levels and COVID-19 severity and mortality [33]. In our study, ICU referral did not significantly differ between COVID-19 cases without any comorbidities and COVID-19 cases with no other comorbidities but having serum 25OHD levels higher than 12 ng/mL. Besides, there was no significant difference between cases with serum 25OHD levels >12 ng/mL and those with 25OHD levels of <12 ng/mL in ICU stay. COVID-19 cases with no comorbidities, who had no vitamin D treatment, and whose serum 25OHD level was <30 ng/mL had the 1.9-fold increased risk of having hospitalization longer than 8 days compared with the COVID-19 cases with comorbidities, whose serum 25OHD level was <30 ng/mL, who had vitamin D treatment. At this point, it is important to note that vitamin D treatment shortened hospital stay even for the COVID-19 cases in our treatment group that had comorbidities. Besides, having vitamin D treatment decreased the mortality rate 2.14 times, even in the presence of comorbidities.

A recent study suggested impaired vitamin D metabolism and elevated PTH levels eight weeks after onset. The study indicated no association between low vitamin D levels and persistent symptom burden, lung function impairment, ongoing inflammation, or more severe CT abnormalities. They suggested that vitamin D deficiency is frequent among COVID-19 patients but not associated with disease outcomes. Cases with severe disease displayed a disturbed parathyroid–vitamin D axis within their recovery phase. [41]. In a study by Mazziotti et al., it was shown that vitamin D deficiency with secondary hyperparathyroidism was associated with acute hypoxemic respiratory failure in COVID19 patients [42]. In our study, PTH levels of COVID-19 cases who did not receive vitamin D supplementation were relatively high. Yet, this level came close to healthy individuals in COVID-19 cases on the 14th day of vitamin D supplementation.

A recent study reported that serum calcium and vitamin D levels in COVID-19 patients were lower than in healthy individuals [43]. Osman et al. showed that hypocalcemic

COVID-19 patients had longer hospitalization duration and higher severity of the disease, yet they could not find a link between vitamin D status and COVID-19 [44]. Our results showed that the Ca^{2+} level of cases was relatively increased on the 7th and 14th day after treatment, yet it was not statistically significant

It is known that vitamin D acts as a regulator of many cytokines in many cell types of the immune system and in many diseases [11,19–21]. Vitamin D enhances innate cellular immunity in part by stimulating many antimicrobial peptides, including human cathelicidin, LL-37, and defensins [45]. In our study, the serum cathelicidin-LL37 level was higher in all case groups compared with controls but was significantly decreased on day 7 and 14 of supplementation compared with non-supplemented cases. Although vitamin D was named as a vitamin, it is rather a secosteroid hormone [10]. Vitamin D can exhibit both anti-inflammatory and pro-inflammatory responses simultaneously, depending on cell, tissue, or microenvironment. This might be a regulatory response of vitamin D to attenuate LL-37 up-regulation in COVID-19 patients.

Vitamin D also regulates the cellular immune response by reducing the cytokine storm stimulated by the innate immune response. As seen in COVID-19, the innate immune response stimulates the release of both pro-inflammatory and anti-inflammatory cytokines in response to viral and bacterial infections [2]. Vitamin D levels are associated with cytokines such as IL-1, IL-6, IL-10, and TNFα [5]; additionally, vitamin D can reduce pro-inflammatory TH1 cytokines such as TNFα and IFNg, and increase anti-inflammatory cytokines released from macrophages [45–47]. In this respect, it is known that it can also regulate the adaptive immune response [14].

IL17 and IL8 are accepted as significant contributors in the pulmonary inflammatory reaction to infectious agents that induce a Th1/Th17 response. These cytokines increase vascular permeability and allow the intense neutrophilic infiltrates to give a response to viral infection. A study indicated the G allele of rs3819025 correlated with higher tissue expression of IL-17A in the COVID-19 cases [46]. In our study, serum IL17 levels of all COVID-19 cases, whether they received vitamin D supplementation or not, remained low compared with controls. A retrospective study investigating cytokine gene expression in COVID-19 patients showed that IL1 β mRNA expression levels were increased in COVID-19 patients compared with healthy individuals [47]. Our results indicated that the IL1β level remained higher in all COVID-19 case groups compared with controls. Although not statistically significant, we observed that the IL6 level on the 14th day was below that of the cases that did not take vitamin D supplements. A systematic review and meta-analysis study reported that elevated IL6 levels are associated with COVID-19 severity [48]. In the study of Li et al., COVID-19 patients had higher IL6 mRNA expression levels compared with healthy individuals [47]. Lakkireddy et al. found that COVID-19 patients with hypovitaminosis D had evaluated IL6 levels and IL6 levels were reduced in patients supplemented with 60,000 IUs/daily of vitamin D for 8–10 days compared with the patients who received standard treatment [49].

IFNg serum levels were found to be decreased in COVID-19 compared with both macrophage activation syndrome (MAS) and secondary hemophagocytic lymphohisti-ocytosis (sHLH), in which cytokine storm is seen [50]. In a study that investigated the expression levels of several cytokine genes in leukocytes of ICU and non-ICU COVID-19 patients, it was shown that IFNg had higher expression levels in non-ICU than in ICU patients [51]. Our data showed IFNg levels were higher than expected in all groups, regardless of their vitamin D supplement status.

NOS1 and S100B were selected as neuronal markers for COVID19 cases. Nitric oxide (NO) functions as an immune mediator and plays an important role in vascular and inflammatory lung diseases [52]. Although a relation was not investigated with neuronal nitric oxide synthase (NOS1), vitamin D was suggested to be the regulator of inducible nitric oxide synthase (NOS2) [53,54] and endothelial nitric oxide synthase (NOS3) [55]. The final products of NO are nitrite and nitrate. The best index of total NO production is accepted as the sum of both nitrite and nitrate (nitrate–nitrite). In our study group, we determined that serum nitrate–nitrite levels, the metabolites of NO, and NOS1 levels were higher in all COVID-19 cases compared with controls. However, the serum 25OHD level and NOS1 level were not correlated in healthy controls but negatively correlated in cases that did not receive supplementation, and positively correlated in cases that received supplementation in our study. Higher serum nitrate levels were also reported in non-surviving COVID-19 patients compared with surviving patients [56]. S100B is a Ca^{+2} binding protein mainly expressed by astrocytes and is used to detect glial activation or death in neurological disorders, or both [57]. Elevated serum levels of the S100B protein were found in COVID-19 patients, reflecting an increased blood–brain barrier permeability [58]. Serum S100B levels were found to be associated with COVID-19 severity [59]. In our study, we found that S100B levels were higher in all COVID-19 cases compared with controls.

In our study, it was observed that ICAM1 levels were higher in COVID-19 cases on the 7th and 14th days of the treatment compared with controls. Moreover, cases on the 14th days of the treatment had higher ICAM1 levels than cases who did not receive supplementation. Although VCAM1 levels were gradually increased in all COVID-19 cases compared with controls, it was not statistically significant. Serum levels of VCAM-1 were found to be higher in COVID-19 patients than in non-COVID-19 patients [60]. Li et al. showed that serum VCAM-1 and ICAM-1 levels were elevated in mild and severe COVID-19 cases compared with healthy subjects [61]. Kessel et al. were found that serum levels of ICAM-1 were increased in COVID-19 patients compared with both (MAS) and (sHLH) patients [50]. In COVID-19-related acute respiratory distress syndrome, plasma ICAM-1 levels were found to be higher in non-survivors than in survivors [62].

The response of vitamin D in individuals with already high vitamin D levels may be more effective than the response of vitamin D, which is increased in a short time with treatment. However, in individuals whose vitamin D level is moved to the normal range by treatment, a longer time may be required to observe the effect of this level on cytokines. This reveals the importance of having normal vitamin D levels for a healthy life.

5. Conclusions

In conclusion, it has been determined that comorbidity is the most important factor in the duration of admission to intensive care unit and hospital stay in the course of COVID-19. It was observed that the length of stay in the ICU was significantly higher in COVID-19 cases without comorbidities, with serum 25OHD levels lower than 12 ng/mL, than in COVID-19 cases with comorbidities. Vitamin D treatment shortened hospital stay in COVID-19 cases even in the existence of comorbidities. Having vitamin D treatment decreased the mortality rate by 2.14 times. It has been determined that vitamin D supplementation is effective on various targeted parameters; therefore, it is an important parameter for the course of COVID-19, and serum vitamin D levels and correlation analyses between these parameters confirm this inference. However, considering the parameters and the chronic characteristics of the disease, it became necessary to examine the long-term effects of vitamin D supplementation on the long-term effects of COVID-19, including full recovery duration and irreversible organ damage. Moreover, it is important to note that further investigations with a high number of healthy individuals and more detailed patient data might widen knowledge on the potential effects of vitamin D.

Author Contributions: Conceptualization, M.S.G., D.G.-A. and E.D. (Erdinç Dursun); methodology, M.S.G., D.K., İ.İ.B., H.V.K., D.G.-A. and E.D. (Erdinç Dursun); validation, D.G.-A. and E.D. (Erdinç Dursun); formal analysis, M.A., D.G.-A. and E.D. (Erdinç Dursun); investigation, M.A., D.G.-A. and E.D. (Erdinç Dursun); resources, M.S.G., D.K., D.G.-A. and E.D. (Erdinç Dursun); data curation, M.S.G., M.A., E.D. (Emre Durcan), Y.Ö., S.Ş., D.K., O.K.N., S.Ü., B.K., İ.İ.B., H.V.K., Ş.B., H.Ö., Z.K., Y.D. and Y.K.; writing—original draft preparation, M.A., D.G.-A. and E.D. (Erdinç Dursun); writing—review and editing, M.S.G., M.A., İ.İ.B., H.V.K., D.G.-A., S.N.K., C.A. and E.D. (Erdinç Dursun); supervision, M.S.G., D.G.-A. and E.D. (Erdinç Dursun); project administration, M.S.G., D.G.-A. and E.D. (Erdinç Dursun); funding acquisition, M.S.G. All authors have read and agreed to the published version of the manuscript.

Funding: This research was funded by the Research Fund of Istanbul University-Cerrahpasa grant number TSG-2020-34998.

Institutional Review Board Statement: The study was conducted according to the guidelines of the Declaration of Helsinki, and approved by, and the study was approved by the Ethics Committee of ISTANBUL UNIVERSITY-CERRAHPASA, and the REPUBLIC OF TURKEY MINISTRY OF HEALTH (Approval Number: Mustafa Sait Gönen-2020-05-06T19_51_05).

Informed Consent Statement: Informed consent was obtained from all subjects involved in the study.

Data Availability Statement: All data generated or analyzed during this study are included in this published article.

Acknowledgments: The study was supported by the Research Fund of Istanbul University-Cerrahpasa (Project No: TSG-2020-34998). The funders had no role in study design, data collection and analysis, decision to publish, or preparation of the manuscript.

Conflicts of Interest: All authors declare that they have no conflict of interest.

References

1. Wan, S.; Yi, Q.; Fan, S.; Lv, J.; Zhang, X.; Guo, L.; Lang, C.; Xiao, Q.; Xiao, K.; Yi, Z.; et al. Relationships among lymphocyte subsets, cytokines, and the pulmonary inflammation index in coronavirus (COVID-19) infected patients. *Br. J. Haematol.* **2020**, *189*, 428–437. [CrossRef] [PubMed]
2. Huang, C.; Wang, Y.; Li, X.; Ren, L.; Zhao, J.; Hu, Y.; Zhang, L.; Fan, G.; Xu, J.; Gu, X.; et al. Clinical features of patients infected with 2019 novel coronavirus in wuhan, china. *Lancet* **2020**, *395*, 497–506. [CrossRef]
3. Cao, W.; Li, T. COVID-19: Towards understanding of pathogenesis. *Cell Res.* **2020**, *30*, 367–369. [CrossRef]
4. Dos Santos, W.G. Natural history of COVID-19 and current knowledge on treatment therapeutic options. *Biomed. Pharmacother.* **2020**, *129*, 110493. [CrossRef] [PubMed]
5. Dursun, E.; Alaylioglu, M.; Bilgic, B.; Hanagasi, H.; Lohmann, E.; Atasoy, I.L.; Candas, E.; Araz, O.S.; Onal, B.; Gurvit, H.; et al. Vitamin d deficiency might pose a greater risk for apo-evarepsilon4 non-carrier alzheimer's disease patients. *Neurol. Sci.* **2016**, *37*, 1633–1643. [CrossRef]
6. Holick, M.F. Vitamin D: A millenium perspective. *J. Cell. Biochem.* **2003**, *88*, 296–307. [CrossRef] [PubMed]
7. Holick, M. Vitamin D: A D-lightful solution for good health. *J. Med. Biochem.* **2012**, *31*, 263–264. [CrossRef]
8. Holick, M.F. Cancer, sunlight and vitamin D. *J. Clin. Transl. Endocrinol.* **2014**, *1*, 179–186. [CrossRef]
9. Visweswaran, R.K.; Lekha, H. Extraskeletal effects and manifestations of Vitamin D deficiency. *Indian J. Endocrinol. Metab.* **2013**, *17*, 602–610. [CrossRef]
10. Gezen-Ak, D.; Dursun, E. Molecular basis of vitamin D action in neurodegeneration: The story of a team perspective. *Hormones* **2018**, *18*, 17–21. [CrossRef]
11. Gezen-Ak, D.; Yilmazer, S.; Dursun, E. Why vitamin d in alzheimer's disease? The hypothesis. *J. Alzheimers Dis.* **2014**, *40*, 257–269. [CrossRef]
12. Annweiler, C.; Dursun, E.; Feron, F.; Gezen-Ak, D.; Kalueff, A.V.; Littlejohns, T.; Llewellyn, D.; Millet, P.; Scott, T.; Tucker, K.L.; et al. Vitamin d and cognition in older adults: Internation-al consensus guidelines. *Geriatr. Psychol. Neuropsychiatr. Vieil.* **2016**, *14*, 265–273. [CrossRef]
13. Dursun, E.; Gezen-Ak, D. Vitamin D basis of Alzheimer's disease: From genetics to biomarkers. *Hormones* **2019**, *18*, 7–15. [CrossRef] [PubMed]
14. Aranow, C. Vitamin D and the Immune System. *J. Investig. Med.* **2011**, *59*, 881–886. [CrossRef] [PubMed]
15. Cannell, J.J.; Vieth, R.; Umhau, J.C.; Holick, M.F.; Grant, W.B.; Madronich, S.; Garland, C.F.; Giovannucci, E. Epidemic influenza and vitamin D. *Epidemiol. Infect.* **2006**, *134*, 1129–1140. [CrossRef] [PubMed]
16. Bodnar, L.M.; Krohn, M.A.; Simhan, H.N. Maternal Vitamin D Deficiency Is Associated with Bacterial Vaginosis in the First Trimester of Pregnancy. *J. Nutr.* **2009**, *139*, 1157–1161. [CrossRef]
17. Villamor, E. A potential role for vitamin d on hiv infection? *Nutr. Rev.* **2006**, *64*, 226–233. [CrossRef]
18. Rodríguez, M.; Daniels, B.; Gunawardene, S.; Robbins, G.K. High Frequency of Vitamin D Deficiency in Ambulatory HIV-Positive Patients. *AIDS Res. Hum. Retrovir.* **2009**, *25*, 9–14. [CrossRef] [PubMed]
19. Joshi, S.; Pantalena, L.C.; Liu, X.K.; Gaffen, S.L.; Liu, H.; Rohowsky-Kochan, C.; Ichiyama, K.; Yoshimura, A.; Steinman, L.; Christakos, S.; et al. 1,25-dihydroxyvitamin d(3) ameliorates th17 autoimmunity via transcriptional modulation of interleukin-17a. *Mol. Cell Biol.* **2011**, *31*, 3653–3669. [CrossRef]
20. Fernandes de Abreu, D.A.; Eyles, D.; Féron, F. Vitamin d, a neuro-immunomodulator: Implications for neurodegenerative and autoimmune diseases. *Psychoneuroendocrinology* **2009**, *34*, 265–277. [CrossRef]
21. Golpour, A.; Bereswill, S.; Heimesaat, M.M. Antimicrobial and immune-modulatory effects of vitamin D provide promising antibiotics-independent approaches to tackle bacterial infections—lessons learnt from a literature survey. *Eur. J. Microbiol. Immunol.* **2019**, *9*, 80–87. [CrossRef] [PubMed]

22. Kumar, R.; Rathi, H.; Haq, A.; Wimalawansa, S.J.; Sharma, A. Putative roles of vitamin D in modulating immune response and immunopathology associated with COVID-19. *Virus Res.* **2021**, *292*, 198235. [CrossRef]

23. Azizi-Soleiman, F.; Vafa, M.; Abiri, B.; Safavi, M. Effects of iron on Vitamin D metabolism: A systematic review. *Int. J. Prev. Med.* **2016**, *7*, 126. [CrossRef]

24. Blanco-Rojo, R.; Pérez-Granados, A.M.; Toxqui, L.; Zazo, P.; De La Piedra, C.; Vaquero, M.P. Relationship between vitamin D deficiency, bone remodelling and iron status in iron-deficient young women consuming an iron-fortified food. *Eur. J. Nutr.* **2012**, *52*, 695–703. [CrossRef]

25. Heldenberg, D.; Tenenbaum, G.; Weisman, Y. Effect of iron on serum 25-hydroxyvitamin D and 24,25-dihydroxyvitamin D concentrations. *Am. J. Clin. Nutr.* **1992**, *56*, 533–536. [CrossRef]

26. Wright, I.; Blanco-Rojo, R.; Fernández, M.C.; Toxqui, L.; Moreno, G.; Pérez-Granados, A.M.; de la Piedra, C.; Remacha, Á.F.; Vaquero, M.P. Bone remodelling is reduced by recovery from iron-deficiency anaemia in premenopausal women. *J. Physiol. Biochem.* **2013**, *69*, 889–896. [CrossRef] [PubMed]

27. Hayden, S.J.; Albert, T.J.; Watkins, T.R.; Swenson, E.R. Anemia in critical illness: Insights into etiology, consequences, and management. *Am. J. Respir. Crit. Care Med.* **2012**, *185*, 1049–1057. [CrossRef] [PubMed]

28. Yang, C.; Liu, Y.; Wan, W. Role and mechanism of vitamin D in sepsis. *Zhonghua Wei Zhong Bing Ji Jiu Yi Xue* **2019**, *31*, 1170–1173.

29. Parekh, D.; Patel, J.; Scott, A.; Lax, S.; Dancer, R.C.A.; D'Souza, V.; Greenwood, H.; Fraser, W.D.; Gao, F.; Sapey, E.; et al. Vitamin D Deficiency in Human and Murine Sepsis. *Crit. Care Med.* **2017**, *45*, 282–289. [CrossRef]

30. Alipio, M.M. *Vitamin D Supplementation Could Possibly Improve Clinical Outcomes of Patients Infected with Coronavirus-2019 (COVID-2019)*; Elsevier BV: Amsterdam, The Netherlands, 2020.

31. Radujkovic, A.; Hippchen, T.; Tiwari-Heckler, S.; Dreher, S.; Boxberger, M.; Merle, U. Vitamin D Deficiency and Outcome of COVID-19 Patients. *Nutrients* **2020**, *12*, 2757. [CrossRef]

32. AlSafar, H.; Grant, W.B.; Hijazi, R.; Uddin, M.; Alkaabi, N.; Tay, G.; Mahboub, B.; Al Anouti, F. COVID-19 disease severity and death in relation to vitamin d status among SARS-CoV-2-positive uae residents. *Nutrients* **2021**, *13*, 1714. [CrossRef]

33. Kazemi, A.; Mohammadi, V.; Aghababaee, S.K.; Golzarand, M.; Clark, C.C.T.; Babajafari, S. Association of vitamin d status with SARS-CoV-2 infection or COVID-19 severity: A systematic review and meta-analysis. *Adv. Nutr.* **2021**, *12*, 1636–1658. [CrossRef]

34. Sanchez-Zuno, G.A.; Gonzalez-Estevez, G.; Matuz-Flores, M.G.; Macedo-Ojeda, G.; Hernandez-Bello, J.; Mora-Mora, J.C.; Perez-Guerrero, E.E.; Garcia-Chagollan, M.; Vega-Magana, N.; Turrubiates-Hernandez, F.J.; et al. Vitamin d levels in COVID-19 outpatients from western mexico: Clinical correlation and effect of its supplementation. *J. Clin. Med.* **2021**, *10*, 2378. [CrossRef]

35. Durlak, J.A. How to Select, Calculate, and Interpret Effect Sizes. *J. Pediatric Psychol.* **2009**, *34*, 917–928. [CrossRef] [PubMed]

36. Gezen-Ak, D.; Alaylıoğlu, M.; Genç, G.; Şengül, B.; Keskin, E.; Sordu, P.; Güleç, Z.E.K.; Apaydın, H.; Bayram-Gürel, Ç.; Ulutin, T.; et al. Altered Transcriptional Profile of Mitochondrial DNA-Encoded OXPHOS Subunits, Mitochondria Quality Control Genes, and Intracellular ATP Levels in Blood Samples of Patients with Parkinson's Disease. *J. Alzheimer's Dis.* **2020**, *74*, 287–307. [CrossRef]

37. Corwin, H.L.; Gettinger, A.; Pearl, R.G.; Fink, M.P.; Levy, M.M.; Abraham, E.; MacIntyre, N.R.; Shabot, M.M.; Duh, M.-S.; Shapiro, M.J. The CRIT Study: Anemia and blood transfusion in the critically ill—Current clinical practice in the United States *. *Crit. Care Med.* **2004**, *32*, 39–52. [CrossRef]

38. Vincent, J.L.; Baron, J.-F.; Reinhart, K.; Gattinoni, L.; Thijs, L.; Webb, A.; Meier-Hellmann, A.; Nollet, G.; Peres-Bota, D. For the ABC Investigators Anemia and Blood Transfusion in Critically Ill Patients. *JAMA* **2002**, *288*, 1499–1507. [CrossRef]

39. Amrein, K.; Schnedl, C.; Holl, A.; Riedl, R.; Christopher, K.B.; Pachler, C.; Urbanic Purkart, T.; Waltensdorfer, A.; Munch, A.; Warnkross, H.; et al. Effect of high-dose vitamin d3 on hospi-tal length of stay in critically ill patients with vitamin d deficiency: The vitdal-icu randomized clinical trial. *JAMA* **2014**, *312*, 1520–1530. [CrossRef]

40. Smith, E.M.; Jones, J.L.; Han, J.E.; Alvarez, J.A.; Sloan, J.H.; Konrad, R.J.; Zughaier, S.M.; Martin, G.S.; Ziegler, T.R.; Tangpricha, V. High-dose vitamin d3 administration is associated with increases in hemoglobin concentrations in mechanically ventilated critically ill adults: A pilot double-blind, randomized, placebo-controlled trial. *J. Parenter Enter. Nutr.* **2018**, *42*, 87–94. [CrossRef] [PubMed]

41. Pizzini, A.; Aichner, M.; Sahanic, S.; Bohm, A.; Egger, A.; Hoermann, G.; Kurz, K.; Widmann, G.; Bellmann-Weiler, R.; Weiss, G.; et al. Impact of vitamin d deficiency on COVID-19-a pro-spective analysis from the covild registry. *Nutrients* **2020**, *12*, 2775. [CrossRef] [PubMed]

42. Mazziotti, G.; Lavezzi, E.; Brunetti, A.; Mirani, M.; Favacchio, G.; Pizzocaro, A.; Sandri, M.T.; Di Pasquale, A.; Voza, A.; Ciccarelli, M.; et al. Vitamin d deficiency, secondary hyperpara-thyroidism and respiratory insufficiency in hospitalized patients with COVID-19. *J. Endocrinol. Investig.* **2021**, *44*, 2285–2293. [CrossRef]

43. Elham, A.S.; Azam, K.; Azam, J.; Mostafa, L.; Nasrin, B.; Marzieh, N. Serum vitamin D, calcium, and zinc levels in patients with COVID-19. *Clin. Nutr. ESPEN* **2021**, *43*, 276–282. [CrossRef] [PubMed]

44. Osman, W.; Al Fahdi, F.; Al Salmi, I.; Al Khalili, H.; Gokhale, A.; Khamis, F. Serum Calcium and Vitamin D levels: Correlation with severity of COVID-19 in hospitalized patients in Royal Hospital, Oman. *Int. J. Infect. Dis.* **2021**, *107*, 153–163. [CrossRef] [PubMed]

45. Grant, W.B.; Lahore, H.; McDonnell, S.L.; Baggerly, C.A.; French, C.B.; Aliano, J.L.; Bhattoa, H.P. Evidence that Vitamin D Supplementation Could Reduce Risk of Influenza and COVID-19 Infections and Deaths. *Nutrients* **2020**, *12*, 988. [CrossRef] [PubMed]

46. Azevedo, M.L.V.; Zanchettin, A.C.; Vaz de Paula, C.B.; Motta Junior, J.D.S.; Malaquias, M.A.S.; Raboni, S.M.; Neto, P.C.; Zeni, R.C.; Prokopenko, A.; Borges, N.H.; et al. Lung neutro-philic recruitment and il-8/il-17a tissue expression in COVID-19. *Front. Immunol.* **2021**, *12*, 656350. [CrossRef] [PubMed]

47. Li, S.; Duan, X.; Li, Y.; Li, M.; Gao, Y.; Li, T.; Li, S.; Tan, L.; Shao, T.; Jeyarajan, A.J.; et al. Differentially expressed immune response genes in COVID-19 patients based on disease severity. *Aging* **2021**, *13*, 9265–9276. [CrossRef]

48. Zawawi, A.; Naser, A.Y.; Alwafi, H.; Minshawi, F. Profile of Circulatory Cytokines and Chemokines in Human Coronaviruses: A Systematic Review and Meta-Analysis. *Front. Immunol.* **2021**, *12*, 666223. [CrossRef]

49. Lakkireddy, M.; Gadiga, S.G.; Malathi, R.D.; Karra, M.L.; Raju, I.P.M.; Chinapaka, S.; Baba, K.S.; Kandakatla, M. Impact of daily high dose oral vitamin d therapy on the inflammatory markers in patients with COVID 19 disease. *Sci. Rep.* **2021**, *11*, 10641. [CrossRef]

50. Kessel, C.; Vollenberg, R.; Masjosthusmann, K.; Hinze, C.; Wittkowski, H.; Debaugnies, F.; Nagant, C.; Corazza, F.; Vely, F.; Kaplanski, G.; et al. Discrimination of COVID-19 from inflam-mation-induced cytokine storm syndromes by disease-related blood biomarkers. *Arthritis Rheumatol.* **2021**, *73*, 1791–1799. [CrossRef]

51. Chen, Z.; Feng, Q.; Zhang, T.; Wang, X. Identification of COVID-19 subtypes based on immunogenomic profiling. *Int. Immunopharmacol.* **2021**, *96*, 107615. [CrossRef]

52. Yamasaki, H. Blood nitrate and nitrite modulating nitric oxide bioavailability: Potential therapeutic functions in COVID-19. *Nitric Oxide* **2020**, *103*, 29–30. [CrossRef]

53. Dursun, E.; Gezen-Ak, D.; Yilmazer, S. A new mechanism for amyloid-beta induction of inos: Vitamin d-vdr pathway disruption. *J. Alzheimer's Dis.* **2013**, *36*, 459–474. [CrossRef] [PubMed]

54. Dursun, E.; Gezen-Ak, D.; Yilmazer, S. The influence of vitamin d treatment on the inducible nitric oxide synthase (inos) expression in primary hippocampal neurons. *Noro. Psikiyatr. Ars.* **2014**, *51*, 163–168. [CrossRef]

55. Andrukhova, O.; Slavic, S.; Zeitz, U.; Riesen, S.C.; Heppelmann, M.S.; Ambrisko, T.D.; Markovic, M.; Kuebler, W.M.; Erben, R.G. Vitamin d is a regulator of endothelial nitric oxide syn-thase and arterial stiffness in mice. *Mol. Endocrinol.* **2014**, *28*, 53–64. [CrossRef] [PubMed]

56. Lorente, L.; Gómez-Bernal, F.; Martín, M.M.; Navarro-Gonzálvez, J.A.; Argueso, M.; Perez, A.; Ramos-Gómez, L.; Solé-Violán, J.; Ramos, J.A.M.Y.; Ojeda, N.; et al. High serum nitrates levels in non-survivor COVID-19 patients. *Med. Intensiva* **2020**. [CrossRef] [PubMed]

57. Yardan, T.; Erenler, A.K.; Baydin, A.; Aydin, K.; Cokluk, C. Usefulness of S100B protein in neurological disorders. *J. Pak. Med. Assoc.* **2011**, *61*, 276–281. [PubMed]

58. Perrin, P.; Collongues, N.; Baloglu, S.; Bedo, D.; Bassand, X.; Lavaux, T.; Gautier-Vargas, G.; Keller, N.; Kremer, S.; Fafi-Kremer, S.; et al. Cytokine release syndrome-associated encephalo-pathy in patients with COVID-19. *Eur. J. Neurol.* **2021**, *28*, 248–258. [CrossRef] [PubMed]

59. Aceti, A.; Margarucci, L.M.; Scaramucci, E.; Orsini, M.; Salerno, G.; Di Sante, G.; Gianfranceschi, G.; Di Liddo, R.; Valeriani, F.; Ria, F.; et al. Serum S100B protein as a marker of severity in COVID-19 patients. *Sci. Rep.* **2020**, *10*, 18665. [CrossRef] [PubMed]

60. Bauer, W.; Ulke, J.; Galtung, N.; Strasser-Marsik, L.C.; Neuwinger, N.; Tauber, R.; Somasundaram, R.; Kappert, K. Role of Cell Adhesion Molecules for Prognosis of Disease Development of Patients with and Without COVID-19 in the Emergency Department. *J. Infect. Dis.* **2021**, *223*, 1497–1499. [CrossRef]

61. Li, L.; Huang, M.; Shen, J.; Wang, Y.; Wang, R.; Yuan, C.; Jiang, L.; Huang, M. Serum Levels of Soluble Platelet Endothelial Cell Adhesion Molecule 1 in COVID-19 Patients Are Associated with Disease Severity. *J. Infect. Dis.* **2021**, *223*, 178–179. [CrossRef]

62. Spadaro, S.; Fogagnolo, A.; Campo, G.; Zucchetti, O.; Verri, M.; Ottaviani, I.; Tunstall, T.; Grasso, S.; Scaramuzzo, V.; Murgolo, F.; et al. Markers of endothelial and epithelial pulmonary injury in mechanically ventilated COVID-19 ICU patients. *Crit. Care* **2021**, *25*, 74. [CrossRef] [PubMed]

 *nutrients*

Article

Assessment of Vitamin D Metabolism in Patients with Cushing's Disease in Response to 150,000 IU Cholecalciferol Treatment

Alexandra Povaliaeva *, Viktor Bogdanov, Ekaterina Pigarova, Artem Zhukov, Larisa Dzeranova, Zhanna Belaya, Liudmila Rozhinskaya, Galina Mel'nichenko and Natalia Mokrysheva

Endocrinology Research Centre, 117292 Moscow, Russia; bogdanov.viktor@endocrincentr.ru (V.B.); pigarova.ekaterina@endocrincentr.ru (E.P.); jukov.artem@endocrincentr.ru (A.Z.); dzeranova.larisa@endocrincentr.ru (L.D.); belaya.zhanna@endocrincentr.ru (Z.B.); rozhinskaya.ludmila@endocrincentr.ru (L.R.); melnichenko.galina@endocrincentr.ru (G.M.); mokrisheva.natalia@endocrincentr.ru (N.M.)
* Correspondence: povalyaeva.alexandra@endocrincentr.ru

Abstract: In this study we aimed to assess vitamin D metabolism in patients with Cushing's disease (CD) compared to healthy individuals in the setting of bolus cholecalciferol treatment. The study group included 30 adults with active CD and the control group included 30 apparently healthy adults with similar age, sex and BMI. All participants received a single dose (150,000 IU) of cholecalciferol aqueous solution orally. Laboratory assessments including serum vitamin D metabolites ($25(OH)D_3$, $25(OH)D_2$, $1,25(OH)_2D_3$, 3-epi-$25(OH)D_3$ and $24,25(OH)_2D_3$), free $25(OH)D$, vitamin D-binding protein (DBP) and parathyroid hormone (PTH) as well as serum and urine biochemical parameters were performed before the intake and on Days 1, 3 and 7 after the administration. All data were analyzed with non-parametric statistics. Patients with CD had similar to healthy controls $25(OH)D_3$ levels ($p > 0.05$) and higher $25(OH)D_3/24,25(OH)_2D_3$ ratios ($p < 0.05$) throughout the study. They also had lower baseline free $25(OH)D$ levels ($p < 0.05$) despite similar DBP levels ($p > 0.05$) and lower albumin levels ($p < 0.05$); 24-h urinary free cortisol showed significant correlation with baseline $25(OH)D_3/24,25(OH)_2D_3$ ratio (r = 0.36, $p < 0.05$). The increase in $25(OH)D_3$ after cholecalciferol intake was similar in obese and non-obese states and lacked correlation with BMI ($p > 0.05$) among patients with CD, as opposed to the control group. Overall, patients with CD have a consistently higher $25(OH)D_3/24,25(OH)_2D_3$ ratio, which is indicative of a decrease in 24-hydroxylase activity. This altered activity of the principal vitamin D catabolism might influence the effectiveness of cholecalciferol treatment. The observed difference in baseline free $25(OH)D$ levels is not entirely clear and requires further study.

Keywords: vitamin D; pituitary ACTH hypersecretion; cholecalciferol; vitamin D-binding protein

1. Introduction

Cushing's disease (CD) is one of the disorders associated with endogenous hypercortisolism and is caused by adrenocorticotropic hormone (ACTH) hyperproduction originating from pituitary adenoma [1]. Skeletal fragility is a frequent complication of endogenous hypercortisolism, and fragility fractures may be the presenting clinical feature of disease. The prevalence of osteoporosis in endogenous hypercortisolism as assessed by dual-energy X-ray absorptiometry (DXA) or incidence of fragility fractures has been reported to be up to 50%. Osteoporosis in CD patients has a complex multifactorial pathogenesis, characterized by a low bone turnover and severe suppression of bone formation [2]. Exogenous glucocorticoids are used in the treatment of a wide range of diseases and it is estimated that 1–2% of the population is receiving long-term glucocorticoid therapy. As a consequence, glucocorticoid-induced osteoporosis is the most common secondary cause of osteoporosis [3].

Citation: Povaliaeva, A.; Bogdanov, V.; Pigarova, E.; Zhukov, A.; Dzeranova, L.; Belaya, Z.; Rozhinskaya, L.; Mel'nichenko, G.; Mokrysheva, N. Assessment of Vitamin D Metabolism in Patients with Cushing's Disease in Response to 150,000 IU Cholecalciferol Treatment. *Nutrients* **2021**, *13*, 4329. https://doi.org/10.3390/nu13124329

Academic Editor: Spyridon N. Karras

Received: 12 November 2021
Accepted: 27 November 2021
Published: 30 November 2021

Publisher's Note: MDPI stays neutral with regard to jurisdictional claims in published maps and institutional affiliations.

Native vitamin D (in particular D_3, or cholecalciferol) and its active metabolites (such as alfacalcidol) are universally considered as the essential components of the osteoporosis management [4,5]. The search for the optimal treatment of bone complications during chronic exposure to glucocorticoid excess provoked the investigation of vitamin D metabolism in this state. Early studies on this topic were focused predominantly on the general vitamin D status (assessed as 25(OH)D level) and on the levels of the active vitamin D metabolite (1,25(OH)$_2$D). These studies showed inconsistent results, reporting that the chronic excess of glucocorticoids decreased [6–9], increased [10–12] or did not change [13–15] the levels of 25(OH)D or 1,25(OH)$_2$D. A likely reason for such inconsistency might have been the high heterogeneity of the studied groups. Some of these studies were performed in humans [6,7,9–13,15] and some in animal models [8,14], and only several of them included subjects with specifically endogenous hypercortisolism [10,12,14,15]. Only two studies assessed both the levels of the active (1,25(OH)$_2$D) and the inactive (24,25(OH)$_2$D) vitamin D metabolites in endogenous hypercortisolism. One of them lacked control group and reported low-normal 24,25(OH)$_2$D levels in patients with Cushing's syndrome [10]. The second study by Corbee et al. reported similar circulating concentrations of 25(OH)D, 1,25(OH)$_2$D and 24,25(OH)$_2$D in studied groups of dogs regardless of either the presence of CD or hypophysectomy status [14].

Several experimental studies were performed to evaluate the impact of glucocorticoid excess on the enzymes involved in vitamin D metabolism. In mouse kidney glucocorticoid treatment increased 24-hydroxylase expression [16] and 24-hydroxylase activity [17]. An increased expression of 24-hydroxylase was also shown in rat osteoblastic and pig renal cell cultures treated with 1,25(OH)$_2$D [18]. Dhawan and Christakos showed that 1,25(OH)$_2$D-induced transcription of 24-hydroxylase was glucocorticoid receptor-dependent [19]. However, some works showed conflicting results. In particular, the steroid and xenobiotic receptor (SXR) which is activated by glucocorticoids [20], repressed 24-hydroxylase expression in human liver and intestine in work by Zhou et al. [21]. Lower 24-hydroxylase expression was observed in the brain and myocardium of glucocorticoid-treated rats [22] as well as in human osteosarcoma cells and human osteoblasts [23].

Nevertheless, based on experimental data, it has been suggested that the acceleration of 25(OH)D catabolism in the presence of glucocorticoid excess may predispose to vitamin D deficiency. Yet, relatively recent meta-analysis of the studies assessing 25(OH)D levels in chronic glucocorticoid users showed that serum 25(OH)D levels in these patients were suboptimal and lower than in healthy controls, but similar to steroid-naive disease controls [24].

Glucocorticoids also affect calcium and phosphorus homeostasis. In particular, they were shown to reduce gastrointestinal absorption by antagonizing vitamin D action (reducing the expression of genes for proteins involved in calcium transport—epithelial Ca channel TRPV6 and calcium-binding protein calbindin-D9K) [25]. Glucocorticoids increased fractional calcium excretion due to mineralocorticoid receptor-mediated action on epithelial sodium channels [26]. Hypercalciuria is highly prevalent in people with CD [27]. These effects might result in a negative calcium balance, although plasma ionized calcium was normal in people and dogs with hypercortisolism compared to control subjects [12,28]. Glucocorticoids also reduced tubular phosphate reabsorption by inhibiting tubular expression of the sodium gradient-dependent phosphate transporter, and induced phosphaturia [29], which was accompanied by phosphate lowering in humans [12].

Overall, current data on vitamin D status in hypercortisolism are conflicting and need clarification. In particular, clinical data on the state of vitamin D metabolism in the state of glucocorticoids excess are quite scarce. Studies were very heterogeneous in design, some lacked a control group, and the absolute majority of the studies were performed before the introduction of vitamin D measurement standardization [30]. Nevertheless, determining the optimal vitamin D treatment regimen in these high-risk patients is fairly relevant.

The aim of this study was to assess vitamin D metabolism in patients with CD compared to healthy individuals particularly in the setting of cholecalciferol treatment.

2. Materials and Methods

2.1. Study Population and Design

The study group included 30 adult patients with CD admitted for inpatient treatment at a tertiary pituitary center. Diagnosis of CD was established in accordance with the federal guidelines [31]. All patients were confirmed to be positive for endogenous hypercortisolism in at least two of the following tests: 24-h urine free cortisol (UFC) greater than the normal range for the assay and/or serum cortisol > 50 nmol/L after the 1-mg overnight dexamethasone suppression test and/or late-night salivary cortisol greater than 9.4 nmol/L). All patients also had morning ACTH $\geq$ 10 pg/mL and pituitary adenoma $\geq$ 6 mm identified by magnetic resonance imaging (MRI) or a positive for CD bilateral inferior petrosal sinus sampling (BIPSS). MRI was performed using a GE Optima MR450w 1.5T with Gadolinium (Boston, MA, USA). BIPSS was performed according to the standard procedure described elsewhere [32,33].

The control group included 30 apparently healthy adult individuals recruited from the staff and the faculty of the facility.

Inclusion criteria were age from 18 to 60 for both groups and the presence of the disease activity for the study group (defined as the presence of endogenous hypercortisolism at the time of participation in the study). Exclusion criteria for both groups were: vitamin D supplementation for 3 months prior to the study; severe obesity (body mass index (BMI) $\geq$ 35 kg/m^2); pregnancy; the presence of granulomatous disease, malabsorption syndrome, liver failure; decreased GFR (less than 60 mL/min per 1.73 m^2); severe hypercalcemia (total serum calcium > 3.0 mmol/L); allergic reactions to vitamin D medications; 25(OH)D level more than 60 ng/mL (determined by immunochemiluminescence analysis). All patients were recruited in the period from October 2019 to April 2021. The study protocol (ClinicalTrials.gov Identifier: NCT04844164) was approved by the Ethics Committee of Endocrinology Research Centre, Moscow, Russia on 10 April 2019 (abstract of record No. 6), all patients signed informed consent to participate in the study.

All participants received standard therapeutic dose (150,000 IU) of an aqueous solution of cholecalciferol (Aquadetrim®, Medana Pharma S.A., Sieradz, Poland) orally as a single dose [34]. Blood and urine samples were obtained before the intake as well as on days 1, 3 and 7 after administration; time points of sample collection were determined based on the authors' previous work evaluating changes in 25(OH)D levels after a therapeutic dose of cholecalciferol [35]. The assessment included serum biochemical parameters (total calcium, albumin, phosphorus, creatinine, magnesium), parathyroid hormone (PTH), vitamin D-binding protein (DBP), vitamin D metabolites (25(OH)D$_3$, 25(OH)D$_2$, 1,25(OH)$_2$D$_3$, 3-epi-25(OH)D$_3$ and 24,25(OH)$_2$D$_3$), free 25(OH)D and urine biochemical parameters (calcium- and phosphorus-creatinine ratios in spot urine).

2.2. Socio–Demographic and Anthropometric Data Collection

At the baseline visit, patients underwent a questionnaire aimed to assess their lifestyle: the presence of unhealthy habits, physical activity level, balanced diet (consumption of dairy products, meat, coffee, soft drinks), exposure to ultraviolet (UV) radiation (solarium and sunscreen usage, traveling south and the number of daytime walks in the sunny weather in the 3 months preceding study participation). Smoking status was classified as current smoker, former smoker and non-smoker; current and former smokers were collectively referred to as total smokers. A unit of alcohol was defined as a glass of wine, a bottle of beer or a shot of spirits, approximating 10–12 g ethanol. Serving of dairy products was defined as 100 g of cottage cheese, 200 mL of milk, 125 g of yogurt or 30 g of cheese. Patients' weight was measured in light indoor clothing with a medical scale to the nearest 100 g, and their height with a wall-mounted stadiometer to the nearest centimeter. BMI was calculated as weight in kilograms divided by height in meters squared.

2.3. Laboratory Measurements

Morning ACTH (reference range 7–66 pg/mL), serum cortisol after a low-dose dexamethasone suppression test (cutoff value for suppression, 50 nmol/L [36]), late-night salivary cortisol (reference range 0.5–9.4 nmol/L [37]) were assayed by electrochemiluminescence assay using a Cobas 6000 Module e601 (Roche, Rotkreuz, Switzerland). The 24-h UFC (reference range 60–413 nmol/24 h) was measured by an immunochemiluminescence assay (extraction with diethyl ether) on a Vitros ECiQ (Ortho Clinical Diagnostics, Raritan, NJ, USA).

Total 25(OH)D levels ($25(OH)D_2$ + $25(OH)D_3$; reference range 30–100 ng/mL) at the baseline visit were determined by the immunochemiluminescence analysis (Liaison, DiaSorin, Saluggia, Italy). PTH levels were evaluated by the electrochemiluminescence immunoassay (ELECSYS, Roche, Basel, Switzerland; reference range for this and subsequent laboratory parameters are given in the Results section for easier reading). Biochemical parameters of blood serum and urine were assessed by the ARCHITECT c8000 analyzer (Abbott, Chicago, IL, USA) using reagents from the same manufacturer according to the standard methods. Serum DBP and free 25(OH)D levels were measured by enzyme-linked immunosorbent assay (ELISA) using commercial kits. The assay used for free 25(OH)D levels assessment (DIAsource, ImmunoAssays S.A., Ottignies-Louvain-la-Neuve, Belgium) has <6.2% intra- and inter-assay coefficient of variation (CV) at levels 5.8–9.6 pg/mL. The assay used for DBP levels assessment (Assaypro, St Charles, MO, USA) has 6.2% average intra-assay CV and 9.9% average inter-assay CV.

The levels of vitamin D metabolites ($25(OH)D_3$, $25(OH)D_2$, $1,25(OH)2D_3$, 3-epi-$25(OH)D_3$ and $24,25(OH)_2D_3$) in serum were determined by ultra-high performance liquid chromatography in combination with tandem mass spectrometry (UPLC-MS/MS) using an in-house developed method, described earlier [38]. With this technique, the laboratory participates in DEQAS quality assurance program (lab code 2388) and the results fall within the target range for the analysis of 25(OH)D and $1,25(OH)_2D$ metabolites in human serum (Supporting Information, Figures S1 and S2). All UPLC-MS/MS measurements were made after the first successful completion (5/5 samples within the target range) of the DEQAS distributions for both analytes simultaneously. Each batch contained control samples (analytes in blank serum) with both high and low analyte concentrations. The samples were barcoded and randomized prior to the measurements to eliminate analyst-related errors.

Serum samples (3 aliquots) collected at each visit were either transferred directly to the laboratory for biochemical analyzes, total 25(OH)D and PTH measurement (1 aliquot) or were stored at −80 °C avoiding repeated freeze-thaw cycles for measurement of DBP, free 25(OH)D and vitamin D metabolites at a later date (2 aliquots).

Albumin-adjusted serum calcium levels were calculated using the formula [39]: total plasma calcium (mmol/L) = measured total plasma calcium (mmol/L) + 0.02 × (40 − measured plasma albumin (g/L)).

Baseline free 25(OH)D levels were also calculated using the formula introduced by Bikle et al. [40,41]. The affinity constant for 25(OH)D and albumin binding (Kalb) used for the calculation was equal 6×10^5 M^{-1}, and affinity constant for 25(OH)D and DBP binding (KDBP) was equal 7×10^8 M^{-1}.

$$\text{Free } 25(OH)D = \frac{\text{total } 25(OH)D}{1 + K_{alb}*\text{albumin} + KDBP * DBP}$$

2.4. Statistical Analysis

Statistical analysis was performed using Statistica version 13.0 (StatSoft, Tulsa, OK, USA). All data were analyzed with non-parametric statistics and expressed as median [interquartile range] unless otherwise specified. Mann-Whitney U-test and Fisher's exact two-tailed test were used for comparisons between two groups. Friedman ANOVA was performed to evaluate changes in indices throughout the study and pairwise comparisons using Wilcoxon test with adjustment for multiple comparisons (Bonferroni) were also made if the Friedman ANOVA was significant. Spearman rank correlation method was used to obtain correlation coefficients among indices. A *p*-value of less than 0.05 was considered statistically significant. When adjusting for multiple comparisons, a *p*-value greater than the significance threshold, but less than 0.05 was considered as a trend towards statistical significance.

3. Results

The groups were similar in terms of age, sex and BMI ($p > 0.05$). Both groups consisted predominantly of young and middle-aged women and the majority of patients were overweight or moderately obese (Table 1). Patients from the study group presented with lower screening levels of total 25(OH)D ($p < 0.05$).

Table 1. General characteristics of the patients at the baseline visits. For detailed description of the data format please refer to the Section 2.

Parameter	Study Group (*n* = 30)	Control Group (*n* = 30)	*p*
Age, years	39.1 [31.2; 48.2]	33.4 [26.5; 42.5]	0.12
Sex (female/male, *n*)	26/4	19/11	0.07
BMI, kg/m²	30.9 [27.1; 31.6]	27.2 [25.4; 30.4]	0.07
25(OH)D total, ng/mL	13.1 [9.6; 17.9]	21.7 [14.4; 28.0]	0.002

The features of the underlying disease course in the study group are listed in Table 2. 15 patients (50%) had diabetes mellitus with an almost compensated state at the time of participation in the study, and 7 patients (23%) reported a history of low-energy fractures.

Table 2. Characteristics of the patients with Cushing's disease (CD) in terms of the underlying disease.

Parameter	Value
24-h UFC, nmoL/24 h	1227 [813; 2970]
Morning ACTH, pg/mL	87 [60; 125]
Diabetes mellitus, *n* (%)	15 (50%)
HbA1c, %	7.8 [7.0; 8.4]
History of low energy fracture, *n* (%)	7 (23%)

The groups did not differ significantly in the reported smoking status, the level of daily physical activity, dietary habits and UV exposure ($p > 0.05$) and although there was a slight difference in alcohol consumption ($p < 0.05$), the absolute values were minor in both groups (Table 3).

Table 3. Questionnaire results.

Parameter	Study Group (*n* = 30)	Control Group (*n* = 30)	*p*
Current smokers, *n* (%)	6 (20%)	13 (43%)	0.09
Total smokers, *n* (%)	10 (33%)	18 (60%)	0.07
Alcohol units, per week	0 [0; 0]	1 [0; 2]	0.007
Exercises lasting more than 30 min, per week	5 [2; 7]	3 [2; 3]	0.09
Dairy products consumption, servings per day	1 [1; 1]	1 [1; 1]	1.0
Meat dishes consumption, portions per week	5 [4; 7]	5 [3; 7]	0.64
Coffee consumption, cups per week	6 [2; 8]	7 [1; 10]	0.4
Soft drinks, mL per week	0 [0; 0]	0 [0; 0]	0.76
Travelers to the south, *n* (%)	3 (10%)	4 (13%)	1.0
Daytime walks in the sunny weather, *n*	7 [0; 20]	4 [1; 11]	0.49
Solarium usage, *n* (%)	0	1 (3%)	1.0

3.1. Baseline Laboratory Evaluation

Detailed results of laboratory studies are presented in Tables 4 and 5.

Table 4. Changes in the levels of the biochemical parameters and parathyroid hormone (PTH) during the study.

Laboratory Parameter	Group	Day 0	Day 1	Day 3	Day 7	p (Friedman ANOVA)	p (Day 0–1)	p (Day 1–3)	p (Day 3–7)	Reference Range
Total calcium, mmol/L	Study	2.39 [2.25; 2.44]	2.37 [2.31; 2.47]	2.35 [2.27; 2.46]	2.39 [2.27; 2.51]	0.89	-	-	-	2.15–2.55
	Control	2.37 [2.31; 2.43]	2.41 [2.36; 2.46]	2.41 [2.37; 2.46]	2.37 [2.34; 2.48]	0.03	0.01	0.47	0.28	
Albumin-adjusted calcium, mmol/L	Study	2.28 [2.21; 2.36]	2.31 [2.23; 2.38]	2.30 [2.22; 2.36]	2.31 [2.23; 2.37]	0.92	-	-	-	2.15–2.55
	Control	2.25 [2.21; 2.31]	2.29 [2.23; 2.33]	2.30 [2.26; 2.35]	2.27 [2.24; 2.34]	0.005	0.01	0.24	0.29	
Phosphorus, mmol/L	Study	1.04 [0.98; 1.13]	1.15 [1.02; 1.19]	1.18 [1.04; 1.23]	1.07 [0.97; 1.19]	0.003	0.006	0.58	0.03	0.74–1.52
	Control	1.10 [1.00; 1.22]	1.15 [1.01; 1.26]	1.19 [1.07; 1.27]	1.15 [1.09; 1.31]	0.06	-	-	-	
PTH, pg/mL	Study	38.9 [33.8; 55.2]	39.5 [29.7; 52.3]	40.1 [31.7; 52.8]	40.1 [30.5; 53.6]	0.6	-	-	-	15–65
	Control	38.6 [31.0; 50.3]	37.0 [28.9; 51.4]	35.8 [28.9; 45.3]	34.3 [25.3; 47.7]	0.03	0.74	0.02	0.93	
Creatinine, μmol/L	Study	67.6 [62.4; 70.1] *	68.9 [63.4; 72.8] *	68.9 [62.1; 72.9]	69.0 [63.3; 72.5]	0.3	-	-	-	63–110 (male) 50–98 (female)
	Control	70.3 [67.4; 78.0]	73.5 [66.9; 79.8]	70.3 [67.1; 79.4]	72.2 [64.0; 83.9]	0.02	0.002	0.20	0.21	
Albumin, g/L	Study	44 [42; 46] *	44 [41; 45] *	44 [40; 47] *	44 [41; 47]	0.3	-	-	-	35–50
	Control	46 [44; 47]	46 [44; 48]	46 [44; 47]	46 [44; 47]	0.48	-	-	-	
Magnesium, mmol/L	Study	0.87 [0.81; 0.92] *	0.86 [0.79; 0.94] *	0.84 [0.78; 0.91] *	0.87 [0.81; 0.93] *	0.27	-	-	-	0.7–1.05
	Control	0.82 [0.76; 0.85]	0.79 [0.77; 0.84]	0.79 [0.76; 0.82]	0.79 [0.75; 0.84]	0.67	-	-	-	
Urine calcium-creatinine ratio, mmol/mmol	Study	0.36 [0.16; 0.49]	0.49 [0.28; 0.63] *	0.37 [0.22; 0.59]	0.49 [0.23; 0.80]	0.05	-	-	-	0.1–0.8
	Control	0.30 [0.13; 0.42]	0.26 [0.21; 0.41]	0.29 [0.21; 0.42]	0.35 [0.20; 0.50]	0.88	-	-	-	
Urine phosphorus-creatinine ratio, mmol/mmol	Study	2.6 [2.0; 3.1] *	2.6 [1.7; 3.1] *	2.6 [2.0; 3.6] *	3.4 [2.4; 4.1] *	0.001	0.51	0.66	0.02	1.4–3.5
	Control	1.8 [1.4; 2.7]	1.7 [0.9; 2.4]	1.6 [1.4; 2.3]	1.7 [1.2; 2.3]	0.09	-	-	-	

* Significant difference in between-group comparison.

Table 5. Changes in the levels of free 25(OH)D, vitamin D-binding protein (DBP) and vitamin D metabolites during the study.

Laboratory Parameter	Group	Day 0	Day 1	Day 3	Day 7	p (Friedman ANOVA)	p (Day 0–1)	p (Day 1–3)	p (Day 3–7)	Reference Range
Free 25(OH)D, pg/mL	Study	4.9 [4.0; 6.1] *	10.7 [8.4; 12.5]	12.9 [11.0; 14.3]	11.4 [10.0; 12.5]	<0.001	<0.001	<0.001	<0.001	2.4–35 [1]
	Control	6.4 [4.1; 7.7]	12.1 [9.5; 15.0]	14.0 [10.3; 18.1]	12.9 [9.4; 15.4]	<0.001	<0.001	<0.001	<0.001	
DBP, mg/L	Study	270 [227; 298]	277 [247; 328]	276 [236; 301] *	252 [206; 281]	0.16	-	-	-	176–623 [1]
	Control	247 [212; 281]	258 [237; 300]	236 [204; 274]	245 [220; 277]	0.31	-	-	-	
25(OH)D$_3$, ng/mL	Study	17.9 [13.0; 24.5]	34.5 [27.6; 38.8]	37.6 [33.2; 45.5]	35.4 [32.1; 42.7]	<0.001	<0.001	<0.001	0.01	$\geq$30 [2]
	Control	19.5 [12.5; 25.7]	31.0 [28.1; 35.0]	33.9 [30.2; 43.1]	34.3 [30.9; 42.9]	<0.001	<0.001	<0.001	0.65	
3-epi-25(OH)D$_3$, ng/mL	Study	0.8 [0.6; 1.1] *	3.0 [2.4; 3.5]	4.2 [3.6; 5.1]	3.1 [2.7; 3.8]	<0.001	<0.001	<0.001	<0.001	not available
	Control	1.4 [0.9; 1.7]	2.7 [2.1; 3.5]	3.9 [3.3; 4.9]	3.6 [3.0; 4.6]	<0.001	<0.001	<0.001	0.003	
1,25(OH)$_2$D$_3$, pg/mL	Study	41 [35; 50]	48 [37; 53]	47 [42; 56]	42 [39; 52]	<0.001	<0.001	0.22	0.02	25–66 [3]
	Control	42 [34; 48]	48 [41; 55]	46 [39; 54]	43 [35; 47]	0.09	-	-	-	
24,25(OH)$_2$D$_3$, ng/mL	Study	1.0 [0.5; 1.4] *	1.4 [0.8; 1.8] *	2.1 [1.6; 2.8]	2.6 [1.8; 3.1]	<0.001	<0.001	<0.001	<0.001	0.5–5.6 [3]
	Control	1.5 [0.9; 2.6]	1.8 [1.2; 2.6]	3.0 [1.9; 3.6]	3.2 [2.1; 4.2]	<0.001	<0.001	<0.001	<0.001	
25(OH)D$_3$/24,25(OH)$_2$D$_3$	Study	19.5 [15.3; 26.8] *	26.8 [21.2; 32.5] *	17.4 [15.9; 21.0] *	13.9 [12.3; 18.8] *	<0.001	<0.001	<0.001	<0.001	7–23 [3]
	Control	12.7 [9.9; 17.0]	18.2 [14.2; 21.3]	13.1 [10.6; 18.2]	12.3 [9.0; 14.9]	<0.001	<0.001	<0.001	<0.001	
25(OH)D$_3$/1,25(OH)$_2$D$_3$	Study	425 [373; 555]	716 [588; 917]	783 [628; 1034]	814 [666; 1044]	<0.001	<0.001	0.08	0.86	not available
	Control	501 [356; 641]	689 [548; 789]	757 [602; 972]	851 [727; 1028]	<0.001	<0.001	0.03	0.05	

* Significant difference in between-group comparison. [1] Reference ranges are specified according to kit manufacturers' recommendations. [2] Reference range is given for total 25(OH)D according to the clinical guidelines [34,42]; the 25(OH)D$_2$ fraction is negligible (<0.5 ng/mL in absolute values) for the purposes of this study. [3] Reference ranges are given according to the literature data [43,44].

Patients with CD had several alterations in biochemical parameters, in particular, lower baseline serum creatinine and albumin levels, while magnesium levels were higher than in the control group ($p < 0.05$). They also had higher levels of urine phosphorus-creatinine ratio ($p < 0.05$). The rest of the studied biochemical parameters did not show significant difference between the groups ($p > 0.05$). 3 patients (10%) from the study group and 5 patients (17%) from the control group had secondary hyperparathyroidism, one patient with CD (3%) was diagnosed with mild primary hyperparathyroidism.

As for the assessment of vitamin D metabolism, unexpectedly the levels of $25(OH)D_3$ occurred to be equal in the groups ($p > 0.05$), with only two patients (7%) from the study group and one patient (3%) from the control group having sufficient vitamin D levels, according to the Endocrine Society and the Russian Association of Endocrinologists guidelines ($\geq$30 ng/mL [34,42]). The levels of the active vitamin D metabolite—$1,25(OH)_2D_3$—were equal between the groups as well ($p > 0.05$), whereas the levels of 3-epi-$25(OH)D_3$ and $24,25(OH)_2D_3$ were lower in CD patients. Further calculation of $25(OH)D_3/24,25(OH)_2D_3$ and $25(OH)D_3/1,25(OH)_2D_3$ ratios corresponded to the observed levels of metabolites: $25(OH)D_3/24,25(OH)_2D_3$ ratio was higher in the study group ($p < 0.05$) assuming lower 24-hydroxylase activity and $25(OH)D_3/1,25(OH)_2D_3$ ratio was equal between the groups ($p > 0.05$).

Levels of free 25(OH)D were lower in CD patients ($p < 0.05$) and the levels of DBP did not differ between the groups ($p > 0.05$). Although calculated free 25(OH)D showed prominent positive correlation with the measured free 25(OH)D in both groups (r = 0.63 in the study group, r = 0.87 in the control group, $p < 0.05$), the association appeared to be weaker in the study group. In the control group, DBP levels correlated with both measured and calculated 25(OH)D levels (r = -0.48, $p < 0.05$ and r = -0.69, $p < 0.05$ respectively), while in patients with CD there was no association with measured free 25(OH)D levels (r = 0.04, $p > 0.05$ and r = -0.50, $p < 0.05$ respectively).

Correlation with 24-h UFC in CD patients was observed for serum albumin level (r = -0.37, $p < 0.05$) and urine calcium-creatinine ratio (r = 0.51, $p < 0.05$) among assessed biochemical parameters, and only with $25(OH)D_3/24,25(OH)_2D_3$ ratio among the parameters of vitamin D metabolism (r = 0.36, $p < 0.05$).

3.2. Laboratory Evaluation after the Intake of Cholecalciferol

All patients from the study group and 28 patients (93%) from the control group completed the study.

The observed baseline differences in biochemical parameters mostly preserved during the follow-up. In the study group there was an increase in serum phosphorus levels by Day 1 ($p = 0.006$) and a tendency to an increase in the urine phosphorus-creatinine ratio by Day 7 ($p = 0.02$). Patients from the control group showed a clinically insignificant increase in serum creatinine levels by Day 1 ($p = 0.002$) and a non-significant trend towards an increase in serum total and albumin-adjusted calcium ($p = 0.01$ for both measurements). No change in PTH levels was observed in patients with CD during the follow-up ($p > 0.05$), while in the control group there was a tendency for PTH to decrease by Day 3 ($p = 0.02$). There were no new cases of hypercalcemia in both groups during the follow-up. One patient from the study group and one patient from the control group had persistently increased urine calcium-creatinine ratio throughout the study. Four patients from the study group (13%) and none from the control group developed hypercalciuria during the follow-up, however these patients had no clinical manifestations during the observation period.

By Day 7, 25 patients (83%) from the study group and 22 patients (79%) reached sufficient $25(OH)D_3$ levels ($\geq$30 ng/mL). Levels of $25(OH)D_3$ continued to increase by Day 3 in both groups ($p < 0.001$), after which tended to decrease in the study group ($p = 0.01$) and remained stable in the control group ($p = 0.65$). The increase in $25(OH)D_3$ after cholecalciferol intake was equal between the groups (18.5 [15.9; 22.5] ng/mL in the study group vs. 16.6 [13.1; 19.8] ng/mL in the control group, $p > 0.05$). In the presence of obesity, $\Delta25(OH)D_3$ was higher in the CD patients than in the control group (18.3 [14.2;

23.0] vs. 12.1 [10.0; 13.1] ng/mL, $p < 0.05$), while in non-obese patients no difference was observed ($p > 0.05$).

Obese and non-obese patients with CD had equal $\Delta25(OH)D_3$ (18.3 [14.2; 23.0] vs. 19.6 [16.0; 21.5] ng/mL, $p > 0.05$), while in the control group it was significantly lower in obese patients (12.1 [10.0; 13.1] vs. 18.3 [15.3; 21.4] ng/mL, $p < 0.05$). BMI showed significant correlation with $\Delta25(OH)D_3$ only in the control group ($r = -0.47$, $p < 0.05$), while in CD patients there was no such association ($r = -0.06$, $p > 0.05$) (Figure 1).

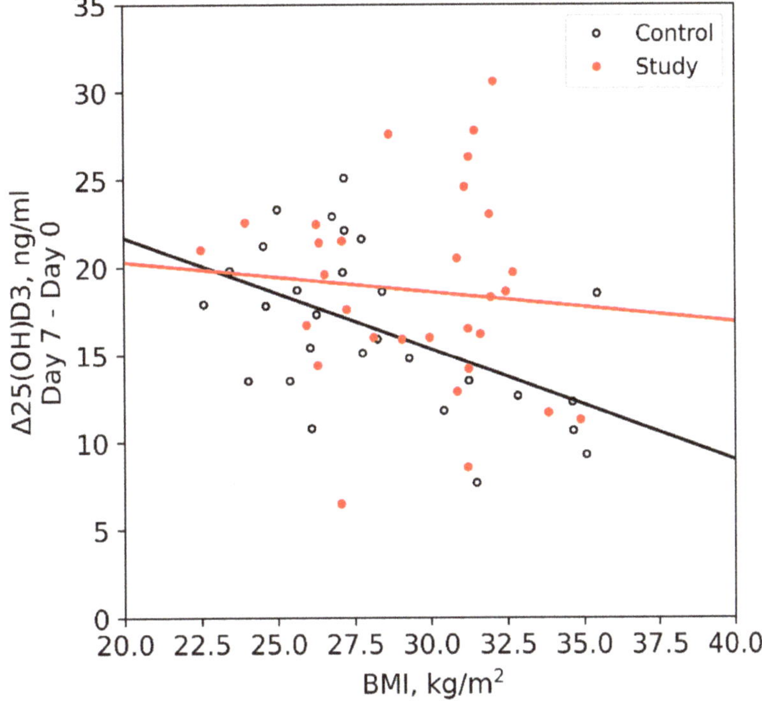

Figure 1. Relationship between $\Delta25(OH)D_3$ and BMI in groups.

$1,25(OH)_2D_3$ levels increased in CD patients by Day 1 and were stable during the follow-up in the control group. The rest of the studied parameters of vitamin D metabolism changed in a similar way between groups: 3-epi-$25(OH)D_3$ levels increased until the Day 3, after which they decreased by the Day 7; $24,25(OH)_2D_3$ levels showed more graduate elevation throughout the follow-up. In both groups $25(OH)D_3/24,25(OH)_2D_3$ ratios increased by Day 1, after which they decreased by Day 7, and $25(OH)D_3/1,25(OH)_2D_3$ ratios increased by Day 1, after which they remained stable. DBP levels didn't change and free 25(OH)D levels showed an increase in both groups during the follow-up. The levels of $25(OH)D_2$ did not exceed 0.5 ng/mL in all examined individuals throughout the study. Among assessed parameters of vitamin D metabolism, higher $25(OH)D_3/24,25(OH)_2D_3$ ratios in the study group was the only difference between the groups which remained significant throughout the observation period ($p < 0.05$) (Figure 2).

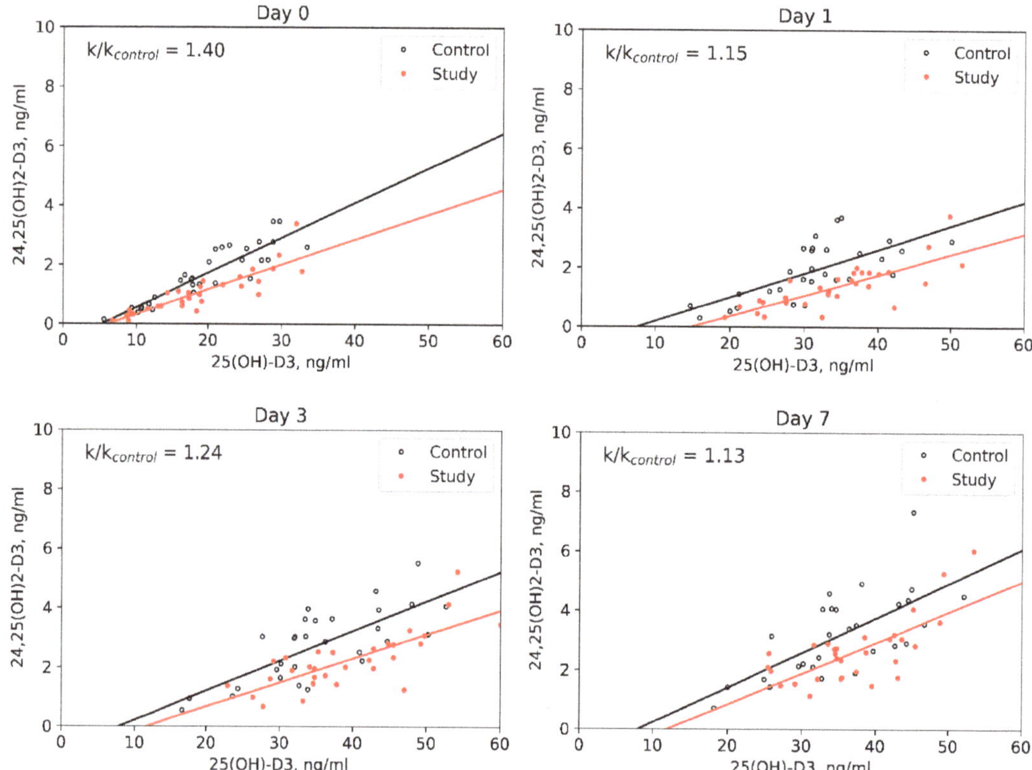

Figure 2. Dynamic evaluation of $25(OH)D_3/24,25(OH)_2D_3$ ratios in groups.

4. Discussion

The main goal of our study was to evaluate the $25(OH)D_3$ levels and its response to the therapeutic dose of cholecalciferol in patients with CD as compared to healthy individuals. We observed no difference in baseline $25(OH)D_3$ assessed by UPLC-MS/MS between groups. Similar to our data were obtained in most studies conducted specifically in the state of endogenous hypercortisolism in humans [12,15] and dogs [14]. The study by Kugai et al. lacked control group and reported plasma levels of $25(OH)D$ corresponding to the vitamin D deficiency in most of the examined patients [10], while in our study only 2/3 of the patients with CD had $25(OH)D$ levels below 20 ng/mL. As for exogenous hypercortisolism, the meta-analysis aimed to explore serum $25(OH)D$ levels in glucocorticoid users showed lower levels than in healthy controls, but similar to steroid-naive disease controls, thus causing concern regarding the influence of the disease status on $25(OH)D$ levels [24]. Somewhat surprisingly, we obtained significantly discordant results in the study group when screening total $25(OH)D$ by ELISA and when measuring baseline $25(OH)D_3$ by UPLC-MS/MS, since the initial difference between the groups revealed by ELISA data with lower total $25(OH)D$ levels in the study group was not replicated by UPLC-MS/MS. It should be noted that our ELISA method did not participate in an external quality control program at the time of the study unlike UPLC-MS/MS; furthermore, a lower analytical performance was previously described for this technique with tendency for low specificity and lower measurement results [45].

When assessing other parameters of vitamin D metabolism, the most significant finding was the higher $25(OH)D_3/24,25(OH)_2D_3$ ratio in CD patients, both initially and during the observation after the intake of the cholecalciferol loading dose, indicating consistently reduced activity of 24-hydroxylase, the main enzyme of vitamin D catabolism. Earlier clinical and experimental studies also suggested altered activity of enzymes of vitamin D metabolism in hypercortisolism. However, these studies were heterogeneous and aimed predominantly at studying the activity of 1α-hydroxylase [7,8,10–12,14], which was not altered in patients with CD as compared to healthy individuals in our study. In the setting of the short-term glucocorticoid administration, Lindgren et al. showed transient increase in $24,25(OH)_2D_3$ levels in rats [8], while in the study of Hahn et al. there was no change in $24,25(OH)_2D_3$ levels [11]. Dogs with CD had similar $24,25(OH)_2D_3$ levels before and after hypophysectomy as well as compared to control dogs [14]. The only study of considerably similar design by Kugai et al. reported low-normal $24,25(OH)_2D_3$ in patients with Cushing's syndrome [10], which is consistent with our result, as well as some experimental works indicative of suppression on CYP24A1 expression by glucocorticoids in human osteoblasts [23], liver and intestine [21] and in rat brain and myocardium [22]. However, in the present work, the activity of 24-hydroxylase in patients with hypercortisolism was for the first time evaluated by calculating the $25(OH)D_3/24,25(OH)_2D_3$ ratio, which has recently emerged as a new tool for vitamin D status assessment [46,47]. Given the correlation of this parameter with laboratory marker of the underlying disease activity (24-h UFC), a direct effect of cortisol overproduction on 24-hydroxylase activity might be assumed. Interestingly, it seems that the decreased activity of 24-hydroxylase observed in CD influenced the effectiveness of cholecalciferol treatment, decreasing the negative effect of obesity, as patients with CD had similar increase in $25(OH)D_3$ in obese and non-obese state and lacked correlation between $\Delta25(OH)D_3$ and BMI, as opposed to the control group. Moreover, the increase in $25(OH)D_3$ in obese patients from the control group was lower not only than in non-obese controls, but also than in obese patients with CD.

Another intriguing finding was lower levels of free 25(OH)D observed in patients with CD despite similar DBP levels and lower albumin levels, which, on the contrary, allows one to expect higher values of free 25(OH)D. Considering the weaker correlation between the measured and calculated free 25(OH)D in patients with CD, as well as the lack of correlation of the measured 25(OH)D with the main transport protein, an altered affinity of DBP might be suspected. One possible explanation is protein glycosylation as a consequence of diabetes mellitus, which was present in half of the patients [38,48,49]. After cholecalciferol intake, which was accompanied by an increase in free 25(OH)D, the differences between the groups were leveled; therefore, another suggested explanation might be competitive binding to the ligand. Since actin binds DBP with high affinity [50] and considering catabolic action of glucocorticoids on muscle tissue [51], actin is a presumable competing ligand candidate. Although this is mostly speculative, as far as the authors are aware, the present work was the first to assess free vitamin D in the glucocorticoid excess, so the described findings require verification of reproducibility and further evaluation.

The obtained discrepancies in the biochemical parameters characterizing calcium and phosphorus metabolism were generally consistent with the data of early studies discussed in the introduction [12,25–29], except for similar to controls serum phosphorus levels and lower prevalence of hypercalciuria. An interesting observation was the complete absence of the PTH decrease in patients with CD after receiving a loading dose of cholecalciferol. The mechanism of this phenomenon is not entirely clear, we tend to agree with the earlier hypothesis that this may be an adaptation to chronic urinary calcium loss [52].

Our research is distinguished by a number of important strengths: a prospective design, substantial sample of patients with CD, accounting for social and behavioral factors affecting vitamin levels D, comprehensive spectrum of vitamin D metabolism parameters investigated and participation in an external quality control program for vitamin D metabolites measurement.

Nevertheless, the study also had several limitations: the amount of dietary vitamin D and phosphorus, as well as possible differences in DBP affinity to vitamin D metabolites due to genetic isoforms of DBP [53] or other possible involved parameters (e.g., fibroblast growth factor-23) were not taken into account. A few patients from both groups received therapy with possible impact on vitamin D and calcium metabolism within 3 months preceding the participation in the study (spironolactone, diuretics, proton pump inhibitors, oral contraceptives, antifungal treatment, antidepressants, barbiturates, antiepileptic drugs). The groups had a trend for differences in sex and BMI ($p = 0.07$ for both parameters). Also, the study lacked a study group of patients with remission of CD to test the hypotheses put forward, however, this is a promising direction for further research.

5. Conclusions

We report that patients with endogenous ACTH-dependent hypercortisolism of pituitary origin have a consistently higher $25(OH)D_3/24,25(OH)_2D_3$ ratio than healthy controls, which is indicative of a decrease in 24-hydroxylase activity. This altered activity of the principal vitamin D catabolism might influence the effectiveness of cholecalciferol treatment. There is also a lack of clarity regarding the lower levels of free 25(OH)D observed in patients with CD, which require further study. To test the proposed hypotheses and to develop specialized clinical guidelines for these patients, longer-term randomized clinical trials are needed.

Supplementary Materials: The following are available online at https://www.mdpi.com/article/10.3390/nu13124329/s1, Method validation against DEQAS, Figure S1: Comparison between DEQAS data for 25(OH)D scheme and our lab results, Figure S2: Comparison between DEQAS data for $1,25(OH)_2D$ scheme and our lab results.

Author Contributions: Conceptualization, L.R., E.P., A.P. and A.Z.; methodology, V.B., Z.B., L.R. and G.M.; formal analysis, A.P.; investigation, A.P., V.B., E.P., L.D. and A.Z.; data curation, A.P. and V.B.; writing—original draft preparation, A.P.; writing—review and editing, V.B., E.P., A.Z., Z.B., L.R.; visualization, V.B.; supervision, L.D., L.R., G.M. and N.M.; project administration, L.R. and N.M.; funding acquisition, L.R. and N.M. All authors have read and agreed to the published version of the manuscript.

Funding: This research was funded by the Russian Science Foundation, grant number 19-15-00243.

Institutional Review Board Statement: This study was performed in line with the principles of the Declaration of Helsinki. Approval was granted by the Ethics Committee of Endocrinology Research Centre, Moscow, Russia on 10 April 2019 (abstract of record No. 6).

Informed Consent Statement: Written informed consent was obtained from all individual participants included in the study.

Data Availability Statement: The datasets generated during and/or analyzed during the current study are available from the corresponding author on reasonable request.

Acknowledgments: We express our deep gratitude to our colleagues: Natalya M. Malysheva, Vitaliy A. Ioutsi, Larisa V. Nikankina for the help with the laboratory research.

Conflicts of Interest: The authors declare no conflict of interest. The funders had no role in the design of the study; in the collection, analyses, or interpretation of data; in the writing of the manuscript, or in the decision to publish the results.

References

1. Nishioka, H.; Yamada, S. Cushing's disease. *J. Clin. Med.* **2019**, *8*, 1951. [CrossRef]
2. Mazziotti, G.; Frara, S.; Giustina, A. Pituitary Diseases and Bone. *Endocr. Rev.* **2018**, *39*, 440–488. [CrossRef] [PubMed]
3. Compston, J. Glucocorticoid-induced osteoporosis: An update. *Endocrine* **2018**, *61*, 7–16. [CrossRef] [PubMed]
4. Buckley, L.; Guyatt, G.; Fink, H.A.; Cannon, M.; Grossman, J.; Hansen, K.E.; Humphrey, M.B.; Lane, N.E.; Magrey, M.; Miller, M. 2017 American College of Rheumatology Guideline for the prevention and treatment of glucocorticoid-induced osteoporosis. *Arthritis Care Res.* **2017**, *69*, 1095–1110. [CrossRef]
5. Belaya, Z.; Rozhinskaya, L.y.; Grebennikova, T.; Kanis, J.; Pigarova, E.; Rodionova, S.; Toroptsova, N.; Nikitinskaya, O.; Skripnikova, I.; Drapkina, O.; et al. Summary of the Draft Federal Clinical Guidelines for Osteoporosis. *Osteoporos. Bone Dis.* **2020**, *23*, 4–21. [CrossRef]
6. Klein, R.G.; Arnaud, S.B.; Gallagher, J.C.; Deluca, H.F.; Riggs, B.L. Intestinal calcium absorption in exogenous hypercortisolism. *J. Clin. Investig.* **1977**, *60*, 253–259. [CrossRef] [PubMed]

7. Seeman, E.; Kumar, R.; Hunder, G.G.; Scott, M.; Iii, H.H.; Riggs, B.L. Production, degradation, and circulating levels of 1,25-dihydroxyvitamin D in health and in chronic glucocorticoid excess. *J. Clin. Investig.* **1980**, *66*, 664–669. [CrossRef]

8. Lindgren, J.U.; Merchant, C.R.; DeLuca, H.F. Effect of 1,25-dihydroxyvitamin D3 on osteopenia induced by prednisolone in adult rats. *Calcif. Tissue Int.* **1982**, *34*, 253–257. [CrossRef]

9. Chaiamnuay, S.; Chailurkit, L.O.; Narongroeknawin, P.; Asavatanabodee, P.; Laohajaroensombat, S.; Chaiamnuay, P. Current daily glucocorticoid use and serum creatinine levels are associated with lower 25(OH) vitamin D levels in thai patients with systemic lupus erythematosus. *J. Clin. Rheumatol.* **2013**, *19*, 121–125. [CrossRef]

10. Kugai, N.; Koide, Y.; Yamashita, K.; Shimauchi, T.; Nagata, N.; Takatani, O. Impaired mineral metabolism in Cushing's syndrome: Parathyroid function, vitamin D metabolites and osteopenia. *Endocrinol. Jpn.* **1986**, *33*, 345–352. [CrossRef]

11. Hahn, T.J.; Halstead, L.R.; Baran, D.T. Effects of short term glucocorticoid administration on intestinal calcium absorption and circulating vitamin D metabolite concentrations in man. *J. Clin. Endocrinol. Metab.* **1981**, *52*, 111–115. [CrossRef]

12. Findling, J.W.; Adams, N.D.; Lemann, J.; Gray, R.W.; Thomas, C.J.; Tyrrell, J.B. Vitamin D metabolites and parathyroid hormone in Cushing's syndrome: Relationship to calcium and phosphorus homeostasis. *J. Clin. Endocrinol. Metab.* **1982**, *54*, 1039–1044. [CrossRef] [PubMed]

13. Slovik, D.M.; Neer, R.M.; Ohman, J.L.; Lowell, F.C.; Clark, M.B.; Segre, G.V.; Potts, J.T., Jr. Parathyroid hormone and 25-hydroxyvitamin D levels in glucocorticoid-treated patients. *Clin. Endocrinol.* **1980**, *12*, 243–248. [CrossRef]

14. Corbee, R.J.; Tryfonidou, M.A.; Meij, B.P.; Kooistra, H.S.; Hazewinkel, H.A.W. Vitamin D status before and after hypophysectomy in dogs with pituitary-dependent hypercortisolism. *Domest. Anim. Endocrinol.* **2012**, *42*, 43–49. [CrossRef] [PubMed]

15. Aloia, J.F.; Roginsky, M.; Ellis, K.; Shukla, K.; Cohn, S. Skeletal metabolism and body composition in Cushing's syndrome. *J. Clin. Endocrinol. Metab.* **1974**, *39*, 981–985. [CrossRef]

16. Van Cromphaut, S.J.; Stockmans, I.; Torrekens, S.; Van Herck, E.; Carmeliet, G.; Bouillon, R. Duodenal calcium absorption in dexamethasone-treated mice: Functional and molecular aspects. *Arch. Biochem. Biophys.* **2007**, *460*, 300–305. [CrossRef] [PubMed]

17. Akeno, N.; Matsunuma, A.; Maeda, T.; Kawane, T.; Horiuchi, N. Regulation of vitamin D-1-hydroxylase and -24-hydroxylase expression by dexamethasone in mouse kidney. *J. Endocrinol.* **2000**, *164*, 339–348. [CrossRef] [PubMed]

18. Kurahashi, I.; Matsunuma, A.; Kawane, T.; Abe, M.; Horiuchi, N. Dexamethasone enhances vitamin D-24-hydroxylase expression in osteoblastic (UMR-106) and renal (LLC-PK 1) cells treated with 1a, 25-dihydroxyvitamin D3. *Endocrine* **2002**, *17*, 109–118. [CrossRef]

19. Dhawan, P.; Christakos, S. Novel regulation of 25-hydroxyvitamin D3 24-hydroxylase (24(OH)ase) transcription by glucocorticoids: Cooperative effects of the glucocorticoid receptor, C/EBPb, and the vitamin D receptor in 24(OH)ase transcription. *J. Cell. Biochem.* **2010**, *110*, 1314–1323. [CrossRef] [PubMed]

20. Luo, G.; Cunningham, M.; Kim, S.; Burn, T.; Lin, J.; Sinz, M.; Hamilton, G.; Rizzo, C.; Jolley, S.; Gilbert, D.; et al. CYP3A4 induction by drugs: Correlation between a pregnane X receptor reporter gene assay and CYP3A4 expression in human hepatocytes. *Drug Metab. Dispos.* **2002**, *30*, 795–804. [CrossRef]

21. Zhou, C.; Assem, M.; Tay, J.C.; Watkins, P.B.; Blumberg, B.; Schuetz, E.G.; Thummel, K.E. Steroid and xenobiotic receptor and vitamin D receptor crosstalk mediates CYP24 expression and drug-induced osteomalacia. *J. Clin. Investig.* **2006**, *116*, 1703–1712. [CrossRef] [PubMed]

22. Jiang, P.; Xue, Y.; Li, H.; Liu, Y. Dysregulation of vitamin D metabolism in the brain and myocardium of rats following prolonged exposure to dexamethasone. *Psychopharmacology* **2014**, *231*, 3445–3451. [CrossRef] [PubMed]

23. Zayny, A.; Almokhtar, M.; Wikvall, K.; Ljunggren, Ö.; Ubhayasekera, K.; Bergquist, J.; Kibar, P.; Norlin, M. Effects of glucocorticoids on vitamin D3-metabolizing 24-hydroxylase (CYP24A1) in Saos-2 cells and primary human osteoblasts. *Mol. Cell. Endocrinol.* **2019**, *496*, 110525. [CrossRef] [PubMed]

24. Davidson, Z.E.; Walker, K.Z.; Truby, H. Do glucocorticosteroids alter vitamin D status? A systematic review with meta-analyses of observational studies. *J. Clin. Endocrinol. Metab.* **2014**, *97*, 738–744. [CrossRef] [PubMed]

25. Huybers, S.; Naber, T.H.J.; Bindels, R.J.M.; Hoenderop, J.G.J. Prednisolone-induced Ca2+ malabsorption is caused by diminished expression of the epithelial Ca2+ channel TRPV6. *Am. J. Physiol.-Gastrointest. Liver Physiol.* **2007**, *292*, 92–97. [CrossRef] [PubMed]

26. Ferrari, P.; Bianchetti, M.G.; Sansonnens, A.; Frey, F.J. Modulation of renal calcium handling by 11β-hydroxysteroid dehydrogenase type 2. *J. Am. Soc. Nephrol.* **2002**, *13*, 2540–2546. [CrossRef]

27. Faggiano, A.; Pivonello, R.; Melis, D.; Filippella, M.; Di Somma, C.; Petretta, M.; Lombardi, G.; Colao, A. Nephrolithiasis in Cushing's disease: Prevalence, etiopathogenesis, and modification after disease cure. *J. Clin. Endocrinol. Metab.* **2003**, *88*, 2076–2080. [CrossRef] [PubMed]

28. Ramsey, I.K.; Tebb, A.; Harris, E.; Evans, H.; Herrtage, M.E. Hyperparathyroidism in dogs with hyperadrenocorticism. *J. Small Anim. Pract.* **2005**, *46*, 531–536. [CrossRef]

29. Freiberg, J.M.; Kinsella, J.; Sacktor, B. Glucocorticoids increase the Na+-H+ exchange and decrease the Na+ gradient-dependent phosphate-uptake systems in renal brush border membrane vesicles. *Proc. Natl. Acad. Sci. USA* **1982**, *79*, 4932–4936. [CrossRef] [PubMed]

30. Sempos, C.T.; Heijboer, A.C.; Bikle, D.D.; Bollerslev, J.; Bouillon, R.; Brannon, P.M.; DeLuca, H.F.; Jones, G.; Munns, C.F.; Bilezikian, J.P.; et al. Vitamin D assays and the definition of hypovitaminosis D: Results from the First International Conference on Controversies in Vitamin D. *Br. J. Clin. Pharmacol.* **2018**, *84*, 2194–2207. [CrossRef]

31. Melnichenko, G.A.; Dedov, I.I.; Belaya, Z.E.; Rozhinskaya, L.Y.; Vagapova, G.R.; Volkova, N.I.; Grigor'ev, A.Y.; Grineva, E.N.; Marova, E.I.; Mkrtumayn, A.M.; et al. Cushing's disease: The clinical features, diagnostics, differential diagnostics, and methods of treatment. *Probl. Endocrinol.* **2015**, *61*, 55–77. [CrossRef]

32. Machado, M.C.; De Sa, S.V.; Domenice, S.; Fragoso, M.C.B.V.; Puglia, P.; Pereira, M.A.A.; De Mendonça, B.B.; Salgado, L.R. The role of desmopressin in bilateral and simultaneous inferior petrosal sinus sampling for differential diagnosis of ACTH-dependent Cushing's syndrome. *Clin. Endocrinol.* **2007**, *66*, 136–142. [CrossRef]

33. Findling, J.W.; Kehoe, M.E.; Raff, H. Identification of patients with Cushing's disease with negative pituitary adrenocorticotropin gradients during inferior petrosal sinus sampling: Prolactin as an index of pituitary venous effluent. *J. Clin. Endocrinol. Metab.* **2004**, *89*, 6005–6009. [CrossRef] [PubMed]

34. Pigarova, E.A.; Rozhinskaya, L.Y.; Belaya, J.E.; Dzeranova, L.K.; Karonova, T.L.; Ilyin, A.V.; Melnichenko, G.A.; Dedov, I.I. Russian Association of Endocrinologists recommendations for diagnosis, treatment and prevention of vitamin D deficiency in adults. *Probl. Endocrinol.* **2016**, *62*, 60–84. [CrossRef]

35. Petrushkina, A.A.; Pigarova, E.A.; Tarasova, T.S.; Rozhinskaya, L.Y. Efficacy and safety of high-dose oral vitamin D supplementation: A pilot study. *Osteoporos. Int.* **2016**, *27*, 512–513. [CrossRef]

36. Pivonello, R.; De Martino, M.C.; De Leo, M.; Lombardi, G.; Colao, A. Cushing's Syndrome. *Endocrinol. Metab. Clin. N. Am.* **2008**, *37*, 135–149. [CrossRef]

37. Belaya, Z.E.; Iljin, A.V.; Melnichenko, G.A.; Rozhinskaya, L.Y.; Dragunova, N.V.; Dzeranova, L.K.; Butrova, S.A.; Troshina, E.A.; Dedov, I.I. Diagnostic performance of late-night salivary cortisol measured by automated electrochemiluminescence immunoassay in obese and overweight patients referred to exclude Cushing's syndrome. *Endocrine* **2012**, *41*, 494–500. [CrossRef]

38. Povaliaeva, A.; Pigarova, E.; Zhukov, A.; Bogdanov, V.; Dzeranova, L.; Mel'nikova, O.; Pekareva, E.; Malysheva, N.; Ioutsi, V.; Nikankina, L.; et al. Evaluation of vitamin D metabolism in patients with type 1 diabetes mellitus in the setting of cholecalciferol treatment. *Nutrients* **2020**, *12*, 3873. [CrossRef] [PubMed]

39. Thode, J.; Juul-Jørgensen, B.; Bhatia, H.M.; Kjaerulf-Nielsen, M.; Bartels, P.D.; Fogh-Andersen, N.; Siggaard-Andersen, O. Comparison of serum total calcium, albumin-corrected total calcium, and ionized calcium in 1213 patients with suspected calcium disorders. *Scand. J. Clin. Lab. Investig.* **1989**, *49*, 217–223. [CrossRef]

40. Bikle, D.D.; Siiteri, P.K.; Ryzen, E.; Haddad, J.G.; Gee, E. Serum protein binding of 1, 25-Dihydroxyvitamin D: A reevaluation by direct measurement of free metabolite levels. *J. Clin. Endocrinol. Metab.* **1985**, *61*, 969–975. [CrossRef]

41. Bikle, D.D.; Gee, E.; Halloran, B.; Kowalski, M.A.N.N.; Ryzen, E.; Haddad, J.G. Assessment of the Free Fraction of 25-Hydroxyvitamin D in Serum and Its Regulation by Albumin and the Vitamin D-Binding Protein. *J. Clin. Endocrinol. Metab.* **1986**, *63*, 954–959. [CrossRef]

42. Holick, M.F.; Binkley, N.C.; Bischoff-Ferrari, H.A.; Gordon, C.M.; Hanley, D.A.; Heaney, R.P.; Murad, M.H.; Weaver, C.M. Evaluation, treatment, and prevention of vitamin D deficiency: An Endocrine Society clinical practice guideline. *J. Clin. Endocrinol. Metab.* **2011**, *96*, 1911–1930. [CrossRef] [PubMed]

43. Dirks, N.F.; Martens, F.; Vanderschueren, D.; Billen, J.; Pauwels, S.; Ackermans, M.T.; Endert, E.; den Heijer, M.; Blankenstein, M.A.; Heijboer, A.C. Determination of human reference values for serum total 1,25-dihydroxyvitamin D using an extensively validated 2D ID-UPLC–MS/MS method. *J. Steroid Biochem. Mol. Biol.* **2016**, *164*, 127–133. [CrossRef] [PubMed]

44. Tang, J.C.Y.; Nicholls, H.; Piec, I.; Washbourne, C.J.; Dutton, J.J.; Jackson, S.; Greeves, J.; Fraser, W.D. Reference intervals for serum 24,25-dihydroxyvitamin D and the ratio with 25-hydroxyvitamin D established using a newly developed LC–MS/MS method. *J. Nutr. Biochem.* **2017**, *46*, 21–29. [CrossRef]

45. Máčová, L.; Bičíková, M. Vitamin D: Current challenges between the laboratory and clinical practice. *Nutrients* **2021**, *13*, 1758. [CrossRef] [PubMed]

46. Ginsberg, C.; Hoofnagle, A.N.; Katz, R.; Hughes-Austin, J.; Miller, L.M.; Becker, J.O.; Kritchevsky, S.B.; Shlipak, M.G.; Sarnak, M.J.; Ix, J.H. The Vitamin D Metabolite Ratio Is Associated With Changes in Bone Density and Fracture Risk in Older Adults. *J. Bone Miner. Res.* **2021**, 1–8. [CrossRef]

47. Cavalier, E.; Huyghebaert, L.; Rousselle, O.; Bekaert, A.C.; Kovacs, S.; Vranken, L.; Peeters, S.; Le Goff, C.; Ladang, A. Simultaneous measurement of 25(OH)-vitamin D and 24,25(OH)2-vitamin D to define cut-offs for CYP24A1 mutation and vitamin D deficiency in a population of 1200 young subjects. *Clin. Chem. Lab. Med.* **2020**, *58*, 197–201. [CrossRef]

48. Rondeau, P.; Bourdon, E. The glycation of albumin: Structural and functional impacts. *Biochimie* **2011**, *93*, 645–658. [CrossRef]

49. Soudahome, A.G.; Catan, A.; Giraud, P.; Kouao, S.A.; Guerin-Dubourg, A.; Debussche, X.; Le Moullec, N.; Bourdon, E.; Bravo, S.B.; Paradela-Dobarro, B.; et al. Glycation of human serum albumin impairs binding to the glucagon-like peptide-1 analogue liraglutide. *J. Biol. Chem.* **2018**, *293*, 4778–4791. [CrossRef]

50. McLeod, J.F.; Kowalski, M.A.; Haddad, J.G. Interactions among serum vitamin D binding protein, monomeric actin, profilin, and profilactin. *J. Biol. Chem.* **1989**, *264*, 1260–1267. [CrossRef]

51. Gupta, Y.; Gupta, A. Glucocorticoid-induced myopathy: Pathophysiology, diagnosis, and treatment. *Indian J. Endocrinol. Metab.* **2013**, *17*, 913. [CrossRef] [PubMed]

52. Smets, P.; Meyer, E.; Maddens, B.; Daminet, S. Cushing's syndrome, glucocorticoids and the kidney. *Gen. Comp. Endocrinol.* **2010**, *169*, 1–10. [CrossRef] [PubMed]

53. Arnaud, J.; Constans, J. Affinity differences for vitamin D metabolites associated with the genetic isoforms of the human serum carrier protein (DBP). *Hum. Genet.* **1993**, *92*, 183–188. [CrossRef] [PubMed]

Article

Upregulation of Irisin and Vitamin D-Binding Protein Concentrations by Increasing Maternal 25-Hydrovitamin D Concentrations in Combination with Specific Genotypes of Vitamin D-Binding Protein Polymorphisms

Spyridon N. Karras [1,*], Erdinç Dursun [2,3], Merve Alaylıoglu [2], Duygu Gezen-Ak [2], Fatme Al Anouti [4], Stefan Pilz [5], Pawel Pludowski [6], Edward Jude [7] and Kalliopi Kotsa [1]

[1] Division of Endocrinology and Metabolism and Diabetes Center, 1st Department of Internal Medicine, Medical School, Aristotle University of Thessaloniki, AHEPA University Hospital, 55535 Thessaloniki, Greece; kalmanthou@yahoo.gr

[2] Brain and Neurodegenerative Disorders Research Laboratories, Department of Medical Biology, Cerrahpasa Faculty of Medicine, Istanbul University-Cerrahpasa, Istanbul 34381, Turkey; erdincdu@gmail.com (E.D.); merve.alaylioglu@hotmail.com (M.A.); duygugezenak@gmail.com (D.G.-A.)

[3] Department of Neuroscience, Institute of Neurological Sciences, Istanbul University-Cerrahpasa, Istanbul 34381, Turkey

[4] Department of Health Sciences, College of Natural and Health Sciences, Zayed University, Abu Dhabi 144534, United Arab Emirates; Fatme.AlAnouti@zu.ac.ae

[5] Division of Endocrinology and Diabetology, Department of Internal Medicine, Medical University of Graz, A-8036 Graz, Austria; stefan.pilz@medunigraz.at

[6] Department of Biochemistry, Radioimmunology and Experimental Medicine, The Children's Memorial Health Institute, 04730 Warsaw, Poland; pludowski@yahoo.com

[7] Department of Endocrinology, Tameside Hospital NHS Foundation Trust, Ashton-under-Lyne OL6 9RW, UK; ejude99@yahoo.co.uk

* Correspondence: karraspiros@yahoo.gr

Citation: Karras, S.N.; Dursun, E.; Alaylıoglu, M.; Gezen-Ak, D.; Al Anouti, F.; Pilz, S.; Pludowski, P.; Jude, E.; Kotsa, K. Upregulation of Irisin and Vitamin D-Binding Protein Concentrations by Increasing Maternal 25-Hydrovitamin D Concentrations in Combination with Specific Genotypes of Vitamin D-Binding Protein Polymorphisms. *Nutrients* 2022, *14*, 90. https://doi.org/10.3390/nu14010090

Academic Editor: Carsten Carlberg

Received: 6 December 2021
Accepted: 23 December 2021
Published: 26 December 2021

Publisher's Note: MDPI stays neutral with regard to jurisdictional claims in published maps and institutional affiliations.

Abstract: Dyshomeostasis of vitamin D-binding protein (VDBP) has been implicated in the pathogenesis of various pregnancy complications, including preeclampsia, preterm birth, gestational diabetes, and adverse metabolic profiles in the offspring. VDBP polymorphisms have been consistently reported to contribute to this intriguing interplay. Until recently, the effects of VDBP polymorphism heterogeneity on maternal and neonatal adipomyokine profiles have not been investigated, specifically after incorporating the different maternal and neonatal 25-hydroxyvitamin D concentration cut-offs at birth. We aimed to investigate the potential effects of maternal and neonatal VDBP polymorphisms on adiponectin, irisin, and VDBP concentrations at birth, according to different cut-offs of vitamin D status, in maternal–neonatal dyads recruited from the sunny region of Northern Greece. We obtained blood samples from 66 mother–child pairs at birth. Results indicated that (i) Neonatal serum biomarkers were not affected by any included neonatal VDBP polymorphism according to different cut-offs of neonatal vitamin D status at birth, (ii) neonatal VDBP concentration was elevated in neonates with maternal rs7041 GG genotype, (iii) maternal 25(OH)D at ≤75 nmol/L resulted in increased concentrations of maternal VBDP and irisin concentrations in women with CC genotype for rs2298850 and rs4588, whereas this effect was also evident for this cut-off for neonatal VDBP concentrations at birth for GC genotype for rs 7041, and (iv) no significant effect of neonatal VDBP polymorphisms was observed on neonatal VDBP, adiponectin, or irisin levels when stratified according to maternal 25(OH)D cut-offs. In conclusion, these findings confirm that among women with the combination of CC genotype for rs2298850 and rs4588, a specific high cut-off of maternal 25(OH)D results in increasing maternal VBDP concentrations, hence providing a mechanistic rationale for aiming for specific cut-offs of vitamin D after supplementation during pregnancy, in daily clinical practice.

Keywords: vitamin D; pregnancy; neonatal health; functional polymorphism

1. Introduction

Vitamin D-binding protein (VDBP) is considered as a crucial factor for vitamin D homeostasis [1,2] since it comprises the major biological parameter regulating the half-life of vitamin D in the systemic circulation. These effects are mediated by both serum VDBP concentrations and VDBP genotype, which affect bioavailable 25-hydroxyvitamin D[25(OH)D] [3,4]. VDBP has been also considered as a potent immuno-regulator implicated in the pathogenesis of autoimmune diseases, either through its effects on vitamin D equilibrium or via direct effects, which are mediated by VDBP concentrations [5].

VDBP dyshomeostasis has been linked to a plethora of complications, including gestational diabetes, preeclampsia, preterm birth, and adverse metabolic profiles in the offspring [6–8]. Our group has previously reported a strong association for VDBP concentrations with adipomyokines, adiponectin, and irisin, which are involved in energy regulation in both mothers and neonates [9]. We also confirmed this association between VDBP and adiponectin in healthy nonpregnant women, with no such association observed among men [10]. Vitamin D receptor and VDBP polymorphisms have been consistently reported to contribute to this intriguing interplay [11–14]. We have recently described that although vitamin D concentrations in the examined neonates were not impacted by maternal VDBP polymorphisms at birth, mothers with CC genotype for rs2298850 and CC genotype for rs4588 manifested higher 25(OH)D concentrations [11]. These effects were evident after adopting conventional international maternal or neonatal cut-offs for 25(OH)D concentrations. VDBP could be also considered as a molecule with significant metabolic functions regulating energy homeostasis. However, the potential effects of VDBP polymorphism heterogeneity on maternal and neonatal adipomyokine profiles remain largely unexplained, specifically after incorporating the effects of maternal and neonatal vitamin D status at birth.

Plasma half-life of 25(OH)D is approximately 3 weeks, according to previous reports [15]. It is a useful biomarker of environmental and physiological determinants of vitamin D status, including dietary and cutaneous synthesis, and is determined by CYP27B1 and CYP24A1 enzyme activity and all factors that influence the delivery and transport of 25(OH)D, including VDBP concentrations and genotypes [15].

Despite the sunny weather in Greece and the other Mediterranean countries, vitamin D deficiency is a major public health burden [16]. Moreover, a systematic review regarding hypovitaminosis D in the Mediterranean region including 2649 pregnant women revealed a prevalence range between 22.7% and 90.3% for vitamin D deficiency [17].We aimed to examine the potential effects of maternal and neonatal VDBP concentrations and polymorphisms on the specific adipomyokines adiponectin and irisin at birth, according to different cut-offs of vitamin D status, in maternal–neonatal dyads from the sunny region of Northern Greece.

2. Methods

2.1. Inclusion and Exclusion Criteria

A cohort of mother–child pairs at birth was included in the study. A detailed description of the enrollment has been previously described [18]. Informed consent was obtained from all mothers. The study was conducted from January 2018 to September 2019 and was granted ethical approval by the Bioethics Committee of the Aristotle University of Thessaloniki, Greece (approval number 1/19-12-2011).

2.2. Demographic and Dietary Data—Biochemical and Hormonal Assays

At enrollment, demographic and social characteristics were recorded. All dietary and demographic data of the cohort and methods of sampling have been reported previously [19]. Concentrations of 25(OH)D2 and 25(OH)D3 were determined using liquid chromatography–tandem mass spectrometry (LC–MS/MS), with lower limits of quantification (LLOQ); 25(OH)D2 (0.5 ng/mL), 25(OH)D3 (0.5 ng/mL), and the combination of the two vitamin D forms, as total 25(OH)D, were provided [20]. VDBP, irisin, and adiponectin were measured with enzyme-linked immunosorbent assay on a Synergy H1 Hybrid reader and Gen5 software (BioTek, Winooski, VT, USA): GC-Globulin/VDBP (AssayPro, St. Charles, MO, USA); irisin (My BioSource, San Diego, CA, USA); adiponectin (R&D Systems, Minneapolis, MN, USA). Intra-assay and inter-assay variance was <8% and <10% for adiponectin and <8% and <10% for irisin, respectively. Detection limits for assays were as follows: 0.098 µg/mL for VDBP, 3.12 ng/mL for irisin, and 0.039 µg/mL for adiponectin. We classified maternal and neonatal vitamin D status at birth according to the following: 25(OH)D $\leq$ 25 nmol/L (deficiency), 25–50 nmol/L (insufficiency), and 25(OH)D $\geq$ 50–75 nmol/L (sufficiency) [21,22].

2.3. VDBP Analysis

DNA isolation was performed by QIAamp DNA Blood Mini Kit (Cat. No. 51304, QIAGEN, Hilden, Germany) according to the manufacturer's protocol. Genotypes of VDBP rs2298850, rs4588, and rs7041 SNPs were determined by LightSNiP assay using simple probes (LightSNiP, TibMolBiol, Berlin, Germany) and LightCycler Fast Start DNA Master HybProbe Kit (Cat. No.12239272001, Roche Diagnostics, Mannheim, Germany). Real-time PCR (RT-PCR) was performed with LightCycler 480 Instrument II (Roche Diagnostics, Mannheim, Germany). Genotyping was performed as previously explained [23].

2.4. UVB Measurements

UVB data for the broad geographical region of Thessaloniki, Greece, were collected as described previously [18]. Mean UVB exposure during the previous 21 days (daily integral) before blood sample collection (estimated mean half-life of vitamin D) was calculated.

2.5. Statistical Analysis

All analyses that involved the distributions of genotypes of VDBP polymorphisms were analyzed with the chi-square (χ2) test, df:2 for genotypes. Significance was also confirmed with Cramer's V/Kendall's tau-c. One-way ANOVA and multiple-comparison tests were run to compare between mean values of the groups, either Tukey HSD or Dunett C, depending on the normality of the data set. Homogeneity of variances was checked with Levene's Test for homogeneity of variances. Data and *p*-values were adjusted as needed by one-way analysis of covariance (ANCOVA) for maternal height (cm), BMI prepregnancy (kg/m^2), BMI terminal (kg/m^2), UVB, and weeks of gestation. All data are presented as mean $\pm$ SD. A *p*-value < 0.05 was considered statistically significant. SPSS version 24.0" software (SPSS, Chicago, IL, USA) was used to perform statistical analysis.

3. Results

3.1. Demographics of the Study Participants

The demographic characteristics of mothers–neonates cohort recruited in the study are shown in Table 1. There were 66 mothers with 28 female and 38 male neonates in total. Mean prepregnancy BMI (kg/m^2) was 24.91 $\pm$ 4.81; while term pregnancy BMI (kg/m^2) was 29.62 $\pm$ 5.80. Daily dietary vitamin D intake during the third trimester was 2.9 $\pm$ 1.2 mcg.

3.2. Distribution of Neonatal Adiponectin, Irisin, and VDBP Concentrations According to Maternal Vitamin D Status or Maternal VDBP Polymorphisms

Distributions of vitamin D status of neonatal adiponectin, irisin, and VDBP concentrations according to maternal vitamin D status are presented in Table 2. Data and *p*-values were adjusted for maternal height (cm), BMI prepregnancy (kg/m^2), BMI at term (terminal has a different connotation) (kg/m^2), UVB exposure, and weeks of gestation. The mean pre-conception BMI was 22.2 $\pm$ 3.3 kg/m^2 (range 16.1–31.6), adjusted BMI was 22.4 $\pm$ 4.3 kg/m^2 (range 13.5–35.5), and duration gestation was 37 to 42 weeks. The frequency distribution for the lower tertile of vitamin D status revealed that 47 mothers–neonates dyads had serum 25(OH)D $\geq$25 nmol/L, with concentrations of 8.97 $\pm$ 13.24 µg/mL and 7.28 $\pm$ 7.73 µg/mL for adiponectin, while 16 dyads had serum 25(OH)D $\leq$25 nmol/L, with concentrations of 20.56 $\pm$ 20.53 µg/mL and 20.95 $\pm$ 21.53 µg/mL for adiponectin. Differences across neonatal adiponectin concentrations according to maternal vitamin D status were remarkably significant (p = 0.048; adjusted = 0.003) with maternal 25(OH)D $\geq$25 nmol/L versus $\leq$25 nmol/L in the lower tertile, while there were no significant differences for the other biomarkers, irisin and VDBP, across the middle and upper tertiles. On the other hand, neonatal VDBP concentration in maternal rs7041 GG genotype was significantly increased compared to GT + TT genotype (GG genotype, n: 17, adjusted mean $\pm$ SD: 495.74 $\pm$ 279.57; GT + TT genotype, n: 47, adjusted mean $\pm$ SD: 295.44 $\pm$ 156.61; adjusted p: 0.02, power: 67%).

Table 1. Demographics of mothers and neonates.

Mothers	
Number (n)	66
Age (years)	31.92 $\pm$ 6.08
Prepregnancy weight (kg)	67.56 $\pm$ 14.54
Term Weight (kg)	85.43 $\pm$ 14.30
Prepregnancy BMI (kg/m^2)	24.91 $\pm$ 4.81
Term BMI (kg/m^2)	29.62 $\pm$ 5.80
Duration of gestation (weeks)	38.80 $\pm$ 1.56
Small for gestational age (SGA,%)	0.04
Appropriate for gestational age (AGA,%)	0.96
Large for gestational age (LGA,%)	0.00
Daily dietary calcium intake during 3rd trimester (mg)	792.5 $\pm$ 334.0
Daily dietary vitamin D intake during 3rd trimester (mcg)	2.9 $\pm$ 1.2
Neonates	
Number (n)	66
Gender; females (n (%))	28 (0.42)
Height (cm)	50.48 $\pm$ 1.96
Weight (g)	3292.12 $\pm$ 414.25

Table 2. Neonatal serum biomarkers according to maternal vitamin D status.

		N	Neonatal VDBP (µg/mL)	N	Neonatal Adiponectin (µg/mL)	N	Neonatal Irisin (ng/mL)
Maternal vitamin D status	<25 nmol/L	17	470.29 $\pm$ 325.47 (426.69 $\pm$ 292.65)	16	20.56 $\pm$ 20.53 (20.95 $\pm$ 21.53)	10	141.03 $\pm$ 90.35
	>25 nmol/L	47	313.63 $\pm$ 176.35 (294.24 $\pm$ 125.52)	47	8.97 $\pm$ 13.24 (7.28 $\pm$ 7.73)	32	184.82 $\pm$ 186.36
	p-value; adjusted p; power		0.074 (0.085) * 41% $^\Phi$		0.048 (0.003) * 88% $^\Phi$		0.32
	<50 nmol/L	42	372.86 $\pm$ 271.59	41	12.00 $\pm$ 15.36	25	157.62 $\pm$ 139.61
	>50 nmol/L	22	321.61 $\pm$ 132.69	22	11.76 $\pm$ 17.67	17	199.06 $\pm$ 206.09
	p-value		0.41		0.96		0.48
	<75 nmol/L	53	365.10 $\pm$ 248.70	52	11.23 $\pm$ 13.99	33	182.44 $\pm$ 172.66
	>75 nmol/L	11	307.78 $\pm$ 136.31	11	15.19 $\pm$ 24.24	9	144.91 $\pm$ 158.43
	p-value		0.46		0.46		0.56

If the Levene's test for equality of variances $p > 0.05$, then equal variances assumed (significant (2-tailed) p-values of t-tests are given). If the Levene's test for equality of variances $p < 0.05$, then equal variances not assumed (significant (2-tailed) p-values of t-tests are given). * The data and p-values adjusted for maternal height (cm), BMI prepregnancy (kg/m^2), BMI terminal (kg/m^2), and weeks of gestation by one-way analysis of covariance (ANCOVA). $^\Phi$ Posthoc power analysis. Abbreviations: VDBP: Vitamin D-binding protein; 25(OH)D: 25-hydroxyvitamin D. Digits in bold refer to significant effects.

3.3. Neonatal Serum Biomarkers according to Neonatal Vitamin D Status at Birth and Neonatal VDBP Polymorphism

Neonatal serum adiponectin, irisin, and VDBP concentrations according to different neonatal vitamin D cut-offs at birth and neonatal VDBP polymorphisms are presented in Table 3. No significant effect was revealed for neonatal biomarkers, according to neonatal vitamin D status at birth and different SNPs and genotypes. The only remarkable effect was evident in neonates with 25(OH)D $\leq$ 75 nmol/L, which demonstrated higher irisin concentrations (233.56 $\pm$ 191.3 ng/mL),when harboring rs2298850 (CG + GG) (p = 0.04). However, results were nonsignificant after adjusting for maternal height (cm), BMI prepregnancy (kg/m^2), BMI at term (kg/m^2), UVB exposure, and weeks of gestation (p = 0.091).

3.4. Maternal Serum Biomarkers according to Maternal Vitamin D Status and Maternal VDBP Polymorphisms

Maternal serum adiponectin, irisin, and VDBP concentrations according to different maternal vitamin D cut-offs during delivery along with maternal VDBP polymorphisms are presented in Table 4. Mothers in the upper tertile of 25(OH)D ($\leq$75 nmol/L),with rs2298850 (CC genotype),manifested higher VDBP concentrations (403.06 $\pm$ 64.72 µg/mL, p = 0.007) after multiple adjustments for maternal height (cm), BMI prepregnancy (kg/m^2), BMI at term (kg/m^2), UVB exposure, and weeks of gestation, compared with women in the middle and lower tertiles of 25(OH)D concentrations. Similar results were obtained for mothers with rs4588 (CC genotype) regarding VDBP concentrations (403.06 $\pm$ 64.72 µg/mL, p = 0.07), whereas the same SNP-genotype pattern revealed a similar effect on maternal irisin concentrations (508.57 $\pm$ 559.87 ng/mL, p = 0.03).

3.5. Neonatal Serum Biomarkers According to Maternal Vitamin D Status and Maternal VDBP Polymorphisms

Neonatal serum adiponectin, irisin, and VDBP concentrations in the context of maternal VDBP polymorphisms and different maternal vitamin D cut-offs at birth are presented in Table 5. We observed that women with rs 7041 (genotype GC) delivered neonates with higher VDBP concentrations, at maternal concentrations of $\leq$50 nmol/L and $\leq$75 nmol/L (524.75 $\pm$ 331.56 µg/mL, p = 0.05 and 526.26 $\pm$ 282.39 µg/mL, p = 0.01, respectively). Significance was more pronounced (p = 0.01) with increasing maternal vitamin D concentrations, including a cut-off of 75 nmol/L.

3.6. Neonatal Serum Biomarkers According to Neonatal Vitamin D Status at Birth and Maternal VDBP Polymorphisms

Neonatal serum adiponectin, irisin, and VDBP concentrations according to different neonatal vitamin D cut-offs at birth and maternal VDBP polymorphisms are presented in Table 6. There were no significant effects of different maternal cut-offs of 25(OH)D according to neonatal vitamin D status and genetic profiles.

Table 3. Neonatal serum biomarkers according to neonatal vitamin D status at birth and neonatal VDBP polymorphisms.

SNP	Neonatal Genotype	N	Neonatal Vitamin D Status at Birth $\leq$ 25 nmol/L			N	Neonatal Vitamin D Status at Birth $\leq$ 50 nmol/L			N	Neonatal Vitamin D Status at Birth $\leq$ 75 nmol/L		
			VDBP (µg/mL)	Adiponectin (µg/mL)	Irisin (ng/mL)		VDBP (µg/mL)	Adiponectin (µg/mL)	Irisin (ng/mL)		VDBP (µg/mL)	Adiponectin (µg/mL)	Irisin (ng/mL)
rs2298850	CC	14	400.07 ± 296.4	8.76 ± 10.6	117.44 ± 105.4	24	406.00 ± 293.6	8.76 ± 9.0	120.35 ± 115.2	29	402.94 ± 276.5	11.02 ± 13.3	126.96 ± 125.9 (152.40 ± 136.5)
	CG + GG	11	389.22 ± 297.3	18.57 ± 20.8	205.54 ± 198.5	26	324.81 ± 207.0	10.44 ± 15.2	202.16 ± 159.2	30	323.37 ± 193.0	10.93 ± 14.4	233.56 ± 191.3 (264.81 ± 214.6) *
	p-value adjusted *p*		0.93	0.19	0.27		0.26	0.64	0.11		0.20	0.98	0.04 0.091 *
rs4588	CC	13	406.69 ± 307.4	9.34 ± 10.8	117.44 ± 105.4	22	402.11 ± 306.5	9.97 ± 9.7	132.64 ± 125.6	27	399.54 ± 286.2	12.17 ± 13.7	136.18 ± 132.6
	CA + AA	12	382.95 ± 284.3	16.98 ± 20.4	117.44 ± 105.4	28	333.66 ± 202.4	9.33 ± 14.5	190.63 ± 157.3	32	331.21 ± 189.7	9.93 ± 13.8	223.31 ± 191.7
	p-value		0.84	0.28	0.27		0.35	0.86	0.27		0.28	0.54	0.11
rs7041	GG	7	407.68 ± 352.0	9.71 ± 14.1	127.62 ± 122.6	14	353.87 ± 260.3	9.19 ± 10.9	133.69 ± 130.5	15	341.49 ± 255.4	10.76 ± 12.1	130.2 ± 123.6
	GT + TT	18	390.47 ± 274.59	14.14 ± 17.0	165.44 ± 166.9	36	367.63 ± 253.8	9.79 ± 13.2	174.39 ± 149.7	44	369.64 ± 235.8	11.05 ± 14.4	194.3 ± 178.8
	p-value		0.89	0.55	0.68		0.86	0.88	0.48		0.70	0.95	0.30

If the Levene's test for equality of variances $p > 0.05$, then equal variances assumed (significant (2-tailed) *p*-values of *t*-tests are given). If the Levene's test for equality of variances $p < 0.05$, then equal variances not assumed (significant (2-tailed) *p*-values of *t*-tests are given). * The data and *p* values adjusted for maternal height (cm), BMI pre-pregnancy (kg/m^2), BMI terminal (kg/m^2), UVB exposure and weeks of gestation by One-way analysis of covariance (ANCOVA). Abbreviations: VDBP: Vitamin D binding protein; 25(OH)D: 25-hydroxy-vitamin D. Digits in bold refer to significant effects.

Table 4. Maternal serum biomarkers according to maternal vitamin D status and maternal VDBP polymorphisms.

SNP	Maternal Genotype	N	VDBP (μg/mL)	Adiponectin (μg/mL)	Irisin (ng/mL)	N	VDBP (μg/mL)	Adiponectin (μg/mL)	Irisin (ng/mL)	N	VDBP (μg/mL)	Adiponectin (μg/mL)	Irisin (ng/mL)
			Maternal Vitamin D Status $\leq$ 25 nmol/L				Maternal Vitamin D Status $\leq$ 50 nmol/L				Maternal Vitamin D Status $\leq$ 75		
rs2298850	CC	7	372.47 ± 79.49	4.74 ± 3.88	304.16 ± 495.99	17	384.16 ± 73.08	4.15 ± 2.92	353.43 ± 414.39	22	394.67 ± 71.29 (403.06 ± 4.72) *	4.81 ± 2.93	452.99 ± 513.74
	CG + GG	10	368.12 ± 6 1.91	5.02 ± 3.22	176.33 ± 134.89	25	365.70 ± 127.39	4.07 ± 2.84	277.50 ± 339.70	33	341.46 ± 121.86 (342.93 ± 64.36) *	3.82 ± 3.23	233.02 ± 298.82
	p-value adjusted *p* power		0.90	0.88	0.52		0.59	0.93	0.55		0.07 0.007 * 80% $^\Phi$	0.26	0.10
rs4588	CC	6	386.40 ± 77.15	5.13 ± 4.10	351.67 ± 525.59	16	390.11 ± 71.09	4.26 ± 2.98	377.31 ± 419.19	21	399.70 ± 68.92 (403.06 ± 64.72) *	4.92 ± 2.95	475.83 ± 517.28 (508.57 ± 559.87) *
	CA + AA	11	360.92 ± 63.40	4.76 ± 3.14	156.68 ± 136.70	26	362.74 ± 125.72	4.00 ± 2.80	265.75 ± 336.06	34	339.92 ± 120.33 (342.93 ± 64.36) *	3.77 ± 3.19	225.89 ± 296.20 (265.22 ± 332.14) *
	p-value adjusted *p* power		0.47	0.84	0.33		0.43	0.78	0.38		0.04 0.007 * 80% $^\Phi$	0.20	0.04 0.03 * 60% $^\Phi$
rs7041	GG	5	383.98 ± 86.01	5.61 ± 4.39	139.86 ± 94.03	11	379.45 ± 80.50	4.68 ± 3.50	215.67 ± 209.68 (198.10 ± 302.52) *	13	391.84 ± 82.84	4.89 ± 3.24	446.47 ± 561.90
	GT + TT	12	364.05 ± 61.42	4.58 ± 3.04	296.01 ± 437.13	31	370.94 ± 117.33	3.88 ± 2.58	340.29 ± 406.95 (178.58 ± 91.93) *	42	353.74 ± 112.89	4.02 ± 3.08	287.0 ± 356.91
	p-value adjusted *p*		0.60	0.59	0.45		0.72	0.10	0.06 0.96 *		0.27	0.38	0.26

If the Levene's test for equality of variances $p > 0.05$, then equal variances assumed (significant (2-tailed) p-values of t-tests are given). If the Levene's test for equality of variances $p < 0.05$, then equal variances not assumed (significant (2-tailed) p-values of t-tests are given). * The data and p-values adjusted for maternal height (cm), BMI prepregnancy (kg/m^2), BMI terminal (kg/m^2), UVB exposure, and weeks of gestation by one-way analysis of covariance (ANCOVA). $^\Phi$ Posthoc power analysis. Abbreviations: VDBP: vitamin D-binding protein; 25(OH)D: 25-hydroxy-vitamin D. Digits in bold refer to significant effects.

Table 5. Neonatal serum biomarkers according to maternal vitamin D status and maternal VDBP polymorphisms.

SNP	Maternal Genotype		Maternal Vitamin D Status ≤ 25 nmol/L				Maternal Vitamin D status ≤ 50 nmol/L				Maternal Vitamin D Status ≤ 75 nmol/L		
		N	Neonatal VDBP (μg/mL)	Neonatal Adiponectin (μg/mL)	Neonatal Irisin (ng/mL)	N	Neonatal VDBP (μg/mL)	Neonatal Adiponectin (μg/mL)	Neonatal Irisin (ng/mL)	N	Neonatal VDBP (μg/mL)	Neonatal Adiponectin (μg/mL)	Neonatal Irisin (ng/mL)
rs2298850	CC	8	550.44 ± 409.76	21.51 ± 20.13	104.42 ± 108.07	18	460.17 ± 349.52	12.55 ± 16.06	131.22 ± 169.93	23	437.54 ± 320.83	11.35 ± 14.59	139.29 ± 173.78
	CG + GG	9	399.04 ± 229.82	19.61 ± 22.28	177.64 ± 57.76	24	307.39 ± 175.37	11.57 ± 15.15	186.23 ± 96.46	30	309.56 ± 159.58	11.13 ± 13.76	223.04 ± 166.43
	p-value		0.38	0.86	0.22		0.10	0.84	0.34		0.09	0.96	0.17
rs4588	CC	7	584.21 ± 430.39	24.42 ± 19.84	104.42 ± 108.07	17	468.76 ± 358.31	13.22 ± 16.29	131.22 ± 169.93	22	443.15 ± 327.22	11.81 ± 14.76	139.29 ± 173.78 (508.57 ± 559.87) *
	CA + AA	10	390.54 ± 218.34	17.56 ± 21.73	177.64 ± 57.76	25	307.65 ± 171.68	11.13 ± 14.97	186.23 ± 96.46	31	309.70 ± 156.90	10.80 ± 13.64	223.04 ± 166.43 (265.22 ± 332.14) *
	p-value		0.30	0.53	0.22		0.10	0.67	0.34		0.09	0.80	0.17
rs7041	GG	5	735.48 ± 420.43	28.19 ± 22.62	55.57 ± 61.63 (67.60 ± 82.02) *	11	580.23 ± 404.75 (524.75 ± 331.56) *	17.00 ± 18.70	126.99 ± 212.45	13	572.62 ± 370.91 (526.26 ± 282.39) *	14.83 ± 17.89	150.78 ± 213.35
	GT + TT	12	359.79 ± 211.0	17.09 ± 19.63	177.66 ± 75.90 (177.66 ± 75.90) *	31	299.28 ± 157.60 (302.78 ± 177.35) *	10.17 ± 13.86	172.04 ± 93.80	40	297.65 ± 145.31 (297.16 ± 161.38) *	10.03 ± 12.48	196.20 ± 155.20
	p-value adjusted p power		0.12	0.33	0.04 ♦		0.04 0.05 * 50% Φ	0.21	0.46		0.02 0.01 * 75% Φ	0.29	0.50

If the Levene's test for equality of variances $p > 0.05$, then equal variances assumed (significant (2-tailed) p-values of t-tests are given). If the Levene's test for equality of variances $p < 0.05$, then equal variances not assumed (significant (2-tailed) p-values of t-tests are given). * The data and p-values adjusted for maternal height (cm), BMI prepregnancy (kg/m^2), BMI terminal (kg/m^2), UVB exposure, and weeks of gestation by one-way analysis of covariance (ANCOVA). ♦ Due to reduced number of cases in GG of rs7041, adjusted p-values could not be calculated. Φ Posthoc power analysis. Abbreviations: VDBP: Vitamin D-binding protein; 25(OH)D: 25-hydroxy-vitamin D. Digits in bold refer to significant effects.

Table 6. Neonatal serum biomarkers according to maternal vitamin D status at birth and neonatal VDBP polymorphisms.

SNP	Neonatal Genotype	N	Maternal Vitamin D Status at Birth < 25 nmol/L			N	Maternal Vitamin D Status at Birth < 50 nmol/L			N	Maternal Vitamin D Status at Birth < 75 nmol/L		
			VDBP (µg/mL)	Adiponectin (µg/mL)	Irisin (ng/mL)		VDBP (µg/mL)	Adiponectin (µg/mL)	Irisin (ng/mL)		VDBP (µg/mL)	Adiponectin (µg/mL)	Irisin (ng/mL)
rs2298850	CC	11	438.79 ± 328.87	17.25 ± 17.82	143.24 ± 108.39	21	429.05 ± 304.32	11.24 ± 14.44	132.62 ± 115.01	27	406.74 ± 281.96	10.38 ± 13.04	132.92 ± 131.37 (164.56 ± 142.93)
	CG + GG	6	528.03 ± 341.32	27.85 ± 26.29	135.87 ± 37.76	21	316.68 ± 227.97	12.80 ± 16.62	189.45 ± 166.11	26	321.85 ± 205.29	12.14 ± 15.16	241.86 ± 200.61 (260.39 ± 239.40)
	p-value adjusted *p*		0.61	0.36	0.91		0.18	0.75	0.32		0.22	0.66	0.07 0.19 * 25% Φ
rs4588	CC	10	451.27 ± 343.90	18.86 ± 17.92	143.24 ± 108.39 (161.87 ± 105.76)	19	426.97 ± 320.20	12.90 ± 15.15	145.80 ± 124.76	25	403.38 ± 292.79	11.58 ± 13.62	143.17 ± 138.10
	CA + AA	7	497.46 ± 321.91	23.40 ± 25.92	135.87 ± 37.76 (135.87 ± 37.76)	23	328.16 ± 221.27	11.22 ± 15.86	172.67 ± 161.55	28	330.91 ± 200.87	10.90 ± 14.58	229.56 ± 201.54
	p-value		0.88	0.15	0.057 ◆		025	0.73	0.64		0.29	0.86	0.16
rs7041	GG	6	412.32 ± 387.53	14.51 ± 14.23	143.15 ± 106.03	10	376.99 ± 296.05	9.50 ± 12.49	159.27 ± 138.64	14	353.88 ± 260.30	9.19 ± 10.90	133.69 ± 130.52
	GT + TT	11	501.91 ± 302.14	24.19 ± 23.48	139.62 ± 89.11	32	371.58 ± 268.52	12.81 ± 16.28	156.98 ± 143.98	39	369.12 ± 247.78	11.98 ± 15.03	200.72 ± 185.13
	p-value		0.60	0.38	0.96		0.96	0.56	0.97		085	0.53	0.33

If the Levene's test for equality of variances $p > 0.05$, then equal variances assumed (significant (2-tailed) *p*-values of *t*-tests are given). If the Levene's test for equality of variances $p < 0.05$, then equal variances not assumed (significant (2-tailed) *p*-values of *t*-tests are given). * The data and *p*-values adjusted for maternal height (cm), BMI prepregnancy (kg/m^2), BMI terminal (kg/m^2), and weeks of gestation by one-way analysis of covariance (ANCOVA). ◆ Due to reduced number of cases in CA + AA of rs4588, adjusted *p*-values could not be calculated. Φ Posthoc power analysis. Abbreviations: VDBP: vitamin D-binding protein; 25(OH)D: 25-hydroxy-vitamin D. Digits in bold refer to significant effects.

4. Discussion

We aimed to investigate interactions between VDBP polymorphisms and adiponectin (adipokine),irisin (myokine), and VDBP concentrations, according to different maternal and neonatal 25(OH)D cut-offs, in mother–neonate pairs at birth. To our knowledge, this is the first study on the effects of different VDBP polymorphisms on the adipo-myokine offspring profiles, in the sunny Mediterranean region of Northern Greece. Our results revealed the following:

(i)   Neonatal serum biomarkers were not affected by any included neonatal VDBP polymorphism according to different cut-offs of neonatal vitamin D status at birth;

(ii)   Neonatal VDBP concentration was increased in neonates with maternal rs7041 GG genotype;

(iii)   Elevated maternal 25(OH)D at $\leq$75 nmol/L resulted in increased concentrations of maternal VBDP and irisin concentrations in women with CC genotype for rs2298850 and rs4588,whereas this effect was also evident for this cut-off for neonatal VDBP concentrations at birth for GC genotype for rs 7041;

(iv)   No significant effect of neonatal VDBP polymorphisms was observed on neonatal VDBP, adiponectin, or irisin levels when stratified according to maternal 25(OH)D cut-offs.

We identified a specific type of functional polymorphism, in relation to vitamin D status, VDBP polymorphisms, and metabolic profiles of future mothers and their neonates. Higher maternal vitamin D status at birth affected concentrations of VDBP and irisin in women and neonates with a specific VDBP SNP-genotype pattern, indicating an intriguing interaction of a modifiable factor (maternal vitamin D status at birth) with a specific genetic background (SNPs for VDBP), resulting in differences in concentrations of metabolites, involved in energy regulation and immune response, such as those described for VDBP and irisin previously [9–12]. Apart from its functions as the primary carrier molecule of vitamin D, VDBP is considered as a critical regulator of the half-life of circulating vitamin D metabolites [1–4]. It is also considered as a potent immunomodulator during pregnancy, at placental and systemic level, inducing systemic and local maternal tolerance to paternal and fetal allo-antigens [7]. VDBP has been implicated in regulating gene expression of certain placental amino-transporters, which might be involved during in utero development in the control of amino acid transfer to the offspring [8]. The exact pathways of the association of VDBP dyshomeostasis and adverse pregnancy and offspring outcomes are still a matter of debate. In this regard, specific VDBP polymorphisms have been also consistently reported to contribute to this intriguing interplay of VDBP biodynamics and pregnancy complications [7,24].

A study among Chinese women showed that the risk allele-A of rs3733359 of VDBP gene was associated with a modest increase of risk for gestational diabetes mellitus (GDM) in the obese subgroup, where other SNPs demonstrated correlations with insulin and glucose homeostasis [25]. Another Chinese group reported that *GC* rs16847024 C > T was significantly associated with an increased risk of GDM, however 25(OH)D concentrations were not evaluated in most women included in the study; nevertheless, genes encoding VDBP were found to be associated with vitamin D status [26]. These interactions, however, should be carefully deciphered, since there are important ethnic variations, which do not necessarily confirm the above findings in other pregnant cohorts. In this regard, a prospective large case-control study from Norway evaluated potential interactions of VDBP and its polymorphisms in pregnant women at 18th week of gestation and after delivery, withT1D risk of offspring [27]. Although higher VDBP concentrations at term were associated with lower risk of T1D in the offspring, no effects of VDBP polymorphisms were evident. Absence or attenuation of the prominent physiological increase of VDBP concentrations during pregnancy has been also reported in women whose offspring later developed T1D from the same region [28]. A recent study from Sweden also investigated associations of gene polymorphisms of vitamin D metabolism with markers of insulin resistance and secretion with regard to the development of GDM. No associations of SNPs for VDBP and postpartum diabetes in women with a history of GDM, after multiple adjustments, were found [29]. It becomes clear that available findings on the field are largely affected by ethnic heterogeneity, highlighting the importance of national or regional data from homogeneous populations.

Similar findings were reported for rs4588 (CC genotype) and rs 7041 (TT genotype) in women with preeclampsia, compared to normotensive pregnant women from South Africa [30]. Results from United States reported that a variant in the *GC* flanking region (rs13150174) and a *GC* missense mutation in rs7041 were correlated with differences in log-transformed 25(OH)D concentration. A meta-analysis conducted by the same authors, also revealed that the minor allele for rs7041 was associated with elevated 25(OH)D concentrations and rs4588 was correlated with reduced 25(OH)D concentrations, among pregnant women [31]. The A-allele of the rs7041 polymorphism of the VDBP gene was also associated with a reduction in circulating 25(OH)D3 (difference in nmol/L) per allele of −5.48, and similar findings were observed for the T-allele of the rs4588 polymorphism (difference in nmol/L) per allele of −6.32 in a pregnant cohort from Northern Europe [28].

In addition, Chinese pregnant women with VDBP Gc-1f and Gc-1s genotypes had elevated plasma 25(OH)D concentrations compared to women with Gc-2 genotypes [32]. Similar results were obtained by a large cohort of 2658 women from the Zhoushan Pregnant Women Cohort study in China. Mutations of rs2298849 and rs7041 on the GC gene were respectively associated with higher 25(OH)D in the first and third trimesters, whereas mutations of seven SNPs (rs1155563, rs16846876, rs17467825, rs2282679, rs2298850, rs3755967, and rs4588) on the GC gene were respectively associated with lower 25(OH)D both in the first and in the third trimester. These effects were modified by season and vitamin D supplementation [33], which was an exclusion condition in our study. Vitamin D-binding protein polymorphisms have been shown to regulate attained 25(OH)D concentrations after the use of supplements during pregnancy. In this regard, a positive association between GC rs2282679 polymorphism and the achieved 25(OH)D status was noted following gestational cholecalciferol supplementation [34]. Recently, a multi-ethnic Asian genome-wide association study analysis pertaining to a birth cohort of three ethnicities identified rs4588 and its defining haplotype as a risk factor for low antenatal and cord blood vitamin D [35]. We have also recently reported that mothers with a CC genotype for rs2298850 and a CC genotype for rs4588 demonstrated higher 25(OH)D concentrations during delivery, confirming these findings in a Southern European pregnant population [11].

Irisin has been involved in the regulation of glucose homeostasis during pregnancy and neonatal body composition [36–38]. Irisin concentrations demonstrated a relationship with anthropometric measurements inappropriate for gestational-age infants, whereas low irisin concentrations in maternal serum were reported in pregnancies that developed preeclampsia and isolated intrauterine growth retardation [37–40]. We have previously described that maternal VDBP concentrations demonstrate a strong positive correlation with maternal adiponectin and irisin concentrations, after adjustments for weeks of gestation, maternal age, and BMI [9,10]. Further investigations are required to decipher the exact dynamic pathways of VDBP, adiponectin, and irisin during pregnancy and their effects on pregnancy complications and offspring body anthropometry.

During pregnancy, the binding affinity of VDBP for vitamin D metabolites is reduced to compensate for the maternal and neonatal higher demand for calcium and elevated VDBP concentrations, with almost two-fold increases between the second and third trimesters during fetal development [6]. Moreover, physiological hemodilution might affect maternal serum 25(OH)D levels due to maternal plasma volume expansion. Inflammation, placental functions, and iron and calcium metabolism also contribute to the peculiarities of vitamin D metabolism during pregnancy. In this regard, free 25(OH)D may be a better indicator compared to total 25(OH)D since it remains comparable to levels reported in nonpregnant women [41]. The analytical significance of several vitamin D metabolites including epimers during pregnancy has been vastly questionable [42]. Recent studies are shedding light on a plausible biologically active role for epimers in vitamin D metabolism and hence its importance upon interpretation of serum 25(OH)D levels based on the measurement assay being used. Specific and accurate assays, namely, LC–MS/MS, separate epimers and thus provide a better diagnostic tool for the measurement of 25(OH)D during pregnancy [43].

5. Strengths and Limitations

To the best of our knowledge, this study is the first to evaluate the contribution of genetic variants of maternal and neonatal VDBP polymorphisms on adiponectin, irisin, and VDBP concentrations at birth, according to different cut-offs of vitamin D status, in maternal–neonatal dyads. Our findings, which were based on the inclusion of both maternal and neonatal polymorphisms in conjunction with the assessment of different 25(OH)D cut-offs, could pave the way for future investigations aiming to examine the potential role of variants of VDBP on maternal–neonatal VDBP and adipokine status. Results could be projected to guide future research for a personalized genotype-based approach that could be particularly valuable for metabolic profiles of future mothers and their offspring.

Despite the strengths of our study, there are a few limitations that should be acknowledged. First, the sample size was limited in its ability to identify additional differences in other maternal–neonatal cut-offs. Second, the causality between the examined correlates could not be confirmed by cross-sectional design. Third, all women were Caucasian, so the findings cannot be confidently projected to other populations. In addition, despite the implications of our results as a personalized genotype-based approach, one major limitation pertains to the practicality of applying such an approach in the current clinical setting in terms of cost-efficiency and feasibility. Hence, to monitor and secure adequate vitamin D status during pregnancy, vitamin D supplementation remains the norm in the clinical practice.

6. Conclusions

The findings in our study emphasize a potential role for VDBP genetic variants, CC genotype for rs2298850 and rs4588, in conjunction with a specific high cut-off of maternal 25(OH)D, in increasing maternal VBDP concentrations, hence providing a mechanistic rationale for aiming for specific cut-offs of vitamin D, after supplementation during pregnancy, in the daily clinical practice.

Author Contributions: S.N.K. designed and conducted the study, interpreted the results, and drafted the original and revised versions. E.D. and D.G.-A. conducted the VDBP polymorphism analysis, statistical analysis, and drafted the original and revised versions. M.A. conducted the VDBP polymorphism analysis. S.P., E.J., F.A.A., P.P. and K.K. contributed to the data interpretation, statistical analysis, and drafting of the original and revised versions. All authors have read and agreed to the published version of the manuscript.

Funding: This study was supported by a grant from National Scholarship Foundation, Greece (no 2019-050-0503-18760).

Institutional Review Board Statement: The study was conducted according to the guidelines of the Declaration of Helsinki, and was approved by the Institutional Review Board (or Ethics Committee) of Aristotle University of Thessaloniki, Greece (Approval number 1/19-12-2011).

Informed Consent Statement: Informed consent was obtained from all subjects involved in the study.

Data Availability Statement: The data presented in this study are available on request from the corresponding author.

Conflicts of Interest: The authors declare no conflict of interest.

References

1. Speeckaert, M.M.; Speeckaert, R.; van Geel, N.; Delanghe, J.R. Vitamin D binding protein: A multifunctional protein of clinical importance. *Adv. Clin. Chem.* **2014**, *63*, 1–57.
2. Haddad, J.G. Plasma vitamin D-binding protein (Gc-globulin): Multiple tasks. *J. Steroid Biochem. Mol. Biol.* **1995**, *53*, 579–582. [CrossRef]

3. Kissmeyer, A.; Mathiasen, I.S.; Latini, S.; Binderup, L. Pharmacokinetic studies of vitamin D analogues: Relationship to vitamin D binding protein (DBP). *Endocrine* **1995**, *3*, 263–266. [CrossRef] [PubMed]

4. Vieth, R.; Kessler, M.J.; Pritzker, K.P. Species differences in the binding kinetics of 25-hydroxyvitamin D3 to vitamin D binding protein. *Can. J. Physiol. Pharmacol.* **1990**, *68*, 1368–1371. [CrossRef]

5. Jassil, N.K.; Sharma, A.; Bikle, D.; Wang, X. Vitamin D binding protein and 25-hydroxyvitamin D levels: Emerging clinical applications. *Endocr. Pract.* **2017**, *23*, 605–613. [CrossRef] [PubMed]

6. Karras, S.N.; Wagner, C.L.; Castracane, V.D. Understanding vitamin D metabolism in pregnancy: From physiology to pathophysiology and clinical outcomes. *Metabolism* **2018**, *86*, 112–123. [CrossRef]

7. Karras, S.N.; Koufakis, T.; Fakhoury, H.; Kotsa, K. Deconvoluting the Biological Roles of Vitamin D-Binding Protein during Pregnancy: A Both Clinical and Theoretical Challenge. *Front. Endocrinol.* **2018**, *9*, 259. [CrossRef]

8. Ma, R.; Gu, Y.; Zhao, S.; Sun, J.; Groome, L.J.; Wang, Y. Expressions of vitamin D metabolic components VDBP, CYP2R1, CYP27B1, CYP24A1, and VDR in placentas from normal and preeclamptic pregnancies. *Am. J. Physiol. Metab.* **2012**, *303*, E928–E935. [CrossRef] [PubMed]

9. Karras, S.N.; Polyzos, S.A.; Newton, D.A.; Wagner, C.L.; Hollis, B.W.; Ouweland, J.V.D.; Dursun, E.; Gezen-Ak, D.; Kotsa, K.; Annweiler, C.; et al. Adiponectin and vitamin D-binding protein are independently associated at birth in both mothers and neonates. *Endocrine* **2018**, *59*, 164–174. [CrossRef]

10. Karras, S.N.; Polyzos, S.A.; Tsekmekidou, X.; Gerou, S.; Gavana, E.; Papageorgiou, V.; Kotsa, K. Adiponectin and vitamin D-binding protein concentrations are independently associated in apparently healthy women but not men: A validation cohort. *Hormones* **2019**, *18*, 99–102. [CrossRef]

11. Karras, S.N.; Dursun, E.; Alaylıoğlu, M.; Gezen-Ak, D.; Annweiler, C.; Al Anouti, F.; Fakhoury, H.M.A.; Bais, A.; Kiortsis, D. Investigating the Role of Functional Polymorphism of Maternal and Neonatal Vitamin D Binding Protein in the Context of 25-Hydroxyvitamin D Cutoffs as Determinants of Maternal-Neonatal Vitamin D Status Profiles in a Sunny Mediterranean Region. *Nutrients* **2021**, *13*, 3082. [CrossRef] [PubMed]

12. Karras, S.N.; Dursun, E.; Alaylıoglu, M.; Gezen-Ak, D.; Annweiler, C.; Skoutas, D.; Evangelidis, D.; Kiortsis, D. Diverse Effects of Combinations of Maternal-Neonatal VDR Polymorphisms and 25-Hydroxyvitamin D Concentrations on Neonatal Birth Anthropometry: Functional Phenocopy Variability Dependence, Highlights the Need for Targeted Maternal 25-Hydroxyvitamin D Cut-Offs during Pregnancy. *Nutrients* **2021**, *13*, 443. [CrossRef]

13. Karras, S.N.; Koufakis, T.; Antonopoulou, V.; Goulis, D.G.; Alaylıoğlu, M.; Dursun, E.; Gezen-Ak, D.; Annweiler, C.; Pilz, S.; Fakhoury, H.; et al. Vitamin D receptor FokI polymorphism is a determinant of both maternal and neonatal vitamin D concentrations at birth. *J. Steroid Biochem. Mol. Biol.* **2019**, *199*, 105568. [CrossRef] [PubMed]

14. Chun, S.K.; Shin, S.; Kim, M.Y.; Joung, H.; Chung, J. Effects of maternal genetic polymorphisms in vitamin D-binding protein and serum 25-hydroxyvitamin D concentration on infant birth weight. *Nutrition* **2017**, *35*, 36–42. [CrossRef] [PubMed]

15. Jones, K.S.; Assar, S.; Harnpanich, D.; Bouillon, R.; Lambrechts, D.; Prentice, A.; Schoenmakers, I. 25(OH)D2 half-life is shorter than 25(OH)D3 half-life and is influenced by DBP concentration and genotype. *J. Clin. Endocrinol. Metab.* **2014**, *99*, 3373–3381. [CrossRef]

16. Lips, P.; Cashman, K.D.; Lamberg-Allardt, C.; Bischoff-Ferrari, H.A.; Obermayer-Pietsch, B.; Bianchi, M.L.; Stepan, J.; El-Hajj, F.G.; Bouillon, R. Current vitamin D status in European and Middle East countries and strategies to prevent vitamin D deficiency: A position statement of the European Calcified Tissue Society. *Eur. J. Endocrinol.* **2019**, *180*, P23–P54. [CrossRef]

17. Karras, S.; Paschou, S.A.; Kandaraki, E.; Anagnostis, P.; Annweiler, C.; Tarlatzis, B.C.; Hollis, B.W.; Grant, W.B.; Goulis, D.G. Hypovitaminosis D in pregnancy in the Mediterranean region: A systematic review. *Eur. J. Clin. Nutr.* **2016**, *70*, 979–986. [CrossRef]

18. Karras, S.N.; Shah, I.; Petroczi, A.; Goulis, D.G.; Bili, H.; Papadopoulou, F.; Harizopoulou, V.; Tarlatzis, B.C.; Naughton, D.P. An observational study reveals that neonatal vitamin D is primarily determined by maternal contributions: Implications of a new assay on the roles of vitamin D forms. *Nutr. J.* **2013**, *12*, 77. [CrossRef] [PubMed]

19. Karras, S.N.; Koufakis, T.; Antonopoulou, V.; Goulis, D.G.; Annweiler, C.; Pilz, S.; Bili, H.; Naughton, D.P.; Shah, I.; Harizopoulou, V.; et al. Characterizing neonatal vitamin D deficiency in the modern era: A maternal-neonatal birth cohort from Southern Europe. *J. Steroid Biochem. Mol. Biol.* **2020**, *198*, 105555. [CrossRef]

20. Shah, I.; James, B.; Barker, J.; Petroczi, A.; Naughton, D.P. Misleading measures in Vitamin D analysis: A novel LC-MS/MS assay to account for epimers and isobars. *Nutr. J.* **2011**, *10*, 46. [CrossRef]

21. Holick, M.F.; Binkley, N.C.; Bischoff-Ferrari, H.A.; Gordon, C.M.; Hanley, D.A.; Heaney, R.P.; Murad, M.H.; Weaver, C.M. Evaluation, treatment, and prevention of vitamin D deficiency: An Endocrine Society clinical practice guideline. *J. Clin. Endocrinol. Metab.* **2011**, *96*, 1911–1930. [CrossRef]

22. Munns, C.F.; Shaw, N.; Kiely, M.; Specker, B.L.; Thacher, T.; Ozono, K.; Michigami, T.; Tiosano, D.; Mughal, M.Z.; Mäkitie, O.; et al. Global Consensus Recommendations on Prevention and Management of Nutritional Rickets. *J. Clin. Endocrinol. Metab.* **2016**, *101*, 394–415. [CrossRef] [PubMed]

23. Gezen-Ak, D.; Dursun, E.; Bilgiç, B.; Hanagasi, H.; Ertan, T.; Gürvit, H.; Emre, M.; Eker, E.; Ulutin, T.; Uysal, O.; et al. Vitamin D receptor gene haplotype is associated with late-onset Alzheimer's disease. *Tohoku J. Exp. Med.* **2012**, *228*, 189–196. [CrossRef] [PubMed]

24. Ganz, A.B.; Park, H.; Malysheva, O.V.; Caudill, M.A. Vitamin D binding protein rs7041 genotype alters vitamin D metabolism in pregnant women. *FASEB J.* **2018**, *32*, 2012–2020. [CrossRef]
25. Wang, Y.; Wang, O.; Li, W.; Ma, L.; Ping, F.; Chen, L.; Nie, M. Variants in vitamin D binding protein gene are associated with gestational diabetes mellitus. *Medicine* **2015**, *94*, e1693. [CrossRef]
26. Shi, A.; Wen, J.; Liu, G.; Liu, H.; Fu, Z.; Zhou, J.; Zhu, Y.; Liu, Y.; Guo, X.; Xu, J. Geneticvariants in vitaminDsignalingpathways and risk of gestationaldiabetesmellitus. *Oncotarget* **2016**, *7*, 67788–67795. [CrossRef] [PubMed]
27. Tapia, G.; Mårild, K.; Dahl, S.R.; Lund-Blix, N.A.; Viken, M.K.; Lie, B.A.; Njølstad, P.R.; Joner, G.; Skrivarhaug, T.; Cohen, A.S.; et al. Maternal and Newborn Vitamin D—Binding Protein, Vitamin D Levels, Vitamin D Receptor Genotype, and Childhood Type 1 Diabetes. *Diabetes Care* **2019**, *42*, 553–559. [CrossRef]
28. Sørensen, I.M.; Joner, G.; Jenum, P.A.; Eskild, A.; Brunborg, C.; Torjesen, P.A.; Stene, L.C. Vitamin D-binding protein and 25-hydroxyvitamin D during pregnancy in mothers whose children later developed type 1 diabetes. *Diabetes Metab. Res. Rev.* **2016**, *32*, 883–890. [CrossRef]
29. Shaat, N.I.; Katsarou, A.; Shahida, B.; Prasad, R.B.; Kristensen, K.; Planck, T. Association between the rs1544410 polymorphism in the vitamin D receptor (VDR) gene and insulin secretion after gestational diabetes mellitus. *PLoS ONE* **2020**, *15*, e0232297. [CrossRef]
30. Naidoo, Y.; Moodley, J.; Ramsuran, V.; Naicke, T. Polymorphisms within vitamin D binding protein gene within a Preeclamptic South African population. *Hypertens. Pregnancy* **2019**, *38*, 260–267. [CrossRef]
31. Baca, K.M.; Govil, M.; Zmuda, J.M.; Simhan, H.N.; Marazita, M.L.; Bodnar, L.M. Vitamin D metabolic loci and vitamin D status in Black and White pregnant women. *Eur. J. Obstet. Gynecol. Reprod. Biol.* **2018**, *220*, 61–68. [CrossRef] [PubMed]
32. Shao, B.; Jiang, S.; Muyiduli, X.; Wang, S.; Mo, M.; Li, M.; Wang, Z.; Yu, Y. Vitamin D pathway gene polymorphisms influenced vitamin D level among pregnant women. *Clin. Nutr.* **2018**, *37*, 2230–2237. [CrossRef]
33. Wu, J.; Shao, B.; Xin, X.; Luo, W.; Mo, M.; Jiang, W.; Si, S.; Wang, S.; Shen, Y.; Yu, Y. Association of vitamin D pathway gene polymorphisms with vitamin D level during pregnancy was modified by season and vitamin D supplement. *Clin. Nutr.* **2021**, *40*, 3650–3660. [CrossRef] [PubMed]
34. Moon, R.J.; Harvey, N.C.; Cooper, C.; D'Angelo, S.; Curtis, E.M.; Crozier, S.R.; Barton, S.; Robinson, S.; Godfrey, K.; Graham, N.J.; et al. Response to antenatal cholecalciferol supplementation is associated with common vitamin D-related genetic variants. *J. Clin. Endocrinol. Metab.* **2017**, *102*, 2941–2949. [CrossRef] [PubMed]
35. Sampathkumar, A.; Tan, K.M.; Chen, L.; Chong, M.F.F.; Yap, F.; Godfrey, K.M.; Chong, Y.S.; Gluckman, P.D.; Ramasamy, A.; Karnani, N. Genetic Link Determining the Maternal-Fetal Circulation of Vitamin D. *Front. Genet.* **2021**, *12*, 721488. [CrossRef]
36. Ökdemir, D.; Hatipoğlu, N.; Kurtoğlu, S.; Siraz, Ü.G.; Akar, H.H.; Muhtaroğlu, S.; Kütük, M.S. The Role of Irisin, Insulin and Leptin in Maternal and Fetal Interaction. *J. Clin. Res. Pediatr. Endocrinol.* **2018**, *10*, 307–315. [CrossRef]
37. Wang, P.; Ma, H.H.; Hou, X.Z.; Song, L.L.; Song, X.L.; Zhang, J.F. Reduced plasma level of irisin in first trimester as a risk factor for the development of gestational diabetes mellitus. *Diabetes Res. Clin. Pract.* **2018**, *142*, 130–138. [CrossRef]
38. Briana, D.D.; Boutsikou, M.; Athanasopoulos, N.; Marmarinos, A.; Gourgiotis, D.; Malamitsi-Puchner, A. Implication of the myokine irisin in maternal energy homeostasis in pregnancies with abnormal fetal growth. *J. Matern. Fetal Neonatal Med.* **2016**, *29*, 3429–3433. [CrossRef]
39. Erol, O.; Erkal, N.; Ellidağ, H.Y.; İsenlik, B.S.; Aydın, Ö.; Derbent, A.U.; Yılmaz, N. Irisin as an early marker for predicting gestational diabetes mellitus: A prospective study. *J. Matern. Fetal Neonatal Med.* **2016**, *29*, 3590–3595. [CrossRef]
40. Ozel, A.; Davutoglu, E.A.; Firat, A.; Erenel, H.; Karslı, M.F.; Korkmaz, S.Ö.; Madazli, R. Maternal serum irisin levels in early and late-onset pre-eclamptic and healthy pregnancies. *J. Obstet. Gynaecol.* **2018**, *38*, 642–646. [CrossRef]
41. Tsuprykov, O.; Buse, C.; Skoblo, R.; Haq, A.; Hocher, B. Reference intervals for measured and calculated free 25-hydroxyvitamin D in normal pregnancy. *J. Steroid Biochem. Mol. Biol.* **2018**, *181*, 80–87. [CrossRef] [PubMed]
42. Karras, S.N.; Kotsa, K.; Angeloudi, E.; Zebekakis, P.; Naughton, D.P. The Road Not So Travelled: Should Measurement of Vitamin D Epimers during Pregnancy Affect Our Clinical Decisions? *Nutrients* **2017**, *9*, 90. [CrossRef] [PubMed]
43. Bikle, D.D.; Schwartz, J. Vitamin D Binding Protein, Total and Free Vitamin D Levels in Different Physiological and Pathophysiological Conditions. *Front. Endocrinol.* **2019**, *10*, 317. [CrossRef] [PubMed]

Article

Vitamin D Intake May Reduce SARS-CoV-2 Infection Morbidity in Health Care Workers

Tatiana L. Karonova [1,*], Alena T. Chernikova [1], Ksenia A. Golovatyuk [1], Ekaterina S. Bykova [1], William B. Grant [2], Olga V. Kalinina [1], Elena N. Grineva [1] and Evgeny V. Shlyakhto [1]

[1] Clinical Endocrinology Laboratory, Department of Endocrinology, Almazov National Medical Research Centre, 194021 Saint Petersburg, Russia; arabicaa@gmail.com (A.T.C.); ksgolovatiuk@gmail.com (K.A.G.); bykova160718@gmail.com (E.S.B.); olgakalinina@mail.ru (O.V.K.); grineva_e@mail.ru (E.N.G.); shlyakhto_ev@almazovcentre.ru (E.V.S.)

[2] Sunlight, Nutrition, and Health Research Center, San Francisco, CA 94164-1603, USA; williamgrant08@comcast.net

* Correspondence: karonova@mai.ru; Tel.: +7-921-310-60-41

Abstract: In the last 2 years, observational studies have shown that a low 25-hydroxyvitamin D (25(OH)D) level affected the severity of infection with the novel coronavirus (COVID-19). This study aimed to analyze the potential effect of vitamin D supplementation in reducing SARS-CoV-2 infection morbidity and severity in health care workers. Of 128 health care workers, 91 (consisting of 38 medical doctors (42%), 38 nurses (42%), and 15 medical attendants (16%)) were randomized into two groups receiving vitamin D supplementation. Participants of group I (n = 45) received water-soluble cholecalciferol at a dose of 50,000 IU/week for 2 consecutive weeks, followed by 5000 IU/day for the rest of the study. Participants of group II (n = 46) received water-soluble cholecalciferol at a dose of 2000 IU/day. For both groups, treatment lasted 3 months. Baseline serum 25(OH)D level in health care workers varied from 3.0 to 65.1 ng/mL (median, 17.7 (interquartile range, 12.2; 24.7) ng/mL). Vitamin D deficiency, insufficiency, and normal vitamin D status were diagnosed in 60%, 30%, and 10%, respectively. Only 78 subjects completed the study. Vitamin D supplementation was associated with an increase in serum 25(OH)D level, but only intake of 5000 IU/day was accompanied by normalization of serum 25(OH)D level, which occurred in 53% of cases. Neither vitamin D intake nor vitamin D deficiency/insufficiency were associated with a decrease in SARS-CoV-2 morbidity (odds ratio = 2.27; 95% confidence interval, 0.72 to 7.12). However, subjects receiving high-dose vitamin D had only asymptomatic SARS-CoV-2 in 10 (26%) cases; at the same time, participants who received 2000 IU/day showed twice as many SARS-CoV-2 cases, with mild clinical features in half of them.

Keywords: COVID-19; SARS-CoV-2; vitamin D; 25(OH)D; health care workers

Citation: Karonova, T.L.; Chernikova, A.T.; Golovatyuk, K.A.; Bykova, E.S.; Grant, W.B.; Kalinina, O.V.; Grineva, E.N.; Shlyakhto, E.V. Vitamin D Intake May Reduce SARS-CoV-2 Infection Morbidity in Health Care Workers. *Nutrients* **2022**, *14*, 505. https://doi.org/10.3390/nu14030505

Academic Editor: Roberto Iacone

Received: 29 December 2021
Accepted: 21 January 2022
Published: 24 January 2022

Publisher's Note: MDPI stays neutral with regard to jurisdictional claims in published maps and institutional affiliations.

1. Introduction

Recent studies showed that vitamin D deficiency and insufficiency are common in the world's general population, including Russia [1,2]. The COVID-19 pandemic that swept the world in 2019 changed our lifestyles, as well as scientific and medical approaches. SARS-CoV-2, a respiratory viral airborne infection with no effective treatment other than applying prevention strategies, had serious damaging health effects [3]. In the last 2 years, researchers have shown that COVID-19 severity may be related to vitamin D status [4]. Considering vitamin D's immunomodulatory properties [5–8], scientists found associations between low 25(OH)D level and COVID-19 severity in observational studies [9–12]. The largest study in the United States reported that the SARS-CoV-2 positivity rate was 6%. It was lowest in subjects with 25(OH)D concentration >55 ng/mL, compared with patients with 25(OH)D of 20 ng/mL, whose rate reached about 11% [13]. However, for hospitalized patients, being aged 50 years or older was a more significant factor than vitamin D status [14]. At the same time, we published data with analysis for more than 300,000 subjects who had a known 25(OH)D level from fall 2019 to fall 2020, and detected that vitamin D deficiency did not increase the rate of positive PCR test to SARS-CoV-2 in the Russian population [15]. One reason for that difference between the United States and Russia could be the high prevalence of vitamin D deficiency in Russia, as well as using different methods for the PCR testing of COVID-19.

Health care workers are now considered a risk group because they are continually exposed to COVID-19. In a study of 120,075 participants, Mutambudzi and colleagues showed that medical support staff had a sevenfold-higher risk of severe COVID-19 infection than other worker groups (relative risk (RR) = 7.43; 95% confidence interval (CI), 5.52 to 10.00) [16]. The same data were found in the United Kingdom, where medical staff had a high risk of COVID-19, especially those with vitamin D deficiency [17].

Vitamin D activates immune cells to produce cathelicidin and defensins, as well as increasing expression of angiotensin-converting enzyme 2, which promotes the binding of the virus in the lung blood vessels [18], leading to reduced survival and replication. As a result, vitamin D supplementation could prevent and treat SARS-CoV-2. Results of the pilot study in Spain, in which hospitalized COVID-19-positive patients took calcifediol, showed that only 1 of 50 treated patients required the intensive-care unit in comparison with 13 of 26 nontreated patients (odds ratio (OR) = 0.02; 95% CI, 0.002 to 0.17) [19]. The bolus vitamin D supplementation also was associated with decreased mortality in hospitalized COVID-19 patients in Turkey [20], and an improved 3-month survival in geriatric patients [21]. This information was confirmed in a meta-analysis that included data from observational and randomized controlled trials that reported reduced severity risk with higher 25(OH)D, and some benefit from vitamin D in treating COVID-19 [22].

Despite many COVID-19 studies having been performed in the last 18 months, we could not find interventional studies among medical staff with the use of vitamin D supplementation to assess its preventive effect on SARS-CoV-2 morbidity. So, this study aimed to analyze the potential effect of vitamin D supplementation in reducing SARS-CoV-2 infection morbidity and severity in health care workers.

2. Materials and Methods

2.1. Patients

The study population included 128 employees of Almazov National Medical Research Centre (111 women and 17 men) who signed an informed consent for participation. They began work in the infectious hospital amid the COVID-19 pandemic, had a negative PCR test, and were off vitamin D or received only preventive doses. This single-center, open-label, randomized, interventional study was performed from October 30 2020 to February 28 2021, with the following inclusion criteria: (i) age 18–65 years, (ii) negative PCR test for SARS-CoV-2, (iii) absence of clinical signs of acute respiratory viral infection (ARVI), (iv) contact with patients with laboratory and/or clinically confirmed SARS-CoV-2 infection. We did not include subjects with a history of intolerance or allergic response to water-soluble cholecalciferol in anamnesis, or those not compliant with the recommendation of the Ministry of Health with regard to personal protective equipment [23]. Exclusion criteria included primary hyperparathyroidism or hypercalcemia of other etiology (including a mutation of 24 hydroxylase); clinically significant gastrointestinal diseases, kidney pathology (estimated glomerular filtration rate less than 45 mL/min/1.73 m^2), and liver diseases that can influence vitamin D absorption and metabolism; a history of granulomatous diseases; a history of oncology diseases (<5 years); intake of glucocorticosteroids or anticonvulsants; and alcohol and drug addiction. Pregnant, breastfeeding women or women planning pregnancies also did not participate. If potential participants had other circumstances that the investigator considered inappropriate, they were excluded. Of note, the general vaccination, including for high-risk groups, was launched only in the end of February to early March 2021, and the first participant was included in the study in

November 2020, when the center started to work with COVID-19 patients. Hence, we did not include vaccinated health care workers at the beginning or throughout the study.

After signing the informed consent and initial physical examination, 23 subjects (18%) were excluded from the survey after the prompt onset of a respiratory tract infection before taking the first dose of cholecalciferol. Eleven employees (9%) withdrew consent soon after signing; and three subjects (2%) were excluded because of a lack of initial laboratory data. Thus, the final survey included data for 91 health care workers. After randomization (random numbers method) at a ratio of 1:1, all participants were divided into two groups. The participants of group I (*n* = 45) received water-soluble cholecalciferol (Aquadetrim, "Akrichin," Staraya Kupavna, Moscow region, Russia) at a dose of 50,000 IU/week for 2 consecutive weeks, followed by 5000 IU/day for the rest of the study. The participants of group II (*n* = 46) received water-soluble cholecalciferol (Aquadetrim) at a dose of 2000 IU/day. Our center was involved in the treatment of COVID-19 patients only for a duration of three months; hence, the health care workers had direct contact with such patients only during this time. Thus, the treatment lasted 3 months for both groups (Figure 1).

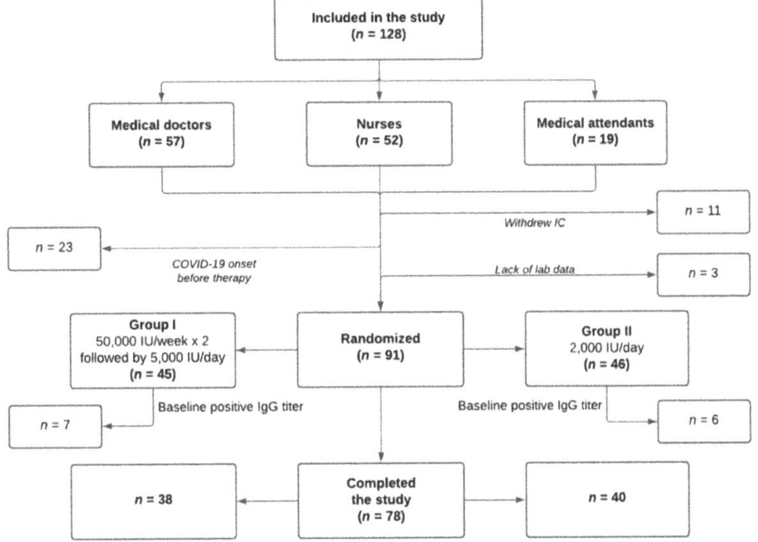

Figure 1. Study design. IC, informed consent; IgG, immunoglobulin G.

2.2. Physical Data

Anthropometric examination included height (centimeters) and weight (kilograms), from which body mass index (BMI) was calculated (kilograms per square meter of body surface area). The participant questionnaire included demographic data, education, medical history and concomitant medication, smoking status, allergies, and vitamin D supplement intake.

2.3. Laboratory Tests

Serum 25(OH)D level was detected by the chemiluminescent immunoassay (Architect i8000; Abbott, Chicago, IL, USA) using laboratory sets and control sera from the manufacturer. Vitamin D deficiency was defined as a serum 25(OH)D level < 20 ng/mL [24].

Testing of immunoglobulin G (IgG) to SARS-CoV-2 was performed with a semiquantitative method by using the enzyme-linked immunosorbent assay on the Bio-Rad 680 microplate reader equipment (Hercules, CA, USA) with the corresponding set SARS-CoV-2-IgG-ELISA-Best (Vector Best; Novosibirsk, Russia). A result was considered negative for positivity index (PI) < 0.8; positive for PI ≥ 1.1; and borderline for 0.8 ≤ PI < 1.1.

In addition, we evaluated biochemical parameters, such as a fasting plasma glucose level and blood lipid profile (Roche Diagnostics GmbH, Mannheim, Germany). Also, 2 weeks after starting the study, participants underwent blood tests to assess serum 25(OH)D (Abbott Architect i8000) and total calcium levels (reference interval, 2.15–2.65 mmol/L; Roche Diagnostics GmbH, Mannheim, Germany).

All blood samples were taken in the morning from the cubital vein, centrifuged, aliquoted, and stored in a freezer at -70 °C before testing.

2.4. Statistical Analysis

For sample calculation, we used Power and Sample Size software [25]. At a 5% significance level and 80% power, the sample size was 72 people (36 per group).

Statistical processing was carried out using SPSS for Windows (ver. 26; IBM, Armonk, NY, USA), with the help of standard methods of variation statistics. Between-group comparison was carried out using the Mann–Whitney criteria for incorrect distribution; results are presented as median and interquartile range, as well as mean and standard deviation for the Student criterion in correct distributed parameters. Associations between quantitative parameters were assessed using Spearman's correlation coefficient. To describe relative risk, we calculated the odds ratio, with a 95% confidence interval calculated using Fisher's exact method. The criterion for the statistical reliability of the obtained results was $p < 0.05$.

3. Results

Of the 128 health care workers who signed the informed consent, 57 were medical doctors (44%), 52 were nurses (41%), and 19 were medical attendants (15%). Baseline serum 25(OH)D level varied from 3.0 to 69.0 ng/mL (mean, 18.5 (interquartile range, 11.9; 26.7) ng/mL). A total of 114 participants presented with baseline 25(OH)D results, with 63 subjects (55%) deficient, 34 subjects (30%) insufficient, and only 17 health care workers (15%) with normal vitamin D status. Medical attendants were diagnosed with vitamin D deficiency more often than medical doctors and nurses: 88%, 46%, and 53%, respectively ($p = 0.001$). The participants with graduate medical education had a higher serum 25(OH)D level (22.1 (16.1; 29.5) ng/mL) than subjects with a secondary medical education (19.3 (10.7; 24.9) ng/mL) or without specialized education (11.1 (9.7; 17.6) ng/mL) ($p = 0.001$; Table 1).

Table 1. Vitamin D status among health care workers ($n = 114$).

Parameter	Medical Doctors $n = 52$	Nurses $n = 45$	Medical Attendants $n = 17$	p
25(OH)D, ng/mL, Me + IQR (25; 75)	22.1 (16.1; 29.5)	19.3 (10.7; 24.9)	11.1 (9.7; 17.6)	0.001
Vitamin D status, n (%)				
Normal	12 (23)	4 (9)	1 (6)	
Insufficiency	16 (31)	17 (38)	1 (6)	0.001
Deficiency	24 (46)	24 (53)	15 (88)	

25(OH)D, 25-hydroxyvitamin D; Me, median; IQR, interquartile range.

After exclusion of subjects infected with COVID-19 or those who withdrew consent before the first dose of vitamin D supplementation, 91 employees (38 medical doctors, 38 nurses, and 15 medical attendants) were randomized. Participants in the groups were comparable and had no significant differences in baseline serum 25(OH)D level, which was 16.9 (11.4; 23.9) for group I, and 18.4 (12.2; 25.1) ng/mL for group II ($p = 0.54$; Table 2).

Table 2. Baseline characteristics of randomized health care workers.

Parameters	Group I $n = 45$	Group II $n = 46$	p
Age, years (mean $\pm$ SD)	35 ± 2	35 ± 2	0.81
Sex, M/F, n (%)	8 (18)/37 (82)	6 (13)/40 (87)	0.53
Education, n (%)			
Graduate medical	15 (33)	23 (50)	
Secondary medical	24 (53)	14 (30)	0.36
Without specialized education	6 (14)	9 (20)	
BMI, kg/m^2, n (%)	24.8 ± 0.8	24.6 ± 0.7	0.98
Normal	25 (55)	29 (63)	
Overweight	12 (27)	10 (22)	0.49
Obese	8 (18)	7 (15)	
FPG, mmol/L	5.3 ± 0.2	5.3 ± 0.2	0.35
TC, mmol/L	5.3 ± 0.2	5.3 ± 0.2	0.95
LDL, mmol/L	2.9 ± 0.2	3.0 ± 0.1	0.46
HDL, mmol/L	1.6 ± 0.1	1.6 ± 0.1	0.44
TG, mmol/L	1.6 ± 0.2	1.6 ± 0.2	0.49
25(OH)D, ng/mL, Me + IQR (25; 75)	16.9 (11.4; 23.9)	18.4 (12.2; 25.1)	0.54
Vitamin D status, n (%)			
Normal	4 (9)	5 (11)	
Insufficiency	12 (27)	15 (33)	0.45
Deficiency	29 (64)	26 (56)	

SD, standard deviation; M, male; F, female; BMI, body mass index; FPG, fasting plasma glucose; TC, total cholesterol; LDL, low-density lipoprotein; HDL, high-density lipoprotein; TG, triglycerides; 25(OH)D, 25-hydroxyvitamin D; Me, median; IQR, interquartile range.

Two weeks after the initiation of vitamin D supplementation, serum 25(OH)D level and total calcium level were measured to control efficacy and safety. So, after 100,000 IU of water-soluble cholecalciferol after 2 weeks, median 25(OH)D level was 32.9 (26.3; 39.6) ng/mL, and was significantly higher than in the initial data ($p = 0.001$). Participants who received 2000 IU had no significant changes in 25(OH)D level (19.3 (14.1; 27.2) ng/mL); $p = 0.08$. Total serum calcium level was within reference values in both groups.

Analyses of IgG to SARS-CoV-2 showed that 13 randomized participants (14%) had initially positive IgG titers, an indicator of past infection—probably asymptomatic. Therefore, data of those participants were excluded from the final analysis.

The analyzable final results included data of 78 employees (34 medical doctors (44%), 33 nurses (42%), and 11 medical attendants (14%)) who had not been exposed to the virus in the past as a result of SARS-CoV-2 (Table 3). The results of their examination showed the absence of significant differences, including the values of the baseline serum 25(OH)D level in group I (18.4 (14.3; 24.5) ng/mL) and group II (18.5 (12.5; 25.0) ng/mL) ($p = 0.94$).

Table 3. Characteristics of health care workers with initially negative IgG titer to SARS-CoV-2.

Parameters	Group I $n = 38$	Group II $n = 40$	p
Age, years (mean $\pm$ SD)	34 $\pm$ 2	36 $\pm$ 2	0.93
Sex, M/F, n (%)	6 (16)/32 (84)	6 (15)/34 (85)	0.92
Education, n (%)			
Graduate medical	15 (39)	19 (48)	
Secondary medical	20 (53)	13 (32)	0.99
Without specialized education	3 (8)	8 (20)	
BMI, kg/m^2, n (%)	24.3 $\pm$ 0.9	24.7 $\pm$ 0.7	0.57
Normal	22 (58)	25 (63)	
Overweight	12 (32)	9 (22)	0.85
Obesity	4 (10)	6 (15)	
FPG, mmol/L	4.9 $\pm$ 0.1	5.4 $\pm$ 0.2	0.12
TC, mmol/L	5.2 $\pm$ 0.2	5.3 $\pm$ 0.2	0.91
LDL, mmol/L	2.8 $\pm$ 0.2	3.0 $\pm$ 0.1	0.44
HDL, mmol/L	1.6 $\pm$ 0.1	1.6 $\pm$ 0.1	0.29
TG, mmol/L	1.5 $\pm$ 0.2	1.6 $\pm$ 0.2	0.81
25(OH)D, ng/mL, Me + IQR (25; 75)	18.4 (14.3; 24.5)	18.5 (12.5; 25.0)	0.94
Vitamin D status, n (%)			
Normal	3 (8)	5 (12)	
Insufficiency	12 (32)	12 (30)	
Deficiency	23 (60)	23 (58)	

SD, standard deviation; M, male; F, female; BMI, body mass index; FPG, fasting plasma glucose; TC, total cholesterol; HDL, high-density lipoprotein; LDL, low-density lipoprotein, TG, triglycerides; Me, median; IQR, interquartile range; 25(OH)D, 25-hydroxyvitamin D.

All participants had an increase in serum 25(OH)D level at the end of the study. Therefore, serum 25(OH)D after 3 months of vitamin D supplementation reached 29.9 (25.2; 42.0) ng/mL in group I and 26.0 (21.3; 30.3) ng/mL in group II ($p = 0.01$), with 53% of participants from group I and 25% from group II reaching normal vitamin D status (Figure 2).

Analysis of positive IgG to SARS-CoV-2 cases among employees showed that 10 (26%) health care workers in group I had a positive PCR test and positive IgG titer, but no clinical features of ARVI. At the same time, 18 (45%) employees in group II had positive results, including 9 (23%) subjects with mild ARVI clinical features, and 9 (23%) subjects with asymptomatic disease. No participants underwent a computed tomography scan owing to asymptomatic or a mild course of COVID-19. Baseline and following vitamin D supplementation serum 25(OH)D level was the same among participants with positive IgG to SARS-CoV-2, and virus-free participants (baseline, 19.3 (12.1; 23.6) and 16.9 (11.8; 24.9) ng/mL, $p = 0.51$; and, at end of the study, 26.4 (20.3; 29.3) and 27.2 (22.2; 36.4) ng/mL, $p = 0.69$, respectively). Assessing the risk of SARS-CoV-2 morbidity depending upon vitamin D status, we found no associations between vitamin D deficiency/insufficiency and increased incidence of viral infection (OR = 2.27; 95% CI, 0.72 to 7.12).

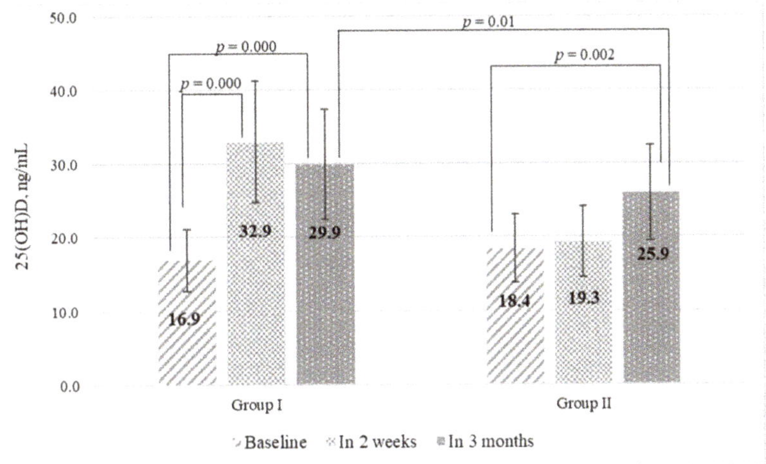

Figure 2. Serum 25(OH)D level before and after different doses of vitamin D supplementation.

4. Discussion

To our knowledge, this is the first randomized interventional trial among health care workers to show that high-dose vitamin D supplementation is safe and effective in achieving normal vitamin D levels, but was not connected to reduced SARS-CoV-2 morbidity. However, intake of 50,000 IU/week twice, followed by 5000 IU/day, seemed to be associated with asymptomatic COVID-19 cases, whereas health care workers receiving 2000 IU/day had a two-fold higher infection that was symptomatic with mild clinical features in half of cases.

Vitamin D is postulated to play an important immunomodulatory role, and deficiency is associated with increased incidence of ARVI, including COVID-19 [26–28]. Our previous results also showed that severe vitamin D deficiency is associated with severity and death in COVID-19 patients [29], and were comparable to recent findings [4,30]. Dissanayake and colleagues, whose meta-analysis included 72 COVID-19 observational and 4 interventional randomized studies, have shown not only correlations between 25(OH)D level and severity or mortality, but also some clinical benefits and improvement in inflammatory markers of vitamin D supplementation in treating COVID-19 [22].

Recent observational studies showed a more frequent vitamin D deficiency among shift workers and newcomers, including health care workers, than day workers [17,31,32], whereas data regarding mortality rate showed a high COVID-19-related mortality among health care workers, as published by the World Health Organization [33]. Thus, a great necessity exists to find new effective measures to prevent SARS-CoV-2 and/or decrease COVID-19 morbidity and severity in medical workers. Taking into account the need to improve preventive actions for medical staff in daily contact with SARS-CoV-2 patients, we developed a hypothesis for this research to assess vitamin D supplementation's effectiveness in preventing COVID-19 among this population.

To reduce the risk of infection, it is recommended that people at risk should rapidly increase 25(OH)D concentrations above 40–60 ng/mL [27,34]. In order to achieve this, patients need to take higher loading vitamin D doses: 100,000–200,000 IU over 8 weeks [34,35]. To maintain that level after the first month, the dose can be decreased to 5000 IU/day [36]. Considering published recommendations for decreasing the morbidity of COVID-19 [27,34,35], we have chosen a high vitamin D supplementation dose of 50,000 IU/week twice for a rapid increase of 25(OH)D level, followed by 5000 IU/day, and compared with the common daily dose used in clinical practice. We showed a good tolerability of the saturating dose of water-soluble cholecalciferol, and a rapid increase in the serum 25(OH)D level to normal values without an increase in total calcium levels in the blood. Those results are comparable to those of previous works [36,37].

Inspection of the population-based study results shows that subjects with cholecalciferol supplementation had a lower risk of SARS-CoV-2 infection (hazard ratio (HR) = 0.95 (95% CI, 0.91 to 0.98); p = 0.004) than deficient unsupplemented subjects. The protective effect was more significant between treated subjects with 25(OH)D > 30 ng/mL and untreated deficient subjects (HR = 0.57 (95% CI, 0.50 to 0.66); p < 0.001) [38]. In our study, no significant differences were evident in morbidity between the comparable groups, and no difference in serum 25(OH)D level emerged between subjects with positive or negative IgG titers despite vitamin D supplementation. That result might be related to our inability to achieve the recommended 25(OH)D level of 40–60 ng/mL. However, subjects receiving a higher dose of cholecalciferol had an asymptomatic course of viral infection. Those differences can be explained by engagement mechanisms as in the cell-bound and adaptive immunity, as well as a protective function on the level of upper-airway mucosa [39].

Possible study limitations include the small sample, absence of lab baseline data of serum 25(OH)D level and IgG before randomization, and short study duration of 3 months. In addition, the study was carried out in a comparative rather than placebo-controlled design. Therefore, conducting more detailed research is necessary to better understand vitamin D's role in preventing SARS-CoV-2 infection.

Author Contributions: E.V.S.: Conceptualization, project administration. T.L.K.: Conceptualization, project administration, study design, writing and editing. A.T.C.: Project administration, data analysis and interpretation, writing and editing. K.A.G.: Patient recruitment, writing and editing. E.S.B.: Data collection, writing. W.B.G.: Study design, conceptualization, review and editing. O.V.K.: Laboratory diagnostic, review and editing. E.N.G.: Conceptualization, project administration. All authors have read and agreed to the published version of the manuscript.

Funding: This study was provided by Almazov National Medical Research Centre, and was financially supported by the Ministry of Science and Higher Education of the Russian Federation (agreement no. 075-15-2020-901).

Institutional Review Board Statement: The study was conducted according to the guidelines of the Declaration of Helsinki, and approved by Almazov National Medical Research Centre Local Ethics Committee, Protocol ID: 0811-20-01C. The study has been registered on clinicaltrials.gov as "COVID-19 Morbidity in Healthcare Workers and Vitamin D Supplementation"; ID: NCT05037253.

Informed Consent Statement: Informed consent was obtained from all subjects involved in the study.

Data Availability Statement: The data generated and analyzed during this study are included in this article and its supplementary information files. More information is available from the corresponding author on reasonable request.

Acknowledgments: We thank study participants for supplying samples to advance our understanding of vitamin D's role in COVID-19 infection.

Conflicts of Interest: The Sunlight, Nutrition, and Health Research Center (W.B.G.) receives funding from Bio-Tech Pharmacal Inc. (Fayetteville, AR, USA). The other authors declare no conflict of interest.

References

1. Zemb, P.; Bergman, P.; Camargo, C.A., Jr.; Cavalier, E.; Cormier, C.; Courbebaisse, M.; Hollis, B.; Joulia, F.; Minisola, S.; Pilz, S.; et al. Vitamin D deficiency and the COVID-19 pandemic. *J. Glob. Antimicrob. Resist.* **2020**, *22*, 133–134. [CrossRef] [PubMed]
2. Karonova, T.; Andreeva, A.; Nikitina, I.; Belyaeva, O.; Mokhova, E.; Galkina, O.; Vasilyeva, E.; Grineva, E. Prevalence of Vitamin D deficiency in the North-West region of Russia: A cross-sectional study. *J. Steroid. Biochem. Mol. Biol.* **2016**, *164*, 230–234. [CrossRef] [PubMed]
3. Wiersinga, W.J.; Rhodesz, A.; Cheng, A.C.; Peacock, S.J.; Prescott, H.C. Pathophysiology, Transmission, Diagnosis and Treatment of Coronavirus Disease 2019 (COVID-19): A Review. *JAMA* **2020**, *324*, 782–793. [CrossRef] [PubMed]
4. Mercola, J.; Grant, W.B.; Wagner, C.L. Evidence Regarding Vitamin D and Risk of COVID-19 and Its Severity. *Nutrients* **2020**, *12*, 3361. [CrossRef]
5. Azrielant, S.; Shoenfeld, Y. Vitamin D and the immune system. *Isr. Med. Assoc. J.* **2017**, *19*, 510–511.

6.   Lemire, J.M.; Adams, J.S.; Kermani-Arab, V.; Bakke, A.C.; Sakai, R.; Jordan, S.C. 1,25-Dihydroxyvitamin D3 suppresses human T helper/inducer lymphocyte activity in vitro. *J. Immunol.* **1985**, *134*, 3032–3035.

7.   Cantorna, M.T.; Snyder, L.; Lin, Y.D.; Yang, L. Vitamin D and 1,25(OH)2D regulation of T cells. *Nutrients* **2015**, *7*, 3011–3021. [CrossRef]

8.   Jeffery, L.E.; Burke, F.; Mura, M.; Zheng, Y.; Qureshi, O.S.; Hewison, M.; Walker, L.S.; Lammas, D.A.; Raza, K.; Sansom, D.M. 1,25-Dihydroxyvitamin D3 and IL-2 combine to inhibit T cell production of inflammatory cytokines and promote development of regulatory T cells expressing CTLA-4 and FoxP3. *J. Immunol.* **2009**, *183*, 5458–5467. [CrossRef]

9.   Panagiotou, G.; Tee, S.A.; Ihsan, Y.; Athar, W.; Marchitelli, G.; Kelly, D.; Boot, C.S.; Stock, N.; Macfarlane, J.; Martineau, A.R.; et al. Low serum 25-hydroxyvitamin D (25[OH]D) levels in patients hospitalized with COVID-19 are associated with greater disease severity. *Clin. Endocrinol.* **2020**, *93*, 508–511. [CrossRef]

10.   Carpagnano, G.E.; Di Lecce, V.; Quaranta, V.N.; Zito, A.; Buonamico, E.; Capozza, E.; Palumbo, A.; Di Gioia, G.; Valerio, V.N.; Resta, O. Vitamin D deficiency as a predictor of poor prognosis in patients with acute respiratory failure due to COVID-19. *J. Endocrinol. Investig.* **2021**, *44*, 765–771. [CrossRef]

11.   Pizzini, A.; Aichner, M.; Sahanic, S.; Bohm, A.; Egger, A.; Hoermann, G.; Kurz, K.; Widmann, G.; Bellmann-Weiler, R.; Weiss, G. Impact of Vitamin D Deficiency on COVID-19—A Prospective Analysis from the CovILD Registry. *Nutrients* **2020**, *12*, 2775. [CrossRef]

12.   Macaya, F.; Espejo Paeres, C.; Valls, A.; Fernandez-Ortiz, A.; Gonzalez Del Castillo, J.; Martin-Sanchez, J.; Runkle, I.; Rubio Herrera, M.A. Interaction between age and vitamin D deficiency in severe COVID-19 infection. *Nutr. Hosp.* **2020**, *37*, 1039–1042. [CrossRef] [PubMed]

13.   Kaufman, H.W.; Niles, J.K.; Kroll, M.H.; Bi, C.; Holick, M.F. SARS-CoV-2 positivity rates associated with circulating 25-hydroxyvitamin D levels. *PLoS ONE* **2020**, *15*, e0239252. [CrossRef] [PubMed]

14.   Merzon, E.; Tworowski, D.; Gorohovski, A.; Vinker, S.; Golan Cohen, A.; Green, I.; Frenkel-Morgenstern, M. Low plasma 25(OH) vitamin D level is associated with increased risk of COVID-19 infection: An Israeli population-based study. *FEBS J.* **2020**, *287*, 3693–3702. [CrossRef] [PubMed]

15.   Karonova, T.L.; Andreeva, A.T.; Golovatuk, K.A.; Bykova, E.S.; Skybo, I.I.; Grineva, E.N.; Shlyakhto, E.V. Inficirovannost SARS-CoV-2 v zavisimosti ot urovnya obespechennosti vitaminom D. *Probl. Endokrinol.* **2021**, *67*, 20–28. [CrossRef]

16.   Mutambudzi, M.; Niedzwiedz, C.; Macdonald, E.B.; Leyland, A.; Mair, F.; Anderson, J.; Celis-Morales, C.; Clelend, J.; Forbes, J.; Gill, J.; et al. Occupation and risk of severe COVID-19: Prospective cohort study of 120,075 UK Biobank participants. *Occup. Environ. Med.* **2021**, *78*, 307–314. [CrossRef]

17.   Faniyi, A.A.; Lugg, S.T.; Faustini, S.E.; Webster, C.; Duffy, J.E.; Hewison, M.; Shields, A.; Nightingale, P.; Richter, A.G.; Thickett, D.R. Vitamin D status and seroconversion for COVID-19 in UK healthcare workers. *Eur. Respir. J.* **2021**, *57*, 2004234. [CrossRef]

18.   Malek Mahdavi, A. A brief review of interplay between vitamin D and angiotensin-converting enzyme 2: Implications for a potential treatment for COVID-19. *Rev. Med. Virol.* **2020**, *30*, 2119. [CrossRef]

19.   Entrenas Castillo, M.; Entrenas Costa, L.M.; Vaquero Barrios, J.M.; Alcalá Díaz, J.F.; López Miranda, J.; Bouillon, R.; Quesada Gomez, J.M. Effect of calcifediol treatment and best available therapy versus best available therapy on intensive care unit admission and mortality among patients hospitalized for COVID-19: A pilot randomized clinical study. *J. Steroid. Biochem. Mol. Biol.* **2020**, *203*, 105751. [CrossRef]

20.   Yildiz, M.; Senel, M.U.; Kavurgaci, S.; Ozturk, F.E.; Ozturk, A. The prognostic significance of vitamin D deficiency in patients with COVID-19 pneumonia. *Bratisl. Lek. Listy* **2021**, *122*, 744–747. [CrossRef]

21.   Annweiler, C.; Beaudenon, M.; Simon, R.; Guenet, M.; Otekpo, M.; Célarier, T.; Gautier, J.; GERIA-COVID Study Group. Vitamin D supplementation prior to or during COVID-19 associated with better 3-month survival in geriatric patients: Extension phase of the GERIA-COVID study. *J. Steroid. Biochem. Mol. Biol.* **2021**, *213*, 105958. [CrossRef] [PubMed]

22.   Dissanayake, H.A.; de Silva, N.L.; Sumanatilleke, M.; de Silva, S.D.N.; Gamage, K.K.K.; Dematapitiya, C.; Kuruppu, D.C.; Ranasinghe, P.; Pathmanathan, S.; Katulanda, P. Prognostic and therapeutic role of vitamin D in COVID-19: Systematic review and meta-analysis. *J. Clin. Endocrinol. Metab.* **2021**, dgab892. [CrossRef] [PubMed]

23.   The prevention, diagnosis and treatment of the new coronavirus infection 2019-nCoV. Temporary guidelines Ministry of Health of the Russian Federation. *PULMONOLOGIYA* **2019**, *29*, 655–672. (In Russian) [CrossRef]

24.   Holick, M.F.; Binkley, N.C.; Bischoff-Ferrari, H.A.; Gordon, C.M.; Hanley, D.A.; Heaney, R.P.; Murad, M.H.; Weaver, C.M. Evaluation, treatment, and prevention of vitamin D deficiency: An Endocrine Society clinical practice guideline. *J. Clin. Endocrinol. Metab.* **2011**, *96*, 1911–1930. [CrossRef]

25.   Sealedenvelope. Available online: https://www.sealedenvelope.com/power/binary-superiority/ (accessed on 8 December 2021).

26.   Turrubiates-Hernández, F.J.; Sánchez-Zuno, G.A.; González-Estevez, G.; Hernández-Bello, J.; Macedo-Ojeda, G.; Muñoz-Valle, J.F. Potential immunomodulatory effects of vitamin D in the prevention of severe coronavirus disease 2019: An ally for Latin America (Review). *Int. J. Mol. Med.* **2021**, *47*, 32. [CrossRef]

27.   Charoenngam, N.; Holick, M.F. Immunologic effects of vitamin D on human health and disease. *Nutrients* **2020**, *12*, 2097. [CrossRef]

28.   Kumar, R.; Rathi, H.; Haq, A.; Wimalawansa, S.J.; Sharma, A. Putative roles of vitamin D in modulating immune response and immunopathology associated with COVID-19. *Virus Res.* **2021**, *292*, 198235. [CrossRef]

29. Karonova, T.L.; Andreeva, A.T.; Golovatuk, K.A.; Bykova, E.S.; Simanenkova, A.V.; Vashukova, M.A.; Grant, W.B.; Shlyakhto, E.V. Low 25(OH)D Level Is Associated with Severe Course and Poor Prognosis in COVID-19. *Nutrients* **2021**, *13*, 3021. [CrossRef]
30. Bychinin, M.V.; Klypa, T.V.; Mandel, I.A.; Andreichenko, S.A.; Baklaushev, V.P.; Yusubalieva, G.M.; Kolyshkina, N.A.; Troitsky, A.V. Low Circulating Vitamin D in Intensive Care Unit-Admitted COVID-19 Patients as a Predictor of Negative Outcomes. *J. Nutr.* **2021**, *151*, 2199–2205. [CrossRef]
31. Sowah, D.; Fan, X.; Dennett, L.; Hagtvedt, R.; Straube, S. Vitamin D levels and deficiency with different occupations: A systematic review. *BMC Public Health* **2017**, *17*, 519. [CrossRef]
32. Sudheesh, S.; Boaz, R.J. Degrees of Deficiency: Doctors and Vitamin D. *Indian J. Community Med.* **2017**, *42*, 53. [CrossRef]
33. World Health Organization. *The Impact of COVID-19 on Health and Care Workers: A Closer Look at Deaths*; World Health Organization: Geneva, Switzerland, 2021; Available online: https://apps.who.int/iris/handle/10665/345300 (accessed on 28 October 2021).
34. Grant, W.B.; Lahore, H.; McDonnell, S.L.; Baggerly, C.A.; Franch, C.B.; Aliano, J.L.; Bhattoa, H.P. Evidence That Vitamin D Supplementation Could Reduce Risk of Influenza and COVID-19 Infections and Deaths. *Nutrients* **2020**, *12*, 988. [CrossRef]
35. Lakkireddy, M.; Gadiga, S.G.; Malathi, R.D.; Karra, M.L.; Raju, I.P.M.; Chinapaka, S.; KSS, S.B.; Kandakatla, M. Impact of Pulse D Therapy on the Inflammatory Markers in Patients with COVID-19. *Sci. Rep.* **2021**, *11*, 1064.
36. Heaney, R.P.; Davies, K.M.; Chen, T.C.; Holick, M.F.; Barger-Lux, M.J. Human serum 25-hydroxycholecalciferol response to extended oral dosing with cholecalciferol. *Am. J. Clin. Nutr.* **2003**, *77*, 204–210. [CrossRef]
37. Gönen, M.S.; Alaylıoğlu, M.; Durcan, E.; Özdemir, Y.; Şahin, S.; Konukoğlu, D.; Nohut, O.K.; Ürkmez, S.; Küçükece, B.; Balkan, İ.İ.; et al. Rapid and Effective Vitamin D Supplementation May Present Better Clinical Outcomes in COVID-19 (SARS-CoV-2) Patients by Altering Serum INOS1, IL1B, IFNg, Cathelicidin-LL37, and ICAM1. *Nutrients* **2021**, *13*, 4047. [CrossRef] [PubMed]
38. Oristrell, J.; Oliva, J.C.; Casado, E.; Subirana, I.; Domínguez, D.; Toloba, A.; Balado, A.; Grau, M. Vitamin D supplementation and COVID-19 risk: A population-based, cohort study. *J. Endocrinol. Investig.* **2021**, *45*, 167–179. [CrossRef] [PubMed]
39. Rondanelli, M.; Miccono, A.; Lamburghini, S.; Avanzato, I.; Riva, A.; Allegrini, P.; Faliva, M.A.; Peroni, G.; Nichetti, M.; Perna, S. Self-Care for Common Colds: The Pivotal Role of Vitamin D, Vitamin C, Zinc, and Echinacea in Three Main Immune Interactive Clusters (Physical Barriers, Innate and Adaptive Immunity) Involved during an Episode of Common Colds-Practical Advice on Dosages and on the Time to Take These Nutrients/Botanicals in order to Prevent or Treat Common Colds. *Evid. Based Complement. Altern. Med.* **2018**, *2018*, 5813095. [CrossRef]

Article

25-Hydroxyvitamin D in Cancer Patients Admitted to Palliative Care: A Post-Hoc Analysis of the Swedish Trial 'Palliative-D'

Maria Helde Frankling [1,2,*], **Caritha Klasson** [1,3] and **Linda Björkhem-Bergman** [1,3]

1 Karolinska Institutet, Department of Neurobiology, Care Sciences and Society (NVS), Division of Clinical Geriatrics, Blickagången 16, Neo Floor 7, SE-141 83 Huddinge, Sweden; caritha.klasson@ki.se (C.K.); linda.bjorkhem-bergman@ki.se (L.B.-B.)
2 Thoracic Oncology Center, Department of Oncology-Pathology, Karolinska Institutet, Karolinska University Hospital, SE-171 76 Stockholm, Sweden
3 Stockholms Sjukhem, Palliative Medicine, Mariebergsgatan 22, SE-112 19 Stockholm, Sweden
* Correspondence: maria.helde.frankling@ki.se

Abstract: The purpose of this study is to explore 25-hydroxyvitamin D (25-OHD) levels in patients with cancer in the palliative phase in relation to season, sex, age, tumor type, colectomy, and survival. To this end, we performed a post-hoc analysis of 'Palliative-D', a randomized placebo-controlled, double-blind trial investigating the effect of daily supplementation with 4000 IU of vitamin D for 12 weeks on pain in patients in palliative cancer care. In the screening cohort ($n = 530$), 10% of patients had 25-OHD levels < 25 nmol/L, 50% < 50, and 84% < 75 nmol/L. Baseline 25-OHD did not differ between seasons or tumor type and was not correlated with survival time. In vitamin D deficient patients supplemented with vitamin D ($n = 67$), 86% reached sufficient levels, i.e., >50 nmol/L, after 12 weeks. An increase in 25-OHD was larger in supplemented women than in men (53 vs. 37 nmol/L, $p = 0.02$) and was not affected by season. In the placebo-group ($n = 83$), decreased levels of 25-OHD levels were noted during the study period for patients recruited during the last quarter of the year. In conclusion, cancer patients in palliative phase have adequate increase in 25-OHD after vitamin D supplementation regardless of season, age, tumor type, or colectomy.

Keywords: vitamin D; cholecalciferol; 25-OHD; vitamin D deficiency; palliative; cancer; latitude; tumor type; season; sex differences

Citation: Helde Frankling, M.; Klasson, C.; Björkhem-Bergman, L. 25-Hydroxyvitamin D in Cancer Patients Admitted to Palliative Care: A Post-Hoc Analysis of the Swedish Trial 'Palliative-D'. *Nutrients* **2022**, *14*, 602. https://doi.org/10.3390/nu14030602

Academic Editors: Spyridon N. Karras and Pawel Pludowski

Received: 10 December 2021
Accepted: 27 January 2022
Published: 29 January 2022

Publisher's Note: MDPI stays neutral with regard to jurisdictional claims in published maps and institutional affiliations.

1. Introduction

Vitamin D is a hormone mainly synthesized in the skin in the presence of sunlight, with 7-deoxycholesterol as a substrate [1]. Smaller amounts of vitamin D are ingested orally, through foodstuffs and supplementation products [2]. Vitamin D is activated in two hydroxylation steps into the active form, 1,25-dihydroxyvitamin D [1]. The active form of vitamin D is the only known ligand to the vitamin D receptor (VDR), a nuclear receptor present in many different cell types [3]. Vitamin D plays an important role in maintaining calcium homeostasis [1], in skeletal health [1], and in the immune system [4,5]. The individual's vitamin D levels is assessed by 25-hydroxyvitamin D (25-OHD), a more stable compound than 1,25(OH)2D [6]. 25-OHD levels below 25 nmol/L constitute severe deficiency and between 25 and 50 nmol/L deficiency [7]. Levels above 50 nmol/L are considered to ensure skeletal health, while 75 nmol/L may be needed for optimal functioning of the immune system [7]. Toxic levels that can cause hypercalcemia and renal failure are identified as levels above 250 nmol/L [7]. Cross sectional data on 25-OHD levels and mortality do however suggest a U-shaped relationship, where levels above 125 nmol/L are not necessarily beneficial for the individual [8].

Mechanistic studies suggesting anti-proliferative and anti-inflammatory effects of vitamin D [4,5,9,10], have spiked interest in epidemiological vitamin D research in cancer patients. Studies on cancer incidence and mortality indicate that vitamin D supplementation may reduce cancer specific, but not overall, mortality [11–15]. Prospective clinical studies investigate the possible potentiating effect of vitamin D on oncologic treatment effect [16,17], as well as its possible role in the management of pain [18–22], fatigue, and quality of life [20,21,23–25]. In a recent US study, 56% of cancer survivors took vitamin D supplementation compared to 37% in the general population [26]. This is a large increase in numbers compared with older cohorts, where fewer than 20% of cancer patients took vitamin D [27].

In Sweden, synthesis of vitamin D cannot take place between October and March ("vitamin D-winter") [28]. Foods are fortified with vitamin D, and risk groups are recommended supplementary vitamin D intake [2]. Still, there is a significant seasonal variation in vitamin D levels [29–31], and 50% of healthy Swedish adults have vitamin D levels below 50 nmol/L during winter [29,31,32]. Although oral vitamin D intake has increased over time [33], vitamin D levels have remained constant [32]. In institutionalized patients in Swedish care homes, most patients were vitamin D deficient [34,35]. In contrast, community-dwelling elderly Swedes have much higher vitamin D levels [36–39]. In Supplementary Materials Table S1, we present cross sectional Swedish studies on 25-OHD levels.

In cross sectional studies on 25-OHD levels in advanced or metastatic cancer patients, 25-OHD levels differ greatly between cohorts [40–46], with no seasonal variation in an Australian cohort [45]. In Table 1, we present an overview of cross-sectional data from cohorts of patients with palliative stage cancer disease.

In the randomized, placebo-controlled, double-blind trial 'Palliative-D', we investigated the effect of 12 weeks of supplementation with 4000 IU vitamin D3 to patients with advanced or metastatic cancer and 25-OHD $\leq$ 50 nmol/L on pain, infections, fatigue, and quality of life (QoL) [20,47]. The mean change in opioid dose (as a proxy for pain) was lower in vitamin D supplemented patients than in controls. Vitamin D treated patients were also less fatigued. There was no difference between groups regarding antibiotic use (as a proxy for infections), or QoL [47].

We have identified a knowledge gap regarding 25-OHD levels in palliative cohorts from Northern latitudes, as well as the effect of vitamin D supplementation in palliative care cohorts with mixed cancer types. In this post-hoc analysis of the randomized, controlled trial (RCT) Palliative-D', the primary aim is to explore 25-OHD levels in relation to season, age, and tumor type in patients with advanced cancer, as well as change in 25-OHD in both untreated and vitamin D supplemented patients. We hypothesize that this severely diseased cohort presents smaller seasonal variations compared to healthier Swedish cohorts due to more time spent indoors and thus experiencing less sun exposure during summer months.

Table 1. Cross-sectional cohorts with measurements of 25-OHD in patients with mixed tumor types in a palliative setting.

Author, Year	Cohort, Location Year	Study Population, Sex, Age	25-OHD, nmol/L Median(Min-Max) or Mean (SD)	Seasonal Variation in 25-OHD	Proportions of Vitamin D Deficient Participants 25-OHD, nmol/L	Vitamin D Supplementation/ Other Comment
Alkan 2019 [40]	Outpatient cancer clinic, Turkey December 2016–May 2018	$n = 706$ 41% men	30.5 (5–241)	Summer: 67% < 50 Autumn: 65% < 50 Winter: 77% < 50 Spring: 78% < 50	Palliative: 76% < 50	Vitamin D supplementation = exclusion criterion
Dev 2011 [42]	Cancer patients, supportive care clinic US 2009–2010	$n = 100$ 68% men median age 60	No information	No information	47% < 50 70% < 75	Deficiency more common in non-whites and females. 15–19% on vitamin D supplementation
Morton 2014 [45]	Oncology/palliative care unit, 76% metastatic disease Australia, 27° S	$n = 100$	Mean 54.6	Higher mean in spring, but only 4 observations	44% < 50 16% < 30 1% < 12.5	No supplements. No association 25-OHD-cancer type. PS associated with 25-OHD.
To 2011 [46]	Inpatient hospice, Australia	$n = 21$, 52% men mean age 69	41 (17–100) 47.5 (23.4)	All measurements during summer	72% < 60	
Edwards 2018 [43]	Cancer patients Texas, US, 2013–2015		No information	No information	49% < 75	No information on metastatic disease
Martinez-Alonso 2016 [44]	Cancer patients (palliative), Spain, March 2013–Aug 2014	$n = 30$, 77% men mean age 63	No information	No information	90% < 75 40% < 20	No supplements. PS and fatigue correlated with 25-OHD
Bergman 2015 [41]	Palliative care Unit Stockholm, 59° N, April 2014–January 2015	$n = 100$ 43% men median age 71	40 (8–154)	No information	65% < 50	Lower 25-OHD in patients who died during follow-up (36 vs. 50, $p = 0.013$)
Wang-Gillam 2008 [48]	Breast cancer, Arkansas, US, 2002–2006	$n = 21$ 100% women	No information	No information	48% < 50 67% < 75	Patients with metastatic disease in a larger cohort
Solomon 2012 [49]	Advanced malignancy and pain Connecticut, US	$n = 260$	No information	No information	21% < 25 43% 25–50 20% 50–75 8% > 75	Poster abstract, no detailed information on supplementation

Abbreviations: 25-OHD: 25-hydroxyvitamin D, °N: degrees North (latitude), PS: Performance Status, °S: degrees South (latitude), SD: Standard deviation, US: United States.

2. Materials and Methods

Patients were all in a palliative phase of their disease trajectory and they were recruited from advanced palliative home care teams in the Stockholm Region (59° N) between November 2017 and March 2020. Vitamin D levels (25-OHD) were assessed as part of the screening procedure in all consenting patients (n = 530), fulfilling inclusion and exclusion criteria. At screening, information on age, sex, and type of cancer was retrieved from medical records [20]. The original study did not comprise assessment of food intake or more specifically vitamin D ingestion.

Patients with 25-OHD ≤50 nmol/L (n = 244) were randomized to study drug, n = 121 to vitamin D3 oil drops (Detremin) 4000 IU/day and n = 123 to placebo [47]. Patients completing all 12 weeks of intervention (n = 150) had their 25-OHD levels measured again at the end of the study (Figure 1). Only 61% of randomized patients could be evaluated after twelve weeks, with clinical deterioration and death due to malignancy causing high attrition rates. In the results section, we present data on the screening cohort (n = 530) and the randomized cohort with two measurements of 25-OHD with a 12-week interval (n = 150) under different subheadings (Figure 1).

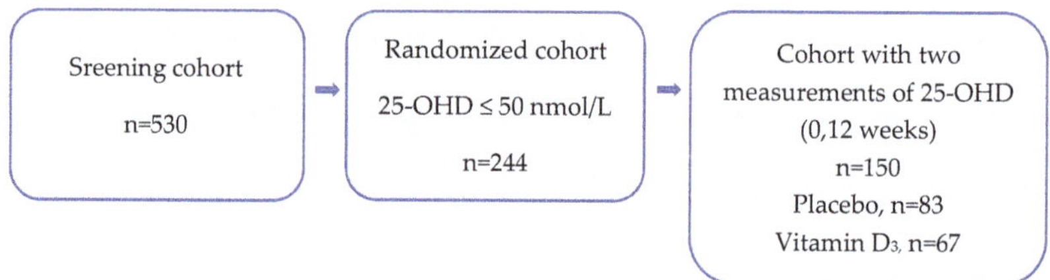

Figure 1. 'Palliative-D' cohorts analyzed regarding 25-OHD levels.

Some data on 25-OHD levels have been presented in previous publications on the studied cohort. In the screening cohort, median 25-OHD was 51 nmol/L (range 8–195) in both men and women [50]. Median baseline values of 25-OHD in randomized patients was 38 nmol/L (IQR 28–45) [47]. In the placebo group, median 25-OHD remained unchanged. In patients supplemented with vitamin D, 25-OHD increased from 36 (±11) to 81 (±26) nmol/L ($p < 0.001$) [47].

As previously reported, the median age in the entire screening cohort was 70 years (IQR 62–77) [50], and the median age was 68 years (IQR 61—75) in patients randomized to study the drug [47]. There were equal numbers of men and women in the screening cohort, with 265 in each group [50]. In randomized patients with two assessments of 25-OHD (n = 150), 49% were men [47]. In both the screening and in the randomized cohorts, colorectal cancer was the most common tumor type, followed by upper gastrointestinal (GI) and lung cancer [47,50]. We did not collect data on physical performance status or socioeconomic factors.

Inclusion criteria allowed for a daily dose of 400 IU vitamin D, and patients were meticulously asked about nutritional supplements during the screening process, so as to avoid recruiting patients who were taking larger than allowed doses of vitamin D [47]. We only recruited patients who planned to spend the next 12 weeks in the Stockholm region, but are aware that a few recruited participants still went on shorter holiday trips during winter months. Compliance was overall good, however 2 patients in the intervention group reported lacking compliance. Compliance is reported in greater detail in the supplementary material of the original publication [47].

Vitamin D levels were assessed as 25-OHD in serum analyzed by chemiluminescence immunoassay (CLIA) on a LIAISON-instrument (DiaSorin Inc, Stillwater, MN, USA) with a detectable range of 7.5 ± 175 nmol/L, CV 2 ± 5% at the Department of Clinical Chemistry, Karolinska University Hospital.

Statistical analysis was performed using Graph-Pad Prism version 8.4.3. Data that do not show Gaussian distribution, medians, IQR, and min-max are presented. For data with Gaussian distribution, we also calculated means and standard deviations (SD). Two tailed significance tests with a significance level of 0.05 were performed with Mann–Whitney U for non-normally distributed data and with Fisher's exact test for normally distributed data. Baseline 25-OHD in relation to tumor type and change in 25-OHD in relation to tumor type in non-supplemented patients were compared using the Kruskal Wallis test. Proportions of categorical variables were compared with Fisher's exact test. An analysis of correlation between 25-OHD and survival was done with simple linear regression.

3. Results

3.1. Baseline Characteristics Not Reported in Previous Publications

In randomized patients, 25-OHD levels ranged from 8–50 nmol/L. Vitamin D levels at screening in patients randomized to intervention were lower in patients who did not complete all 12 weeks, compared to those who did (median 25-OHD 34 vs. 39 nmol/L), however the difference was not statistically significant ($p = 0.075$).

3.2. Cutoff Levels for Vitamin D Deficiency

3.2.1. Screening Cohort

In the screening cohort, 10% of patients had 25-OHD levels < 25 nmol/L, 50% < 50 nmol/L, and 84% < 75 nmol/L. Two percent of screened patients had 25-OHD above 125, with the highest individual value at 195 nmol/L. Two patients in the screening cohort had undetectable vitamin D levels at screening (<8 nmol/L), one of them being a patient with breast cancer who died shortly after inclusion. The other patient with unmeasurable 25-OHD also had a diagnosis of breast cancer, and was randomized to vitamin D supplementation 4000 IU/day in September and increased 25-OHD to 130 nmol/L after 12 weeks of follow up (the largest individual increase in the intervention group).

3.2.2. Randomized Cohort

In patients who received vitamin D supplementation for 12 weeks ($n = 67$), 13% had baseline values < 25 nmol/L and 15/67 patients with initial values below 50 nmol/L reached levels above 100 nmol/L after 12 weeks (median increase in this subpopulation was 79 nmol/L, IQR 64–94). In contrast, 9/67 vitamin D supplemented patients remained vitamin D deficient. In this group, 8/9 patients were male and 7/9 had gastrointestinal tumors. One patient with a gastrointestinal neuroendocrine tumor (GI-NET) and one with pancreatic malignancy dropped their 25-OHD-levels with two units during follow-up. Very small increases in 25-OHD (2–5 nmol/L) were seen in three colorectal cancer patients, two of which had undergone total or partial colectomy. Still, median 25-OHD levels in all patients with gastrointestinal cancer were not significantly lower than other tumor groups. Change in 25-OHD in relation to cutoff values for vitamin D deficiency are presented in Supplementary Materials Table S2.

3.3. 25-OHD in Relation to Season

3.3.1. Screening Cohort ($n = 530$)

Baseline 25-OHD did not differ significantly between months or quarters of the year or summer (April–September) versus winter season (October–March) in the screening cohort (Figure 2, Supplementary Materials Table S3). There were differences in the number of patients screened each month, with the lowest numbers in June ($n = 12$) and July ($n = 16$), and the highest numbers in November ($n = 92$), March ($n = 68$), and January ($n = 67$). Only 10 percent of patients were recruited during the summer months June, July, and August (52/530).

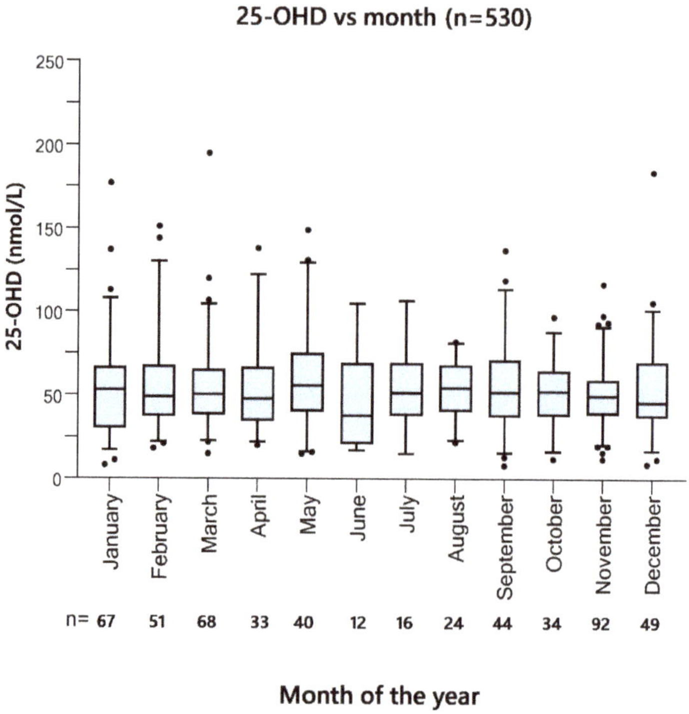

Figure 2. Cross-sectional 25-OHD in relation to screening month. Median 25-OHD values in nmol/L in the screening cohort (*n* = 530) of the 'Palliative-D' study. Boxes show interquartile range whiskers 5/95 percentiles and dots outliers. There were no statistically significant differences between groups (Mann-Whitney U).

3.3.2. Randomized Cohort (*n* = 150)

In patients receiving placebo, the difference in median change in 25-OHD over 12 weeks for patients recruited during the first quarter of the year (3 nmol/L) was significantly higher than in those recruited during Q4 (-3 nmol/L p = 0.003) (Supplementary Materials Table S4). In patients supplemented with 4000 IU vitamin D/day, median change in 25-OHD for patients recruited during the first two quarters of the year was 44 and 47 nmol/L respectively, and 36.5 and 37 nmol/L in Q3 and Q4, however differences between time periods were not significant (Supplementary Materials Table S4).

3.4. Change in 25-OHD in Relation to Sex, Randomized Cohort (n = 150)

In patients receiving placebo, there was no difference in median change in 25-OHD over 12 weeks between men and women (data not shown). In patients supplemented with 4000 IU vitamin D for 12 weeks (*n* = 67), a median increase in 25-OHD for men was 37 nmol/L and for women, it was 53 nmol/L (confidence interval, CI, for difference between groups -26 to -2, p = 0.02). The difference between groups was mainly due to large increases in 25-OHD in a small number of women, i.e., outliers.

3.5. 25-OHD in Relation to Cancer Type

3.5.1. Screening Cohort (*n* = 530)

In the screening cohort, there were no significant differences in 25-OHD between patients with breast, colorectal, lung, gynecological, prostate cancer, upper gastrointestinal (GI) cancer, or "other", a group in which tumor types with fewer observations were pooled (cancer of unknown primary, tumors of the central nervous system, head & neck cancer, hematological malignancy, malignant melanoma, sarcoma, and urinary tract tumors) (Figure 3).

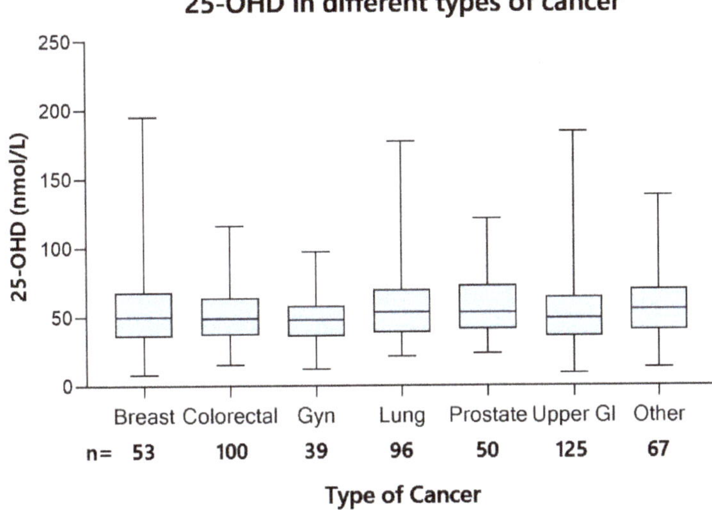

Figure 3. Cross-sectional 25-OHD in relation to tumor type ($n = 530$). Median 25-hydroxyvitamin D (25-OHD) values in nmol/L in patients with different types of cancer from the screening cohort of the 'Palliative-D'-study. Boxes show interquartile range and whiskers min-max values. There were no statistically significant differences between groups (Kruskal Wallis).

3.5.2. Randomized Cohort ($n = 150$)

A change in 25-OHD over time in patients receiving placebo did not vary between cancer types ($p = 0.56$). In vitamin D supplemented patients, the median change in 25-OHD was largest in patients with gynecological tumors ($n = 7$, median 64, IQR 54–86) and lowest in patients with prostate cancer ($n = 7$, median 26, IQR 21–58). Due to few observations, we did not perform a significance test for change in 25-OHD in supplemented patients across all tumor types. When comparing median change in 25-OHD between the two largest supplemented groups, colorectal cancer and upper GI-cancer, results were very similar (data not shown). When looking at individual values, all vitamin D supplemented patients with very small changes in 25-OHD after 12 weeks had GI-tumors.

Median 25-OHD values in nmol/L in the screening cohort ($n = 530$) of the 'Palliative-D' study included the interquartile range (boxes) and min-max values (whiskers). Comparisons between groups was performed with Mann–Whitney U, and no significant difference between the types of cancer was observed.

3.6. 25-OHD in Relation to Age, Screening Cohort ($n = 530$)

In the screening cohort, median vitamin D levels in patients aged 70–79 years old ($n = 206$) was significantly higher compared to the rest of the screening cohort (56 vs. 51, 95% CI of difference 1–8, $p = 0.02$). In 60–69-year-old patients ($n = 145$), median 25-OHD-levels were instead lower than in other age groups (45 vs. 51 nmol/L, 95% CI of difference 2–10, $p = 0.005$). All other comparisons between age groups and the remaining cohort were not significant. However, when comparing age groups with each other, we also noted that the small group of younger patients (<39 years, $n = 7$) had lower 25-OHD levels than elderly patients. Patients 60–69 years of age had significantly lower 25-OHD levels compared to those who were older, as seen in Figure 4.

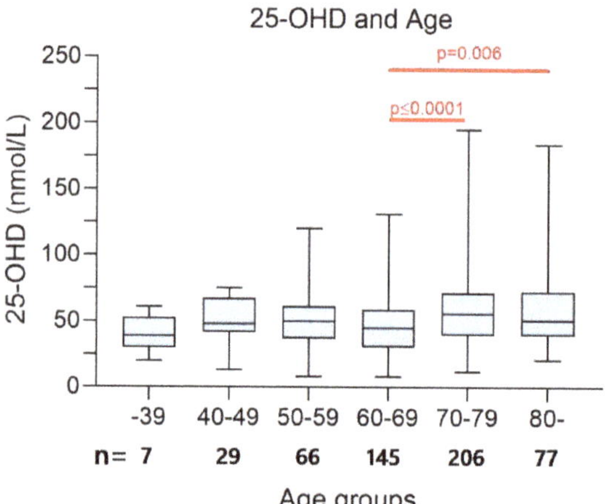

Figure 4. Cross-sectional 25-OHD in relation to age (*n* = 530). Median 25-hydroxyvitamin D (25-OHD) values in nmol/L in the screening cohort (*n* = 530) of the 'Palliative-D' study, included the interquartile range and min-max values. Comparisons between groups was performed with Mann–Whitney U. In the 60–69 years age group, the median 25-OHD was lower than in the 70–79 years age group (45 vs. 56 nmol/L, 95% CI −15 to −5, *p* < 0.0001, and in the 80+ years age group (45 vs. 51 nmol/L, 95% CI −15 to −2, *p* = 0.006).

3.7. 25-OHD in Relation to Survival

3.7.1. Screening Cohort (*n* = 530)

In patients in the screening cohort who were deceased by 9 June 2021 (*n* = 440), 25-OHD at screening did not correlate with survival time (*p* = 0.159). In Figure 5, we present median 25-OHD in patients with a survival time of less than 1 month, 1–3, 3–6, 6–12, and more than 12 months between survival time periods. In Figure 6, we have plotted 25-OHD values versus survival in days. In patients who survived for less than a month after screening (*n* = 44), median 25-OHD was 41.5 and in patients surviving longer than a year (*n* = 59), 56 nmol/L respectively (*p* = 0.11).

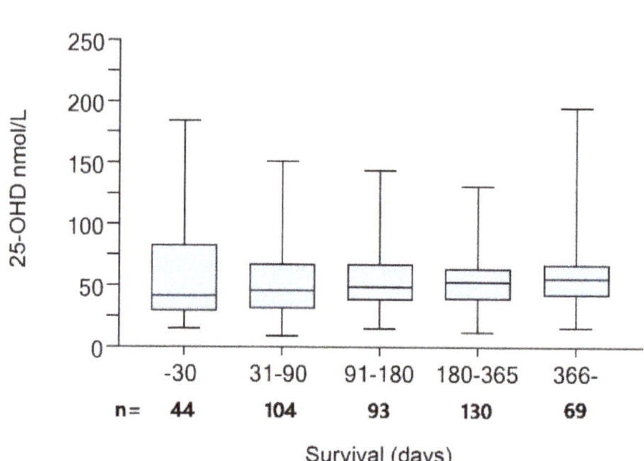

Figure 5. Median 25-hydroxyvitamin D (25-OH) values in nmol/L in deceased patients from the screening cohort (*n* = 440) of the 'Palliative-D' study included the interquartile range and min-max values. Comparisons between groups was performed with Mann–Whitney U and there were no statistically significant differences between groups.

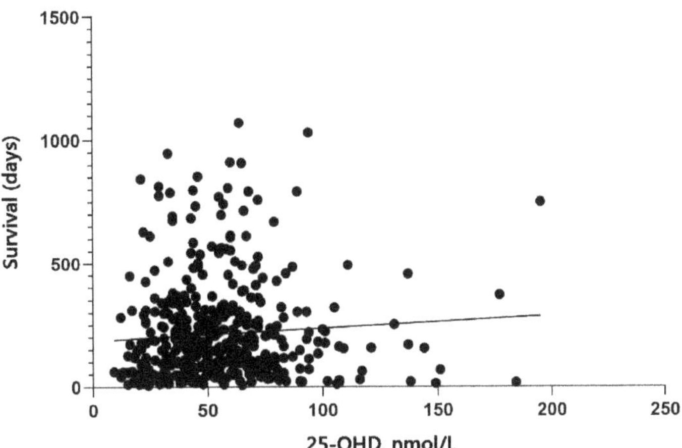

Figure 6. Median 25-hydroxyvitamin D (25-OHD) values in nmol/L in deceased patients from the screening cohort (*n* = 440) of the 'Palliative-D' study plotted against survival time in days.

3.7.2. Randomized Cohort (*n* = 150)

There was no correlation between 25-OHD and survival after 12 weeks of vitamin D supplementation (*n* = 112, *p* = 0.40).

3.8. 25-OHD in Colectomized Patients/Patients with Short Bowel Syndrome
3.8.1. Screening Cohort (*n* = 530)

There was no difference in median 25-OHD at screening between colectomized patients (*n* = 60) and the rest of the screening cohort (*n* = 470, Figure 7). There were six patients with short bowel syndrome in the screening cohort. In this group, median 25-OHD was 42.5 (range 21–137). Due to the small number of observations, we could not make a comparison of 25-OHD between groups.

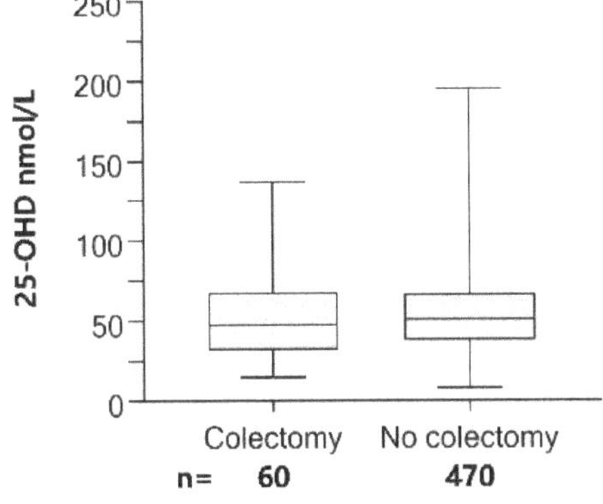

Figure 7. Cross-sectional 25-OHD in colectomized patients. Median 25-hydroxyvitamin D (25-OHD) values in nmol/L in colectomized and non-colectomized patients from the screening cohort of the 'Palliative-D'-study. Boxes show interquartile range and whiskers min-max values. There was no statistically significant difference between groups (Fischer's exact test).

3.8.2. Randomized Cohort (*n* = 150)

One patient with short bowel syndrome was supplemented with vitamin D and increased 25-OHD levels from 40 to 98 nmol/L.

4. Discussion

In this explorative post-hoc analysis of 25-OHD levels in the 'Palliative-D' cohort of severely diseased patients with cancer, 50 percent of patients were vitamin D deficient, and 84% had 25-OHD values below the proposed desired level of 75 nmol/L. Cross-sectional 25-OHD-levels did not vary with time of the year. As shown in the original publication, there was no difference in cross-sectional 25-OHD between men and women, but women had a significantly larger increase in 25-OHD when supplemented with 4000 IU of vitamin D$_3$ for 12 weeks. In our material, 60–70-year-old patients had significantly lower 25-OHD levels and 70–80-year-old patients had higher levels compared to other age groups. There was no association between tumor type and 25-OHD. Colectomized patients did not exhibit lower 25-OHD levels than non-colectomized patients. Almost one in four patients supplemented with vitamin D for 12 weeks increased their 25-OHD levels with more than 50 nmol/L. In patients who did not reach vitamin D levels above 50 nmol/L, a majority were male patients with gastrointestinal tumors. There was no association between baseline 25-OHD and survival.

In comparison with other Swedish cohorts with cross-sectional data on 25-OHD, our cohort of severely diseased patients had higher 25-OHD values compared to comparatively older Swedish nursing home residents [34,39,51], however values were lower than in healthy elderly [39]. We also observed significantly less seasonal variation in 25-OHD levels compared to healthy (and younger) Swedish cohorts [29,31]. We suggest that as patients with advanced cancer spend less time outside, consequently their 25-OHD levels rely less on sun exposure.

As seen in Table 1, 25-OHD levels in palliative cohorts with mixed tumor types vary greatly [40–43,46,48,49]. Our cohort is the yet largest to be studied. Patients screened in 'Palliative-D' had higher 25-OHD levels compared to recently studied cohorts in Spain and Turkey [24,40], well in line with Australian and US experiences [43,45,46,48,49]. The fact that 25-OHD-levels did not differ between tumor types is consistent with previous findings [45].

In our material, women increased their 25-OHD-levels more than men when supplemented with vitamin D, and more specifically a small number of women had very large increases in 25-OHD. We do not know whether this is due to sex differences in vitamin D uptake and metabolization, whether these individuals took more vitamin D supplementation than prescribed in the study, or otherwise changed their lifestyle and eating habits to increase 25-OHD.

In the multivariate analysis of 'Palliative-D', colectomy and cancer type did not affect results regarding pain, infections, and quality of life [47]. However, all vitamin D supplemented patients with very small changes in 25-OHD after 12 weeks had GI-tumors. This indicates that at least in some GI-cancer patients, reduced uptake of vitamin D may be an issue, as seen in other studies on patients with GI cancer [52].

We consider the fact that vitamin D supplementation is recommended in all citizens aged 75 years and above in Sweden plays a role in the higher levels of 25-OHD observed in the elderly in our cohort [2]. In the small group of young patients, 25-OHD levels were low. We do not interpret this as a difference between groups attributable to age itself. Rather, there were individuals in this group with very long disease trajectories and many lines of palliative oncological treatment.

We noted that 10 patients had very high levels of 25-OHD at screening, although ongoing vitamin D supplementation was an exclusion criterion in the 'Palliative-D' study. These patients were recruited to the study during all four quarters of the year. To us it seems unlikely that levels above 150 nmol/L are attained solely through sun exposure, and we suspect that these patients were taking vitamin D supplementation. Use of vitamin D supplementation is presently very common in cancer patients [26]. In Sweden, we do not routinely screen for vitamin D deficiency, and possibly some patients enrolled in the study mainly to have their vitamin D status checked.

Several patients had high 25-OHD levels at screening, and a few also reached levels above 150 nmol/L after 12 weeks of supplementation. The safety, over time, of such high levels is debated [8,53].

A strength of this study is the size of the screening cohort and the fact that we included many different types of cancer types. Limitations include the fact that the subgroups are small, especially when analyzing patients followed for 12 weeks, and this makes comparisons across subgroups less reliable. Furthermore, only 10% of patients were recruited to the screening cohort during the months of June, July, and August. We did not collect data on dietary intake and did not assess levels of the parathyroid hormone (PTH). However, dietary intake could not account for the large increases of 25-OHD in supplemented patients.

5. Conclusions

Levels of 25-OHD in palliative cancer patients in northern latitudes have less seasonal variation than healthy populations from the same latitudes. The type of cancer does not predict vitamin D levels in a palliative setting. There is large inter-individual variation in cross-sectional 25-OHD levels, indicating that some patients take larger doses of vitamin D supplementation. Even in a severely diseased population, patients respond well to vitamin D supplementation with adequate increase in 25-OHD levels, regardless of season, age, tumor type, or colectomy, however the increase may be more pronounced in women than in men.

Supplementary Materials: The following are available online at https://www.mdpi.com/article/10.3390/nu14030602/s1, Table S1: Cross sectional cohorts with measurements of 25-OHD in Sweden, Table S2: Baseline levels and change in 25-OHD (nmol/L) after 12 weeks of supplementation of vitamin D3 4000 IU/day in relation to cutoff levels in a cohort of patients with cancer in palliative phase, Table S3: 25-OHD in the screening cohort (*n* = 530) in relation to season, Table S4: Change in 25-OHD over 12 weeks in relation to season in vitamin D supplemented patients and in patients receiving placebo.

Author Contributions: Conceptualization, M.H.F. and L.B.-B.; methodology, M.H.F. and L.B.-B.; validation, M.H.F. and L.B.-B.; formal analysis, M.H.F. and L.B.-B.; investigation, M.H.F., C.K. and L.B.-B.; resources, L.B.-B.; data curation, M.H.F. and L.B.-B.; writing—original draft preparation, M.H.F.; writing—review and editing, M.H.F., C.K. and L.B.-B.; visualization, M.H.F. and L.B.-B.; supervision, L.B.-B.; project administration, M.H.F., C.K. and L.B.-B.; funding acquisition, L.B.-B. All authors have read and agreed to the published version of the manuscript.

Funding: This research was funded by grants from Region Stockholm (clinical research appointment), Swedish Cancer Society, CIMED and Stockholms Sjukhem Foundation. The funders of this trial took no part in study design, data collection, data interpretation, writing, or reviewing of the manuscript.

Institutional Review Board Statement: The "Palliative-D" study was conducted according to the guidelines of the Declaration of Helsinki and was approved by the Regional Ethics Committee in Stockholm (Dnr 2017/405-31); date of approval 7 April 2017.

Informed Consent Statement: Written informed consent was obtained from all subjects involved in the original study before any study-related procedure was performed.

Data Availability Statement: Raw data from the "Palliative-D" study is available from the corresponding author upon request.

Acknowledgments: The authors want to express their sincere gratitude to all of the patients who participated in the "Palliative-D" study. The authors also want to thank all of the staff at ASIH Stockholm Södra, ASIH Stockholm Norr, and Stockholms Sjukhem for their kind help in performing "Palliative-D". We especially want to thank study nurses Carina Sandberg and Tina Nyman Olsson and study physician Marie Nordström for their skillful work.

Conflicts of Interest: The authors declare no conflict of interest.

References

1. Holick, M.F. Vitamin D deficiency. *N. Engl. J. Med.* **2007**, *357*, 266–281. [CrossRef] [PubMed]
2. Itkonen, S.T.; Andersen, R.; Bjork, A.K.; Brugard Konde, A.; Eneroth, H.; Erkkola, M.; Holvik, K.; Madar, A.A.; Meyer, H.E.; Tetens, I.; et al. Vitamin D status and current policies to achieve adequate vitamin D intake in the Nordic countries. *Scand. J. Public Health* **2020**, *49*, 616–622. [CrossRef] [PubMed]
3. Christakos, S.; Dhawan, P.; Verstuyf, A.; Verlinden, L.; Carmeliet, G. Vitamin D: Metabolism, Molecular Mechanism of Action, and Pleiotropic Effects. *Physiol. Rev.* **2016**, *96*, 365–408. [CrossRef] [PubMed]
4. Charoenngam, N.; Holick, M.F. Immunologic Effects of Vitamin D on Human Health and Disease. *Nutrients* **2020**, *12*, 2097. [CrossRef] [PubMed]
5. Hewison, M. Vitamin D and immune function: An overview. *Proc. Nutr. Soc.* **2012**, *71*, 50–61. [CrossRef] [PubMed]
6. Holick, M.F. Vitamin D status: Measurement, interpretation, and clinical application. *Ann. Epidemiol.* **2009**, *19*, 73–78. [CrossRef]
7. Ross, A.C.; Manson, J.E.; Abrams, S.A.; Aloia, J.F.; Brannon, P.M.; Clinton, S.K.; Durazo-Arvizu, R.A.; Gallagher, J.C.; Gallo, R.L.; Jones, G.; et al. The 2011 report on dietary reference intakes for calcium and vitamin D from the Institute of Medicine: What clinicians need to know. *J. Clin. Endocrinol. Metab.* **2011**, *96*, 53–58. [CrossRef]
8. Durup, D.; Jorgensen, H.L.; Christensen, J.; Tjonneland, A.; Olsen, A.; Halkjaer, J.; Lind, B.; Heegaard, A.M.; Schwarz, P. A Reverse J-Shaped Association Between Serum 25-Hydroxyvitamin D and Cardiovascular Disease Mortality: The CopD Study. *J. Clin. Endocrinol. Metab.* **2015**, *100*, 2339–2346. [CrossRef]
9. Bandera Merchan, B.; Morcillo, S.; Martin-Nuñez, G.; Tinahones, F.J.; Macías-González, M. The role of vitamin D and VDR in carcinogenesis: Through epidemiology and basic sciences. *J. Steroid Biochem. Mol. Biol.* **2017**, *167*, 203–218. [CrossRef]
10. Negri, M.; Gentile, A.; de Angelis, C.; Montò, T.; Patalano, R.; Colao, A.; Pivonello, R.; Pivonello, C. Vitamin D-Induced Molecular Mechanisms to Potentiate Cancer Therapy and to Reverse Drug-Resistance in Cancer Cells. *Nutrients* **2020**, *12*, 1798. [CrossRef]
11. Akutsu, T.; Kitamura, H.; Himeiwa, S.; Kitada, S.; Akasu, T.; Urashima, M. Vitamin D and Cancer Survival: Does Vitamin D Supplementation Improve the Survival of Patients with Cancer? *Curr. Oncol. Rep.* **2020**, *22*, 62. [CrossRef] [PubMed]
12. Chandler, P.D.; Chen, W.Y.; Ajala, O.N.; Hazra, A.; Cook, N.; Bubes, V.; Lee, I.M.; Giovannucci, E.L.; Willett, W.; Buring, J.E.; et al. Effect of Vitamin D3 Supplements on Development of Advanced Cancer: A Secondary Analysis of the VITAL Randomized Clinical Trial. *JAMA Netw. Open* **2020**, *3*, e2025850. [CrossRef] [PubMed]
13. Han, J.; Guo, X.; Yu, X.; Liu, S.; Cui, X.; Zhang, B.; Liang, H. 25-Hydroxyvitamin D and Total Cancer Incidence and Mortality: A Meta-Analysis of Prospective Cohort Studies. *Nutrients* **2019**, *11*, 2295. [CrossRef] [PubMed]
14. Manson, J.E.; Cook, N.R.; Lee, I.M.; Christen, W.; Bassuk, S.S.; Mora, S.; Gibson, H.; Gordon, D.; Copeland, T.; D'Agostino, D.; et al. Vitamin D Supplements and Prevention of Cancer and Cardiovascular Disease. *N. Engl. J. Med.* **2019**, *380*, 33–44. [CrossRef] [PubMed]
15. Zhang, Y.; Fang, F.; Tang, J.; Jia, L.; Feng, Y.; Xu, P.; Faramand, A. Association between vitamin D supplementation and mortality: Systematic review and meta-analysis. *BMJ* **2019**, *366*, l4673. [CrossRef] [PubMed]
16. Ng, K.; Nimeiri, H.S.; McCleary, N.J.; Abrams, T.A.; Yurgelun, M.B.; Cleary, J.M.; Rubinson, D.A.; Schrag, D.; Miksad, R.; Bullock, A.J.; et al. Effect of High-Dose vs. Standard-Dose Vitamin D3 Supplementation on Progression-Free Survival Among Patients With Advanced or Metastatic Colorectal Cancer: The SUNSHINE Randomized Clinical Trial. *JAMA* **2019**, *321*, 1370–1379. [CrossRef] [PubMed]
17. Urashima, M.; Ohdaira, H.; Akutsu, T.; Okada, S.; Yoshida, M.; Kitajima, M.; Suzuki, Y. Effect of Vitamin D Supplementation on Relapse-Free Survival Among Patients With Digestive Tract Cancers: The AMATERASU Randomized Clinical Trial. *JAMA* **2019**, *321*, 1361–1369. [CrossRef] [PubMed]
18. Amir, E.; Simmons, C.E.; Freedman, O.C.; Dranitsaris, G.; Cole, D.E.; Vieth, R.; Ooi, W.S.; Clemons, M. A phase 2 trial exploring the effects of high-dose (10,000 IU/day) vitamin D(3) in breast cancer patients with bone metastases. *Cancer* **2010**, *116*, 284–291. [CrossRef]
19. Beer, T.M.; Garzotto, M.; Katovic, N.M. High-dose calcitriol and carboplatin in metastatic androgen-independent prostate cancer. *Am. J. Clin. Oncol.* **2004**, *27*, 535–541. [CrossRef]
20. Helde-Frankling, M.; Bergqvist, J.; Klasson, C.; Nordstrom, M.; Hoijer, J.; Bergman, P.; Bjorkhem-Bergman, L. Vitamin D supplementation to palliative cancer patients: Protocol of a double-blind, randomised controlled trial 'Palliative-D'. *BMJ Supportive Palliat. Care* **2017**, *7*, 458–463. [CrossRef] [PubMed]
21. Helde-Frankling, M.; Hoijer, J.; Bergqvist, J.; Bjorkhem-Bergman, L. Vitamin D supplementation to palliative cancer patients shows positive effects on pain and infections-Results from a matched case-control study. *PLoS ONE* **2017**, *12*, e0184208. [CrossRef] [PubMed]
22. Van Veldhuizen, P.J.; Taylor, S.A.; Williamson, S.; Drees, B.M. Treatment of vitamin D deficiency in patients with metastatic prostate cancer may improve bone pain and muscle strength. *J. Urol.* **2000**, *163*, 187–190. [CrossRef]
23. Anand, A.; Singh, S.; Sonkar, A.A.; Husain, N.; Singh, K.R.; Singh, S.; Kushwaha, J.K. Expression of vitamin D receptor and vitamin D status in patients with oral neoplasms and effect of vitamin D supplementation on quality of life in advanced cancer treatment. *Contemp. Oncol.* **2017**, *21*, 145–151. [CrossRef]

24. Martinez-Alonso, M.; Dusso, A.; Ariza, G.; Nabal, M. The effect on quality of life of vitamin D administration for advanced cancer treatment (VIDAFACT study): Protocol of a randomised controlled trial. *BMJ Open* **2014**, *4*, e006128. [CrossRef] [PubMed]
25. Schöttker, B.; Kuznia, S.; Laetsch, D.C.; Czock, D.; Kopp-Schneider, A.; Caspari, R.; Brenner, H. Protocol of the VICTORIA study: Personalized vitamin D supplementation for reducing or preventing fatigue and enhancing quality of life of patients with colorectal tumor-randomized intervention trial. *BMC Cancer* **2020**, *20*, 739. [CrossRef] [PubMed]
26. Du, M.; Luo, H.; Blumberg, J.B.; Rogers, G.; Chen, F.; Ruan, M.; Shan, Z.; Biever, E.; Zhang, F.F. Dietary Supplement Use among Adult Cancer Survivors in the United States. *J. Nutr.* **2020**, *150*, 1499–1508. [CrossRef]
27. Velicer, C.M.; Ulrich, C.M. Vitamin and mineral supplement use among US adults after cancer diagnosis: A systematic review. *J. Clin. Oncol. Off. J. Am. Soc. Clin. Oncol.* **2008**, *26*, 665–673. [CrossRef]
28. O'Neill, C.M.; Kazantzidis, A.; Ryan, M.J.; Barber, N.; Sempos, C.T.; Durazo-Arvizu, R.A.; Jorde, R.; Grimnes, G.; Eiriksdottir, G.; Gudnason, V.; et al. Seasonal Changes in Vitamin D-Effective UVB Availability in Europe and Associations with Population Serum 25-Hydroxyvitamin D. *Nutrients* **2016**, *8*, 533. [CrossRef]
29. Klingberg, E.; Oleröd, G.; Konar, J.; Petzold, M.; Hammarsten, O. Seasonal variations in serum 25-hydroxy vitamin D levels in a Swedish cohort. *Endocrine* **2015**, *49*, 800–808. [CrossRef]
30. Lundström, P.; Caidahl, K.; Eriksson, M.J.; Fritz, T.; Krook, A.; Zierath, J.R.; Rickenlund, A. Changes in Vitamin D Status in Overweight Middle-Aged Adults with or without Impaired Glucose Metabolism in Two Consecutive Nordic Summers. *J. Nutr. Metab.* **2019**, *2019*, 1840374. [CrossRef]
31. Nälsén, C.; Becker, W.; Pearson, M.; Ridefelt, P.; Lindroos, A.K.; Kotova, N.; Mattisson, I. Vitamin D status in children and adults in Sweden: Dietary intake and 25-hydroxyvitamin D concentrations in children aged 10–12 years and adults aged 18–80 years. *J. Nutr. Sci.* **2020**, *9*, e47. [CrossRef] [PubMed]
32. Summerhays, E.; Eliasson, M.; Lundqvist, R.; Söderberg, S.; Zeller, T.; Oskarsson, V. Time trends of vitamin D concentrations in northern Sweden between 1986 and 2014: A population-based cross-sectional study. *Eur. J. Nutr.* **2020**, *59*, 3037–3044. [CrossRef] [PubMed]
33. Samuelsson, J.; Rothenberg, E.; Lissner, L.; Eiben, G.; Zettergren, A.; Skoog, I. Time trends in nutrient intake and dietary patterns among five birth cohorts of 70-year-olds examined 1971–2016: Results from the Gothenburg H70 birth cohort studies, Sweden. *Nutr. J.* **2019**, *18*, 66. [CrossRef] [PubMed]
34. Arnljots, R.; Thorn, J.; Elm, M.; Moore, M.; Sundvall, P.D. Vitamin D deficiency was common among nursing home residents and associated with dementia: A cross sectional study of 545 Swedish nursing home residents. *BMC Geriatr.* **2017**, *17*, 229. [CrossRef] [PubMed]
35. Samefors, M.; Tengblad, A.; Östgren, C.J. Sunlight Exposure and Vitamin D Levels in Older People- An Intervention Study in Swedish Nursing Homes. *J. Nutr. Health Aging* **2020**, *24*, 1047–1052. [CrossRef] [PubMed]
36. Björk, A.; Ribom, E.; Johansson, G.; Scragg, R.; Mellström, D.; Grundberg, E.; Ohlsson, C.; Karlsson, M.; Ljunggren, Ö.; Kindmark, A. Variations in the vitamin D receptor gene are not associated with measures of muscle strength, physical performance, or falls in elderly men. Data from MrOS Sweden. *J. Steroid Biochem. Mol. Biol.* **2019**, *187*, 160–165. [CrossRef]
37. Buchebner, D.; Bartosch, P.; Malmgren, L.; McGuigan, F.E.; Gerdhem, P.; Akesson, K.E. Association Between Vitamin D, Frailty, and Progression of Frailty in Community-Dwelling Older Women. *J. Clin. Endocrinol. Metab.* **2019**, *104*, 6139–6147. [CrossRef]
38. Samefors, M.; Scragg, R.; Länne, T.; Nyström, F.H.; Östgren, C.J. Association between serum 25(OH)D(3) and cardiovascular morbidity and mortality in people with Type 2 diabetes: A community-based cohort study. *Diabet. Med. J. Br. Diabet. Assoc.* **2017**, *34*, 372–379. [CrossRef]
39. Carlsson, M.; Wanby, P.; Brudin, L.; Lexne, E.; Mathold, K.; Nobin, R.; Ericson, L.; Nordqvist, O.; Petersson, G. Older Swedish Adults with High Self-Perceived Health Show Optimal 25-Hydroxyvitamin D Levels Whereas Vitamin D Status Is Low in Patients with High Disease Burden. *Nutrients* **2016**, *8*, 717. [CrossRef]
40. Alkan, A.; Köksoy, E.B. Vitamin D deficiency in cancer patients and predictors for screening (D-ONC study). *Curr. Probl. Cancer* **2019**, *43*, 421–428. [CrossRef]
41. Bergman, P.; Sperneder, S.; Hoijer, J.; Bergqvist, J.; Bjorkhem-Bergman, L. Low vitamin D levels are associated with higher opioid dose in palliative cancer patients–results from an observational study in Sweden. *PLoS ONE* **2015**, *10*, e0128223. [CrossRef] [PubMed]
42. Dev, R.; Del Fabbro, E.; Schwartz, G.G.; Hui, D.; Palla, S.L.; Gutierrez, N.; Bruera, E. Preliminary report: Vitamin D deficiency in advanced cancer patients with symptoms of fatigue or anorexia. *Oncologist* **2011**, *16*, 1637–1641. [CrossRef] [PubMed]
43. Edwards, B.J.; Sun, M.; Zhang, X.; Holmes, H.M.; Song, J.; Khalil, P.; Karuturi, M.; Shah, J.B.; Dinney, C.P.; Gagel, R.F.; et al. Fractures frequently occur in older cancer patients: The MD Anderson Cancer Center experience. *Supportive Care Cancer* **2018**, *26*, 1561–1568. [CrossRef] [PubMed]
44. Martinez-Alonso, M.; Dusso, A.; Ariza, G.; Nabal, M. Vitamin D deficiency and its association with fatigue and quality of life in advanced cancer patients under palliative care: A cross-sectional study. *Palliat. Med.* **2016**, *30*, 89–96. [CrossRef]
45. Morton, A.; Hardy, J.; Morton, A.; Tapuni, A.; Anderson, H.; Kingi, N.; Shannon, C. Vitamin D deficiency in patients with malignancy in Brisbane. *Supportive Care Cancer* **2014**, *22*, 2223–2227. [CrossRef]
46. To, T. Vitamin D deficiency in an Australian inpatient hospice population. *J. Pain Symptom Manag.* **2011**, *41*, e1–e2. [CrossRef]

47. Helde Frankling, M.; Klasson, C.; Sandberg, C.; Nordström, M.; Warnqvist, A.; Bergqvist, J.; Bergman, P.; Björkhem-Bergman, L. 'Palliative-D'-Vitamin D Supplementation to Palliative Cancer Patients: A Double Blind, Randomized Placebo-Controlled Multicenter Trial. *Cancers* **2021**, *13*, 3707. [CrossRef]
48. Wang-Gillam, A.; Miles, D.A.; Hutchins, L.F. Evaluation of vitamin D deficiency in breast cancer patients on bisphosphonates. *Oncologist* **2008**, *13*, 821–827. [CrossRef]
49. Solomon, L.; Blatt, L. Vitamin D Deficiency as a Risk Factor for Pain: Prevalence in the Setting of Advanced Malignancy (413-C). *J. Pain Symptom Manag.* **2012**, *43*, 379–380. [CrossRef]
50. Klasson, C.; Helde-Frankling, M.; Sandberg, C.; Nordström, M.; Lundh-Hagelin, C.; Björkhem-Bergman, L. Vitamin D and Fatigue in Palliative Cancer: A Cross-Sectional Study of Sex Difference in Baseline Data from the Palliative D Cohort. *J. Palliat. Med.* **2020**, *24*, 433–437. [CrossRef]
51. Samefors, M.; Östgren, C.J.; Mölstad, S.; Lannering, C.; Midlöv, P.; Tengblad, A. Vitamin D deficiency in elderly people in Swedish nursing homes is associated with increased mortality. *Eur. J. Endocrinol.* **2014**, *170*, 667–675. [CrossRef] [PubMed]
52. Savoie, M.B.; Paciorek, A.; Zhang, L.; Van Blarigan, E.L.; Sommovilla, N.; Abrams, D.; Atreya, C.E.; Bergsland, E.K.; Chern, H.; Kelley, R.K.; et al. Vitamin D Levels in Patients with Colorectal Cancer Before and After Treatment Initiation. *J. Gastrointest. Cancer* **2019**, *50*, 769–779. [CrossRef] [PubMed]
53. Rizzoli, R. Vitamin D supplementation: Upper limit for safety revisited? *Aging Clin. Exp. Res.* **2021**, *33*, 19–24. [CrossRef] [PubMed]

MDPI

St. Alban-Anlage 66

4052 Basel

Switzerland

Tel. +41 61 683 77 34

Fax +41 61 302 89 18

www.mdpi.com

Nutrients Editorial Office

E-mail: nutrients@mdpi.com

www.mdpi.com/journal/nutrients